ROCKET PROPULSION ELEMENTS

Rocket Propulsion Elements

Ninth Edition

GEORGE P. SUTTON

Acknowledged expert on rocket propulsion
Formerly Executive Director of Engineering at Rocketdyne
(now Aerojet Rocketdyne)
Formerly Laboratory Associate at Lawrence Livermore National Laboratory

OSCAR BIBLARZ

Professor Emeritus
Department of Mechanical and Aerospace Engineering
Naval Postgraduate School

Copyright © 2017 by John Wiley & Sons, Inc. All rights reserved.

Published by John Wiley & Sons, Inc., Hoboken, New Jersey.

Published simultaneously in Canada.

No part of this publication may be reproduced, stored in a retrieval system, or transmitted in any form or by any means, electronic, mechanical, photocopying, recording, scanning, or otherwise, except as permitted under Section 107 or 108 of the 1976 United States Copyright Act, without either the prior written permission of the Publisher, or authorization through payment of the appropriate per-copy fee to the Copyright Clearance Center, 222 Rosewood Drive, Danvers, MA 01923, (978) 750-8400, fax (978) 646-8600, or on the web at www.copyright.com. Requests to the Publisher for permission should be addressed to the Permissions Department, John Wiley & Sons, Inc., 111 River Street, Hoboken, NJ 07030, (201) 748-6011, fax (201) 748-6008, or online at www.wiley.com/go/permissions.

Limit of Liability/Disclaimer of Warranty: While the publisher and author have used their best efforts in preparing this book, they make no representations or warranties with the respect to the accuracy or completeness of the contents of this book and specifically disclaim any implied warranties of merchantability or fitness for a particular purpose. No warranty may be created or extended by sales representatives or written sales materials. The advice and strategies contained herein may not be suitable for your situation. You should consult with a professional where appropriate. Neither the publisher nor the author shall be liable for damages arising herefrom.

For general information about our other products and services, please contact our Customer Care Department within the United States at (800) 762-2974, outside the United States at (317) 572-3993 or fax (317) 572-4002.

Wiley publishes in a variety of print and electronic formats and by print-on-demand. Some material included with standard print versions of this book may not be included in e-books or in print-on-demand. If this book refers to media such as a CD or DVD that is not included in the version you purchased, you may download this material at http://booksupport.wiley.com. For more information about Wiley products, visit www.wiley.com.

Library of Congress Cataloging-in-Publication Data is Available

ISBN 9781118753651 (Hardcover)
ISBN 9781118753880 (ePDF)
ISBN 9781118753910 (ePub)

Cover design: Wiley
Cover image: SpaceX

This is a photograph of the rocket propulsion system at the aft end of the recoverable booster stage of the Falcon 9 Space Launch Vehicle. This propulsion system has nine Merlin liquid propellant rocket engines, but only eight of these can be seen in this view. The total take-off thrust at sea level is approximately 1.3 million pounds of thrust force and at orbit altitude (in a vacuum) it is about 1.5 million pounds of thrust. Propellants are liquid oxygen and RP-1 kerosene. More information about this multiple rocket engine propulsion system can be found in Chapter 11 Section 2 and more information about RP-1 kerosene can be found in Chapter 7. The Falcon space vehicle and the Merlin rocket engines are designed, developed, manufactured, and operated by Space Exploration Technologies Corporation, better known as SpaceX, of Hawthorne, California.

This book is printed on acid-free paper. ♾

Printed in the United States of America

V10007084_122618

CONTENTS

3 Nozzle Theory and Thermodynamic Relations   45

4 Flight Performance   99

8 Thrust Chambers

9 Liquid Propellant Combustion and Its Stability

PREFACE

The rocket propulsion business in the United States of America appears to be changing. In the past, and also currently, the business has been planned, financed, and coordinated mostly by the Department of Defense and NASA. Government funding, government test or launch facilities, and other government support was provided. As it happens in all fields old-time companies have changed ownership, some have been sold or merged, some went out of business, some reduced the number of employees, and other companies have entered the field. New privately financed companies have sprung up and have developed their own rocket propulsion systems and flight vehicles as well as their own test, manufacturing, and launch facilities. These new companies have received some government contracts. Several privately owned companies have developed on their own useful space vehicles and rocket propulsion systems that were not originally in the government's plan. Although business climate changes noticeably influence rocket activities, it is not the purpose of this book to describe such business effects, but to present rocket propulsion principles and to give recent information and data on technical and engineering aspects of rocket propulsion systems.

All aerospace developments are aimed either at better performance, or higher reliability, or lower cost. In the past, when developing or modifying a rocket propulsion system for space applications, the emphasis has been primarily on very high reliability and, to a lesser extent, on high performance and low cost. Each of the hundreds of components of a propulsion system has to do its job reliably and without failure during operation. Indeed, the reliability of space launches has greatly improved world wide. In recent years emphasis has been placed primarily on cost reduction, but with continuing lower priority efforts to further improve performance and reliability. Therefore, this Ninth Edition has a new section and table on cost reduction of rocket propulsion systems. Also, in this book environmental compatibility is considered to be part of reliability

This Ninth Edition is organized into the same 21 chapters and subjects, as in the Eighth Edition, except that some aspects are treated in more detail. The names of the 21 chapters can be found in the Table of Contents. There are some changes, additions, improvements, and deletions in every chapter. A few problems have printed answers so students or other readers can self-check their solutions.

About half of this new edition is devoted to chemical rocket propulsion (solid propellant motors, liquid propellant rocket engines, and hybrid rocket propulsion systems). The largest number of individual rocket propulsion systems (currently in use, on stand-by, or in production) are solid propellant rocket motors; they vary in size, complexity, and duration; most systems are for military or defense applications. The next largest number in production or currently in use for space flight or missile defense are liquid propellant rocket engines; they vary widely in size, thrust or duration. Many people in aerospace consider this rocket propulsion technology to be mature. Enough technical information is available from public sources and from skilled personnel so that any new or modified rocket propulsion system can be developed with some confidence.

There have been several new applications (different flight vehicles, different missions) using existing or modified rocket propulsion systems. Several of these new applications are mentioned in this book.

Compared to the prior edition this new edition has less information or data of recently retired rocket engines, such as the engines for the Space Shuttle (retired in 2011) or Energiya; these have been replaced with facts from rocket propulsion systems that are likely to be in production for a long time. This new edition gives data on several rocket propulsion systems that are currently in production; examples are the RD-68 and the Russian RD-191 engines. Relatively little discussion of current research and developments is contained in this Ninth Edition; this is because it is not known when any particular development will lead to a better propulsion system, a better material of construction, a better propellant, or a better method of analysis, even if it appears to be promising at the present time. It is unfortunate that a majority of Research and Development programs do not lead to production applications.

Subjects new to the book include the Life of Liquid Propellant Thrust Chambers, a powerful new solid propellant explosive ingredient and two sections on variable thrust rocket propulsion. The discussion of dinitrogen oxide propellant is new, and additions were made to the write-ups of hydrogen peroxide and methane. Several different liquid propellant rocket engines are shown as examples of different engine types. The rocket propulsion system of the MESSENGER space probe is described as an example of a multiple thruster pressure feed system; its flow diagram replaces the Eighth Edition's one for the Space Shuttle. The Russian RD-191 engine (for the Angara series of launch vehicles) serves as an example for a high performance staged combustion engine cycle. The RD-68A presently has the highest thrust of any liquid oxygen/liquid hydrogen engine and it is an example of an advanced gas generator engine cycle. The RD-0124 illustrates an upper stage rocket engine with four thrust chambers and a single turbopump. Currently, a new manufacturing process known as Additive Manufacturing is being investigated for replacing parts or components of existing liquid propellant rocket engines.

The Ninth Edition also has the following other subjects, which are new to this book: upper stages with all electric propulsion, a dual inlet liquid propellant centrifugal pump for better cavitation resistance, topping-off cryogenic propellant tanks just prior to launching, benefits of pulsing of small thrusters, avoiding carbon containing deposits in the passages of liquid propellant cooling jackets, and a two-kilowatt arcjet. Since it is unlikely that nuclear power rocket propulsion systems development will again be undertaken in the next decade or that gelled propellants or aerospike nozzles will enter into production anytime soon, these three topics have largely been deleted from the new edition.

All Problems and Examples have been reviewed. Some have been modified, and some are new. A few of the problems which were deemed hard to solve have been deleted. The index at the end of the book has been expanded, making it somewhat easier to find specific topics in the book.

Since its first edition in 1949 this book has been a most popular and authoritative work in rocket propulsion and has been acquired by at least 77,000 students and professionals in more than 35 countries. It has been used as a text in graduate and undergraduate courses at about 55 universities. It is the longest living aerospace book ever, having been in print continuously for 67 years. It is cited in two prestigious professional awards of the American Institute of Aeronautics and Astronautics. Earlier editions have been translated into Russian, Chinese, and Japanese. The authors have given lectures and three-day courses using this book as a text in colleges, companies and Government establishments. In one company all new engineers are given a copy of this book and asked to study it.

As mentioned in prior editions, the reader should be very aware of the hazards of propellants, such as spills, fires, explosions, or health impairments. The authors and the publisher recommend that readers of this book do not work with hazardous propellant materials or handle them without an exhaustive study of the hazards, the behavior, and properties of each propellant, and without rigorous safety training, including becoming familiar with protective equipment. People have been killed, when they failed to do this. Safety training and propellant information is given routinely to employees of organizations in this business. With proper precautions and careful design, all propellants can be handled safely. Neither the authors nor the publisher assume any responsibility for actions on rocket propulsion taken by the reader, either directly or indirectly. The information presented in this book is insufficient and inadequate for conducting propellant experiments or rocket propulsion operations.

This book and its prior editions use both the English Engineering (EE) system of units (foot, pound) and the SI (Système International) or metric system of units (m, kg), because most drawings and measurements of components and subassemblies of chemical rocket propulsion systems, much of the rocket propulsion design and most of the manufacturing is still done in EE units, Some colleges and research organizations in the United States, and most propulsion organizations in other countries use the SI system of units. This dual set of units is used, even though the United States has been committed to switch to SI units.

Indeed the authors gratefully acknowledge the good help and information obtained from experts in specific areas of propulsion. James H. Morehart, The Aerospace Corporation, (information on various rocket engines and propellants) 2005 to 2015; Jeffrey S. Kincaid, Vice President (retired), Aerojet Rocketdyne, Canoga Park, CA (RS-68 engine data and figures, various propulsion data) 2012 to 2915; Roger Berenson, Engine Program Chief Engineer, Aerojet Rocketdyne, Canoga Park, CA, (RS-68 and RS-25 engine and general propulsion data) 2015; Mathew Rottmund, United Launch Alliance, Centennial, CO. (launch vehicle propulsion issues), 2014 to 2015; Olwen M. Morgan (retired), Marketing Manager, Aerojet Rocketdyne, Redmond, WA, (MESSENGER space probe; monopropellants); 2013 to 2016; Dieter M. Zube, Aerojet Rocketdyne, Redmond (view and data on hydrazine arcjet); 2013-2015; Jeffrey D. Haynes, Manager, Aerojet Rocketdyne, (additive manufacturing information), 2015; Leonard H. Caveny, Consultant, Fort Washington, MD, (solid propellant rocket motors); Russell A. Ellis, Consultant, (solid propellant rocket motors); 2015; David K. McGrath, Director Systems Engineering, Orbital ATK, Missile Defense and Controls, Elkton, MD, (solid propellant rocket motors); 2014 to 2015; Eckart W. Schmidt, Consultant for Hazardous Materials, Bellevue, WA, (Hydrazine and liquid propellants), 2013 to 2015; Michael J. Patterson, Senior Technologist, In-Space Propulsion, NASA Glenn Research Center, Cleveland, OH (electric propulsion information), 2014; Rao Manepalli, Deptford, NJ, formerly with Indian Space Research Organization (rocket propulsion systems information); 2011 to 2013; Dan Adamski, Aerojet Rocketdyne, (RS-68 flowsheet), 2014; Frederick S. Simmons (retired), The Aerospace Corporation (review of Chapter 20); 2015 to 2016.

The authors have made an effort to verify and/or validate all information in this ninth edition. If the reader finds any errors or important omissions in the text of this edition we would appreciate bringing them to our attention so that we may evaluate them for possible inclusion in subsequent printings.

George P. Sutton
Los Angeles, California

Oscar Biblarz,
Monterey, California

ROCKET PROPULSION ELEMENTS

CHAPTER 1

CLASSIFICATION

In general terms, propulsion is the act of changing the motion of a body with respect to an inertial reference frame. Propulsion systems provide forces that either move bodies initially at rest or change their velocity or that overcome retarding forces when bodies are propelled through a viscous medium. The word *propulsion* comes from the Latin *propulsus,* which is the past participle of the verb *propellere,* meaning "to drive away." *Jet propulsion* is a type of motion whereby a reaction force is imparted to a vehicle by the momentum of ejected matter.

Rocket propulsion is a class of jet propulsion that produces thrust by ejecting matter, called the working fluid or *propellant,* stored entirely in the flying vehicle. *Duct propulsion* is another class of jet propulsion and it includes turbojets and ramjets; these engines are more commonly called air-breathing engines. Duct propulsion devices mostly utilize their surrounding medium as the propellant, energized by its combustion with the vehicle's stored fuel. Combinations of rockets and duct propulsion devices have been attractive for some applications, and one is briefly described in this chapter.

The *energy source* most commonly used in rocket propulsion is *chemical combustion.* Energy can also be supplied by *solar radiation* and by a *nuclear reactor.* Accordingly, the various propulsion devices in use can be divided into *chemical propulsion, nuclear propulsion,* and *solar propulsion.* Table 1–1 lists many important propulsion concepts according to their energy source and type of propellant. Radiant energy may originate from sources other than the sun and theoretically includes the transmission of energy by ground-based microwaves and laser beams. Nuclear energy originates in transformations of mass within atomic nuclei and is generated by either fission or fusion. Energy sources are central to rocket performance and several kinds, both within and external to the vehicle, have been investigated. The useful energy

TABLE 1–1. Energy Sources and Propellants for Various Propulsion Concepts

Propulsion Device	Energy Source[a]			Propellant or Working Fluid
	Chemical	Nuclear	Solar	
Turbojet	D/P			Fuel + air
Turbo–ramjet	TFD			Fuel + air
Ramjet (hydrocarbon fuel)	D/P	TFD		Fuel + air
Ramjet (H$_2$ cooled)	TFD			Hydrogen + air
Rocket (chemical)	D/P	TFD		Stored propellant
Ducted rocket	TFD			Stored solid fuel + surrounding air
Electric rocket	D/P		D/P	Stored propellant
Nuclear fission rocket		TFD		Stored H$_2$
Solar-heated rocket			TFD	Stored H$_2$
Photon rocket (big light bulb)		TFND		Photon ejection (no stored propellant)
Solar sail			TFD	Photon reflection (no stored propellant)

[a]D/P developed and/or considered practical; TFD, technical feasibility has been demonstrated, but development is incomplete; TFND, technical feasibility has not yet been demonstrated.

input modes in rocket propulsion systems are either heat or electricity. Useful output thrust comes from the kinetic energy of the ejected matter and from the propellant pressure on inner chamber walls and at the nozzle exit; thus, rocket propulsion systems primarily convert input energies into the kinetic energy of the exhausted gas. The ejected mass can be in a solid, liquid, or gaseous state. Often, combinations of two or more phases are ejected. At very high temperatures, ejected matter can also be in a plasma state, which is an electrically conducting gas.

1.1. DUCT JET PROPULSION

This class, commonly called *air-breathing engines*, comprises devices which entrain and energize air flow inside a duct. They use atmospheric oxygen to burn fuel stored in the flight vehicle. This class includes turbojets, turbofans, ramjets, and pulsejets. These are mentioned here primarily to provide a basis for comparison with rocket propulsion and as background for combined rocket–duct engines, which are mentioned later. Table 1–2 compares several performance characteristics of specific chemical rockets with those of typical turbojets and ramjets. A high specific impulse (which is a measure of performance to be defined later) relates directly to long-flight ranges and thus indicates the superior range capability of air-breathing engines over chemical rocket propulsion systems at relatively low earth altitudes. However, the uniqueness of rocket propulsion systems (for example, high thrust to weight, high thrust to frontal area, and thrust nearly independent of altitude) enables flight in rarefied air and exclusively in space environments.

TABLE 1–2. Comparison of Several Characteristics of a Typical Chemical Rocket Propulsion System and Two-Duct Propulsion Systems

Feature	Chemical Rocket Engine or Rocket Motor	Turbojet Engine	Ramjet Engine
Thrust-to-weight ratio, typical	75:1	5:1, turbojet and afterburner	7:1 at Mach 3 at 30,000 ft
Specific fuel consumption (pounds of propellant or fuel per hour per pound of thrust)[a]	8–14	0.5–1.5	2.3–3.5
Specific thrust (pounds of thrust per square foot frontal area)[b]	5000–25,000	2500 (low Mach[c] numbers at sea level)	2700 (Mach 2 at sea level)
Specific impulse, typical[d] (thrust force per unit propellant or fuel weight flow per second)	270 sec	1600 sec	1400 sec
Thrust change with altitude	Slight increase	Decreases	Decreases
Thrust vs. flight speed	Nearly constant	Increases with speed	Increases with speed
Thrust vs. air temperature	Constant	Decreases with temperature	Decreases with temperature
Flight speed vs. exhaust velocity	Unrelated, flight speed can be greater	Flight speed always less than exhaust velocity	Flight speed always less than exhaust velocity
Altitude limitation	None; suited to space travel	14,000–17,000 m	20,000 m at Mach 3 30,000 m at Mach 5 45,000 m at Mach 12

[a]Multiply by 0.102 to convert to kg/(hr-N).
[b]Multiply by 47.9 to convert to N/m^2.
[c]Mach number is the ratio of gas speed to the local speed of sound (see Eq. 3–22).
[d]*Specific impulse* is a performance parameter defined in Chapter 2.

The *turbojet engine* is the most common of ducted engines. Figure 1–1 shows its basic elements.

For supersonic flight speeds above Mach 2, the *ramjet engine* (a pure duct engine) becomes possible for flights within the atmosphere. Compression is purely gas dynamic and thrust is produced by increasing the momentum of the subsonic compressed air as it passes through the ramjet, basically as is accomplished in the turbojet and turbofan engines but without any compressor or turbine hardware. Figure 1–2 shows the basic components of a ramjet. Ramjets with subsonic combustion and hydrocarbon fuels have an upper speed limit of approximately Mach 5; hydrogen fuel, with hydrogen cooling, raises this to at least Mach 16. Ramjets with supersonic combustion, known as *scramjets*, have flown in experimental vehicles. All ramjets

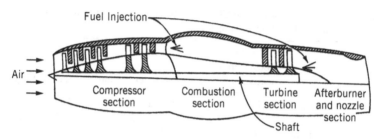

FIGURE 1–1. Simplified schematic diagram of a turbojet engine.

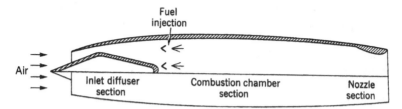

FIGURE 1–2. Simplified diagram of a ramjet with a supersonic inlet (a converging/diverging flow passage).

must depend on rocket or aircraft boosters for initial acceleration to supersonic conditions and operating altitudes, and on oblique shocks to compress and decelerate the entrance air. Applications of ramjets with subsonic combustion include shipboard- and ground-launched antiaircraft missiles. Studies of a hydrogen-fueled ramjet for hypersonic aircraft looked promising, but as of this writing they have not been properly demonstrated; one supersonic flight vehicle concept combines a ramjet-driven high-speed airplane and a one- or two-stage rocket booster for driving the vehicle to its operating altitude and speed; it can travel at speeds up to a Mach number of 25 at altitudes of up to 50,000 m.

No truly new or significant rocket technology concepts have been implemented in recent years, reflecting a certain maturity in this field. Only a few new applications for proven concepts have been found, and those that have reached production are included in this edition. The culmination of research and development efforts in rocket propulsion often involves adaptations of new approaches, designs, materials, as well as novel fabrication processes, cost, and/or schedule reductions to new applications.

1.2. ROCKET PROPULSION

Rocket propulsion systems may be classified in a number ways, for example, according to energy source type (chemical, nuclear, or solar) or by their basic function (booster stage, sustainer or upper stages, attitude control, orbit station keeping, etc.)

or by the type of vehicle they propel (aircraft, missile, assisted takeoff, space vehicle, etc.) or by their size, type of propellant, type of construction, and/or by the number of rocket propulsion units used in a given vehicle.

Another useful way to classify rockets is by the method of producing thrust. The thermodynamic expansion of a gas in a supersonic nozzle is utilized in most common rocket propulsion concepts. The internal energy of the propellant is converted into exhaust kinetic energy, and thrust is also produced by the pressure on surfaces exposed to the exhaust gases, as will be shown later. This same thermodynamic theory and the same generic equipment (i.e., a chamber plus a nozzle) is used for jet propulsion, rocket propulsion, nuclear propulsion, laser-thermal and solar-thermal propulsion, and in some types of electrical propulsion. Totally different methods of producing thrust are used in nonthermal types of electric propulsion. As described below, these electric systems use magnetic and/or electric fields to accelerate electrically charged atoms or molecules at very low gas densities. It is also possible to obtain very small accelerations by taking advantage of the difference in gravitational attraction as a function of earth altitude, but this method is not treated in this book.

The Chinese developed and used solid propellant in rocket missiles over 800 years ago, and military "bombardment rockets" were used frequently in the eighteenth and nineteenth centuries. However, the most significant developments of rocket propulsion took place in the twentieth century. Early pioneers included the Russian Konstantin E. Ziolkowsky, who is credited with the fundamental rocket flight equation and his 1903 proposals to build rocket vehicles. Robert H. Goddard, an American, is credited with the first flight using a liquid propellant rocket engine in 1926. For the history of rockets, see Refs. 1–1 to 1–7.

Chemical Rocket Propulsion

Energy from the combustion reaction of chemical propellants, usually a fuel and an oxidizer, in a high-pressure chamber goes into heating reaction product gases to high temperatures (typically 2500 to 4100 °C or 4500 to 7400 °F). These gases are subsequently expanded in a supersonic nozzle and accelerated to high velocities (1800 to 4300 m/sec or 5900 to 14,100 ft/sec). Since such gas temperatures are about twice the melting point of steel, it is necessary to cool or insulate all the surfaces and structures that are exposed to the hot gases. According to the physical state of the stored propellant, there are several different classes of chemical rocket propulsion devices.

Liquid propellant rocket engines use propellants stored as liquids that are fed under pressure from tanks into a *thrust chamber.** A typical pressure-fed liquid propellant rocket engine system is schematically shown in Fig. 1–3. The *bipropellant* consists of a liquid oxidizer (e.g., liquid oxygen) and a liquid fuel (e.g., kerosene). A *monopropellant* is a single liquid that decomposes into hot gases when properly catalyzed.

---

*The term *thrust chamber*, used for the assembly of the injector, nozzle, and chamber, is preferred by several official agencies and therefore has been used in this book. For small spacecraft control rockets the term *thruster* (a small thrust chamber) is commonly used, and this term will be used in some sections of this book.

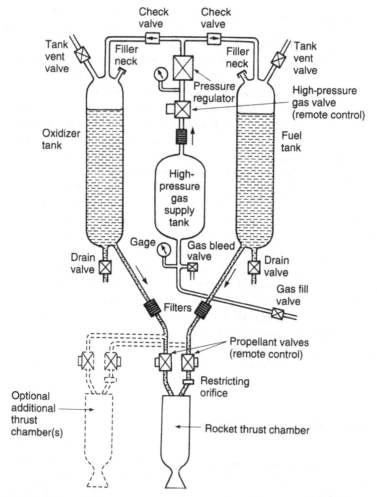

FIGURE 1–3. Schematic flow diagram of a liquid propellant rocket engine with a gas pressure feed system. The dashed lines show a second thrust chamber, but some engines have more than a dozen thrust chambers supplied by the same feed system. Also shown are components needed for start and stop, controlling tank pressure, filling propellants and pressurizing gas, draining or flushing out remaining propellants, tank pressure relief or venting, and several sensors.

Gas pressure feed systems are used mostly on low-thrust, low-total-energy propulsion systems, such as those used for attitude control of flying vehicles, often with more than one thrust chamber per engine. The larger bipropellant rocket engines use one or more turbopump-fed liquids as shown in Fig. 1–4. Pump-fed liquid rocket systems are most common in applications needing larger amounts of propellant and higher thrust, such as those in space launch vehicles. See Refs. 1–1 to 1–6.

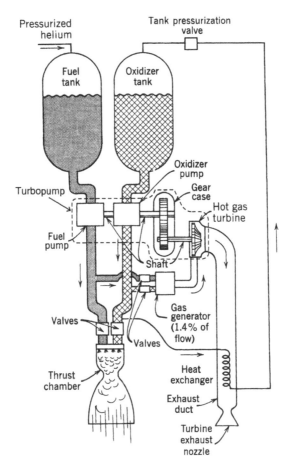

FIGURE 1–4. Simplified schematic diagram of a liquid propellant rocket engine with one type of turbopump feed system and a separate gas generator, which generates "warm" gas for driving the turbine. Not shown are components necessary for controlling the operation, filling, venting, draining, or flushing out propellants, filters, pilot valves, or sensors. This turbopump assembly consists of two propellant pumps, a gear case, and a high-speed turbine. Many turbopumps have no gear case.

Chemical propellants react to form hot gases inside the *thrust chamber* and proceed in turn to be accelerated through a supersonic nozzle from which they are ejected at a high velocity, thereby imparting momentum to the vehicle. Supersonic nozzles consist of a converging section, a constriction or throat, and a conical or bell-shaped diverging section, as further described in the next two chapters.

Some liquid rocket engines permit repetitive operation and can be started and shut off at will. If the thrust chamber is provided with adequate cooling capacity, it is possible to run liquid rockets for hours, depending only on the propellant supply. A liquid rocket propulsion system requires several precision valves and some have

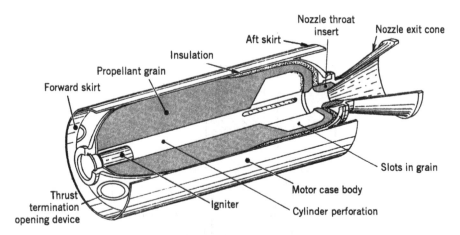

FIGURE 1–5. Simplified perspective three-quarter section view of a typical solid propellant rocket motor with the propellant grain bonded to the case and to the insulation layer, and with a conical exhaust nozzle. The cylindrical case with its forward and aft hemispherical domes forms a pressure vessel containing the combustion chamber pressure. Adapted with permission from Ref. 12–1.

complex feed mechanisms that include propellant pumps, turbines and gas generators. All have propellant-pressurizing devices and relatively intricate combustion/thrust chambers.

In *solid propellant rocket motors** the ingredients to be burned are already stored within a combustion chamber or *case* (see Fig. 1–5). The solid propellant (or charge) is called the *grain*, and it contains all the chemical elements for complete burning. Once ignited, it is designed to burn smoothly at a predetermined rate on all the exposed internal grain surfaces. In the figure, initial burning takes place at the internal surfaces of the cylinder perforation and at the four slots. The internal cavity expands as propellant is burned and consumed. The resulting hot gases flow through the supersonic nozzle to impart thrust. Once ignited, motor combustion is designed to proceed in an orderly manner until essentially all the propellant has been consumed. There are no feed systems or valves. See Refs. 1–6 to 1–9.

Liquid and solid propellants, together with the propulsion systems that use them, are discussed in Chapters 6 to 11 and 12 to 15, respectively. Rocket propulsion selection systems, for both liquid and solid propellants, are compared in Chapter 19.

Gaseous propellant rocket engines use a stored high-pressure gas, such as air, nitrogen, or helium, as working fluid. Such stored gases require relatively heavy tanks. These cold gas thrusters were used in many early space vehicles for low-thrust maneuvers and for attitude-control systems, and some are still used today. Heating the gas with electrical energy or by the combustion of certain monopropellants in a chamber

---

*Historically, the word *engine* is used for a liquid propellant rocket propulsion system and the word *motor* is used for solid propellant rocket propulsion. They were developed originally by different groups.

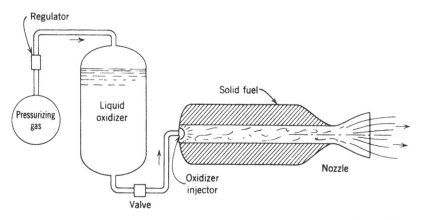

FIGURE 1–6. Schematic diagram of a typical hybrid rocket engine. The relative positions of the oxidizer tank, high-pressure gas tank, and the fuel chamber with its nozzle depend on the particular vehicle design.

improves their performance, and this has often been called *warm gas propellant rocket propulsion*. Chapter 7 reviews gaseous propellants.

Hybrid propellant rocket propulsion systems use both liquid and solid propellant storage. For example, if a liquid oxidizing agent is injected into a combustion chamber filled with a solid carbonaceous fuel grain, the chemical reaction produces hot combustion gases (see Fig. 1–6). They are described further in Chapter 16. Several have flown successfully.

Combinations of Ducted Jet Engines and Rocket Engines

The Tomahawk surface-to-surface missile uses a sequenced two-stage *propulsion system*. The solid propellant rocket booster lifts the missile away from its launch platform and is discarded after its operation. A small turbojet engine then sustains their low-level flight at nearly constant subsonic speed toward the target.

Ducted rocket propulsion systems, sometimes called *air-augmented rocket propulsion systems*, combine the principles of rocket and ramjet engines; they give higher performance (specific impulse) than chemical rocket engines but can only operate within the earth's atmosphere. Usually, the term *air-augmented rocket* denotes mixing of air with the rocket exhaust (made fuel rich for afterburning) in proportions that enable the propulsion device to retain those characteristics that typify rocket engines, for example, high static thrust and high thrust-to-weight ratio. In contrast, the ducted rocket is often like a ramjet in that it must be boosted to operating speed and uses the rocket components more as a fuel-rich gas generator (liquid or solid).

The action of rocket propulsion systems and ramjets can be combined. An example of these two are propulsion systems operating in sequence and then in tandem and yet utilizing a common combustion chamber volume, as shown in Fig. 1–7. Such a low-volume configuration, known as an *integral rocket–ramjet*, has been attractive

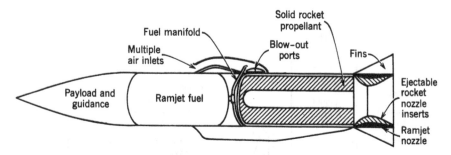

FIGURE 1–7. Simplified diagram of an air-launched missile with integral rocket–ramjet propulsion. After the solid propellant has been consumed in boosting the vehicle to flight speed, the rocket combustion chamber becomes the ramjet combustion chamber with air burning the ramjet liquid fuel. Igniter and steering mechanisms are not shown.

for air-launched missiles using ramjet propulsion. The transition from rocket engine to ramjet requires enlarging the exhaust nozzle throat (usually by ejecting rocket nozzle parts), opening the ramjet air inlet–combustion chamber interface, and following these two events with a normal ramjet starting sequence.

A *solid fuel ramjet* uses grains of solid fuel that gasify or ablate and then react with air. Good combustion efficiencies have been achieved with a patented boron-containing solid fuel fabricated into grains similar to a solid propellant rocket motor and burning in a manner similar to a hybrid rocket propulsion system.

Nuclear Rocket Engines

These are basically a type of liquid propellant rocket engine where the power input comes from a single nuclear reactor and not from any chemical combustion. During the 1960s an experimental rocket engine with a nuclear fission graphite reactor was built and ground tested with liquid hydrogen as the propellant. It delivered an equivalent altitude specific impulse (this performance parameter is explained in Chapter 2) of 848 sec, a thrust of over 40,000 lbf at a nuclear reactor power level of 4100 MW with a hydrogen temperature of 2500 K. No further ground tests of nuclear fission rocket engines have been undertaken.

Public concerns about any ground and/or flight accident with the inadvertent spreading of radioactive materials in the Earth's environment have caused the termination of nuclear rocket engine work. It is unlikely that nuclear rocket engines will be developed in the next few decades and therefore no further discussion is given in here. Our Eighth Edition has additional information and references on nuclear propulsion.

Electric Rocket Propulsion

Electric propulsion has been attractive because of its comparatively high performance, producing desired amounts thrust with moderately low propellant utilization, but they are limited to relatively low thrusts by existing electrical power supplies. This type of propulsion is much too low for earth launches and atmospheric fight because it requires rather massive and relatively inefficient power sources (but in spacecraft they can often be shared with other subsystems). Unlike chemical propulsion, electric propulsion utilizes energy sources (nuclear, solar radiation, or batteries) not contained in the propellant being utilized. The thrust is usually quite low, levels typical of orbit maintenance (0.005 to 1 N). In order to accomplish significant increases in vehicle velocity, it is necessary to apply such low thrusts (and their small accelerations) during times considerably longer than with chemical propulsion, some for months and even years (see Ref. 1–10 and Chapter 17).

Of the three basic electric types, *electrothermal* rocket propulsion most resembles the previously discussed liquid-propellant chemical rocket units; a propellant is heated electrically (with solid resistors called resistojets or electric arcs called arcjets), and the hot gas is then thermodynamically expanded through a supersonic nozzle (see Fig. 1–8). These electrothermal thrusters typically have thrust ranges of 0.01 to 0.5 N, with exhaust velocities of 1000 to 7800 m/sec; they use ammonium, hydrogen, nitrogen, or hydrazinede composition products as the propellant.

The two other types—the *electrostatic* or *ion propulsion* thrusters and the *electromagnetic* or *magnetoplasma* thrusters—accomplish propulsion by different means, and no thermodynamic gas expansion in a nozzle is necessary. Both work only in a vacuum. In an ion thruster (see Fig. 1–9) the working fluid (typically, xenon) is ionized by stripping off electrons, and then the heavy ions are accelerated to very high velocities (2000 to 60,000 m/sec) by means of electrostatic fields. The ions are subsequently electrically neutralized by combining them with emitted electrons to prevent the buildup of a "space charge" on the vehicle.

In electromagnetic thrusters, a plasma (an energized gaseous mixture of ions, electrons, and neutral particles) is accelerated by the interaction between electric currents and perpendicular magnetic fields and this plasma is then ejected at high velocities (1000 to 75,000 m/sec). There are many different types and geometries.

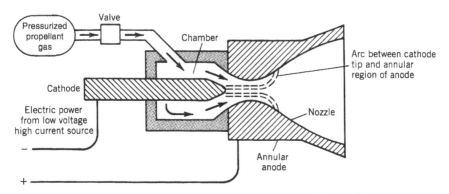

FIGURE 1–8. Simplified schematic diagram of arcjet thruster. The arc plasma temperature is very high (perhaps 15,000 K) and the anode, cathode, and chamber will get hot (1000 K) due to heat transfer.

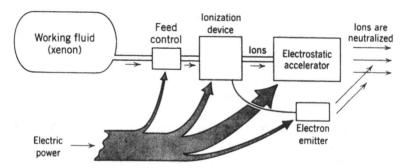

FIGURE 1–9. Simplified schematic diagram of a typical ion thruster, showing the approximate distribution of the electric power by the width of the arrow.

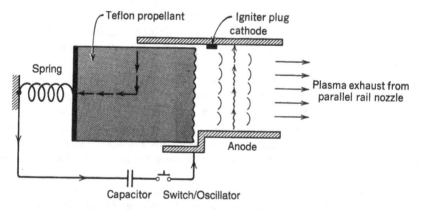

FIGURE 1–10. Simplified diagram of a pulsed plasma thruster with self-induced magnetic acceleration. When the capacitor is discharged, an arc is struck at the left side of the rails. The high arc current closes the loop, thus inducing a magnetic field. The action of the current and the magnetic field causes the plasma to be accelerated at right angles to both the magnetic field and the current in the direction of the rails. Each time the arc is created, a small amount of solid propellant (Teflon) is vaporized and converted to a small plasma cloud, which (when ejected) gives a small pulse of thrust. Actual units can operate with many pulses per second.

The Hall-effect thruster, a relatively new entry, may also be considered as electrostatic (see Chapter 17). A simply configured, pulsed electrical thruster with a solid (stored) propellant is shown in Fig. 1–10; it has been used for spacecraft attitude control (Ref. 1–10).

Other Rocket Propulsion Concepts

One concept is the *solar thermal rocket;* it has large-diameter optics to concentrate the sun's radiation (e.g., with lightweight precise parabolic mirrors or Fresnel lenses) onto a receiver or optical cavity, see Ref. 1–11. Figure 1–11 shows one embodiment

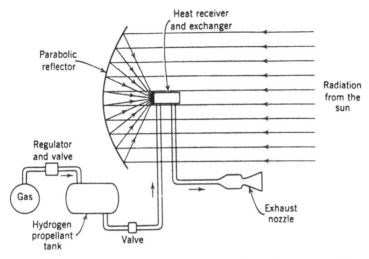

FIGURE 1–11. Schematic diagram of a solar thermal rocket concept.

and some data is given in Table 2–1. The receiver is made of high-temperature metal (such as tungsten or rhenium) and has a heat exchanger that heats the working fluid, usually hydrogen, up to perhaps 2500 °C; the hot gas is then exhausted through one or more nozzles. The large reflective mirror must be pointed toward the sun, and this requires orientation adjustments if the spacecraft orbits around the Earth or other planets. Performance can be two to three times higher than that of chemical rockets and thrust levels are low (1 to 10 N). Since large lightweight optical elements cannot withstand drag forces without deformation, such units are deployed outside the atmosphere. Contamination is negligible, but storage or refueling of liquid hydrogen is a challenge. Problems being investigated include rigid, lightweight mirror or lens structures; operational lifetimes; and minimizing hydrogen evaporation and heat losses to other spacecraft components. An experimental solar-thermal rocket propulsion system flew in a satellite in 2012. It has not been approved for a production application (as of 2015) to the authors' knowledge.

The *solar sail* is another concept. It is basically a large photon reflecting surface. The power source for the solar sail is external to the vehicle (see Ref. 1–12), but the vehicle can only move away from the sun. Concepts for transmitting radiation energy (by lasers or microwaves) from ground stations on Earth to satellites have been proposed but not yet tested.

International Rocket Propulsion Effort

Active development and/or production of rocket propulsion systems have been under way in more than 30 different countries. A few foreign rocket units are mentioned in this book together with their characteristics and with references to the international rocket literature. Although most of the data in this book are taken from U.S. rocket experiences, this is not intended to minimize the significance of foreign achievements.

At the time of this writing, the only joint major international program has been the *International Space Station* (ISS), a multiyear cooperative effort with major contributions from the United States and Russia and active participation by several other nations. This manned orbital space station is used for conducting experiments and observations on a number of research projects. See Ref. 1–13.

1.3. APPLICATIONS OF ROCKET PROPULSION

Because rocket propulsion can reach performances unequaled by other prime movers, it has its own field of applications and does not usually compete with other propulsion devices. Selection of the best rocket propulsion system type and design for any given application is a complex process involving many factors, including system performance, reliability, cost, propulsion system size, and compatibility, as described in Chapter 19. Examples of important applications are given below and some are discussed further in Chapter 4.

Space Launch Vehicles

Since 1957 there have been numerous space launch attempts with a better than 95% success record. *Space launch vehicles* or *space boosters* can be broadly classified as expendable or recoverable/reusable, by the type of propellant (storable or cryogenic liquid or solid propellants), number of stages (single-stage, two-stage, etc.), size/mass of payloads or vehicles, and as manned or unmanned. There are many different *missions* and *payloads* for space launch vehicles. Discussed below are the following categories: commercial missions (e.g., communications), military missions (e.g., reconnaissance), nonmilitary missions (e.g., weather observation), and space exploration missions (e.g., flights to the planets).

Each space launch has a specific space flight objective, such as an Earth orbit or a moon landing. See Ref. 1–14. It uses between two and five stages, each with its own propulsion systems and each usually fired sequentially after the lower stage is expended. Selection for the number of stages is based on the specific space trajectory, the number and types of maneuvers, the energy content of a unit mass of the propellant, the payload size, as well as other factors. The initial stage, usually called the booster stage, is the largest; this stage is then separated from the ascending vehicle before the second-stage propulsion system is ignited and operated. As explained in Chapter 4, adding extra stages may permit significant increases in the payload (such as more scientific instruments or more communications gear).

Each stage of a multistage launch vehicle is essentially a complete vehicle in itself and carries its own propellant, its own rocket propulsion system or systems, and its own control system. Once the propellant of a given stage is expended, its remaining mass (including empty tanks, cases, structure, instruments, etc.) is no longer useful to succeeding stages. By dropping off this mass, it is possible to accelerate the final stage with its payload to a higher terminal velocity than would be attained if multiple staging were not used. Both solid propellant and liquid propellant rocket propulsion systems have been amply utilized in low Earth orbit missions.

Figure 1–12 shows the Delta IV HEAVY lift space launch vehicle at takeoff. Its propellants are liquid oxygen/liquid hydrogen (LOX/LH$_2$) in all its main engines. Its booster engine, the Aerojet Rocketdyne RS-68A, is shown in Figs. 6–9a and 6–9b and data is in Table 11–2; its second-stage engine, the Aerojet Rocketdyne RL-10B-2 LOX/LH$_2$ (24,750 lbf thrust) is shown in Fig. 8–17 and data is in Table 8–1. The two liquid propellant strap-on booster pods (with the same booster engine) are removed for launching smaller payloads. Figure 1–13 shows the Atlas V space launch vehicle. Its booster engine is the Russian (Energomash) RD-180, it has Aerojet solid propellant strap-on boosters, and the upper stage engine is the Aerojet Rocketdyne RL 10A-4-2 LOX/LH$_2$ engine. The Russian (Energomash) LOX/kerosene RD-180 engine is shown in Ref. 1–2 as its Figure 7.10–11 and data is in its Table 7.10–2. In both of these launch vehicles the payload is carried on top of the second stage, which has its own propulsion set of small thrusters. Table 1–3 gives data for the larger propulsion systems in these two U.S. launch vehicles. Not shown in Table 1–3 are two additional stage separation systems for the Delta IV HEAVY space launch vehicle. One consists of a set of small solid propellant rocket motors that are installed in the two outboard boosters; one quarter of these motors are installed under each of the nozzle housings and one quarter under each nose fairing at the outboard booster stages. Their purpose is to move two outboard boosters (just after thrust termination) from the center or core booster (which continues to operate) and thus prevent any collision. These separation solid propellant motor boosters have a relatively high thrust of very short duration.

Also not listed in Table 1–3 is an attitude control system for the second stage of the Delta IV HEAVY. It has 12 small restartable monopropellant hydrazine thrusters that provide pitch, yaw, and roll control forces to this upper stage during the powered flight (from the RL-10B-2) and during the unpowered portion of this upper stage. These thrusters and the separation motors are not evident in Fig. 1–12.

The U.S. *Space Shuttle*, which was retired in 2011, provided the first *reusable spacecraft* that could glide and land on a runway. Figure 1–14 shows the basic configuration of the Space Shuttle at launch, which consisted of two stages, the booster, the orbiter stage, and an external tank. It shows all the 67 rocket propulsion systems of the shuttle. These consisted of 3 main engines (LOX/LH$_2$ of 470,000 lbf vacuum thrust each, see Chapter 7 for chemical nomenclature), 2 orbital maneuvering engines (N$_2$O$_4$/MMH of 6,000 vacuum thrust each), 38 reaction control primary thrusters (N$_2$O$_4$/MMH of 870 lbf vacuum thrust each), 6 reaction control Vernier thrusters (N$_2$O$_4$/MMH of 25 lbf vacuum thrust each), 2 large segmented booster solid propellant motors (composite solid propellant of 3.3×10^6 lbf thrust at sea level each), and 16 stage separation rocket motors (with composite solid propellants of 22,000 lbf vacuum thrust each, operating for 0.65 sec); they were activated after thrust termination of the boosters in order to separate them from the external tank. The orbiter was the reusable vehicle, a combination space launch vehicle, spacecraft, and glider for landing. The two solid propellant strap-on rocket motors were then

FIGURE 1–12. Delta IV HEAVY lift space launch vehicle. The center liquid propellant booster stage has a Pratt & Whitney Rocketdyne RS-68A rocket engine (LOX/LH$_2$). The two strap-on stages each use the same engine. See Figs. 6–9a and 6–9b. Courtesy Aerojet-Rocketdyne.

FIGURE 1–13. Atlas V space launch vehicle with three (or five) strap-on stages using Aerojet solid rocket motors and a central Energomash (Russia) RD-180 liquid propellant booster rocket engine running on LOX/kerosene. See Table 1–3 for key parameters. Courtesy United Launch Alliance.

TABLE 1–3. Major Propulsion Systems for Two U.S. Launch Vehicles

Vehicle	Propulsion System Designation (Propellant)	Stage	No. of Propulsion Systems per Stage	Thrust (lbf/kN) per Engine/Motor	Specific Impulse (sec)	Mixture Ratio, Oxidizer to Fuel Flow	Chamber Pressure (psia)	Nozzle Exit Area Ratio	Inert Engine Mass (lbm/kg)
DELTA IV HEAVY	RS-68A (LOX/LH$_2$)	1	1 or 3	797,000/3548[a]	411[a]	5.97	1557	21.5:1	14,770/6,699
	RL 10B-2 (LOX/LH$_2$)	2	1	702,000/3123[b,c] 24,750/0.110[a]	362[b] 465.5[a]	5.88	633	285:1	664 lbm
ATLAS V	Solid Booster	0	Between 1[d] and 5	287,346/1.878[b] each	279.3[b]	N/A	3722	16:1[c]	102,800 lbm (loaded)
	RD-180[d] (LOX/Kerosene)	1	1	933,400/4.151[a]	310.7[b]	2.72	610	36.4:1	12,081/5,480
	RD 10A-4-2 (LOX/LH$_2$)	2	1 or 2	860,200/3.820[b,c] 22,300/99.19[a]	337.6[a] 450.5[a]	4.9–5.8		84.1:1	5330 kg 370/168

[a] Vacuum value.
[b] Sea-level value.
[c] At ignition.
[d] Russian RD-180 engine has 2 gimbal-mounted thrust chambers.

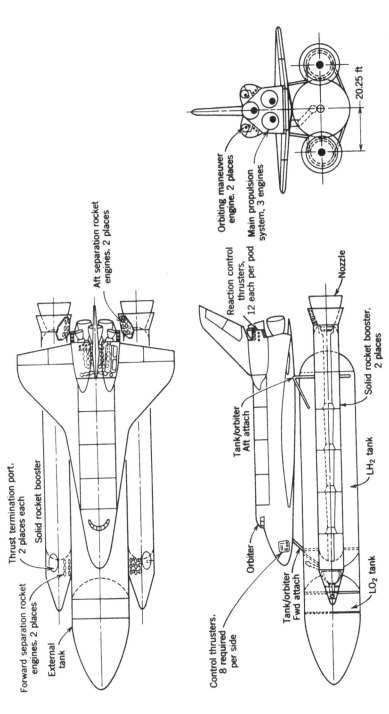

FIGURE 1–14. Simplified diagrams of the original Space Shuttle vehicle. The shuttle orbiter—the delta-winged vehicle about the size of a medium-range jet liner—was a reusable, cargo-carrying, spacecraft–airplane combination that took off vertically and landed horizontally as a glider. Each shuttle orbiter was designed for a minimum of 100 missions and could carry as much as 65,000 lb of payload to a low Earth orbit, and a crew of up to four members. It could return up to 25,000 lb of payload back to earth. NASA's new manned Space Launch System plans using a modified Space Shuttle engine (see Fig. 6–1) designated as the RS-25.

the largest in existence; they were equipped with parachutes for sea recovery of the burned-out motors. The large LO_2/LH_2 external tank was jettisoned and expended just before orbit insertion (see Ref. 1–15). The Space Shuttle accomplished both civilian and military missions, placing astronauts and satellites in orbit, undertaking scientific exploration, supplying the International Space Station, and repairing, servicing, and retrieving satellites. Today, the retired orbiters (the winged vehicles of the Space Shuttle) are on museum displays. At the time of this writing (2015), the National Aeronautics and Space Administration (NASA) had awarded the initial research and development (R&D) contracts in several critical areas for a new large manned space flight vehicle identified as Orion and as SLS (Space Launch System).

A new generation of manned space flight vehicles is currently being developed by several entrepreneurial U.S. and foreign companies, some with private capital. They are aimed at future commercial markets that include sending tourists into space for-ays. All are based on reusable spacecraft, some with reusable launch vehicles, some with vertical and some with horizontal takeoff or landing. A number of suborbital flights have already been accomplished by test pilots using new winged vehicles. It is too early to predict which of these organizations will be successful in commercializing manned space flight. Some new companies have already developed structures and engines for practical "low-cost" launch vehicles; these are, at the time of this writing, being routinely used for transporting supplies to the International Space Station (which is in a low Earth orbit). To bring down costs, the first stage of future launch vehicles will need to be recoverable and reusable.

A single-stage-to-orbit vehicle, an attractive concept because it avoids the costs and complexities of staging, would be expected to have improved reliability (simpler structures, fewer components). However, its envisioned payload has been too small for economic operation. To date efforts at developing a rocket-propelled single-stage-to-orbit vehicle have not been successful.

Spacecraft

Depending on their mission, *spacecraft* can be categorized as Earth satellites, lunar, interplanetary, and trans-solar types, and/or as manned and unmanned spacecraft. Reference 1–16 lists over 20,000 satellites and categorizes them as satellites for communications, weather, navigation, scientific exploration, deep space probes, observation (including radar surveillance), reconnaissance, and other applications. Rocket propulsion is needed for both *primary propulsion* (i.e., acceleration along the flight path, such as for ascents, orbit insertion, or orbit change maneuvers) and for secondary propulsion functions in these vehicles. Some of the *secondary propulsion* functions are attitude control, spin control, momentum wheel and gyro unloading, rendezvous in space, stage separation, and for the settling of liquids in tanks. Spacecraft need a variety of different rocket propulsion systems and some thrusters can be very small. For spacecraft attitude control about three perpendicular axes, each in two rotational directions, the system must allow the application of pure torque for six modes of angular freedom, thus requiring a minimum of 12 thrusters. Some missions require as few as 4 to 6 thrusters, whereas the more complex manned spacecraft have 20 to 50 thrusters in all of their stages. Often, the small *attitude control rockets* must give pulses or short bursts of thrust, necessitating thousands of restarts. See Section 6.7 and Ref. 1–17.

A majority of spacecraft and space launch vehicles use liquid propellant engines for principal propulsion because of their better performance. Liquid propellants are used as both primary and secondary propulsion systems. A few vehicles have used solid propellant rocket motors for booster stages and some for orbit injection. Some spacecraft operate successfully with electrical propulsion for attitude control. Recently electrical propulsion (EP) systems have also been used for some primary and secondary spacecraft propulsion missions in long-duration space flights. Because of their low thrust, space operations/maneuvers with EP require relatively longer times to reach desired velocity increases. For example, transfer from a low Earth orbit to a geosynchronous orbit may take as long as two or three months compared to a few hours with chemical propulsion systems. See Chapter 17. Designs utilizing all-electric propulsion systems for upper stages are a relatively recent trend.

Micropropulsion is a new designation for thrust levels applicable to small spacecraft of less than 100 kg, or 220 lbm. See Ref. 1–18. It encompasses a variety of different propulsion concepts, such as certain very low thrust liquid mono- and bipropellant rocket engines, small gaseous propellant rocket engines, several types of electrical propulsion systems, and emerging advanced versions of these. Many are based on recent fabrication techniques for very small components (valves, thrusters, switches, insulators, or sensors) by micromachining and electromechanical processes.

Military and Other Applications

Military applications can be classified as shown in Table 1–4. Rocket propulsion for new U.S. missiles is presently based almost exclusively on solid propellant rocket motors. These can be *strategic missiles*, such as long-range ballistic missiles (800 to 9000 km range), which are aimed at military targets within an enemy country, or *tactical missiles*, which are intended to support or defend military ground forces, aircraft, or navy ships. Table 1–5 shows some preferred rocket propulsion systems for selected applications.

The term *surface launch* can mean launching from the ground, the ocean surface (from a ship), or from underneath the sea (submarine launch). Some tactical missiles, such as the air-to-surface short-range attack missile (SRAM), have a two-pulse solid propellant motor, where two separate insulated grains of different solid properties are in the same motor case; the time interval before starting the second pulse can be timed to control the flight path or speed profile. Many countries now have tactical missiles in their military inventories, and many of these countries have the capability to produce their own vehicles and their rocket propulsion systems.

Applications that were popular 40 to 70 years ago but are no longer active include liquid propellant rocket engines for propelling military fighter aircraft to altitude, assisted takeoff rocket engines and rocket motors, and superperformance rocket engines for augmenting the thrust of an aircraft jet engine.

TABLE 1–4. Selected Military Applications Using Rocket Propulsion Systems

Category	Vehicle/System	Comments/Examples
Military Satellites	Reconnaissance/Observation	Rely on existing vehicles, such as Delta IV or Delta II and Atlas V
	Secure communications	
Strategic Weapons	Early warning of ICBM launches	
	ICBM—silo launched	Minuteman III; Trident; SPRMs; 5000 to 9000 km range;
	ICBM—submarine launched	Tomahawk with SPRM booster and turbofan engine
	Cruise missile (subsonic flight)	
Surface-to-Surface Tactical Weapons	Intermediate range ballistic missile	Pershing II (2 stages)
	Battlefield support (very short range)	Guided or unguided
	Ship launched to ship or shore	All use SPRM and single-stage,
	Small shoulder-fired missile	Redeye is a surface-to-air shoulder-fired missile
	Wire-guided small missile	
Surface-to-Air and Surface-to-Incoming-Missile Tactical weapons	Local area defense (e.g., airfield)	Standard missile, Patriot missile,
	Large area defense	Multistage vehicle with SPRM
	Battlefield support	booster and LPRE divert top stage
	Shoulder-fired small missile	Guided or unguided SPRM, Redeye
	Ship defense	Unwinding wire to control flight path of local battlefield missile
Air-to-Surface Missiles	Short range	All use SPRMs
	Long range	Larger guided missiles
Air-to-Air Missiles	Carried under aircraft wings	Guided, SPRM, Phoenix, Sparrow
	Antitank missiles	With armor piercing warhead
	Antisubmarine missiles	Hawk, TOW, SPRM
	Anti-radar missiles	Subroc, SPRM
Specialized Weapons or Devices	Torpedo propulsion	Homing air launched SPRM
	Rocket-assisted artillery	Gas generator, SPRM
	Aircraft pilot seat ejection	Up to $20,000 g_0$ in gun barrel, increased range
		Emergency maneuver, SPRM

ICBM, intercontinental ballistic missile; LPRE, liquid propellant rocket engine; SPRM, solid propellant rocket motor.

TABLE 1–5. Selected Examples of Propulsion Characteristics of Rocket Applications

Application	Type of Propellant	Thrust Profile	Typical Firing Duration	Maximum Acceleration[a]
Large space launch vehicle	Liquid or cryogenic liquid	Nearly constant thrust 100,000 to 3,300,000 lbf	2–8 min	1.2–6 g_0
Strap-on booster	Solid or liquid		½ to 2 min	1.2 to 3 g_0
Spent strap-on stage separation	4 to 8 SPRMs	10,000 to 20,000 lbf each	Less than 1 sec	N. A.
Antiaircraft or antimissile-missile	Solid, some with liquid terminal divert stage	High thrust boost, decreasing thrust sustain phase; high thrust divert	2–75 sec each	5 to 20 g_0, but can be up to 100 g_0
Spacecraft orbit maneuvers and/or maintenance	Storable liquid or cryogenic liquid; electric propulsion	Multiple restarts in space; can be pulsed	Up to 10 min cumulative duration	0.2–6 g_0
Air-launched guided missile	Solid	High thrust boost phase with low thrust or decreasing thrust for sustain phase; sometimes 2 pulses	Boost: 2–5 sec / Sustain: 10–30 sec	Up to 25 g_0
Battlefield support—surface launched	Solid	Decreasing thrust	Up to 2 min each stage	Up to 10 g_0
Rocket-assisted projectile, gun launched	Solid	Constant or decreasing thrust	A few sec	Up to 20,000 g_0 in gun barrel
Spacecraft attitude control—large vehicles	Storable liquid (monopropellant or bipropellant); electric propulsion; xenon	Many restarts (up to several thousands); pulsing	Up to several hours cumulative duration	Less than 0.1 g_0
Spacecraft attitude control—small vehicle	Electric propulsion; Cold or warm gas or storable liquid.	Same	Up to several hours cumulative	Same
Reusable main engines for Space Shuttle	Cryogenic liquid (O_2/H_2)	Variable thrust, many flights with same engine	8 min, over 7 hr cumulative in several missions	
Lunar landing	Storable bipropellant	10:1 thrust variation	4 min	Several g_0
Weather sounding rocket	Solid	Single burn period—often decreasing thrust	5–30 sec	Up to 15 g_0
Antitank	Solid	Single burn period	0.2–3 sec	Up to 20 g_0

[a] g_0 is acceleration of gravity at the Earth's surface = 9.8066 m/sec^2 or 32.17 ft/sec^2.

Other applications of rocket propulsion systems to space operations include communication satellites—these have been successfully deployed for many years providing relays of telephone and television signals between Earth stations; this application is managed and operated by commercial organizations. Other countries also have their own satellite-based communications. The U.S. government sponsored programs that include weather satellites and the now ubiquitous Global Positioning System (GPS). Examples of space exploration include missions to the planets and/or into deep space, such as the Voyager mission and the MESSENGER program. A good number of these missions and probes were developed by NASA's Jet Propulsion Laboratory; they use the multiple thrusters from a monopropellant hydrazine reaction control system. Another application is for suborbital winged space vehicles for space tourism; as of 2014, two such vehicles (privately financed) have been flown in experimental versions (see Fig. 16–3).

REFERENCES

1–1. E. C. Goddard and G. E. Pendray (Eds), *The Papers of Robert H. Goddard*, three volumes, McGraw-Hill Book Company, 1970, 1707 pages. It includes the pioneering treatise "A Method of Reaching Extreme Altitudes" originally published as Smithsonian Miscellaneous Collections, Vol. 71, No. 2, 1919.

1–2. G. P. Sutton, *History of Liquid Propellant Rocket Engines*, published by AIAA, 2006, 911 pages.

1–3. B. N. Yur'yev (Ed), *Collected Works of K. E. Tsiolkowski*, Vols. 1–3, USSR Academy of Sciences, 1951; also NASA Technical Translation F-236, April 1965.

1–4. H. Oberth, *Die Rakete zu den Planetenräumen* (By Rocket to Planetary Space), R. Oldenburg, Munich, 1923 (in German), a classical text.

1–5. W. von Braun and F. Ordway, *History of Rocketry and Space Travel*, 3rd ed., Thomas Y. Crowell, New York, 1974.

1–6. L. H. Caveny, R. L. Geisler, R. A. Ellis, and T. L. Moore, "Solid Enabling Technologies and Milestones in the USA," *Journal of Propulsion and Power*, Vol. 19, No. 6, Nov.–Dec. 2003, AIAA, pp. 1038–1066.

1–7. A. M. Lipanov, "Historical Survey of Solid Propellant Rocket Development in Russia," *Journal of Propulsion and Power*, Vol. 19, No. 6. Nov.–Dec. 2003, pp. 1063–1088.

1–8. AGARD Lecture Series 150, *Design Methods in Solid Rocket Motors*, AGARD/NATO, Paris, April 1988.

1–9. A. Davenas, *Solid Rocket Propulsion Technology*, Pergamon Press, London (originally published in French), revised edition 1996.

1–10. C. Zakrzwski et al., "On-Orbit Testing of the EO–1 Pulsed Plasma Thruster." AIAA2002–3973, Reston, VA, 2002. http://eo1.gsfc.nasa.gov/new/validationReport/ Technology/Documents/Summaries/08-PPT_Rev-0.pdf

1–11. T. Nakamura et al., "Solar Thermal Propulsion for Small Spacecraft—Engineering System Development and Evaluation," Report PSI-SR-1228, July 2005.

1–12. T. Svitek et al., "Solar Sails as Orbit Transfer Vehicle—Solar Sail Concept Study," Phase II Report, AIAA paper 83–1347, 1983.

1–13. NASA *International Space Station* (A resource on the ISS by NASA; includes operational use, wide range of background material, archives, image gallery and planned missions). www.nasa.gov/station-34k

1–14. S. J. Isakowitz, J. B. Hopkins, and J. P. Hopkins, *International Reference Guide to Space Launch Systems*, 4th ed., AIAA, 2004, 596 pages.

1–15. National Aeronautics and Space Administration (NASA), *National Space Transportation System Reference*, Vol. 1, *Systems and Facilities*, U.S. Government Printing Office, Washington, DC, June 1988; a description of the Space Shuttle.

1–16. A. R. Curtis (Ed), *Space Satellite Handbook*, 3rd ed., Gulf Publishing Company, Houston, TX, 1994, 346 pages.

1–17. G. P. Sutton, "History of Small Liquid Propellant Thrusters," presented at the 52nd JANNAF Propulsion Meeting, May 2004, Las Vegas, NE, published by the Chemical Propulsion Information Analysis Center, Columbia, Maryland, June 2004.

1–18. M. M. Micci and A. D. Ketsdever, *Micropropulsion for Small Spacecraft*, Progress in Aeronautics and Astronautics Series, Vol. 187, AIAA, 2000, 477 pages.

CHAPTER 2

DEFINITIONS AND FUNDAMENTALS

This chapter deals with the definitions and basic relations for the propulsive force, exhaust velocity, and efficiencies related to creating and converting energy; comparisons of various propulsion systems and the simultaneous performance of multiple propulsion systems are also presented. The basic principles of rocket propulsion are essentially those of mechanics, thermodynamics, and chemistry. Propulsion is achieved by applying a force to a vehicle, that is, accelerating it or, alternatively, maintaining a given velocity against a resisting force. The propulsive force derives from momentum changes that originate from ejecting propellant at high velocities, and the equations in this chapter apply to all such systems. Symbols used in all equations are defined at the end of the chapter. Wherever possible, the American Standard letter symbols for rocket propulsion (as given in Ref. 2–1) are used.

2.1. DEFINITIONS

The *total impulse* I_t is found from the thrust force F (which may vary with time) integrated over the time of its application t:

$$I_t = \int_0^t F\,dt \qquad (2-1)$$

For constant thrust with negligibly short start and stop transients, this reduces to

$$I_t = Ft \qquad (2-2)$$

Total impulse I_t is essentially proportional to the total energy released by or into all the propellant utilized by the propulsion system.

The *specific impulse* I_s represents the thrust per unit propellant "weight" flow rate. It is an important figure of merit of the performance of any rocket propulsion system, a concept similar to miles per gallon parameter as applied to automobiles. A higher number often indicates better performance. Values of I_s are given in many chapters of this book and the concept of optimum specific impulse for a particular mission is introduced later. If the total propellant mass flow rate is $\dot{m}$ and the standard acceleration of gravity g_0 (with an *average* value at the Earth's sea level of 9.8066 m/sec^2 or 32.174 ft/sec^2), then

$$I_s = \frac{\int_0^t F\,dt}{g_0 \int_0^t \dot{m}\,dt} \tag{2-3}$$

This equation will give a time-averaged specific impulse value in units of "seconds" for any rocket propulsion system and is particularly useful when thrust varies with time. During transient conditions (during start or thrust buildup or shutdown periods, or during a change of flow or thrust levels) values of I_s may be obtained by either the integral above or by using average values for F and $\dot{m}$ for short time intervals. As written below, m_p represents the total effective propellant mass expelled.

$$I_s = I_t/(m_p g_0) \tag{2-4}$$

In Chapters 3, 12, and 17, we present further discussions of the specific impulse concept. For constant propellant mass flow $\dot{m}$, constant thrust F, and negligible start or stop transients Eq. 2–3 simplifies,

$$I_s = F/(\dot{m} g_0) = F/\dot{w} = I_t/w \tag{2-5}$$

At or near the Earth's surface, the product $m_p g_0$ is the effective propellant weight w, and its corresponding weight flow rate given by $\dot{w}$. But for space or in satellite outer orbits, mass that has been multiplied by an "arbitrary constant," namely, g_0 does not represent the weight. In the *Système International* (SI) or metric system of units, I_s is in "seconds." In the United States today, we still use the English Engineering (EE) system of units (foot, pound, second) for many chemical propulsion engineering, manufacturing, and test descriptions. In many past and some current U.S. publications, data, and contracts, the specific impulse has units of thrust (lbf) divided by weight flow rate of the propellants (lbf/sec), also yielding the unit of seconds; thus, the numerical values for I_s are the same in the EE and the SI systems. Note, however, that this unit for I_s does not indicate a measure of elapsed time but the thrust force per unit "weight flow rate." In this book, we use the symbol I_s exclusively for the specific impulse, as listed in Ref. 2–1. For solid propellants and other rocket propulsion systems, the symbol I_{sp} is more commonly used to represent specific impulse, as listed in Ref. 2–2.

In actual rocket nozzles, the exhaust velocity is not really uniform over the entire exit cross section and such velocity profiles are difficult to measure accurately. A uniform axial velocity c is assumed for all calculations which employ one-dimensional problem descriptions. This *effective exhaust velocity* c represents an average or mass-equivalent velocity at which propellant is being ejected from the rocket vehicle. It is defined as

$$c = I_s g_0 = F/\dot{m} \qquad (2-6)$$

It is given either in meters per second or feet per second. Since c and I_s only differ by a constant (g_0), either one can be used as a measure of rocket performance. In the Russian literature c is used in lieu of I_s.

In solid propellant rockets, it is difficult to measure propellant flow rate accurately. Therefore, in ground tests, the specific impulse is often calculated from total impulse and the propellant weight (using the difference between initial and final rocket motor weights and Eq. 2–5). In turn, the total impulse is obtained from the integral of the measured thrust with time, using Eq. 2–1. In liquid propellant units, it is possible to measure thrust and instantaneous propellant flow rate and thus Eq. 2–3 is used for the calculation of specific impulse. Equation 2–4 allows yet another interpretation for specific impulse, namely, the amount of total impulse imparted to a vehicle per total sea-level weight of propellant expended.

The term *specific propellant consumption* corresponds to the reciprocal of the specific impulse and is not commonly used in rocket propulsion. It is used in automotive and air-breathing duct propulsion systems. Typical values are listed in Table 1–2.

The mass ratio **MR** of the total vehicle or of a particular vehicle stage or of the propulsion system itself is defined to be the final mass m_f divided by the mass before rocket operation, m_0. Here, m_f consists of the mass of the vehicle or stage after the rocket has ceased to operate when all the useful propellant mass m_p has been consumed and ejected. The various terms are depicted in Fig. 4–1.

$$\mathbf{MR} = m_f/m_0 \qquad (2-7)$$

This equation applies to either a single or to a multistage vehicle; for the latter, the overall mass ratio is the product of the individual vehicle stage mass ratios. The final vehicle mass m_f has to include components such as guidance devices, navigation gear, the payload (e.g., scientific instruments or military warheads), flight control systems, communication devices, power supplies, tank structures, residual propellants, along with all the propulsion hardware. In some vehicles it may also include wings, fins, a crew, life support systems, reentry shields, landing gears, and the like. Typical values of **MR** can range from 60% for some tactical missiles down to 10% for some unmanned launch vehicle stages. This mass ratio is an important parameter for analyzing flight performance, as explained in Chapter 4. When the **MR** applies only to a single lower stage, then all its upper stages become part of its "payload." It is important to specify when the **MR** applies either to a multiple-stage vehicle, to a single stage, or to a particular propulsion system.

The *propellant mass fraction* ζ indicates the ratio of the useful propellant mass m_p to the initial mass m_0. It may apply to a vehicle, or a single stage, or to an entire rocket propulsion system.

$$\zeta = m_p/m_0 \tag{2-8}$$

$$\zeta = (m_0 - m_f)/m_0 = m_p/(m_p + m_f) \tag{2-9}$$

$$m_0 = m_p + m_f \tag{2-10}$$

Like the mass ratio **MR**, the propellant fraction ζ is used to describe a rocket propulsion system; its values will differ when applied to an entire vehicle, single or multistage. For a rocket propulsion system, the initial or loaded mass m_0 consists of the inert propulsion mass (the hardware necessary to burn and store the propellant) and the effective propellant mass. It would exclude masses of nonpropulsive components, such as payload or guidance devices (see Fig. 4–1). For example, in liquid propellant rocket engines the final or inert propulsion mass m_f would include propellant tanks, their feed and empty pressurization system (a turbopump and/or gas pressure system), one or more thrust chambers, various piping, fittings and valves, engine mounts or engine structures, filters, and some sensors. Any *residual or unusable remaining propellant* is normally considered to be part of the final inert mass m_f, as in this book. However, some rocket propulsion manufacturers and some literature assign residuals to be part of the propellant mass m_p. When applied to an entire rocket propulsion system, the value of the propellant mass fraction ζ indicates the quality of the design; a value of, say, 0.91 means that only 9% of the mass is inert rocket hardware, and this small fraction is needed to contain, feed, and burn the substantially larger mass of propellant; high values of ζ are desirable.

The *impulse-to-weight* ratio of a complete propulsion system is defined as the total impulse I_t divided by the initial (propellant-loaded) vehicle sea-level weight w_0. A high value indicates an efficient design. Under our assumptions of constant thrust and negligible start and stop transients, it can be expressed as

$$\frac{I_t}{w_0} = \frac{I_t}{(m_f + m_p)g_0} = \frac{I_s}{m_f/m_p + 1} \tag{2-11}$$

The *thrust-to-weight* ratio F/w_0 expresses the acceleration (in multiples of the Earth's surface acceleration of gravity, g_0) that an engine is capable of giving to its own loaded propulsion system mass. Values of F/w_0 are given in Table 2–1. The *thrust-to-weight ratio* is useful in comparing different types of rocket propulsion systems and/or in identifying launch capabilities. For constant thrust the maximum value of the thrust-to-weight ratio, or maximum acceleration, invariably occurs just before thrust termination (i.e., burnout) because the vehicle's mass has been diminished by the mass of useful propellant.

TABLE 2–1. Ranges of Typical Performance Parameters for Various Rocket Propulsion Systems

Engine Type	Specific Impulse[a] (sec)	Maximum Temperature (°C)	Thrust-to-Weight Ratio[b]	Propulsion Duration	Specific Power[c] (kW/kg)	Typical Working Fluid	Status of Technology
Chemical—solid or liquid bipropellant, or hybrid	200–468	2500–4100	10^{-2}–100	Seconds to a few minutes	10^{-1}–10^3	Liquid or solid propellants	Flight proven
Liquid monopropellant	194–223	600–800	10^{-1}–10^{-2}	Seconds to minutes	0.02–200	N_2H_4	Flight proven
Resistojet	150–300	2900	10^{-2}–10^{-4}	Days	10^{-3}–10^{-1}	H_2, N_2H_4	Flight proven
Arc heating—electrothermal	280–800	20,000	10^{-4}–10^{-2}	Days	10^{-3}–1	N_2H_4, H_2, NH_3	Flight proven
Electromagnetic including pulsed plasma (PP)	700–2500	—	10^{-6}–10^{-4}	Weeks	10^{-3}–1	H_2 Solid for PP	Flight proven
Hall effect	1220–2150	—	10^{-4}	Weeks	10^{-1}–5×10^{-1}	Xenon	Flight proven
Ion—electrostatic	1310–7650	—	10^{-6}–10^{-4}	Months, years	10^{-3}–1	Xenon	Flight proven
Solar heating	400–700	1300	10^{-3}–10^{-2}	Days	10^{-2}–1	H_2	In development

[a] At $p_1 = 1000$ psia and optimum gas expansion at sea level ($p_2 = p_3 = 14.7$ psia).
[b] Ratio of thrust force to full propulsion system sea level weight (with propellants, but without payload).
[c] Kinetic power per unit exhaust mass flow.

Example 2–1.   A rocket vehicle has the following characteristics:

Initial mass	200 kg
Mass after rocket operation	130 kg
Payload, nonpropulsive structure, etc.	110 kg
Rocket operation duration	3.0 sec
Average specific impulse of propellant	240 sec

Determine the mass ratios for the vehicle and its propulsive unit, the propellant mass fraction for the vehicle and for its propulsive unit, the effective exhaust velocity, and the total impulse. Also, sensitive electronic equipment in the payload would limit the maximum acceleration to 35 g_0's. Is this exceeded during the flight? Assume constant thrust and neglect the short start and stop transients as well the force of gravity (which is being balanced by the lift of aerodynamic wings).

SOLUTION.   The difference between initial mass and that after rocket operation is the mass of the expended propellant, namely, 70 kg. The mass ratio (Eq. 2–8) for the vehicle is therefore **MR** $= m_f/m_0 = 130/200 = 0.65$, and for the propulsion unit it is $(130 - 110)/(200 - 110) = 0.222$; note that these are different. The propellant mass fraction of the propulsion system is (Eq. 2–8)

$$\zeta = 70/(200 - 110) = 0.778$$

This mass fraction is acceptable for tactical missiles but might be only "fair" for spacecraft. The effective exhaust velocity c is proportional to the specific impulse, Eq. 2–6, $c = 240 \times 9.81 = 2354.4$ m/sec. The propellant mass flow rate is $\dot{m} = 70/3 = 23.3$ kg/sec and the thrust becomes (Eq. 2–6)

$$F = I_s g_0 = 23.3 \times 2354.4 = 54{,}936 \text{ N}$$

Now the total impulse can be calculated, $I_t = 54{,}936 \times 3 = 164{,}808$ N- sec. The impulse-to-weight ratio (Eq. 2–11) for the propulsion system at the Earth's surface is $I_t/w_0 = 164{,}808/[(200 - 110)9.81] = 187$ sec (this might only reflect a "fair design" because a "good design" would approach a value closer to the specific impulse, $I_s = 240$ sec).

For a horizontal trajectory, the maximum acceleration is found at the end of the thrusting schedule, just before shutdown (because while the thrust is unchanged the vehicle mass is now at its minimum value of 130 kg).

$$\text{Final acceleration} = F/m = 54{,}936/130 = 422.58 \text{ m/sec}^2$$

This value represents 43.08 g_0's, at the Earth's surface and it exceeds the stated 35 g_0 limit of the equipment. This can be remedied by changing to a lower thrust along with a longer flight time.

2.2. THRUST

Thrust is the force produced by the rocket propulsion system acting at the vehicle's center of mass. It is a reaction force, experienced by vehicle's structure from the ejection of propellant at high velocities (the same phenomenon that pushes a garden hose backward or makes a gun recoil). Momentum is a vector quantity defined as the product of mass times its vector velocity. In rocket propulsion, relatively small amounts of propellant mass *carried within* the vehicle are ejected at high velocities.

The thrust, due to a change in momentum, is shown below (a derivation can be found in earlier editions of this book). Here, the exit gas velocity is assumed constant, uniform, and purely axial, and when the mass flow rate is constant, thrust is itself constant. Often, this idealized thrust can actually be close to the actual thrust.

$$F = \frac{d(mv_2)}{dt} = \dot{m}v_2 \text{ at sea level} = \frac{\dot{w}}{g_0}v_2 \qquad (2\text{–}12)$$

But this force represents the total propulsive force only when the nozzle exit pressure equals the ambient pressure.

The pressure of the surrounding fluid (i.e., the local atmosphere) gives rise to the second component of thrust. Figure 2–1 shows a schematic of an external uniform pressure acting on the outer surfaces of a rocket chamber as well as the changing gas pressures inside of a typical thermal rocket propulsion system. The direction and length of the arrows indicates the relative magnitude of the pressure forces. Axial thrust can be determined by integrating all the pressures acting on areas that have a projection on the plane normal to the nozzle axis. Any forces acting radially outward may be appreciable but do not contribute to the thrust when rocket chambers are axially symmetric.

For a fixed nozzle geometry, changes in ambient pressure due to variations in altitude during flight result in imbalances between the atmospheric pressure p_3 and the propellant gas pressure p_2 at the exit plane of the nozzle. For steady operation in a

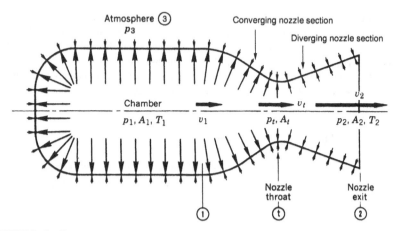

FIGURE 2–1. Gas pressures on the chamber and nozzle-interior walls are not uniform. The internal pressure (indicated by length of arrows) is highest in the chamber (p_1) and decreases steadily in the nozzle until it reaches the nozzle exit pressure p_2. The external or atmospheric pressure p_3 is uniform. At the throat the pressure is p_t. The four subscripts (shown inside circles) are used to identify quantities such as A, v, T, and p at those specific locations. The centerline horizontal arrows denote relative velocities.

homogeneous atmosphere, the total thrust can be shown to equal (this equation is derived using the *control volume* approach in gas dynamics, see Refs. 2–3 and 2–4):

$$F = \dot{m}v_2 + (p_2 - p_3)A_2 \qquad (2\text{--}13)$$

The first term is the *momentum thrust* given by the product of the propellant mass flow rate and its exhaust velocity relative to the vehicle. The second term represents the *pressure thrust*, consisting of the product of the cross-sectional area at the nozzle exit A_2 (where the exhaust jet leaves the vehicle) and the difference between the gas pressure at the exit and the ambient fluid pressure (p_2 may differ from p_3 only in supersonic nozzle exhausts). When the exit gas pressure is less than the surrounding fluid pressure, the pressure thrust is negative. Because this condition gives a lower thrust and is undesirable for other reasons (discussed in Chapter 3), rocket nozzles are usually designed so that their exhaust pressures equal or are slightly higher than the ambient pressure.

When the ambient or atmospheric pressure equals the exhaust pressure, the pressure term is zero and the thrust is the same as in Eq. 2–12. In the vacuum of space $p_3 = 0$ and the pressure thrust becomes a maximum,

$$F = \dot{m}v_2 + p_2 A_2 \qquad (2\text{--}14)$$

Most nozzles are have their area ratio A_2/A_t designed so that their exhaust pressure will equal the surrounding air pressure (i.e., $p_2 = p_3$) somewhere at or above sea level. For any fixed nozzle configuration this can occur only at one altitude, and this location is referred to as nozzle operation at its *optimum expansion ratio*; this case is treated in Chapter 3.

Equation 2–13 shows that thrust is independent of a rocket unit's flight velocity. Because changes in ambient pressure affect the pressure thrust, rocket thrust varies noticeably with altitude. Since atmospheric pressure decreases with increasing altitude, thrust and specific impulse will increase as the vehicle reaches higher altitudes. On Earth, this increase in pressure thrust due to altitude changes can amount to between 10 and 30% of the sea-level thrust; the fact that as flights gain altitude, atmospheric pressure continuously diminishes causing the thrust to increase is a unique feature of rocket propulsion systems. Table 8–1 shows the sea level and high-altitude thrust for several liquid rocket engines. Appendix 2 gives the properties of the *standard atmosphere* (the ambient pressure, p_3) as a function of altitude.

2.3. EXHAUST VELOCITY

The *effective exhaust velocity* as defined by Eq. 2–6 applies to all mass-expulsion thrusters. From Eq. 2–13 and for constant propellant mass flow it can be modified to give the equation below. As before, g_0 is a constant whose numerical value equals the average acceleration of gravity at sea level and does not vary with altitude.

$$c = v_2 + (p_2 - p_3)A_2/\dot{m} = I_s g_0 \qquad (2\text{--}15)$$

In Eq. 2–6 the value of c may be determined from measurements of thrust and propellant flow. When $p_2 = p_3$, the value of the effective exhaust velocity c equals v_2, the average actual nozzle exhaust velocity of the propellant gases. But even when $p_2 \neq p_3$ and $c \neq v_2$, the second term of the right-hand side of Eq. 2–15 usually remains small in relation to v_2; thus, the effective exhaust velocity always stays relatively close in value to the actual exhaust velocity. At the Earth's surface and when $c = v_2$, the thrust, from Eq. 2–13, can be simply written as (the second version below is less restricted)

$$F = (\dot{w}/g_0)v_2 = \dot{m}c \qquad (2-16)$$

The *characteristic velocity* c^*, pronounced "cee-star," is a term frequently used in the rocket propulsion literature. It is defined as

$$c^* = p_1 A_t / \dot{m} \qquad (2-17)$$

The characteristic velocity c^*, though not a physical velocity, is used for comparing the relative performance of different chemical rocket propulsion system designs and propellants; it may be readily determined from measurements of $\dot{m}$, p_1, and A_t. Being essentially independent of nozzle characteristics, c^* may also be related to the efficiency of the combustion process. However, the specific impulse I_s and the effective exhaust velocity c remain functions of nozzle geometry (such as the nozzle area ratio A_2/A_t, as shown in Chapter 3). Some typical values of I_s and c^* are given in Tables 5–5 and 5–6.

Example 2–2.   The following measurements were made in a sea-level test of a solid propellant rocket motor (all cross sections are circular and unchanging):

Burn duration	40 sec
Initial propulsion system mass	1210 kg
Mass of rocket motor after test	215 kg
Sea-level thrust	62,250 N
Chamber pressure	7.00 MPa
Nozzle exit pressure	70.0 kPa
Nozzle throat diameter	8.55 cm
Nozzle exit diameter	27.03 cm

Determine $\dot{m}$, v_2, c^*, and c at sea level. Also, determine the *pressure thrust* (Eq. 2–13) and the specific impulse at sea level, 1000 m, and 25,000 m altitude. Assume that the momentum thrust is invariant during the rocket ascent, and that start and stop transients can be neglected.

SOLUTION.   The cross-sectional areas corresponding to the given diameters are $A_t = 0.00574$ and $A_2 = 0.0574\,\text{m}^2$ (i.e., an area ratio of 10). The steady-state mass flow rate for all altitudes is

$$\dot{m} = (1210 - 215)/40 = 24.88 \text{ kg/sec}$$

The desired c^* (all altitudes) and c at sea level follow from Eqs. 2–17 and 2–6:

$$c^* = p_1 A_t / \dot{m} = 7.00 \times 10^6 \times 0.00574/24.88 = 1615 \text{ m/sec}$$

$$c = F/\dot{m} = 62,250/24.88 = 2502 \text{ m/sec}$$

The *pressure thrust* at sea level is negative:

$$(p_2 - p_3)A_2 = (0.070 - 0.1013) \times 10^6 \times 0.0574 = -1797 \text{ N}$$

so the nozzle exit velocity v_2 becomes (for all altitudes), from Eq. 2–13,

$$v_2 = (62{,}250 + 1797)/24.88 = 2574 \text{ m/sec}$$

The remaining answers are shown in the table below. We obtain the ambient pressures from Appendix 2.

Altitude	p_3 (kPa)	Pressure Thrust (N)	I_s (sec)
Sea level	101.32	−1797	255
1000 m	89.88	−1141	258
25,000 m	2.55	3871	278

Further examination of the standard atmosphere table in Appendix 2 reveals that the *pressure thrust* becomes zero at just under 3000 m.

2.4. ENERGY AND EFFICIENCIES

Although efficiencies are not commonly used directly in designing rocket propulsion systems, they permit an understanding of the energy balance in these systems. Their definitions arbitrarily depend on the losses considered and any consistent set of efficiencies, such as the set presented in this section, is satisfactory in evaluating energy losses. As stated previously, two types of energy conversion processes occur in all propulsion systems, namely, the production of energy, which is most often the conversion of stored energy into available energy and, subsequently, its conversion to the form with which a reaction thrust can be obtained. The kinetic energy of ejected matter is the main useful form of energy for propulsion. The *power of the jet* P_{jet} is the time derivative of this energy, and for a constant gas exit velocity ($v_2 \approx c$ and at sea level), this is a function of I_s and F:

$$P_{jet} = \frac{1}{2}\dot{m}v_2^2 = \frac{1}{2}\dot{w}g_0 I_s^2 = \frac{1}{2}Fg_0 I_s = \frac{1}{2}Fv_2 \tag{2-18}$$

The term *specific power* is sometimes used as a measure of the utility of the mass of the propulsion system, including its power source; it equals the jet power divided by the loaded propulsion system mass, P_{jet}/m_0. For electrical propulsion systems that need to include a massive, relatively inefficient energy source, specific power can be much lower than that for chemical rockets. The source of energy input to a rocket propulsion system is different in different thruster types; for chemical rockets this energy is created solely by combustion. The maximum energy available in chemical propellants is their heat of combustion per unit of propellant mass Q_R; the *power input to a chemical engine* is given by

$$P_{chem} = \dot{m}Q_R J \tag{2-19}$$

where J is a units-conversion constant (see Appendix 1). A significant portion of the energy may leave the nozzle as residual enthalpy in the exhaust gases and is unavailable for conversion into kinetic energy. This is analogous to the energy lost in the hot exhaust gases of internal combustion engines.

The *combustion efficiency* η_{comb} for chemical rockets is the ratio of the actual energy released to the ideal heat of reaction per unit of propellant mass and represents a measure of the source efficiency. Its value can be high (nearly 94 to 99%). When the power input P_{chem} is multiplied by the combustion efficiency, it becomes the power available to the propulsive device, where it is then converted into the kinetic power of the exhaust jet. In electric propulsion the analogous efficiency is the power conversion efficiency; with solar cells this efficiency has a relatively low value because it depends on the efficiency of converting solar radiation energy into electric power (presently between 10 and 30%).

The *power transmitted to the vehicle* at any one instant of time is defined in terms of the product of the thrust of the propulsion system F and the *vehicle velocity* u:

$$P_{vehicle} = Fu \qquad (2-20)$$

The *internal efficiency* of a rocket propulsion system reflects the effectiveness of converting the system's input energy to the propulsion device into kinetic energy of ejected matter; for example, for a chemical unit it is the ratio of the kinetic power of the ejected gases expressed by Eq. 2–18 divided by the power input of the chemical reaction, Eq. 2–19. The energy balance diagram for a chemical rocket (Fig. 2–2) shows typical losses. The internal efficiency η_{int} may be expressed as

$$\eta_{int} = \frac{\text{kinetic power in jet}}{\text{available chemical power}} = \frac{\frac{1}{2}\dot{m}v^2}{\eta_{comb}P_{chem}} \qquad (2-21)$$

Any object moving through a fluid medium affects the fluid (i.e., stirs it) in ways that may hinder its motion and/or require extra energy expenditures. This is one consequence of skin friction, which can be substantial. The *propulsive efficiency* η_p (Fig. 2–3) reflects this energy cost for rocket vehicles. The equation that determines how much exhaust kinetic energy is useful for propelling a rocket vehicle is defined as

$$\eta_p = \frac{\text{vehicle power}}{\text{vehicle power} + \text{residual kinetic jet power}}$$

$$= \frac{Fu}{Fu + \frac{1}{2}\dot{m}(c-u)^2} = \frac{2u/c}{1+(u/c)^2} \qquad (2-22)$$

where F is the thrust, u the absolute vehicle velocity, c the effective rocket exhaust velocity with respect to the vehicle, $\dot{m}$ the propellant mass flow rate, and η_p the desired propulsive efficiency. This propulsive efficiency becomes one when the forward vehicle velocity is exactly equal to the effective exhaust velocity; here any residual kinetic energy becomes zero and the exhaust gases effectively stand still in space.

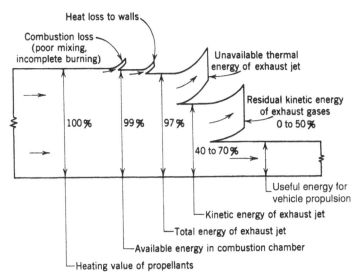

FIGURE 2–2. Typical energy distribution diagram for a chemical rocket.

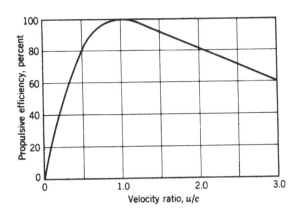

FIGURE 2–3. Propulsive efficiency at varying velocities.

While it is desirable to use energy economically and thus have high efficiencies, there is also the need for minimizing the expenditure of ejected mass, which in many cases is more important than minimizing the energy. With nuclear-reactor energy and some solar-energy sources, for example, there is an almost unlimited amount of heat energy available; yet the vehicle can carry only a limited amount of propellant. For a given thrust, economy of mass expenditures in the working fluid can be obtained when the exhaust velocity is high. Because the specific impulse is proportional to the exhaust velocity, it therefore measures this propellant mass economy.

2.5. MULTIPLE PROPULSION SYSTEMS

The relationships below are used for determining the total or overall (subscript oa) thrust and the overall mass flow of propellants for a group of propulsion systems (two or more) firing in parallel (i.e., in the same direction at the same time). These relationships apply to liquid propellant rocket engines, solid propellant rocket motors, electrical propulsion systems, hybrid propulsion systems, and any combinations of these. Many space launch vehicles and larger missiles use multiple propulsion systems. The Space Shuttle, for example, had three large liquid engines and two large solid motors firing jointly at liftoff.

Overall thrust, F_{oa}, is needed for determining the vehicle's flight path and overall mass flow rate, $\dot{m}_{oa}$, is needed for determining the vehicle's mass; together they determine the overall specific impulse, $(I_s)_{oa}$:

$$F_{oa} = \sum F = F_1 + F_2 + F_3 + \ldots \qquad (2\text{-}23)$$

$$\dot{m}_{oa} = \sum \dot{m} = \dot{m}_1 + \dot{m}_2 + \dot{m}_3 + \ldots \qquad (2\text{-}24)$$

$$(I_s)_{oa} = F_{oa}/(g_0 \dot{m}_{oa}) \qquad (2\text{-}25)$$

For liquid propellant rocket engines with a turbopump and a gas generator, there is a separate turbine outlet flow that is usually dumped overboard through a pipe and a nozzle (see Fig. 1–4), which needs to be included in the equations above and this is treated in Examples 2–3 and 11–1.

Example 2–3. The MA-3 multiple liquid engine for the Atlas 2 missile had two booster engines ($F = 165,000$ lbf at sea level and $\dot{m} = 667$ lbm/sec each). The turbine exhaust gas of both boosters adds an extra 2300 lbf of thrust at a propellant flow rate of about 33 lbm/sec. These booster engines are dropped from the vehicle after about 145 sec operation. The central sustainer rocket engine, which also starts at liftoff, operates for a total 300 sec. It has a sea-level thrust of 57,000 lbf and altitude thrust of 70,000 lbf both at a mass flow of 270.5 lbm/sec. Its turbine exhaust gases are aspirated into the nozzle of the engine and do not directly contribute to the thrust. There are also two small vernier engines used for roll control of the vehicle during the sustainer-only portion of the flight. They each have an altitude thrust of 415 lbf with a propellant flow of 2.13 lbm/sec. Determine the overall thrust F_{oa} and the overall mass flow rate $\dot{m}_{oa}$ at liftoff and at altitude, when all the nozzles are pointing vertically down.

SOLUTION. Use Eqs. 2–23 and 2–24. The thrust from the turbine exhaust must be included in Eq. 2–23. At sea-level,

$$F_{oa} = 165,000 \times 2 + 2300 + 57,000 = 389,300 \text{ lbf}$$

$$\dot{m}_{oa} = 667 \times 2 + 33 + 270.5 = 1637.5 \text{ lbm/sec}$$

At altitude (vacuum),

$$F_{oa} = 70,000 + 2 \times 415 = 70,830 \text{ lbf}$$

$$\dot{m}_{oa} = 270.5 + 2 \times 2.13 = 274.76 \text{ lbm/sec}$$

2.6. TYPICAL PERFORMANCE VALUES

Typical values of representative performance parameters for different types of rocket propulsion are given in Table 2–1 and in Fig. 2–4.

Chemical rocket propulsion systems have relatively low values of specific impulse, relatively light machinery (i.e., low engine total mass), with very high thrust capabilities, and therefore can provide high acceleration and high specific power. At the other extreme, ion-propulsion thrusters have a very high specific impulse, but they carry massive electrical power systems to deliver the power necessary for such high ejection velocities. The very low acceleration potential for electrical propulsion units (and for others using solar energy) usually requires long accelerating periods and thus these systems are best suited for missions where powered flight times can be long. The low thrust magnitudes of electrical systems also imply that they are not useful in fields of strong gravitational gradients (e.g., for Earth takeoff or landing) but are best suited for flight missions in space.

Performance of rocket propulsion systems also depends on their application. Typical applications are shown in several chapters of this book (see Index).

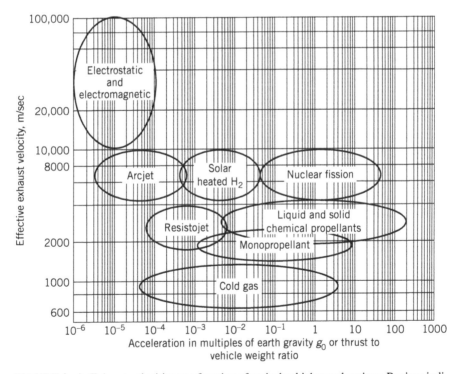

FIGURE 2–4. Exhaust velocities as a function of typical vehicle accelerations. Regions indicate approximate performance values for different types of propulsion systems. The mass of the vehicle includes the propulsion system, but the payload is assumed to be zero. Nuclear fission is shown only for comparison as it is not being pursued.

Chemical systems (solid and liquid propellant rockets) are presently the most developed and are widely used for many different vehicle applications. They are described in Chapters 5 to 16. Electrical propulsion has also been in operation in many space flight applications (see Chapter 17). Other types are still in their exploratory or development phase.

The accelerations shown in Figure 2–4 are directly related to the magnitudes of the thrust that is applied by the various rocket propulsion systems depicted. The effective exhaust velocities are related the required jet power per unit thrust (see Eq. 2–18) and the resulting input power (which is proportional to the square of the velocity and depends on the internal efficiency, Eq. 2–21) would turn out to be very high for electric propulsion systems unless the thrust itself is reduced to comparatively low values. As of 2015, electrical power levels of about 100 kW are expected to be available in space.

In Figure 2–4, hybrid rocket propulsion systems (see Chapter 16) are not shown separately because they are included as part of liquid and solid chemical propellants. Compressed (cold) gases stored at ambient temperatures have been used for many years for roll control in larger vehicles, for complete attitude control of smaller flight vehicles, and in rocket toys. Nuclear fission rocket propulsion systems as shown in Figure 2–4 represent analytical estimates; their development has been stopped. Hall and ion thrusters are part of the electrostatic and electromagnetic entry in this figure; they have flown successfully in many space applications (see Chapter 17).

2.7. VARIABLE THRUST

Most operational rocket propulsion systems have essentially constant propellant mass flow producing constant thrust or slightly increasing thrust with altitude. Only some flight missions require large thrust changes during flight; Table 2–2 shows several applications; the ones that require randomly variable thrust use predominantly liquid propellant rocket engines. Some applications require high thrust during a short initial period followed by a pre-programmed low thrust for the main flight portion (typically 20 to 35% of full thrust); these use predominantly solid propellant rocket motors. Section 8.8 describes how liquid propellant rocket engines can be designed and controlled for randomly variable thrust. Section 12.3 explains how the grain of solid rocket motors can be designed to give predetermined thrust changes.

Some solid and liquid propellant experimental propulsion systems have used variable nozzle throat areas (achieved with a variable position "tapered pintle" at the nozzle throat) and one experimental version has flown. To date, there has been no published information on production and implementation of such systems.

TABLE 2–2. Applications of Variable Thrust

Application	Type*	L/S*	Comment
1. Vertical ascent through atmosphere of large booster stage	AB	L	Reduced thrust avoids excessive aerodynamic pressure on vehicle
2. Short range tactical surface-to-surface missile	B	S	100 % initial thrust, 20 to 35% thrust for sustaining portion of flight
3. Tactical surface-to-air defensive missile	B	S	Same as # 2
4. Aircraft pilot emergency seat ejection capsule	B	S	Rapid ejection to get away from aircraft to go to higher altitude to deploy parachute
5. Soft landing on planet or moon, "retro-firing"	A	L	Thrust can be reduced by a factor of 10 with automatic landing controls
6. Top stage of multistage area defense missile against attacking ballistic missile	A or B	L, S	Axial thrust, side thrust, and attitude control thrust to home in on predicted vehicle impact point
7. Sounding rocket or weather rocket (vertical ascent)	B	S	Programmed two-thrust levels for many, but not all such rockets

*A   Random variable thrust.
*B   Preprogrammed (decreasing) thrust profile.
*L   Liquid propellant rocket engine.
*S   Solid propellant rocket motor.

SYMBOLS

A area, m^2 (ft^2)

A_t nozzle throat area, m^2 (ft^2)

A_2 exit area of nozzle, m^2 (ft^2)

c effective exhaust velocity, m/sec (ft/sec)

c^* characteristic velocity, m/sec (ft/sec)

E energy, J (ft-lbf)

F thrust force, N (lbf)

F_{oa} overall force, N (lbf)

g_0 average sea-level acceleration of gravity, 9.81 m/sec^2 (32.2 ft/sec^2), [at equator 9.781, at poles 9.833 m/sec^2]

I_s specific impulse, sec

$(I_s)_{oa}$ overall specific impulse, sec

I_t impulse or total impulse, N-sec (lbf-sec)

J conversion factor or mechanical equivalent of heat, 4.184 J/cal or 1055 J/Btu or 778 ft-lbf/Btu

m	mass, kg (slugs, 1 slug = mass of a 32.174 lb-weight at sea level)
m_{oa}	overall mass, kg
$\dot{m}$	mass flow rate, kg/sec (lbm/sec)
m_f	final mass (after rocket propellant is ejected), kg (lbm or slugs)
m_p	propellant mass, kg (lbm or slugs)
m_0	initial mass (before rocket propellant is ejected), kg (lbm or slugs)
MR	mass ratio (m_f/m_0)
p	pressure, pascal [Pa] or N/m^2 (lbf/ft^2)
p_3	ambient or atmospheric pressure, Pa (lbf/ft^2)
p_2	rocket gas pressure at nozzle exit, Pa (lbf/ft^2)
p_1	chamber pressure, Pa (lbf/ft^2)
P	power, J/sec (ft-lbf/sec)
P_s	specific power, J/sec-kg (ft-lbf/sec-lbm)
Q_R	heat of reaction per unit propellant, J/kg (Btu/lbm)
t	time, sec
u	vehicle velocity, m/sec (ft/sec)
v_2	gas velocity leaving the nozzle, m/sec (ft/sec)
w	weight, N or kg-m/sec^2 (lbf)
$\dot{w}$	weight flow rate, N/sec (lbf/sec)
w_0	initial weight, N or kg-m/sec^2 (lbf)

Greek Letters

ζ	propellant mass fraction
η	efficiency
η_{comb}	combustion efficiency
η_{int}	internal efficiency
η_p	propulsive efficiency

PROBLEMS

When solving problems, three appendixes (see end of book) may be helpful:

Appendix 1. Conversion Factors and Constants
Appendix 2. Properties of the Earth's Standard Atmosphere
Appendix 3. Summary of Key Equations

1. A jet of water hits a stationary flat plate in the manner shown below.

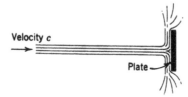

a. If 50 kg per minute flows at an absolute velocity of 200 m/sec, what will be the force on the plate?

b. What will this force be when the plate moves in the flow direction at $u = 50$ km/h? Explain your methodology.

Answers: 167 N; 144 N.

2. The following data are given for a certain rocket unit: thrust, 8896 N; propellant consumption, 3.867 kg/sec; velocity of vehicle, 400 m/sec; energy content of propellant, 6.911 MJ/kg. Assume 100% combustion efficiency.

Determine **(a)** the effective velocity; **(b)** the kinetic jet energy rate per unit flow of propellant; **(c)** the internal efficiency; **(d)** the propulsive efficiency; **(e)** the overall efficiency; **(f)** the specific impulse; **(g)** the specific propellant consumption.

Answers: **(a)** 2300 m/sec; **(b)** 2.645 MJ/kg; **(c)** 38.3%; **(d)** 33.7%; **(e)** 13.3%; **(f)** 234.7 sec; **(g)** 0.00426 sec^{-1}.

3. A certain rocket engine (flying horizontally) has an effective exhaust velocity of 7000 ft/sec; it consumes 280 lbm/sec of propellant mass, and liberates 2400 Btu/lbm. The unit operates for 65 sec. Construct a set of curves plotting the propulsive, internal, and overall efficiencies versus the velocity ratio u/c ($0 < u/c < 1.0$). The rated flight velocity equals 5000 ft/sec. Calculate **(a)** the specific impulse; **(b)** the total impulse; **(c)** the mass of propellants required; **(d)** the volume that the propellants occupy if their average specific gravity is 0.925. Neglect gravity and drag.

Answer: **(a)** 217.4 sec; **(b)** 3,960,000 lbf-sec; **(c)** 18,200 lbm; **(d)** 315 ft^3.

4. For the rocket in Problem 2, calculate the specific power, assuming a propulsion system dry mass of 80 kg and a duration of 3 min.

5. A Russian rocket engine (RD-110 with LOX-kerosene) consists of four thrust chambers supplied by a single turbopump. The exhaust from the turbine of the turbopump then is ducted to four vernier nozzles (which can be rotated to provide some control of the flight path). Using the information below, determine the thrust and mass flow rate of the four vernier gas nozzles. For individual thrust chambers (vacuum):

$$F = 73.14\,\text{kN}, c = 2857\,\text{m/sec}$$

For overall engine with verniers (vacuum):

$$F = 297.93\,\text{kN}, c = 2845\,\text{m/sec}$$

Answers: 5.37 kN, 2.32 kg/sec.

6. A certain rocket engine has a specific impulse of 250 sec. What range of vehicle velocities (u, in units of ft/sec) would keep the propulsive efficiencies at or greater than 80%. Also, how could rocket–vehicle staging be used to maintain these high propulsive efficiencies for the range of vehicle velocities encountered during launch?

Answers: 4021 to 16,085 ft/sec; design upper stages with increasing I_s.

7. For a solid propellant rocket motor with a sea-level thrust of 207,000 lbf, determine: (a) the (constant) propellant mass flow rate $\dot{m}$ and the specific impulse I_s at sea level, (b) the altitude for optimum nozzle expansion as well as the thrust and specific impulse at this

optimum condition and (c) at vacuum conditions. The initial total mass of the rocket motor is 50,000 lbm and its propellant mass fraction is 0.90. The residual propellant (called slivers, combustion stops when the chamber pressure falls below a deflagration limit) amounts to 3 % of the burnt. The burn time is 50 seconds; the nozzle throat area (A_t) is 164.2 in.^{2} and its area ratio (A_2/A_t) is 10. The chamber pressure (p_1) is 780 psia and the pressure ratio (p_1/p_2) across the nozzle may be taken as 90.0. Neglect any start/stop transients and use the information in Appendix 2.

Answers: (a) $\dot{m}$ = 873 lbm/sec, 237 sec., (b) F = 216,900 lbf, I_s = 248.5 sec., (c) F = 231,000 lbf, I_s = 295 sec.

8. During the boost phase of the Atlas V, the RD-180 engine operates together with three solid propellant rocket motors (SRBs) for the initial stage. For the remaining thrust time, the RD-180 operates alone. Using the information given in Table 1–3, calculate the *overall effective exhaust velocity* for the vehicle during the initial combined thrust operation.

Answer: 309 sec.

9. Using the values given in Table 2–1, choose three propulsion systems and calculate the total impulse for a fixed propellant mass of 20 kg.

10. Using the MA-3 rocket engine information given in Example 2–3, calculate the overall specific impulse at sea level and at altitude, and compare these with I_s values for the individual booster engines, the sustainer engine, and the individual vernier engines.

Answers: $(I_s)_{oa}$ = 238 sec (SL) and 258 sec (altitude)

11. Determine the mass ratio **MR** and the mass of propellant used to produce thrust for a solid propellant rocket motor that has an inert mass of 82.0 kg. The motor mass becomes 824.5 kg after loading the propellant. For safety reasons, the igniter is not installed until shortly before motor operation; this igniter has a mass of 5.50 kg of which 3.50 kg is igniter propellant. Upon inspection after firing, the motor is found to have some unburned residual propellant and a motor mass of 106.0 kg.

Answers: **MR** = 0.1277, propellant burned = 720.5 kg.

REFERENCES

2–1. "American National Standard Letter Symbols for Rocket Propulsion," *ASME Publication Y 10.14*, 1959.

2–2. *"Solid Propulsion Nomenclature Guide,"* CPIA Publication 80, Chemical Propulsion Information Agency, Johns Hopkins University, Laurel, MD, May 1965, 18 pages.

2–3. P. G. Hill and C. R. Peterson, *Mechanics and Thermodynamics of Propulsion*, Addison-Wesley, Reading, MA, 1992. [Paperback edition, 2009]

2–4. R. D. Zucker and O. Biblarz, *Fundamentals of Gas Dynamics*, 2nd ed., John Wiley & Sons, Hoboken, NJ, 2002.

CHAPTER 3

NOZZLE THEORY AND THERMODYNAMIC RELATIONS

In rocket propulsion systems the mathematical tools needed to calculate performance and to determine several key design parameters involve the principles from gas dynamics and thermodynamics that describe processes inside a rocket nozzle and its chamber. These relations are also used for evaluating and comparing the performance between different rocket systems since with them we can predict operating parameters for any system that uses the thermodynamic gas expansion in a supersonic nozzle; they allow the determination of nozzle size and generic shape for any given performance requirement. This theory applies to chemical rocket propulsion systems (liquid and solid and hybrid propellant types), nuclear rockets, solar-heated and resistance or arc-heated electrical rocket systems, and all propulsion systems that use gas expansion as the mechanism for ejecting matter at high velocities.

Fundamental thermodynamic relations are introduced and explained in this chapter. By using these equations, the reader can gain a basic understanding of the thermodynamic processes involved in high-temperature and/or pressure gas expansions. Some knowledge of both elementary thermodynamics and fluid mechanics on the part of the reader is assumed (see Refs. 3–1 to 3–3). This chapter also addresses different nozzle configurations, nonoptimum performance, energy losses, nozzle alignment, variable thrust, and four different alternate nozzle performance parameters.

3.1. IDEAL ROCKET PROPULSION SYSTEMS

The concept of an ideal rocket propulsion system is useful because the relevant basic thermodynamic principles can be expressed with relatively simple mathematical relationships, as shown in subsequent sections of this chapter. These equations describe quasi-one-dimensional nozzle flows, which represent an idealization and simplification of the full two- or three-dimensional equations of real aerothermo-chemical behavior. However, within the assumptions stated below, these descriptions are very adequate for obtaining useful solutions to many rocket propulsion systems and for preliminary design tasks. In chemical rocket propulsion, measured actual performances turn out to be usually between 1 and 6% below the calculated ideal values. In designing new rocket propulsion systems, it has become accepted practice to use such ideal rocket parameters, which can then be modified by appropriate corrections, such as those discussed in Section 3.5. An *ideal rocket* propulsion unit is defined as one for which the following assumptions are valid:

1. The working fluid (which usually consists of chemical reaction products) is *homogeneous in composition.*

2. All the species of the working fluid are treated as *gaseous.* Any condensed phases (liquid or solid) add a negligible amount to the total mass.

3. The working fluid obeys the *perfect gas law.*

4. There is *no heat transfer* across any and all gas-enclosure walls; therefore, the flow is *adiabatic.*

5. There is no appreciable *wall friction* and all *boundary layer* effects may be neglected.

6. There are no *shock waves* or other *discontinuities* within the nozzle flow.

7. The *propellant flow rate* is *steady* and *constant.* The expansion of the working fluid is uniform and steady, without gas pulsations or significant turbulence.

8. *Transient effects* (i.e., start-up and shutdown) are of such short duration that may they be neglected.

9. All exhaust gases leaving the rocket nozzles travel with a *velocity parallel to the nozzle axis.*

10. The gas velocity, pressure, temperature, and density are all *uniform* across any section normal to the nozzle axis.

11. *Chemical equilibrium* is established within the preceding combustion chamber and gas composition does not change in the nozzle (i.e., frozen composition flow).

12. Ordinary propellants are stored at ambient temperatures. Cryogenic propellants are at their boiling points.

These assumptions permit the derivation of the relatively compact, quasi-one-dimensional set of equations shown in this chapter. Later in this book we present more sophisticated theories and/or introduce correction factors for several of the items on the above list, which then allow for more accurate determinations. The next paragraph explains why the above assumptions normally cause only small errors.

For liquid bipropellant rockets, the idealized situation postulates an injection system in which the fuel and oxidizer mix perfectly so that a homogeneous working medium results; a good rocket injector can closely approach this condition. For solid propellant rocket units, the propellant must essentially be homogeneous and uniform and the burning rate must be steady. For solar-heated or arc-heated propulsion systems, it must be assumed that the hot gases can attain a uniform temperature at any cross section and that the flow is steady. Because chamber temperatures are typically high (2500 to 3600 K for common propellants), all gases are well above their respective saturation conditions and do follow closely the perfect gas law. Assumptions 4, 5, and 6 above allow the use of *isentropic expansion* relations within the rocket nozzle, thereby describing the maximum conversion from heat and pressure to kinetic energy of the jet (this also implies that the nozzle flow is thermodynamically reversible). Wall friction losses are difficult to determine accurately, but they are usually negligible when the inside walls are smooth. Except for very small chambers, the heat losses to the walls of the rocket are usually less than 1% (occasionally up to 2%) of the total energy and can therefore be neglected. Short-term fluctuations in propellant flow rates and pressures are typically less than 5% of their steady value, small enough to be neglected. In well-designed supersonic nozzles, the conversion of thermal and/or pressure energy into directed kinetic energy of the exhaust gases may proceed smoothly and without normal shocks or discontinuities—thus, flow expansion losses are generally small.

Some rocket companies and/or some authors do not include all or the same 12 items listed above in their definition of an ideal rocket. For example, instead of assumption 9 (all nozzle exit velocity is axially directed), they use a conical exit nozzle with a 15° half-angle as their base configuration for the ideal nozzle; this item accounts for the divergence losses, a topic later described in this chapter (using the correction factor λ).

3.2. SUMMARY OF THERMODYNAMIC RELATIONS

In this section we briefly review some of the basic relationships needed for the development of the nozzle flow equations. Rigorous derivations and discussions of these relations can be found in many thermodynamics or fluid dynamics texts, such as Refs. 3–1 to 3–3.

The principle of *conservation of energy* may be readily applied to the adiabatic, no shaft-work processes inside the nozzle. Furthermore, in the absence of shocks or friction, flow entropy changes are zero. The concept of *enthalpy* is most useful in flow systems; the enthalpy h comprises the *internal thermal energy* plus the *flow work* (or work performed by the gas of velocity v in crossing a boundary). For ideal gases the enthalpy can conveniently be expressed as the product of the specific heat c_p times the absolute temperature T (c_p is the specific heat at constant pressure, defined as the partial derivative of the enthalpy with respect to temperature at constant pressure). Under the above assumptions, the total or stagnation enthalpy per unit mass h_0 remains constant in nozzle flows, that is,

$$h_0 = h + v^2/2J = \text{constant} \tag{3-1}$$

Other stagnation conditions are introduced later, below Eq. 3–7. The symbol J is the mechanical equivalent of heat which is utilized only when thermal units (i.e., the Btu and calorie) are mixed with mechanical units (i.e., the ft-lbf and the joule). In SI units (kg, m, sec) the value of J is one. In the EE (English Engineering) system of units the value of the constant J is given in Appendix 1. Conservation of energy applied to isentropic flows between any two nozzle axial sections x and y shows that the decrease in static enthalpy (or thermodynamic content of the flow) appears as an increase of kinetic energy since any changes in potential energy may be neglected.

$$h_x - h_y = \frac{1}{2}(v_y^2 - v_x^2)/J = c_p(T_x - T_y) \tag{3–2}$$

The principle of *conservatism of mass* in steady flow, for passages with a single inlet and single outlet, is expressed by equating the mass flow rate $\dot{m}$ at any section x to that at any other section y; this is known as the mathematical form of the continuity equation. Written in terms of the cross-sectional area A, the velocity v, and the "specific volume" V (i.e., the volume divided by the mass within), at any section

$$\dot{m}_x = \dot{m}_y \equiv \dot{m} = Av/V \tag{3–3}$$

The *perfect gas law* can be written as (at an arbitrary location x)

$$p_x V_x = RT_x \tag{3–4}$$

where the gas constant R is found from the universal gas constant R' divided by the molecular mass $\mathfrak{M}$ of the flowing gas mixture. The molecular volume at standard conditions becomes $22.41\,\mathrm{m^3/kg - mol}$ or $\mathrm{ft^3/lb - mol}$ and it relates to a value of $R' = 8314.3\,\mathrm{J/kg\text{-}mol\text{-}K}$ or $1544\,\mathrm{ft\text{-}lbf/lb\text{-}mol\text{-}°R}$. We often find Eq. 3–3 written in terms of density ρ which is the reciprocal of the specific volume V. The specific heat at constant pressure c_p, the specific heat at constant volume c_v, and their ratio k are constant for perfect gases over a wide range of temperatures and are related as follows:

$$k = c_p/c_v \tag{3–5a}$$

$$c_p - c_v = R/J \tag{3–5b}$$

$$c_p = kR/(k-1)J \tag{3–6}$$

For any *isentropic flow process*, the following relations may be shown to hold between any two nozzle sections x and y:

$$T_x/T_y = (p_x/p_y)^{(k-1)/k} = (V_y/V_x)^{k-1} \tag{3–7}$$

During an isentropic expansion the pressure drops substantially, the absolute temperature drops somewhat less, and the specific volume increases. When flows are stopped isentropically the prevailing conditions are known as *stagnation conditions* which are designated by the subscript 0. Sometimes the word "total" is used instead of stagnation. As can be seen from Eq. 3–1 the stagnation enthalpy consists of the sum of the static or local enthalpy and the fluid kinetic energy. The absolute stagnation temperature T_0 is found from this energy equation as

$$T_0 = T + v^2/(2c_pJ) \tag{3–8}$$

where T is the static absolute fluid temperature. In adiabatic flows, the stagnation temperature remains constant. A useful relationship between the stagnation pressure and the local pressure in isentropic flows can be found from the previous two equations:

$$p_0/p = [1 + v^2/(2c_pJT)]^{k/(k-1)} = (V/V_0)^k \tag{3–9}$$

When local velocities are close to zero, the corresponding local temperatures and pressures approach the stagnation pressure and stagnation temperature. Inside combustion chambers, where gas velocities are typically small, the local combustion pressure essentially equals the stagnation pressure. Now, the *velocity of sound a* also known as the acoustic velocity in perfect gases is independent of pressure. It is defined as

$$a = \sqrt{kRT} \tag{3–10}$$

In the EE system the units of R must be corrected and a conversion constant $g_c \equiv g_0$ must be added – Equation 3–10 becomes $\sqrt{g_c kRT}$. This correction factor allows for the proper velocity units. The *Mach number M* is a dimensionless flow parameter and is used to locally define the ratio of the flow velocity v to the local acoustic velocity a:

$$M = v/a = v/\sqrt{kRT} \tag{3–11}$$

Hence, Mach numbers less than one correspond to subsonic flows and greater than one to supersonic flows. Flows moving at precisely the velocity of sound would have Mach numbers equal to one. It is shown later that at the throat of all one-dimensional supersonic nozzles the Mach number must be equal to one. The relation between stagnation temperature and Mach number may now be written from Eqs. 3–2, 3–7, and 3–10 as

$$T_0 = T\left[1 + \frac{1}{2}(k-1)M^2\right] \tag{3–12}$$

or

$$M = \sqrt{\frac{2}{k-1}\left(\frac{T_0}{T} - 1\right)}$$

T_0 and p_0 designate the temperature and pressure stagnation values. Unlike the temperature, the stagnation pressure during an adiabatic nozzle expansion only remains constant for totally isentropic flows (i.e., no losses of any kind). It may be computed from

$$p_0 = p\left[1 + \frac{1}{2}(k-1)M^2\right]^{k/(k-1)} \tag{3–13}$$

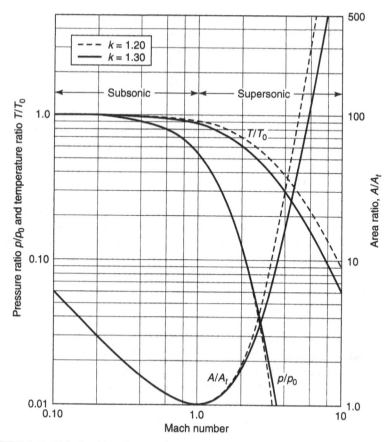

FIGURE 3–1. Relationship of area ratio, pressure ratio, and temperature ratio as functions of Mach number in a converging/diverging nozzle depicted for the subsonic and supersonic nozzle regions.

The nozzle *area ratio* for isentropic flow may now be expressed in terms of Mach numbers for two arbitrary locations x and y within the nozzle. Such a relationship is plotted in Fig. 3–1 for $M_x = 1.0$, where $A_x = A_t$ the throat or minimum area, along with corresponding ratios for T/T_0 and p/p_0. In general,

$$\frac{A_y}{A_x} = \frac{M_x}{M_y} \sqrt{\left\{ \frac{1 + [(k-1)/2]M_y^2}{1 + [(k-1)/2]M_x^2} \right\}^{(k+1)/(k-1)}} \qquad (3\text{–}14)$$

As can be seen from Fig. 3–1, for subsonic flows any chamber contraction (from station $y = 1$) or nozzle entrance ratio A_1/A_t can remain small, with values approaching 3 to 6 depending on flow Mach number, and the passage is convergent. There are no noticeable effects from variations of k. In solid rocket motors the chamber area

A_1 refers to the flow passage or port cavity in the virgin grain. With supersonic flow the nozzle section diverges and the area ratio enlarges rather quickly; area ratios are significantly influenced by the value of k. At $M = 4$ the exit area ratio A_2/A_t ranges between 15 and 30, depending on the value of k indicated. On the other hand, pressure ratios are less sensitive to k but temperature ratios show more variation.

The average *molecular mass* $\mathfrak{M}$ of a mixture of gases is the sum of all chemical species' n_i in kg-mols multiplied by their molecular mass $(n_i\mathfrak{M}_i)$ and then divided by the sum of all molar masses. This is further discussed in Chapter 5, Eq. 5–5. The symbol $\mathfrak{M}$ has been used to avoid confusion with M for the Mach number. In the older literature $\mathfrak{M}$ is called the molecular weight.

Example 3–1.   An "ideal rocket engine" is designed to operate at sea level using a propellant whose products of combustion have a specific heat ratio k of 1.3. Determine the required chamber pressure if the exit Mach number is 2.52. Also determine the nozzle area ratio between the throat and exit.

SOLUTION. For "optimum expansion" the nozzle exit pressure must equal the local atmospheric pressure, namely, 0.1013 MPa. If the chamber velocity may be neglected, then the ideal chamber pressure is the total stagnation pressure, which can be found from Eq. 3–13 as

$$p_0 = p\left[1 + \frac{1}{2}(k-1)M^2\right]^{k/(k-1)}$$

$$= 0.1013[1 + 0.15 \times 2.52^2]^{1.3/0.3} = 1.84 \text{ MPa}$$

The ideal nozzle area ratio A_2/A_t is determined from Eq. 3–14 setting $M_t = 1.0$ at the throat (see also Fig. 3–1):

$$\frac{A_2}{A_t} = \frac{1}{M_2}\left[\frac{1 + [(k-1)/2]M_2^2}{(k+1)/2}\right]^{(k+1)/2(k-1)}$$

$$= \frac{1}{2.52}\left[\frac{1 + 0.15 \times 2.52^2}{1.15}\right]^{2.3/0.6} = 3.02$$

Note that *ideal* implies no losses, whereas *optimum* is a separate concept reflecting the best calculated performance at a particular set of given pressures. *Optimum* performance is often taken as the design condition and it occurs when $p_2 = p_3$ as will be shown in the section on the *thrust coefficient* (the peak of the curves in Figs. 3–6 and 3–7 for fixed p_1/p_3).

3.3. ISENTROPIC FLOW THROUGH NOZZLES

Within converging–diverging nozzles a large fraction of the thermal energy of the gases flowing from the chamber is converted into kinetic energy. As will be shown, gas pressures and temperatures may drop dramatically and gas velocities can reach values in excess of 2 miles per second. This expansion is taken as a reversible or isentropic flow process in the analyses described here. When a nozzle's inner wall has a flow obstruction or a wall protrusion (a piece of weld splatter or slag), then the gas kinetic energy is locally converted back into thermal energy essentially recovering the stagnation temperature and stagnation pressure in the chamber. Since this would quickly lead to local material overheating and wall failure, all nozzle inner walls must be smooth without any protrusions. Stagnation conditions can also occur at the leading edge of a jet vane (described in Chapter 18) or at the tip of a gas sampling tube inserted into the flow.

Velocity

The nozzle exit velocity (at $y = 2$) v_2 can be solved for from Eq. 3–2:

$$v_2 = \sqrt{2J(h_1 - h_2) + v_1^2} \tag{3-15a}$$

As stated, this relation strictly applies when there are no heat losses. This equation also holds between any two locations within the nozzle, but hereafter subscripts 1 and 2 will only designate nozzle inlet and exit conditions. For constant k the above expression may be rewritten with the aid of Eqs. 3–6 and 3–7:

$$v_2 = \sqrt{\frac{2k}{k-1} RT_1 \left[1 - \left(\frac{p_2}{p_1} \right)^{(k-1)/k} \right] + v_1^2} \tag{3-15b}$$

When the chamber cross section is large compared the nozzle throat, the chamber velocity or nozzle entrance velocity is comparatively small and the term v_1^2 may be neglected above. The chamber temperature T_1, which is located at the nozzle inlet, under isentropic conditions differs little from the stagnation temperature or (for chemical rocket propulsion) from the combustion temperature T_0. Hence we arrive at important equivalent simplified expressions of the velocity v_2, ones commonly used for analysis:

$$v_2 = \sqrt{\frac{2k}{k-1} RT_1 \left[1 - \left(\frac{p_2}{p_1} \right)^{(k-1)/k} \right]}$$

$$= \sqrt{\frac{2k}{k-1} \frac{R'T_0}{\mathfrak{M}} \left[1 - \left(\frac{p_2}{p_1} \right)^{(k-1)/k} \right]} \tag{3-16}$$

As can be seen above, the gas velocity exhausting from an ideal nozzle is a function of the prevailing *pressure ratio* p_1/p_2, the ratio of specific heats k, and the absolute temperature at the nozzle inlet T_1, as well as of the gas constant R. Because the gas constant for any particular gas is inversely proportional to the molecular mass $\mathfrak{M}$, exhaust velocities or their corresponding specific impulses strongly depend on the ratio of the absolute nozzle entrance temperature (which is close to T_0) divided by the average molecular mass of the exhaust gases, as is shown in Fig. 3–2. The fraction

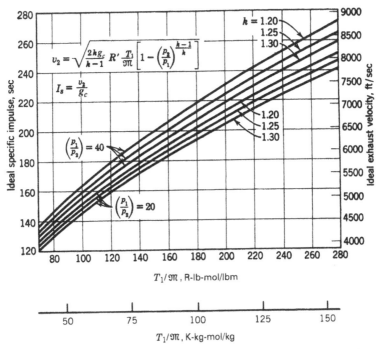

FIGURE 3–2. Specific impulse and exhaust velocity of an ideal rocket propulsion unit at optimum nozzle expansion as functions of the absolute chamber gas temperature T_1 and the molecular mass $\mathfrak{M}$ for several values of k and p_1/p_2.

$T_0/\mathfrak{M}$ plays an important role in optimizing the mixture ratio (oxidizer to fuel flow) in chemical rocket propulsion defined in Eq. 6–1.

Equations 2–13 and 2–16 give relations between the thrust F and the velocities v_2 and c, and the corresponding specific impulse I_s which is plotted in Fig. 3–2 for two pressure ratios and three values of k. Equation 3–16 clearly shows that any increase in entrance gas temperatures (arising from increases in chamber energy releases) and/or any decrease of propellant molecular mass (commonly achieved from using low molecular mass gas mixtures rich in hydrogen) will enhance the ratio $T_0/\mathfrak{M}$; that is, they will increase the specific impulse I_s through increases in the exhaust velocity v_2 and, thus, the performance of the rocket vehicle. Influences of pressure ratio across the nozzle p_1/p_2 and of specific heat ratio k are less pronounced. As can be seen from Fig. 3–2, performance does increase with an increase of the pressure ratio; this ratio increases when the value of the chamber pressure p_1 rises and/or when the exit pressure p_2 decreases, corresponding to high altitude designs. The small influence of k values is fortuitous because low molecular masses, found in many diatomic and monatomic gases, correspond to the higher values of k.

Values of the pressure ratio must be standardized when comparing specific impulses from one rocket propulsion system to another or for evaluating the influence of various design parameters. Presently, a chamber pressure of 1000 psia (6.894 MPa) and an exit pressure of 1 atm (0.1013 MPa) are used as standard reference for such comparisons, or $p_1/p_2 = 68.06$.

For *optimum expansion* $p_2 = p_3$ and the effective exhaust velocity c (Eq. 2–15) and the ideal rocket exhaust velocity v_2 become equal, namely,

$$v_2 = (c)_{opt} \tag{3-17}$$

Thus, here c can be substituted for v_2 in Eqs. 3–15 and 3–16. For a fixed nozzle exit area ratio, and constant chamber pressure, this optimum condition occurs only at the particular altitude where the ambient pressure p_3 just equals the nozzle exhaust pressure p_2. At all other altitudes $c \neq v_2$.

The maximum theoretical value of nozzle outlet velocity is reached with an infinite nozzle expansion (when exhausting into a vacuum):

$$(v_2)_{max} = \sqrt{2kRT_0/(k-1)} \tag{3-18}$$

This maximum theoretical exhaust velocity is finite, even though the pressure ratio is not, because it corresponds to finite thermal energy contents in the fluid. In reality, such an expansion cannot occur because, among other things, the temperature of most gas species will eventually fall below their liquefaction or their freezing points; thus, they cease to be a gas and can no longer contribute to the expansion or to any further velocity increases.

Example 3–2.   A rocket propulsion system operates near sea level with a chamber pressure of $p_1 = 2.068$ MPa or 300 psia, a chamber temperature of 2222 K, and a propellant consumption of $\dot{m} = 1.0$ kg/sec. Take $k = 1.30$ and R = 345.7 J/kg – K. Calculate the ideal thrust and the ideal specific impulse. Also plot the cross-sectional area A, the local velocity v, the specific volume V, the absolute temperature T, and the local Mach number M with respect to pressure along the nozzle.

SOLUTION.   Assume that the operation of this rocket propulsion system is at an optimum for expansion to sea-level pressure $p_3 = 0.1013$ MPa so that $p_2/p_1 = 0.049$. in. From Eq. 3–16 the effective exhaust velocity is calculated to be 1827 m/sec since here $v_2 = c$. Hence,

$$I_s = c/g_0 = 1827/9.81 = 186 \text{ sec} \quad \text{from Eq. 2-5} \quad \text{and}$$

$$F = \dot{m}c = 1.0 \times 1827 = 1827 \text{ N} \quad \text{from Eq. 2-16}$$

In this example, we denote the nozzle axial location x as either the nozzle entrance (location 1 in Fig. 2–1) condition or the throat section (location t in Fig. 2–1) and keep the axial location y as a variable which spans the nozzle ranging from $1 \rightarrow 2$ (exit) in Fig. 2–1.

The calculation of the velocity follows from Eq. 3–16 with p_y set equal to the exit pressure p_3, the local atmospheric pressure or 0.01013 MPa.

$$v_2 = \sqrt{\frac{2k}{k-1} RT_1 \left[1 - \left(\frac{p_y}{p_1}\right)^{(k-1)/k} \right]}$$

$$= \sqrt{\frac{2 \times 1.30 \times 345.7 \times 2222}{1.30 - 1} \left[1 - \left(\frac{0.1013}{2.068}\right)^{0.2308} \right]} = 1827 \text{ m/sec}$$

The cross-sectional area can be obtained from Eq. 3–3 and the Mach number from Eq. 3–11. By stepwise varying the pressure from 2.068 to 0/1013 MPa, one can arrive at the information shown in Fig. 3–3; it is a tedious process. A faster way is to modify the relevant equations in terms of p_y so that they can be plotted directly and this is shown below.

The initial specific volume V_1 is calculated from the equation of state of a perfect gas (Eq. 3–4),

$$V_1 = RT_1/p_1 = 345.7 \times 2222/(2.068 \times 10^6) = 0.3714 \text{ m}^3/\text{kg}$$

Now we can write equations for the specific volume and the temperature (Eq. 3–7) as a function of the pressure p_y

$$V_y = V_1(p_1/p_y)^{1/k} = 0.3714(2.068/p_y)^{0.769} \text{ (m}^3/\text{kg)}$$

$$T_y = T_1(p_y/p_1)^{(k-1)/k} = 2222(p_y/2.068)^{0.231} \text{ (K)}$$

These can be plotted by inserting p_y values between 2.068 and 0.1013 in MPa. The calculations for the velocity follow from Eq. 3–16,

$$v_y = 2580[1 - (p_y/2.068)^{0.231}]^{1/2} \text{ (m/sec)}$$

The cross-sectional area is found from Eq. 3–3 and the relations above:

$$A_y = \dot{m} \, V_y/v_y = (1.0 \times 0.3714/2580) \times (2.068/p_y)^{0.769}[1 - (p_y/2.068)^{0.231}]^{-1/2}$$

$$= 1.44 \times 10^{-4} \times (2.068/p_y)^{0.769}[1 - (p_y/2.068)^{0.231}]^{-1/2} \text{ (m}^2)$$

Finally the Mach number is obtained using Eq. 3–11 and the relations above:

$$M_y = v_y/[kRT_y]^{1/2}$$

$$= [2580/(1.30 \times 345.7 \times 2222)^{1/2}][1 - (p_y/2.068)^{0.231}]^{1/2}(p_y/2.068)^{-0.231}$$

$$= 2.582[1 - (p_y/2.068)^{0.231}]^{1/2}(p_y/2.068)^{-0.231}$$

With the equations indicated above and graphing software such as MATLAB or MAPLE or other, the desired plots may be obtained. Figure 3–3 shows semiquantitative plots of key parameter variations and should only be used to exhibit trends. Note that in this figure the pressures are shown as decreasing to the right, representing a nozzle flow from left to right (also its units are in psia where the throat pressure is 164.4 psia and the exit pressure would be 14.7 psia, just to the left of zero).

A number of interesting deductions can be made from this example. Very *high gas velocities* (over 1 km/sec) can be obtained in rocket nozzles. The *temperature drop* of the combustion gases flowing through a rocket nozzle can be appreciable. In the example above the temperature changed 1115 °C in a relatively short distance.

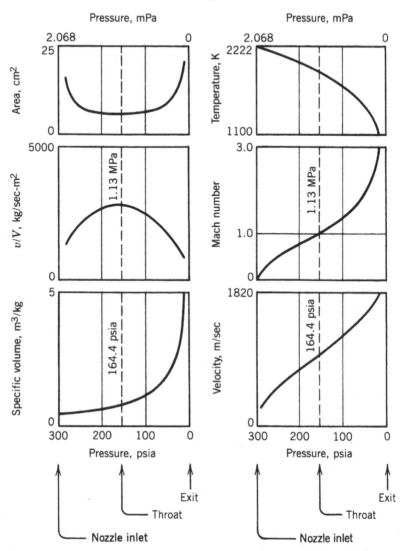

FIGURE 3–3. Typical variations of cross-sectional area, temperature, specific volume, Mach number, and velocity with pressure in a rocket nozzle (from Example 3–2).

This should not be surprising, for the increase in the kinetic energy of the gases comes from a decrease of the enthalpy, which in turn is roughly proportional to the decrease in temperature. Because the exhaust gases are still relatively hot (1107 K) when leaving the nozzle, they contain considerable thermal energy which is not available for conversion because it is seldom realistic to match exit and ambient temperatures, (see Fig. 2–2).

Nozzle Flow and Throat Condition

Supersonic nozzles (often called De Laval nozzles after their inventor) always consist of a convergent section leading to a *minimum* area followed by a divergent section. From the continuity equation, this area is inversely proportional to the ratio v/V. This quantity has also been shown in Fig. 3–3, where there is a *maximum* in the curve of v/V because at first the velocity increases at a greater rate than the specific volume; however, in the divergent section, the specific volume increases faster.

The minimum nozzle cross-sectional area is commonly referred to as the *throat area* A_t. The ratio of the nozzle exit area A_2 to the throat area A_t is called the *nozzle expansion area ratio* and is designated here by the Greek letter ϵ. It represents an important nozzle parameter:

$$\epsilon = A_2/A_t \tag{3-19}$$

The maximum gas flow rate per unit area occurs at the throat where a unique gas pressure ratio exists, which is only a function of the ratio of specific heats k. This pressure ratio is found by setting $M = 1$ in Eq. 3–13:

$$p_t/p_1 = [2/(k+1)]^{k/(k-1)} \tag{3-20}$$

The throat pressure p_t for which an isentropic mass flow rate attains its maximum is called the *critical pressure*. Typical values of the above critical pressure ratio range between 0.53 and 0.57 of the nozzle inlet pressure. The flow through a specified rocket nozzle with a given inlet condition is less than the maximum if the pressure ratio is larger than that given by Eq. 3–20. However, note that this ratio is not the value across the entire nozzle and that the maximum flow or choking condition (explained below) is always established internally at the throat and not at the exit plane. The nozzle inlet pressure is usually very close to the chamber stagnation pressure, except in narrow combustion chambers where there is an appreciable drop in pressure from the injector region to the nozzle entrance region. This is discussed in Section 3.5. At the location of critical pressure, namely the throat, the Mach number is always one and the values of the specific volume and temperature can be obtained from Eqs. 3–7 and 3–12:

$$V_t = V_1[(k+1)/2]^{1/(k-1)} \tag{3-21}$$

$$T_t = 2T_1/(k+1) \tag{3-22}$$

In Eq. 3–22 the nozzle inlet temperature T_1 is typically very close to the combustion chamber temperature and hence close to the nozzle flow stagnation temperature T_0. At the throat there can only be a small variation for these properties. Take, for example, a gas with $k = 1.2$; the critical pressure ratio is about 0.56 (which means that p_t is a little more than half of the chamber pressure p_1; the temperature drops only slightly ($T_t = 0.91T_1$), and the specific volume expands by over 60% ($V_t = 1.61V_1$). Now, from Eqs. 3–15, 3–20, and 3–22, the *critical or throat velocity* v_t is obtained:

$$v_t = \sqrt{\frac{2k}{k+1}\,RT_1} = \sqrt{kRT_t} = a_t \tag{3-23}$$

The first version of this equation permits the critical velocity to be calculated directly from the nozzle inlet conditions and the second one gives v_t when throat temperature is known. The throat is the only nozzle section where the flow velocity is also the *local sonic velocity* ($a_t = v_t$ because $M_t = 1.00$). The inlet flow from the chamber is subsonic and downstream of the nozzle throat it is supersonic. The divergent portion of the nozzle provides for further decreases in pressure and increases in velocity under supersonic conditions. If the nozzle is cut off at the throat section, the exit gas velocity is sonic and the flow rate per unit area will remain its maximum. Sonic and supersonic flow conditions can only be attained when the critical pressure prevails at the throat, that is, when p_2/p_1 is equal to or less than the quantity defined by Eq. 3–20. There are, therefore, three basically different types of nozzles: subsonic, sonic, and supersonic, and these are described in Table 3–1.

The supersonic nozzle is the only one of interest for rocket propulsion systems. It achieves a high degree of conversion of enthalpy to kinetic energy. The ratio between the inlet and exit pressures in all rocket propulsion systems must be designed sufficiently large to induce supersonic flow. Only when the absolute chamber pressure drops below approximately 1.78 atm will there be subsonic flow in the divergent portion of any nozzle during sea-level operation. This pressure condition actually occurs for very short times during start and stop transients.

The velocity of sound, a, is the propagation speed of an elastic pressure wave within the medium, sound being an infinitesimal pressure wave. If, therefore, sonic velocity is reached at any location within a steady flow system, it is impossible for pressure disturbances to travel upstream past that location. Thus, any partial obstruction or disturbance of the flow downstream of a nozzle throat running critical has no influence on the throat or upstream of it, provided that the disturbance does not raise the downstream pressure above its critical value. It is not possible to increase the sonic velocity at the throat or the flow rate in a fixed configuration nozzle by further lowering the exit pressure or even evacuating the exhaust section. This important condition has been described as *choking* the flow, which is always established at the throat (and not at the nozzle exit or any other plane). *Choked mass flow* rates through the critical section of a supersonic nozzle may be derived from Eqs. 3–3, 3–21, and 3–23.

TABLE 3–1. Nozzle Types

	Subsonic	Sonic	Supersonic
Throat velocity	$v_1 < a_t$	$v_t = a_t$	$v_t = a_t$
Exit velocity	$v_2 < a_2$	$v_2 = v_t$	$v_2 > v_t$
Mach number	$M_2 < 1$	$M_2 = M_t = 1.0$	$M_2 > 1$
Pressure ratio	$\dfrac{p_1}{p_2} < \left(\dfrac{k+1}{2}\right)^{k/(k-1)}$	$\dfrac{p_1}{p_2} = \dfrac{p_1}{p_t} = \left(\dfrac{k+1}{2}\right)^{k/(k-1)}$	$\dfrac{p_1}{p_2} > \left(\dfrac{k+1}{2}\right)^{k(k-1)}$
Shape			

From continuity, $\dot{m}$ always equals to the mass flow at any other cross section within the nozzle for steady flow.

$$\dot{m} = \frac{A_t v_t}{V_t} = A_t p_1 k \frac{\sqrt{[2/(k+1)]^{(k+1)/(k-1)}}}{\sqrt{kRT_1}} \tag{3-24}$$

The (chocked) mass flow through a rocket nozzle is therefore proportional to the throat area A_t and the chamber (stagnation) pressure p_1; it is also inversely proportional to the square root of $T/\mathfrak{M}$ as well as a function of other gas properties. For supersonic nozzles the *ratio between the throat and any downstream area* at which a pressure p_y prevails can be expressed as a function of pressure ratios and the ratio of specific heats, by using Eqs. 3–4, 3–16, 3–21, and 3–23, as follows:

$$\frac{A_t}{A_y} = \frac{V_t v_y}{V_y v_t} = \left(\frac{k+1}{2}\right)^{1/(k-1)} \left(\frac{p_y}{p_1}\right)^{1/k} \sqrt{\frac{k+1}{k-1}\left[1 - \left(\frac{p_y}{p_1}\right)^{(k-1)/k}\right]} \tag{3-25}$$

When $p_y = p_2$, then $A_y/A_t = A_2/A_t = \epsilon$ and Eq. 3–25 shows the inverse nozzle exit area expansion ratio. For low-altitude operation (sea level to about 10,000 m) nozzle area ratios are typically between 3 and 30, depending on chamber pressure, propellant combinations, and vehicle envelope constraints. For high altitudes (100 km or higher) area ratios are typically between 40 and 200, but there have been some as high as 400. Similarly, an expression for the ratio of the velocity at any point downstream of the throat with a pressure p_y to the throat velocity may be written from Eqs. 3–15 and 3–23 as:

$$\frac{v_y}{v_t} = \sqrt{\frac{k+1}{k-1}\left[1 - \left(\frac{p_y}{p_1}\right)^{(k-1)/k}\right]} \tag{3-26}$$

Note that Eq. 3–26 is dimensionless but does *not* represent a Mach number. Equations 3–25 and 3–26 permit direct determination of the velocity ratio or the area ratio for any given pressure ratio, and vice versa, in ideal rocket nozzles. They both are plotted in Fig. 3–4. When $p_y = p_2$, Eq. 3–26 describes the velocity ratio between the nozzle exit area and the throat section. When the exit pressure coincides with the atmospheric pressure ($p_2 = p_3$, see Fig. 2–1), these equations apply to optimum nozzle expansions. For rockets that operate at high altitudes, it can be shown that not much additional exhaust velocity may be gained by increasing the area ratio above 1000. In addition, design difficulties and a heavy inert nozzle mass have made applications above area ratios of about 350 marginal.

Appendix 2 is a table of several properties of the Earth's atmosphere with established standard values. It gives ambient pressure for different altitudes. These properties do vary somewhat from day to day (primarily because of solar activity) and between hemispheres. For example, the density of the atmosphere at altitudes between 200 and 3000 km can change by more than an order of magnitude, affecting satellite drag estimates.

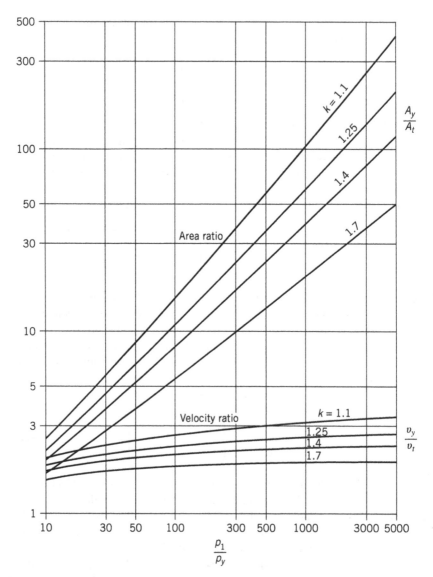

FIGURE 3–4. Area and velocity ratios as function of pressure ratio for the diverging section of a supersonic nozzle.

Example 3–3. Design an ideal nozzle for a rocket propulsion system that operates at 25 km altitude and delivers 5000 N thrust with a chamber pressure of 2.039 MPa and a chamber temperature of 2800 K. For this particular propellant, $k = 1.20$ and $R = 360$ J/kg-K. Determine the throat area, exit area, throat temperature, and exit velocity for optimum expansion.

SOLUTION. At 25 km altitude the atmospheric pressure equals 2.549 kPa (in Appendix 2 the ratio shown is 0.025158, which must be multiplied by the pressure at sea level, 0.1013 MPa).

The controlling pressure ratio is

$$p_2/p_1 = 0.002549/2.039 = 0.00125 = 1/800$$

The area ratio may now be obtained directly from Eq. 3–25 (or Fig. 3–4) as

$$\frac{A_t}{A_2} = \left(\frac{k+1}{2}\right)^{1/(k-1)} \left(\frac{p_2}{p_1}\right)^{1/k} \sqrt{\frac{k+1}{k-1}\left[1 - \left(\frac{p_2}{p_1}\right)^{(k-1)/k}\right]}$$

$$= (1.10)^5 (0.00125)^{0.833} \sqrt{11.0[1 - (0.00125)^{0.167}]}$$

$$= 1.67 \times 10^{-2} = 1/60$$

The temperature at the throat may be found directly from Eq. 3–22:

$$T_t = 2T_1/(k+1) = 2 \times 2800/2.2 = 2545 \text{ K}$$

The ideal exit velocity is found from Eq. 3–16 as

$$v_2 = \sqrt{\frac{2k}{k-1} RT_1 \left[1 - \left(\frac{p_2}{p_1}\right)^{(k-1)/k}\right]}$$

$$= \sqrt{1.21 \times 10^7 [1 - (0.00125)^{0.167}]} = 2851 \text{ m/sec}$$

Next, we need to calculate the throat area, which can be found via the mass flow rate, Eq. 3–24, $A_t = \dot{m} V_t/v_t$. To arrive at the mass flow rate explicitly we note that $v_2 = c$ because $p_2 = p_3$ (Eq. 2–15) and from Eq. 2–16

$$\dot{m} = F/c = 5000/2851 = 1.754 \text{ kg/sec}$$

$$A_t = \frac{\dot{m}}{p_1} \sqrt{\frac{RT_1}{k[2/(k+1)]^{(k+1)/(k-1)}}}$$

$$= \frac{1.754}{2.039 \times 10^6} \sqrt{\frac{360 \times 2800}{1.2[2/2.2]^{11}}} = 13.32 \text{ cm}^2$$

And thus the ideal exit area becomes $A_2 = 60 \times 13.32 = 799 \text{ cm}^2$. The designer would next have to choose an actual nozzle configuration and modify the area ratio to account for nonideal effects, and to choose materials together with cooling methods that can accommodate the high chamber and throat temperatures.

Thrust and Thrust Coefficient

As already stated, the efflux of propellant gases (i.e., their momentum flowing out) causes thrust—a reaction force on the rocket structure. Because this flow is supersonic, the pressure at the exit plane of the nozzle may differ from the ambient pressure, and a pressure component in the thrust equation adds (or subtracts) to the momentum thrust as given by Eq. 2–13, which is repeated here:

$$F = \dot{m}v_2 + (p_2 - p_3)A_2$$

Maximum thrust for any given nozzle operation is found in a vacuum where $p_3 = 0$. Between sea level and the vacuum of space, Eq. 2–13 gives the variation of thrust with altitude, applying the properties of the atmosphere as listed in Appendix 2. A typical variation of thrust with altitude from a chocked nozzle would be represented by a smooth increase curving to an asymptotic value (see Example 2–2). To modify values calculated for the optimum operating condition ($p_2 = p_3$), for other conditions of p_1, k, and ϵ, the following expressions may be used. For the thrust,

$$F = F_{\text{opt}} + p_1 A_t \left(\frac{p_2}{p_1} - \frac{p_3}{p_1} \right) \epsilon \qquad (3\text{–}27)$$

For the specific impulse, using Eqs. 2–5, 2–17, and 2–13,

$$I_s = (I_s)_{\text{opt}} + \frac{c^* \epsilon}{g_0} \left(\frac{p_2}{p_1} - \frac{p_3}{p_1} \right) \qquad (3\text{–}28)$$

When, for example, the specific impulse at a new exit pressure p_2 corresponding to a new area ratio A_2/A_t is to be calculated, the above relations may be used.

Equation 2–13 can be modified by substituting v_2, v_t, and V_t from Eqs. 3–16, 3–21, and 3–23:

$$F = \frac{A_t v_t v_2}{V_t} + (p_2 - p_3) A_2$$

$$= A_t p_1 \sqrt{ \frac{2k^2}{k-1} \left(\frac{2}{k+1} \right)^{(k+1)/(k-1)} \left[1 - \left(\frac{p_2}{p_1} \right)^{(k-1)/k} \right] } + (p_2 - p_3) A_2 \qquad (3\text{–}29)$$

The first version of this equation is general and applies to the gas expansion of all rocket propulsion systems, the second form applies to an ideal rocket propulsion system with k being constant throughout the expansion process. This equation shows that thrust is proportional to the throat area A_t and the chamber pressure (or the nozzle inlet pressure) p_1 and is a function of the pressure ratio across the nozzle p_1/p_2, the specific heat ratio k, and of the pressure thrust. Equation 3–29 is called the ideal thrust equation. A thrust coefficient C_F may now defined as the thrust divided by the chamber pressure p_1 and the throat area A_t. Equation 3–30 then results:

$$C_F = \frac{F}{p_1 A_t} = \frac{v_2^2 A_2}{p_1 A_t V_2} + \frac{p_2 A_2}{p_1 A_t} - \frac{p_3 A_2}{p_1 A_t}$$

$$= \sqrt{ \frac{2k^2}{k-1} \left(\frac{2}{k+1} \right)^{(k+1)/(k-1)} \left[1 - \left(\frac{p_2}{p_1} \right)^{(k-1)/k} \right] } + \frac{p_2 - p_3}{p_1} \frac{A_2}{A_t} \qquad (3\text{–}30)$$

The thrust coefficient C_F is dimensionless. It is a key parameter for analysis and depends on the gas property k, the nozzle area ratio ϵ, and the pressure ratio across the nozzle p_1/p_2, but is not directly dependent on chamber temperature. For any fixed ratio p_1/p_3, the thrust coefficient C_F and the thrust F have a peak when $p_2 = p_3$ as area ratio $\epsilon = A_2/A_t$ varies. This peak value is known as the *optimum thrust coefficient* and is an important criterion in nozzle designs. The introduction of the thrust coefficient permits the following form of Eq. 3–29:

$$F = C_F A_t p_1 \qquad (3\text{--}31)$$

Equation 3–31 may be solved for C_F to experimentally determine thrust coefficient from measured values of chamber pressure, throat diameter, and thrust. Even though the thrust coefficient is a function of chamber pressure, it is not simply proportional to p_1, as can be seen from Eq. 3–30. However, it is directly proportional to throat area. Thrust coefficient has values ranging from just under 1.0 to roughly 2.0. Thrust coefficient may be thought of as representing the amplification of thrust due to the gas expansion in the supersonic nozzle as compared to the thrust that would be exerted if the chamber pressure acted over the throat area only. It is a convenient parameter for visualizing effects of chamber pressure and/or altitude variations in a given nozzle configuration, or for modifying sea-level results to flight altitude conditions.

Figure 3–5 shows variations of the *optimum expansion thrust coefficient* ($p_2 = p_3$) with pressure ratio p_1/p_2, for different values of k and area ratio ϵ. Complete thrust coefficient variations for $k = 1.20$ and 1.30 are plotted in Figs. 3–6 and 3–7 as a function of pressure ratio p_1/p_3 and area ratio ϵ. These sets of curves are very useful in examining various nozzle problems for they permit the evaluation of under- and overexpanded nozzle operation, as explained below. The values given in these figures are ideal and do not consider such losses as divergence, friction, or internal expansion waves.

When p_1/p_3 becomes very large (e.g., expansions into near vacuum), the thrust coefficient approaches an asymptotic maximum as shown for $k = 1.20$ and 1.30 in Figs. 3–6 and 3–7. These figures also give values of C_F for some mismatched nozzle conditions ($p_2 \neq p_3$), provided the nozzle is flowing full at all times (the working fluid does not separate or break away from the nozzle walls). Flow separation is discussed later in this section.

Characteristic Velocity and Specific Impulse

The *characteristic velocity c^** was defined by Eq. 2–17. From Eqs. 3–24 and 3–31, it can be shown that

$$c^* = \frac{p_1 A_t}{\dot{m}} = \frac{I_s g_0}{C_F} = \frac{c}{C_F} = \frac{\sqrt{k R T_1}}{k \sqrt{[2/(k+1)]^{(k+1)/(k-1)}}} \qquad (3\text{--}32)$$

This velocity is only a function of propellant characteristics and combustion chamber properties, independent of nozzle characteristics. Thus, it can be used as a figure of merit when comparing different propellant combinations for combustion chamber performance. The first version of this equation applies in general and allows determination of c^* from experimental values of $\dot{m}$, p_1, and A_t. The last version gives the ideal

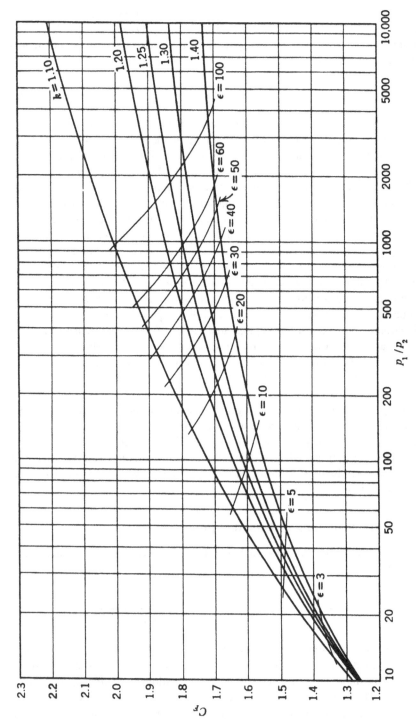

FIGURE 3–5. Thrust coefficient C_F as a function of pressure ratio, nozzle area ratio, and specific heat ratio for optimum expansion conditions $(p_2 = p_3)$.

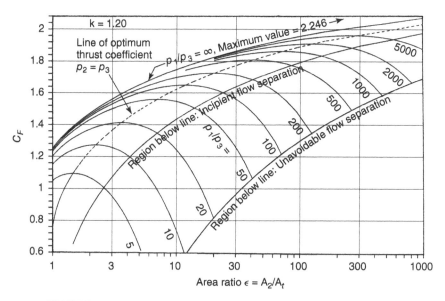

FIGURE 3–6. Thrust coefficient C_F versus nozzle area ratio for $k = 1.20$.

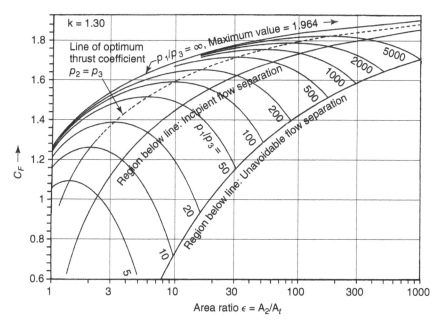

FIGURE 3–7. Thrust coefficient C_F versus nozzle area ratio for $k = 1.30$.

value of c^* as a function of working gas properties, namely k, chamber temperature, and effective molecular mass $\mathfrak{M}$, determined from theory as presented in Chapter 5; some values of c^* are shown in Tables 5–5 and 5–6.

The term c^*-*efficiency* can be used to express the degree of completion of chemical energy releases in the generation of high-temperature, high-pressure gases in combustion chambers. It is the ratio of the actual value of c^*, as determined from measurements (the first part of Eq. 3–32), to the theoretical value (last part of Eq. 3–32) and has typical values between 92 and 99.5%.

Combining now Eqs. 3–31 and 3–32 allows to express the thrust itself as the mass flow rate ($\dot{m}$) times a function of combustion chamber parameters (c^*) times a function of nozzle expansion parameters (C_F), namely,

$$F = \dot{m}c^*C_F \qquad (3\text{--}33)$$

Some authors use a term called the *discharge coefficient* C_D, which is merely the reciprocal of c^*. Both C_D and the characteristic exhaust velocity c^* can be used only with chemical rocket propulsion systems.

The influence of *variations in the specific heat ratio k* on various quantities (such as c, c^*A_2/A_t, v_2/v_t, or I_s) is not as large as those from changes in chamber temperature, pressure ratio, or molecular mass. Nevertheless, they are a noticeable factor, as can be seen by examining Figs. 3–2 and 3–4 to 3–7. The value of k is 1.67 for monatomic gases such as helium and argon, 1.4 for cold diatomic gases such as hydrogen, oxygen, and nitrogen, and for triatomic and beyond it varies between 1.1 and 1.3 (methane is 1.11 and ammonia and carbon dioxide 1.33), see Table 7–3. In general, the more complex the molecule the lower the value of k; this is also true for molecules at high temperatures when their vibrational modes have been activated. Average values of k and $\mathfrak{M}$ for typical rocket exhaust gases with several constituents depend strongly on the composition of the products of combustion (chemical constituents and concentrations), as explained in Chapter 5. Values of k and $\mathfrak{M}$ are given in Tables 5–4, 5–5, and 5–6.

Example 3–4.   What is the percentage variation in thrust between sea level and 10 km for a launch vehicle whose rocket propulsion system operates with a chamber pressure of 20 atm and has a nozzle expansion area ratio ϵ of 6? (Use $k = 1.30$.) Also, at what altitude is the performance of this rocket optimum?

SOLUTION.   According to Eq. 3–33, the only component of the thrust that depends on the ambient pressure is C_F, the thrust coefficient. This coefficient can be found from Eq. 3–30 or from Fig. 3–7 by following a vertical line corresponding to $A_2/A_t = 6.0$. We first have to find the ratio p_1/p_3 at sea level and at 10 km. Solving Eq. 3–25 for $p_2/p_1 = 0.0198$, or $p_2 = 0.396$ atm, and using values from Appendix 2 we have

Sea level: $p_1/p_3 = 20/1.0 = 20 \quad C_F = 1.33$

10 km altitude: $p_1/p_3 = 20/0.26151 = 76.5 \quad C_F = 1.56$

Thrust increase $= (1.56 - 1.33)/1.33 = 17.3\%$

Note that performance at optimum nozzle expansion ($p_2 = p_3 = 0.396$ atm.) this nozzle delivers a thrust that is somewhat below the maximum. From Fig. 3–7 we find that for $\epsilon = 6$ the optimum $C_F = 1.52$, which corresponds to 7.2 km elevation, whereas the maximum value is $C_F = 1.63$ in a vacuum.

Under- and Overexpanded Nozzles

An *underexpanded nozzle* discharges the gases at an exit pressure greater than the external pressure because its exit area is too small for an optimum expansion. Gas expansion is therefore incomplete within the nozzle, and further expansion will take place outside of the nozzle exit because the nozzle exit pressure is higher than the local atmospheric pressure.

In an *overexpanded nozzle* the gas exits at lower pressure than the atmosphere as it has a discharge area too large for optimum. The phenomenon of overexpansion inside a supersonic nozzle is indicated schematically in Fig. 3–8, from typical early pressure measurements of superheated steam along the nozzle axis and different back pressures or pressure ratios. Curve *AB* shows the variation of pressure with

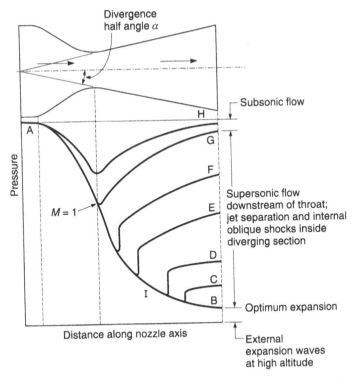

FIGURE 3–8. Distribution of pressures in a converging–diverging nozzle for different flow conditions. Inlet pressure is the same, but exit pressure changes. Based on early experimental data.

optimum back pressure fully corresponding to the nozzle area ratio. Curves AC to AH show pressure variations along the axis for increasingly higher external pressures. The expansion within the nozzle proceeds normally along its initial portion but, for example, at any location after I on curve AD the pressure is lower than the exit pressure and a sudden rise in pressure takes place that is accompanied by *separation* of the flow from the walls (separation is described later in this chapter).

Separation behavior in nozzles is deeply influenced by the presence of compression waves or shock waves inside the diverging nozzle section, which are strong discontinuities and exist only in supersonic flow. The sudden pressure rise in curve ID represents such a compression wave. Expansion waves (below point B), also strictly supersonic phenomena, act to match flows from the nozzle exit to lower ambient pressures. Compression and expansion waves are further described in Chapter 20.

These different possible flow conditions in supersonic nozzles may be stated as follows:

1. When the external pressure p_3 is below the nozzle exit pressure p_2, the nozzle will flow full but there will form external expansion waves at its exit (i.e., underexpansion). The expansion of the gas inside the nozzle is incomplete and the values of C_F and I_s will be less than at optimum expansion.

2. For external pressures p_3 somewhat higher than the nozzle exit pressure p_2, the nozzle will continue to flow full. This will occur until p_2 drops to a value between about 10 and 40% of p_3. The expansion is somewhat inefficient and C_F and I_s will have lower values than an optimum nozzle would have. Weak oblique shock waves will develop outside the nozzle exit section.

3. At higher external pressures, flow separation will begin to take place inside the divergent portion of the nozzle. The diameter of the exiting supersonic jet will be smaller than the nozzle exit diameter (with steady flow, separation remains typically axially symmetric). Figures 3–9 and 3–10 show diagrams for separated flows. The axial location of the separation plane depends on both the local pressure and the wall contour. The separation point travels upstream with increasing external pressure. At the nozzle exit, separated flow remains supersonic in the center portion but is surrounded by an annular-shaped section of subsonic flow. There is a discontinuity at the separation location and the thrust is reduced compared to an optimum expansion nozzle at the available area ratio. Shock waves may exist outside the nozzle in the plume.

4. For nozzles in which the exit pressure is just below the value of the inlet pressure (i.e., curve AH in Fig. 3–8), the pressure ratio is below the critical pressure ratio (as defined by Eq. 3–20) and subsonic flow would prevail throughout the entire nozzle. This condition normally occurs in rocket nozzles for a short time during the start and stop transients.

Methods for estimating pressure at the location of the separation plane inside the diverging section of a supersonic nozzle have traditionally been empirical (see Ref. 3–4). Reference 3–5 describes a variety of nozzles, their behavior, and methods used to estimate the location and pressure at separation. Actual values of pressure for

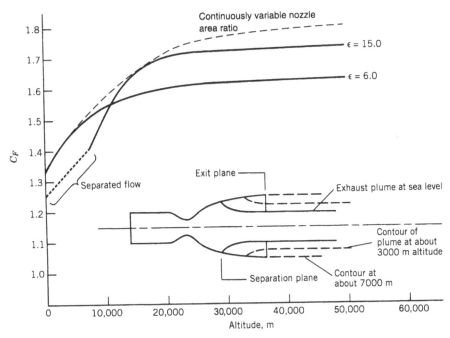

FIGURE 3–9. Thrust coefficient C_F for two nozzles of different area ratios. One has jet separation below about 7000 m altitude. The fully expanded exhaust plume at higher altitudes is not shown in the sketch.

the overexpanded and underexpanded regimes described above are functions of the specific heat ratio and the area ratio (see Ref. 3–1).

The axial thrust direction is not usually altered by separation because steady flows tend to separate uniformly over the divergent portion of a nozzle, but during start and stop transients this separation may not be axially symmetric and may momentarily produce large side forces at the nozzle walls. During normal sea-level transients in large rocket nozzles (before the chamber pressure reaches its full value), some momentary flow oscillations and nonsymmetric separation of the jet may occur while in overexpanded flow operation. The magnitude and direction of these transient side forces may change rapidly and erratically. Because these resulting side forces can be large, they have caused failures of nozzle exit cone structures and thrust vector control gimbal actuators. References 3–5, 3–6, and 3–7 discuss techniques for estimating these side forces.

When flow separates, as it does in a highly overexpanded nozzle, the thrust coefficient C_F can be estimated only when the location of separation in the nozzle is known. The C_F thus determined relates to an equivalent smaller nozzle with an exit area equal to that at the separation plane. The effect of separation is move the effective value of ϵ in Figs. 3–6 and 3–7 to the left of the physical or design value. Thus, with separated gas flows, a nozzle designed for high altitude (large value of ϵ) could have a larger

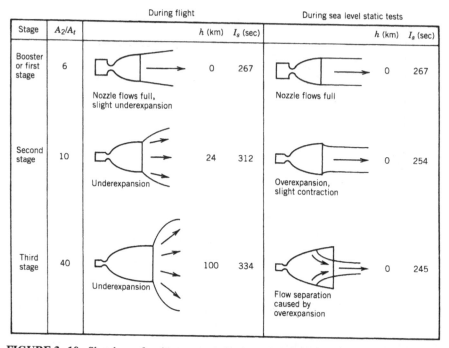

Stage	A_2/A_t	During flight		During sea level static tests	
		h (km)	I_s (sec)	h (km)	I_s (sec)
Booster or first stage	6	Nozzle flows full, slight underexpansion — 0	267	Nozzle flows full — 0	267
Second stage	10	Underexpansion — 24	312	Overexpansion, slight contraction — 0	254
Third stage	40	Underexpansion — 100	334	Flow separation caused by overexpansion — 0	245

FIGURE 3–10. Sketches of exhaust gas behavior of three typical rocket nozzles of a three-stage launch vehicle. The first vehicle stage has the biggest chamber and the highest thrust but the lowest nozzle exit area ratio, and the top or third stage usually has the lowest thrust but the highest nozzle exit area ratio.

thrust at sea level than expected; in this case, separation may actually be desirable. But with continuous separated flow operation a large and usually heavy portion of the nozzle is not utilized making it bulkier and longer than necessary. Such added engine mass and size decrease flight performance. Designers therefore usually select an area ratio that will not cause separation.

Because of potentially uneven flow separation and its destructive side loads, sea-level static tests of an upper stage or of a space propulsion system with high area ratios are usually avoided because they result in overexpanded nozzle operation; instead, tests are done with a sea-level substitute nozzle (of a much smaller area ratio) often called a "stub nozzle." However, actual or simulated altitude testing (in an altitude test facility, see Chapter 21) is done with nozzles having the correct area ratio. One solution that avoids separation at low altitudes and has high values of C_F at high altitudes would consist of a nozzle that changes area ratio in flight. This is discussed at the end of this section.

For most atmospheric flights, rocket systems have to operate over a range of altitudes; for a fixed chamber pressure this implies a range of nozzle pressure ratios. The most desirable nozzle for such an application is not necessarily one that gives optimum nozzle gas expansion, but one that gives the best vehicle overall flight performance (say, total impulse, or range, or payload); this can often be related to a time average over the powered flight trajectory.

Example 3–5.   Using the data from Example 3–4 ($p_1 = 20$ atm, $k = 1.30$), consider two nozzle area ratios, $\epsilon = 6.0$ and $\epsilon = 15.0$. Compare the performance of these two as a function of altitude by plotting C_F up to 50 km. Assume no shocks inside the nozzle.

SOLUTION. The procedures are the same as those in Example 3–4 and the resulting plot has been shown in Fig. 3–9. Values of C_F can be obtained by following vertical lines for $\epsilon = 6.0$ and 15.0 in Fig. 3–7 corresponding to the different pressure ratios at increasing altitudes. Alternatively, Eq. 3–30 may be used for greater accuracy. The lower area ratio gives better performance at the lower altitudes, but above about 10 km the larger area ratio gives higher values of C_F. Also note that in Fig. 3–9 that the upper dashed line represents what a continuously variable nozzle area ratio could accomplish by matching p_2 to p_3 at all altitudes.

The optimum pressure ratio always occurs when $p_1/p_3 = p_1/p_2$. For the $\epsilon = 15.0$ case, from Fig. 3–5 or 3–7, this ratio is about 180; here $p_3 = 20/180 = 0.111$ atm corresponding to about 15.5 km altitude. Below this altitude such a nozzle is overexpanded. At sea level, where $p_1/p_3 = 20$, separation could occur as indicated in Fig. 3–7. For similarly shaped nozzles, it is estimated that sea-level separation begins to take place at a cross section where the local internal pressure is between 10 and 40% of p_3 or below 0.4 atm. Upon separation, this nozzle would not flow full downstream beyond an area ratio of about 6 or 7 and the gas jet would occupy only the central portion of the nozzle exit (see Figs. 3–9 and 3–10). Weak oblique shock waves and jet contractions would then raise the exhaust jet's pressure to match the value of the surrounding atmosphere. If the jet does not separate, it would reach an exit pressure of 0.111 atm, but this is often an unstable condition. As the vehicle gains altitude, the separation plane would gradually move downstream until, at an altitude of about 7000 m, the exhaust gases could occupy the full exit diverging section.

Figure 3–10 shows behavior comparisons of altitude and sea-level ground-tests of three nozzles and their plumes at different area ratios for a typical three-stage satellite launch vehicle. When fired at sea-level conditions, the nozzle of the third stage with the highest area ratio will experience flow separation and suffer a major performance loss; the second stage will flow full but the external plume will contract; since $p_2 < p_3$, there is a loss in I_s and F. There is no effect on the first stage nozzle. When the second and third stages actually operate at their proper altitude, they would experience no separation.

Example 3–6.   A rocket engine test near sea level gives the following data: thrust $F = 53{,}000$ lbf, propellant mass flow rate $\dot{m} = 208$ lbm/sec, nozzle exit area ratio $A_2/A_t = 10.0$, actual local atmospheric pressure at test station $p_3 = 13.8$ psia, and chamber pressure $p_1 = 620$ psia. The test engineer also knows that for the same flow rate (and mixture ratio) the theoretical specific impulse is 289 sec at the standard reference conditions of $p_1 = 1000$ psia and $p_3 = 14.7$ psia, and that $k = 1.20$. For many propellants we may assume that the combustion temperature T_1 and k do not vary significantly with chamber pressure. Compare the test performance of this rocket with its equivalent at sea level, standard, and vacuum performance.

SOLUTION. The pressure ratio for the test condition is $620/13.8 = 44.93$; if the test had been conducted at 14.7 psia, this pressure ratio would have been $620/14.7 = 42.18$; for the standard reference conditions the pressure ratio is $1000/14.7 = 68.03$.

Since this nozzle is operating supersonically, we find, from Eqs. 2–5, 3–24, and 3–31, that specific impulse ratios depend only on ratios of the thrust coefficient, that is, $(I_s)_b = (I_s)_a[(C_F)_b/(C_F)_a]$. From Fig. 3–6 or from Eq. 3–30 one obtains $C_F = 1.48$ for the test condition where $\epsilon = 10$, $k = 1.20$, and $p_1/p_3 = 44.93$. The corresponding specific impulse is, from Eq. 2–5, $I_s = 53000/208 = 255$ sec.

The following table shows all results:

	Test Results	Test for Sea Level	Standard Condition	Vacuum Condition
p_1 (psia)	620	620	1000	620
p_3 (psia)	13.8	14.7	14.7	0
p_1/p_3	44.93	42.18	68.03	∞
F (lbf)	53,000	52,480	55,702	60,690
I_s (sec)	255	252.5	268	292
C_F	1.52	1.505	1.60	1.74

Figures 3–9 and 3–10 suggest that a continuous optimum performance for an ascending (e.g., launch) rocket vehicle would require a "rubber-like" diverging section that would lengthen and enlarge so that the nozzle exit area would increase as the ambient pressure is reduced. Such design would then allow the rocket vehicle to attain its best performance at all altitudes as it ascends. As yet we have not achieved such flexible, stretchable mechanical nozzle designs with full altitude compensation similar to "stretching rubber." However, there are a number of practical nozzle configurations that can be used to alter nozzle shape with altitude and obtain peak performance at different altitudes. These are discussed in the next section.

Influence of Chamber Geometry

When the chamber has a cross section that is about four times larger than the throat area $(A_1/A_t > 4)$, the chamber velocity v_1, may be neglected, as was mentioned in introducing Eqs. 3–15 and 3–16. However, some vehicle space- or weight-constraints for liquid propellant engines often require smaller thrust chamber areas, and for solid propellant motors grain design considerations may lead to small void volumes or small perforations or port areas. In these cases, v_1's contribution to performance can no longer be neglected. Gases in the chamber expand as heat is being added. The energy expended in accelerating these expanding gases within the chamber will also cause a pressure drop, and processes in the chamber can be adiabatic (no heat transfer) but not isentropic. Losses are a maximum when the chamber diameter equals the nozzle throat diameter, which means that there is no converging nozzle section. This condition is called a *throatless rocket motor* and is found in a few tactical missile booster applications, where there is a premium on minimum inert mass and length. Here, flight performance improvements due to inert mass savings have supposedly outweighed the nozzle performance loss of a throatless motor.

Because of the significant pressure drops within narrow chambers, chamber pressures are lower at the nozzle entrance than they would be with a larger A_1/A_t.

TABLE 3–2. Estimated Losses for Small-Diameter Chambers

Chamber-to-Throat Area Ratio	Throat Pressure (%)	Thrust Reduction (%)	Specific Impulse Reduction (%)
∞	100	0	0
3.5	99	1.5	0.31
2.0	96	5.0	0.55
1.0	81	19.5	1.34

$k = 1.20$; $p_1/p_2 = 1000$.

This causes a small loss in thrust and specific impulse. The theory of such losses is given in the Second and Third Editions of this book, and some results are listed in Table 3–2.

3.4. NOZZLE CONFIGURATIONS

Several different proven nozzle configurations are available today. This section describes their geometries and performance. Other chapters in this book (6, 8, 12, 15, and 18) discuss nozzle materials, heat transfer, and their applications with mention of certain requirements, design, construction, and thrust vector control. Nozzles and chambers are typically circular in cross section and must have a converging section, a throat or minimum cross section, and a diverging section. Nozzles are apparent in Figs. 1–4, 1–5, 1–8, 2–1, 3–10, as well as in Figs. 3–11, 3–13, 11–1, 12–1 to 12–3, and 15–6 and 15–8. Refs. 3–4 and 3–8 describe several other nozzle configurations.

The *converging nozzle section* shape between the chamber and the throat does not significantly affect nozzle performance. Its subsonic flow can easily be turned with very low pressure drops and almost any given radius, cone angle, wall contour curve, or nozzle inlet shape is satisfactory. A few small attitude control thrust chambers have had their nozzle turn at 90° from the combustion chamber axis without measurable performance loss. The *throat contour* itself is also not very critical to performance, and any smooth curved shape is usually acceptable. Since pressure gradients are high in these two regions, the flow will always adhere to the walls. Principal differences in nozzle configurations are found in the diverging supersonic-flow section, as described below. In general, internal wall surfaces throughout the nozzle should be smooth and reflective to minimize friction, radiation absorption, and convective heat transfer, which is enhanced by surface roughness. Gaps, holes, sharp edges, and/or protrusions must be avoided.

Six different nozzle configuration types have been considered and they are depicted in Fig. 3–11. The first three sketches show the more common conical and bell-shaped nozzles and they are further described in this chapter. The last three have a center body inside the nozzle in order to provide altitude compensation (this means that outside the nozzle the hot-gas boundaries can grow as the outside pressure decreases allowing a more optimum expansion with altitude). Although these last three have been thoroughly ground tested, to date none of them has flown in a

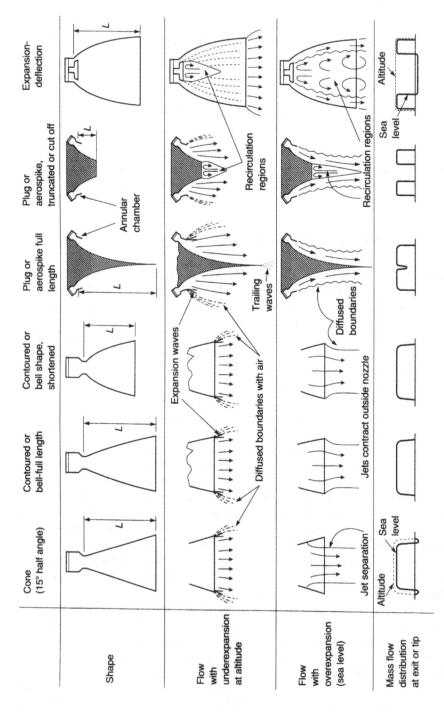

FIGURE 3–11. Simplified diagrams of different generic nozzle configurations and their flow effects.

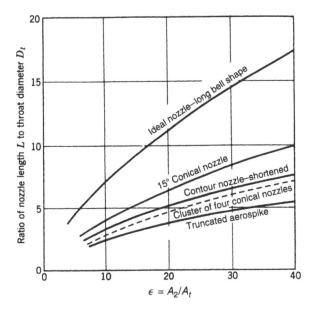

FIGURE 3–12. Length comparison of several types of nozzles. Taken in part from Ref. 3–9.

production space launch vehicle (see Eighth edition). The lengths of several nozzle types are compared in Fig. 3–12. The objectives for a good nozzle configuration are to obtain the highest practical I_s minimizing inert nozzle mass and conserving length (shorter nozzles can reduce vehicle length, vehicle structure, and vehicle inert mass).

Cone- and Bell-Shaped Nozzles

The *conical nozzle* is the oldest and perhaps the simplest configuration. It is relatively easy to fabricate and is still used today in many small nozzles. A theoretical correction factor λ must be applied to the nozzle exit momentum in any ideal rocket propulsion system using a conical nozzle. This factor is the ratio between the momentum of the gases exhausting with a finite nozzle angle 2α and the momentum of an ideal nozzle with all gases flowing in the axial direction:

$$\lambda = \tfrac{1}{2}\,(1 + \cos \alpha) \tag{3–34}$$

For ideal rockets $\lambda = 1.0$. For a rocket nozzle with a divergence cone angle of 30° (half angle $\alpha = 15°$), the exit momentum and therefore the axial exhaust velocity will be 98.3% of the velocity calculated by Eq. 3–15b. Note that the correction factor λ only applies to the first term (momentum thrust) in Eqs. 2–13, 3–29, and 3–30 and not to the second term (pressure thrust).

A half angle value of 15° has become the unofficial standard of reference for comparing correction factors or energy losses or lengths from different diverging conical nozzle contours.

Small nozzle divergence angles may allow most of the momentum to remain axial and thus produce high specific impulses, but they result long nozzles introducing performance penalties in the rocket propulsion system and the vehicle mass, amounting to small losses. Larger divergence angles give shorter, lightweight designs, but their performance may become unacceptably low. Depending on the specific application and flight path, there is an optimum conical nozzle divergence (typically between 12 and 18° half angle).

Bell-shaped or *contour nozzles* (see Figs. 3–11 and 3–13) are possibly the most common nozzle shapes used today. They have a high angle expansion section (20 to 50°) immediately downstream of the nozzle throat followed by a gradual reversal of nozzle contour slope so that at the nozzle exit the divergence angle is small, usually less than a 10° half angle. It is possible to have large divergence angles immediately behind the throat (of 20 to 50°) because high relative local pressures, large pressure gradients, and rapid expansions of the working fluid do not allow separation in this region (unless there are discontinuities in the nozzle contour). Gas expansion in supersonic bell nozzles is more efficient than in simple straight cones of similar area ratio and length because wall contours can be designed to minimize losses, as explained later in this section.

Changing the direction of supersonic flows with an expanding wall geometry requires *expansion waves*. These occur at thin regions within the flow, where the flow velocity increases as it slightly changes its direction, and where the pressure and temperature drop. These wave surfaces are at oblique angles to the flow. Beyond the throat, the gas crosses a series of expansion waves with essentially no loss of energy. In the bell-shaped configuration shown in Fig. 3–13 these expansions occur internally in the flow between the throat and the inflection location I; here the area is steadily increasing (like a flare on a trumpet). The contour angle θ_i is a maximum at this inflection location. Between the inflection point I and the nozzle exit E the flow area increases at a diminishing rate. Here, the nozzle wall contour is different and the rate of change in cross-sectional area per unit length is decreasing. One purpose of this last nozzle segment is to reduce the nozzle exit angle θ_e and thus to reduce flow divergence losses as the gas leaves the nozzle exit plane. The angle at the exit θ_e is small, usually less than 10°. The difference between θ_i and θ_e is called the turn-back angle. As the gas flow is turned in the opposite direction (between points I and E) oblique compression waves occur. These compression waves are thin surfaces where the flow undergoes a mild shock as it is turned back and its velocity slightly reduced. Each of these multiple compression waves causes a small energy loss. By carefully determining the wall contour (through an analysis termed "method of characteristics"), it is possible to balance the oblique expansion waves with the oblique compression waves and minimize these energy losses. Analyses leading to bell-shaped nozzle contours are presented in Chapter 20.33 of Ref. 3–3 and also in Refs. 3–8 to 3–11. Most rocket organizations have developed their own computer codes for such work. It has been found that the radius of curvature or contour shape at the throat region noticeably influences the contours in diverging bell-shaped nozzle sections.

The length of a bell nozzle is usually given as fractions of the length of a reference conical nozzle with a 15° half angle. An 80% bell nozzle has a length (distance

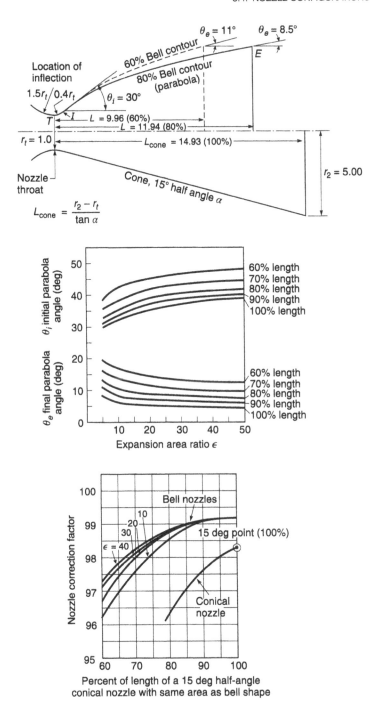

FIGURE 3–13. Top sketch shows comparison of nozzle inner diverging wall surfaces for a 15° conical nozzle, an 80% length bell nozzle, a 60% length bell nozzle, all at an area ratio of 25. The lengths are expressed in multiples of the throat radius r_t. The middle set of curves shows the initial angle θ_i and the exit angle θ_e for bell nozzles as functions of the nozzle area ratio and percent length. The bottom curves show the nozzle losses in terms of a correction factor.

between throat plane and exit plane) that is 20% shorter than a comparable 15° cone of the same area ratio. Reference 3–9 shows the original method of characteristics as initially applied to determine bell nozzle contours and later to shortened bell nozzles; a parabola has been used as a good approximation for the bell-shaped contour curve (Ref. 3–3, Section 20.33), and actual parabolas have been used in some nozzle designs. The top part of Fig. 3–13 shows a parabolic contour that is tangent (θ_i) at the inflection point I and has an exit angle (θ_e) at point E, and that its length L that has to be corrected for the curve TI. These conditions allow the parabola to be determined either from simple geometric analysis or from a drawing. A throat approach radius of $1.5r_t$ and a throat expansion radius of $0.4r_t$ were used here. If somewhat different radii had been used, the results would have been only slightly different. The middle set of curves gives the relation between length, area ratio, and the two angles of the bell contour. The bottom set of curves gives the correction factors, equivalent to the λ-factor for conical nozzles, which are to be applied to the thrust coefficient or the exhaust velocity, provided the nozzles operate at optimum expansion, that is, $p_2 = p_3$.

Table 3–3 shows data for parabolas developed from Fig. 3–13 in order to allow the reader to apply this method and check results. The table shows two shortened bell nozzles and a conical nozzle, each for three area ratios. A 15° half angle cone is given as reference. It can be seen that as the length decreases, the losses are higher for the shorter length and slightly higher for the smaller nozzle area ratios. A 1% improvement in the correction factor gives about 1% greater specific impulse (or thrust), and this difference can be significant in many applications. A reduced length is an important benefit usually reflected in improvements of the vehicle mass ratio. Table 3–3 and Fig. 3–13 show that bell nozzles (75 to 85% the length of cones) can be slightly more efficient than a longer 15° conical nozzle (100% length) at the same area ratio. For shorter nozzles (below 70% equivalent length), the energy losses due

TABLE 3–3. Data on Several Bell-Shaped Nozzles

Nozzle Exit Area Ratio	10	25	50
Cone (15° Half Angle)			
Length (100%)[a]	8.07	14.93	22.66
Correction factor λ	0.9829	0.9829	0.9829
80% Bell Contour			
Length[a]	6.45	11.94	18.12
Correction factor λ	0.985	0.987	0.988
Approximate half angle at inflection point and exit (degrees)	25/10	30/8	32/7.5
60% Bell Contour			
Length[a]	4.84	9.96	13.59
Correction factor λ	0.961	0.968	0.974
Approximate half angle at inflection point and exit (degrees)	32.5/14	36/17	39/18

[a]The length is given in dimensionless form as a multiple of the throat radius.

to internal oblique shock waves can become substantial so that such short nozzles are not commonly used today.

For solid propellant rocket motor exhausts containing small solid particles in the gas (usually aluminum oxide), and for the exhaust of certain gelled liquid propellants, solid particles impinge against the nozzle wall in the reversing curvature section between *I* and *E* in Fig. 3–13. While the gas can be turned by oblique waves to have less divergence, any particles (particularly the larger particles) will tend to move in straight lines and hit the walls at high velocities. The resulting abrasion and erosion of the nozzle wall can be severe, especially with commonly used ablative and graphite materials. This abrasion by hot particles increases with turn-back angle. Only when the turn-back angle and thus also the inflection angle θ_i are reduced such erosion can become acceptable. Typical solid rocket motors flying today have values of inflection angles between 20 and 26° and turn-back angles of 10 to 15° with different wall contours in their divergent section. In comparison, current liquid rocket engines without entrained particles have inflection angles between 27 and 50° and turn-back angles of between 15 and 30°. Therefore, performance enhancements resulting from the use of bell-shaped nozzles (with high correction factor values) are somewhat lower in solid rocket motors with solid particles in the exhaust.

An ideal (minimum loss through oblique waves) bell-shaped nozzle is relatively long, equivalent to a conical nozzle of perhaps 10 to 12°, as seen in Fig. 3–11. Because long nozzles are heavy, in applications where vehicle mass is critical, somewhat shortened bell nozzles are preferred.

Two-Step Nozzles. Two-position nozzles (with different expansion area ratio sections) can give better performance than conventional nozzles with a fixed single area ratio. This is evident in Fig. 3–9; the lower area ratio nozzle ($\epsilon = 6.0$) performs best at low altitudes and the higher area ratio nozzle performs best at higher altitudes. If these two nozzles could somehow be mechanically combined, the resulting two-position nozzle would perform closer to a nozzle that adjusts continuously to the optimum area ratio, as shown by the thin dashed curve. When integrated over flight time, the extra performance has a noticeable payoff for high-velocity missions, such as Earth-orbit injection and deep space missions. Several bell-shaped nozzle concepts have evolved that achieve maximum performance at more than a single altitude. Figure 3–14 shows three different two-step nozzle concepts, having an initial low area ratio A_2/A_t for operation at or near the Earth's surface and a larger second area ratio that improves performance at high altitudes. See Ref. 3–4.

Extendible Nozzles. The *extendible nozzle* is now relatively common for top stages of multistage vehicles. Its nozzle extension is moved into operating position after a lower stage has finished working and has dropped off. These propulsion systems only operate in space with the nozzle extended. Extendible nozzles require actuators, a power supply, mechanisms for moving the extension into position during flight, and fastening and sealing devices. They have successfully flown in several solid rocket motor nozzles and liquid engine applications, where they were deployed

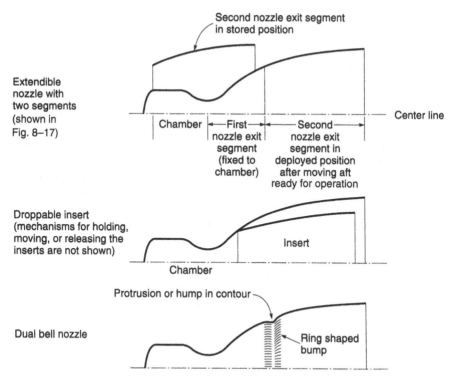

FIGURE 3–14. Simplified diagrams of three altitude-compensating two-step nozzle concepts. See Ref. 3–4.

prior to ignition. See Fig. 8–17. Key concerns here are reliable, rugged mechanisms to move the extension into position, the hot gas seal between the nozzle sections, and the extra weights involved. Its principal merit is the short nozzle length while stored during operation of the lower stages, which reduces vehicle length and inert mass. There have been versions of this concept with three nozzle segments; one is shown in Fig. 12–3.

The *droppable insert concept* shown in Fig. 3–14 avoids any moving mechanisms and gas seals but has a potential stagnation temperature problem at the joint of the two segments. To date, it has not flown in U.S. production vehicles. See Refs. 3–4 and 3–12.

The *dual-bell nozzle concept* uses two shortened bell nozzles combined into one with a "ring-shaped bump" or inflection point between them, as shown in Fig. 3–14. During ascent, it functions first with the lower area ratio, with flow separation occurring at the bump. As altitude increases, the flow expands further and attaches itself downstream of this inflection point, filling the full nozzle exit section and operating with the higher area ratio for higher performance. There is a small performance penalty for a contour with a circular bump and some concerns about heat transfer. To date, there has been little flight experience with this concept.

Multiple Nozzles. Whenever possible, reducing nozzle length by replacing a single large nozzle with a cluster of smaller nozzles on liquid engines or solid motor units (with the same total thrust) will reduce vehicle length and often vehicle structure and inert mass. Quadruple thrust chamber arrangements have been used effectively in several U.S. and many large Russian space launch vehicles and missiles. Multiple nozzles can be designed to provide thrust vector control (see Figure 18–9). As indicated in Fig. 3–12, the nozzle length of such a cluster can be about 30% shorter than a single nozzle of an equivalent larger thrust chamber. But vehicle diameter at the cluster nozzle exit location needs to be somewhat larger, vehicle drag can be somewhat higher, and there is additional engine complexity often with more engine mass.

3.5. REAL NOZZLES

The assumptions listed in Section 3.1 are approximations that allow relatively simple mathematical relations for the analysis of real rocket nozzles. With most of these assumptions it is possible either (1) to use an empirical correction factor (based on experimental data) or (2) to develop or use more accurate algorithms, which represent a better simulation of energy losses, physical or chemical phenomena, and contain more complex theoretical descriptions. Some of these more accurate approaches are briefly covered in this section.

Compared to an ideal nozzle, real nozzles have losses and some of the flow internal energy is unavailable for conversion into kinetic energy of the exhaust. Principal losses are listed below and several of these are then discussed in more detail.

1. The *divergence of the flow* in the nozzle exit sections is a loss that varies as a function of the cosine of the divergence angle, as shown by Eq. 3–34 for conical nozzles. These losses can be reduced with bell-shaped nozzle contours.

2. Small chamber or port area cross sections relative to the throat area or *low nozzle contraction ratios* A_1/A_t cause pressure losses in the chamber and slightly reduce the thrust and exhaust velocity. See Table 3–2.

3. The lower velocities at the wall *boundarylayers* reduce the effective average exhaust velocity by 0.5 to 1.5%.

4. *Solid particles* and/or *liquid droplets* in the gas may cause losses of perhaps up to 5% (depending on particle size, shape and percent solids), as described below.

5. *Unsteady combustion* and/or flow oscillations may result in small losses.

6. *Chemical reactions within nozzle flows* change gas composition and gas properties and gas temperatures, amounting to typically a 0.5% loss. See Chapter 5.

7. Chamber pressures and overall performance are lower during *transient operations*, for example, during start, stop, or pulsing.

8. Any gradual *erosion* of the *throat region* increases its diameter by perhaps 1 to 6% during operation with uncooled nozzle materials (such as fiber-reinforced plastics or carbon). In turn, this will reduce the chamber pressure and thrust

by about 1 to 6%. Such throat area enlargements cause a slight reduction in specific impulse, usually less than 0.7%.

9. *Nonuniform gas compositions* may reduce performance (due to incomplete mixing or incomplete combustion).

10. *Real gas properties* may noticeably modify gas composition, that is, actual values of k and $\mathfrak{M}$ cause a small loss in performance, about 0.2 to 0.7%.

11. Operation at nonoptimum altitudes reduces thrust and specific impulse for nozzles of fixed area ratios. There is no loss if the vehicle flies only at the altitude designed for optimum nozzle expansion ($p_2 = p_3$). If it flies at higher or lower altitudes, then the losses (during those portions of the flight) can amount up to 10% of thrust compared to a nozzle with altitude compensation, as can be seen in Figs. 3–6 and 3–7. These conditions may also reduce overall vehicle flight performance between 1 and 5%.

Boundary Layers

Real nozzles always develop thin viscous *boundary layers* adjacent to their inner walls, where gas velocities are much lower than in the free-stream region. An enlarged depiction of a boundary layer is shown in Fig. 3–15. Immediately next to the wall the flow velocity is zero; beyond the wall the boundary layer may be considered as being built up of successive annular-shaped thin layers of increasing velocity until the free-stream value is reached. The low-velocity flow region closest to the wall is always laminar and subsonic, but at the higher-velocity regions of the boundary layer the flow becomes supersonic and can be turbulent. Local temperatures at some locations in the boundary layer can be noticeably higher than the free-stream temperature because of the conversion of kinetic energy into thermal energy that occurs as the velocity is locally slowed down and by heat created by viscous effects. The flow layer immediately next to the wall will be cooler because of heat transfer to the wall. These gaseous boundary layers have a profound effect on the overall heat transfer to nozzle and chamber walls. They also affect rocket performance, particularly in applications with very small nozzles and also with relatively long nozzles with high nozzle area ratios, where a comparatively high proportion of the total mass flow (2 to 25%) can be in the lower-velocity region of the boundary layer. High gradients in pressure, temperature, or density and changes in local velocity (direction and magnitude) also influence boundary layers. Flow separation in nozzles always originates inside the boundary layers. Scaling laws for boundary layer phenomena have not been reliable to date.

Theoretical approaches to boundary layer effects can be found in Chapters 26 to 28 of Ref. 3–1 and in Ref. 3–18. Even though theoretical descriptions of boundary layers in rocket nozzles are yet unsatisfactory, their overall effects on performance can be small—for most rocket nozzles with unseparated flow, boundary layer derived losses of specific impulse seldom exceeds 1%.

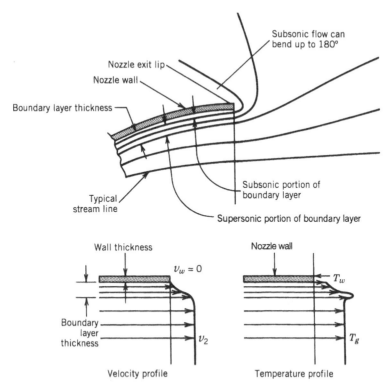

FIGURE 3–15. Flow conditions at a nozzle exit lip at high altitudes, showing streamline, boundary layer, velocity, and temperature profiles.

Multiphase Flow

In some propulsion systems, the gaseous working fluid contains small liquid droplets and/or solid particles entrained in the flow. They may heat the gas during nozzle expansion. This heating, for example, occurs with solid propellants (see Section 13–4) or in some gelled liquid propellants, which contain aluminum powder because small oxide particles form in the exhaust. This can also occur with iron oxide catalysts, or solid propellants containing aluminum, boron, or zirconium.

In general, if the particles are relatively small (typically with diameters of 0.005 mm or less), they will have almost the same velocity as the gas and will be in thermal equilibrium with the nozzle gas flow. Thus, even if some kinetic energy is transferred to accelerate the embedded particles, the flow gains thermal energy from them. As particle diameters become larger, the mass (and thus the inertia) of each particle increases as the cube of its diameter; however, the drag or entrainment force increases only as the square of the diameter. Larger particles therefore lag the gas movement and do not transfer heat to the gas as readily as do smaller particles. Larger particles have a lower momentum than an equivalent mass agglomeration of smaller particles and reach the nozzle exit at higher temperatures because they give up less of their thermal energy.

It is possible to derive simple equations for correcting performance (I_s, c, or c^*) as shown below and as given in Refs. 3–13 and 3–14. These formulas are based on the following assumptions: specific heats of the gases and the particles are constant throughout the nozzle flow, particles are small enough to move at essentially the same velocity as the gas and are in thermal equilibrium with it, and that particles do not exchange mass with the gas (no vaporization or condensation). Expansion and acceleration occur only in the gas, and the volume occupied by the particles is negligibly small compared to the overall gas volume. When the amount of particles is small, the energy needed to accelerate these particles may also be neglected. There are also no chemical reactions.

The enthalpy h, the specific volume V, the specific heat ratio k, and the gas constant R can be expressed as functions of a particle fraction β, which represents the mass of particles (liquid and/or solid) divided by the total mass. The specific heats at constant pressure and volume are denoted as c_p and c_v. Using the subscripts g and s to refer to gaseous or solid states, the following relationships then apply:

$$h = (1 - \beta)(c_p)_g T + \beta c_s T \tag{3–35}$$

$$V = V_g(1 - \beta) \tag{3–36}$$

$$p = R_g T / V_g \tag{3–37}$$

$$R = (1 - \beta)R_g \tag{3–38}$$

$$k = \frac{(1 - \beta)c_p + \beta c_s}{(1 - \beta)c_v + \beta c_s} \tag{3–39}$$

Performance parameters for nozzle flows from solid-propellant combustion products that contain small particles or small droplets in the exhaust gas are briefly discussed at the end of this section.

Equations 3–35 to 3–39 are then used with the formulas for simple one-dimensional nozzle flow, such as Eq. 2–15, 3–15, or 3–32. Values for specific impulse or characteristic velocity will decrease with β, while the particle percent increases. For very small particles (less than 0.01 mm in diameter) and small values of β (less than 6%) losses in specific impulse are often below 2%. For larger particles (over 0.015 mm diameter) and larger values of β this theory does not apply—the specific impulse can become 10 to 20% less than the value without any flow lag. Actual particle sizes and their distribution depend on the particular propellant combustion products, the actual particle composition, and the specific rocket propulsion system; these parameters usually have to be measured (see Chapters 13 and 20). Thus, adding a metal, such as aluminum, to a solid propellant will increase performance only when the additional heat release can sufficiently increase the combustion temperature T_0 so that it more than offsets any decrease caused by the nonexpanding mass portion of the particles in the exhaust.

With very high area ratio nozzles and low nozzle exit pressures (at high altitudes or the vacuum of space) some gaseous propellant ingredients may condense; when temperatures drop sharply in the nozzle, condensation of H_2O, CO_2, or NH_3 are known to occur. This causes a decrease in the gas flow per unit area while transfer of the latent heat of particle vaporization to the remaining gas takes place. It is also possible to form a solid phase and precipitate fine particles of snow (H_2O) or frozen fogs of other species. Overall effect on performance is small only if the particle sizes are small and the percent of precipitate mass is moderate.

Other Phenomena and Losses

The *combustion process* can never be totally steady. Low- and high-frequency oscillations in chamber pressure of up to perhaps 5% of rated values are considered to be smooth-burning and sufficiently steady flow. Gas properties (k, $\mathfrak{M}$, c_p) and flow properties (v, V, T, p, etc.) will also oscillate with time and will not necessarily be uniform across the flow channel. We therefore deal with "average" values of these properties only, but it is not always clear what kind of time averages are appropriate. Energy losses due to nonuniform unsteady burning are difficult to assess theoretically; for smooth-burning rocket systems these losses are neglected, but they may become significant for the larger-amplitude oscillations.

Gas composition may change somewhat in the nozzle when chemical reactions are occurring in the flowing gas, and the assumption of a uniform or "chemically-frozen-composition" gas flow may not be fully valid for all chemical systems. A more sophisticated approach for determining performance with changing composition and changing gas properties is described in Chapter 5. Also, any thermal energy that exits the nozzle ($\dot{m}\, c_p\, T_2$) is unavailable for conversion to useful propulsive (kinetic) energy, as shown in Fig. 2–2. The only way to decrease this loss is to reduce the nozzle exit temperature T_2 (with a larger nozzle area ratio), but most often this amounts to other sizable losses.

When operating durations are multiple and short (as with antitank rockets or pulsed attitude control rockets which start and stop repeatedly), start and stop *transients* become a significant fraction of total operating time. During these transient periods the average thrust, chamber pressure, and specific impulse will be lower than during steady operating conditions. This situation may be analyzed with a step-by-step process; for example, during start-up the amount of propellant flowing in the chamber has to equal the flow of gas through the nozzle plus the amount of gas needed to fill the chamber to a higher pressure; alternatively, an empirical curve of chamber pressure versus time, if available, could be used as the basis of such calculations. Transition times can be negligibly short in small, low-thrust propulsion systems, perhaps a few milliseconds, but they become longer (several seconds) for large propulsion systems.

Performance Correction Factors

In addition to the theoretical cone- and bell-shaped nozzle correction factors discussed in Section 3.4, we discuss here another set of empirical correction factors. These factors represent a variety of nonideal phenomena (such as friction, imperfect mixing and/or combustion, heat transfer, chemical nonequilibrium, and two- and three-dimensional effects) that are unavoidably present; refer to Fig. 2–2 and Ref. 3–4. Correction factors are arbitrarily defined in rocket analysis and may differ from the more conventional efficiencies—there are no "universal performance correction factors." For specific propulsion systems, where accurately measured data are available, they allow for simple predictions of actual performance. For example, a velocity correction factor of 0.942 means that the velocity or the actual specific impulse is about 94% of theoretical (a commonly used value might be closer to 0.92).

Corrections factors are used by engineers to determine performance ahead of testing and for preliminary designs, informal proposals, and health monitoring systems. In all these, a set of *accepted* or *nominal values* is needed to estimate performance together with some useful formula recipes. For accurate calculations, the industry relies extensively on computer programs which are largely proprietary.

Correction factors are ratios of measured or *actual* (subscript "a") to formulated or *ideal* (subscript "i") values. In the ordinary testing of rocket propulsion systems, the combustion chamber pressure, the propellant mass flow rates, the thrust force, and the throat and exit areas are typically measured. These measurements yield two direct ratios, namely, the *thrust correction factor* ($\zeta_F = F_a/F_i$) and the *discharge correction factor* ($\zeta_d = \dot{m}_a/\dot{m}_i$) and, using measured chamber pressures, the product $[(p_1)_a(A_t)_a]$ that has units of force. This product enters into the formulation of two other correction factors discussed below. The *nozzle area ratio* $[\epsilon_a = (A_2)_a/(A_t)_a]$ in real nozzles is found from measurements and this ratio may differ from the ideal or calculated value as will be shown in Example 3–7. In liquid propellant rocket engines, the propellant mass flow rate $\dot{m}_a$ is measured during ground tests using calibrated flowmeters, but it is practically impossible to directly measure mass flow rates in solid propellant motors. Only an effective $\dot{m}_a$ is determined from the initial and final weights at ground testing. Therefore, the discharge correction factor ζ_d for solid propellant rocket motors represents only an averaged value.

The thrust correction factor (ζ_F) is found from the ratio of thrust measurements with the corresponding ideal values given by Eq. 3–29. The discharge correction factor (ζ_d) can be determined from the ratio of mass-flow-rate measurements with the corresponding theoretical values of Eq. 3–24. Unlike incompressible flows, in rocket propulsion systems the value of ζ_d is always somewhat larger than 1.0 (up to 1.15) because actual flow rates may exceed ideal flow rates for the following reasons:

1. Incomplete combustion (a lower combustion temperature), which results in increases of the exhaust gas densities.
2. Cooling at the walls that reduces boundary layer temperatures and thus the average gas temperature, especially in small thrust chambers.
3. Changes in the specific heat ratio and molecular mass in an actual nozzle that affect the flow rate and thus the discharge correction factor (see Eq. 3–24).

The c^* *efficiency* or ζ_{c^*} *correction factor* represents a combined effectiveness of the combustion chamber and the injector design. It can be determined from the ratio of measured values of $[(p_1)_a(A_t)_a]/\dot{m}_a$ (from Eq. 2–17) with the corresponding ideal value of the right-hand side of Eq. 3–32. In well-designed combustion chambers, the value of ζ_{c^*} correction factor is over 95%. The ζ_{C_F} *correction factor*, also known as the C_F *efficiency*, represents the effectiveness of the nozzle design at its operating conditions. It can be determined using measured values of $F_a/[(p_1)_a(A_t)_a]$ (from Eq. 3–31) with the corresponding ideal value of Eq. 3–30. In well-designed nozzles, the value of the ζ_{C_F} correction factor is above 90%.

An *effective exhaust velocity correction factor* $[\zeta_v \equiv (F_a/\dot{m}_a)/c_i]$ may now be introduced using that velocity's definition given in Eq. 2–16 (or from Eq. 3–32, $c = c^* C_F$) as

$$\zeta_v = \zeta_F/\zeta_d = \zeta_{c^*}\zeta_{C_F} \qquad (3\text{–}40)$$

This further suggests a correction factors equation, a form equivalent to Eq. 3–33, written as $\zeta_F = \zeta_d\zeta_{c^*}\zeta_{C_F}$, which has meaning only within existing experimental uncertainties and implies that in steady flow all four correction factors cannot be arbitrarily determined but must be so related.

Additional useful relations may also be written. The actual specific impulse, from Eq. 2–5, may now be calculated from

$$(I_s)_a = (I_s)_i(\zeta_v) \qquad (3\text{–}41)$$

The thermodynamic *nozzle efficiency* η_n which is traditionally defined as the ratio of ideal to the actual enthalpy changes (see Eqs. 3–15a and 3–16) under a given pressure ratio becomes

$$\eta_n = (\Delta h)_a/(\Delta h)_i = \tfrac{1}{2}(v^2_2)_a/\tfrac{1}{2}(v^2_2)_i \approx (\zeta_v)^2 \qquad (3\text{–}42)$$

The approximate sign above becomes an equality when $p_2 = p_3$. This nozzle efficiency will always have a value less than 1.0 and represents losses inside the nozzle (Ref. 3–3). Small nozzles being used for *micropropulsion* can have quite low η_n's because of their relatively large frictional effects which depend in the surface-to-volume ratio (some cross sections being rectangular thus having additional sharp-corner losses).

Under strictly *chemically-frozen-composition* flow assumptions, the right-hand side of Eq. 3–32 leads to a useful relation between the ideal and actual stagnation temperatures [here $T_1 \equiv (T_0)_i$]:

$$(T_0)_a/(T_0)_i \approx (\zeta_{c^*})^2 \qquad (3\text{–}43)$$

With conditions at the inlet of a supersonic nozzle readily calculable, one task for the analyst is to determine the actual throat area required to pass a specified mass flow rate of gaseous propellant and also to determine the area and the flow properties at the nozzle exit plane. Across the nozzle heat losses modify the local stagnation temperatures and, together with friction, the stagnation pressures (both of which begin as $(p_0)_1 = p_1$ and T_1 in the combustion chamber). Because ideal formulations are based on the *local stagnation* values at any given cross section, the challenge is to work with the appropriate flow assumptions as discussed below.

When the applicable correction factors values are available, the overall ratio of the product of the stagnation pressure with the throat area may be inferred using

$$[(A_t)_a(p_1)_a]/[(A_t)_i(p_1)_i] = \zeta_F/\zeta_{C_F} \text{ or } \zeta_d\zeta_{c*} \qquad (3-44)$$

In order to arrive at an actual nozzle throat area from such available parameters, additional information is needed. The key question is how to relate stagnation pressure ratios to their corresponding temperature ratios across the nozzle. When known, a *polytropic index "n"* provides a relation between gas properties (Ref. 3–3); otherwise, we may apply the isentropic relation given as Eq. 3–7 because nozzle efficiencies are typically high (around 90%). Real nozzle flows are not isentropic but *the net entropy* change along the nozzle may be insignificant, though noticeable decreases of the stagnation pressure and temperature do occur as Example 3–7 will show next. When dealing with *real nozzles* with given expansion ratios, it becomes necessary to increase somewhat the stagnation pressure beyond $(p_1)_i$ to achieve any previously defined ideal nozzle performance.

Example 3–7. Design a rocket propulsion nozzle to conform to the following conditions:

Chamber pressure	20.4 atm = 2.067 MPa
Atmospheric pressure	1.0 atm
Mean molecular mass of gases	21.87 kg/kg-mol
Specific heat ratio	1.23
Ideal specific impulse or c_i	230 sec or 2255.5 m/sec (at operating condition)
Desired thrust	1300 N
Chamber temperature	2861 K

Determine the actual values of exhaust velocity, specific impulse, and nozzle throat and exit areas. Also calculate the discharge correction factor and nozzle efficiency implied by the following correction factors that apply: $\zeta_F = 0.96$, $\zeta_{c*} = 0.98$, and $\zeta_{C_F} = 0.97$.

SOLUTION. The theoretical thrust coefficient can be calculated from Eq. 3–30. For optimum operation, $p_2 = p_3$. By substituting $k = 1.23$ and $p_1/p_2 = 20.4$, the thrust coefficient is found as $(C_F)_i = 1.405$ and $(A_t)_i = 4.48 \text{ cm}^2$. These calculated values may be checked by interpolating between the values shown in Fig. 3–5.

An *actual effective exhaust velocity* c_a is not measured, but for this problem it may be calculated from

$$c_a = (c_i)(\zeta_v) = (2255.5)(0.98 \times 0.97) = 2144 \text{ m/sec}$$

The actual specific impulse is $(I_s)_a = 2144/9.81 = 219$ sec and the discharge correction factor from Eq. 3–40

$$\zeta_d = \zeta_F/\zeta_{c*}\zeta_{C_F} = 0.96/(0.98 \times 0.97) = 1.01$$

Note that this value is slightly greater than 1.0. Since for this problem, $v_2 = c$, the nozzle efficiency may be found from $\eta_n = (\zeta_v)^2 = 90\%$.

An approximate *exit stagnation temperature* may be calculated using Eq. 3–43, $(T_0)_a \approx T_1(\zeta_{c*})^2 = 2748$ K. Next, an estimate of the stagnation pressure at the throat is needed, but our correction factors only apply to the overall nozzle. Because the nozzle efficiency is only 10% less than ideal and the overall change of stagnation temperature is less than 4%, a sufficient approximation is to assign half of the *stagnation temperature drop* to the region upstream of the throat, that is, $(T_0)_t \approx 2804$ K. Thus,

$$(p_0)_t \approx p_1[2804/2861]^{(k/k-1)} = 1.86 \text{ MPa}$$

An actual throat area estimate may now be computed from Eq. 3–44:

$$(A_t)_a/(A_t)_i = (\zeta_F/\zeta_{CF})(p_1/(p_0)_t) \approx 1.10$$

As expected, the actual throat area is larger than the ideal for the more realistic flow conditions (by about 10%). The actual exit area may be obtained next. Going back to the mass flow rate:

$$\dot{m} = \zeta_F F/c_a = 0.582 \text{ kg/sec} = p_2 A_2 c/RT_2$$

Now, $p_2 = p_3$ as given and T_2 may be found approximately from

$$T_2 \approx 2748 \times (0.1013/1.67)^{(0.229/1.229)} = 1631 \text{ K}$$

where $(p_0)_a \approx 1.67$ MPa would be an actual stagnation pressure applicable at the exit of the real nozzle. Finally, the exit area and the area ratio are found to be $(A_2)_a \approx 16.61$ cm^2 and $\epsilon_a \approx 3.37$, which may be compared with the ideal values from Fig. 3–4.

Four Performance Parameters

When using or quoting values of thrust, specific impulse, propellant flow, and other performance parameters one must be careful to specify or as bare minimum qualify the conditions under which any particular number applies. There are at least four sets of performance parameters, and they are different in concept and value, even when referring to the same rocket propulsion system. Parameters such as F, I_s, c, v_2, and/or $\dot{m}$ should be accompanied by a clear definition of the conditions under which they apply. Not all the items below apply to every parameter.

(a) Chamber pressure; also, for slender chambers, the location where this pressure prevails or is measured (e.g., at nozzle entrance).

(b) Ambient pressure or altitude or space (vacuum).

(c) Nozzle expansion area ratio and whether this is an optimum.

(d) Nozzle shape and exit angle (see Fig. 3–11 and Eq. 3–34).

(e) Propellants, their composition and mixture ratio.

(f) Initial ambient temperature of propellants, prior to start.

1. *Theoretical performance values* as defined in Chapters 2, 3, and 5 generally apply only to ideal rockets. Conditions for ideal nozzles are given in Section 3.1. The theoretical performance of an ideal nozzle per se is not used, but one that includes corrections is used instead; analyses of one or more losses may be included as well as

values for the correction factors described in the previous section and these will yield a lower performance. Most of these modifications are two dimensional and correct for chemical reactions in the nozzle using real gas properties, and most correct for exit divergence. Many also correct for one or more of the other losses mentioned above. For example, computer programs for solid propellant motor nozzles can include losses for throat erosion and multiphase flow; for liquid propellant engines they may include two or more concentric zones, each at different mixture ratios and thus with different gas properties. Nozzle wall contour analyses with expansion and compression waves may apply finite element analysis and/or a method of characteristics approach. Some of the more sophisticated programs include analyses of viscous boundary layer effects and heat transfer to the walls. Typically, these computer simulations are based on finite element descriptions using the basic Navier–Stokes relationships. Most organizations doing nozzle design have their own computer programs, often different programs for different nozzle designs, different thrust levels, or operating durations. Many also use simpler, one-dimensional computer programs, which may include one or more correction factors; these programs are used frequently for preliminary estimates, informal proposals, or evaluation of ground test results.

2. *Delivered*, that is, *actually measured, performance values* are obtained from static tests or flight tests of full-scale propulsion systems. Again, the conditions should be defined (e.g., p_1, A_2/A_t, p_3, etc.) and the measured values should be corrected for instrument deviations, errors, or calibration constants. Flight test data need to be corrected for aerodynamic effects, such as drag. Often, empirical values of the thrust correction factor, the velocity correction factor, and the mass discharge flow correction factors are used to convert theoretical values from item 1 above to approximate actual values, something that is often satisfactory for preliminary estimates. Sometimes subscale thrust chambers are used in the development of new large thrust chambers and then scale factors are applied to correct the measured data to full-scale values.

3. *Performance values at standard conditions* are the corrected values of items 1 and 2 above. These standard conditions are generally rigidly specified by the customer reflecting commonly accepted industrial practice. Usually, they refer to conditions that allow evaluation or comparison with reference values and often they refer to conditions that can be easily measured and/or corrected. For example, to allow for a proper comparison of specific impulse for several propellants or rocket propulsion systems, the values are often corrected to the following standard conditions (see Example 3–6):

(a) $p_1 = 1000$ psia (6.894 MPa) or other agreed-upon value.

(b) $p_2 = p_3 = 14.69$ psia (0.10132 MPa) sea level.

(c) Nozzle exit area ratio, optimum, $p_2 = p_3$.

(d) Nozzle divergence half angle $\alpha = 15°$ for conical nozzles, or other agreed-upon value.

(e) Specific propellant, its design mixture ratio, and/or propellant composition and maximum allowed impurities.

(f) Propellant initial ambient temperature: 21°C (sometimes 20 or 25°C) or boiling temperature, if cryogenic.

A rocket propulsion system is generally designed, built, tested, and delivered in accordance with certain predetermined requirements or customer-approved *specifications* found in formal documents (often called the *rocket engine or rocket motor specifications*). These may define the performance as shown above along with detailing many other requirements. More discussion of these specifications is given as a part of the selection process for propulsion systems in Chapter 19.

4. Rocket manufacturers are often required by their customers to deliver rocket propulsion systems with a *guaranteed minimum performance*, such as minimum F and/or minimum I_s. The determination of such values can be based on a nominal value (items 1, 2, or 3 above) diminished by all likely losses, including changes in chamber pressure due to variation of pressure drops in injector or pipelines, losses due to nozzle surface roughness, propellant initial ambient temperatures, ambient pressure variations, and manufacturing variations from rocket to rocket (e.g., in grain volume, nozzle dimensions, or pump impeller diameters, etc.). Minimum values can be determined by probabilistic evaluations of these losses, which are then usually validated by actual full-scale static and flights tests.

3.6. NOZZLE ALIGNMENT

When the thrust vector direction of the propulsion system does not go through the center of mass of a flying vehicle, a turning moment exists that will rotate the vehicle while thrust is being applied. Predetermined controlled turning moments may be desirable and necessary for turning or for attitude control of the vehicle and can be accomplished by thrust vector deflections, aerodynamic fins, or by separate attitude control rocket engines. However, any turning becomes undesirable when its magnitude or direction is not known—this happens when a fixed nozzle in a major propulsion system has its thrust axis misaligned. A large high-thrust booster rocket system, even if misaligned by a very small angle (less than 0.50°), can cause major and upsetting turning moments during firing. If not corrected or compensated, such a small misalignment can cause the flight vehicle to tumble and/or deviate from the intended flight path. For such turning moments not to exceed a vehicle's compensating attitude control capability, it is necessary to align very accurately the nozzle axis of each and all propulsion systems that contain fixed (nongimbaled) nozzles. Normally, the geometric axis of the nozzle diverging exit surface geometry is taken to be the thrust axis. Special alignment fixtures are needed to orient the nozzle axis to be within less than ± 0.25° of the intended line though the vehicle's center of gravity and to position the center of a large nozzle throat on the vehicle's centerline, say within 1 or 2 mm. See Ref. 3–15.

There are other types of misalignments: (1) irregularities in the nozzle geometry (out of round, protuberances, or unsymmetrical roughness in the surface); (2) transient flow misalignments during start or stop; (3) uneven deflection of the propulsion system or vehicle structure under load; and (4) irregularities in the gas flow (faulty injector, uneven burning rate in solid propellants). For simple unguided rocket vehicles, it has been customary to rotate (spin) the vehicle to prevent any existing misalignments from being in one direction only, evening out any misalignments during powered flight.

In the cramped volume of spacecraft or launch vehicles, it is sometimes not possible to accommodate the full length of a large-area-ratio nozzle within the available vehicle envelope. Nozzles of attitude control thrusters are sometimes cut off at an angle at the vehicle surface, which allows a more compact installation. Figure 3–16 shows a diagram of two (out of four) roll control thrusters whose nozzle exit conforms to the vehicle contour. The thrust direction of such a *scarfed nozzle* is no longer along the nozzle axis centerline, as it is with fully symmetrical nozzles, and the nozzle exit flow will not be axisymmetric. Reference 3–16 shows how to estimate the performance and thrust direction of scarfed nozzles.

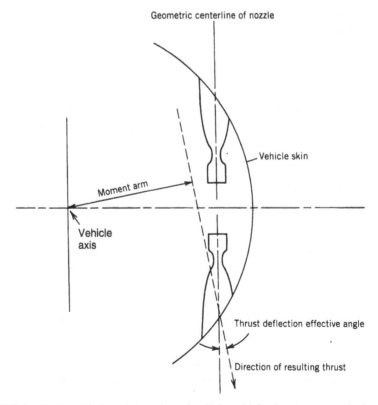

FIGURE 3–16. Simplified partial section of a flight vehicle showing two attitude control thrusters with scarfed nozzles to fit a cylindrical vehicle envelope.

SYMBOLS

A	area, m^2 (ft^2)
c	effective exhaust velocity, m/sec (ft/sec)
c_p	specific heat at constant pressure, J/kg-K (Btu/lbm-°R)
c_s	specific heat of solid phase, J/kg-K (Btu/lbm-°R)
c_v	specific heat at constant volume, J/kg-K (Btu/lbm-°R)
c^*	characteristic velocity, m/sec (ft/sec)
C_F	thrust coefficient
C_D	discharge coefficient ($1/c^*$), sec/m (sec/ft)
D	diameter, m (ft)
F	thrust, N (lbf)
g_0	standard sea-level gravitational acceleration, 9.80665 m/sec^2 (32.174 ft/sec^2)
h	enthalpy per unit mass, J/kg (Btu/lbm) or altitude, km
I_s	specific impulse, sec
J	mechanical equivalent of heat; $J = 4.186$ J/cal in SI units or
	1 Btu $= 777.9$ ft-lbf
k	specific heat ratio
L	length of nozzle or liquid level distance above thrust chamber, m (ft)
$\dot{m}$	propellant mass flow rate, kg/sec (lbm/sec)
M	Mach number
$\mathfrak{M}$	molecular mass, kg/kg-mol (lbm/lb-mol)
MR	mass ratio
n_i	molar fraction of species i
p	pressure, N/m^2 (lbf/ft^2 or lbf/in.^{2})
R	gas constant per unit weight, J/kg-K (ft-lbf/lbm-°R) ($R = R'/\mathfrak{M}$)
R'	universal gas constant, 8314.3 J/kg mol-K (1544 ft-lb/lb mol-°R)
T	absolute temperature, K (°R)
v	velocity, m/sec (ft/sec)
V	specific volume, m^3/kg(ft^3/lbm)
$\dot{w}$	propellant weight flow rate, N/sec (lbf/sec)

Greek Letters

α	half angle of divergent conical nozzle section
β	mass fraction of solid particles
ϵ	nozzle expansion area ratio A_2/A_t
η_n	nozzle efficiency
ζ_{C_F}	thrust coefficient correction factor
ζ_{c^*}	c^* correction factor
ζ_d	discharge or mass flow correction factor
ζ_F	thrust correction factor
ζ_v	velocity correction factor
λ	divergence angle correction factor for conical nozzle exit

Subscripts

a	actual
g	gas
i	ideal, or a particular species in a mixture
max	maximum
opt	optimum nozzle expansion
s	solid
sep	point of separation
t	nozzle throat
x	any direction or section within rocket nozzle
y	any direction or section within rocket nozzle
0	stagnation or impact condition
1	nozzle inlet or combustion chamber
2	nozzle exit
3	atmospheric or ambient

PROBLEMS

1. Certain experimental results indicate that the propellant gases from a liquid oxygen–gasoline reaction have a mean molecular mass of 23.2 kg/kg-mol and a specific heat ratio of 1.22. Compute the specific heats at constant pressure and at constant volume, assuming perfect-gas relations apply.

2. The actual conditions for an optimum expansion nozzle operating at sea level are given below. Calculate v_2, T_2, and C_F. Use $k = 1.30$ and the following parameters: $\dot{m} = 3.7$ kg/sec; $p_1 = 2.1$ MPa; $T_1 = 2585$ K; $\mathfrak{M} = 18.0$ kg/kg-mol

3. A certain nozzle expands a gas under isentropic conditions. Its chamber or nozzle entry velocity equals 90 m/sec, its final velocity 1500 m/sec. What is the change in enthalpy of the gas? What percentage of error is introduced if the initial velocity is neglected?

4. Nitrogen ($k = 1.38$, molecular mass $= 28.00$ kg/kg-mol) flows at a Mach number of 2.73 and 500 °C. What are its local and acoustic velocities?

5. The following data are given for an optimum rocket propulsion system:

Average molecular mass	24 kg/kg-mol
Chamber pressure	2.533 MPa
External pressure	0.090 MPa
Chamber temperature	2900 K
Throat area	0.00050 m²
Specific heat ratio	1.30

Determine (a) throat velocity; (b) specific volume at throat; (c) propellant flow and specific impulse; (d) thrust; (e) Mach number at the throat.

6. Determine the ideal thrust coefficient for Problem 5 by two methods.

7. A certain ideal rocket with a nozzle area ratio of 2.3 and a throat area of 5 in.^{2} delivers gases at $k = 1.30$ and $R = 66$ ft-lbf/lbm-°R at a chamber pressure of 300 psia and a constant chamber temperature of 5300 °R against a back atmospheric pressure of 10 psia. By means of an appropriate valve arrangement, it is possible to throttle the propellant flow to the thrust chamber. Calculate and plot against pressure the following quantities for 300, 200, and 100 psia chamber pressure: **(a)** pressure ratio between chamber and atmosphere; **(b)** effective exhaust velocity for area ratio involved; **(c)** ideal exhaust velocity for optimum and actual area ratio; **(d)** propellant flow; **(e)** thrust; **(f)** specific impulse; **(g)** exit pressure; **(h)** exit temperature.

8. For an ideal rocket with a characteristic velocity $c^* = 1500$ m/sec, a nozzle throat diameter of 20 cm, a thrust coefficient of 1.38, and a mass flow rate of 40 kg/sec, compute the chamber pressure, the thrust, and the specific impulse.

9. For the rocket propulsion unit given in Example 3–2 compute the new exhaust velocity if the nozzle is cut off, decreasing the exit area by 50%. Estimate the losses in kinetic energy and thrust and express them as a percentage of the original kinetic energy and the original thrust.

10. What is the maximum velocity if the nozzle in Example 3–2 was designed to expand into a vacuum? If the expansion area ratio was 2000?

11. Construction of a variable-area conventional axisymmetric nozzle has often been considered to operate a rocket thrust chamber at the optimum expansion ratio at any altitude. Because of the enormous difficulties of such a mechanical device, it has never been successfully realized. However, assuming that such a mechanism could eventually be constructed, what would have to be the variation of the area ratio with altitude (plot up to 50 km) if such a rocket had a chamber pressure of 20 atm? Assume that $k = 1.20$.

12. Design a supersonic nozzle to operate at 10 km altitude with an area ratio of 8.0. For the hot gas take $T_0 = 3000$ K, $R = 378$ J/kg-K, and $k = 1.3$. Determine the exit Mach number, exit velocity, and exit temperature, as well as the chamber pressure. If this chamber pressure is doubled, what happens to the thrust and the exit velocity? Assume no change in gas properties. How close to optimum nozzle expansion is this nozzle?

13. The German World War II A-4 propulsion system had a sea-level thrust of 25,400 kg and a chamber pressure of 1.5 Mpa. If the exit pressure is 0.084 MPa and the exit diameter 740 mm, what would be its thrust at 25,000 m?

14. Derive Eq. 3–34. (*Hint*: Assume that all the mass flow originates at the apex of the cone.) Calculate the nozzle angle correction factor for a conical nozzle whose divergence half angle is 13°.

15. Assuming that the thrust correction factor is 0.985 and the discharge correction factor is 1.050 in Example 3–2, determine **(a)** the actual thrust; **(b)** the actual exhaust velocity; **(c)** the actual specific impulse; **(d)** the velocity correction factor.

16. An ideal rocket has the following characteristics:

Chamber pressure	27.2 atm
Nozzle exit pressure	3 psia
Specific heat ratio	1.20
Average molecular mass	21.0 lbm/lb-mol
Chamber temperature	4200 °F

Determine the critical pressure ratio, the gas velocity at the throat, the expansion area ratio, and the theoretical nozzle exit velocity.

Answers: 0.5645; 3470 ft/sec; 14.8; and 8570 ft/sec.

17. For an ideal rocket with a characteristic velocity c^* of 1220 m/sec, a mass flow rate of 73.0 kg/sec, a thrust coefficient of 1.50, and a nozzle throat area of 0.0248 m², compute the effective exhaust velocity, the thrust, the chamber pressure, and the specific impulse.

Answer: 1830 m/sec; 133,560 N; 3.590×10^6 N/m²; 186.7 sec.

18. Derive Eqs. 3–24 and 3–25.

19. An upper stage of a launch vehicle propulsion unit fails to meet expectations during sea-level testing. This unit consists of a chamber at 4.052 MPa feeding hot propellant to a supersonic nozzle of area ratio $\epsilon = 20$. The local atmospheric pressure at the design condition is 20 kPa. The propellant has a $k = 1.2$ and the throat diameter of the nozzle is 9 cm.

    a. Calculate the ideal thrust at the design condition.

    b. Calculate the ideal thrust at the sea-level condition.

    c. State the most likely source of the observed nonideal behavior.

Answer: (a) 44.4 kN, (b) 34.1 kN, (c) separation in the nozzle.

20. Assuming ideal flow within some given propulsion unit:

    a. State all necessary conditions (realistic or not) for

$$c^* = c = v_2$$

    b. Do the above conditions result in an *optimum* thrust for a given p_1/p_3?

    c. For a launch vehicle designed to operate at some intermediate Earth altitude, sketch (in absolute or relative values) how c^*, c, and v_2 would vary with altitude.

21. A rocket nozzle has been designed with $A_t = 19.2$ in.² and $A_2 = 267$ in.² to *operate optimally* at $p_3 = 4$ psia and produce 18,100 lbf of *ideal thrust* with a chamber pressure of 570 psia. It will use the proven design of a previously built combustion chamber that operates at $T_1 = 6000\,°R$ with $k = 1.25$ and $R = 68.75$ ft-lbf/lbm°R, with a c^*-efficiency of 95%. But test measurements on this thrust system, at the stated pressure conditions, yield a thrust of only 16,300 lbf when the measured flow rate is 2.02 lbm/sec. Find the applicable correction factors ($\zeta_F, \zeta_d, \zeta_{C_F}$) and the actual specific impulse *assuming frozen flow* throughout.

Answers: $\zeta_F = 0.90$; $\zeta_d = 1.02$; $\zeta_{C_F} = 0.929$; $(I_s)_a = 250$ sec.

22. The reason optimum thrust coefficient (as shown on Figs. 3–6 and 3–7) exists is that as the nozzle area ratio increases with fixed p_1/p_3 and k, the pressure thrust in Eq. 3–30 changes sign at $p_2 = p_3$. Using $k = 1.3$ and $p_1/p_3 = 50$, show that as p_2/p_1 drops with increasing ϵ, the term $1.964[1-(p_2/p_1)^{0.231}]^{0.5}$ increases more slowly than the (negative) term $[1/50-p_2/p_1]\,\epsilon$ increases (after the peak, where $\epsilon \approx 7$). (Hint: use Eq. 3–25.)

REFERENCES

3–1. A. H. Shapiro, *The Dynamics and Thermodynamics of Compressible Fluid Flow*, Vols. 1 and 2, Ronald Press Company, New York, 1953; and M. J. Zucrow and J. D. Hoffman, *Gas Dynamics*, Vols. I and II, John Wiley & Sons, New York, 1976 (has section on nozzle analysis by method of characteristics).

3–2. M. J. Moran and H. N. Shapiro, *Fundamentals of Engineering Thermodynamics*, 3rd ed., John Wiley & Sons, New York, 1996; also additional text, 1997.

3–3. R. D. Zucker and O. Biblarz, *Fundamentals of Gas Dynamics*, 2nd ed., John Wiley & Sons, Hoboken, NJ, 2002.

3–4. R. Stark, "Flow Separation in Rocket Nozzles – an Overview", AIAA paper 2013-3849, July 2013.

3–5. G. Hagemann, H. Immich, T. Nguyen, and G. E. Dummov, "Rocket Engine Nozzle Concepts," Chapter 12 of *Liquid Rocket Thrust Chambers: Aspects of Modeling, Analysis and Design*, V. Yang, M. Habiballah, J. Hulka, and M. Popp (Eds.), Progress in Astronautics and Aeronautics, Vol. 200, AIAA, 2004.

3–6. P. Vuillermoz, C. Weiland, G. Hagemann, B. Aupoix, H. Grosdemange, and M. Bigert, "Nozzle Design Optimization," Chapter 13 of *Liquid Rocket Thrust Chambers: Aspects of Modeling, Analysis and Design*, V. Yang, M. Habiballah, J. Hulka, and M. Popp (Eds), *Progress in Astronautics and Aeronautics*, Vol. 200, AIAA, 2004.

3–7. "Liquid Rocket Engine Nozzles," NASA SP-8120, 1976.

3–8. J. A. Muss, T. V. Nguyen, E. J. Reske, and D. M. McDaniels, "Altitude Compensating Nozzle Concepts for RLV," AIAA Paper 97-3222, July 1997.

3–9. G. V. R. Rao, Recent Developments in Rocket Nozzle Configurations, *ARS Journal*, Vol. 31, No. 11, November 1961, pp. 1488–1494; and G. V. R. Rao, Exhaust Nozzle Contour for Optimum Thrust, *Jet Propulsion*, Vol. 28, June 1958, pp. 377–382.

3–10. J. M. Farley and C. E. Campbell, "Performance of Several Method-of-Characteristics Exhaust Nozzles," *NASA TN D-293*, October 1960.

3–11. J. D. Hoffman, Design of Compressed Truncated Perfect Nozzles, *Journal of Propulsion and Power*, Vol. 3, No. 2, March–April 1987, pp. 150–156.

3–12. G. P. Sutton, Stepped Nozzle, U.S. Patent 5,779,151, 1998; M. Ferlin, "Assessment and benchmarking of extendible nozzle systems in liquid propulsion," AIAA Paper 2012-4163, July/August 2012.

3–13. F. A. Williams, M. Barrère, and N. C. Huang, "Fundamental Aspects of Solid Propellant Rockets," *AGARDograph 116*, Advisory Group for Aerospace Research and Development, NATO, October 1969, 783 pages.

3–14. M. Barrère, A. Jaumotte, B. Fraeijs de Veubeke, and J. Vandenkerckhove, *Rocket Propulsion*, Elsevier Publishing Company, Amsterdam, 1960.

3–15. R. N. Knauber, "Thrust Misalignments of Fixed Nozzle Solid Rocket Motors," AIAA Paper 92-2873, 1992.

3–16. J. S. Lilley, "The Design and Optimization of Propulsion Systems Employing Scarfed Nozzles," *Journal of Spacecraft and Rockets*, Vol. 23, No. 6, November–December 1986, pp. 597–604; and J. S. Lilley, "Experimental Validation of a Performance Model for Scarfed Nozzles," *Journal of Spacecraft and Rockets*, Vol. 24, No. 5, September–October 1987, pp. 474–480.

CHAPTER 4

FLIGHT PERFORMANCE

This chapter serves as an introduction to the flight performance of rocket-propelled vehicles such as spacecraft, space launch vehicles, and missiles or projectiles. It presents these subjects from a rocket propulsion point of view. Propulsion systems deliver forces to a flight vehicle and cause it to accelerate (or at times decelerate), overcome drag forces, or to change flight direction. Some propulsion systems also provide torques to the flight vehicles for rotation or other maneuvers. Flight missions can be classified into several flight regimes: (1) flight within the Earth's atmosphere (e.g., air-to-surface missiles, surface-to-surface short-range missiles, surface-to-air missiles, air-to-air missiles, assisted takeoff units, sounding rockets, or aircraft rocket propulsion systems), see Refs. 4–1 and 4–2; (2) near space environment (e.g., Earth satellites, orbital space stations, or long-range ballistic missiles), see Refs. 4–3 to 4–9; (3) lunar and planetary flights (with or without landing or Earth return), see Refs. 4–5 to 4–12; and (4) deep space exploration and sun escape. Each of these is discussed in this chapter except for the operation of very low thrust units which is treated in Chapter 17. We begin with a basic one-dimensional analysis of space flight and then consider some more the complex fight path scenarios for various flying rocket-propelled vehicles. Conversion factors, atmospheric properties, and a summary of key equations can be found in the appendices.

4.1. GRAVITY-FREE DRAG-FREE SPACE FLIGHT

This simplified rocket flight analysis applies to outer space environments, far enough from any star, where there is no atmosphere (thus no drag) and essentially no significant gravitational attraction. The flight direction is the same as the thrust direction (along the axis of the nozzle, namely, a linear acceleration path); the propellant mass flow is $\dot{m}$ and the propulsive thrust F remains constant for the propellant burning time duration t_p. The thrust force F has been defined in Eq. 2–16. For constant propellant flow, the flow rate $\dot{m}$ becomes m_p/t_p, where m_p is the total usable propellant mass. From Newton's second law and for an instantaneous vehicle mass m with flight velocity u,

$$F = m\,du/dt \tag{4–1}$$

For any rocket where start and shutdown durations may be neglected, the instantaneous mass of the vehicle m may be expressed as a function of time t using the initial mass of the full vehicle m_0, the initial propellant mass m_p, the time at power cutoff t_p as follows:

$$m = m_0 - \frac{m_p}{t_p}t = m_0\left(1 - \frac{m_p}{m_0}\frac{t}{t_p}\right) \tag{4–2}$$

$$= m_0\left(1 - \zeta\frac{t}{t_p}\right) = m_0\left[1 - (1 - \mathbf{MR})\frac{t}{t_p}\right] \tag{4–3}$$

Equation 4–3 expresses the vehicle mass in a form useful for trajectory calculations. The total vehicle mass ratio $\mathbf{MR}$ and the propellant mass fraction ζ have been defined in Eqs. 2–7 and 2–8 (see section 4–7 for an extension to multistage vehicles). They are related by

$$\zeta = 1 - \mathbf{MR} = 1 - m_f/m_0 = m_p/m_0 \tag{4–4}$$

A depiction of the relevant masses is shown in Fig. 4–1. The initial mass at takeoff m_0 equals the sum of the useful propellant mass m_p plus the empty or final vehicle mass m_f; in turn m_f equals the sum of the inert masses of the engine system (such as nozzles, tanks, cases, or unused residual propellant) plus the guidance, control, electronics, and related equipment and the payload. After thrust termination, any *residual propellant* in the propulsion system is considered to be a part of the final engine mass. This includes any liquid propellant remaining in tanks after operation, trapped in pipe pockets, valve cavities, and pumps or wetting the tank and pipe walls. For solid propellant rocket motors it is the remaining unburned solid propellant, also called "slivers," and sometimes also unburned insulation.

For constant propellant mass flow $\dot{m}$ and finite propellant burning time t_b the total useful propellant mass m_p is simply $\dot{m}t_p$ and the instantaneous vehicle mass $m = m_0 - \dot{m}t$. Equation 4–1 may be written as

$$du = (F/m)dt = (c\dot{m}/m)dt$$

$$= \frac{(c\dot{m})dt}{m_0 - m_p t/t_p} = \frac{c(m_p/t_p)dt}{m_0(1 - m_p t/m_0 t_p)} = c\frac{d(\zeta t/t_p)}{1 - \zeta t/t_p}$$

When start and shutdown periods consume relatively little propellant, they may be neglected. Integration leads to the maximum vehicle velocity u_p at propellant burnout that can be attained in a gravity-free vacuum environment. Depending on the frame of

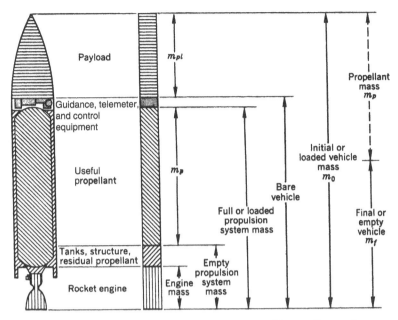

FIGURE 4–1. Definitions of various vehicle masses. For solid propellant rocket motors the words "rocket engine, tanks, and structures" are to be replaced by "rocket motor nozzle, thermal insulation, and case." m_p includes only the usable propellant mass; residuals are part of the empty mass.

reference, u_0 will not necessarily be zero, and the result is often written as a velocity increment Δu:

$$\Delta u = -c \ln(1 - \zeta) + u_0 = c \ln(m_0/m_f) + u_0 \qquad (4\text{–}5)$$

However, when the initial velocity u_0 may be taken as zero, the velocity at thrust termination u_p becomes

$$u_p \equiv \Delta u = -c \ln(1 - \zeta) = -c \ln[m_0/(m_0 - m_p)]$$
$$= -c \ln \mathbf{MR} = c \ln(1/\mathbf{MR}) = c \ln(m_0/m_f) \qquad (4\text{–}6)$$

The symbol "ln" stands for the natural logarithm. Thus, u_p is the maximum velocity increment Δu that can be obtained in a gravity-free vacuum with constant propellant flow, starting from rest. The effect of variations in c, I_s, and ζ on this flight velocity increment is shown in Fig. 4–2. An alternate way to write Eq. 4–6 using "e," the base of the natural logarithm is

$$e^{\Delta u/c} = 1/\mathbf{MR} = m_0/m_f \qquad (4\text{–}7)$$

The concept of maximum attainable flight velocity increment Δu in a gravity-free vacuum is valuable for understanding the influence of the basic parameters involved. It is used in comparing one propulsion system on a vehicle or one flight mission with another, as well as in comparing proposed upgrades or possible design improvements.

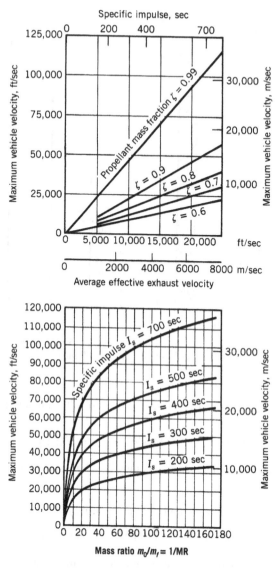

FIGURE 4–2. Maximum vehicle flight velocity in gravitationless, drag-free space for different mass ratios and specific impulses (plot of Eq. 4–6). Single-stage vehicles can have values of **MR** up to about 20 and multistage vehicles can exceed 200.

From Eq. 4–6 it can be seen that the vehicle's *propellant mass fraction* has a logarithmic effect on the vehicle velocity. By increasing this ratio from 0.80 to 0.90, u_p is increased by 43%. A mass fraction of 0.80 indicates that only 20% of the total vehicle mass is needed for structure, skin, payload, propulsion hardware, radios, guidance system, aerodynamic lifting surfaces, and so on; the remaining 80% is useful propellant. To exceed 0.85 requires a careful design; mass fraction ratios approaching 0.95

appear to be the probable practical upper limit for single-stage vehicles with currently known materials (when the mass fraction is 0.90, then $\mathbf{MR} = 0.1$ or $1/\mathbf{MR} = 10.0$). This noticeable influence of mass fraction or mass ratio on the velocity at power cut-off, and therefore also on vehicle range, is fundamental and applies to most types of rocket-powered vehicles. For this reason, high importance is placed on saving inert mass on each and every vehicle component, including the propulsion system.

Equation 4–6 can be solved for the effective propellant mass m_p required to achieve a desired velocity increment for a given initial takeoff mass or a final shutdown mass of the vehicle. The final mass consists of the payload, the structural mass of the vehicle, the empty propulsion system mass (that includes residual propellant), plus a small additional mass for guidance, communications, and control devices. Here $m_p = m_0 - m_f$:

$$m_p = m_f(e^{\Delta u/c} - 1) = m_0(1 - e^{(-\Delta u/c)}) \tag{4-8}$$

In a gravity-free environment, the flight velocity increment Δu or u_p is proportional to the effective exhaust velocity c and, therefore, to the specific impulse (see Eq. 2–6). Thus, any improvement in I_s (such as better propellants, more favorable nozzle area ratios, or higher chamber pressures) reflects itself in improved vehicle performance, provided that such an improvement does not also necessitate an excessive increase in rocket propulsion system inert mass, which would lead to a decrease in the effective propellant fraction. Figure 4–3 shows how payload fraction (m_{pl}/m_0) varies as a function of the ratio ($\Delta u/I_s g_0$) for several values of propellant stage-mass fraction ζ_i, as calculated from the equations introduced above for single-staged vehicles.

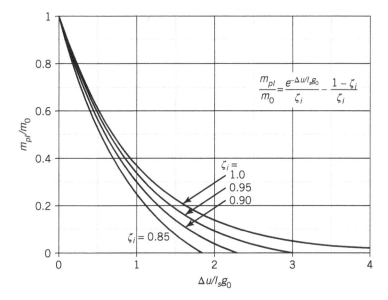

FIGURE 4–3. Maximum payload fraction in gravitationless, drag-free space, single-stage vehicle flight for different stage-mass fractions, velocity increments and specific impulses. For each stage $m_0 = m_i + m_{pl}$ and Eq. 2–8 becomes $\zeta_i = m_p/m_i$.

Figure 4–3 complements Fig. 4–2 showing m_{pl}/m_0 as a function of both $\Delta u/I_s g_0$ and ζ_i (Example 4–3 treats ζ_i for multi-staged vehicles). Note that for any given $\zeta_i < 1.0$ certain payload fractions become unavailable as values of $\Delta u/I_s g_0$ increase. This is evident for the three curves that terminate at zero payload fraction.

4.2. FORCES ACTING ON A VEHICLE IN THE ATMOSPHERE

The external forces commonly acting on vehicles flying in the Earth's atmosphere consist of thrust, aerodynamic forces, and gravitational attractions. Other forces, such as wind or solar radiation pressure, are usually small and generally can be neglected in many calculations.

Thrust is the force produced by the vehicle's power plant, such as a propeller or a rocket propulsion system. It most often acts in the direction of the axis of the power plant, that is, along the propeller shaft axis or the rocket nozzle axis. The thrust force of a rocket propulsion system with constant mass flow has been formulated in Chapter 2 as a function of the effective exhaust velocity c and the propellant flow rate $\dot{m}$. In many rockets the mass rate of propellant consumption $\dot{m}$ is essentially constant, and starting and stopping transients may be neglected. Therefore, the thrust as given from Eq. 2–6 (or Eqs. 2–13 and 2–15) may be written with $\dot{m}$ as m_p/t_p

$$F = c\dot{m} = cm_p/t_p \qquad (4-9)$$

As explained in Chapter 3, the value of the effective exhaust velocity c (or the specific impulse I_s) depends on both nozzle area ratio and nozzle exhaust pressure. However, as Earth altitude increases, values of c change only by a relatively small factor bounded between about 1.2 and 1.8 (with the higher values applicable in the vacuum of space).

There are two relevant *aerodynamic forces* in the atmosphere. The *drag D* acts in a direction opposite to the flight path and is due to resistance to the body's motion by the surrounding fluid. The *lift L* is the aerodynamic force acting in a direction normal to the flight path. They are both given as functions of the vehicle's flight speed u, the mass density ρ of the atmosphere in which it moves, and a *characteristic surface area A*:

$$L = C_L \tfrac{1}{2}\rho A u^2 \qquad (4-10)$$

$$D = C_D \tfrac{1}{2}\rho A u^2 \qquad (4-11)$$

where C_L and C_D are lift and drag coefficients, respectively. For airplanes and winged missiles the area A is the wing area. For wingless missiles or space launch vehicles it is the maximum cross-sectional area normal to the missile axis. These lift and drag coefficients are primarily functions of the vehicle configuration, flight Mach number, and angle of attack—the angle between the vehicle axis (or the wing plane) and the flight direction. At low flight speeds, Mach number effects may be neglected, and the drag and lift coefficients are only functions of the angle of attack. A typical variation of the drag and lift coefficients for a supersonic missile is shown in Fig. 4–4. Values for these coefficients reach a maximum near a Mach number of unity. For wingless vehicles the effective angle of attack α is usually very small ($0 < \alpha < 1°$). The "standard density" and other properties of the Earth's atmosphere are listed in Appendix 2, but note that the local density can vary from day to day by a factor up to 2 (for altitudes from 300 to 1200 km) depending on solar activity and night-to-day

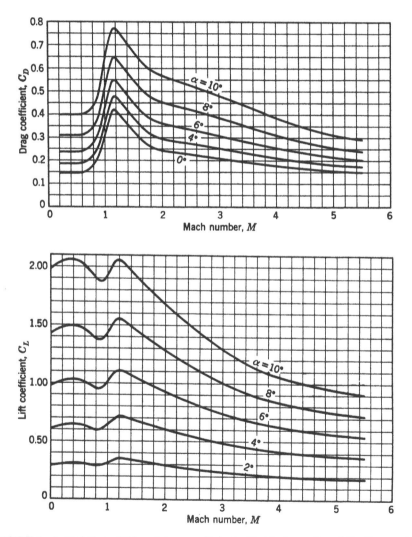

FIGURE 4–4. Variation of lift and drag coefficient with Mach number of the German V-2 missile at several angles of attack α based on body cross-sectional area with jet off and without exhaust plume effects. The Mach number shown, defined in Eq. 3–11, relates to the external flow.

temperature variations. This introduces a major uncertainty in calculations of lift and drag. Aerodynamic forces are also affected by the flow and the pressure distribution of the rocket exhaust gases, as explained in Chapter 20.

A vehicle's flight regime in the neighborhood of Mach 1 is called the transonic flight region. Here strong unsteady aerodynamic forces often develop (due to shock-induced buffeting), which are noticeable in the steep rise and subsequent decrease of the coefficients as shown in Fig. 4–4. In some cases a vehicle's maximum load capabilities during transonic flight have been exceeded leading to structural failures.

For properly designed space launch vehicles and ballistic missiles the integrated drag losses, when expressed in terms of Δu, are typically 5 to 10% of the final ideal vehicle velocity. These relatively low values result from air densities (and thus dynamic pressures), being low at high altitudes where velocities are high, and being high at altitudes where vehicle velocities are low.

Gravitational attraction is exerted upon any flying space vehicle by all planets, stars, and moons. Gravity forces pull the vehicle in the direction of the center of mass of the attracting body. Within the immediate vicinity of the Earth, the attraction of other planets and celestial bodies (like our sun and moon) can be negligibly small compared to the Earth's gravitational force. This force or gravitational pull makes up the object's *weight*.

When Earth gravity variations with the geographical features and the oblate shape are neglected, the acceleration of gravity g varies inversely as the square of the distance from the Earth's center. If R_0 is the mean radius at the (spherical) Earth's surface at which g_0 is the acceleration of gravity, by Newton's law of gravitation, g changes with altitude (h) as

$$g = g_0(R_0/R)^2$$
$$= g_0[R_0/(R_0 + h)]^2 \qquad (4-12)$$

At the equator the spherical Earth's radius is 6378.388 km and the standard value of g_0 is 9.80665 m/sec². At distances as far away as the moon, the Earth's gravity acceleration is only about $3.3 \times 10^{-4} g_0$. For the most accurate analyses, the value of g will vary locally with the Earth's bulge at the equator, with the height of nearby mountains, and with the local difference of the Earth's density at specific regions.

4.3. BASIC RELATIONS OF MOTION

For any vehicle that flies within the Earth's proximity, the gravitational attractions from all other heavenly bodies are neglected. Assume next that the vehicle is moving in rectilinear equilibrium flight and that all control forces, lateral forces, and moments that tend to turn the vehicle are zero. The resulting trajectory is two-dimensional and is contained in a fixed plane. Assume further that the vehicle has wings inclined to the flight path at an angle of attack α providing lift in a direction normal to the flight path. The direction of flight need not coincide with the direction of thrust as shown schematically in Figure 4–5.

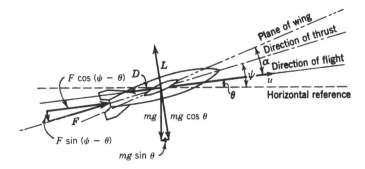

FIGURE 4-5. Two-dimensional free-body force diagram for flying vehicle with wings and fins.

Let θ be the angle of the flight path with the horizontal and ψ the angle of the direction of thrust with the horizontal. Along the flight path direction, the product of the mass and the acceleration has to equal the vector sum of the propulsive, aerodynamic, and gravitational forces:

$$m(du/dt) = F \cos(\psi - \theta) - D - mg \sin\theta \qquad (4-13)$$

The acceleration perpendicular to the flight path is $u(d\theta/dt)$; for a given value of u and at the instantaneous Earth radius R (from the Earth's center) of the flight path, it becomes u^2/R. The equation of motion in a direction normal to the flight velocity is

$$mu(d\theta/dt) = F \sin(\psi - \theta) + L - mg \cos\theta \qquad (4-14)$$

By substituting from Eqs. 4-10 and 4-11, these two basic equations can be solved for the accelerations as follows:

$$\frac{du}{dt} = \frac{F}{m} \cos(\psi - \theta) - \frac{C_D}{2m} \rho u^2 A - g \sin\theta \qquad (4-15)$$

$$u\frac{d\theta}{dt} = \frac{F}{m} \sin(\psi - \theta) + \frac{C_L}{2m} \rho u^2 A - g \cos\theta \qquad (4-16)$$

No general solution can be given to these equations, since t_p, u, C_D, C_L, ρ, θ, and/or ψ may vary independently with time, mission profile, and/or altitude. Also, C_D and C_L are functions of velocity or Mach number. In more sophisticated analyses, other factors may also be considered, such as any propellant used for nonpropulsive purposes (e.g., attitude control or flight stability). See Refs. 4-1, 4-8, 4-11, and 4-12 for background material on flight performance in some of these flight regimes. Because rocket propulsion systems are usually tailored to fit specific flight missions, different flight performance parameters are maximized (or optimized) for different rocket flight missions or flight regimes such as Δu, range, orbit height and shape, time-to-target, or altitude.

For actual trajectory analyses, navigation computation, space flight path determination, or missile-firing tables, the above two-dimensional simplified theory is not sufficiently accurate; perturbation effects, such as those listed in Section 4.4, must then be considered in addition to drag and gravity, and computer modelling is necessary to handle such complex relations. Suitable divisions of the trajectory into small elements and step-by-step or numerical integrations to define a trajectory are usually indicated. More generally, three-body theories include the gravitational attraction among three masses (for example, the Earth, the moon, and the space vehicle) and this is considered necessary in many space flight problems (see Refs. 4–2 to 4–5). The form and the solution to the given equations become further complicated when propellant flow and thrust are not constant and when the flight path is three dimensional.

For each mission or flight one can obtain actual histories of velocities and distances traveled and thus complete trajectories when integrating Eqs. 4–15 and 4–16. More general cases require six equations: three for translation along each of three perpendicular axes and three for rotation about these axes. The choice of coordinate systems of reference points can simplify the mathematical solutions (see Refs. 4–3 and 4–5) but there are always a number of trade-offs in selecting the best trajectory for the flight of a rocket vehicle. For example, for a fixed thrust the trade-off is between burn time, drag, payload, maximum velocity, and maximum altitude (or range). Reference 4–2 describes the trade-offs between payload, maximum altitude, and flight stability for sounding rockets.

Equations 4–15 and 4–16 may be further simplified for various special applications, as shown below; results of such calculations for velocity, altitude, or range using the above two basic equations are often adequate for rough design estimates. A form of these equations is also useful for determining the actual thrust or actual specific impulse during vehicle flight from accurately observed trajectory data, such as from optical or radar tracking data. Vehicle acceleration (du/dt) is essentially proportional to net thrust and, by making assumptions or measurements of propellant flow (which usually varies in a predetermined manner) and from analyses of aerodynamic forces, it is possible to determine a rocket propulsion system's actual thrust under flight conditions.

Equations 4–15 and 4–16 simplify for wingless rocket projectiles, space launch vehicles, or missiles with constant thrust and propellant flow. In Fig. 4–6 the flight direction θ is the same as the thrust direction and any lift forces for a symmetrical, wingless, stably flying vehicle are neglected at zero angle of attack. For a two-dimensional trajectory in a single plane (no wind forces) and a stationary Earth, the acceleration in the direction of flight is given below, where t_p is the operating or burn time of the propellant and ζ is the propellant mass fraction:

$$\frac{du}{dt} = \frac{c\zeta/t_p}{1 - \zeta t/t_p} - g\sin\theta - \frac{C_D \frac{1}{2}\rho u^2 A/m_0}{1 - \zeta t/t_p} \qquad (4\text{--}17)$$

The force vector diagram in Fig. 4–6 also shows the net force (the addition of thrust, drag, and gravity vectors) to be at an angle to the flight path, which will therefore

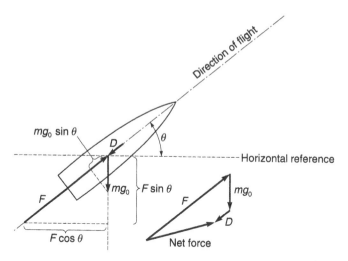

FIGURE 4-6. Simplified free-body force diagram for a vehicle without wings or fins. The force vector diagram shows the net force on the vehicle. All forces act through the vehicle's center of gravity.

be curved. These types of diagram form the basis for iterative trajectory numerical solutions.

All further relationships in this section correspond to two-dimensional flight paths, ones that lie in a single plane. If maneuvers out of that plane take place (e.g., due to solar attraction, thrust misalignment, or wind), then another set of equations will be needed—it requires both force and energy to push a vehicle out of its flight plane; Reference 4–1 describes equations for the motion of rocket projectiles in the atmosphere in three dimensions. Trajectories must be calculated accurately in order to reach any intended flight objective and today these are done exclusively with the aid of computers. Several *computer programs* for analyzing flight trajectories exist (which are maintained by aerospace companies and/or government agencies). Some are two-dimensional, relatively simple, and used for making preliminary estimates or comparisons of alternative flight paths, alternative vehicle designs, or alternative propulsion schemes. Several use a stationary flat Earth, while others use a rotating curved Earth. Three-dimensional programs used for more accurate flight path analyses may include some or all significant perturbations, orbit plane changes, or flying at angles of attack. Reference 4–4 explains their nature and complexity.

When the flight trajectory is vertical (as for a sounding rocket), then Eq. 4–17 becomes

$$\frac{du}{dt} = \frac{c\zeta/t_p}{1 - \zeta t/t_p} - g - \frac{C_D \frac{1}{2}\rho u^2 A/m_0}{1 - \zeta t/t_p} \qquad (4-18)$$

The velocity at the end of burning can be found by integrating between the limits of $t = 0$ and $t = t_p$ where $u = u_0$ and then $u = u_p$. The first two terms can readily be integrated. The last term is of significance only if the vehicle spends a considerable portion of time within the lower atmosphere. It may be integrated graphically or numerically, and its value can be designated by the term $BC_D A/m_0$, where

$$B \equiv \int_0^{t_p} \frac{\frac{1}{2}\rho u^2}{1 - \zeta t/t_p} dt$$

The cutoff velocity or velocity at the end of propellant burning u_p then becomes

$$u_p = -\bar{c} \ln(1 - \zeta) - \bar{g}t_p - \frac{BC_D A}{m_0} + u_0 \qquad (4-19)$$

where u_0 is an initial velocity such as may be given by a booster, $\bar{g}$ is an average gravitational attraction evaluated with respect to time and altitude from Eq. 4–12, and $\bar{c}$ is a time average of the effective exhaust velocity, which also depends on altitude. For nonvertical flight paths, the gravity loss becomes a function of the angle between the flight direction and the local horizontal; more specifically, the gravity loss is then given by the integral $\int (g \sin \theta) dt$.

When aerodynamic forces within the atmosphere may be neglected (or for vacuum operation) and when no booster or means for attaining an initial velocity ($u_0 = 0$) are present, the velocity at the end of the burning reached with a vertically ascending trajectory becomes simply

$$u_p = -\bar{c} \ln(1 - \zeta) - \bar{g}t_p$$
$$= -\bar{c} \ln \mathbf{MR} - \bar{g}t_p = \bar{c} \ln(1/\mathbf{MR}) - \bar{g}t_p \qquad (4-20)$$

The first term on the right side is usually the largest and is identical to Eq. 4–6. It is directly proportional to the effective rocket exhaust velocity and very sensitive to changes in the mass ratio. The second term is related to the Earth's gravity and is always negative during ascent, but its magnitude can be small when the burn time t_p is short or when flight is taking place at high orbits or in space where $\bar{g}$ is comparatively small.

For the simplified case given in Eq. 4–20 the net initial acceleration a_0 for vertical takeoff at sea level is

$$a_0 = (F_0 g_0/w_0) - g_0 \qquad (4-21)$$
$$a_0/g_0 = (F_0/w_0) - 1 \qquad (4-22)$$

where a_0/g_0 is the *initial takeoff acceleration* in multiples of the sea-level gravitational acceleration g_0, and F_0/w_0 is the thrust-to-weight ratio at takeoff. For large surface-launched vehicles, this initial-thrust-to-initial-weight ratio typically has values between 1.2 and 2.2; for small missiles (air-to-air, air-to-surface, and surface-to-air types) this ratio is usually larger, sometimes even as high as 50 or 100. *The final or terminal acceleration a_f of a vehicle in vertical Earth ascent usually occurs just before the rocket engine is shut off and/or before the useable propellant* is completely consumed. If drag is neglected, then the final acceleration a_f acting on the final mass m_f becomes

$$a_f = (F_f/m_f) - g \qquad (4\text{–}23)$$

This equation applies when the powered flight path traverses a substantial range of altitude with g decreasing according to Eq. 4–12. For rocket propulsion systems with constant propellant flow, the final acceleration is usually also the maximum acceleration because the vehicle mass being accelerated is a minimum just before propellant termination, and for ascending rockets thrust usually increases with altitude. When this terminal acceleration is found to be too large (e.g., causes overstressing of the structure, thus necessitating an increase in structure mass), then the thrust must be reduced by redesign to a lower value for the last portion of the burning period. In manned flights, maximum accelerations are limited to the maximum g-loading that can be withstood by the crew.

Example 4–1. A simple single-stage rocket for a rescue flare has the following characteristics. Its flight path nomenclature is shown in the accompanying sketch.

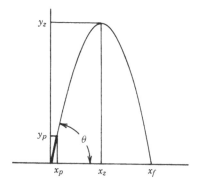

Launch weight (w_0)	4.0 lbf
Useful propellant weight (w_p)	0.4 lbf
Effective specific impulse	120 sec
Launch angle θ (relative to horizontal)	80°
Burn time t_p (with constant thrust)	1.0 sec

The heavy line in the ascending trajectory designates the powered portion of the flight.

Drag may be neglected since flight velocities are low. Assume that the acceleration of gravity is unchanged from its sea-level value g_0, which then makes the propellant mass numerically equal to the propellant weight in the EE system, or 0.4 lbm. Also assume that start and stop transients are short and can be neglected.

Solve for the initial and final acceleration during powered flight, the maximum trajectory height and the time to reach maximum height, the range or horizontal distance to impact, and the angle at propulsion cutoff.

SOLUTION. We will divide the flight path into three portions: the powered flight for 1 sec, the unpowered ascent after cutoff, and the free-fall descent. The thrust is obtained from Eq. 2–5:

$$F = I_s w_p/t_p = 120 \times 0.4/1.0 = 48 \text{ lbf}$$

The initial accelerations along the x-horizontal and y-vertical directions are, from Eq. 4–22,

$$(a_0)_y = g_0[(F \sin\theta/w) - 1] = 32.17[(48/4.0)\sin 80° - 1] = 348 \text{ ft/sec}^2$$

$$(a_0)_x = g_0(F \cos\theta/w) = 32.17(48/4.0)\cos 80° = 67.03 \text{ ft/sec}^2$$

At thrust termination the final flight acceleration becomes

$$a_0 = \sqrt{(a_0)_x^2 + (a_0)_y^2} = 354.4 \text{ ft/sec}^2$$

The vertical and horizontal components of the velocity u_p at the end of powered flight can be obtained from Eq. 4–20. Note that the vehicle mass has been diminished by the propellant that has been consumed:

$$(u_p)_y = c \ln(m_0/m_f) \sin \theta - g_0 t_p = 32.17 \times 120 \times \ln(4/3.6) \times 0.984 - 32.17 \times 1.0$$

$$= 368 \text{ m/sec}$$

$$(u_p)_x = c \ln(m_0/m_f) \cos \theta = 32.17 \times 120 \times \ln(4/3.6) \times 0.1736 = 70.6 \text{ m/sec}$$

The effective exhaust velocity $c = I_s g_0$ (Eq. 2–6). The trajectory angle with the horizontal at rocket cutoff in dragless flight is

$$\tan^{-1}(368/70.6) = 79.1°$$

The final acceleration is found, using Eq. 4–22 with the final mass, as $a_f = 400 \text{ m/sec}^2$.

For the powered flight, the coordinates at propulsion burnout y_p and x_p can be calculated from the time integration of their respective velocities, using Eq. 2–6. The results are

$$y_p = ct_p[1 - \ln(m_0/m_f)/(m_0/m_f - 1)]\sin \theta - 1/2 g_0 t_p^2$$

$$= 32.17 \times 120[1 - \ln(4/3.6)/(4/3.6 - 1)] \times 0.984 - \tfrac{1}{2} \times 32.17 \times (1.0)^2 = 181 \text{ ft}$$

$$x_p = ct_p[1 - \ln(m_0/m_f)/(m_0/m_f - 1)]\cos \theta$$

$$= 32.17 \times 120[1 - \ln(4/3.6)/(4/3.6 - 1)] \times 0.173 = 34.7 \text{ ft}$$

The unpowered part of the trajectory reaches zero vertical velocity at its zenith. The height gained in unpowered free flight may be obtained by equating the vertical kinetic energy at power cutoff to its equivalent potential energy, $g_0(y_z - y_p) = \tfrac{1}{2}(u_p)_y^2$
so that

$$(y_z - y_p) = \tfrac{1}{2}(u_p)_y^2/g_o = \tfrac{1}{2}(368)^2/32.17 = 2105 \text{ ft}$$

The maximum height or zenith location thus becomes $y_z = 2105 + 181 = 2259$ ft. What remains now is to solve the free-flight portion of vertical descent. The time for descent from the zenith is $t_z = \sqrt{2y_z/g_0} = 11.85$ sec and the final vertical or impact vertical velocity $(u_f)y - g_0 t_z = 381$ ft/sec.

The total horizontal range to the zenith is the sum of the powered and free-flight contributions. During free flight, the horizontal velocity remains unchanged at 70.6 ft/sec because there are no accelerations (i.e., no drag, wind, or gravity component). We now need to find the free-flight time from burnout to the zenith, which is $t = (u_p)y/g_0 = 11.4$ sec. The total free-flight time becomes $t_{ff} = 11.4 + 11.85 = 23.25$ sec.

Now, the horizontal or total range becomes $\Delta x = 34.7 + 70.6 \times 23.25 = 1676$ ft.

The impact angle would be around 79°. If drag had been included, solving this problem would have required information on the drag coefficient (C_D) and a numerical solution using Eq. 4–18. All resulting velocities and distances would turn out somewhat lower in value. A set of flight trajectories for sounding rockets is given in Ref. 4–3.

4.4. SPACE FLIGHT

Newton's law of gravitation defines the gravitational attraction force F_g between two bodies in space as:

$$F_g = Gm_1m_2/R^2 = \mu m_2/R^2 \qquad (4\text{--}24)$$

Here G is the universal gravity constant ($G = 6.674 \times 10^{-11} \text{m}^3/\text{kg} - \text{sec}^2$), m_1 and m_2 are the masses of the two attracting bodies (such as the Earth and the moon, or the Earth and a spacecraft, or the sun and a planet), and R is the distance between their centers of mass. The Earth's gravitational constant μ is the product of Newton's universal constant G and the mass of the Earth, m_1 ($5.974 \times 10^{24}\text{kg}$). It is $\mu = 3.987 \times 10^{14}\text{m}^3/\text{sec}^2$.

Rockets offer a means for escaping the Earth's pull for lunar and interplanetary travel, for escaping our solar system, and for launching stationary or moving platforms in space. The flight velocity required to escape from the Earth can be found by equating the kinetic energy of the moving body to the work necessary to overcome gravity, neglecting the rotation of the Earth and the attraction of other celestial bodies, namely,

$$\tfrac{1}{2}mu^2 = m \int g dR$$

By substituting for g from Eq. 4–12 and neglecting air friction, the following relation for the Earth's escape velocity v_e is obtained:

$$v_e = R_0\sqrt{\frac{2g_0}{R_0 + h}} = \sqrt{\frac{2\mu}{R}} \qquad (4\text{--}25)$$

Here, R_0 is the effective Earth mean radius (6374.2 km), h is the circular orbit's altitude above sea level, and g_0 is the acceleration of gravity at the Earth's surface (9.806 m/sec). The satellite flight radius R as measured from the earth's center is $R = R_0 + h$. The velocity of escape from the Earth's surface is 11,179 m/sec or 36,676 ft/sec and does not vary appreciably within the Earth's atmosphere, as shown by Fig. 4–7. Escape velocities for surface launches are given in Table 4–1 from the sun, the planets, and the moon. Launching from the Earth's surface at the escape velocity is not practical because as such vehicle ascends through the Earth's atmosphere, it is subject to severe aerodynamic heating and dynamic pressures. A practical launch vehicle has to traverse the lower atmosphere at relatively low velocities and then accelerate to high velocities beyond the dense atmosphere. For example, during a portion of the Space Shuttle's ascent, its main engines were actually throttled to a lower thrust. Alternatively, an Earth escape vehicle may be launched above the denser atmosphere from an orbiting space station or from an orbiting launch vehicle's upper stage.

Any rocket or spaceship may become an Earth *satellite*, revolving around the Earth in a fashion similar to that of the moon. Low Earth orbits, typically below 500 km altitude, are designated as LEO. Most orbit altitudes are above the Earth's lower atmosphere because this minimizes the energy expended to overcome the drag that

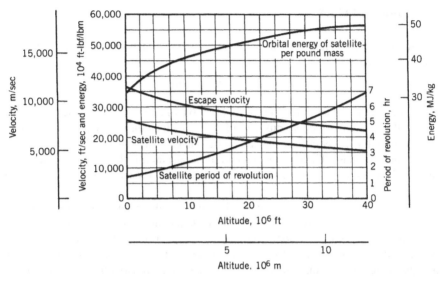

FIGURE 4–7. Orbital energy, orbital velocity, period of revolution, and Earth escape velocity for a space vehicle as a function of altitude for circular satellite orbits. Values are based on a spherical Earth and neglect the Earth's oblate shape, its rotation, and atmospheric drag.

continuously brings the vehicle's orbit closer to the Earth. However, radiation effects in the Van Allen belt on human beings and on sensitive equipment at times often necessitate the selection of Earth orbits at low altitudes.

For a satellite's circular trajectory, the velocity must be sufficient so that its centrifugal force precisely balances the Earth's gravitational attraction:

$$mu_s^2/R = mg$$

For a circular orbit, the satellite velocity u_s is found by using Eq. 4–12,

$$u_s = R_0\sqrt{g_0/(R_0 + h)} = \sqrt{\mu/R} \qquad (4\text{–}26)$$

which is smaller than the escape velocity by a factor of $\sqrt{2}$. The period τ in seconds for one revolution in a circular orbit relative to a stationary Earth is

$$\tau = 2\pi(R_0 + h)/u_s = 2\pi(R_0 + h)^{3/2}/(R_0\sqrt{g_0}) \qquad (4\text{–}27)$$

Neglecting drag, the energy E necessary to bring a unit of mass into a circular satellite orbit consists of its kinetic and potential energy, namely,

$$E = \tfrac{1}{2}u_s^2 + \int_{R_0}^{R} g\,dR$$

$$= \tfrac{1}{2}R_0^2\frac{g_0}{R_0+h} + \int_{R_0}^{R} g_0\frac{R_0^2}{R^2}dR = \tfrac{1}{2}R_0g_0\frac{R_0+2h}{R_0+h} \qquad (4\text{–}28)$$

TABLE 4–1. Characteristic Data for Several Heavenly Bodies

Name	Mean Radius of Orbit (million km)	Period of Revolution		Mean Diameter (km)	Relative Mass (Earth = 1.0)[a]	Specific Gravity	Acceleration of Gravity at Surface (m/sec^2)	Escape Velocity at Surface (m/sec)
Sun	—	—		1,393,000	332,950	1.41	273.4	616,000
Moon	0.383	27.3	days	3475	0.012	3.34	1.58	2380
Mercury	57.87	87.97	days	4670	0.06	5.5	3.67	4200
Venus	108.1	224.70	days	12,400	0.86	5.3	8.67	10,300
Earth	149.6	365.256	days	12,742	1.00^3	5.52	9.806	11,179
Mars	227.7	686.98	days	6760	0.15	3.95	3.749	6400
Jupiter	777.8	11.86	year	143,000	318.4	1.33	26.0	59,700
Saturn	1486	29.46	year	121,000	95.2	0.69	11.4	35,400
Uranus	2869	84.0	year	47,100	17.0	1.7	10.9	22,400
Neptune	4475	164.8	year	50,700	17.2	1.8	11.9	31,000
Pluto	5899	248	year	2368	0.00218	1.44	0.658	1229

[a]Earth mass is 5.976 × 10^{24} kg.

Source: In part from Refs 4–3 and 4–4.

Escape velocity, satellite velocity, satellite period, and satellite orbital energy are all shown as functions of altitude in Fig. 4–7.

A satellite moving around the Earth at an altitude of 300 miles or 482.8 km has a velocity u_s of about 7375 m/sec or 24,200 ft/sec, circles a stationary Earth in $\tau = 1.63$ hr.; ideally it requires an energy of 3.35×10^7 J to place 1 kg of spaceship mass into its orbit. An equatorial satellite in a circular orbit at an altitude of 6.611 Earth radii (about 26,200 miles, 42,200 km, or 22,700 nautical miles) has a period of revolution of exactly 24 hr. It will, therefore, appear stationary to an observer on Earth. This is known as a *synchronous* satellite in *geo synchronous Earth orbit*, usually abbreviated as GEO. This orbit is used extensively for communications satellite and Earth observation applications. In the part of Section 4.7 on launch vehicles, we describe how the payload of any given space vehicle diminishes as the orbit's circular altitude is increased and as the inclination (angle between orbit plane and Earth equatorial plane) is changed. See Refs. 4–3, 4–4, 4–5, 4–6 and 4–9.

Elliptical Orbits

The circular orbit described above is a special case of the more general elliptical orbit shown in Fig. 4–8; here, the Earth (or any other heavenly body around which another body is moving) is located at one of the focal points of this ellipse. The relevant equations of motion come from Kepler's laws and elliptical orbits may be described as follows, when expressed in polar coordinates:

$$u = \left[\mu \left(\frac{2}{R} - \frac{1}{a} \right) \right]^{1/2} \qquad (4\text{--}29)$$

where u is the velocity of the body in the elliptical orbit, R is the instantaneous radius from the center of the Earth (a vector quantity, which changes direction as well as magnitude), a is the major axis of the ellipse, and μ is the Earth's gravitational constant, 3.986×10^{14} m^3/sec^2. These symbols are defined in Fig. 4–8. From Eq. 4–29 it can be seen that the velocity has its maximum value u_p when the moving body comes closest to its focal point at its orbit's *perigee* and the minimum value u_a at its *apogee*. By substituting for R in Eq. 4–29, and by defining the ellipse's shape factor e as the *eccentricity of the ellipse*, $e = \sqrt{a^2 - b^2} / a$, the apogee and perigee velocities can be expressed as

$$u_a = \sqrt{\frac{\mu(1-e)}{a(1+e)}} \qquad (4\text{--}30)$$

$$u_p = \sqrt{\frac{\mu(1+e)}{a(1-e)}} \qquad (4\text{--}31)$$

Another property of an elliptical orbit is that the product of velocity and instantaneous radius remains constant for any location x or y on the ellipse, namely, $u_x R_x = u_y R_y = uR$. The exact path that a satellite takes depends on the velocity (magnitude and vector orientation) with which it is started or was injected into orbit.

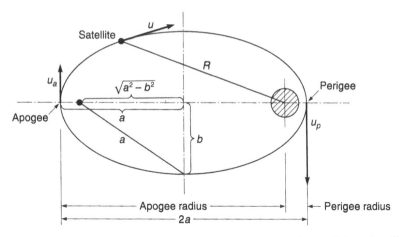

FIGURE 4–8. Elliptical orbit; the attracting body is at one of the focal points of the ellipse.

For *interplanetary transfers*, an ideal mission can be achieved with minimum energy with a simple transfer ellipse, as suggested originally by Hohmann (see Ref. 4–7). Assuming that planetary orbits about the sun are circular and coplanar, Hohmann demonstrated that the path of minimum energy is an ellipse tangent to both planetary orbits as shown in Fig. 4–9. This operation requires a velocity increment (of relatively high thrust) at the initiation (planet A at time t_1) and another at termination (planet B at time t_2): both increments equal the velocity differences between the respective circular planetary velocities and the perigee and apogee velocities which define the transfer ellipse. Thrust levels at the beginning and end maneuvers of the Hohmann ellipse must be high enough to amount to a short operating time and an acceleration of at least 0.01 g_0, but preferably more. Note that because electrical propulsion accelerations are much lower, amounting

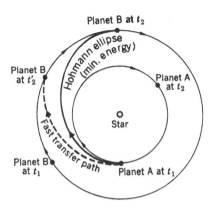

FIGURE 4–9. Schematic of interplanetary transfer paths. These same transfer maneuvers apply when going from a low-altitude Earth satellite orbit to a higher orbit.

to about $10^{-5}g_0$, and operating times longer, weeks or months, the best transfer trajectories in electrical propulsion turn out to be much different from Hohmann ellipses; these are described in Chapter 17.

Departure dates or the *relative positions of the launch planet and the target planet* in planetary transfer missions become critical, because the spacecraft must meet with the target planet when it arrives at the target orbit. Transfer times $(t_2 - t_1)$ for Hohmann-ellipse flights starting on Earth are about 116 hours to go to the moon and about 259 days to Mars. If faster flight paths (shorter transfer times) are desired (see dashed lines in Fig. 4–9), they will require more energy than those with a Hohmann transfer ellipse. This means a larger vehicle with more propellant and a larger propulsion system, or a higher total impulse. There always is a *time window* for launching a spacecraft that will make for a successful rendezvous. For Mars missions an Earth-launched spacecraft may have a launch time window of more than two months. Hohmann transfer ellipses or faster transfer paths apply not only to planetary flight but also to Earth satellites when they go from one circular orbit to another (but within the same plane). Also, if one spacecraft goes to rendezvous with another spacecraft in a different orbit, the two have to be in the proper predetermined positions prior to any thrust application to simultaneously reach their rendezvous location.

When the launch orbit (or launch planet) is not in the same plane as the target orbit, then additional energy will be needed for applying thrust in directions normal to the launch orbit plane. More information can be found in Refs. 4–3, 4–4, 4–6, and 4–10.

Example 4–2. A satellite is launched from a circular equatorial parking orbit at an altitude of 160 km into a coplanar circular synchronous orbit by using a Hohmann transfer ellipse. Assume a homogeneous spherical Earth with a radius of 6371 km. Determine the velocity increments for entering the transfer ellipse and for achieving the synchronous orbit at 42,200 km altitude. See Fig. 4–9 for the terminology of the orbits.

SOLUTION. The two orbits are $R_A = 6.531 \times 10^6$ m; $R_B = 48.571 \times 10^6$ m. The major axis a_{te} of the transfer ellipse is

$$a_{te} = \tfrac{1}{2}(R_A + R_B) = 27.551 \times 10^6 \text{m}$$

The orbit velocities of the two satellites are

$$u_A = \sqrt{\mu/R_A} = [3.986005 \times 10^{14}/6.571 \times 10^6]^{1/2} = 7788 \text{ m/sec}$$

$$u_B = \sqrt{\mu/R_B} = 2864.7 \text{ m/sec}$$

The velocities needed to enter and exit the transfer ellipse are

$$(u_{te})_A = \sqrt{\mu}[(2/R_A) - (1/a)]^{1/2} = 10,337 \text{ m/sec}$$

$$(u_{te})_B = \sqrt{\mu}[(2/R_B) - (1/a)]^{1/2} = 1394 \text{ m/sec}$$

The changes in velocity going from parking orbit to ellipse Δu_A and from ellipse to final orbit Δu_B are

$$\Delta u_A = |(u_{te})_A - u_A| = 2549 \text{ m/sec}$$

$$\Delta u_B = |u_B - (u_{te})_B| = 1471 \text{ m/sec}$$

The total velocity change for the transfer maneuvers is

$$\Delta u_{total} = \Delta u_A + \Delta u_B = 4020 \text{ m/sec}$$

Figure 4–10 shows the elliptical transfer trajectory of a ballistic missile or a satellite launch or an ascent vehicle. During the initial powered flight the trajectory angle is adjusted by signals from the guidance system and torques from the reaction control system to an angle that will allow the vehicle to reach the apogee of its elliptical path at exactly the desired altitude. An *orbit injection velocity* increase of the space vehicle is now applied by a chemical propulsion system at this apogee, which causes the vehicle to change from an elliptical transfer flight path to a circular-orbit flight path. The horizontal arrow symbolizes this velocity increase. For an ideal satellite the simplified theory assumes that an orbit injection maneuver is essentially an instantaneous application of the total impulse when the ballistic elliptic trajectory reaches its

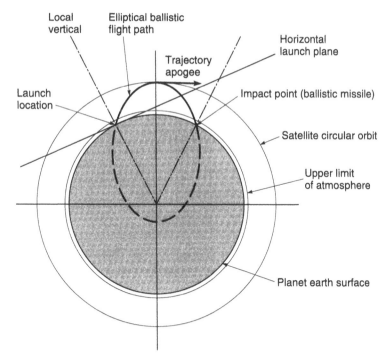

FIGURE 4–10. Long-range ballistic missiles follow an elliptical free flight trajectory, which is nearly drag free, with the Earth's center as one of the focal points. The surface launch is usually vertically up (not shown here) but the flight path is quickly tilted during the early powered flight to enter into an elliptic trajectory. The ballistic range is the arc distance on the Earth's surface. The same elliptical flight path can be used by launch vehicles for satellites; another powered flight period occurs (called orbit injection) just as the vehicle reaches its elliptical apogee (as indicated by the arrow), causing the vehicle to enter an orbit.

apogee or zenith. In reality, the rocket propulsion system for orbit injection operates over a finite time, during which gravity losses and changes in altitude occur.

Deep Space

Lunar and *interplanetary* missions may include circumnavigation, landing, and return flights. The energy necessary to escape from the Earth may be calculated as $\frac{1}{2}mv_e^2$ from Eq. 4–25 as 6.26×10^7 J/kg, which is more than that required to launch an Earth satellite. The gravitational attraction of various heavenly bodies and their respective escape velocities depends on their mass and diameter; approximate values are listed in Table 4–1. An idealized diagram of an interplanetary landing mission is shown in Fig. 4–11.

Escape from the solar system requires approximately 5.03×10^8 J/kg which is eight times as much energy as is required for escape from the Earth. Technology exists today to send small, unmanned probes away from the sun into outer space, but before any mission to the nearest star can be achieved some very long-duration, novel, rocket propulsion system must be introduced. The ideal trajectory for a spacecraft to escape from the sun is either a parabola (minimum energy) or a hyperbola. See Refs. 4–6 and 4–10.

The *Voyager 2 Spacecraft*, developed by NASA's Jet Propulsion Laboratory, was the first man-made object to escape from the solar system and enter interplanetary space. It was launched on August 20, 1977 for exploring the outer planets (flybys of Jupiter, Saturn, Neptune, and Uranus) and then leaving the solar system. It was not expected that Voyager 2 would continue to be operational for over 37 years. A three-axis stabilization system with gyroscopic and celestial reference instruments

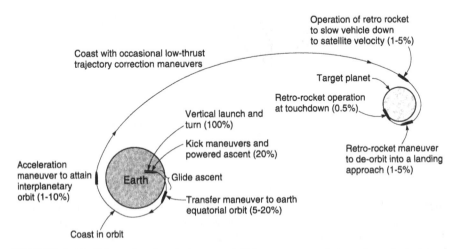

FIGURE 4–11. Schematic of typical powered flight maneuvers during a hypothetical two-dimensional interplanetary mission with a landing (not drawn to scale). The numbers indicate typical thrust magnitudes of the maneuvers in percent of launch takeoff thrust. Heavier lines show powered flight path segments.

is needed to provide a signal that periodically operates its rocket propulsion system which consists of a gas pressurized feed system and 16 small hydrazine monopropellant thrusters, 8 of which remain working to keep a 12-foot-diameter antenna pointed to Earth. Voyager 2 has been powered by three radioisotope thermoelectric generators, which collectively delivered 420 electrical watts at launch (Reference: http://en.wikipedia.org/wiki/Voyager2).

Perturbations

This section gives a brief discussion of forces and torques that cause perturbations and/or deviations from intended space flight paths or satellite's flight orbits. For a more detailed treatment of flight paths and their perturbations, see Refs. 4–3, 4–4, and 4–13. A system that measures the satellite's position and its deviation from the intended flight path is required to determine the needed periodic correction maneuvers in order to apply corrective forces and/or torques. It is called an *orbit maintenance system;* it corrects the perturbed or altered orbit by periodically applying small rocket propulsion forces in predetermined directions. Typically, these corrections are performed by a set of small reaction control thrusters that provide predetermined total impulses in desired directions. These corrections are needed throughout the life of any spacecraft (for 1 to 20 years and sometimes more) to overcome the effects of disturbances so as to maintain the intended flight regime.

Perturbations may be categorized as short term and long term. Daily or orbital period oscillating forces are called *diurnal,* and those with long periods are called *secular.*

High-altitude Earth satellites (36,000 km and higher) experience perturbing forces primarily as gravitational pulls from the sun and the moon, with these forces acting in different directions as the satellite flies around the Earth. Such third-body effects can increase or decrease the velocity magnitude and change the satellite's direction. In extreme cases the satellite may come close enough to the third body, such as a planet or one of its moons, to undergo what is called a *hyperbolic maneuver* (caused by the attraction of that heavenly body) that will radically change its trajectory. Such encounters have been used to increase or decrease the satellite's energy and intentionally change the velocity and the shape of the orbit.

Medium- and low-altitude satellites (500 to 35,000 km) experience perturbations because of the Earth's oblateness. The Earth bulges at the equator, and its cross section through the poles is not entirely circular. Depending on the inclination of the orbital plane to the Earth equator and the altitude of the satellite orbit, two perturbations result: (1) the regression of the nodes and (2) a shifting of the apsides line (major axis). Regression of the nodes is shown in Fig. 4–12 as a rotation of the plane of the orbit in space, and it can be as high as 9° per day at relatively low altitudes. Theoretically, regression does not occur in truly equatorial orbits.

Figure 4–13 shows an exaggerated shift of the apsidal line, with the center of the Earth remaining as a focus point. This perturbation may be visualized as the movement of the prescribed elliptical orbit in a fixed plane. Obviously, both apogee and perigee points change in position, the rate of change being a function of the satellite

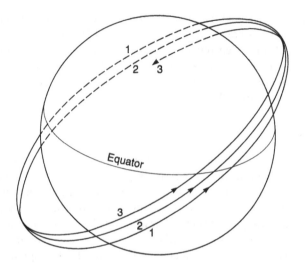

FIGURE 4–12. The regression of nodes is shown as a rotation of the plane of the orbit. The movement direction will be opposite to the east–west components of the Earth's satellite motion.

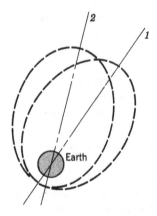

FIGURE 4–13. Shifting of the apsidal line of an elliptic orbit from position 1 to 2 because of the oblateness of the Earth.

altitude and plane inclination angle. At an apogee altitude of 1000 nautical miles (n.m.) and a perigee of 100 n.m. in an equatorial orbit, the apsidal drift is approximately 10° per day.

Satellites of modern design, with irregular shapes due to protruding antennas, solar arrays, or other asymmetrical appendages, experience torques and forces that tend to perturb the satellite's position and orbit throughout its orbital life. Principal torques and forces result from the following factors:

1. *Aerodynamic drag.* This factor is significant at orbital altitudes below 500 km and is usually assumed to cease at 800 km above the Earth. Reference 4–8 gives

a detailed discussion of aerodynamic drag, which, in addition to affecting the attitude of unsymmetrical vehicles, causes a change in elliptical orbits known as apsidal drift, a decrease in the major axis, and a decrease in eccentricity of orbits about the Earth. See Refs. 4–6, 4–8, 4–12, and 4–13.

2. *Solar radiation.* This factor dominates at high altitudes (above 800 km) and is due to impingement of solar photons upon satellite surfaces. The solar radiation pressure p (N/m^2) on a given surface of the satellite in the vicinity of the Earth exposed to the sun can be determined from

$$p = 4.5 \times 10^{-6} \cos\theta [(1 - k_s)\cos\theta + 0.67\, k_d] \qquad (4\text{--}32)$$

where θ is the angle (in degrees) between the incident radiation vector and the normal to the surface and k_s and k_d are the specular and diffuse coefficients of reflectivity. Typical values are 0.9 and 0.5, respectively, for k_s and k_d on the body and antenna and 0.25 and 0.01, respectively, for k_s and k_d with solar array surfaces. Radiation intensity varies as the square of the distance from the sun (see Refs. 4–4 and 4–14). The torque T on the vehicle is given by $T = pAl$, where A is the projected area normal to the flight direction (or normal to the sun's rays) and l is the offset distance between the spacecraft's center of gravity and the center of solar pressure. For a nonsymmetrical satellite with a large solar panel on one side, solar radiation will cause a small torque that will rotate the vehicle.

3. *Gravity gradients.* Gravitational torques in spacecraft result from variations in the gravitational force on the distributed mass of a spacecraft. Determination of this torque requires knowledge of the gravitational field and the distribution of spacecraft mass. This torque decreases as a function of the orbit radius and increases with the offset distances of masses within the spacecraft (including booms and appendages); it is most significant in large spacecraft or in space stations operating in relatively low orbits (see Refs. 4–4 and 4–15).

4. *Magnetic field.* The Earth's magnetic field and any magnetic moment within the satellite can interact to produce torque. The Earth's magnetic field precesses about the Earth's axis but is very weak (0.63 and 0.31 gauss at poles and equator, respectively). This field is continually fluctuating in direction and intensity because of magnetic storms and other influences. Since the field strength decreases with $1/R^3$ with the orbital altitude, magnetic field forces are often neglected in the preliminary analysis of satellite flight paths (see Ref. 4–16).

5. *Internal accelerations.* Deployment of solar array panels, the shifting of liquid propellants, the movement of astronauts or other masses within the satellite, or the "unloading" of reaction wheels may produce noticeable torques and forces.

6. For precise low Earth orbits the *oblateness* of the Earth (diameter at equator being slightly larger than diameter between poles), high mountains, or Earth surface areas of different densities will cause perturbations of these orbits.

We can categorize satellite propulsion needs according to their function as listed in Table 4–2, which shows the total impulse "budget" applicable to a typical

TABLE 4–2. Typical Propulsion Functions and Approximate Total Impulse Needs of a 2000-lbm Geosynchronous Satellite with a Seven-Year Life

Function	Total Impulse (N-sec)
Acquisition of orbit	20,000
Attitude control (rotation)	4000
Station keeping, E–W	13,000
Station keeping, N–S	270,000
Repositioning (Δu, 200 ft/sec)	53,000
Control apsidal drift (third-body attraction)	445,000
Deorbit	12,700
Total	817,700

high-altitude, elliptic orbit satellite. The control system designer often distinguishes two different kinds of station-keeping orbit corrections needed to keep the satellite in a synchronous position. The east–west correction refers to a correction that moves the point at which a satellite orbit intersects the Earth's equatorial plane in an east or west direction; it usually corrects forces caused largely by the oblateness of the Earth. The north–south correction counteracts forces usually connected with the third-body effects of the sun and the moon.

For many satellite missions any gradual changes in orbit caused by perturbation forces are of little concern. However, in certain missions it is necessary to compensate for these perturbing forces and maintain the satellite in a specific orbit at a particular position in that orbit. For example, synchronous communications satellites in a Geosynchronous Earth Orbit, or GEO, need to maintain their position and their orbit so as to be able to (1) keep covering a specific area of the Earth or communicate with the same Earth stations within its line of sight and (2) not become a hazard to other satellites in this densely populated synchronous equatorial orbit. Another example is Low Earth Orbit or LEO communications satellites system with several coordinated satellites; here at least one satellite has to be in a position to receive and transmit radio-frequency (RF) signals to specific locations on the Earth. The orbits and relative positions of several satellites with respect to each other also need to be controlled and maintained (see Ref. 4–3).

Orbit maintenance requires applying small correcting forces and torques periodically to compensate for perturbation effects; for GEO this happens every few months. Typical velocity increments for the orbit maintenance of synchronous satellites require a Δu between 10 and 50 m/sec per year. For a satellite mass of about 2000 kg a 50-m/sec correction for a 10-year orbit life would need a total impulse of about 100,000 N-sec, which corresponds to a chemical propellant mass of 400 to 500 kg (about a quarter of the satellite mass) when done with small monopropellant or bipropellant thrusters. It would require much less propellant if electrical propulsion were to be used, but for some spacecraft the inert mass of the power supply and mission duration might represent a substantial increase . See Refs. 4–6, 4–13, and 4–14.

Mission Velocity

A convenient way to describe the magnitude of the energy requirement for a space mission is to use the concept of the *mission velocity*. It is the sum of all the flight velocity increments needed (in all the vehicle's stages) to attain the mission objective even though these increments are provided by different propulsion systems and their thrusts may be in different directions. In the sketch of a planetary landing mission of Fig. 4–11, it is the sum of all the Δu velocity increments shown by the heavy lines (rocket-powered flight segments) of the trajectories. Even through some velocity increments might be achieved by retro-action (a negative propulsion force to decelerate the flight velocity), all these maneuvers require energy and their absolute magnitude is counted in the mission velocity. The initial velocity from the Earth's rotation (464 m/sec at the equator and 408 m/sec at a launch station at 28.5° latitude) does not need to be provided by the vehicle's propulsion systems. For example, the required mission velocity for launching at Cape Kennedy, bringing the space vehicle into an orbit at 110 km, staying in orbit for a while, and then entering a deorbit maneuver has the Δu components shown in Table 4–3.

The required mission velocity is the sum of the absolute values of all translation velocity increments that have forces going through the center of gravity of the vehicle (including turning maneuvers) during the mission flight. It is the hypothetical velocity that would be attained by the vehicle in a gravity-free vacuum, if all the propulsive energy of the momentum-adding thrust chambers in all stages were to be applied in the same direction. This theoretical value is useful for comparing one flight vehicle design with another and as an indicator of total mission energy.

The required mission velocity must equal to the "supplied" mission velocity, that is, the sum of all the velocity increments provided by the propulsion systems during each of the various vehicle stages. The total velocity increment that was "supplied" by the Shuttle's propulsion systems for the Shuttle mission (solid rocket motor strap-on boosters, main engines and, for orbit injection, also the increment from the orbital maneuvering system—all shown in Fig. 1–14) had to equal or exceed 9347 m/sec. When the reaction control system propellant and an uncertainty factor are added, this value would have needed to exceed 9621 m/sec. With chemical propulsion systems and a single stage, we can achieve space mission velocities of 4000 to 13,000 m/sec,

TABLE 4–3. Typical Estimated Space Shuttle Incremental Flight Velocity Breakdown for Flight to Low Earth Orbit and Return

Ideal satellite velocity	7790 m/sec
Δu to overcome gravity losses	1220 m/sec
Δu to turn the flight path from the vertical	360 m/sec
Δu to counteract aerodynamic drag	118 m/sec
Orbit injection	145 m/sec
Deorbit maneuver to reenter atmosphere and aerodynamic braking	60 m/sec
Correction maneuvers and velocity adjustments	62 m/sec
Initial velocity provided by the Earth's rotation at 28.5° latitude	−408 m/sec
Total required mission velocity	9347 m/sec

depending on the payload, mass ratio, vehicle design, and propellant. With two stages they can be between perhaps 12,000 and 22,000 m/sec.

Rotational maneuvers (to be described later) do not change the flight velocity and some analysts do not add them to the mission velocity requirements. Also, maintaining a satellite in orbit against long-term perturbing forces (see prior section) is often not counted as part of the mission velocity. However, designers need to provide additional propulsion capabilities for these purposes. These are often separate propulsion systems, called *reaction control systems*.

Typical vehicle velocities required for various interplanetary missions have been estimated as shown in Table 4–4. By starting interplanetary journeys from a space station, considerable savings in vehicle velocity can be achieved, namely, the velocity necessary to attain the Earth-circling satellite orbit. As space flight objectives become more ambitious, mission velocities increase. For a given single or multistage vehicle it is possible to increase the vehicle's terminal velocity, but usually only at the expense of payload. Table 4–5 shows some typical ranges of payload values for a

TABLE 4–4. Approximate Vehicle Mission Velocities for Typical Space and Interplanetary Missions

Mission	Ideal Velocity (km/sec)	Approximate Actual Velocity (km/sec)
Satellite orbit around Earth (no return)	7.9–10	9.1–12.5
Escape from Earth (no return)	11.2	12.9
Escape from moon	2.3	2.6
Earth to moon (soft landing on moon, no return)	13.1	15.2
Earth to Mars (soft landing)	17.5	20
Earth to Venus (soft landing)	22	25
Earth to moon (landing on moon and return to Earth[a])	15.9	17.7
Earth to Mars (landing on Mars and return to Earth[a])	22.9	27

[a] Assumes air braking within atmospheres.

TABLE 4–5. Approximate Relative Payload-Mission Comparison Chart for Typical Multistage Rocket Vehicles Using Chemical Propulsion Systems

Mission	Relative Payload[a] (%)
Earth satellite	100
Earth escape	35–45
Earth 24-hr orbit	10–25
Moon landing (hard)	35–45
Moon landing (soft)	10–20
Moon circumnavigation (single fly-by)	30–42
Moon satellite	20–30
Moon landing and return	1–4
Moon satellite and return	8–15
Mars flyby	20–30
Mars satellite	10–18
Mars landing	0.5–3

[a] 300 nautical miles (555.6 km) Earth orbit is 100% reference.

given multistage vehicle as a percentage of a payload for a relatively simple Earth orbit. Thus, a vehicle capable for putting a substantial payload into a near-Earth orbit can only land a very small fraction of this payload on the moon, since it has to have additional upper stages, which displace payload mass. Therefore, much larger vehicles are required for space flights with high mission velocities when compared to a vehicle of less mission velocity but identical payload. The values listed in Tables 4–4 and 4–5 are only approximate because they depend on specific vehicle design features, the propellants used, exact knowledge of the trajectory–time relation, and other factors that are beyond the scope of this abbreviated treatment.

4.5. SPACE FLIGHT MANEUVERS

Ordinarily, all propulsion operations are controlled (started, monitored, and stopped) through the vehicle's guidance and control system. The following types of *space flight maneuvers and vehicle accelerations* utilize rocket propulsion:

a. The propulsion systems in the *first stage* and *strap-on booster stage* add momentum during launch and ascent of the space vehicles. They require high or medium level thrust of limited durations (typically 0.7 to 8 min). To date all Earth-launch operations have used chemical propulsion systems and these constitute the major mass portion of the space vehicle; they are discussed further in the next section.

b. *Orbit injection* is a powered maneuver that adds velocity to the top stage (or payload stage) of a launch vehicle, just as it reaches the apogee of its ascending elliptical trajectory. Figure 4–10 shows the ascending portion of such elliptical flight path which the vehicle follows up to its apogee. The horizontal arrow in the figure symbolizes the application of thrust in a direction such as to inject the vehicle into an Earth-centered coplanar orbit. This is performed by the main propulsion system in the top stage of the launch vehicle. For injection into a low Earth orbit, thrust levels are typically between 200 and 45,000 N or 50 and 11,000 lbf, depending on the payload size, transfer time, and specific orbit.

c. A space vehicle's *transfer from one orbit to another* (coplanar) orbit has two stages of rocket propulsion operation. One operates at the beginning of the transfer maneuver from the launching orbit and the other at the arrival at the destination orbit. The Hohmann ellipse shown in Figure 4–9 gives the most energy efficient transfer (i.e., the smallest amount of propellant usage). Reaction control systems are used here to orient the transfer thrust chamber into a proper direction. If the new orbit is higher, then thrusts are applied in the flight direction. If the new orbit is at a lower altitude, then thrusts must be applied in a direction opposite to the flight velocity vector. Transfer orbits can also be achieved with a very low thrust level (0.001 to 1 N) using an electric propulsion system, but flight paths will be very different (multiloop spirals) and transfer durations will be much longer, (see Chapter 17). Similar maneuvers are also performed with lunar or interplanetary flight missions, such as the planetary landing mission shown schematically in Fig. 4–11.

d. *Velocity vector adjustment and minor in-flight correction maneuvers* for both translation and rotation are usually performed with low-thrust, short-duration, intermittent (pulsing) operations using a reaction control system with multiple small liquid propellant thrusters. Vernier rockets on ballistic missiles are used to accurately calibrate its terminal velocity vector for improved target accuracy. Reaction control rocket systems on a space launch vehicle will allow for accurate orbit-injection adjustment maneuvers after it has been placed into orbit by its other, less accurate, propulsion systems. Midcourse guidance-directed *correction maneuvers* for the trajectories of deep space vehicles similarly fall into this category. Propulsion systems for *orbit maintenance maneuvers*, also called *station-keeping maneuvers* (for overcoming perturbing forces), that keep spacecraft in their intended orbit and orbital position are also considered to be in this category.

e. *Reentry and landing maneuvers* may take several forms. If landing occurs on a planet that has an atmosphere, then drag in that atmosphere will heat and slow down the reentering vehicle. For multiple elliptical orbits, drag will progressively reduce the perigee altitude and the perigee velocity at every orbit. Landing at a precise or preplanned location requires a particular velocity vector at a predetermined altitude and distance from the landing site. The vehicle has to be rotated into a proper position and orientation so as to use its decelerators and heat shields correctly. Precise velocity magnitudes and directions prior to entering any denser atmosphere are critical for minimizing heat transfer (usually to a vehicle's heat shield) and to achieve touchdown at the intended landing site or, in the case of ballistic missiles, the intended target and this commonly requires a relatively minor maneuver (of low total impulse). If there is very little or no atmosphere (for instance, landing on the moon or Mercury), then an amount of reverse thrust has to be applied during descent and at touchdown. Such rocket propulsion system needs to have a variable thrust capability to assure a soft landing and to compensate for the decrease in vehicle mass as propellant is consumed during descent. The U.S. lunar landing rocket engine, for example, had a 10 to 1 thrust variation.

f. *Rendezvous and docking* between two space vehicles involve both rotational and translational maneuvers with small reaction control thrusters. Docking (sometimes called lock-on) is the linking up of two spacecraft and requires a gradual gentle approach (low thrust, pulsing-mode thrusters) so as not to cause spacecraft damage.

g. *Simple rotational maneuvers* rotate the vehicle on command into a specific angular position in order to orient/point a telescope or other instrument (a solar panel, or an antenna for purposes of observation, navigation, communication, or solar power receptor). Such a maneuver is also used to keep the orientation of a satellite in a specific direction; for example, if an antenna needs to be continuously pointed at the center of the Earth, then the satellite needs to be rotated around its own axis once every satellite revolution. Rotation is also used to point nozzles in the primary propulsion system into their intended direction just prior to their operation. The reaction control system can also provide pulsed thrusting; it has been used for achieving flight stability, and/or for correcting angular oscillations that would otherwise increase drag or cause tumbling of the vehicle. Spinning or rolling a vehicle about its axis will not only improve flight stability but also average out thrust vector misalignments. Chemical multi-thruster reaction-control systems are used when rotation needs to be performed quickly. When rotational changes may be done over long periods of time, electrical propulsion systems (operating at a higher specific impulse) with multiple thrusters are often preferred.

h. *Any change of flight trajectory plane* requires the application of a thrust force (through the vehicle's center of gravity) in a direction normal to the original flight-path plane. This is usually performed by a propulsion system that has been rotated (by its reaction control system) into the proper nozzle orientation. Such maneuvers are done to change a satellite orbit's plane or when traveling to a planet, such as Mars, whose orbit is inclined to the plane of the Earth's orbit.

i. *Deorbiting and disposal of used or spent rocket stages* and/or *spacecraft* is an important requirement of increasing consequences for removing space debris and clutter. Spent spacecraft must not become a hazard to other spacecraft or in the event of reentry threaten population centers. Relatively small thrusts are used to drop the vehicle to a low enough elliptical orbit so that atmospheric drag will slow down the vehicle at the lower elevations. In the denser regions of the atmosphere during reentering, expended vehicles will typically break up and/or overheat (and burn up). Ground based energetic lasers and in-space debris removers are also being considered. International efforts will be required to solve this fast growing "space junk" problem.

j. *Emergency or alternative mission.* When there is a malfunction in a spacecraft and it is decided to abort the mission, such as a premature quick return to the Earth without pursuing the originally intended mission, then specifically suitable rocket engines can be used for such alternate missions. For example, the main rocket engine in the Apollo lunar mission service module was normally used for retroaction to attain a lunar orbit and for return from lunar orbit to the Earth; it could also have been used for emergency separation of the payload from the launch vehicle and for unusual midcourse corrections during translunar coast, enabling an emergency Earth return.

Table 4–6 lists all maneuvers that have just been described, together with some others, and shows the various types of rocket propulsion system (as introduced in Chapter 1) that have been used for each of these maneuvers. The table omits several propulsion systems, such as solar thermal or nuclear rocket propulsion, because these have not yet flown in routine space missions. One of the three propulsion systems on the right of Table 4–6 is electrical propulsion which has relatively high specific impulse (see Table 2–1), and this makes it very attractive for deep space missions and for certain station-keeping jobs (orbit maintenance). However, electrical thrusters perform best when applied to missions where sufficiently long thrust action times for

TABLE 4–6. Types of Rocket Propulsion Systems Commonly Used for Different Flight Maneuvers or Application

Flight Maneuvers and Applications ↓ / Propulsion System →	Liquid Propellant Rocket Engines			Solid Propellant Rocket Motors		Electrical Propulsion		
	High Thrust, Liquid Propellant Rocket Engine, with Turbopump	Medium to Low Thrust, Liquid Propellant Rocket Engine	Pulsing Liquid Propellant, Multiple Small Thrusters	Large Solid Propellant Rocket Motor, Often Segmented	Medium to Small Solid Propellant Motors	Arcjet, Resistojet	Ion Propulsion, Electromagnetic Propulsion	Pulsed Plasma Jet
Launch vehicle booster	××			××				
Strap-on motor/engine	××			××				
Upper stages of launch vehicle	××	××		×	××			
Satellite orbit injection and transfer orbits		××			××	×	×	
Flight velocity adjustments, flight path corrections, orbit changes		×	××			×	×	
Orbit/position maintenance, rotation of spacecraft			××			×	×	×
Docking of two spacecraft			××					
Reentry and landing, emergency maneuvers		×	×		×			
Deorbit		×	×		×	×		
Deep space, sun escape		×	×				×	
Tactical missiles					××			
Strategic missiles	×	×	×	××	××			
Missile defense			×	××	××			
Artillery shell boost					××			

Legend: × = in use: ×× = preferred for use in recent years.

reaching the desired vehicle velocity or rotation positions are available because of their very relatively small accelerations.

During high-speed atmospheric reentry vehicles encounter extremely high heating loads. In the Apollo program a heavy thermal insulation layer was located at the bottom of the Apollo Crew Capsule and in the Space Shuttle Orbiter low-conductivity, lightweight bricks on the wings provided thermal protection to the vehicle and crew. An alternate method in multi-engine main rocket propulsion vehicles is to reduce high Earth reentry velocities by reversing or retro directing the thrust of some of the engines. This requires to turn the vehicle around in space by 180° (usually by means of several attitude control thrusters) prior to the return maneuver and then firing the necessary portion of the main propulsion system. An example of this method can be found in the Falcon 9 Space Vehicle booster stage reentry, the lower portion of which is shown in the front cover of this book. The aim here is simply to recover and reuse this stage. All nine (9) Merlin liquid propellant rocket engines are needed during ascent to orbit but only 3 of these are sufficient for the retro-slowdown maneuver during reentry, and only the central engine need be operated during the final vertical landing maneuver. Before reusing and relaunching, the recovered booster stage with its multiple rocket engines is refurbished—all residual propellant is removed and the unit is cleaned, flushed, and purged with hot dry air. Upon inspection, further maintenance may be performed as needed. Since the booster stage is usually the most expensive stage, this reuse will allow some cost reduction (if used often enough).

Reaction Control System

All functions of a reaction control system have been described in the previous section on flight maneuvers; they are used for the maneuvers identified by paragraphs d, f, and h. In some vehicle designs they are also used for tasks described in b and c, and parts of e and g, if the thrust levels are low.

A *reaction control system* (RCS), often also called an *auxiliary rocket propulsion system*, is needed to provide trajectory corrections (small Δu additions) as well as for correcting rotational or attitude positions in almost all spacecraft and all major launch vehicles. If mostly rotational maneuvers are made, the RCS has been called an *attitude control system* (but this nomenclature is not consistent throughout the industry or the literature).

An RCS is usually incorporated into the payload stage and into each of the stages of a multiple-stage vehicle. In some missions and designs the RCS is only built into the uppermost stage; it operates throughout the flight and provides needed control torques and forces for all the stages. In large vehicle stages, thrust levels of multiple thrusters of an RCS are correspondingly large (500 to 15,000 lbf), and for terminal stages in small satellites they can be small (0.01 to 10.0 lbf) and may be pulsed. Liquid propellant rocket engines with multiple thrusters are presently used in nearly all launch vehicles and in most spacecraft. Cold gas systems were used exclusively with early spacecraft. In the last two decades, an increasing number of electrical propulsion systems are being used, primarily on spacecraft (see Chapter 17). The life of an RCS may be short (when used on an individual vehicle stage) or it may be used throughout the mission duration (some more than 10 years) when part of an orbiting spacecraft.

Vehicle attitude has to be controlled about three mutually perpendicular axes, each with two degrees of freedom (clockwise and counterclockwise rotation), giving a total of six degrees of freedom. *Pitch* control raises or lowers the nose of the vehicle, *yaw* torques induce motion to the right or left side, and *roll* torques will rotate the vehicle about its axis, either clockwise or counterclockwise. In order to apply a pure torque it is necessary to use two thrust chambers of equal thrust and equal start and

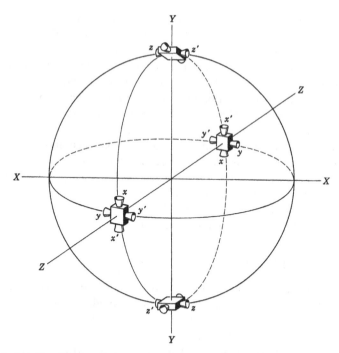

FIGURE 4–14. Simplified attitude control system diagram for spacecraft. It requires 12 thrusters (identified as x, y, z pairs) to allow the application of pure torques about three perpendicular axes. The four unlabeled thrusters, shown at the top and bottom clusters, would be needed for translation maneuvers along the Z axis.

stop times, placed equidistant from the center of mass. Figure 4–14 shows a simple spherical spacecraft attitude control system; thrusters $x - x$ or $x' - x'$ apply torques that rotate about the X axis. There should be a minimum of 12 thrusters in such a system, but some spacecraft with geometrical or other limitations on the placement of these nozzles or with provisions for redundancy may actually have more than 12. The same system can, by operating different sets of nozzles, also provide translation forces; for example, if one each of the (opposite) thrust units x and x' were operated simultaneously, the resulting forces would propel the vehicle in the direction of the Y axis. With clever designs it is possible to use fewer thrusters, but they will usually not provide a pure torque.

An RCS usually contains the following major subsystems: (1) sensing devices for determining the attitude, velocity, and/or position of the vehicle with respect to some reference direction at any given time, such as provided by gyroscopes, star-trackers, or radio beacons; (2) a control-command system that compares the actual space and rotary position with the desired or programmed position and issues command signals to change the vehicle's position within a desired time period; and (3) devices for changing the angular position, such as a set of high-speed gyroscopic wheels or sets of RCS or small attitude control thrusters. See Refs. 4–13 and 4–14.

Precise attitude angular corrections can also be achieved by the use of inertial or high-speed rotating reaction wheels, which apply torques when their rotational speed is increased or decreased. While these wheels are quite simple and effective, the total amount of angular momentum change they can supply is limited. By using pairs of supplementary attitude control thrust rocket units, it is possible to unload or even respin each wheel so it can continue to supply small angular position corrections as needed.

The torque T of a pair of thrust chambers of thrust F and separation distance l provides the vehicle with an angular or rotational moment of inertia M_a an angular acceleration of magnitude α:

$$T = Fl = M_a \alpha \qquad (4-33)$$

For a cylinder of radius r and of equally distributed mass the rotational moment of inertia is $M_a = \frac{1}{2}mr^2$ and for a homogeneous sphere it is $M_a = \frac{2}{5}mr^2$. The largest possible practical value of moment arm l will minimize thrust and propellant requirements. If the angular acceleration is constant over a time t, the vehicle will move at an angular speed ω and through a displacement angle θ, namely,

$$\omega = \alpha t \quad \text{and} \quad \theta = \frac{1}{2}\alpha t^2 \qquad (4-34)$$

Commonly, the control system senses a small angular disturbance and then commands an appropriate correction. For detection of angular position changes by an accurate sensor, it is usually necessary for the vehicle to undergo a slight angular displacement. Care must be taken to avoid overcorrection and hunting of the vehicle's position by the control system. This is one of the reasons many spacecraft use extremely short multiple pulses (0.010 to 0.040 sec per pulse) and low thrust (0.01 to 100 N) (see Refs. 4-11, 4-13, and 4-14).

Reaction control systems may be characterized by the magnitude of the total impulse and by the number, thrust level, and direction of the thrusters and their duty cycles. The *duty cycle* refers to the number of thrust pulses, their operating times, times between thrust applications, and timing of short operations during the mission operating period. For any given thruster, a 30% duty cycle would mean an average active cumulative thrust period of 30% during the propulsion system's flight duration. These propulsion parameters can be determined from the mission, the guidance and control approach, the desired accuracy, flight stability, the likely thrust misalignments of the main propulsion systems, the three-dimensional flight path variations, the perturbations to the trajectory, and several other factors. Some of these parameters can often be difficult to determine.

4.6. EFFECT OF PROPULSION SYSTEM ON VEHICLE PERFORMANCE

This section gives several methods for improving flight vehicle performance and most of enhancements listed are directly influenced by the flight mission and by the selection or design of the vehicle and the propulsion system. Only a few flight vehicle performance improvements do not depend on the propulsion system. Most of those listed below apply to all missions, but some are peculiar to some missions only.

1. The *effective exhaust velocity c* or its equivalent the *specific impulse I_s* usually have a direct effect on the vehicle's overall performance. The vehicle's final velocity increment Δu may be increased by a higher I_s. This can be done by using more energetic chemical propellants (see Chapters 7 and 12), by higher chamber pressures and, for upper stages operating at high altitudes, also by larger nozzle area ratios provided that all these do not appreciably increase the vehicle's inert mass. With electrical propulsion a higher available I_s can enhance vehicle performance but, as explained in Chapter 17, their very low thrusts do limit their use to certain space missions.

2. The vehicle mass ratio's (m_0/m_f) logarithmic effect can be increased in several ways. One way is by reducing the final mass m_f (which consists of the inert hardware and payload plus the nonusable, residual propellant mass). Reducing the inert mass implies lighter structures, smaller payloads, lighter guidance/control devices, and/or less unavailable residual propellant; this often means going to stronger structural materials at higher stresses, more efficient power supplies, or smaller electronic packages. During design, there is always great emphasis on reducing all hardware and residual propellant masses to their practical minima. Another way is to increase the initial vehicle mass, and use higher thrust and more propellant, but with accompanying small increases in the structure or inert propulsion system mass.

3. Reducing the burning time (i.e., increasing the thrust level) will reduce the gravitational loss in some applications. However, higher accelerations usually require more structural and propulsion system mass, which in turn cause the mass ratio to be less favorable.

4. Atmospheric *drag*, which can be considered as negative thrust, can be reduced in at least four ways. Drag has several components: (a) Form drag depends on the flight vehicle's aerodynamic shape; slender pointed noses or sharp, thin leading edges on fins or wings have less drag than stubby, blunt shapes. (b) A vehicle with a small cross-sectional area has less drag; propulsion designs that can be packaged in long, thin shapes will be preferred. (c) Drag is proportional to the cross-sectional or frontal vehicle area; higher propellant densities will decrease propellant volume and therefore will allow smaller cross sections. (d) Skin drag is caused by the friction of the gas flowing over all vehicle's outer surfaces; smooth contours and polished surfaces are preferred; skin drag also depends on higher propellant densities because they require smaller volumes and thus lower surface areas. (e) Base drag is the fourth component; it is a function of the ambient pressure acting over the surface of the vehicle's base or rear section; it is influenced by the nozzle exit gas pressure and turbulence, any discharges of turbine exhaust gases, and by the geometry of the vehicle base design. These are discussed further in Chapter 20.

5. The length of the propulsion nozzle is often a significant contributor to the over-all vehicle or stage length. As described in Chapter 3, for each mission there is an optimum nozzle contour and length, which can be determined by trade-off analyses. A shorter nozzle length or multiple nozzles on the same propulsion system may allow a somewhat shorter vehicle; in many designs this implies a somewhat lighter vehicle structure and a slightly better vehicle mass ratio.

6. The final vehicle velocity at propulsion termination can be increased by increasing the initial velocity u_0. By launching a satellite in an eastward direction the rotational speed of the Earth adds to the final satellite orbital velocity. This Earth tangential velocity is as high as 464 m/sec or 1523 ft/sec at the equator; Sea Launch, a commercial enterprise, launched from a ship at the equator to take full advantage of this velocity increment. For an easterly launch at John F. Kennedy Space Center (latitude of 28.5° north) this extra velocity is less, about 408 m/sec or 1340 ft/sec. Conversely, a westerly satellite launch has a negative initial velocity and thus requires higher-velocity increments. Another way to increase u is to launch a spacecraft or an air-to-surface missile from a satellite or an aircraft, which imparts its initial vehicle velocity vector and allows launching in the desired direction. An example is the Pegasus three-stage space vehicle, which is launched from an airplane.

7. For vehicles that fly within the atmosphere it is possible to increase their range when aerodynamic lift is used to counteract gravity and reduce gravity losses. Using a set of wings and flying at an angle of attack increases the lift, but also adds to the drag. Vehicle lift may also be used to increase the maneuverability and trajectory flexibility.

8. When the vehicle's flight velocity u is close to the rocket's effective exhaust velocity c, the propulsive efficiency is highest (Eq. 2–22 or Fig. 2–3) and more of the rocket exhaust gas energy is transformed into the vehicle's flight energy. When $u = c$ this propulsive efficiency reaches 100%. Trajectories where u is close in value to c for a major portion of the flight therefore would need less propellant.

9. When a mission is changed during flight, the liquid propellant of a particular rocket engine (in a multistage vehicle) may not be fully used and such unused propellant is then available to be transferred within the vehicle to augment the propellant of a different rocket engine system unit. An example was the transfer of storable liquid propellant in the Space Shuttle from the Orbital Maneuvering System (OMS) to the Reaction Control System (RCS – 14 small bipropellant thrusters). This transfer allowed for additional orbit maintenance operations and more time in orbit.

Several of these influencing parameters can be optimized. Therefore, for every mission or flight application there is an optimum propulsion system design and the propulsion parameters that define the optimum condition are dependent on vehicle or flight parameters.

4.7. FLIGHT VEHICLES

As previously stated, a vast majority of rocket-propelled vehicles use a relatively simple single-stage design and commonly employ solid propellant rocket motors. Most are used in military applications as described in the next section. In this section we discuss the more sophisticated multistage space launch vehicles and mention others, such as large ballistic missiles (often called strategic missiles) and some sounding rockets. All these require some intelligence acquisition for their guidance and must include navigation-system hardware.

A *single stage to orbit* vehicle (e.g., to LEO) is very limited in the payload it can carry and this concept has been only of research interest. Figures 4–2 and 4–3 indicate that a high-performance single-stage vehicle with a propellant fraction of 0.95 and an average I_s of 400 sec may achieve an ideal terminal velocity of nearly 12,000 m/sec without payload. If the analysis includes drag and gravity forces, a correspondingly higher value of I_s would be needed. Accounting for maneuvers in the trajectory and an attitude control system, depending on design it is likely that a single stage's payload would remain at about 1.0% of the gross takeoff mass. For typical larger payload percentages and particularly for more ambitious missions, we use vehicles with two or more stages as described below.

Multistage Vehicles

Multistep or *multistage rocket vehicles* permit higher vehicle velocities, more payload for space vehicles, larger area coverage for defensive missiles, and improved performance for long-range ballistic missiles or area defense missiles. After the useful propellant has been consumed in a particular stage, the remaining empty mass of that expended stage is dropped from the vehicle and the operation of the propulsion system of the next step or stage is started. The last or top stage, which is usually the smallest, carries the payload. Separating the empty mass of expended stages from the remainder of the vehicle avoids additional energy expenditures. As the number of stages is increased, the initial takeoff mass can decrease, but the relative gains in lowering initial mass become less and less with each additional stage. Moreover, adding stages increases the required physical mechanisms increasing vehicle complexity and total mass. The most economically useful number of stages is usually between two and six, depending on the mission. See Example 4–3. Several different multistage launch vehicle configurations have been used successfully, and four are shown in Fig. 4–15. Most vehicles are launched vertically, but a few have been otherwise launched from airplanes, such as the three-stage Pegasus space vehicle.

Even though it represents only a very small portion of the initial mass, the payload of a multistage rocket is roughly proportional its takeoff mass. If a payload of 50 kg requires a 6000-kg multistage rocket, a 500-kg payload would require approximately a 60,000-kg rocket unit with identical number of stages and similar configuration, using the same payload fraction and the same propellants. When the operation of an upper stage is started immediately after thrust termination of a lower stage, then the total ideal velocity of a multistage vehicle with purely series-stage (or tandem) arrangement is simply the sum of the individual stage velocity increments.

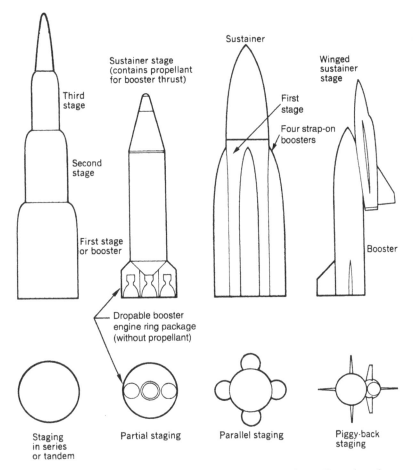

FIGURE 4–15. Schematic renditions of four different geometric configurations for assembling individual stages into a launch vehicle.

For n stages in series as shown in the first sketch of Fig. 4–15 the final velocity increment Δu_f is

$$\Delta u_f = \sum_1^n \Delta u = \Delta u_1 + \Delta u_2 + \Delta u_3 + \cdots \qquad (4\text{--}35)$$

In vertical atmospheric flight, the individual velocity increments are each given by Eq. 4–19. For the simplified case of a vacuum flight in a gravity-free field this equation may be expressed as (applying Eq. 4–6)

$$\Delta u_f = c_1 \ln(1/\mathbf{MR}_1) + c_2 \ln(1/\mathbf{MR}_2) + c_3 \ln(1/\mathbf{MR}_3) + \cdots \qquad (4\text{--}36)$$

This equation defines the maximum velocity that an ideal multistage vehicle in a tandem configuration may attain. It assumes a space environment (no drag or gravity) and that the upper stage propulsion system starts at full thrust exactly when the lower stage stops (no time delays) without any of the common thrust declines at shutdown In atmospheric trajectories because the actual stage separation process has some unavoidably small delays (with very low or no thrust), the individual velocity increments in Eq. 4–35 need to be determined by integrating Eqs. 4–15 and 4–16 which are more general and consider drag and gravity losses; this is discussed in the next subsection. Other losses or trajectory perturbations can also be included as mentioned earlier in this chapter, but such an approach requires numerical solutions.

For two- or three-stage vehicles the overall vehicle mass ratio (initial mass at take-off to final mass of last stage) can reach values of over 100. Figure 4–2 may be separated into regions applicable to each stage, such as single-stage vehicles (with $1/MR \leq 95$) and tandem or multistage vehicles (with $1/MR$ up to and beyond 180). Equation 4–36 does not apply to the parallel, partial or piggy back staging identified in Fig. 4–15. For such stages where more than one propulsion system is operating at the same time and producing thrust in the same direction, the overall specific impulse, overall propellant mass flow and overall thrust and mass flow are given by Eqs. 2–23 to 2–25 or 11–1 to 11–3.

The first sketch in Fig. 4–15 depicts a common configuration where the stages are stacked vertically on top of each other, as in the Minuteman long-range missile or the Russian Zenit (or Zenith) launch vehicle. Partial staging was used on the early versions of the U.S. Atlas vehicle; it allowed all engines to be started together, thus avoiding an altitude start for the sustainer engine, which was unproven in those early days; the two Atlas booster engines arranged in a doughnut-shaped assembly were dropped off in flight. The third sketch in Fig. 4–15 has two or more separate booster "strap-on" stages attached to the bottom stage of a vertical configuration (they can be either solid or liquid propellants) and this allows for increases in vehicle capability. The piggyback configuration concept on the right was used on the Space Shuttle—its two large solid rocket motor boosters are shown in Fig. 1–14.

Stage Separation

It takes a finite time for the thrust termination of a lower stage propulsion system to go to essentially zero (typically 1 to 3 sec for large thruster chambers and for small thrusters as brief as 1 msec). In some multistage flight vehicles (with stage separation devices), there can be further delays (about 4 to 10 sec) to achieve a respectable separation distance between stages before firing the upper stage propulsion system can be initiated. This is needed in order to prevent any blow-back or damage from hot flames onto the upper stage. Also, upper stage engine start-ups are not instantaneous but require one or more seconds in larger rocket propulsion systems. During these several-second cumulative delays, the Earth's gravity pull acts to diminish the vehicle's upward velocity, causing a reduction of the flight velocity by perhaps 20 to 500 ft/sec (7 to 160 m/sec). A scheme called *hot staging* has been introduced to diminish this velocity loss and shorten staging time intervals—the upper stage propulsion system is actually started at a low but increasing thrust before the lower stage propulsion system has been fully shut off or well before it reaches essentially zero thrust; special flame-resistant ducts are placed in the interstage structure to allow the hot exhaust gases from the upper stage engine to be symmetrically deflected and safely discharged prior to and immediately after the actual separation of the stages. Because this improves flight performance, hot staging schemes have been used in large multistage vehicles such as the Titan II in the United States and in certain Chinese and Russian launch vehicles.

For multistage vehicles the stage mass ratios, thrust levels, propulsion durations, and location or travel of the center of gravity of the stages are all usually optimized, often using complex trajectory computer programs. High specific impulse chemical rocket engines (e.g., those using hydrogen–oxygen propellants) are normally employed in upper stages of space launch vehicles because here any small increase in specific impulse can be usually more effective than in the lower stages.

Example 4–3. A two-stage exploration vehicle is launched from a high-orbit satellite into a gravity-free vacuum trajectory. The following notation is used and explained in the accompanying diagram as well as in Fig. 4–1:

m_0 = initial mass of vehicle (or stage) at launch

m_p = useful propellant mass of stage

m_i = initial mass of stage(s)

m_f = final mass after rocket operation; it includes the empty propulsion system with its residual propellant, vehicle structures plus propulsion system with control, guidance and payload masses

m_{pl} = payload mass such as scientific instruments; it can include guidance, control and communications equipment, antennas, scientific instruments, military equipment, research apparatus, power supply, solar panels, sensors, etc.

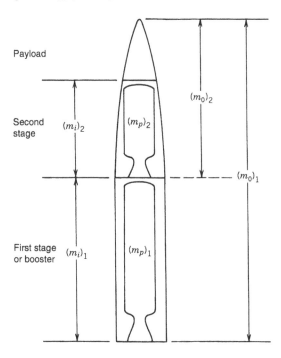

Subscripts 1 and 2 refer to the first and second stages. The following data are given:

Flight velocity increment in gravity-free vacuum	4700 m/sec
Specific impulse (each stage)	310 sec
Initial takeoff launch vehicle mass	4500 kg
Propellant mass fraction, ζ_i (each stage)	0.88

Assume that the same propellant is used for both stages and that there is no "stage separation delay."

Determine the payload for the following cases:

1. When the two propulsion system or stage masses are equal $[(m_i)_1 = (m_i)_2]$.
2. When the mass ratios of the two stages are equal $[(m_f/m_0)_1 = (m_f/m_0)_2]$.

SOLUTION. The following relationships apply to both cases. The takeoff mass or launch mass can be divided into three parts, namely, the two propulsion stages and the payload.

$$(m_0) = (m_i)_1 + (m_i)_2 + (m)_{pl} = 4,500 \text{ kg}$$

The propellants represent 88% of their propulsion system mass, so that

$$(m_p)_1 = 0.88(m_i)_1 \quad \text{and} \quad (m_p)_2 = 0.88(m_i)_2$$

The nozzle exit area ratio and the chamber pressure are also the same in both stages. Thus, the exhaust velocities are the same,

$$c_1 = c_2 = c = I_s g_0 = 310 \times 9.807 = 3040 \text{ m/sec}$$

Case 1. The stage masses and sizes are the same, or

$$(m_i)_1 = (m_i)_2 = m_i$$

Equation 4–36 can be rewritten as

$$e^{\Delta u/c} = (1/\mathbf{MR}_1)(1/\mathbf{MR}_2) = \frac{(m_0)_1}{(m_0)_1 - (m_p)_1} \frac{(m_0)_2}{(m_0)_2 - (m_p)_2}$$

$$e^{4700/3040} = 4.693 = \frac{4500}{4500 - 0.88m_i} \frac{4500 - m_i}{4500 - 1.88m_i}$$

The quadratic equation above can be solved to yield $m_i = 1925.6$ kg. Here, the payload becomes $m_{pl} = m_0 - 2m_i = 4500 - 2 \times 1925.6 = 649$ kg.

Case 2. The mass ratios for the two stages are equal, or

$$e^{\Delta u/c} = 4.693 = 1/\mathbf{MR}_2 \times 1/\mathbf{MR}_2 = \left[\frac{(m_0)_1}{(m_0)_1 - (m_p)_1}\right]^2 = \left[\frac{4500}{4500 - (m_p)_1}\right]^2$$

which yields $(m_p)_1 = 2422.8$ kg so that $(m_i)_1 = 2422.8/0.88 = 2753$ kg. Solving now for $(m_p)_2$

$$e^{\Delta u/c} = 4.693 = \left[\frac{(m_0)_2}{(m_0)_2 - (m_p)_2}\right]^2 = \left[\frac{1747}{1747 - 0.88(m_i)_2}\right]^2$$

Giving $(m_i)_2 = 1068.8$ and thus $m_{pl} = 4500 - 2763 - 1068.8 = 678.2$ kg.

The result from Case (2) is a payload higher than in Case (1). The payload is thus shown to be about 30 kg higher when the mass ratios of the stages are equal, and this conclusion applies to other stage pairs in space flight. For a single stage with the same take off mass and same performance the payload would have been 476 kg or about 70% of the two stage payload (see Problem 4–11).

If a three-stage vehicle had been used in Example 4–3 instead two, the theoretical payload increase will add only about 8 or 10%. A fourth stage gives an even smaller theoretical improvement; it would add only 3 to 5% to the payload. Hence, the potential amount of performance improvement diminishes with each added stage. Moreover, each additional stage means extra complexity in an actual vehicle (such as a reliable separation mechanism, an interstage structure, more propulsion systems, joints or couplings in connecting pipes and cables, etc.), requires additional inert mass (increasing the mass ratio **MR**), and compromises the overall reliability. Therefore, the minimum number of stages that will meet the payload and Δu requirements is usually selected.

The flight paths taken by the vehicles in the two simplified cases of Example 4–3 have to be different because their time of flight and the acceleration histories are different. One conclusion from this example that applies to all multistage rocket-propelled vehicles is that for each mission there is an optimum number of stages, an optimum distribution of the mass between the stages, and usually also an optimum flight path for each design, where key vehicle parameters such as payload, velocity increment, or range are maximized.

Launch Vehicles

The *first or lowest stage*, often called a *booster stage*, is usually the largest and requires the largest thrust and largest total impulse. For Earth surface launches, all stages presently use chemical propulsion to achieve desired thrust-to-weight ratios. Thrust magnitudes decrease with each subsequent stage, also known as the *upper* or *sustainer stages*. Thrust requirements depend on the total mass of the vehicle, which in turn depend on the mass of the payload and on the mission. Typical configurations are shown in the sketches of Fig. 4–15.

Many launch vehicles with large payloads have between one and six large *strap-on stages*, also called "zero stages or half stages." These augment the thrust of the booster stage; all units are usually started at the same time. A schematic diagram is shown as the parallel staging sketch in Fig. 4–15. Solid propellant strap-on stages are common, such as the Atlas V shown in Fig. 1–13 or the Space Shuttle shown in Fig. 1–14. These strap-on stages are usually smaller in size than their equivalent liquid propellant units (due to higher propellant density) and have less drag but may produce a very toxic exhaust. Liquid propellant strap-on stages are used in the Delta IV Heavy lift launch vehicle (see Fig. 1–12), and have been used in the first Soviet ICBM (intercontinental ballistic missile, circa 1950) and in several Soviet/Russian space launch vehicles. Most deliver higher specific impulse than their solid propellant counterparts and enhance vehicle performance but require propellant filling at the launch site.

There is some variety in existing launch vehicles. The smaller ones are for low payloads and low orbits; the larger ones usually have more stages, are heavier, and have larger payloads or higher mission velocities. Vehicle cost increases with the number of stages and with initial vehicle launch mass. Once a particular launch vehicle has proven to be reliable, it is often modified and uprated to allow for improvements in its capability or mission flexibility. Each stage of a space launch vehicle can have several rocket engines, each for specific missions or maneuvers. The Space Shuttle system, shown in Fig. 1–14, had 67 different rocket propulsion systems. In most cases each rocket engine was used for a single maneuver, but in some cases the same engine could be used for more than one specific purpose; the small reaction control thrusters in the Shuttle, for example, served to give attitude control (pitch, yaw, and roll) during orbit insertion and reentry, for counteracting internal shifting of masses (astronaut movement, extendible arm), small trajectory corrections, minor flight path adjustments, docking, and for the precise pointing of scientific instruments.

The *spacecraft* is that portion of a launch vehicle that carries the payload. It is the only part of the vehicle that goes into orbit or into deep space and/or returns to Earth. Final major space maneuvers, such as orbit injection or planetary landing, often require substantial velocity increments; the propulsion system, which provides the force for such maneuvers, may be integrated with the spacecraft or may be part of a discardable stage, just below the spacecraft. Several of the maneuvers described in Section 4.5 may often be accomplished by propulsion systems located in two different stages of a multistage vehicle. The selection of the most desirable propulsion systems, together with the decision of which of the several propulsion systems will perform specific maneuvers, will depend on optimizing performance, cost, reliability, schedule, and mission flexibility as further described in Chapter 19.

When a space vehicle is launched from the Earth's surface into orbit, it flies through three distinct trajectory phases: (1) Most are usually launched vertically and then undergo a turning maneuver while under rocket power to point the flight velocity vector into the desired direction; (2) the vehicle then follows a free-flight (unpowered) ballistic trajectory (usually elliptical), up to its apex; finally, (3) satellites would need an extra push from a rocket propulsion system to add enough total impulse or energy to accelerate to orbital velocity. This last maneuver is also known as *orbit insertion* or sometimes as a *kick maneuver*. During the initial powered flight, the trajectory angle and the thrust cutoff velocity of the last stage are adjusted by the guidance system to a velocity vector in space that will allow the vehicle to reach the apogee of its elliptic path exactly at the desired orbit altitude. As shown in Fig. 4–10, a *multistage ballistic missile* follows the same two ascent flight phases mentioned above, but it then continues its elliptical ballistic trajectory going down to its target.

Historically, launch vehicles have been successfully modified, enlarged, and improved in performance. Newer versions retain most of the old, proven, reliable components, materials, and subsystems. This reduces development efforts and costs. Upgrading a vehicle allows for an increase in mission energy (i.e., more ambitious missions) or payload or both. Typically, this is done by one or more of following types of improvement: increasing the mass of propellant without an undue increase in tank or case mass; uprating the thrust and strengthening the engine; increasing the specific impulse; or adding successively more or bigger strap-on boosters. Upgrading also usually requires a strengthening of the structure to accept higher loads.

Figure 4–16 shows effects of orbit inclination and altitude on payload capability of the Pegasus (a relatively small, airplane-launched space launch vehicle). Inclination is the angle between the equatorial plane of the Earth and the trajectory; an equatorial orbit has zero inclination and a polar orbit has 90° inclination. Since the Earth's rotation gives the vehicle an initial velocity, launching from the equator in an eastward direction will give the highest payload. For the same orbit altitudes other trajectory inclinations have lower payloads. For the same inclinations, payload decreases with orbit altitude since more energy has to be expended to overcome

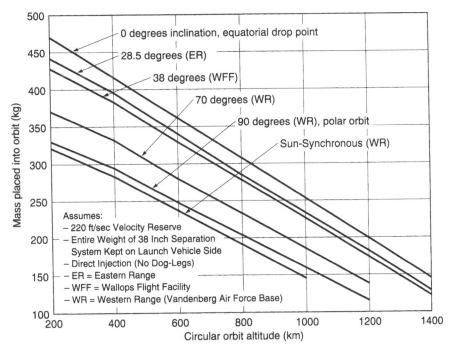

FIGURE 4–16. Changes of payload with circular orbit altitude and orbit inclination for early version of the airplane-launched Pegasus launch vehicle. This is a relatively simple three-stage launch vehicle of 50 in. diameter driven by a solid propellant rocket motor in each stage. Courtesy Orbital ATK

gravitational attraction. The figure shows that a practical payload becomes too small for orbits higher than about 1200 km. To lift heavier payloads and to go to higher orbits requires a larger launch vehicle than Pegasus vehicle. Figure 4–16 is based on the assumption of a particular payload separation length (38 in.) and a specific Δu vehicle velocity reserve (220 ft./sec) for variables such as the normal changes in atmospheric density (which can double the drag) or the mass tolerances of the propulsion systems. Similar curves are produced by the makers of other launch vehicles.

The Space Shuttle achieved its maximum payload when launched due east into an orbit with 28.5° inclination from Kennedy Space Flight Center in Florida, namely about 56,000 lbm (or 25,402 kg) at a 100-nautical-mile (185-km) orbit altitude. Such payload decreased by about 100 lbm (45.4 kg) for every nautical mile increase in altitude. When the inclination is 57°, the payload diminishes to about 42,000 lbm (or 19,051 kg). If launched in a southerly direction from Vandenberg Air Force Base on the U.S. West Coast in a 98° inclination into a circular, nearly polar orbit, the payload will be only about 30,600 lbm or 13,880 kg.

4.8. MILITARY MISSILES

A majority of all rocket propulsion systems built today is used for military purposes. There is a large variety of missiles, projectiles and military missions and therefore many different propulsion systems, all using chemical propulsion systems. They range from simple, small, unguided, fin-stabilized, single-stage rocket projectiles (used in air-to-surface missions and surface-to-surface bombardment) up to complex, sophisticated, expensive, long-range, multistage ballistic missiles (intended for faraway military or strategic targets). The term *surface* means not only land surface (ground launch or ground target) but also ocean surface (ship launched) and below ocean surface (submarine launched). A *tactical missile* can be used for attacking or defending ground troops, nearby military or strategic installations, military aircraft, short-range missiles, and/or antitank missiles. Armed forces also use *military satellites* for missions such as reconnaissance, early warning of impending attack, secure communication (including command and control) and to accurately locate particular items on the Earth's surface (latitude and longitude).

Strategic missiles with a range of 3000 km or more have traditionally been two- or three-stage surface-to-surface rocket-propelled missiles. Early designs used liquid propellant rocket engines and some are still in service in certain countries. Beginning about 50 years ago, newer strategic missiles have used solid propellant rocket motors by the United States and France. Both types usually also have a liquid propellant reaction-control-system (RCS) for accurately adjusting the final payload flight velocity (in magnitude, direction, and position in space) at the cutoff of the propulsion system of the last stage. Solid propellant RCS versions also exist (see Figs. 12–27 and 12–28). Flight analyses and ballistic trajectories of long-range missiles are similar in many ways to those described for space launch vehicles in this chapter.

Solid propellant rocket motors are preferred for most tactical missile missions because they allow relatively simple logistics and can be launched quickly. Furthermore, solid propellants don't spill and have long storage times (see Tables 19–1 and 19–3). Cryogenic propellants are not suitable for military missiles. If altitudes are low and flight durations are long, such as with a cruise missile, an air-breathing jet engine and a vehicle that provides lift will usually be more effective than a long-duration rocket. However, a large solid propellant rocket motor may still be used as a booster to launch the cruise missile and bring it up to speed.

Liquid propellant rocket engines have recently been used for upper stages in two-stage anti-aircraft missiles and ballistic defense missiles because they can be pulsed for different durations and randomly throttled. For each application, optima can be found for total impulse, thrust and thrust-time profile, nozzle configuration (single or multiple nozzles, with or without thrust vector control, with optimal area ratios), chamber pressure, and some favored liquid or solid propellant grain configuration. Low-exhaust plume gas radiation emissions in the visible, infrared, and/or ultraviolet spectrum and certain safety features (making the system insensitive to energy stimuli) become very important in some of the tactical missile applications; these are discussed in Chapters 13 and 20.

Short-range, uncontrolled, unguided, single-stage rocket vehicles, such as military rocket projectiles (ground and air launched) and rescue rockets, can be quite simple in design. The applicable equations of motion are derived in Section 4.3, and a detailed analysis is given in Ref. 4–1.

Unguided military rocket-propelled missiles are currently produced in larger numbers than any other category of rocket-propelled vehicles. In the past, 2.75-in. diameter, folding fin unguided solid propellant rocket missiles were produced in the U.S. in quantities of about 250,000 per year. Guided missiles for anti-aircraft, antitank, or infantry support have been produced in annual quantities of over a thousand.

Because these rocket projectiles are essentially unguided missiles, the accuracy of hitting a target depends on the initial aiming and the dispersion induced by uneven drag, wind forces, oscillations, and misalignment of nozzles, body, or fins. Deviations from the intended trajectory are amplified if the projectile is moving at low initial velocities, because the aerodynamic stability of a projectile with fins decreases at low flight speeds. When projectiles are launched from an aircraft at a relatively high initial velocity, or when projectiles are given stability by spinning them on their axis, their accuracy of reaching a target is increased 2- to 10-fold, compared to simple fin-stabilized rockets launched from rest.

In guided *air-to-air* and *surface-to-air rocket-propelled missiles* the time of flight to a given target, usually called the *time to target* t_t, is an important flight performance parameter. With the aid of Fig. 4–17 it can be derived in a simplified form by considering the distance traversed by the rocket (called the range) to be the integrated area underneath the velocity–time curve. Simplifications here include the assumptions of no drag, no gravity effect, horizontal flight, relatively small distances traversed during powered flight compared to total range, and linear increases in velocity during powered flight:

$$t_t = \frac{S + \frac{1}{2}u_p t_p}{u_0 + u_p} \tag{4–37}$$

Here, S is the flight vehicle's range to target corresponding to the integrated area under the velocity–time curve, and u_p is the velocity increase of the rocket during

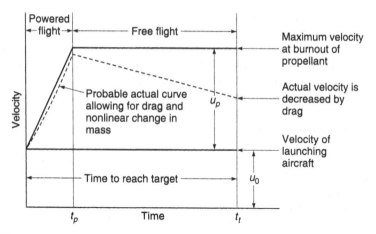

FIGURE 4–17. Simplified trajectory for an unguided, nonmaneuvering, air-launched rocket projectile. Solid line shows ideal flight velocity without drag or gravity and the dashed curve shows a likely actual flight. Powered flight path is assumed to be a straight line, which introduces only a small error.

powered flight (up to the time of burnout or propellant termination). The time of rocket operation is t_p and u_0 is the initial velocity of the launching aircraft. For the same flight time, the range of the actual vehicle velocity (dashed line) is less than for the dragless vehicle. For more accuracy, the velocity increase u_p as given by Eq. 4–19 may be used. More accurate values are also calculated through step-by-step trajectory analyses including the effects of drag and gravity from Eq. 4–17.

In unguided air-to-air or air-to-surface rocket-powered projectiles, target aiming is principally done by orienting and flying the launching aircraft into the direction of the target. A relatively simple solid propellant rocket motor is the most common propulsion choice. In guided missiles, such as air-to-air, air-to-ground, or ground-to-air, the flight path to target is controlled and can be achieved by moving aerodynamic control surfaces and/or propulsion systems, which may be pulsed and/or throttled to a lower thrust. As the guidance system and the target seeker system of a guided missile senses and tracks the flight path of a flying target, a computer calculates a predicted impact point, and the missile's flight control changes the flight path of the guided missile to achieve impact with the intended target. The control system may command the propulsion system to operate or fire selected liquid propellant thrusters from an engine with multiple thrusters (or to selectively provide thrust through multiple nozzles with hot-gas shutoff valves in solid motors). A similar set of events can occur in a defensive ground-to-incoming-ballistic-missile scenario. This requires propulsion systems capable of pulsing or repeated starts, possibly with some throttling and side forces. Rocket engines with such capabilities can be seen in Figs. 6–14, 12–27, and 12–28.

In both unguided projectiles and guided missiles, the hit probability increases as the *time to target* t_t is reduced. In any particular air-to-air combat situation, the effectiveness of the rocket projectile varies approximately inversely as the cube of the time to target. Best results (e.g., best hit probability) are usually achieved when the time to target is as small as practically possible.

Any analysis of missile and propulsion configuration that gives the minimum time to target over all likely flight scenarios can be complicated. The following rocket propulsion features and parameters will help to reduce the time to target but their effectiveness will depend on the specific mission, range, guidance and control system, thrust profile, and the particular flight conditions.

1. *High initial thrust* or *high initial acceleration* for the missile to quickly reach a high-initial-powered flight velocity. See Fig. 12–19.

2. Application of a subsequent *lower thrust* to counteract drag and gravity losses and thus maintain the high flight velocity. This can be done with a single rocket propulsion system that gives a short high initial thrust followed by a smaller (10 to 25%) sustaining thrust of longer duration.

3. For the higher supersonic flight speeds, a *two-stage missile* can be more effective. Here, the first stage is dropped off after its propellant has been consumed, thus reducing the inert mass of the next stage and improving its mass ratio and thus its flight velocity increase.

4. If the target is highly maneuverable and if the closing velocity between missile and target is large, it may be necessary not only to provide an axial thrust but also to apply large *side forces* or side accelerations to a defensive missile. This can be accomplished either by aerodynamic forces (lifting surfaces or flying at an angle of attack) or by multiple-nozzle propulsion systems with variable or pulsing thrusts; the rocket engine then would have an axial thruster and one or more side thrusters. The side thrusters have to be so located that all the thrust forces are essentially directed through the center of gravity of the vehicle in order to minimize turning moments. Thrusters that provide the side accelerations have also been called *divert* thrusters, since they divert the vehicle in a direction normal to the axis of the vehicle.

5. *Drag losses* can be reduced when the missile has a large L/D ratio (or a small cross-sectional area) and when the propellant density is high, allowing a smaller missile volume. Drag forces are highest when missiles travel at low altitudes and high speeds. Long and thin propulsion system configurations and high-density propellants help to reduce drag.

One unique military application is the *rocket-assisted gun-launched projectile* for attaining longer artillery ranges. Their small rocket motors located at the bottom of gun projectiles must withstand the very high accelerations in the gun barrel (5000 to 10,000 g_0's are typical). These have been in production.

4.9. FLIGHT STABILITY

Stability of a vehicle is achieved when it does not randomly rotate or oscillate during flight. Unstable flights are undesirable because pitch or yaw oscillations increase drag (flying at an angle of attack most of the time) and cause problems with instruments and sensors (target seekers, horizon scanners, sun sensors, or radar). Instability often leads to tumbling (uncontrolled turning) of vehicles, which often results in missing orbit insertion, missing targets, and/or the sloshing liquid propellant in tanks.

Stability may be built in by proper design so that the flying vehicle will be inherently stable, or stability may be obtained by appropriate controls, such as using the aerodynamic control surfaces on airplanes, reaction control systems, or hinged multiple rocket nozzles.

Flight stability exists when the overturning moments (e.g., those due to a wind gust, thrust misalignment, or wing misalignment) are smaller than the stabilizing moments induced by thrust vector controls or by aerodynamic control surfaces. In unguided vehicles, such as low-altitude rocket projectiles, stability of flight in a rectilinear motion is achieved by giving large stability margins to the vehicle using tail fins and by locating the center of gravity ahead of the center of aerodynamic pressure. In a vehicle with an active stability control system, a nearly neutral inherent stability is desired, so that the applied control forces are small, thus requiring small control devices, small RCS thrusters, small actuating mechanisms, and structural mass. Neutral stability is achieved by locating aerodynamic surfaces and the mass distribution of the components within the vehicle in such a manner that the center of gravity is only slightly above the center of aerodynamic pressure. Because aerodynamic moments change with Mach number, the center of pressure does not necessarily stay fixed during accelerating flight but shifts usually along the vehicle axis. The center of gravity also changes its position as propellant is consumed because the vehicle mass decreases. Thus, it is usually difficult to achieve neutral missile stability at all altitudes, speeds, and flight conditions.

Stability considerations affect rocket propulsion system design in several ways. By careful nozzle design and careful installation it is possible to minimize thrust misalignments and thus to minimize undesirable torques on the vehicle and to reduce the reaction control propellant consumption. It is also possible to exercise considerable control over the travel of the center of gravity by judicious design. In liquid propellant rockets, special design provisions, special tank shapes, and a careful selection of tank location in the vehicle afford this possibility. By using nozzles at the end of a blast tube, as shown in Fig. 15–6, it is possible to place the mass of solid propellants close to the vehicle's center of gravity. Attitude control liquid propellant engines with multiple thrusters have been used satisfactorily to obtain control moments for turning vehicles in desired ways, as described in Section 4.5 and in Chapter 6.

Unguided rocket projectiles and missiles are often rotated or "rolled" with inclined aerodynamic fins or inclined multiple rocket exhaust gas nozzles to improve their flight stability and accuracy. This is similar to the rotation given to bullets by spiral-grooved gun barrels. Such *spin stability* is achieved in part by gyroscopic effects, where an inclination of the spin axis is resisted by torques. Centrifugal effects, however, cause problems in emptying liquid propellant tanks and produce extra stresses on solid propellant grains. In some applications a low-speed roll is applied not for spin stability but to assure that any effects of thrust vector deviations or aerodynamic vehicle shape misalignments are minimized and canceled out.

SYMBOLS

a	major axis of ellipse, m, or acceleration, m/sec^2 (ft/sec^2)
A	area, m^2 (ft^2)
b	minor axis of ellipse, m
B	numerical value of drag integral
c	effective exhaust velocity, m/sec (ft/sec)
$\bar{c}$	average effective exhaust velocity, m/sec
C_D	drag coefficient
C_L	lift coefficient
D	drag force, N (lbf)
e	eccentricity of ellipse, $e = \sqrt{1 - b^2/a^2}$
E	energy, J
F	thrust force, N (lbf)
F_f	final thrust, N
F_g	gravitational attraction force, N
F_0	initial thrust force, N
g	local gravitational acceleration, m/sec^2
g_0	gravitational acceleration at sea level, 9.8066 m/sec^2
$\bar{g}$	average gravitational attraction, m/sec^2
G	universal or Newton's gravity constant, 6.674×10^{-11} m^3/kg $-$ sec^2
h	altitude, m (ft)
h_p	altitude of rocket at power cutoff, m
I_s	specific impulse, sec
k_d	diffuse coefficient of reflectivity
k_s	specular coefficient of reflectivity
l	distance of moment arm, m
L	lift force, N (lbf)
m	instantaneous vehicle mass, kg (lbm)
m_f	final vehicle/stage mass after rocket operation, kg
m_p	useful propellant mass, kg
m_0	initial vehicle launching mass, prior to rocket operation, kg
$\dot{m}$	mass flow rate of propellant, kg/sec
M_a	angular moment of inertia, kg-m^2
MR	mass ratio of vehicle $= m_f/m_0$
n	number of stages
p	pressure, N/m^2 or Pa (psi)
r	radius, m, or distance between the centers of two attracting masses, m
R	instantaneous radius from vehicle to center of Earth, m
R_0	effective mean Earth radius, 6.3742×10^6 m
S	range, m
t	time, sec
t_p	time from launching to power cutoff or time from propulsion start to thrust termination, sec

t_t	time to target, sec
T	torque, N-m (ft-lbf)
u	vehicle flight velocity, m/sec (ft/sec)
u_a	orbital velocity at apogee, m/sec
u_p	velocity at power cutoff, m/sec, or orbital velocity at perigee, m/sec
u_0	initial or launching velocity, m/sec
v_e	escape velocity, m/sec
w	weight, N (lbf)
x, y	arbitrary points on an elliptical orbit

Greek Letters

α	angle of attack, deg or rad, or angular acceleration, angle/sec^2
ζ, ζ_i	propellant mass fraction ($\zeta = m_p/m_0$)
θ	angle between flight direction and horizontal, or angle of incident radiation, deg or rad
μ	gravity constant for Earth, 3.986×10^{14} m^3/sec^2
ρ	mass density, kg/m^3
τ	period of revolution of satellite, sec
ψ	angle of thrust direction with horizontal
ω	angular speed, deg/sec (rad/sec)

Subscripts

e	escape condition
f	final condition at rocket thrust termination
i	initial condition
max	maximum
p	condition at power cutoff or propulsion termination or propellant related
pl	payload
s	satellite
z	zenith
0	initial condition or takeoff condition

PROBLEMS

1. For a vehicle in gravitationless space, determine the mass ratio necessary to boost the vehicle velocity by **(a)** 1600 m/sec and **(b)** 3400 m/sec; the effective exhaust velocity is 2000 m/sec. If the initial total vehicle mass is 4000 kg, what are the corresponding propellant masses?

   Answer: **(a)** 2204 kg.

2. Determine the burnout velocity and burnout altitude for a dragless projectile with the following parameters for a simplified vertical trajectory: $\bar{c} = 2209$ m/sec; $m_p/m_0 = 0.57$; $t_p = 5.0$ sec; $m_p/m_0 = 0.57$; $t_p = 5.0$ sec; and $u_0 = 0$; $h_0 = 0$. Select a relatively small diameter missile with L/D of 10 and an average vehicle density of 1200 kg/m^3.

3. Assuming that the projectile of Problem 2 has a drag coefficient essentially similar to the 0° curve in Fig. 4–4, redetermine the answers above and the approximate percentage errors in u_p and h_p. Use a numerical method.

4. A research space vehicle in gravity-free and drag-free space launches a smaller space-craft into a meteor shower region. The 2-kg sensitive instrument package of this space-craft (25 kg total mass) limits the maximum acceleration to no more than 50 m/sec². It is launched by a solid propellant rocket motor ($I_s = 260$ sec and $\zeta = 0.88$). Assume instantaneous start and stop of the rocket motor.

   (a) Determine the minimum allowable burn time, assuming steady constant propellant mass flow;

   (b) Determine the maximum velocity relative to the launch vehicle.

   (c) Solve for (a) and (b) when half of the total impulse is delivered at the previous propellant mass flow rate, with the other half at 20% of this mass flow rate.

5. For a satellite cruising in a circular orbit at an altitude of 500 km, determine the period of revolution, the flight speed, and the energy expended to bring a unit mass into this orbit. *Answer:* 1.58 hr, 7613 m/sec, 33.5 MJ/kg.

6. A large ballistic rocket vehicle has the following characteristics: propellant mass flow rate: 12 slugs/sec (1 slug = 32.2 lbm = 14.6 kg); nozzle exit velocity: 7100 ft/sec; nozzle exit pressure: 5 psia (assume no separation); atmospheric pressure: 14.7 psia (sea level); takeoff weight: 12.0 tons (1 ton = 2000 lbf); burning time: 50 sec; nozzle exit area: 400 in.² Determine (a) the sea-level thrust; (b) the sea-level effective exhaust velocity; (c) the initial thrust-to-weight ratio; (d) the initial acceleration; (e) the mass inverse ratio m_0/m_f. *Answers* : 81,320 lbf; 6775 ft./sec; 3.38; 2.38 g_0.

7. In Problem 6 compute the altitude and missile velocity at the time of power plant cutoff, neglecting atmospheric drag and assuming a simple vertical trajectory.

8. A spherical satellite has 12 identical monopropellant thrust chambers for attitude control with the following performance characteristics: thrust (each unit): 5 lbf; I_s (steady state or more than 2 sec): 240 sec; I_s (pulsing duration 20 msec): 150 sec; I_s (pulsing duration 100 msec): 200 sec; satellite weight: 3500 lbf; satellite diameter: 8 ft; satellite internal density distribution is essentially uniform; disturbing torques, Y and Z axes: 0.00005 ft-lbf average; disturbing torque, for X axis: 0.001 ft-lbf average; distance between thrust chamber axes: 8 ft; maximum allowable satellite pointing position error: ±1°. Time interval between pulses is 0.030 sec.

   (a) What would be the maximum and minimum vehicle angular drift per hour if no correction torque were applied? *Answers*: 4.65 and 18.65 rad/hr.

   (b) What is the frequency of pulsing action (i.e., how often does an engine pair operate?) at 20-msec, 100-msec, and 2-sec pulses in order to correct for angular drift? Discuss which pulsing mode is best and which is impractical.

9. For an ideal multistage launch vehicle with several stages in series, discuss the following: (a) the effect on the ideal mission velocity if the second and third stages are not started immediately but are each allowed to coast for a short period after shutoff and separation of the expended stage before rocket engine start of the next stage; (b) the effect on the mission

velocity if an engine malfunctions and delivers a few percent less than the intended thrust but for a longer duration and essentially the full total impulse of that stage.

10. Given a cylindrically shaped space vehicle ($D = 1.0$ m, height is 1.0 m, average density is 1.1 g/cm^3) with a flat solar cell panel on an arm (mass of 32 kg, effective moment arm is 1.5 m, effective average area facing normally toward sun is 0.6 m^2) in a set of essentially frictionless bearings and in a low circular orbit at 160 km altitude with sunlight being received, on the average, about 50% of the time:

   (a) Compute the maximum solar pressure-caused torque and the angular displacement this would cause during 1 day if not corrected.

   (b) Using data from the atmospheric table in Appendix 2 and an arbitrary average drag coefficient of 1.0 for both the body and the flat plate, compute the drag force.

   (c) Using stored, high-pressure air at 5000 psia initial pressure as the propellant for attitude control, design an attitude control system to periodically correct for these disturbances (F, I_s, t, etc.). State all assumptions.

11. Determine the payload for a single-stage vehicle in Example 4–3. Using results from this example compare your answer with a two-and three-stage vehicle.

12. An Earth satellite is in an elliptical orbit with the perigee at 600 km altitude and an eccentricity of $e = 0.866$. Determine the parameters of the new satellite trajectory, if a rocket propulsion system is fired in the direction of flight giving an incremental velocity of 200 m/sec when (a) fired at apogee, (b) fired at perigee, and (c) fired at perigee, but in the opposite direction, reducing the velocity.

13. A sounding rocket (75 kg mass, 0.25 m diameter) is speeding vertically upward at an altitude of 5000 m and a velocity of 700 m/sec. What is the deceleration in multiples of g_0 due to gravity and drag? (Use C_D from Fig. 4–4 and Appendix 2.)

14. Derive Eq. 4–37; state all your assumptions.

15. Calculate the change of payload relative to the overall mass (i.e., the payload fraction) for a single stage rocket propelled vehicle when replacing $I_s = 292$ sec and propellant mass fraction (for the propulsive stage) 0.95 with a new engine of 455 sec and 0.90. Assume that the total mass, the structure and any other fixed masses remain unchanged. Take $\Delta u = 5000$ m/sec.

REFERENCES

4–1. J. B. Rosser, R. R. Newton, and G. L. Gross, *Mathematical Theory of Rocket Flight*, McGraw-Hill Book Company, New York, 1947; or F. R. Gantmakher and L. M. Levin, *The Flight of Uncontrolled Rockets*, Macmillan, New York, 1964.

4–2. R. S. Wolf, "Development of a Handbook for Astrobee F (Sounding Rocket) Flight Performance Predictions," *Journal of Spacecraft and Rockets*, Vol. 24, No. 1, January–February 1987, pp. 5–6.

4–3. *Orbital Flight Handbook*, NASA SP33, 1963, Part 1: Basic Techniques and Data. Part 2: Mission Sequencing Problems. Part 3: Requirements.

4–4. V. A. Chobotov (Ed) *Orbital Mechanics*, 3rd ed., Educational Series, AIAA, Reston, VA., 2002; and T. Logsdon, *Orbital Mechanics and Application*, John Wiley & Sons, New York, October 1997.

4–5. J. W. Cornelisse, H. F. R. Schöyer, and K. F. Wakker, *Rocket Propulsion and Space Flight Dynamics*, Pitman Publishing, London, 1979.

4–6. J. P. Vinti, G. J. Der, and L. Bonavito, *Orbits and Celestial Mechanics*, Vol. 177 of Progress in Aeronautics and Astronautics Series, AIAA, Reston, VA, 1998, 409 pages.

4–7. W. Hohmann, *Die Erreichbarkeit der Himmelskörper* (Accessibility of Celestial Bodies), Oldenburg, Munich, 1925.

4–8. W. J. Larson and J. R. Wertz, *Space Mission Analysis and Design*, 3rd ed., published jointly by Microcosm, Inc. and Kluwer Academic Press, 1999.

4–9. J. J. Pocha, *An Introduction to Mission Design for Geostationary Satellites*, Kluwer Academic Publishers, Hingham, MA, 1987, 222 pages.

4–10. M. H. Kaplan, *Orbital Spacecraft Dynamics and Control*, John Wiley & Sons, New York, 1976.

4–11. R. W. Humble, G. N. Henry, and W. J. Larson, *Space Propulsion Analysis and Design*, McGraw-Hill, New York, 1995, 748 pages.

4–12. J. P. Vinti, G. J. Der, and L. Bonavito, *Orbit and Celestial Mechanics*, Vol. 177 of Progress in Astronautics and Aeronautics Series, AIAA, Reston, VA, 1998, 409 pages.

4–13. C-C. G. Chao, *Applied Orbit Perturbations and Maintenance*, Aerospace Press, 2005, 264 pages.

4–14. "Spacecraft Aerodynamic Torques," *NASA SP 8058*, January 1971 (N 71-25935).

4–15. "Spacecraft Radiation Torques," *NASA SP 8027*, October 1969 (N 71-24312).

4–16. "Spacecraft Gravitational Torques," *NASA SP 8024*, May 1964 (N 70-23418).

4–17. "Spacecraft Magnetic Torques," *NASA SP 8018*, March 1969 (N 69-30339).

CHAPTER 5

CHEMICAL ROCKET PROPELLANT PERFORMANCE ANALYSIS

This chapter is an introduction to the theory, performance, and description of the most useful gas parameters in chemical rocket propulsion systems. It identifies relevant chemical fundamentals, basic analytical approaches, and key equations. While not presenting complete and complex analytical results to actual propulsion systems, it does list references for the required detailed analyses. Also included are tables of calculated results for several common liquid propellant combinations and select solid propellants, including exhaust gas composition for some of these. Furthermore, it describes how various key parameters influence both performance and exhaust gas composition.

In Chapter 3, simplified one-dimensional performance relations were developed that require knowledge of the composition of rocket propulsion gases and some properties of propellant reaction products, such as their chamber temperature T_1, average molecular mass $\mathfrak{M}$, and specific heat ratios or enthalpy change across the nozzle $(h_1 - h_2)$. This chapter presents several theoretical approaches to determine these relevant thermochemical properties for any given composition or propellant mixture, chamber pressure, and nozzle area ratio or exit pressure. Such information then allows the determination of performance parameters, including theoretical exhaust velocities and specific impulses in chemical rockets.

By knowing temperature, pressure, and composition, it is possible to calculate other combustion gas properties. This information also permits the analysis and selection of materials for chamber and nozzle structures. Heat transfer analyses require knowledge of specific heats, thermal conductivities, and viscosities for the gaseous mixture. The calculated exhaust gas composition also forms the basis for estimating environmental effects, such as the potential spreading of toxic clouds near a launch site, as discussed in Chapter 21. Exhaust gas parameters are needed for the analysis of exhaust plumes (Chapter 20), which include plume profiles and mixing effects external to the nozzle.

The advent of digital computers has made possible solving large sets of equations involving mass balances and energy balances together with the thermodynamics and chemical equilibria of complex systems for a variety of propellant ingredients. In this chapter, we discuss sufficient theoretical analysis background material so the reader can understand the thermodynamic and chemical basis for computer programs in use today. We do not introduce any specific computer programs but do identify relevant phenomena and chemical reactions.

The reader is referred to Refs. 5–1 through 5–5 (and 5–16) for much of the general chemical and thermodynamic background and principles. For detailed descriptions of the properties of each possible reactant and reaction products, consult Refs. 5–6 through 5–15.

All analytical work requires some simplifying assumptions and as more phenomena are understood and mathematically simulated analytical approaches and resulting computer codes become more realistic (but also more complex). The assumptions made in Section 3.1 for an ideal rocket are valid only for quasi-one-dimensional flows. However, more sophisticated approaches make several of these assumptions unnecessary. Analytical descriptions together with their assumptions are commonly divided into two separate parts:

1. The *combustion process* is the first part. It normally occurs at essentially constant pressure (isobaric) in the combustion chamber and the resulting gases follow Dalton's law, which is discussed in this chapter. Chemical reactions do occur very rapidly during propellant combustion. The chamber volume is assumed to be large enough and the residence time long enough for attaining chemical equilibrium within the chamber.

2. The *nozzle gas expansion process* constitutes the second set of calculations. The equilibrated gas combustion products from the chamber then enter a supersonic nozzle where they undergo an adiabatic expansion without further chemical reactions. The gas entropy is assumed constant during reversible nozzle gas expansions, although in real nozzles it increases slightly.

The principal chemical reactions commonly taken into account occur only inside the combustion chamber of a liquid propellant rocket engine or inside the grain cavity of a solid propellant rocket motor (usually within a short distance from the burning surface). Chamber combustion analyses are discussed further in Chapters 9 and 14. In reality, however, some chemical reactions also occur in the nozzle as the gases expand; with such shifting equilibrium the composition of the flowing reaction products may noticeably change inside the nozzle, as described later in this chapter. A further set of chemical reactions can occur in the exhaust plume outside the nozzle, as described in Chapter 20—many of the same basic thermochemical approaches described in this chapter may also be applied to exhaust plumes.

From Eqs. 2–6 and 3–33, we see that $c = g_0 I_s = C_F c^*$ represent a rocket's performance. As discussed in Chapters 2 and 3, the thrust coefficient depends almost entirely on the nozzle gas expansion process and the characteristic velocity depends almost exclusively on the effects of combustion. In this chapter, we will focus on the key parameters that make up c^*.

5.1. BACKGROUND AND FUNDAMENTALS

The description of chemical reactions between one or more fuels with one or more oxidizing reactants forms the basis for chemical rocket propulsion combustion analysis. The heat liberated in such reactions transforms the propellants into hot gaseous products, which are subsequently expanded in a nozzle to produce thrust.

The propellants or stored chemical reactants can initially be either liquid or solid and sometimes also gaseous (such as hydrogen heated in the engine's cooling jackets). Reaction products are predominantly gaseous, but for some propellants one or more reactant species may partly remain in the solid or liquid phase. For example, with aluminized solid propellants, the chamber reaction gases contain liquid aluminum oxide and the colder gases in the nozzle exhaust may contain solid, condensed aluminum oxide particles. For some chemical species, therefore, analysis must consider all three phases and energy changes involved in their phase transitions. When the amount of solid or liquid species in the exhaust is negligibly small and the particles themselves are small, perfect gas descriptions introduce only minor errors.

It is often necessary to accurately know the chemical composition of the propellants and their relative proportion. In liquid propellants, this means knowing the mixture ratio and all major propellant impurities; in gelled or slurried liquid propellants knowledge of suspended or dissolved solid materials is needed; and in solid propellants this means knowledge of all ingredients, their proportions, impurities, and phase (some ingredients, such as plasticizers, are stored in a liquid state).

Dalton's law may be applied to the resulting combustion gases. It states that a mixture of gases at equilibrium exerts a pressure that is the sum of all the partial pressures of its individual constituents, each acting at a common total chamber volume and temperature. The subscripts a, b, c, and so on below refer to individual gas constituents:

$$p = p_a + p_b + p_c + \cdots \tag{5-1}$$

$$T = T_a = T_b = T_c = \cdots \tag{5-2}$$

The perfect gas equation of state $pV = RT$ accurately represents high-temperature gases. For the j-th chemical species, V_j is the "specific volume" (or reaction-chamber volume per unit component mass) and R_j is the gas constant for that species, which is obtained by dividing the universal gas constant R' (8314.3 J/kg-mol-K) by the species molecular mass $\mathfrak{M}_j$ (called molecular weight in the earlier literature). Using Eq. 5–1 we may now write the total pressure in a mixture of chemical species as

$$p = R_a T/V_a + R_b T/V_b + R_c T/V_c + \cdots \equiv R'T/(\mathfrak{M}V_{\text{mix}}) \tag{5-3}$$

The volumetric proportions for each gas species in a gas mixture are determined from their *molar concentrations*, n_j, expressed as kg-mol for a particular species j per kg of mixture. If n is the total number of kg-mol of all species per kilogram of uniform gas mixture, then the *mol fraction* X_j becomes

$$X_j = \frac{n_j}{n} \quad \text{since} \quad n = \sum_{j=1}^{m} n_j \tag{5-4}$$

where n_j is the kg-mol of species j per kilogram of mixture and the index m represents the total number of different gaseous species in the equilibrium combustion gas mixture. In Eq. 5-3, the effective average molecular mass $\mathfrak{M}$ for a gas mixture becomes

$$\mathfrak{M} = \frac{\sum_{j=1}^{m} n_j \mathfrak{M}_j}{\sum_{j=1}^{m} n_j} \tag{5-5}$$

When there are ℓ possible species which chemically coexist in a mixture and of these m are gaseous, then $\ell - m$ represents the number of condensed species. The *molar specific heat* for a gas mixture at constant pressure C_p can be determined from the individual gas molar fractions n_j and their molar specific heats as shown by Eq. 5-6. The specific heat ratio k for the perfect gas mixture is shown in Eq. 5-7:

$$(C_p)_{\text{mix}} = \frac{\sum_{j=1}^{m} n_j (C_p)_j}{\sum_{j=1}^{m} n_j} \tag{5-6}$$

$$k_{\text{mix}} = \frac{(C_p)_{\text{mix}}}{(C_p)_{\text{mix}} - R'} \tag{5-7}$$

When a chemical reaction goes to completion, that is, when all of reactants are consumed and transformed into products, the reactants appear in *stoichiometric* proportions. For example, consider the gaseous reaction:

$$H_2(g) + \frac{1}{2}O_2(g) \rightarrow H_2O(g) \tag{5-8}$$

All the hydrogen and oxygen are fully consumed to form the single product—water vapor—without any reactant residue. It requires 1 mol of the H_2 and $\frac{1}{2}$ mol of the O_2 to obtain 1 mol of H_2O. On a mass basis, this *stoichiometric mixture* requires of 16.0 kg of O_2 and 2 kg of H_2, which are in the "stoichiometric mixture mass ratio" of 8:1. The release of energy per unit mass of propellant mixture and the combustion temperature are always highest at or near the stoichiometric condition.

It is usually not advantageous in rocket propulsion systems to operate with the oxidizer and fuel at their stoichiometric mixture ratio. Instead, they tend to operate fuel rich because this allows low molecular mass molecules such as hydrogen to remain unreacted; this reduces the average molecular mass of the reaction products, which in turn increases their exhaust velocity (see Eq. 3-16) provided other factors are comparable. For rockets using H_2 and O_2 propellants, the best operating mixture mass ratio for high-performance rocket engines ranges between 4.5 and 6.0 (fuel rich) because here the drop in combustion temperature (T_0) remains small even though there is unreacted $H_2(g)$ in the exhaust.

Equation 5–8 is actually a reversible chemical reaction; by adding energy to the $H_2O(g)$ the reaction can be made to go backward to recreate $H_2(g)$ and $O_2(g)$ at high temperature and the arrow in the equation would be reversed. The decomposition of solid propellants, identified with an (s), into reaction product gases involves irreversible chemical reactions, as is the burning of liquid propellants, denoted with an (l), to create gases. However, reactions among gaseous combustion product may be reversible.

Chemical equilibrium occurs in reversible chemical reactions when the rate of product formation exactly equals the reverse reaction (one forming reactants from products). Once this equilibrium is reached, no further changes in concentration take place. In Equation 5–8 all three gases would be present in relative proportions that would depend on the pressure, temperature, and equilibrium state of the mixture.

The *heat of formation* $\Delta_f H^0$ (also called *enthalpy of formation*) is the energy released (or absorbed), or the value of enthalpy change, when 1 mol of a chemical compound is formed from its constituent atoms or elements at 1 bar (100,000 Pa) and isothermally at 298.15 K or 25°C. The symbol Δ implies an energy change. The subscript f refers to "formation" and the superscript 0 means that each product or reactant substance is at its "thermodynamic standard state" and at the reference pressure and temperature. By convention, heats of formation of the elements in gaseous form (e.g., H_2, O_2, Ar, Xe, etc.) are set to zero at standard temperature and pressure. Typical values of $\Delta_f H^0$ and other properties are given in Table 5–1 for selected species. When heat is absorbed in the formation a chemical compound, the given $\Delta_f H^0$ has a positive value. Earlier published tables show values given at a temperature of 273.15 K and a slightly higher standard reference pressure of 1 atm (101, 325 Pa) than Table 5–1.

The *heat of reaction* $\Delta_r H^0$ can be negative or positive, depending on whether the reaction is *exothermic* or *endothermic*. The heat of reaction at other than standard reference conditions has to be corrected in accordance with corresponding changes in the enthalpy. Also, when a species changes from one state to another (e.g., liquid becomes gas), it may lose or gain energy. In most rocket propulsion computations, the heat of reaction is determined for a constant-pressure combustion process. In general, the heat of reaction can be determined from sums of the heats of formation of products and reactants, namely,

$$\Delta_r H^0 = \sum [n_j(\Delta_f H^0)_j]_{\text{products}} - \sum [n_j(\Delta_f H^0)_j]_{\text{reactants}} \qquad (5\text{--}9)$$

Here, n_j is the molar concentration of each particular species j. In a typical rocket propellant, there are a number of different chemical reactions going on simultaneously; Eq. 5–9 provides the heat of reaction for all of these simultaneous reactions. For data on heats of formation and heats of reaction, see Refs. 5–7 through 5–13 and 5–15.

Various thermodynamic criteria for representing necessary and sufficient conditions for stable equilibria were first advanced by J. W. Gibbs and are based on

TABLE 5–1. Chemical Thermodynamic Properties of Selected Substances at 298.15 K (25°C) and 0.1 MPa (1 bar)

Substance	Phase[a]	Molar Mass (g/mol)	$\Delta_f H^0$ (kJ/mol)	$\Delta_f G^0$ (kJ/mol)	S^0 (J/mol-K)	C_p (J/mol-K)
Al (crystal)	s	29.9815	0	0	28.275	24.204
Al_2O_3	l	101.9612	−1620.567	−1532.025	67.298	79.015
C (graphite)	s	12.011	0	0	5.740	8.517
CH_4	g	16.0476	−74.873	−50.768	186.251	35.639
CO	g	28.0106	−110.527	−137.163	197.653	29.142
CO_2	g	44.010	−393.522	−394.389	213.795	37.129
H_2	g	2.01583	0	0	130.680	28.836
HCl	g	36.4610	−92.312	−95.300	186.901	29.136
HF	g	20.0063	−272.546	−274.646	172.780	29.138
H_2O	l	18.01528	−285.830	−237.141	69.950	75.351
H_2O	g	18.01528	−241.826	−228.582	188.834	33.590
N_2H_4	l	32.0453	+50.434	149.440	121.544	98.666
N_2H_4	g	32.0453	+95.353	+159.232	238.719	50.813
NH_4ClO_4	s	117.485	−295.767	−88.607	184.180	128.072
N_2O_4	l	92.011	−19.564	+97.521	209.198	142.509
N_2O_4	g	92.011	9.079	97.787	304.376	77.256
NO_2	g	46.0055	33.095	51.258	240.034	36.974
HNO_3	g	63.0128	−134.306	−73.941	266.400	53.326
N_2	g	28.0134	0	0	191.609	29.125
O_2	g	31.9988	0	0	205.147	29.376
NH_3	g	17.0305	−45.898	−16.367	192.774	35.652

[a]s = solid, l = liquid, g = gas. When species are listed twice, as liquid and gas, their existence is due to evaporation or condensation.
The molar mass can be in g/g-mol or kg/kg-mol; C_p is given in J/g-mol-K or kJ/kg-mol-K, for J/kg-mol-K multiply tabulated values by 10^3.
Source: Refs. 5–8 and 5–9.

minimizing the system's energy. The *Gibbs free energy G* (often called the *chemical potential*) is a convenient function or "property of state" for the chemical system describing its thermodynamic potential, and it is directly related to the constituents' internal energy U, pressure p, molar specific volume V, enthalpy h, temperature T, and entropy S. For any single species j the free energy is defined below as G_j; it can be determined for specific thermodynamic conditions, for mixtures of gases as well as for individual gas species:

$$G_j = U_j + p_j V_j - T_j S_j = h_j - T_j S_j \tag{5-10}$$

For most materials used as rocket propellants, the Gibbs free energy has been determined and tabulated as a function of temperature. It can then be corrected for pressure. Typical G_j's units are J/kg-mol. For a series of different species the mixture or total free energy G is

$$G = \sum_{j=1}^{m} n_j G_j \tag{5-11}$$

For perfect gases, the *free energy* is only a function of temperature and pressure. It represents another defined property, just as the enthalpy or the density; only two such independent properties are required to characterize the state of a single-species gas. The free energy may be thought of as a tendency or driving force for a chemical substance to enter into a chemical (or physical) change. Only differences in chemical potential can be measured. When the chemical potential of the reactants is higher than that of their likely products, a chemical reaction can occur and the gas composition may change. The change in free energy ΔG for reactions at constant temperature and pressure is the value of the chemical potential of the products less that of the reactants:

$$\Delta G = \sum_{j=1}^{m} [n_j (\Delta_f G^0)_j]_{products} - \sum_{j=1}^{r} [n_j (\Delta_f G^0)_j]_{reactants} \tag{5-12}$$

Here, the index m accounts for the number of gas species in the combustion products, the summation index r accounts for the number of gas species in the reactants, and the ΔG represents the maximum energy that can be "freed to do work on an open system" (i.e., one where mass enters and leaves the system). At equilibrium the free energy is a minimum—at its minimum any small change in mixture fractions causes negligible change in ΔG, and the free energies of the products and the reactants are essentially equal; here

$$d\Delta G / dn = 0 \quad \text{at equilibrium} \tag{5-13}$$

and a curve of molar concentration n versus ΔG would display a minimum.

If the reacting propellants are liquid or solid materials, energy will be needed to change phase and/or vaporize them or to break them down into other (gaseous) species. This energy has to be subtracted from that available to heat the gases from the reference temperature to the combustion temperature. Therefore, values of ΔH^0 and ΔG^0 for liquid and solid species are considerably different from those for the same species initially in a gaseous state. The standard *free energy of formation* $\Delta_f G^0$ is the increment in free energy associated with the reaction forming a given compound or species from its elements at their reference state. Table 5–2 gives values of $\Delta_f H^0$ and $\Delta_f G^0$ and other properties for carbon monoxide as a function of temperature. Similar data for other species can be obtained from Refs. 5–7 and 5–13. The *entropy* is another thermodynamic property of matter that is relative, which means that it is determined only as a change in entropy. In the analysis of isentropic nozzle flow, for example, it is assumed that the entropy remains constant. For a perfect gas, the change of entropy is given by

$$dS = \frac{dU}{T} + \frac{p\,dV}{T} = C_p \frac{dT}{T} - R \frac{dp}{p} \tag{5-14}$$

TABLE 5–2. Variation of Thermochemical Data with Temperature for Carbon Monoxide (CO) as an Ideal Gas

Temp. (K)	C_p^0	S^0	$H^0 - H^0(T)$ (kJ/mol)	$\Delta_f H^0$ (kJ/mol)	$\Delta_f H^0$ (kJ/mol)
	(J/mol-K)				
0	0	0	−8.671	−113.805	−113.805
298.15	29.142	197.653	0	110.527	−137.163
500	29.794	212.831	5.931	−110.003	−155.414
1000	33.183	234.538	21.690	−111.983	−200.275
1500	35.217	248.426	38.850	−115.229	−243.740
2000	36.250	258.714	56.744	−118.896	−286.034
2500	36.838	266.854	74.985	−122.994	−327.356
3000	37.217	273.605	93.504	−127.457	−367.816
3500	37.493	279.364	112.185	−132.313	−407.497
4000	37.715	284.386	130.989	−137.537	−446.457

To change units to J/kg-mol multiply tabulated values by 10^3.
Source: Refs. 5–8 and 5–9.

and for constant C_p the corresponding integral becomes

$$S - S^0 = C_p \ln \frac{T}{T_0} - R \ln \frac{p}{p_0} \tag{5-15}$$

where the subscript "zero" applies to the reference state. For mixtures, the total entropy becomes

$$S = \sum_{j=1}^{m} S_j n_j \tag{5-16}$$

Here, the entropy S_j is in J/kg-mol-K. The entropy for each gaseous species is

$$S_j = (S_T^0)_j - R \ln \frac{n_j}{n} - R \ln p \tag{5-17}$$

For solid and liquid species, the last two terms are zero. Here, (S_T^0) refers to the standard state entropy at a temperature T. Typical values for entropy are listed in Tables 5–1 and 5–2.

5.2. ANALYSIS OF CHAMBER OR MOTOR CASE CONDITIONS

The objective here is to determine the theoretical combustion temperature and the theoretical composition of the reaction products, which in turn will allow the determination of the physical properties of the combustion gases (C_p, k, ρ, or other). Before we can perform such analyses, some basic information has to be known or postulated (e.g., propellants, their ingredients and proportions, desired chamber pressure, and all likely reaction products). Although combustion processes really consist of a series of different chemical reactions that occur almost simultaneously and include the break-down of chemical compounds into intermediate and subsequently into final products, here we are only concerned with initial and final conditions, before and after com-bustion. We will mention several approaches to analyzing chamber conditions; in this section we first give some definitions of key terms and introduce relevant principles.

The first principle concerns the *conservation of energy*. The heat created by the combustion is set equal to the heat necessary to adiabatically raise the resulting gases to their final combustion temperature. The heat of reaction of the combustion $\Delta_r H$ has to equal the enthalpy change ΔH of the reaction product gases. *Energy balances* may be thought of as a two-step process: the chemical reaction process occurs instan-taneously but isothermally at the reference temperature, and then the resulting energy release heats the gases from this reference temperature to the final combustion tem-perature. The heat of reaction, Equation 5–9, becomes

$$\Delta_r H = \sum_1^m n_j \int_{T_{\text{ref}}}^{T_1} C_{pj} dT = \sum_1^m n_j \Delta h_j |_{T_{\text{ref}}}^{T_1} \qquad (5\text{–}18)$$

Here, the Δh_j, the increase in enthalpy for each species, is multiplied by its molar concentration n_j and C_{pj} is the species molar specific heat at constant pressure.

The second principle is the *conservation of mass*. The mass of any atomic species present in the reactants before the chemical reaction must equal that of the same species in the products. This can be better illustrated with a more general case of the reaction shown in Equation 5–8 when the reactants are not in stoichiometric proportion.

The combustion of hydrogen with oxygen is used below as an example. It may yield six possible products: water, hydrogen, oxygen, hydroxyl, atomic oxygen, and atomic hydrogen. Here, all reactants and products are gaseous. Theoretically, there could be two additional products: ozone O_3 and hydrogen peroxide H_2O_2; however, these are unstable compounds that do not exist for long at high temperatures and can be ignored. In chemical notation the mass balance may be stated as

$$a H_2 + b O_2 \rightarrow n_{H_2O} H_2O + n_{H_2} H_2 + n_{O_2} O_2 + n_O O + n_H H + n_{OH} OH \qquad (5\text{–}19)$$

The left side shows the condition before the reaction and the right side the condition after. Since H_2 and O_2 are found on both sides, it means that not all of these species are consumed and a portion, namely, n_{H_2} and n_{O_2}, will remain unreacted. At any particular temperature and pressure, the molar concentrations on the right side will remain fixed when chemical equilibrium prevails. Here, a, b, n_{H_2O}, n_{H_2}, n_{O_2}, n_O, n_H, and n_{OH} are the respective molar concentrations of these substances before and after the reaction, these are expressed in kg-mol per kilogram of propellant reaction prod-ucts or of mixture; initial proportions of a and b are usually known. The number of kg-mol per kilogram of mixture of each element can be established from this initial mix of oxidizer and fuel ingredients. For the hydrogen–oxygen relation above, the mass balances would be

$$\text{for hydrogen}: 2a = 2n_{H_2O} + 2n_{H_2} + n_H + n_{OH}$$

$$\text{for oxygen}: 2b = n_{H_2O} + 2n_{O_2} + n_O + n_{OH} \tag{5-20}$$

The *mass balances* of Eq. 5–20 provide two more equations for this reaction (one for each atomic species) in addition to the energy balance equation. There are six unknown product percentages and an unknown combustion or equilibrium temperature. However, three equations can only solve for three unknowns, say the combustion temperature and the molar fractions of two of the species. When, for example, it is known that the initial mass mixture ratio of b/a is fuel rich, so that the combustion temperature will be relatively low, the percentage of remaining O_2 and the percentage of the dissociation products (O, H, and OH) would all be very low and may be neglected. Thus, n_O, n_H, n_{OH}, and n_{O_2} are set to be zero. The solution requires knowledge of the enthalpy change of each of the species, and that information can be obtained from existing tables, such as Table 5–2 or Refs. 5–8 and 5–9.

In more general form, the mass for any given element must be the same before and after the reaction. The number of kg-mol of a given element per kilogram of reactants and product is equal, or their difference is zero. For each atomic species, such as the H or the O in Eq. 5–20,

$$\left[\sum_{j=1}^{m} a_{ij}n_j \right]_{\text{products}} - \left[\sum_{j=1}^{r} a_{ij}n_j \right]_{\text{reactants}} = 0 \tag{5-21}$$

Here, the atomic coefficients a_{ij} are the number of kilogram atoms of element i per kg-mol of species j, and m and r are indices as defined above. The average molecular mass for the products, using Eqs. 5–5 and 5–19, becomes

$$\mathfrak{M} = \frac{2n_{H_2} + 32n_{O_2} + 18n_{H_2O} + 16n_O + n_H + 17n_{OH}}{n_{H_2} + n_{O_2} + n_{H_2O} + n_O + n_H + n_{OH}} \tag{5-22}$$

The computational approach used in Ref. 5–13 is the one commonly used today for thermochemical analyses. It relies on the minimization of the Gibbs free energy and on the mass balance and energy balance equations. As was indicated in Eqs. 5–12 and 5–13, the change in the Gibbs free energy function is zero at equilibrium; here, the chemical potential of the gaseous propellants has to equal that of the gaseous reaction products, which is Eq. 5–12:

$$\Delta G = \sum (n_j \Delta G_j)_{\text{products}} - \sum (n_j \Delta G_j)_{\text{reactants}} = 0 \tag{5-23}$$

To assist in solving this equation a "Lagrangian multiplier," a factor representing the degree of the completion of the reaction, is often used. An alternative older method in solving for gas composition, temperature, and gas properties is to use the energy balance (Eq. 5–18) together with several mass balances (Eq. 5–21) and certain *equilibrium constant* relationships (see for example Ref. 5–16).

After assuming a chamber pressure and setting up the energy balance, mass balances, and equilibrium relations, another method of solving all the equations is to estimate a combustion temperature and then solve for the various values of n_j. Then, check to see if a balance has been achieved between the heat of reaction $\Delta_r H^0$ and the heat absorbed by the gases, $H_T^0 - H_0^0$, going from the reference temperature to the combustion temperature. If they do not balance, the value of the combustion temperature is iterated until there is convergence and the energy balances.

The *energy release efficiency*, sometimes called the *combustion efficiency*, can be defined here as the ratio of the actual change in enthalpy per unit propellant mixture to the calculated change in enthalpy necessary to transform the reactants from the initial conditions to the products at the chamber temperature and pressure. The actual enthalpy change is evaluated when the initial propellant conditions and the actual compositions and the temperatures of the combustion gases are measured. Measurements of combustion temperature and gas composition are difficult to perform accurately, and combustion efficiency is therefore only experimentally evaluated in rare instances (such as in some R & D programs). Combustion efficiencies in liquid propellant rocket thrust chambers also depend on the method of injection and mixing and increases with increasing combustion temperature. In solid propellants the combustion efficiency becomes a function of grain design, propellant composition, and degree of uniform mixing among the several solid constituents. In well-designed rocket propulsion systems, actual measurements yield energy release efficiencies from 94 to 99%. These high efficiencies indicate that combustion is essentially complete, that is, that negligible amounts of unreacted propellant remain and that chemical equilibrium is indeed closely established.

The number of compounds or species in combustion exhausts can be large, up to 40 or more with solid propellants or with liquid propellants that have certain additives. The number of nearly simultaneous chemical reactions that take place may easily exceed 150. Fortunately, many of these chemical species are present only in relatively small amounts and may usually be neglected.

Example 5–1.   Hydrogen peroxide is used both as a monopropellant and as an oxidizer in bipropellant systems (see Chapter 7). It is stored in liquid form and available in various degrees of dilution with liquid water. For rocket applications, concentrations (70 to 98+%), known as high-test peroxide (HTP) are used. For a monopropellant application, calculate the *adiabatic flame or dissociation temperature* as a function of water content based on an initial mixture temperature of 298.15 K (the standard condition).

SOLUTION. In this application, one kilogram hydrogen peroxide dissociates while passing through a catalyst and releases energy, which only goes to increase the propellant temperature in the absence of any heat transfer losses. But some of this heat will be required to evaporate the diluent water. The mass balance, Eq. 5–21, is satisfied by 2 mol of hydrogen peroxide in

n mols of liquid water, producing $n + 2$ mol of water vapor plus 1 mol of oxygen gas. Since the reaction goes to completion, no equilibrium constant is needed:

$$2H_2O_2(l) + nH_2O(l) \rightarrow (n + 2)H_2O(g) + O_2(g)$$

The symbols (l) and (g) refer to the liquid state and the gaseous state, respectively. The heats of formation from the standard state $\Delta_f H^0$ and molar specific heats C_p are shown below (see Table 5–1 and other common sources such as the NIST Chemistry Web-Book, http://webbook .nist.gov/chemistry/). For these calculations the heat of mixing may be ignored.

Species	$\Delta_f H^0$ (kJ/kg-mol)	C_p (J/kg-mol-K)	$\mathfrak{M}$, (kg/kg-mol)
$H_2O_2(l)$	−187.69		34.015
$H_2O(l)$	−285.83		18.015
$H_2O(g)$	−241.83	0.03359	18.015
$O_2(g)$	0	0.02938	31.999

The energy balance, Eq. 5–9, for 2 mol of decomposing hydrogen peroxide becomes

$$\Delta_r H^0 = [n\Delta_f H^0]_{H_2O} - [n\Delta_f H^0]_{H_2O_2}$$
$$= 2 \times (-241.83) - 2 \times (-187.69)$$
$$= -108.28 \text{ kJ}$$

The reaction is exothermic but, as stated, some of this energy is used up in vaporizing the diluent liquid water, namely, $285.83 - 241.83 = 44.0$ kJ/kg-mol (at standard conditions). The net available heat release thus becomes $108.28 - 44.0n$ (kJ). In order to calculate the *adiabatic temperature*, we assume ideal-gas heating at constant pressure, Eq. 5–18 (values for the molar specific heats are from Table 5–1 and are taken as constant):

$$\int (n_{H_2O} C_{pH_2O} + n_{O_2} C_{pO_2}) dT = [(2 + n)C_{pH_2O} + C_{pO_2}] \Delta T = 108.28 - 44.0n$$

It will be more convenient to give results in terms of z, a *mass fraction* of the diluent water in the original mixture, and for this molecular masses need to be inserted (also shown in Table 5–1):

$$z = m_{H_2O}/(m_{H_2O_2} + m_{H_2O}) = n\mathfrak{M}_{H_2O}/(2\mathfrak{M}_{H_2O_2} + n\mathfrak{M}_{H_2O})$$
$$= 18.015n/(2 \times 34.015 + 18.015n)$$

Now solve for n, $n = 3.78z/(1 - z)$ $n = 3.78z/(1 - z)$, and substitute it in the relation for the temperature. The resulting value (the adiabatic temperature, T_{ad}) may then be plotted (see figure next page) in terms of the mass fraction z with the initial temperature as given:

$$T_{ad} = 298.15 + \frac{108.28 - 44 \times 3.78z/(1 - z)}{0.03359 \times [2 + 3.78z/(1 - z)] + 0.02838}$$

The figure also displays the values of c^* (where $T_1 = T_{ad}$), $\mathfrak{M}$, and k, which are calculated according to Eqs. 3–32 and 5–5 and 5–7.

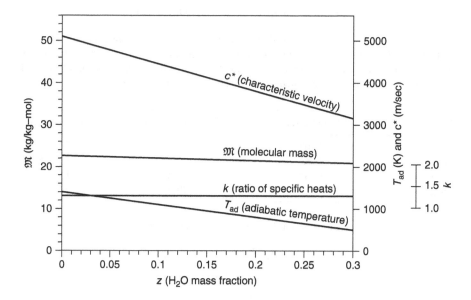

5.3. ANALYSIS OF NOZZLE EXPANSION PROCESSES

There are several methods for analyzing the nozzle flow, depending on the chemical equilibrium assumptions made, nozzle expansion particulates, and/or energy losses. Several are outlined in Table 5–3.

Once the gases reach a supersonic nozzle, they experience an adiabatic, reversible expansion process which is accompanied by substantial drops in temperature and pressure, reflecting the conversion of thermal energy into kinetic energy. Several increasingly more complicated methods have been used for analyzing nozzle processes. For the simplest case, frozen (composition) equilibrium and one-dimensional flow, the state of the gas throughout expansion in the nozzle is fixed by the entropy of the system, which is considered to be invariant as the pressure is reduced. All assumptions listed in Chapter 3 for ideal rockets would be valid here. Again, effects of friction, divergence angle, heat losses, shock waves, and nonequilibrium are neglected in the simplest cases but are considered for the more sophisticated solutions. Any condensed (liquid or solid) phases present are similarly assumed to have zero volume and to be in kinetic as well as thermal equilibrium with the gas flow. This implies that particles and/or droplets are very small in size, move at the same velocity as the gas stream, and have the same temperature as the gas everywhere in the nozzle.

Chemical composition during nozzle expansion may be treated analytically in the following ways:

1. When the expansion is sufficiently rapid, composition may be assumed as invariant throughout the nozzle, that is, there are no chemical reactions or phase changes and the reaction products composition at the nozzle exit are identical to those of the chamber. Such composition results are known as *frozen equilibrium* rocket performance. This approach is the simplest, but tends to underestimate the system's performance typically by 1 to 4%.

2. Instantaneous chemical equilibrium among all molecular species may be significant in some cases under the continuously variable pressure and temperature conditions of the nozzle expansion process. Here, product compositions do shift because the chemical reactions and phase change equilibria taking place between gaseous and condensed phases in all exhaust gas species are fast compared to their nozzle transit time. The composition results so calculated are called *shifting equilibrium* performance. Here, gas composition mass fractions are different at the chamber and nozzle exits. This method usually overstates real performance values, such as c^* or I_s, typically by 1 to 4%. Such analysis is more complex and many more equations are needed.

3. Even though the chemical reactions may occur rapidly, they do require some finite time. Reaction rates for specific reactions are often estimates; these rates are a function of temperature, the magnitude of deviation from the equilibrium molar composition, and the nature of the chemicals or reactions involved. Values of T, c^*, or I_s in most types of actual flow analyses usually fall between those of frozen and instantaneously shifting equilibria. This approach is seldom used because of the lack of good data on reaction rates with multiple simultaneous chemical reactions.

The simplest nozzle flow analysis is also one dimensional, which means that all velocities and temperatures or pressures are equal at any normal cross section of an axisymmetric nozzle. This is often satisfactory for preliminary estimates. In two-dimensional analyses, the resulting velocity, temperature, density, and/or Mach number do not have a flat profile varying somewhat over the cross sections. For nozzle shapes that are not bodies of revolution (e.g., rectangular, scarfed, or elliptic), three-dimensional analyses need to be performed.

When solid particles or liquid droplets are present in the nozzle flow and when the particles are larger than about 0.1 μm average diameter, there will be both a thermal lag and a velocity lag. Solid particles or liquid droplets cannot expand as a gas; their temperature decrease depends on how they lose energy by convection and/or radiation, and their velocity depends on the drag forces exerted on the particle. Larger-diameter droplets or particles are not accelerated as rapidly as smaller ones and their flow velocities are lower than that of those of the accelerating gases. Also, these particulates remain hotter than the gas and provide heat to it. While particles

TABLE 5–3. Typical Steps and Alternatives in the Analysis of Rocket Thermochemical Processes in Nozzles

Step	Process	Method/Implication/Assumption
Nozzle inlet condition	Same as chamber exit; need to know $T_1, p_1, v_1, H, c^*, \rho_1$, etc.	For simpler analyses assume the flow to be uniformly mixed and steady.
Nozzle expansion	An adiabatic process, where flow is accelerated and thermal energy is converted into kinetic energy. Temperature and pressure drop drastically. Several different analyses have been used with different specific effects. Can use one-, two-, or three-dimensional flow pattern. The number of species can be small (Eq. 5–19 has 6) or large (Table 5–8 has 30).	1. Simplest method is inviscid isentropic expansion flow. 2. Include internal weak shock waves; no longer a truly isentropic process. 3. If solid particles are present, they will create drag, thermal lag, and a hotter exhaust gas. Must assume an average particle size and optical surface properties of the particulates. Flow is no longer isentropic. 4. Include viscous boundary layer effects and/or nonuniform velocity profile.

Often a simple single correction factor is used with one-dimensional analyses to modify nozzle exit condition for items 2, 3, and/or 4 above. Computational fluid dynamic codes with finite element analyses have been used with two- and three-dimensional nozzle flow.

Chemical equilibrium during nozzle expansion	Due to rapid decrease in T and p, the equilibrium composition can change from that in the chamber. The four processes listed in the next column allow progressively more realistic simulation and require more sophisticated techniques.	1. Frozen equilibrium; no change in gas composition; usually gives low performance. 2. Shifting equilibrium or instantaneous change in composition; usually overstates the performance slightly. 3. Use reaction time rate analysis to estimate the time to reach equilibrium for each chemical reaction; some rate constants are not well known; analysis is more complex. 4. Use different equilibrium analysis for boundary layer and main inviscid flow; will have nonuniform gas temperature, composition, and velocity profiles.

TABLE 5–3. (*Continued*)

Step	Process	Method/Implication/Assumption
Heat release in nozzle	Recombination of dissociated molecules (e.g., H + H = H$_2$) and exothermic reactions due to changes in equilibrium composition cause internal heating of the expanding gases. Particulates release heat to the gas.	Heat released in subsonic portion of nozzle will increase the exit velocity. Heating in the supersonic flow portion of nozzle can increase the exit temperature but reduce the exit Mach number.
Nozzle shape and size	Can use straight cone, bell-shaped, or other nozzle contour; bell can give slightly lower losses. Make correction for divergence losses and nonuniformity of velocity profile.	Must know or assume a particular nozzle configuration. Calculate bell contour by method of characteristics. Use Eq. 3–34 for divergence losses in conical nozzle. Most analysis programs are one- or two-dimensional. Unsymmetrical nonround nozzles may need three-dimensional analysis.
Gas properties	The relationships governing the behavior of the gases apply to both nozzle and chamber conditions. As gases cool in expansion, some species may condense.	Either use perfect gas laws or, if some of the gas species come close to being condensed, use real gas properties.
Nozzle exit conditions	Will depend on the assumptions made above for chemical equilibrium, nozzle expansion, and nozzle shape/contour. Assume no jet separation. Determine velocity profile and the pressure profile at the nozzle exit plane. If pressure is not uniform across a section it will have some cross flow.	Need to know the nozzle area ratio or nozzle pressure ratio. For quasi-one-dimensional and uniform nozzle flow, see Eqs. 3–25 and 3–26. If v_2 is not constant over the exit area, determine effective average values of v_2 and p_2. Then calculate profiles of T, ρ, etc. For nonuniform velocity profile, the solution requires an iterative approach. Can calculate the gas conditions (T, p, etc.) at any point in the nozzle.
Calculate specific impulse	Can be determined for different altitudes, pressure ratios, mixture ratios, nozzle area ratios, etc.	Can be determined for average values of v_2, p_2, and p_3 based on Eq. 2–6 or 2–13.

contribute to the momentum of the exhaust mass, they are not as efficient as an all-gaseous flow. For composite solid propellants with aluminum oxide particles in the exhaust gas, losses due to particles could typically amount to 1 to 3%. Analyses of two- or three-phase flows require assumptions about the nongaseous amounts from knowledge of the sizes (diameters), size distributions, shapes (usually assumed as spherical), optical surface properties (for determining the emission/absorption or scattering of radiant energy), and their condensation or freezing temperatures. Some of these parameters are seldom well known. Performance estimates of flows with particles are treated in Section 3.5.

The viscous boundary layers adjacent to nozzle walls have velocities substantially lower than those of the inviscid free stream. This viscous drag near the walls actually causes a conversion of kinetic energy into thermal energy, and thus some parts of the boundary layer can be hotter than the local free-stream static temperature. A diagram of a two-dimensional boundary layer is shown in Figure 3–15. With turbulent flows, this boundary layer can be relatively thick in small nozzles. Boundary layers also depend on the axial pressure gradient in the nozzle, nozzle geometry (particularly at the throat region), surface roughness, and/or the heat losses to the nozzle walls. The layers immediately adjacent to the nozzle walls always remain laminar and subsonic. Presently, boundary layer analyses with unsteady flow are only approximations, but are expected to improve as our understanding of relevant phenomena grows and as computational fluid dynamics (CFD) techniques improve. The net effect of such viscous layers appears as nonuniform velocity and temperature profiles, irreversible heating (and therefore increases in entropy), and minor reductions (usually less than 5%) of the kinetic exhaust energy for well-designed systems.

At high combustion temperatures some portion of the gaseous molecules dissociate (splitting into simpler species); in this *dissociation* process, some energy is absorbed by the flow; this reduces the stagnation temperature of the flow within the nozzle even if some of this energy may be released back during reassociation (at the lower pressures and temperatures in the nozzle).

For propellants that yield only gaseous products, extra energy is released in the nozzle, primarily from the recombination of free-radical and atomic species, which become unstable as the temperature decreases in the nozzle expansion process. Some propellant products include species that may actually condense as the temperature drops in the nozzle. If the heat release upon condensation is large, the difference between frozen and shifting equilibrium calculations can be substantial.

In the simplest approach, the exit temperature T_2 is calculated for an isentropic process (with frozen equilibrium). This determines the temperature at the exit and thus the gas conditions at the exit. From the corresponding change in enthalpy, it is then possible to obtain the exhaust velocity and the specific impulse. When nozzle flow is not really isentropic because the expansion process is only partly reversible, it is necessary to include losses due to friction, shock waves, turbulence, and so on. This results in a somewhat higher average nozzle exit temperature and a slight decrease in I_s. A possible set of steps used in such nozzle analysis is given in Table 5–3.

When the contraction between the combustion chamber (or port area) and the throat area is small ($A_p/A_t \leq 3$), acceleration of the gases in the chamber causes an appreciable drop in the effective chamber pressure at the nozzle entrance. This pressure loss in the chamber causes a slight reduction of the values of c and I_s. The analysis of this chamber configuration is treated in Ref. 5–14 and some data are shown in Table 3–2.

5.4. COMPUTER-ASSISTED ANALYSIS

At the present time, all analyses discussed in this chapter are carried out with computer software. Most are based on minimizing the free energy. This is a simpler approach than relying on equilibrium constants, which was common some years ago. Once the values of n_j and T_1 are determined, it is possible to calculate the molecular mass of the gas mixture (Eq. 5–5), the average molar specific heats C_p by Eq. 5–6, and the specific heat ratio k from Eq. 5–7. This then characterizes the thermodynamic state conditions leaving the combustion chamber. With these data we may calculate c^*, R, and other gas-mixture parameters at the combustion chamber exit. From the process of nozzle expansion, as formulated in computer codes, we can then calculate performance (such as I_s, c, or A_2/A_t) and gas conditions in the nozzle; these calculations may include several of the *correction factors* mentioned in Chapter 3 for more realistic results. Programs exist for one-, two-, and three-dimensional flow patterns.

More sophisticated solutions may include a supplementary analysis of combustion chamber conditions when the chamber velocities are high (see Ref. 5–14), boundary layer analyses, heat transfer analyses, and/or two-dimensional axisymmetric flow models with nonuniform flow properties across nozzle cross sections. Time-dependent chemical reactions in the chamber, which are usually neglected, may be analyzed by estimating the time rates at which the reactions occur. This is described in Ref. 5–3.

A commonly used computer program, based on equilibrium compositions, has been developed at the NASA Glenn Laboratories and is known as the *NASA CEA* code (Chemical Equilibrium with Applications). It is described in Ref. 5–13, Vols. 1 and 2, and is available for download (http://www.grc.nasa.gov/WWW/CEAWeb/ ceaguiDownload-win.htm). Key assumptions in this program are one-dimensional forms of the continuity, energy and momentum equations, negligible velocity at the forward end of the combustion chamber, isentropic expansion in the nozzle, ideal gas behavior, and chemical equilibrium in the combustion chamber. It includes options for frozen flow and for narrow chambers (for liquid propellant combustion) or port areas with small cross sections (for solid propellant grains), where the chamber flow velocities are relatively high, resulting in noticeable pressure losses and slight losses in performance. NASA's CEA code has become part of a commercially available code named Cequel^{TM}, which also extends the code's original capabilities.

Other relatively common computer codes used in the United States for analyzing converging–diverging nozzle flows include:

> *ODE (one-dimensional equilibrium code)*, which features instantaneous chemical reactions (shifting equilibrium) and includes all gaseous constituents.

ODK (*one-dimensional kinetics*), which incorporates finite chemical reaction rates for temperature-dependent composition changes in the flow direction with uniform flow properties across any nozzle section. It is used as a module in more complex codes but has no provision for embedded particles.

TDK (*two-dimensional kinetic code*), which incorporates finite kinetic chemical reaction rates and radial variation in flow properties. It has no provision for embedded particles.

VIPERP (*viscous interaction performance evaluation routine for two-phased flows*), a parabolized Navier–Stokes code for internal two-phase nozzle flows with turbulent and nonequilibrium reacting gases. It can be used with embedded solid particles but requires data (or assumptions) on the amount of solids, particle size distribution, or their shape (see, e.g., pp. 503 to 505 in the Seventh Edition of this book).

More information on these computer codes may be obtained from the appropriate government offices and/or from private companies (who actually run the necessary codes for their customers). Many of the more sophisticated codes are proprietary to propulsion organizations or otherwise restricted and not publicly available.

5.5. RESULTS OF THERMOCHEMICAL CALCULATIONS

Extensive computer generated results are available in the literature and only a few samples are indicated here to illustrate effects typical to the variations of key parameters. In general, high specific impulse or high values of c^* can be obtained when the average molecular mass of the reaction products is low (usually this implies formulations rich in hydrogen) and/or when the available chemical energy (heat of reaction) is large, which means high combustion temperatures (see Eqs. 3–16 and 3–32).

Table 5–4 shows computed results for a liquid oxygen, liquid hydrogen thrust chamber taken from Ref. 5–13. It shows *shifting equilibrium* results in the nozzle flow. The narrow chamber has a cross section that is only a little larger than the throat area. The large pressure drop in the chamber (approximately 126 psi) is due to the energy needed to accelerate the gas, as discussed in Section 3.3 and Table 3–2.

The above calculated values of specific impulse will be higher than those obtained from firing actual propellants in rocket units. In practice, it has been found that the experimental values can be lower than those calculated for shifting equilibrium by up to 12%. Because nozzle inefficiencies as explained in Chapter 3 must be considered, only a portion of this correction (perhaps 1 to 4%) is due to combustion inefficiencies.

Much input data for rocket-propulsion-system computer programs (such as the physical and chemical properties of various propellant species used in this chapter) are based on experiments that are more than 25 years old. A few of those have newly revised values, but the differences are believed to be relatively small.

Figures 5–1 through 5–6 indicate calculated results for the liquid propellant combination, liquid oxygen-RP-1 (Rocket Propellant #1). These data are taken from Refs. 5–7 and 5–8. RP-1 is a narrow-cut hydrocarbon similar to kerosene with

TABLE 5–4. Calculated Parameters for a Liquid Oxygen and Liquid Hydrogen Rocket Engine with Four Different Nozzle Expansions

Chamber pressure at injector 773.3 psia or 53.317 bar; $c^* = 2332.1$ m/sec; shifting equilibrium nozzle flow mixture ratio $O_2/H_2 = 5.551$; chamber to throat area ratio $A_1/A_t = 1.580$.

Location	Injector face	Comb. end	Throat	Exit I	Exit II	Exit III	Exit IV
					Parameters		
p_{inj}/p	1.00	1.195	1.886	10.000	100.000	282.15	709.71
T (K)	3389	3346	3184	2569	1786	1468	1219
$\mathfrak{M}$ (molec. mass)	12.7	12.7	12.8	13.1	13.2	13.2	13.2
k (spec. heat ratio)	1.14	1.14	1.15	1.17	1.22	1.24	1.26
C_p (spec. heat, kJ/kg-K)	8.284	8.250	7.530	4.986	3.457	3.224	3.042
M (Mach number)	0.00	0.413	1.000	2.105	3.289	3.848	4.379
A_2/A_t	1.580[a]	1.580[a]	1.000	2.227	11.52	25.00	50.00
c (m/sec)	NA	NA	2879[b]	3485	4150	4348	4487
v_2 (m/sec)	NA	NA	1537[b]	2922	3859	4124	4309
					Mol fractions of gas mixture		
H	0.03390	0.03336	0.02747	0.00893	0.00024	0.00002	0.00000
HO_2	0.00002	0.00001	0.00001	0.00000	0.00000	0.00000	0.00000
H_2	0.29410	0.29384	0.29358	0.29659	0.30037	0.30050	0.30052
H_2O	0.63643	0.63858	0.65337	0.68952	0.69935	0.69948	0.69948
H_2O_2	0.00001	0.00001	0.00000	0.00000	0.00000	0.00000	0.00000
O	0.00214	0.00204	0.00130	0.00009	0.00000	0.00000	0.00000
OH	0.03162	0.03045	0.02314	0.00477	0.00004	0.00000	0.00000
O_2	0.00179	0.00172	0.00113	0.00009	0.00000	0.00000	0.00000

[a] Chamber contraction ratio A_1/A_r.
[b] If cut off at throat.
[c] is the effective exhaust velocity in a vacuum.
v_2 is the nozzle exit velocity at optimum nozzle expansion.
NA means not applicable.

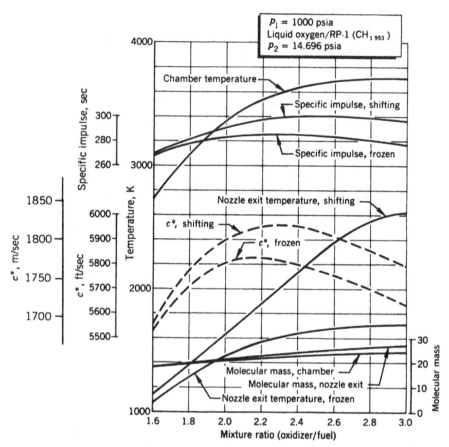

FIGURE 5–1. Calculated performance of liquid oxygen and hydrocarbon combustion as a function of mixture mass ratio.

an average of 1.953 g-atoms of hydrogen for each g-atom of carbon; thus, it has a nominal formula of $CH_{1.953}$. These calculations are for a chamber pressure of 1000 psia. Most of the curves are for optimum area ratio expansion to atmospheric pressure, namely 1 atm or 14.696 psia, and for a limited range of oxidizer-to-fuel mixture mass (not mol) ratios.

For maximum specific impulse, Figs. 5–1 and 5–4 show an *optimum mixture ratio* of approximately 2.3 (kg/sec of oxidizer flow divided by kg/sec of fuel flow) for frozen equilibrium expansion and 2.5 for shifting equilibrium, with the gases expanding to sea-level pressure. The maximum values of c^* occur at slightly different mixture ratios. These optimum mixture ratios are not at the value for highest temperature, which is usually fairly close to stoichiometric. The stoichiometric mixture ratio

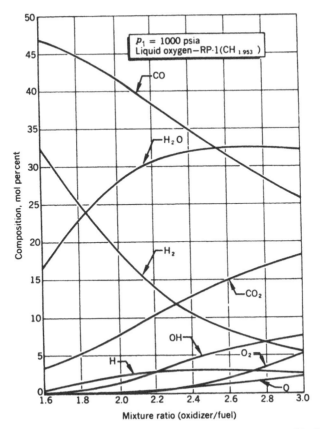

FIGURE 5–2. Calculated chamber gas composition for liquid oxygen and hydrocarbon fuel as a function of mixture ratio. Aggressive gases, such as O_2, O, or OH, can cause reactions with the wall materials in the chamber and the nozzle.

is more than 3.0 where much of the carbon is burned to CO_2 and almost all of the hydrogen to H_2O.

Because shifting equilibrium makes more enthalpy available for conversion to kinetic energy, it gives higher values of performance (higher I_s or c^*) and higher values of nozzle exit temperature for the same exit pressure (see Fig. 5–1). The influence of mixture ratio on *chamber gas composition* is evident from Fig. 5–2. A comparison with Fig. 5–3 indicates marked changes in the gas composition as the gases are expanded under *shifting equilibrium* conditions. The influence of the degree of expansion, or of nozzle exit pressure on gas composition is shown in Fig. 5–6 as well as in Table 5–4. As gases expand to higher area ratios and lower exit pressures (or higher pressure ratios) system performance increases; however, the relative increase diminishes as the pressure ratio is further increased (see Figs. 5–5 and 5–6).

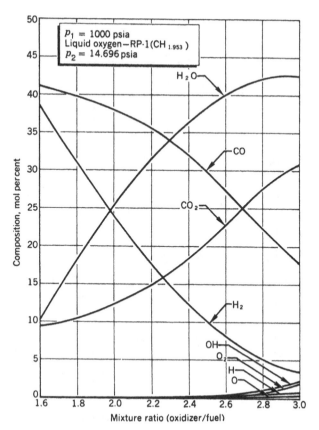

FIGURE 5–3. Calculated nozzle exit gas composition for shifting equilibrium conditions as a function of mixture mass ratio. Breakdown into O, OH, or H *and free* O_2 occurs only at the higher temperatures or higher mixture ratios.

The *dissociation* of gas molecules absorbs considerable energy and decreases the combustion temperature, which in turn reduces the specific impulse. Dissociation of reaction products increases as chamber temperature rises, and decreases with increasing chamber pressure. Atoms or radicals such as monatomic O or H and OH are formed, as can be seen from Fig. 5–2; some unreacted O_2 also remains at the higher mixture ratios and very high combustion temperatures. As gases cool in the nozzle expansion, the dissociated species tend to recombine and release heat into the flowing gases. As can be seen from Fig. 5–3, only a small percentage of dissociated species persists at the nozzle exit and only at the high mixture ratios, where the exit temperature is relatively high. (See Fig. 5–1 for exit temperatures with shifting equilibria). Heat release in supersonic flows actually reduces the Mach number.

Results of thermochemical calculations for several different liquid and solid propellant combinations are given in Tables 5–5 and 5–6. For the liquid propellant

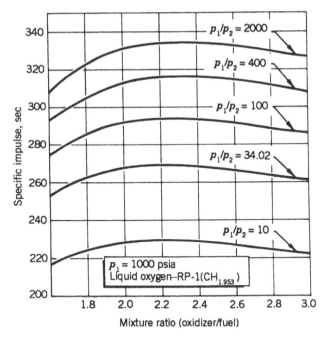

FIGURE 5–4. Variation of theoretical specific impulse with mixture mass ratio and pressure ratio, calculated for frozen equilibrium.

combinations, the listed mixture ratios are optimum and their performance is a maximum. For solid propellants, practical considerations (such as propellant physical properties, e.g., insufficient binder) do not always permit the development of satisfactory propellant grains where ingredients are mixed to optimum performance proportions; therefore values listed for solid propellants in Table 5–6 correspond in part to practical formulations with reasonable physical and ballistic properties.

Calculated results obtained from Ref. 5–13 are presented in Tables 5–7 through 5–9 for a solid propellant to indicate typical variations in performance or gas composition. This particular propellant consists of 60% ammonium perchlorate (NH_4ClO_4), 20% pure aluminum powder, and 20% of an organic polymer of a given chemical composition, namely, $C_{3.1}ON_{0.84}H_{5.8}$. Table 5–7 shows the variation of several performance parameters with different chamber pressures expanding to atmospheric pressure. The area ratios listed are optimum for this expansion with shifting equilibrium. The exit enthalpy, exit entropy, thrust coefficient, and the specific impulse also reflect shifting equilibrium conditions. The characteristic velocity c^* and the chamber molecular mass are functions of chamber conditions only. Table 5–8 shows the variation of gas composition with chamber pressure; here, some reaction products are in the liquid phase, such as Al_2O_3. Table 5–9 shows the variation of nozzle exit characteristics and composition for shifting equilibria as a function of exit pressure

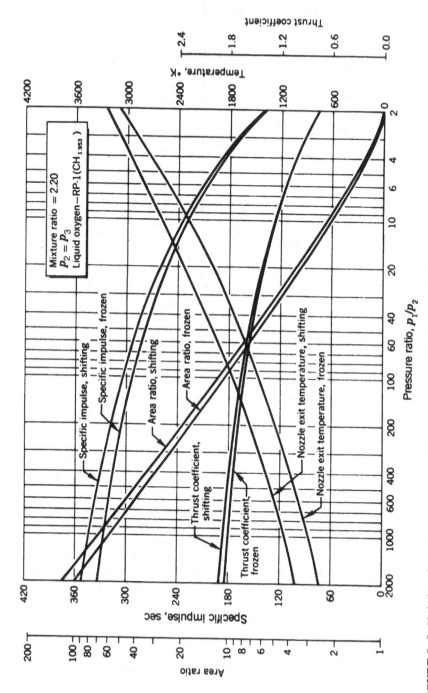

FIGURE 5–5. Variation of calculated parameters with pressure ratio for liquid oxygen–hydrocarbon propellant at a mixture ratio of 2.20. An increase in pressure ratio may be due to an increase in chamber pressure, a decrease of nozzle exit pressure (larger area ratio and higher altitude), or both.

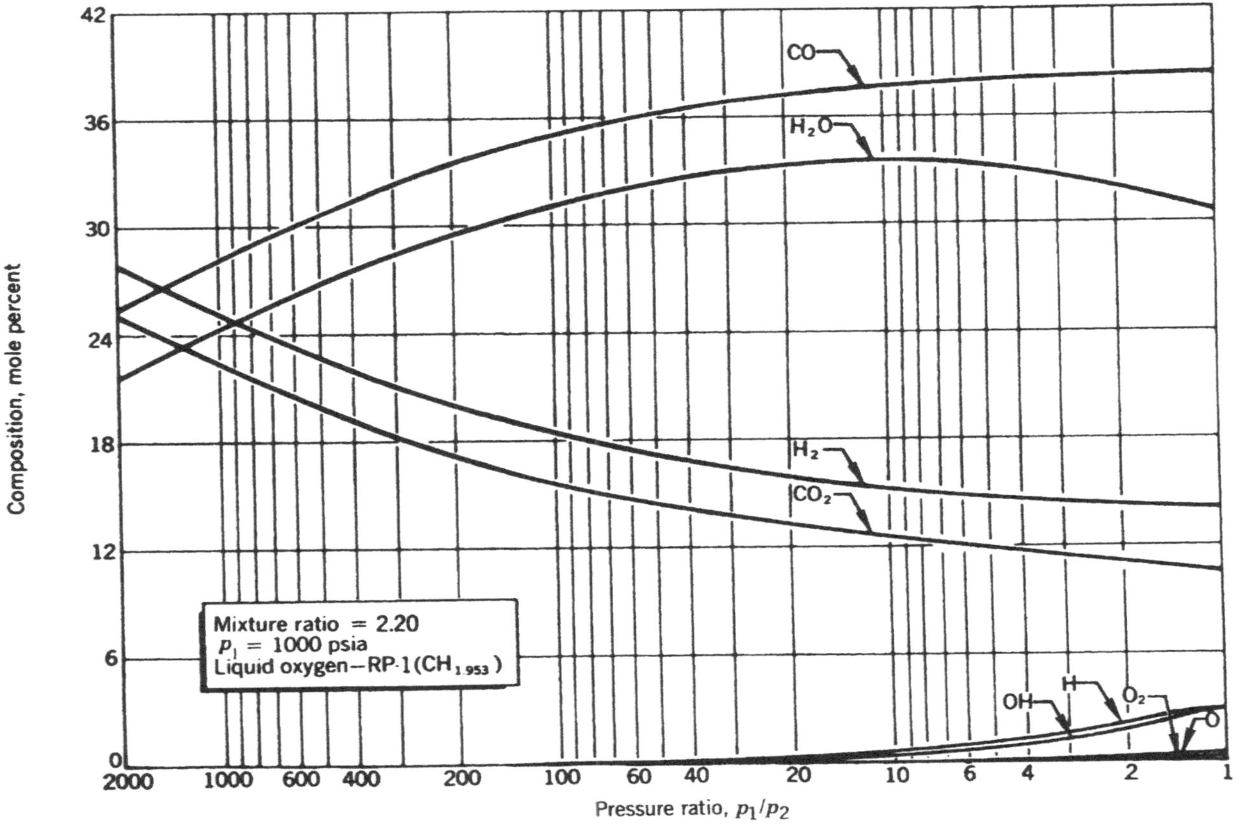
42
36
30
24
18
12
6
0
Composition, mole percent
CO
H₂O
H₂
CO₂
OH
H
O₂
O
Mixture ratio = 2.20
p_1 = 1000 psia
Liquid oxygen—RP-1(CH$_{1.953}$)
2000
1000
600
400
200
100
60
40
20
10
6
4
2
1
Pressure ratio, p_1/p_2

FIGURE 5–6. Variation of exhaust gas composition at nozzle exit with pressure ratio at a fixed mixture ratio and for shifting equilibrium. For frozen equilibrium compositions would be the same as those of the chamber, as shown in Fig. 5–2.

TABLE 5–5. Theoretical Chamber Performance of Liquid Rocket Propellant Combinations

Oxidizer	Fuel	Mixture Ratio		Average Specific Gravity	Chamber Temp.(K)	Chamber c^* (m/sec)	$\mathfrak{M}$ (kg/mol)	I_s (sec)		k
		By Mass	By Volume					Shifting	Frozen	
Oxygen	Methane	3.20	1.19	0.81	3526	1835	20.3		296	1.20
		3.00	1.11	0.80	3526	1853		311		
	Hydrazine	0.74	0.66	1.06	3285	1871	18.3		301	1.25
		0.90	0.80	1.07	3404	1892	19.3	313		
	Hydrogen	3.40	0.21	0.26	2959	2428	8.9		386	1.26
		4.02	0.25	0.28	2999	2432	10.0	389.5		
	RP-1	2.24	1.59	1.01	3571	1774	21.9		285.4	1.24
		2.56	1.82	1.02	3677	1800	23.3	300		
	UDMH	1.39	0.96	0.96	3542	1835	19.8		295	1.25
		1.65	1.14	0.98	3594	1864	21.3	310		
Fluorine	Hydrazine	1.83	1.22	1.29	4553	2128	18.5	334		1.33
		2.30	1.54	1.31	4713	2208	19.4		365	
	Hydrogen	4.54	0.21	0.33	3080	2534	8.9		389	1.33
		7.60	0.35	0.45	3900	2549	11.8	410		
Nitrogen tetroxide	Hydrazine	1.08	0.75	1.20	3258	1765	19.5		283	1.26
		1.34	0.93	1.22	3152	1782	20.9	292		
	50% UDMH	1.62	1.01	1.18	3242	1652	21.0		278	1.24
	50% hydrazine	2.00	1.24	1.21	3372	1711	22.6	289		
	RP-1	3.4	1.05	1.23	3290		24.1	289	297	1.23
	MMH	2.15	1.30	1.20	3396	1747	22.3	289		1.22
Red fuming nitric acid	RP-1	4.1	2.12	1.35	3175	1591	21.7		278	1.23
		4.8	2.48	1.33	3230	1594	24.6		258	1.22
	50% UDMH	1.73	1.00	1.23	2997	1609	25.8	269	272	
	50% hydrazine	2.20	1.26	1.27	3172	1682	20.6			
Hydrogen peroxide (90%)	RP-1	7.0	4.01	1.29	2760	1701	22.4	279	297	1.19

Notes:

Combustion chamber pressure—1000 psia (6895 kN/m²); nozzle exit pressure—14.7 psia (1 atm); optimum expansion.

Adiabatic combustion and isentropic expansion of ideal gases.

The specific gravity at the boiling point has been used for those oxidizers or fuels that boil below 20 °C at 1 atm pressure, see Eq. 7–1.

Mixture ratios are for approximate maximum values of I_s.

TABLE 5–6. Theoretical Performance of Typical Solid Rocket Propellant Combinations

Oxidizer	Fuel	ρ_b (g/cm^3)^a	T_1 (K)	c^* (m/sec)^b	$\mathfrak{M}$, (kg/mol)	I_s (sec)^b	k
Ammonium nitrate	11% binder and 7% additives	1.51	1282	1209	20.1	192	1.26
Ammonium perchlorate 78–66%	18% organic polymer binder and 4–20% aluminum	1.69	2816	1590	25.0	262	1.21
Ammonium perchlorate 84–68%	12% polymer binder and 4–20% aluminum	1.74	3371	1577	29.3	266	1.17

^{a}Density of solid propellant, see Eq. 12–1.
^{b}Conditions for I_s and c^*: Combustion chamber pressure: 1000 psia; nozzle exit pressure: 14.7 psia; optimum nozzle expansion ratio; frozen equilibrium.

or pressure ratio for a fixed value of chamber pressure. Table 5–9 shows how composition shifts during expansion in the nozzle and indicates several species present in the chamber that do not appear at the nozzle exit. These three tables show computer results—some thermodynamic properties of the reactants and reaction products probably do not warrant the indicated high accuracy of five significant figures. In the analysis for chemical ingredients of this solid propellant, approximately 76 additional reaction products have been considered in addition to the major product species. These include, for example, CN, CH, CCl, Cl, NO, and so on. Their calculated mol fractions are very small, and therefore they may be neglected and are not included in Table 5–8 or 5–9.

Such calculated results are useful in estimating performance (I_s, c^*, C_F, ϵ, etc.) for particular chamber and nozzle exit pressures, and knowledge of gas composition, as indicated in the previous figures and tables, permits more detailed estimates of other design parameters, such as convective properties for heat transfer determination, radiation characteristics of the flame inside and outside the thrust chambers, and acoustic characteristics of the gases. Some performance data relevant to hybrid propellants are presented in Chapter 16.

The thermochemical analyses found in this chapter can also be applied to gas generators; results (such as the gas temperature T_1, specific heat c_p, specific heat ratio k, or composition) are used for estimating turbine inlet conditions or turbine power. In gas generators and preburners for staged combustion cycle rocket engines (explained in Section 6.6) gas temperatures are much lower, to avoid damage to the turbine blades. Typically, combustion reaction gases are at 800 to 1200 K, which is lower than the gas temperature in the thrust chamber (2900 to 3600 K). Examples are listed in Table 5–10 for a chamber pressure of 1000 psia. Some gaseous species will not be present (such as atomic oxygen or hydroxyl), and often real gas properties will need to be used because some of these gases do not behave as a perfect gas at these lower temperatures.

TABLE 5–7. Variation of Calculated Performance Parameters for an Aluminized Ammonium Perchlorate Composite Propellant as a Function of Chamber Pressure for Expansion to Sea Level (1 atm) with Shifting Equilibrium

	1500	1000	750	500	200
Chamber pressure (psia)	1500	1000	750	500	200
Chamber pressure (atm) to sea-level pressure ratio p_1/p_2	102.07	68.046	51.034	34.023	13.609
Chamber temperature (K)	3346.9	3322.7	3304.2	3276.6	3207.7
Nozzle exit temperature (K)	2007.7	2135.6	2226.8	2327.0	2433.6
Chamber enthalpy (cal/g)	−572.17	−572.17	−572.17	−572.17	−572.17
Exit enthalpy (cal/g)	−1382.19	−1325.15	−1282.42	−1219.8	−1071.2
Entropy (cal/g-K)	2.1826	2.2101	2.2297	2.2574	2.320
Chamber molecular mass (kg/mol)	29.303	29.215	29.149	29.050	28.908
Exit molecular mass (kg/mol)	29.879	29.853	29.820	29.763	29.668
Exit Mach number	3.20	3.00	2.86	2.89	2.32
Specific heat ratio – chamber, k	1.1369	1.1351	1.1337	1.1318	1.1272
Specific impulse, vacuum (sec)	287.4	280.1	274.6	265.7	242.4
Specific impulse, sea-level expansion (sec)	265.5	256.0	248.6	237.3	208.4
Characteristic velocity, c^* (m/sec)	1532	1529	1527	1525	1517
Nozzle area ratio[a], A_2/A_t	14.297	10.541	8.507	8.531	6.300
Thrust coefficient[a], C_F	1.700	1.641	1.596	1.597	1.529

[a]At optimum expansion.
Source: From Ref. 5–13.

TABLE 5-8. Mol Fraction Variation of Chamber Gas Composition with Combustion Chamber Pressure for an Aluminum Containing Composite Solid Propellant

Pressure p_1 (psia)	1500	1000	750	500	200
Pressure (atm) or press. ratio to sea level	102.07	68.046	51.034	34.023	13.609
Ingredient					
Al	0.00007	0.00009	0.00010	0.00012	0.00018
AlCl	0.00454	0.00499	0.00530	0.00572	0.00655
$AlCl_2$	0.00181	0.00167	0.00157	0.00142	0.00112
$AlCl_3$	0.00029	0.00023	0.00019	0.00015	0.00009
AlH	0.00002	0.00002	0.00002	0.00002	0.00002
AlO	0.00007	0.00009	0.00011	0.00013	0.00019
AlOCl	0.00086	0.00095	0.00102	0.00112	0.00132
AlOH	0.00029	0.00032	0.00034	0.00036	0.00041
AlO_2H	0.00024	0.00026	0.00028	0.00031	0.00036
Al_2O	0.00003	0.00004	0.00004	0.00005	0.00006
Al_2O_3 (solid)	0.00000	0.00000	0.00000	0.00000	0.00000
Al_2O_3 (liquid)	0.09425	0.09378	0.09343	0.09293	0.09178
CO	0.22434	0.22374	0.22328	0.22259	0.22085
COCl	0.00001	0.00001	0.00001	0.00001	0.00000
CO_2	0.00785	0.00790	0.00793	0.00799	0.00810
Cl	0.00541	0.00620	0.00681	0.00772	0.01002
Cl_2	0.00001	0.00001	0.00001	0.00001	0.00001
H	0.02197	0.02525	0.02776	0.03157	0.04125
HCl	0.12021	0.11900	0.11808	0.11668	0.11321
HCN	0.00003	0.00002	0.00001	0.00001	0.00000
HCO	0.00003	0.00002	0.00002	0.00002	0.00001
H_2	0.32599	0.32380	0.32215	0.31968	0.31362
H_2O	0.08960	0.08937	0.08916	0.08886	0.08787
NH_2	0.00001	0.00001	0.00001	0.00000	0.00000
NH_3	0.00004	0.00003	0.00002	0.00001	0.00001
NO	0.00019	0.00021	0.00023	0.00025	0.00030
N_2	0.09910	0.09886	0.09867	0.09839	0.09767
O	0.00010	0.00014	0.00016	0.00021	0.00036
OH	0.00262	0.00297	0.00324	0.00364	0.00458
O_2	0.00001	0.00001	0.00002	0.00002	0.00004

Source: From Ref. 5-13.

TABLE 5–9. Calculated Variation of Thermodynamic Properties and Exit Gas Composition for an Aluminized Perchlorate Composite Propellant with $p_1 = 1500$ psia and Various Exit Pressures at Shifting Equilibrium and Optimum Expansion

	Chamber	Throat	Nozzle Exit				
Pressure (atm)	102.07	58.860	2.000	1.000	0.5103	0.2552	0.1276
Pressure (MPa)	10.556	5.964	0.2064	0.1032	0.0527	0.0264	0.0132
Nozzle area ratio	>0.2	1.000	3.471	14.297	23.972	41.111	70.888
Temperature (K)	3346.9	3147.3	2228.5	2007.7	1806.9	1616.4	1443.1
Ratio chamber pressure/local pressure	1.000	1.7341	51.034	102.07	200.00	400.00	800.00
Molecular mass (kg/mol)	29.303	29.453	29.843	29.879	29.894	29.899	29.900
Composition (mol%)							
Al	0.00007	0.00003	0.00000	0.00000	0.00000	0.00000	0.00000
AlCl	0.00454	0.00284	0.00014	0.00008	0.00000	0.00000	0.00000
AlCl$_2$	0.00181	0.00120	0.00002	0.00000	0.00000	0.00000	0.00000
AlCl$_3$	0.00029	0.00023	0.00002	0.00000	0.00000	0.00000	0.00000
AlOCl	0.00086	0.00055	0.00001	0.00000	0.00000	0.00000	0.00000
AlOH	0.00029	0.00016	0.00000	0.00000	0.00000	0.00000	0.00000
AlO$_2$H	0.00024	0.00013	0.00000	0.00000	0.00000	0.00000	0.00000
Al$_2$O	0.00003	0.00001	0.00000	0.00000	0.00000	0.00000	0.00000
Al$_2$O$_3$ (solid)	0.00000	0.00000	0.00000	0.00000	0.00000	0.00000	0.00000
Al$_2$O$_3$ (liquid)	0.09425	0.09608	0.09955	0.09969	0.09974	0.09976	0.09976
CO	0.22434	0.22511	0.22553	0.22416	0.22008	0.21824	0.21671
CO$_2$	0.00785	0.00787	0.00994	0.01126	0.01220	0.01548	0.01885
Cl	0.00541	0.00441	0.00074	0.00028	0.00009	0.00002	0.00000
H	0.02197	0.01722	0.00258	0.00095	0.00030	0.00007	0.00001
HCl	0.12021	0.12505	0.13635	0.13707	0.13734	0.13743	0.13746
H$_2$	0.32599	0.33067	0.34403	0.34630	0.34842	0.35288	0.35442
H$_2$O	0.08960	0.08704	0.08091	0.07967	0.07796	0.07551	0.07214
NO	0.00019	0.00011	0.00001	0.00000	0.00000	0.00000	0.00000
N$_2$	0.09910	0.09950	0.10048	0.10058	0.10063	0.10064	0.10065
O	0.00010	0.00005	0.00000	0.00000	0.00000	0.00000	0.00000
OH	0.00262	0.00172	0.00009	0.00005	0.00002	0.00000	0.00000

Source: From Ref. 5–13.

TABLE 5–10. Typical Gas Characteristics for Fuel-Rich Liquid Propellant Gas Generators

Propellant	T_1(K)	k	Gas Constant R (ft-lbf/lbm-°R)	Oxidizer-to-Fuel Mass Ratio	Specific Heat c_p (kcal/kg-K)
Liquid oxygen and	900	1.370	421	0.919	1.99
liquid hydrogen	1050	1.357	375	1.065	1.85
	1200	1.338	347	1.208	1.78
Liquid oxygen and	900	1.101	45.5	0.322	0.639
kerosene	1050	1.127	55.3	0.423	0.654
	1200	1.148	64.0	0.516	0.662
Dinitrogen tetroxide	1050	1.420	87.8	0.126	0.386
and dimethyl	1200	1.420	99.9	0.274	0.434
hydrazine					

SYMBOLS

Symbols referring to chemical elements, compounds, or mathematical operators are not included in this list.

a or b	number of kilogram atoms
A_t	throat area, m^2
A_p	port area, m^2
c^*	characteristic velocity, m/sec
c_p	specific heat per unit mass at constant pressure, J/kg-K
C_p	molar specific heat at constant pressure of gas mixture, J/kg-mol-K
g_0	acceleration of gravity at sea level, 9.8066 m/sec^2
G	Gibbs free energy for a propellant combustion gas mixture, J/kg
$\Delta_f G^0$	change in free energy of formation at 298.15 K and 1 bar
G_j	free energy for a particular species j, J/kg
ΔH	overall enthalpy change, J/kg or J/kg-mol
ΔH_j	enthalpy change for a particular species j, J/kg
$\Delta_r H^0$	heat of reaction at reference 298.15 K and 1 bar, J
$\Delta_f H^0$	heat of formation at reference 298.15 K and 1 bar, J/kg
h_j	enthalpy for a particular species, J/kg or J/kg-mol
I_s	specific impulse, sec
k	specific heat ratio
ℓ	total number of given chemical species in a mixture
m	number of gaseous species, also total number of products
$\dot{m}$	mass flow rate, kg/sec
$\mathfrak{M}$	molecular mass of gas mixture, kg/kg-mol (lbm/lb-mol)
n	total number of mols per unit mass (kg-mol/kg or mol) of mixture
n_j	mols of species j, kg-mol/kg or mol
p	pressure of gas mixture, N/m^2

r	total number of reactants
R	gas constant, J/kg-K
R'	universal gas constant, 8314.3 J/kg mol-K
S	entropy, J/kg mol-K
T	absolute temperature, K
T_{ad}	adiabatic temperature, K
U	internal energy, J/kg-mol
v	gas velocity, m/sec
V	specific volume, m^3/kg
X_j	mol fraction of species j

Greek Letters

ρ	density, kg/m^3

Subscripts

a, b	molar fractions of reactant species A or B
c, d	molar fractions of product species C or D
i	atomic or molecular species in a specific propellant
j	constituent or species in reactants or products
mix	mixture of gases
ref	at reference condition (also superscript 0)
1	chamber condition
2	nozzle exit condition
3	ambient atmospheric condition

PROBLEMS

1. Explain the physical and/or chemical reasons for the maximum value of specific impulse at a particular flow mixture ratio of oxidizer to fuel.

2. Explain why, in Table 5–8, the relative proportion of monatomic hydrogen and monatomic oxygen changes markedly with different chamber pressures and exit pressures.

3. This chapter contains several charts for the performance of liquid oxygen and RP-1 hydrocarbon fuel. If by mistake a new shipment of cryogenic oxidizer contains at least 15% liquid nitrogen discuss what general trends should be expected in results from its testing in performance values, likely composition of the exhaust gases under chamber and nozzle conditions, and find the new optimum mixture ratio.

4. A mixture of perfect gases consists of 3 kg of carbon monoxide and 1.5 kg of nitrogen at a pressure of 0.1 MPa and a temperature of 298.15 K. Using Table 5–1, find **(a)** the effective molecular mass of the mixture, **(b)** its gas constant, **(c)** specific heat ratio, **(d)** partial pressures, and **(e)** density.
Answer: **(a)** 28 kg/kg-mol, **(b)** 297 J/kg-K, **(c)** 1.40, **(d)** 0.0666 and 0.0333 MPa, **(e)** 1.13 kg/m^3.

5. Using information from Table 5–2, plot the value of the specific heat ratio for carbon monoxide (CO) as a function of temperature. Notice the trend of this curve, which is typical of the temperature behavior of other diatomic gases.
   Answer: $k = 1.28$ at 3500 K, 1.30 at 2000 K, 1.39 at 500 K.

6. Modify and tabulate two entries in Table 5–5 for operation in the vacuum of space, namely oxygen/hydrogen and nitrogen tetroxide/hydrazine. Assume the data in the table represent the design condition.

7. Various experiments have been conducted with a liquid monopropellant called nitromethane (CH_3NO_2), which can be decomposed into gaseous reaction products. Determine the values of T, $\mathfrak{M}$, k, c^*, C_F, and I_s using the water–gas equilibrium conditions. Assume no dissociations and no O_2.
   Answer: 2470 K, 20.3 kg/kg-mol, 1.25, 1527 m/sec, 1.57, 244 sec.

8. The figures in this chapter show several parameters and gas compositions of liquid oxygen burning with RP-1, which is a kerosene-type material. For a mixture ratio of 2.0, use the given compositions to verify the molecular mass in the chamber and the specific impulse (frozen equilibrium flow in nozzle) in Fig. 5–1.

REFERENCES

5–1. F. Van Zeggeren and S. H. Storey, *The Computation of Chemical Equilibria*, Cambridge University Press, Cambridge, England, 1970.

5–2. S. S. Penner, *Thermodynamics for Scientists and Engineers*, Addison-Wesley, Reading, MA, 1968.

5–3. S. I. Sandler, *Chemical and Engineering Thermodynamics*, John Wiley & Sons, New York, 1999.

5–4. R. H. Dittman and M. W. Zemansky, *Heat and Thermodynamics*, 7th ed., McGraw-Hill, New York, 1996.

5–5. K. Denbigh, *The Principles of Chemical Equilibrium*, 4th ed., Cambridge University Press, Cambridge, England, 1981.

5–6. K. K. Kuo, *Principles of Combustion*, 2nd ed., John Wiley & Sons, Hoboken, NJ, 2005.

5–7. *JANAF Thermochemical Tables*, Dow Chemical Company, Midland, MI, Series A (June 1963) through Series E (January 1967).

5–8. M. W. Chase, C. A. Davies, J. R. Downey, D. J. Frurip, R. A. McDonald, and A. N. Syverud, *JANAF Thermochemical Tables*, 3rd ed., Part I, *Journal of Physical and Chemical Reference Data*, Vol. 14, Supplement 1, American Chemical Society, American Institute of Physics, and National Bureau of Standards, 1985.

5–9. D. D. Wagman et al., "The NBS Tables of Chemical Thermodynamic Properties," *Journal of Physical and Chemical Reference Data*, Vol. 11, Supplement 2, American Chemical Society, American Institute of Physics, and National Bureau of Standards, 1982.

5–10. J. B. Pedley, R. D. Naylor, and S. P. Kirby, *Thermochemical Data of Organic Compounds*, 2nd ed., Chapman & Hall, London, 1986, xii + 792 pages; ISBN: 9780412271007.

5–11. B. J. McBride, S. Gordon, and M. Reno, "Thermodynamic Data for Fifty Reference Elements," *NASA Technical Paper 3287*, January 1993. Also NASA/TP-3287/REV1; NASA NTRS Doc. ID 20010021116; http://hdl.handle.net/2060/20010021116

5–12. B. J. McBride and S. Gordon, "Computer Program for Calculating and Fitting Thermodynamic Functions," *NASA Reference Publication 1271*, November 1992; http://hdl.handle.net/2060/19930003779; http://hdl.handle.net/2060/19880011868

5–13. S. Gordon and B. J. McBride, "Computer Program for Calculation of Complex Chemical Equilibrium Compositions and Applications, Vol. 1: Analysis" (October 1994), http://hdl.handle.net/2060/19950013764; and "Vol. 2: User Manual and Program Description" (June 1996), *NASA Reference Publication 1311*.

5–14. S. Gordon and B. J. McBride, "Finite Area Combustor Theoretical Rocket Performance," *NASA TM* 100785, April 1988; http://hdl.handle.net/2060/19880011868.

5–15. D. R. Stull, E. F. Westrum, and G. C. Sinke, "The Chemical Thermodynamics of Organic Compounds," John Wiley & Sons, New York, 1969, xvii + 865 pages; ISBN: 9780471834908.

5–16. P. G. Hill and C. R. Peterson, "Mechanics and Thermodynamics of Propulsion," 2nd ed., Addison-Wesley/Prentice Hall, Reading, MA, 1992, xi + 754 pages; ISBN: 9780201146592.

CHAPTER 6

LIQUID PROPELLANT ROCKET ENGINE FUNDAMENTALS

This chapter presents an overview of liquid propellant chemical rocket engines. It is the first of six chapters devoted to this subject. It identifies types of liquid rocket engines, their key components, different propellants, and tank configurations. It also discusses two types of propellant feed systems, engine cycles, propellant tanks, their pressurization subsystems, engine controls, valves, piping, and structure. Chapter 7 covers liquid propellants in more detail. Chapter 8 describes thrust chambers (and their nozzles), small thrusters, and heat transfer. Chapter 9 is about the combustion process and Chapter 10 discusses turbopumps. Chapter 11 presents engine design, engine controls, propellant budgets, engine balance and calibration, and overall engine systems.

In this book, *liquid propellant rocket propulsion systems* consist of a *rocket engine* and a set of *tanks* for storing and supplying propellants. They have all the hardware components and the propellants necessary for their operation, that is, for producing thrust. See Ref. 6–1. The *rocket engine* consists of one or more *thrust chambers*, a *feed mechanism* for supplying the propellants from their tanks to the thrust chamber(s), a *power source* to furnish the energy for the feed mechanism, suitable *plumbing* or *piping* to transfer the liquid propellants under pressure, a *structure* to transmit the thrust force, and *control devices* (including valves) to start and stop and sometimes also to vary the propellant flow and thus the thrust. Liquid propellants are either expelled from their tanks by a high-pressure gas or they are delivered by pumps to the thrust chambers. Figure 6–1 shows the Space Shuttle Main Engine (SSME), which was retired in 2011—at the time of this writing, the RS-25 engine, essentially identical, is being developed (Refs. 6–2 and 6–3) for the initial flights in NASA's Space Launch System (SLS) missions. The RS-25 has a somewhat higher thrust capability (512,000 lbf) but is a simplified nonreusable version

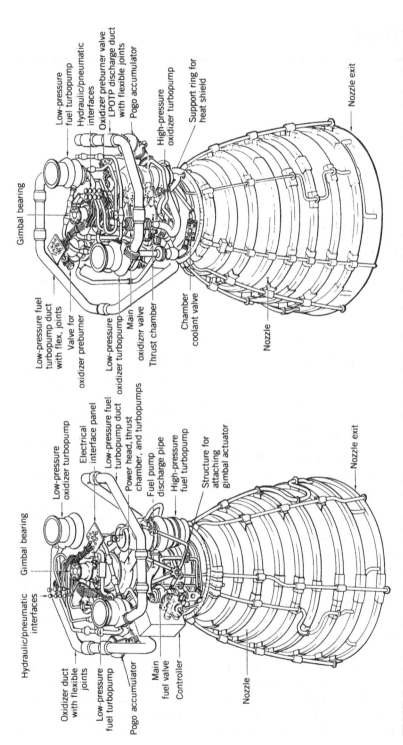

FIGURE 6–1. Two views of the man-rated, throttleable, reusable Space Shuttle main engine. This engine, now modified and designated as the RS-25, will support initial missions in NASA's Space Launch System (SLS). It uses liquid oxygen and liquid hydrogen as propellants, and its vacuum thrust is 512,000 lbf. Courtesy of Aerojet Rocketdyne and NASA

190

of the SSME. It has two low-pressure-rise booster turbopumps in addition to the two main high-pressure, high-speed pumps—the booster-pump discharge pressure avoids cavitation at the main pump impellers (see Section 10.5). The large propellant tanks are pressurized by small flows of gasified oxygen and of gasified hydrogen, respectively. Tank pressurization is discussed in Section 6.5.

In some applications rocket engines may also include a *thrust vector control* system (for changing the thrust vector direction; see Chapter 18), a random *variable thrust* feature (see Section 8.5), an engine *condition monitoring* or engine *health monitoring* subsystem (see Section 11.4), and various *instrumentation/measuring* devices (see Chapter 21). The liquid propellant storage *tanks* and the subsystem for *pressurizing the tanks with gas* are discussed in this chapter and are considered in this book to be part of the *rocket propulsion system.**

The design of any propulsion system is tailored to fit a specific *mission requirement* as explained in Chapter 19. These requirements are usually stated in terms of their application, such as anti-aircraft missile or second stage of a space launch vehicle, flight velocity increment, flight path and flight maneuvers, launch sites, minimum life (in storage or in orbit), or number of vehicles to be delivered. Such requirements usually also include constraints of the inert engine mass, cost, or safety provisions. Other criteria, constraints and the selection process are given in Chapter 19.

From *mission requirements* and their *definition* one can derive *propulsion system* and *engine requirements*, which include the thrust–time profile, minimum specific impulse, number of thrust chambers, total impulse, number of restarts (if any), likely propellants, and/or constraints of engine masses or engine sizes. Some engine parameters, such as thrust, chamber pressure, mixture ratio, engine mass, or nozzle exit area ratio, may be analytically optimized for a specific mission. Other engine parameters can be selected based on experience and/or design studies, including the feed system, engine components arrangement, engine cycle, thrust modulation, and alternate methods of thrust vector control. Two or more preliminary or conceptual designs may be compared for the purpose of selecting a propulsion design for the mission under consideration.

Tables 1–3, 11–2, and 11–3 present typical data for selected rocket engines. Many different types of rocket engines have been studied, built, and flown, ranging in thrust size from less than 0.01 lbf to over 1.75 million pounds (0.044 N to 7.7 MN),

---

*The responsibilities for the design, development, fabrication, and operation of a propulsion system are usually shared between a rocket engine organization and a flight vehicle organization. However, responsibility allocations for components or subsystems have been neither rigid nor consistent in the literature or in actual industry practice. For example, some vehicle design/development organizations have considered the tanks and parts of the engine structure to be really a part of their vehicle. The tank pressurization system has in various scenarios been considered to be part of either the engine, the propulsion system, or the vehicle. For some reaction control systems the vehicle developer often assumed the responsibility for the propulsion systems and obtained only the small thrusters and their small propellant valves from a rocket engine company. In some programs, such as the peacekeeper missile fourth stage, the rocket engine developer has developed not only the engine, but also much of the vehicle stage with its propellant tanks and pressurization system.

with one-time operation or multiple starts (some small thrusters have over 150,000 restarts), with or without thrust modulation (called throttling), single use or reusable, arranged as single engines, or in clusters of multiple units.

One way to categorize liquid propellant rocket engines is described in Table 6–1. There are two categories, namely those used for *boosting* a payload and imparting a significant velocity increase to it and those for *auxiliary propulsion* used in *trajectory adjustments* and *attitude control*. Liquid propellant rocket engine systems are *classified* in several other ways. They can be *reusable* (like the Space Shuttle main engine or a rocket engine for quick ascent or maneuvers of fighter aircraft) or suitable for a *single flight* only (as the engines in expendable launch vehicles), and they can be *restartable*, like a reaction control engine, or *single firing*, as in space launch vehicle boosters. They can also be categorized by their *propellants, application*, or *stage*, such as an upper stage or booster stage, their *thrust level*, and by the *feed system type* (pressurized or turbopump).

The *thrust chamber* or *thruster* includes the combustion device where liquid propellants are metered, injected, atomized, mixed, and then burned to form hot gaseous reaction products, which in turn are accelerated and ejected at high velocities to impart thrust. The thrust chamber has three major parts: an *injector*, a *combustion chamber*, and a *nozzle*. In regeneratively *cooled thrust chambers*, one of the propellants (usually the fuel) is circulated through cooling jackets or a special cooling passage to absorb the heat transfer from the hot reaction gases to the thrust chamber walls (see Figs. 8–2 and 8–9). A *radiation-cooled* thrust chamber uses high-temperature materials, such as niobium metal, which can radiate away all their excess heat. There are also *uncooled* or *heat-absorbing* thrust chambers, such as those using *ablative* materials. Thrust chambers are discussed in Chapter 8.

There are two types of feed systems used for liquid propellant rocket engines: one that uses pumps for moving propellants from their vehicle's storage tanks to the thrust chamber and the other that uses a high-pressure gas for expelling their propellants from their tanks. These are discussed further in Sections 6.3, 6.4 and 6.6.

Solid propellants are covered in Chapters 12 to 15. Tables 19–1 to 19–4 compare advantages and disadvantages of liquid propellant rocket engines and solid propellant rocket motors. Hybrid propulsion is discussed in Chapter 16.

6.1. TYPES OF PROPELLANTS

Propellants, the working substances of rocket engines, constitute the fluid that undergoes chemical and thermodynamic changes. The term *liquid propellant* embraces all the various propellants stored as liquids and may be one of the following (all these are described in Chapter 7):

1. Oxidizer (liquid oxygen, nitric acid, nitrogen tetroxide, etc.).
2. Fuel (kerosene, alcohol, liquid hydrogen, etc.).

TABLE 6–1. Typical Characteristics of Two Categories of Liquid Propellant Rocket Engines

Purpose/Feature	Boost Propulsion	Auxiliary Propulsion
Mission	Impart significant velocity to propel a vehicle along its flight path	Attitude control, minor space maneuvers, trajectory corrections, orbit maintenance
Applications	Booster stage and upper stages of launch vehicles, large missiles	Spacecraft, satellites, top stage of antiballistic missile, space rendezvous
Total impulse	High	Low
Number of thrust chambers per engine	Usually 1; sometimes 4, 3, or 2	Between 4 and 24
Thrust level per thrust chamber	High; 4500 N up to 7,900,000 N or 1000–1,770,000 lbf	Small; 0.001 up to 4500 N, a few go up to 1000 lbf
Feed system, typical	Mostly turbopump type; occasionally pressurized feed system for smaller thrusts	Pressurized feed system with high-pressure gas supply
Tank pressure range	0.138–0.379 MPa or 20–55 psi	0.689–17.23 MPa or 100–2500 psi
Most common cooling method	Propellant cooled	Radiation cooled
Propellants (see next section)	Cryogenic and storable liquids	Storable liquids, monopropellants, and/or stored cold gas
Chamber pressure	2.4–21 MPa or 350–3600 psi	0.14–2.1 MPa or 20–400 psi
Number of starts during a single mission	Usually no restart; sometimes one, but up to four in some cases	Several thousand starts for some space missions
Cumulative duration of firing	Up to a few minutes	Up to several hours
Shortest firing duration	Typically 5–40 sec	0.02 sec typical for pulsing small thrusters
Time elapsed to reach full thrust	Up to several seconds	Usually very fast, 0.004–0.080 sec
Life in space	Hours, days, or months	Up to 15 years or more in space

3. Chemical compound (or mixtures of oxidizer and fuel ingredients) capable of self-decomposition, such as hydrazine.
4. Any of the above, but with a gelling agent (these have yet to be approved for production).

A *bipropellant* consists of two separate liquid propellants, an oxidizer and a fuel. They are the most common type. They are stored separately and are mixed inside the combustion chamber (see definition of the mixture ratio below). A *hypergolic* bipropellant combination *self-ignites* upon contact between the oxidizer and the liquid fuel. A *nonhypergolic* bipropellant combination needs energy to start combusting (e.g., heat from an electric discharge) and such engines need an ignition system.

A *monopropellant* may contain an oxidizing agent and combustible matter in a single liquid substance. It may be a stored mixture of several compounds or it may be a homogeneous material, such as hydrogen peroxide or hydrazine. Monopropellants are stable at ambient storage conditions but decompose and yield hot combustion gases when heated or catalyzed in a chamber.

A *cold gas propellant* (e.g., helium, argon, or gaseous nitrogen) is stored at ambient temperatures but at relatively high pressures; it gives a comparatively low performance but allows a simple system, and is usually very reliable. They have been used for roll control and attitude control.

A *cryogenic propellant* is a liquefied gas at lower than ambient temperatures, such as liquid oxygen (-183 °C) or liquid hydrogen (-253 °C). Provisions for venting the storage tank and minimizing vaporization losses are necessary with this type.

Storable propellants (e.g., nitric acid or gasoline) are liquid at ambient temperatures and at modest pressures and can be stored for long periods in sealed tanks. *Space-storable propellants* remain liquid in the space environment; their storability depends on the specific tank design, thermal conditions, and tank pressures. An example is ammonia.

A *gelled propellant* is a thixotropic liquid with a gelling additive. It behaves in storage as a jelly or thick paint (it will not spill or leak readily) but can flow under pressure and will burn, thus being safer in some respects. Gelled propellants have been used in a few experimental rocket engines but, to date, gelled propellants have not been in production (see Eighth edition of this book).

Hybrid propellants usually have a liquid oxidizer and a solid fuel. These are discussed in Chapter 16.

For bipropellants, the propellant *mixture ratio* represents the ratio at which the oxidizer and fuel flows are mixed and react in the chamber to give the hot flow of gases. The mixture ratio r is defined as the ratio of the oxidizer mass flow rate $\dot{m}_o$ to the fuel mass flow rate $\dot{m}_f$ or

$$r = \dot{m}_o/\dot{m}_f \qquad (6-1)$$

As explained in Chapter 5, this mixture ratio affects the composition and temperature of the combustion products. It is usually chosen to give a maximum value of specific impulse (or the ratio $T_1/\mathfrak{M}$, where T_1 is the absolute combustion temperature and $\mathfrak{M}$ is the average molecular mass of the reaction gases, see Eq. 3–16 and Fig. 3–2). For a given thrust F and a given effective exhaust velocity c, the total propellant flow rate $\dot{m}$ is given by Eq. 2–6; namely, $\dot{m} = F/c$. Actual relationships between $\dot{m}, \dot{m}_o, \dot{m}_f$, and r are as follows:

$$\dot{m} = \dot{m}_o + \dot{m}_f \qquad (6-2)$$

$$\dot{m}_o = r\dot{m}/(r + 1) \qquad (6-3)$$

$$\dot{m}_f = \dot{m}/(r + 1) \qquad (6-4)$$

The above four equations are often valid when w and $\dot{w}$ (weight and weight flow rate) are substituted for m and $\dot{m}$. Calculated performance values for a number of different propellant combinations are given for specific mixture ratios in Table 5–5. Physical properties and a discussion of several common liquid propellants together with their safety concerns are described in Chapter 7.

Example 6–1. A liquid oxygen–liquid hydrogen rocket thrust chamber that produces 10,000-lbf thrust, operates at a chamber pressure of 1000 psia, a mixture ratio of 3.40, has exhaust products with a mean molecular mass $\mathfrak{M}$ of 8.90 lbm/lb-mol, combustion temperature T_1 of 4380°F, and specific heat ratio of 1.26. Determine the nozzle throat area, nozzle exit area for optimum operation at an altitude where $p_3 = p_2 = 1.58$ psia, the propellant sea-level weight and the volume flow rates, and the total propellant requirements for 2.5 min of operation. For this problem, assume that the actual specific impulse I_s is 97% of theoretical and that the thrust coefficient C_F is 98% of the ideal value.

SOLUTION. The exhaust velocity for an optimum nozzle is determined from Eq. 3–16, but with a correction factor of g_0 for the English Engineering system:

$$c = v_2 = \sqrt{\frac{2g_0 k}{k-1} \frac{R'T_1}{\mathfrak{M}} \left[1 - \left(\frac{p_2}{p_1} \right)^{(k-1)/k} \right]}$$

$$= \sqrt{\frac{2 \times 32.2 \times 1.26}{0.26} \frac{1544 \times 4840}{8.9} (1 - 0.00158^{0.206})} = 13{,}890 \text{ ft/sec}$$

The theoretical specific impulse is c/g_0, or in this optimum expansion case v_2/g_0 or $13{,}890/32.2 = 431$ sec. The actual specific impulse then becomes $0.97 \times 431 = 418$ sec. The theoretical or ideal thrust coefficient can be found from Eq. 3–30 or from Fig. 3–5 ($p_2 = p_3$) and for a pressure ratio $p_1/p_2 = 633$ to be $C_F = 1.76$. The actual thrust coefficient is slightly less, or 98%, so $C_F = 1.72$. The throat area required may be found from Eq. 3–31.

$$A_t = F/(C_F p_1) = 10{,}000/(1.72 \times 1000) = 5.80 \text{ in.}^2 \text{ (2.71 in. diameter)}$$

The optimum area ratio can be found from Eq. 3–25 or Fig. 3–4 to be 42. The exit area is $5.80 \times 42 = 244$ in.^{2} (17.6 in. diameter). At sea level, the weight density of oxygen is 71.1 lbf/ft^3 and that of hydrogen 4.4 lbf/ft^3 (see Fig. 7–1 or Table 7–1). The propellant weight flow rates (Eqs. 2–5, 6–3, and 6–4) are at sea level:

$$\dot{w} = F/I_s = 10{,}000/418 = 24.0 \text{ lbf/sec}$$

$$\dot{w}_o = \dot{w}r/(r+1) = 24.0 \times 3.40/4.40 = 18.55 \text{ lbf/sec}$$

$$\dot{w}_f = \dot{w}/(r+1) = 24/4.40 = 5.45 \text{ lbf/sec}$$

The volume flow rates are determined from their liquid weight densities (or specific gravities) and the calculated sea-level weight flow rates:

$$\dot{V}_o = \dot{w}_o/\rho_o = 18.55/71.1 = 0.261 \text{ ft}^3/\text{sec}$$

$$\dot{V}_f = \dot{w}_f/\rho_f = 5.45/4.4 = 1.24 \text{ ft}^3/\text{sec}$$

For 150 sec operation (arbitrarily allowing the equivalent of two additional seconds for start and stop transients and unavailable residual propellant), the weights and volumes of required propellant are

$$w_o = 18.55 \times 152 = 2820 \text{ lbf of oxygen}$$

$$w_f = 5.45 \times 152 = 828 \text{ lbf of hydrogen}$$

$$V_o = 0.261 \times 152 = 39.7 \text{ ft}^3 \text{ of oxygen}$$

$$V_f = 1.24 \times 152 = 188.5 \text{ ft}^3 \text{ of hydrogen}$$

Note that, because of the low-density fuel, the volume flow rate and therefore the tank volume of hydrogen is nearly five times as large compared to that of the oxidizer. Hydrogen is unique in that its liquid state has an extremely low density.

6.2. PROPELLANT TANKS

In liquid bipropellant rocket engine systems the propellants are stored in separate oxidizer and fuel tanks within the flying vehicle. Monopropellant rocket engine systems have, by definition, only one propellant tank. There are usually also one or more high-pressure auxiliary gas tanks, the gas being used to pressurize the propellant tanks. However, as will be discussed in Section 6.5, there are also tank pressurization schemes using heated gas from the engine voiding the need for extra heavy, high-pressure gas storage tanks. Tanks can be arranged in a variety of ways, and tank design, shape, and location can be used to apply some control over the change in the location of the vehicle's center of gravity. Typical arrangements are shown in Fig. 6–2 (concepts for positive expulsion are shown later on Fig. 6–4). Because propellant tanks also have to fly, their mass cannot be neglected and tank materials can be highly stressed. Common tank materials are aluminum, stainless steel, titanium, alloy steels, and fiber-reinforced plastics (with an impervious thin inner liner of metal to prevent leakage through the pores of the fiber-reinforced walls). Chapter 8 of Ref. 6–1 describes the design of propellant tanks.

Any gas volume above the propellant in sealed tanks is called the *ullage*. It is a necessary space that allows for thermal expansion of the propellant liquids, for the accumulation of gases that were originally dissolved in the propellant or, with some propellants, for gaseous products from any slow chemical reactions within the propellant during storage. Depending on the storage temperature, range, the propellants' coefficient of thermal expansion, and on the particular application, ullage volumes usually range between 3 and 10% of tank volumes. Once loaded, ullage volume (and, if not vented, also pressure) will change as the bulk temperature and density of the stored propellant varies.

The *expulsion efficiency* of a tank and/or propellant piping system is the amount of propellant that can be expelled or available for propulsion divided by the total amount of propellant initially present. Typical values are 97 to 99.7%. Here losses consist of unavailable propellants left in tanks after rocket operation, trapped in grooves or corners of pipes, fittings, filters, and valves, or wetting the walls, retained

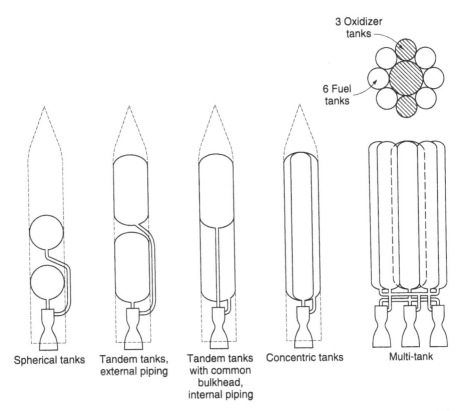

FIGURE 6–2. Simplified sketches of typical tank arrangements for large turbopump-fed liquid bipropellant rocket engines.

by surface tension, or caught in instrument taps. Such *residual propellant* is not available for combustion and must be treated as inert mass, causing the vehicle mass ratio to decrease slightly. In the design of tanks and piping systems, an effort is made to minimize such residual propellant.

An optimum shape for propellant tanks (and also for gas pressurizing tanks) is spherical because for a given volume it results in a tank with the least mass. Small spherical tanks are often used with reaction control engine systems, where they can be packaged with other vehicle equipment. Unfortunately, larger spheres, which would be needed for the principal propulsion systems, do not fill up all the available space in a flight vehicle. Therefore, larger tanks are often made integral with the vehicle fuselage or wing. Most are cylindrical with half ellipses at the ends, but they can also be irregular in shape. A more detailed discussion of tank pressurization is given in Section 6.5.

Cryogenic propellants cool the tank wall temperature far below the ambient temperature. This causes condensation of moisture from air surrounding the tank's exposed sides and usually also formation of ice during and prior to launch. Ice formation is undesirable because it increases the vehicle inert mass. Also, as pieces of ice are shaken off or tend to break off during the initial flight, they can damage the vehicle; in one notable example, pieces of ice from the Space Shuttle's cryogenic tank hit the Orbiter vehicle.

For extended storage periods, cryogenic tanks are usually thermally insulated; any porous external insulation layers have to be sealed to prevent moisture from being condensed inside the insulation. With liquid hydrogen it is even possible to liquefy or solidify the ambient air on the outside of the fuel tank. Even with the best thermal insulation and low-conductivity structural tank supports, it has not been possible to prevent the continuous evaporation of cryogenic fluids, and therefore they cannot be kept in a vehicle for more than perhaps a few days without refilling. For vehicles that need to be stored after filling or to operate for longer periods, a storable propellant combination is preferred.

Prior to loading any cold cryogenic propellant into a flight tank, it is necessary to remove or evacuate air from the tank and propellant passages to avoid forming solid air particles and ice from any existing moisture. These frozen particles would plug up injection holes, cause valves to freeze shut, and/or prevent valves from being fully closed. Tanks, piping, and valves also need to be chilled or cooled down before they can contain cryogenic liquids without excessive bubbling. This is usually done by admitting an initial amount of cryogenic liquid to absorb the heat from the relatively warm hardware prior to engine start. During this initial cool-down, the propellant is vaporized and vented through appropriate vent valves and is not available for propulsion.

When a tank or any segment of piping containing low-temperature cryogenic liquid is sealed for an extended period of time, ambient heat and ambient-temperature surrounding hardware will cause evaporation, and this will greatly raise the pressure, which may exceed the strength of the sealed container; controlled self-pressurization can be difficult to achieve. Uncontrolled self-pressurization will cause failure, usually as a major leak or even a tank explosion. All cryogenic tanks and piping systems are therefore vented during countdown on the launch pad, equipped with pressure safety devices (such as burst diaphragms or relief valves), and evaporated propellant is allowed to escape from its container. For long-term storage of cryogenic propellants in space (or on the ground) some form of a powered refrigeration system is needed to recondense the vapors and minimize evaporation losses. Cryogenic propellant tanks are usually refilled or topped off just before launch to replace the evaporated and vented cool-down propellant. When such a tank is pressurized, just before launch, the boiling point is slightly raised and the cryogenic liquid can better absorb any heat being transferred to it during the several minutes of rocket firing.

There are several categories of tanks in liquid propellant propulsion systems. With few exceptions, the relevant pressure values are listed below.

1. For *pressurized feed systems,* the *propellant tanks* typically operate at an average pressure between 1.3 and 9 MPa or about 200 to 1800 lbf/in.^{2} Such tanks have thick walls and are heavy.

2. For *high-pressure stored gases* (used to expel the propellants), the tank pressures need to be much higher, typically between 6.9 and 69 MPa or 1000 to 10,000 lbf/in.^{2} These tanks are usually spherical for minimum inert mass. Several small spherical tanks can be connected together. In some vehicles, the smaller high-pressure gas tanks are placed within the liquid propellant tanks.

3. For *turbopump feed systems,* it is necessary to pressurize the propellant tanks slightly (to suppress pump cavitation as explained in Sections 10.3 and 10.4) to average values of between 0.07 and 0.34 MPa or 10 to 50 lbf/in.^{2} These low pressures allow thin tank walls, and therefore turbopump feed systems have relatively low inert tank mass.

During flight, liquid propellant tanks can be difficult to empty under side accelerations, zero-g, or negative-g conditions. Special devices and special types of tanks are needed to operate under these conditions. Some of the effects that have to be overcome are described below.

Oscillations and side accelerations of vehicles in flight may cause *sloshing* of the stored liquid, very similar to a glass of water that is being jiggled. In anti-aircraft missiles, for example, side accelerations can be large and may initiate severe sloshing. Typical analysis of sloshing can be found in Refs. 6–4 and 6–5. When the tank is partly empty, sloshing can uncover a tank's outlet and allow gas bubbles to enter into the propellant tank discharge line. These bubbles may cause major combustion problems in the thrust chambers; the aspiration of bubbles or the uncovering of tank outlets by liquids therefore must be avoided. Sloshing also causes irregular shifts in the vehicle's center of gravity making flight control difficult.

Vortexing also allows gas to enter the tank outlet pipe; this phenomenon is similar to the Coriolis force effects in bathtubs being emptied and can be augmented by vehicle spins or rotations in fight. A series of internal baffles can be used to reduce the magnitude of sloshing and vortexing in tanks with modest side accelerations. A positive expulsion mechanism described below can prevent gas from entering the propellant piping under multidirectional major accelerations or spinning (centrifugal) acceleration. Both the vortexing and sloshing can also greatly increase unavailable or residual propellants and thus some reduction in vehicle performance.

In the gravity-free environment of space, stored liquids will float around in a partly emptied tank and may not always cover the tank outlet, thus allowing gas to enter the tank outlet or discharge pipe. Figure 6–3 shows how gas bubbles have no particular orientation. Various devices have been developed to solve this problem: namely, *positive expulsion devices* and *surface tension devices*. The positive expulsion tank design may include movable pistons, inflatable flexible bladders, or thin movable and flexible metal diaphragms. Surface tension devices (such as 200-mesh screens) rely on surface tension forces to keep the outlet covered with liquid. See Ref. 6–5. Alternatively, a small acceleration may be applied in a zero-g space environment (using supplementary thrusters) in order to orient the liquid propellant in the tank.

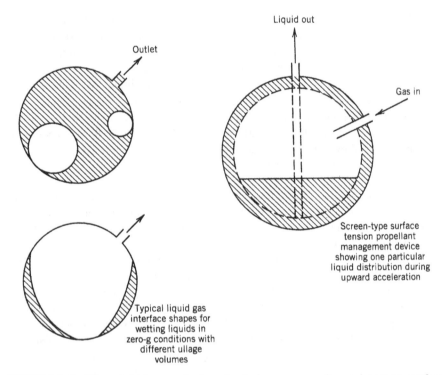

FIGURE 6–3. Ullage bubbles can float randomly in a zero-gravity environment; surface tension device are needed to keep tank outlets covered with liquid.

Several basic types of *positive expulsion devices* have been used successfully in propellant tanks with pressurized feed systems. They are compared in Table 6–2 and shown in Fig. 6–4 for simple tanks. These devices mechanically separate the pressurizing gas from the liquid propellant in the propellant tank. Separation is useful for these reasons:

1. It prevents pressurizing gas from dissolving in the propellant and propellant vapors from mixing with the gases. Any dissolved pressurizing gas dilutes the propellant, reduces its density as well as its specific impulse, and makes the pressurization inefficient.

2. It allows moderately hot and reactive gases (such as generated by gas generators) to be used for pressurization, thus permitting a reduction in the pressurizing gas system mass and volume. Mechanical separation prevents chemical reactions between hot gases and propellants, prevents any gas from being dissolved in the propellant, prevents propellant vapors from diffusing into the (unheated) pressurant lines and freeze, and reduces the heat transfer to the liquid.

TABLE 6–2. Comparison of Propellant Positive Expulsion Methods for Spacecraft Hydrazine Tanks

Selection Criteria	Single Elastomeric Diaphragm (Hemispherical)	Inflatable Dual Elastomeric Bladder (Spherical)	Foldable Metallic Diaphragm (Hemispherical or cylindrical)	Piston or Bellows	Rolling Diaphragm	Surface Tension Screens
Application history	Extensive	Extensive	Limited	Extensive in high acceleration vehicles	Modest	Extensive
Weight (normalized)	1.0	1.1	1.15	1.2	1.0	0.9
Expulsion efficiency	Excellent	Very good	Good	Excellent	Very good	Good or fair
Maximum side acceleration	Low	Low	Medium	High	Medium	Lowest
Control of center of gravity	Poor	Limited	Good	Excellent	Good	Poor
Long service life	Excellent	Excellent	Excellent	Very good	Good	Excellent
Preflight check	Leak test	Leak test	Leak test	Leak test	Leak test	None
Disadvantages	Chemical deterioration	Chemical deterioration; fits only into a few tank geometries	High-pressure drop; fits only certain tank geometries; high weight	Potential seal failure; critical tolerances on piston seal; heavy	Weld inspection is difficult; adhesive (for bonding to wall) can deteriorate; bellows have high residuals	Limited to low accelerations

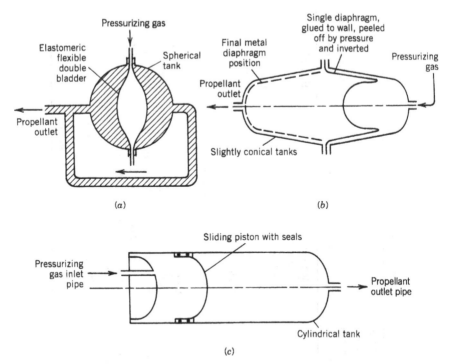

FIGURE 6–4. Sketches of three concepts of propellant tanks with positive expulsion: (a) inflatable dual bladder; (b) rolling, peeling diaphragm; (c) sliding piston. As the propellant volume expands or contracts with changes in ambient temperature, the piston or diaphragm will also move slightly and the ullage volume will change during storage.

3. In some cases, the ullage of tanks containing toxic liquid propellant must be vented without spilling any toxic liquid propellant or its vapor. For example, in servicing a reusable rocket, the tank pressure needs to be relieved without venting or spilling any potentially hazardous materials.

A *piston expulsion* device permits the center of gravity (CG) to be accurately controlled, making its location known during engine operation. This is important in rockets with high side accelerations such as anti-aircraft missiles or space defense interceptor missiles, where the thrust vector needs to go through the vehicle's CG—if the CG is not well known, unpredictable turning moments may be imposed on the vehicle. A piston also prevents sloshing and/or vortexing.

Surface tension devices use capillary attraction for supplying liquid propellant to the tank outlet pipe. These devices (see Fig. 6–3) are often made of very fine (300 mesh) stainless steel wire woven into a screen and formed into tunnels or other shapes (see Ref. 6–5). These screens are located near the tank outlet and, in some tanks, tubular galleries are used designed to connect various parts of the tank volume to the outlet pipe sump. These devices work best in relatively low-acceleration environments, where surface tension forces can overcome the inertia forces.

The combination of surface tension screens, baffles, sumps, and traps is called a *propellant management device*. Although not shown in any detail, they are included inside the propellant tanks of Figs. 6–3 and 6–14.

During gravity-free flights, suddenly accelerated from relatively large thrusts, high forces can be imposed on tanks and thus on the vehicle by strong liquid sloshing motions or by sudden changes in position of the liquid in a partly empty tank. The resulting forces will depend on the tank geometry, baffles, ullage volume and its initial propellant location, and on the acceleration magnitude and direction. Such forces can be large and may cause tank failure.

6.3. PROPELLANT FEED SYSTEMS

Propellant feed systems have two principal functions: (1) to raise the pressure of the propellants and (2) to supply them at design mass flow rates to one or more thrust chambers. The energy for these functions comes either from a high-pressure gas, centrifugal pumps, or a combination of the two. The selection of a particular feed system and its components is governed primarily by the rocket application, duration, number or type of thrust chambers, past experience, mission and by the general requirements of simplicity of design, ease of manufacture, low cost, and minimum inert mass. A classification of several of the more important types of feed system is shown in Fig. 6–5, and some types are discussed in more detail in other parts of this book. All feed systems consist of piping, a series of valves, provisions for filling and usually also for removing (draining and flushing) the liquid propellants, filters, and control devices to initiate, stop, and regulate their flow and operation. See Ref. 6–1.

In general, *gas pressure feed systems* give superior performance to turbopump systems when the vehicle's total impulse or propellant mass is relatively low, the chamber pressure is relatively low, the engine thrust-to-weight ratio is low (usually less than 0.6), and when there are repeated short-duration thrust pulses; here, the usually heavy-walled propellant tanks and the pressurizing gas constitute the major inert mass of the engine system. In a *turbopump feed system*, propellant tank pressures are much lower (by a factor of 10 to 40) and thus vehicles' tank masses are much lower (by the same factor). Turbopump systems usually give superior performance when the vehicle's total impulse is relatively large, the chamber pressure is high, and the mission velocity is high.

Local Pressures and Flows

Key parameters for any feed system's description in liquid propellant rocket engines involve flow magnitudes (oxidizer and fuel flow including subsystems and thrust chamber flow passages) together with local pressures (pressurizing gas subsystems). An inspection of the flow diagram of a relatively simple rocket engine with a pressurized feed system, similar to Fig. 1–3, shows that the gas flow splits into two branches, and the propellant flow splits into pipes leading to each of the thrust chambers. The highest pressure resides in the high-pressure gas supply tank. The pressure drops along the pressurizing gas subsystem (pipes, valves, regulator) and then drops

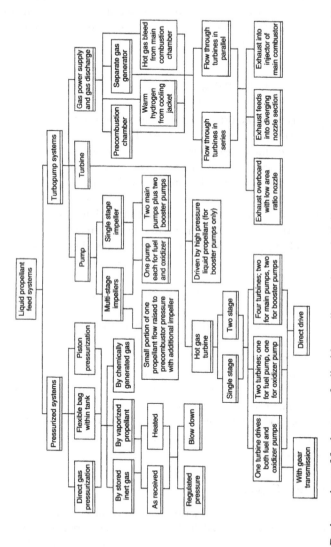

FIGURE 6–5. Design options of feed systems for liquid propellant rocket engines. The more common types are designated with a double line at the bottom of their boxes.

further in the liquid propellant flow subsystems (more pipes, valves, filters, injector, or cooling jacket) as the liquid propellants flow into the thrust chamber, where they burn at the chamber pressure. As shown in Chapter 3, the gas pressure always reaches a minimum at the nozzle exit. It is possible to predict relevant pressure drops and flow distributions when enough information is known about all components and their flow passages. If such analysis can be validated by data from previous pertinent tests, or prior proven rocket engine flights, it will have a higher confidence factor. Most rocket engine design organizations have developed software for estimating pressures and flows at different parts of an engine.

Knowing local flows and the local pressures is important for the following reasons:

1. Such information is used in the *stress analysis* and sometimes in the thermal analysis of related components and subsystems.

2. They are needed for *rocket engine calibrations* so that engines will operate at the intended mixture ratio, chamber pressure, and/or thrust (such tasks are accomplished by using control devices or simple orifices to adjust the pressures). Calibration also requires the proper balance of flows and pressures. For a feed system with one or more turbopumps, such as shown in Fig. 1–4, one needs to also include the rise in pressure as propellants flow through a pump. Furthermore, feed system analyses and calibration with turbopumps become more complex when there are other combustion devices (gas generators or preburners). A more detailed discussion of engine system calibration is given in Section 11.5.

3. The measurement of actual flows and key local pressures during engine ground tests (or actual flight operation) and subsequent comparison with predicted values makes it possible to identify discrepancies between practice and theory. Such discrepancies can yield clues to possible malfunctions inadequate designs and/or fabrications, all of which may then be possibly corrected. If actual values can be compared with analysis in real time, some experimental test hardware may be saved from self-destruction. This can also be the basis for a real-time engine health monitoring system, which is discussed in Sections 11.5 and 21.3.

Sometimes analyses are done for strictly transient conditions, such as starting or shutdown, or during changes in thrust (throttling). Transient analyses must also provide information for the filling of empty propellant flow passages at different propellant temperatures, for water hammer, valve reaction time, and the like.

6.4. GAS PRESSURE FEED SYSTEMS

One of the simplest and most common means of pressurizing liquid propellants and force them out of their respective tanks is to use a high-pressure gas. Rocket engines with pressurized gas feed systems can be very reliable. References 6–1, 6–6, and 6–7 give additional information. A rocket engine with a gas-pressurized feed system was the first to be tested and flown (1926). There are two common types of pressurized feed systems both still used often today.

The first uses a *gas pressure regulator* in the gas feed line with the engine operating at essentially constant tank pressure and nearly constant thrust. This is shown schematically in Fig. 1–3 and consists of a high-pressure gas tank, a gas starting valve, a gas pressure regulator, propellant tanks, propellant valves, and feed lines. Additional components, such as filling and draining provisions, check valves, filters, flexible elastic bladders for separating the liquid from the pressurizing gas, and pressure sensors or gauges, are also often incorporated. After all tanks are filled, the high-pressure gas valve in Fig. 1–3 is remotely actuated and admits gas through the pressure regulator at a constant pressure to the propellant tanks. Check valves prevent mixing of the oxidizer with the fuel, particularly when the unit is not in an upright position. Propellants are fed to the thrust chamber by opening appropriate valves. When the propellants are completely consumed, the pressurizing gas can also be used to scavenge and clean lines and valves of much of the liquid propellant residue. Any variations in this system, such as the combination of several valves into one or the elimination and addition of certain components, depend on the application. If a unit is to be used and flown repeatedly, such as a space-maneuver rocket, it may include several additional features such as a thrust-regulating device and a tank level gauge; these will not be found in expendable, single-shot units, which may not even have a tank-drainage provision.

The second common type of gas pressure feed system is called a *blow-down feed system*. It is shown in Fig. 6–6 and discussed in Refs. 6–7 and 6–8. Here, the

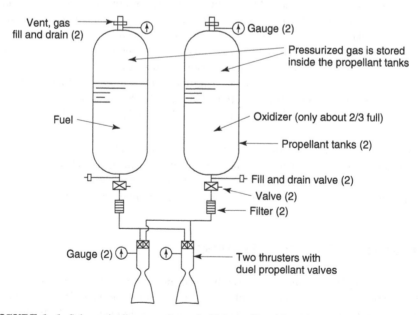

FIGURE 6–6. Schematic diagram of a typical bipropellant blow-down pressurized gas feed system with two thrusters.

propellant tanks are larger because they store not only the propellants but also the pressurizing gas at an initial maximum propellant tank pressure. There is no separate high-pressure gas tank and no pressure regulator. The expansion of the gas already in the tanks provides for the expulsion of the propellants. Blow-down systems can be lighter than a regulated pressure system, but gas temperatures, pressures and the resulting thrust all steadily decrease as propellants are consumed. A comparison of these two common types is shown in Table 6–3.

Different bipropellant pressurization concepts are evaluated in Refs. 6–1, 6–6, 6–7, and 6–8. Table 6–4 lists various optional features aimed at satisfying particular design goals. Many of these features also apply to pump-fed systems, which are discussed in Section 6.6. Individual engine feed systems have some but certainly not all of the features listed in Table 6–4. Monopropellants gas pressure feed systems are simpler since there is only one propellant, thus reducing the number of pipes, valves, and tanks.

An example of a multi-thruster liquid propellant rocket engine with an intricate gas pressurized feed system is the MESSENGER (MErcury Surface, Space ENvironment, GEochemistry, and Ranging), Refs. 6–9, 6–10, planetary space probe. Launched on August 3, 2004, this propulsion system enabled two flybys of Venus, and three flybys of the planet Mercury, before going into orbit around Mercury

TABLE 6–3. Comparison of Two Types of Gas Pressurization Systems

Type	Regulated Pressure	Blowdown
Pressure/thrust	Stays essentially constant[a]	Decreases as propellant is consumed
Gas storage	In separate high-pressure tanks	Gas is stored inside propellant at tank pressure with large ullage volume (30–60%)
Required components	Needs regulator, filter, gas valves, and gas tank	Larger, heavier propellant tanks
Advantages	Nearly constant-pressure feed gives essentially constant propellant flow and approximately constant thrust, I_s and r	Simpler system. Less gas required. Can be less inert mass.
	Better control of mixture ratio	No high-pressure gas tank
Disadvantages	Slightly more complex	Thrust decreases with burn duration
	Regulator introduces a small pressure drop	Somewhat higher residual propellant due to less accurate mixture ratio control
	Gas stored under high pressure, often for a long time	Thruster must operate and be stable over wide range of thrust values and modest range of mixture ratios
	Requires more pressurizing gas	Propellants stored under pressure; slightly lower I_s toward end of burning time

[a]See section 6.9 for valves and pressure regulators

TABLE 6–4. Typical Features of Liquid Propellant Feed Systems

---

Enhance Safety

Check valves to prevent backflow of propellant into the gas tank and inadvertent mixing of propellants inside flow passages.

Pressurizing gas should be inert, clean, and insoluble in propellant.

Burst diaphragms or isolation valves to isolate the propellants in their tanks and positively prevent leakage into the thrust chamber or into the other propellant tank during storage.

Isolation valves to shut off a section of a system that has a leak or malfunction.

Sniff devices to detect leak of hazardous vapor in some vehicle compartments or ground test areas.

Features that prevent an unsafe condition to occur or persist and shut down the engine safely, such as relief valves or relief burst diaphragms (to prevent tank overpressurization), or a vibration monitor to shut off operation in the case of combustion instability.

Provide Control

Valves to control pressurization and flow to the thrust chambers (start/stop/throttle).

Sensors to measure temperatures, pressures, valve positions, thrust, etc., and computers to monitor/analyze/record system status. Compare measured values with analytical estimates, issue command signals, and correct if sensed condition is outside predetermined limits.

Manned vehicle can require a system status display and command signal override.

Fault detection, identification, and automatic remedy, such as shutoff isolation valves in compartment in case of fire, leak, or disabled thruster.

Control thrust (throttle valve) to fit a desired thrust–time flight profile.

Enhance Reliability and Life

Fewest practical number of components/subassemblies.

Filters to catch dirt in propellant lines, which could prevent valves from closing or might cause small injector holes to be plugged or might cause bearings to gall.

Duplication of key components, such as redundant small thrusters, regulators, or check valves. If malfunction is sensed, then often remedial action is to shut down the malfunctioning component and activate the redundant spare component.

Heaters to prevent freezing of moisture or low-melting-point propellant.

Long storage life—use propellants with little or no chemical deterioration and no reaction with wall materials, valves, pipes, gaskets, pressurizing gas, etc.

Operate over the life of the mission and its duty cycle, including long-term storage.

Provide for Reusability, If Required

Provisions to drain propellants or pressurants remaining after operation.

Provision for cleaning, purging, flushing, and drying the feed system and refilling propellants and pressurizing gas.

Devices to check functioning of key components prior to next operation.

Features to allow checking of engine calibration and leak testing after operation.

Access for inspection devices and for visual inspection of internal surfaces or components for damage or failure. Look for cracks in nozzle throat inner walls and in turbine blade roots.

Enable Effective Propellant Utilization

High tank expulsion efficiency with minimum residual, unavailable propellant.

Lowest possible ambient temperature variation and/or matched propellant property variation with temperature so as to minimize mixture ratio change and residual propellant.

Alternatively, measure remaining propellant in tanks (using special gauges) and automatically adjust mixture ratio (throttling) to minimize residual propellant.

Minimize pockets in the piping and valves that cannot be readily drained or flushed.

on March 13, 2011. This mission ended in April 2015 when the probe ran out of fuel for the attitude control system that kept the antenna pointed to Earth. A flow diagram is shown in Figure 6–7 and selected data is in Table 6–5 (Refs. 6–9, 6–10). The main bipropellant thruster is identified in the diagram as the LVA (large velocity adjustment)—with nominally 145 lbf thrust; it provided the changes in the flight path; four of the larger monopropellant thrusters (nominally 5 lbf thrust each) provided the pitch and yaw control and four of the 12 smaller monopropellant thrusters provided the roll control (nominally 1 lbf of thrust each) during LVA operation. All 12 small thrusters provided vehicle attitude control whenever needed. As can be seen in Figure 4–14, it requires a minimum of 12 small thrusters to achieve pure torques for vehicle rotation about three mutually perpendicular axes. For very precise rotary orientation, the MESSENGER spacecraft also had a reaction wheel system (see Reaction Control System in Section 4.5). As seen by the dotted lines in Fig. 6–7, there were four modules, each with one MR-106E and one MR-111C, all pointing in the same axial direction as the LVA (also called Leros 1b).

All monopropellant thrusters were radiation cooled and operated in either steady state or pulse modes (as explained below), and most had redundant valve seats for protection against leakage and an inlet filter to keep particulate away from the valve seats. The propulsion system of Fig.6–7 shows two hydrazine fuel tanks (FT#1 and FT#2), one oxidizer tank (OT), one high-pressure helium (GHe) tank, redundant check valves, filters, latch valves,* pyrotechnic isolation valves, service valves, redundant temperature sensors and redundant pressure sensors (redundant here means if one malfunctions, its reading can be ignored—this improves reliability). There is also a small, spherical, positive expulsion auxiliary fuel tank (AFT) with a flexible diaphragm separating the helium gas and the propellants. It was used to start four of the small monopropellant thrusters (firing in a direction parallel to the LVA thruster) in the gravity-free space environment. Their acceleration would cause the liquids in the three main tanks to be oriented and cover the respective tank outlets with propellant. This maneuver avoided letting helium gas from the tanks enter the thruster. It is known that gas bubbles being supplied to an operating thruster may cause uncontrolled and irregular thrust and in bipropellants, bubbles can also initiate destructive combustion instability, so any inadvertent feeding of gas to the thrusters must be avoided. The auxiliary tank was refilled with propellant from the main tanks during some of the axial maneuvers which enabled propellant availability for propellant settling just before the next maneuvers.

Not shown in Fig. 6–7 is the electric subsystem (heaters, switches, valve position indicators, controllers, etc.). Electrical heaters were provided on all components and propellant lines which were exposed to hydrazine to prevent freezing in the space environment. Care was taken to make sure locations where hydrazine vapor might migrate were also heated. A catalyst bed temperature of at least 250 °F allowed faster catalyst reactions and provided longer catalyst bed lifetimes.

---

*A "latch valve" requires power only to open and close it. Its valve seat is "latched" or locked in either the open or closed position at all other times, and the valve does not require power to maintain it in either position.

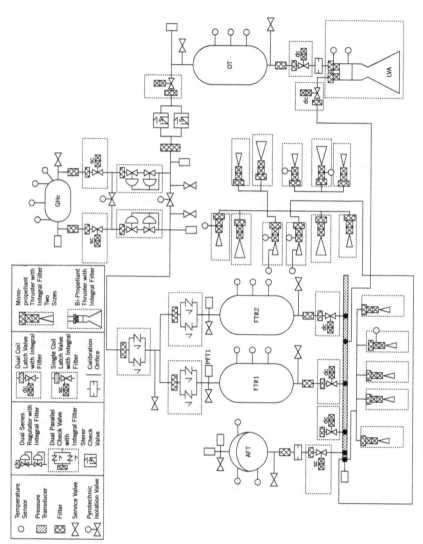

FIGURE 6–7. This schematic of the Liquid Rocket Propulsion System for the MESSENGER Space Probe shows the designed component redundancy and duplication that enhances engine system reliability for many years of use in space. Some diagram abbreviations defined in the text. Courtesy of Aerojet Rocketdyne and NASA

TABLE 6–5. Selected Data on the Gas Pressurized Propulsion System of the MESSENGER Space Probe

Designation	Leros-1b Main Thrust Chamber (M006)	MR-106E Attitude Control Thrusters	MR-111C Attitude Control Thrusters
Propellant	Hydrazine/N_2O_4	Hydrazine	Hydrazine
Number of units	1	4	12
Thrust chamber cooling	Fuel film cooled and radiation cooled	Radiation cooled	Radiation cooled
Thrust (per nozzle) (lbf)	145	6.9-2.6; Nominal 5 lbf	1.2-0.3; nominal 1.0 lbf
Feed pressure (psia)	300-218 psia	350-100	450-50
Chamber pressure, nominal (psia)	125 nominal	180-65	200-35
Vacuum specific impulse (lbf-sec/lbm)	318	235-229	229-215
Nozzle area ratio	150:1	60:1	74:1
Mixture ratio (oxidizer flow/fuel flow mass ratio)	0.85 +/– 0.02	N/A	N/A
Minimum burn time (sec)	N/A	0.016	0.015
Maximum burn time (sec)	2520 (single burn) 20,500 (cumulative)	2000 (single burn) 4670 (cumulative)	5000 (single burn)
Number of starts, cumulative	70	>1000	>1500
System Parameters			
Oxidizer mass (lbm)	510		
Fuel mass (lbm)	805		
Pressurant (helium) mass (lbm)	5.4		
Propulsion system dry mass (lbm)	180		

Data from Refs. 6–9, 6–10

Some of the MESSENGER vehicle maneuvers require axial thrust, which is lower than that provided by the LVA thruster (145 lbf nominal). This can be achieved by operating the four MR-106E thrusters (nominally 5 lbf each for a total of 20 lbf of thrust) together with the four MR-111C (nominally 1 lbf of thrust each) in the same module for pitch and yaw control. The MESSENGER rocket propulsion system has series-redundant gas pressure regulators for each propellant and operates most of the time with a nearly constant flow gas pressure feed system; however, for parts of the operation it also operates in a blow-down pressurization mode. The small auxiliary fuel tank can be recharged with hydrazine from the main fuel tanks during flight.

Pulsing of thrusters is common in small monopropellant units. A short thrust period (typically 15 to 100 milliseconds) is followed by a period of no thrust (typically 15 to 300 milliseconds), and this can be frequently repeated. An ideal simple thrust profile is shown in Problem 6–3. One merit of pulsing the thrust is the ease for changing the total impulse to fit particular flight maneuvers by changing the number of pulses; alternatively one can change the pulse duration or change the time between pulses. Also pulsing valves are less costly than throttling valves. The disadvantage of pulsing, when compared to continuous nonpulsing operations of the same peak thrust, is a decrease of specific impulse and that requires more propellant usage and lengthens the operation duration. A cluster of thrusters (e.g., the four MR-106L thrusters in Fig. 6–7) can also be "off-pulsed" to balance the spacecraft. In this case, the thruster with the lowest thrust (or with the least influence on the spacecraft's position) will fire at steady state and the other thrusters will turn off for varying amounts of time in order to guide the spacecraft in the commanded direction.

Compartmented propellant metal tanks with anti-slosh and anti-vortex baffles, sumps, and a surface tension propellant retention device allow propellant to be delivered independent of the propellant load, the orientation, or the acceleration environment (some of the time in zero-*g*). Gauges in each tank allow a determination of the amount of propellant remaining, and they also indicate leaks. Safety features can include sniff lines at each propellant valve actuator to sense leakage. With cryogenic propellants electrical heaters are provided at propellant valves, certain lines, and injectors to prevent fuel freezing or moisture turning into ice.

Some pressure feed systems can be prefilled with storable propellant and pressurizing agent at the factory and stored in readiness for operation. Compared to solid propellant rocket units, these prepackaged liquid propellant pressurized feed systems offer advantages in long-term storability, random restarts, and better resistance to transportation vibration or shock.

Thrust levels in rocket propulsion systems are a function of propellant flow magnitudes which in pressurized gas feed systems are governed by the gas pressure regulator settings. The propellant *mixture ratio* in this type of feed system is a function of the hydraulic resistance of the liquid propellant lines, cooling jacket, and injector, and the mixture ratio is usually adjusted by means of variable restrictors or fixed orifices. Further discussion of the adjusting of thrust and mixture ratio can be found in Sections 11.5 and 11.6 and in Example 11–2.

6.5. TANK PRESSURIZATION

As stated earlier, the objective of *feed systems* is to move propellants under pressure from propellant tanks to thrust chamber(s). The *tank pressurization system* is that part of the feed system that provides such a propellant expellant gas. See Refs. 6–1, 6–6, 6–7, and 6–10. As was described in Section 6.3, there are two types: (1) in a pressurized gas feed system, a relatively high-pressure gas displaces the propellants from the tanks, and (2) in a pumped feed system (described in the next section) the main energy for feeding the propellants comes from one or more pumps. It requires lower gas pressures in the tanks to move the propellants to the pump inlet, and it helps to avoid pump cavitation.

There are several *sources of pressurizing gas* used in tank pressurization systems.

1. *High-pressure inert gases* stored at ambient temperature are the most common. Typical gases are helium, nitrogen, and air. Table 6–3 shows a comparison of the regulated pressure system (see Fig. 1–3) and the blow-down system (see Fig. 6–6). This is discussed further in this section. When gases expand adiabatically, their temperature drops.

2. *Heated high-pressure inert gases* (typically 200 to 800 °F or 93 to 427 °C) reduce the amount of required gas and thus the inert mass of the pressurizing system. Examples are gases heated by a heat exchanger with the warm exhaust from a gas generator or a turbine or with electrical heaters inside the gas tank.

3. *Gases created by a chemical reaction* using either liquid bipropellants or a monopropellant, or alternatively a solid propellant, all at mixture ratios or compositions that result in "warm gas." Uncooled hardware can be used. The term *warm gas* (say 400 to 1600 °F or 204 to 871 °C) distinguishes such gas from the "hot gas" (4000 to 6000 °F, or 2204 to 3319 °C) in the main combustion chamber. Chemically generated warm gases usually result in tank pressurization systems lighter than heated inert gas systems, particularly for high total impulse applications. These gases may also come from two separate small gas generators for tank pressurization—one produces fuel-rich gas for pressurizing the fuel tank and the other feeds oxidizer-rich pressurizing gas into the oxidizer tank. Such a scheme was first used in the United States on the Bomarc rocket engine around 1952 and in the Russian RD-253 around 1961. Another common warm-gas scheme has been to bleed a small amount of warm gas from the engine's gas generator or preburner. If such a gas is fuel rich, then it can only be used for pressurizing the fuel tank, and it may need to be cooled. Chemical reaction gases are typically at 1200 to 1700 °F or 649 to 921 °C, a range within the gas temperatures allowable for most alloy turbines and steel tanks. Catalyzed decomposed hydrazine products have successfully pressurized liquid hydrazine tanks. With aluminum propellant tanks it is often necessary to cool the warm gas further; many aluminum alloys melt around 1100 °F (593 °C). This gas cooling has been accomplished using a heat exchanger with one of the propellants. Solid propellant gas generators have been used in experimental liquid propellant rocket engines, but to date none are known to have been adapted for a production flight vehicle. One clever system developed in the former Soviet Union uses two rocket engines operating simultaneously on the same vehicle. The larger engine has bleeds gases from an oxidizer-rich preburner to pressurize a common oxidizer tank. The second engine feeds four smaller hinge-mounted vernier thrust chamber used for attitude control and also for extra thrust; their fuel-rich gas generator has a bleed of gas for pressurizing the common fuel tank. Ref. 6–11.

4. *Evaporated flow of a small portion of a cryogenic liquid propellant*, usually liquid hydrogen or liquid oxygen, by applying heat from a thrust chamber cooling

jacket or from turbine exhaust gases (with heat exchangers), and then using a part or all of this evaporated flow for tank pressurization. Orifices or pressure regulators may be needed for attaining the desired tank pressure and mass flow rates. This scheme was used for the fuel tank of the Space Shuttle Main Engine (now the RS-25 engine), and the tap-off stub for pressurizing gas can be seen later in Fig. 6–11 at the turbine exhaust manifold of the fuel booster pump. The oxygen tank was pressurized by gasified liquid oxygen, which was tapped off the discharge side of the main oxygen pump and heated in a heat exchanger around the turbine of the main oxygen pump, as shown in Figure 6–11.

5. *Direct injection* of a small stream of hypergolic fuel into the main oxidizer tank and a small flow of hypergolic oxidizer into the fuel tank has been tried by several countries but with limited success. It is really another form of chemical gas generation.

6. *Self-pressurization of cryogenic propellants* by evaporation is feasible but can be difficult to control. Experience here is limited.

In order to design or analyze any pressurizing system it is necessary to have relevant information about the tank and the engine. This can include basic engine parameters, such as propellant flow, thrust, duration, pulse width, propellant tank volume, percent ullage of tank volume, storage temperature range, propellant and pressurizing gas properties, propellant tank pressure, gas tank pressure, and/or amount of unavailable residual propellant. For many of these items, nominal, maximum, and minimum values may be needed.

Factors Influencing the Required Mass of Pressurizing Gas

A key task in the analysis and design of tank pressurization systems is determining the required mass of pressurizing gas. Many different factors influence this mass, and some of them can be quite intricate, as shown in Ref. 6–1. The gaseous mass may be estimated using simplifying assumptions, but it is always more accurate when based on actual test results and/or data from similar proven pressurizing systems. Following are some *influencing factors* that should be accounted for:

1. *Evaporation of propellant* at the interface between the pressurizing gas and the liquid propellant is a key phenomenon. The evaporated propellant dilutes the gas and changes its expansion properties. This change depends on the temperature difference between the gas and the liquid, sloshing, vapor pressure of the propellant, turbulence, and local gas impingement velocities. Furthermore, any propellant film on those portions of the tank walls and baffles that are above the liquid level will also evaporate. "Warm gases" (e.g., a bleed from a gas generator) will heat the top layer of the liquid propellant and increase propellant evaporation reducing its local density. Because with cryogenic propellants the gas is always warmer than the liquid, as the gas cools, more liquid evaporates.

2. The temperature of those propellant tank walls which form part of exterior vehicle surfaces exposed to the atmosphere are affected by *aerodynamic heating*,

which may vary during flight. Such heating can increase the gas temperature as well as the liquid propellant temperature, augmenting liquid evaporation.

3. The *solubility of a gas in a liquid* is affected by temperature and pressure. For example, because nitrogen gas is soluble in liquid oxygen, it requires approximately four times as much nitrogen gas to pressurize liquid oxygen as an equivalent volume of water. This dilutes the oxygen and causes a small performance loss. Helium is a preferred pressurant gas because it is least soluble in liquid oxygen.

4. *Condensation* of certain gaseous species can dilute the propellant. The water vapor in warm gases is an example. Condensation can also occur on the exposed wetted inner walls of propellant tanks, and this requires more pressurizing gas.

5. *Changes in the gas temperature* may take place during operation. Compressed gases undergoing an adiabatic expansion can cause noticeable gas cooling; temperatures as low as 160 to 200 K (-228 to -100 °F) have been recorded with helium. A cold gas will absorb heat from the propellants and the engine hardware. The particular nature of the expansion process will depend largely on the time of rocket operation. For large liquid propellant rocket engines, which run only for a few minutes, the expansion process will be close to adiabatic, which means little heat transfer from the hardware to the gas. For satellites, which stay in orbits for years and where the thrusters operate only occasionally and for short operating periods, heat will be transferred from the vehicle hardware to the gas, and expansion will be close to isothermal (no change in temperature).

6. *Chemical reactions* in some species of pressurizing gas with the liquid propellant have occurred, some that can generate heat or increase the pressure. Inert gases, such as helium, undergo no chemical reactions with propellants.

7. *Turbulence, impingement, and irregular flow distributions* of the entering gas will increase the heat transfer between the liquid and the gas. Depending on the temperature difference, it can cause additional heating or cooling of the top liquid layers.

8. *Vigorous sloshing* can quickly change the gas temperature. In some of the experimental flights of the Bomarc missile, side accelerations induced sloshing, which caused a sudden cooling of the warm tank pressurizing gas and resulted in a sudden reduction of tank pressure and propellant flow. See Refs. 6–4 and 6–5.

9. In many rocket engines, a portion of the pressurizing gas is used for purposes other than tank pressurization, such as actuation of valves or controls. The amount required, once determined, must be added to the total gas mass needed.

Simplified Analysis for the Mass of Pressurizing Gas

This section describes the case of a pressurizing system using a compressed gas stored initially in a separate tank at ambient temperatures. It is the first category of pressurizing systems discussed above and perhaps the most common type. When the tank has some insulation and the operation of the rocket engines is relatively short (1 or 2 minutes), the expansion process in the gas tank is close to adiabatic (i.e., no heat transfer to or from the system hardware and the gas expansion in the storage tank noticeable drops the gas temperature and changes its density). At the other extreme is the isothermal expansion; a considerably slower process requiring longer times for temperature equilibration, an example being many short operations in a multi-year orbit maintenance. We may assume here perfect gas behavior, where perfect gas formulations apply (see Chapters 3 and 5). Furthermore, we assume no evaporation of the liquid propellant (only valid for propellants with low vapor pressure), an inert pressurizing gas that does not dissolve in the liquid propellant, and that no propellant sloshing or vortexing occurs.

Let the initial condition in the gas tank be given by subscript 0, the final gas conditions in the gas tank by subscript g, and the gas in the propellant tank by subscript p; final pressures may differ because in practice $p_g \geq p_p$ to account for valve, piping, and regulator pressure drops. The relevant equations include mass continuity and the perfect equation of state as shown below:

$$m_0 = m_g + m_p \quad \text{and} \quad pV = mRT \tag{6-5}$$

Here, V represents the gas volume (m^3 or ft^3). When the end points of the process of propellant displacement are isothermal (a relatively slow process), then the temperatures before and after will be the same (and heat will be drawn from the environment). Substituting the gas law into the mass balance and solving for V_0, after T_0 equilibrates,

$$\frac{p_0 V_0}{RT_0} = \frac{p_g V_0}{RT_0} + \frac{p_p V_p}{RT_0} \quad \text{or} \quad V_0 = \frac{p_p}{p_0 - p_g} V_p \tag{6-6}$$

Now, returning to Eqs. 6–5 we arrive at the total gas mass m_0,

$$\text{Isothermal}: m_0 = \frac{p_0}{RT_0} \frac{p_p}{p_0 - p_g} V_p = \frac{p_p V_p}{RT_0} \left(\frac{1}{1 - p_g/p_0} \right) \tag{6-7}$$

In real pressurizations, end conditions should fall somewhere between isothermal and adiabatic. In thermodynamics, a reversible-adiabatic process is isentropic so that here we may speak of a reversible "polytropic expansion" of a perfect gas, where the polytropic exponent lies between 1.0 and k. Example 6–2 provides a set of tank volume and total mass estimates using either helium or nitrogen as the pressurizing gas.

Example 6–2. Compare helium to nitrogen gas when used for pressurizing a propellant tank with 250 kg of 90% hydrogen peroxide by estimating their required mass and volume. The pressurizing tank is initially at a gas pressure p_0 of 14 MPa and the required propellant final tank pressure p_p is 3.40 MPa. The density of this liquid propellant is 1388 kg/m³ and the ambient temperature is 298 K. Assume $p_g \approx p_p$ together with ideal flow conditions so that we may use well-known expressions for isentropic and isothermal expansions.

SOLUTION. Since the vapor pressure of 90% hydrogen peroxide propellant is low, the amount of liquid propellant that would evaporate can be neglected and the assumption of no evaporation applies. The propellant volume to be displaced is 250/1388 = 0.180 m³.

The expansion will likely be polytropic, but we will only estimate two related limits. Under our assumptions above, the fully isothermal case in Eq. 6–7 becomes simply $pV = $ constant, and the isentropic case is known from thermodynamics to be $pV^k = $ constant, where k is the gas specific heat ratio. For helium we use $R = 2079$ J/kg-K and $k = 1.68$, and for nitrogen $R = 296.7$ J/kg-K and $k = 1.40$. Solving for the storage volume,

$$\text{Isothermal: } p_0V_0 = p_p(V_0 + 0.180) \quad \text{or} \quad V_0 = \frac{0.180p_p}{p_0 - p_p} = 0.0577 \text{ m}^3$$

In the isothermal case, tank volumes are the same for both helium and nitrogen. Now in a reversible adiabatic process (isentropic, see Eq. 3–7),

$$p_0V_0^k = p_p(V_0 + 0.180)^k \quad \text{or} \quad \frac{V_0 + 0.180}{V_0} = (4.118)^{1/k}$$

Isentropic: Helium $1.334V_0 = 0.180$ or $V_0 = 0.135$ m^3

Nitrogen $1.748V_0 = 0.180$ or $V_0 = 0.103$ m^3

Here, the ideal tank volume for nitrogen is somewhat smaller. Next, the masses are obtained from the perfect gas law as follows:

Helium $m_0 = \dfrac{p_0V_0}{RT_0} = \dfrac{14 \times 10^6 V_0}{2079 \times 298} = 3.05$ kg isentropic and 1.30 kg isothermal

Nitrogen $m_0 = \dfrac{p_0V_0}{RT_0} = \dfrac{14 \times 10^6 V_0}{296.7 \times 298} = 16.35$ kg isentropic and 9.16 kg isothermal

For both helium and nitrogen, the ideal adiabatic process requires a larger mass and volume than the fully isothermal process, but the former can be much faster. Because of the stated assumptions, we only arrive at theoretical (i.e., minimum) mass estimates but the comparisons remain valid. Clearly, the smaller volume advantage of nitrogen is offset by its much larger mass disadvantage—resulting from their molecular mass difference; helium is about seven times lighter than nitrogen and thus more desirable. Nitrogen is also prone to condensation under strong adiabatic expansions, whereas helium is not.

6.6. TURBOPUMP FEED SYSTEMS AND ENGINE CYCLES

The principal components of a rocket engine with one type of turbopump system are shown in the simplified diagram of Fig. 1–4. Here, the propellants are pressurized by means of separate *pumps*, which in turn are driven by one or more *turbines*. These turbines derive their power from the expansion of high enthalpy gases.

Figures 10–1, 10–2, and 10–3 show turbopump examples, and Chapter 10 is devoted exclusively to this topic. The turbopump is a high-precision, accurately balanced piece of high-shaft-speed (rpm) rotating machinery. It consists usually of one or two centrifugal pump(s) and a turbine. Its high-speed, high-load bearings support the shaft(s) on which the pump impeller(s) and turbine disk are mounted. It has shaft seals to prevent propellant leakage and also to prevent the two propellants from mixing with each other inside the turbopump. Some turbopumps also have a gear transmission, which allows the pumps or turbine to rotate at different, usually more efficient, shaft speeds. Chapter 10 describes the design of turbopumps and their major components, several arrangements of the key components, and alternate configurations. Starting turbopump feed systems usually takes longer than pressurized feed systems, because it takes some time for the rotating components (pumps, turbines) to accelerate to operating shaft speeds. Starting is discussed in Sections 7.1 and 11.4.

Engines with turbopumps are preferred for the booster and sustainer stages of space launch vehicles, long-range missiles, and in the past also for aircraft performance augmentation. These feed systems include the tanks, and they are usually lighter than other types for such high-thrust, long-duration applications (the inert hardware mass of the rocket engine without tanks is essentially independent of duration). From turbopump feed system options, such as depicted in Chapter 10, the designer can select the most suitable concept for any particular application.

In summary, turbopump feed systems are usually preferred when the engine has a relatively high total impulse, which usually means high thrust and/or very long cumulative firing duration. Pressurized feed systems are best for rocket engines with relatively low total impulse, that is, low thrust and/or many multiple starts.

Engine Cycles

All liquid propellant rocket engines with a turbopump feed system operate with one of several engine cycles. The three most common are shown in Fig. 6–8, and they have flown many times. Reference 6–12 shows variations of these three cycles and other engine cycles, some of which have not yet flown and some have not yet even been built.

An *engine cycle* for turbopump-fed engines consists of (1) specific propellant flow paths through the major engine components, (2) a method for providing the hot gas to one or more turbines, and (3) a method for handling the turbine exhaust gases. There are two types of operating cycle, *open* and *closed cycles. Open* denotes that the working fluid coming from the turbine is discharged into the nozzle exit section of the thrust chamber at a location in the expanding section far downstream of the nozzle throat as shown schematically in Fig. 6–8 (and discussed in Table 6–6), or is discharged overboard, usually after having been expanded in a separate nozzle of its own as in Figs. 1–4 and in 6–9a and 6–9b. In *closed cycles* or *topping cycles* all the working fluid from the turbine is injected into the rocket engine combustion chamber to use more efficiently its remaining energy. Here, the turbine exhaust gas is fed into the injector of the thrust chamber and is expanded through the full pressure ratio of the main thrust chamber nozzle, thus giving a somewhat higher performance than the open cycles, where these exhaust gases expand only through a relatively small pressure ratio.

Table 6–6 shows key parameters for each of the three common cycles, and it describes the differences between them. The schematic diagrams of Fig. 6–8 show each cycle with a separate turbopump for fuel and for oxidizer. However, arrangements where the fuel and oxidizer pump are driven by the same turbine are also

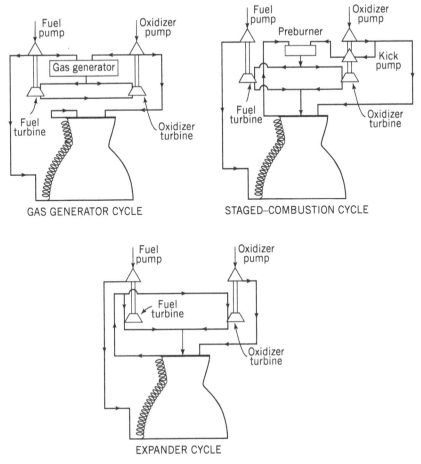

FIGURE 6–8. Simplified diagrams of three common engine cycles for liquid propellant rocket engines. The spirals are a symbol for an axisymmetric cooling jacket where heat is absorbed.

common because sometimes such schemes reduce the hardware mass, volume, and cost. The "best" cycle needs to be selected on the basis of mission, suitability of existing engines, and criteria established for each particular vehicle. There is an optimum chamber pressure and an optimum mixture ratio for each application and for each engine cycle, both of which depend in part on optimization factors such as maximum range, lowest cost, and/or highest payload.

The *gas generator cycle* has been the most commonly used. Compared to other engine cycles, its rocket engines are relatively simple, pressures are usually lower, and generally they have a lower inert mass and engine cost. However, its performance (specific impulse) is less by a few percentage points than the other two cycles. Such performance is nevertheless adequate for many space flight and military missions.

TABLE 6–6. Qualitative Characteristics for Three Different Engine Cycles

Engine Cycle	Gas Generator Cycle	Expander Cycle	Staged Combustion Cycle
Engine specific impulse, as % of gas gen. cycle	Set to 100%	102–106%	102–108%
Turbine flow, as % of total propellant flow	1.5–7%	75–96% of the fuel flow or 12–20% of total	60–80%
Typical pressure drop, across turbine as % of chamber pressure	50–90	5–30	60–100
Propellant type	All types	Cryogenic fuel for cooling	All types
Pump discharge pressure, % of chamber pressure	135–180	150–200	170–250
Turbine exhaust gas	Dumped overboard through a separate nozzle or aspirated into main nozzle exit section	Fed into main thrust chamber injector	Fed into main thrust chamber injector
Relative inert mass of engine	Relatively low	Higher	Highest
Thrust control, typical	Regulate flow and/or mixture ratio in gas generator	Regulate bypass of some gasified fuel flow around turbine	Regulate preburner mixture ratio and propellant flows
Maximum pressure in feed system	Relatively low	Higher	Highest
First ground tests	Goddard, USA, 1934	Aerojet-Rocketdyne, 1960	RNII[a], Russia, 1958
First flight operation	Hellmuth Walter Comp., Germany, 1939	Aerojet-Rocketdyne, 1963	Korolev Design Bureau, 1961

[a]Rossiyskiy Naucho Isseldovatelskiy Institut (Reaction Propulsion Research Institute).

In gas generator cycles, the turbine inlet gas comes from a separate gas generator, whose propellants can be supplied from two separate pressurized propellant tanks or can be drawn off main propellant pump discharges. Some early engines also used a separate monopropellant for creating the generator gas; the German V-2 missile engine used hydrogen peroxide decomposed by a catalyst. Typically, turbine exhaust gases are discharged overboard through one or two separate uncooled ducts and small low-area-ratio nozzles (at relatively low specific impulse), as shown schematically in Fig. 1–4 and in the Vulcain engine or RS-68 engine (Figs. 6–9a and 6–9b and listed in Table 11–2). Alternatively, this turbine exhaust can be aspirated into the main flow through multiple openings in the diverging nozzle section, as shown schematically in Fig. 6–8 for a gas generator engine cycle. This turbine exhaust gas then can protect the diverging walls near the nozzle exit from high temperatures. Both methods can only provide small amounts of additional thrust. The gas generator mixture ratio is usually fuel rich so that the gas temperatures are low enough (typically 900 to 1350 K) to allow the use of uncooled turbine blades and uncooled nozzle exit segments.

The liquid-oxygen/liquid-hydrogen RS-68 rocket engine, shown in Figs. 6–9a and 6–9b, is an example of an engine operating on a gas generator cycle. See Refs. 6–13, 6–14, and 6–15. Three RS-68 engines are used on the Delta IV Heavy launch vehicle; one propels the first or center stage of the vehicle, and the others propel each of the two strap-on outboard stages, as seen in Fig. 1–12. Data from this engine is under Fig. 6–9a and in Table 11–2. With a gas generator cycle the specific impulse of the thrust chamber by itself is always somewhat higher (by one-half to 4%) than the specific impulse of the engine. This difference is due to the small turbine exhaust flow with its very low specific impulse, making the overall thrust of the rocket engine always just a little larger in gas generator cycles. This engine is started by flowing helium (from a ground-based tank) through the gas generator pumps and turbines. The helium flow also purges any air initially in the engine passages and thus prevents any freezing of air and/or moisture. Each engine has two separate turbopumps (see Chapter 10) to raise the pressure and control the flow of propellants that feed the thrust chamber (see Chapter 8). Not shown in the RS-68 flow diagram are (a) thrust vector control components, which change angular thrust directions, such as a gimbal mounting block on top of the injector and two hydraulic actuators (see Chapter 18), (b) the ignition system (see Section 8.6), (c) electrical subsystems (wires, sensors, switches) and (d) the separate power supply for providing a high-pressure hydraulic fluid. This pressurized hydraulic fluid energizes (through two actuators) the engine's angular motion and the movable roll control nozzle (which uses exhaust gases from the fuel turbine), it is also used for operating and throttling the four principal valves. Flexible joints in the high-pressure propellant ducts and the exhaust ducts are both needed to allow for angular motion of the engine and for thermal growth. There is usually an intentional leak of propellant at the rotary seals of the turbopumps and at the stems of major valves, and the diagram shows drains for the safe discharge of such leaks; these small leaks allow for seal cooling and lubrication. The heat exchanger shown in the flow diagram gasifies a small flow of liquid oxygen used to pressurize the liquid oxygen tank during flight at low tank pressures. The hydrogen tank is pressurized by

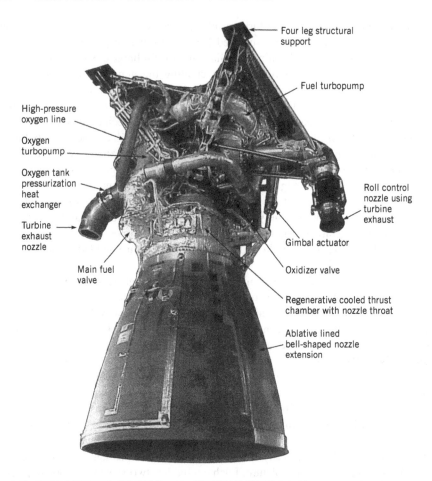

Parameter	Thrust chamber	Engine
Specific impulse at sea level (max.), sec	363	357
Specific impulse in vacuum (max.), sec	415	409
Thrust, at sea level, lbf	640,700	656,000
Thrust in vacuum lbf	732,400	751,000
Mixture ratio	6.74	6.0

FIGURE 6–9a. Large RS-68 rocket engine with a gas generator cycle. For engine data, see Table 11–2. Courtesy of Aerojet Rocketdyne

a small gaseous hydrogen flow from the exit of thrust chamber cooling jacket after reducing its pressure. The nozzle exit section of the thrust chamber is uncooled and internally lined with an ablative high-temperature material; the combustion chamber and nozzle throat section are regeneratively cooled by liquid hydrogen in the cooling jacket (see heat transfer in Chapter 8). Just before start, the engine's liquid propellant

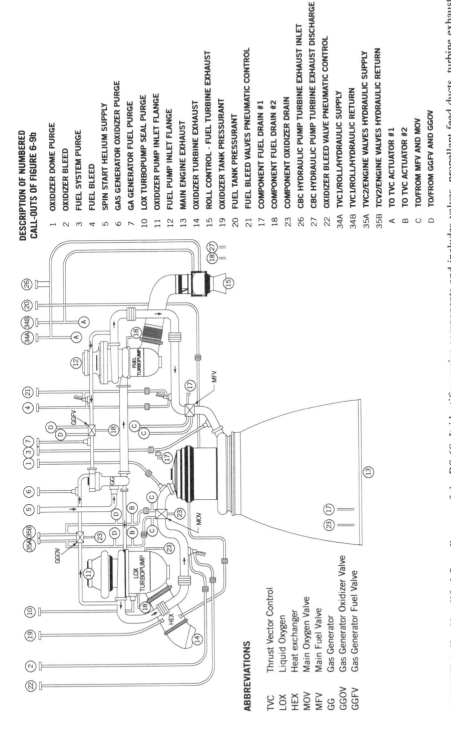

DESCRIPTION OF NUMBERED CALL-OUTS OF FIGURE 6-9b

1 OXIDIZER DOME PURGE
2 OXIDIZER BLEED
3 FUEL SYSTEM PURGE
4 FUEL BLEED
5 SPIN START HELIUM SUPPLY
6 GAS GENERATOR OXIDIZER PURGE
7 GA GENERATOR FUEL PURGE
10 LOX TURBOPUMP SEAL PURGE
11 OXIDIZER PUMP INLET FLANGE
12 FUEL PUMP INLET FLANGE
13 MAIN ENGINE EXHAUST
14 OXIDIZER TURBINE EXHAUST
15 ROLL CONTROL - FUEL TURBINE EXHAUST
19 OXIDIZER TANK PRESSURANT
20 FUEL TANK PRESSURANT
21 FUEL BLEED VALVES PNEUMATIC CONTROL
17 COMPONENT FUEL DRAIN #1
18 COMPONENT FUEL DRAIN #2
23 COMPONENT OXIDIZER DRAIN
26 CBC HYDRAULIC PUMP TURBINE EXHAUST INLET
27 CBC HYDRAULIC PUMP TURBINE EXHAUST DISCHARGE
22 OXIDIZER BLEED VALVE PNEUMATIC CONTROL
34A TVC1/ROLL/HYDRAULIC SUPPLY
34B TVC1/ROLL/HYDRAULIC RETURN
35A TVC2/ENGINE VALVES HYDRAULIC SUPPLY
35B TCV2/ENGINE VALVES HYDRAULIC RETURN

A TO TVC ACTUATOR #1
B TO TVC ACTUATOR #2
C TO/FROM MFV AND MOV
D TO/FROM GGFV AND GGOV

ABBREVIATIONS

TVC Thrust Vector Control
LOX Liquid Oxygen
HEX Heat exchanger
MOV Main Oxygen Valve
MFV Main Fuel Valve
GG Gas Generator
GGOV Gas Generator Oxidizer Valve
GGFV Gas Generator Fuel Valve

FIGURE 6–9b. Simplified flow diagram of the **RS-68**. It identifies major components and includes valves, propellant feed ducts, turbine exhaust ducts, small sized tubing for drains, purges, hydraulic controls, and it shows the pipe for helium spin-up of the turbines for starting. The circled numbers are explained on that page. Some of the small tubing is not shown in full length; only the first and last few inches are shown. Courtesy of Aerojet Rocketdyne

passages are brought to a very low temperature by periodically bleeding cryogenic propellant—down to the main propellant valves. Such drastic cooling is simultaneous with the loading and pressurization of propellants into the vehicle. Thrust in the RS-68 can be throttled to 60% of full value and this is needed during ascent to reduce acceleration and avoid high aerodynamic pressures on the vehicle at certain altitudes. See Refs. 6–11 and 6–12.

The RS-68 engine has recently been replaced with a simplified version (RS-68A) having a somewhat higher thrust and chamber pressure. Some of its parameters are: thrust 797,000 lbf (vacuum) and 702,000 lbf (sea level), mixture ratio 5.97, chamber pressure 1,557 psia, specific impulse 411 sec (vacuum) and 362 sec (sea level), nozzle exit area ratio 21.5, and engine weight a sea level 14,770 lbf.

The *expander cycle* and the *staged combustion cycle* are both closed cycles, and any small improvement they offer makes a substantial difference in payloads for flight missions with high mission velocities. Alternatively, they allow somewhat smaller flight vehicles. However, these engines are usually more complex, heavier, and more expensive.

A flow diagram of an *expander cycle* is shown in Fig. 6–10. The preferred fuel for this cycle is cryogenic hydrogen. It is evaporated, heated, and then fed to low-pressure-ratio turbines after having passed through the engine's cooling jackets. Part of the coolant, perhaps 5 to 15%, bypasses the turbine (as shown in Fig. 6–10) and

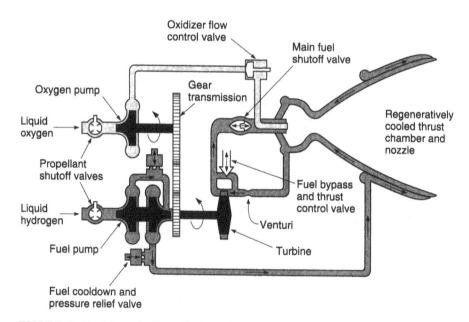

FIGURE 6–10. Schematic flow diagram of the RL10B-2 upper-stage rocket engine is an example of an expander engine cycle. Courtesy of Aerojet Rocketdyne

rejoins the turbine exhaust flow before the entire coolant flow is sent to the injector or into the engine combustion chamber. Advantages of the expander cycle include good specific impulse, no gas generator, and a relatively low engine mass. In an expander cycle, all the propellants are fully burned in the engine combustion chamber and efficiently expanded in the exhaust nozzle of the thrust chamber.

An expander cycle is used in the RL10 hydrogen/oxygen rocket engine and different versions of this engine have flown successfully in the upper stages of several space launch vehicles. Data on the RL10 engines are given in Tables 6–6, 8–1, and 11–2, Section 6.8 and Ref. 6–16. A modification of this engine, the RL10B-2 with an extendible nozzle skirt, can be seen in Fig. 8–17. The turbine drives a single-stage liquid oxygen pump (through a gear case) and a direct drive for a two-stage liquid hydrogen pump. Cooling down of the hardware to cryogenic temperatures is accomplished by flowing (prior to engine start) cold propellant through the engine by opening "cool-down valves." Pipes for discharging the cooling propellants overboard are not shown here but can be seen in Fig. 8–17. Thrust is regulated by controlling the flow of hydrogen gas to the turbine, bypassing the turbine to maintain constant chamber pressure. Helium is used for a power boost, actuating several of the larger valves through solenoid-operated pilot valves.

In *staged combustion cycles*, the coolant flow path through the cooling jacket is the same as that of the expander cycle. But here, a preburner (a high-pressure gas generator) burns all the fuel with part of the oxidizer to provide high-enthalpy gas to the turbines. The total turbine exhaust gas flow is then injected into the main combustion chamber where it burns with the remaining oxidizer. This cycle lends itself to high-chamber-pressure operations that permit a relatively small thrust chamber size. The extra pressure drop in the preburner and turbines causes the pump discharge pressures of both the fuel and the oxidizer to be higher than with open cycles, requiring heavier and more complex pumps, turbines, and piping. Turbine flows can be relatively high and turbine pressure drops low, when compared to other cycles. Staged combustion cycles give high specific impulses, but their engines are more complex and massive. A variation of the staged combustion cycle was used in the Space Shuttle main engine, now modified to the RS-25, as shown in Figs. 6–1 and 6–11. This engine actually used two separate preburner chambers, each mounted directly on a separate main turbopump. In addition, there were two more low-speed, low-power turbopumps for providing boost pressures to the main pumps, but their turbines were not driven by combustion gases; instead, high-pressure liquid oxygen drove one booster pump and evaporated hydrogen drove the other.

The Russians have flown more than 10 different staged-combustion cycle rocket engines, all of which use an oxidizer-rich preburner. The Japanese and also the Chinese have also flown their own oxidizer-rich designs of this cycle. The United States' version of this cycle with a fuel-rich preburner was found in the Space Shuttle Main Engine (or RD-25 engine). If there is a leak in the oxidizer-rich gas portion of the engine, the leaking gas will not ignite with air unlike any hot fuel-rich leaks that will readily ignite with air causing engine fires. There are some disadvantages in oxidizer-rich systems related to pump power—because there is more oxidizer

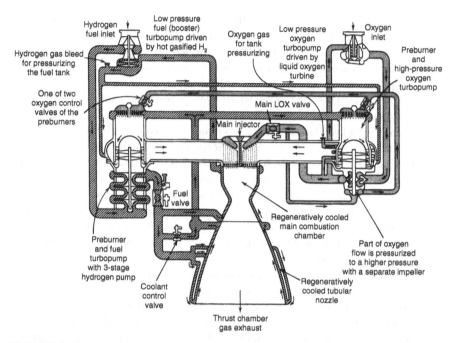

FIGURE 6–11. Flow diagram illustrating the staged combustion cycle of both the Space Shuttle Main Engine (SSME) and the RS-25 engine which use liquid oxygen and liquid hydrogen fuel. Courtesy of Aerojet Rocketdyne and NASA

pressure being handled, higher pump powers are required at the preburner or turbine inlets than with fuel-rich systems where all the fuel is pumped to the higher pressures (pump efficiencies do improve at the larger oxidizer flows and higher pressures). A fuel-rich staged combustion engine cycle (the RS-25) can be seen in Fig. 6–11, a simplified diagram of which can be found in Fig. 6–8; an oxidizer-rich cycle is shown later in Fig. 6–13 (the RD-191).

The Russian RD-191 *liquid propellant rocket engine* (developed and built by NPO Energomash) is another example of a staged combustion cycle (Ref. 6–11 and 6–18). One, three, or five of these engines are used to power the first and strap-on stages of several versions of the new Angara fleet with different sizes of Russian space launch vehicles. See Figs. 6–12 and 6–13, Table 6–7, and Refs. 6–11 and 6–17. This engine is a single thrust chamber derivative of both the RD-170 (with four thrust chambers, which is no longer in production) and the RD-180 (two thrust chambers, used in the Atlas 5). It has an oxidizer-rich preburner, a two-axis gimbal mount, an essentially identical thrust chamber to its predecessors, and a turbopump with a 2-stage turbine, an oxidizer pump, a fuel pump and a second fuel pump feeding the high-pressure preburner. The more powerful main turbopump has one liquid oxygen pump and two fuel (kerosene) pumps, one of which is a smaller

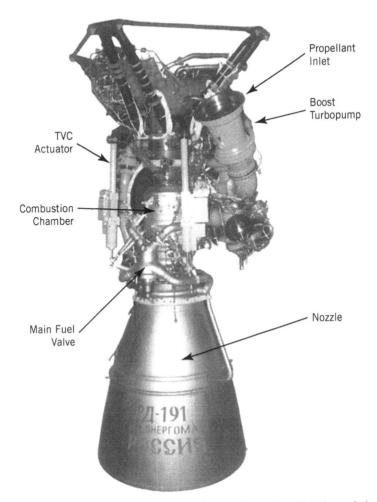

FIGURE 6–12. Russian RD-191 liquid propellant rocket engine. This is a relatively new engine that operates on a staged combustion cycle. From NPO Energomash, Khimki, Russia

kick pump (small flow second stage – higher discharge pressure). Two lower-speed booster turbopumps (one for the liquid oxygen and one for the kerosene) slightly raise the propellant pressure to avoid cavitation in the impellers of the main pumps. See Section 10.5. A portion of the warm turbine exhaust gas is used in a heat exchanger to heat and expand the helium that is used to pressurize the propellant tanks in the flight vehicle. Not shown in Fig. 6–13 is a bleed for a portion of the high-pressure fuel (from the pump discharge) that drives two hydraulic actuators of the gimbal-mounted thrust chamber for thrust vector control. In Section 10.8 there is a discussion of fuel rich and oxidizer rich gases that come from preburners.

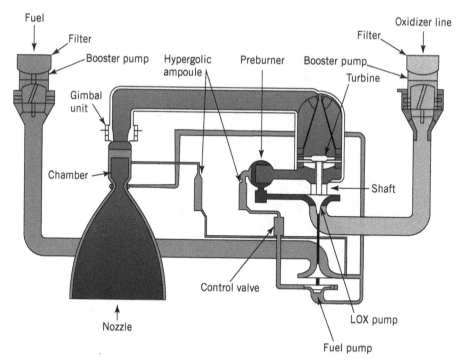

FIGURE 6–13. Simplified RD-191 flow diagram. See Figure 8–11 for a sectioned view of the thrust chamber and Figure 8–4 12 for a sectioned view of the injector From NPO Energomash, Khimki, Russia

TABLE 6–7. Selected Performance and Operational Characteristics of the RD-191 Engine

Number of thrust chambers per engine	1
Thrust, sea level, and vacuum, kN	1921 and 2084
Specific impulse, sea level, and vacuum, sec	310.8 and 337.9
Mixture ratio,[a] oxygen/kerosene flow	2.63 ± 7%
Chamber pressure (at injector face), MPa	25.813
Engine, dry mass, kg	2200
Engine mass with propellants, kg	2430
Thrust chamber internal diameter, mm	380
Throat diameter, mm	235.5
Nozzle exit area ratio	36.87
Engine, maximum-height and diameter, mm	3780 and 1930
In-flight operating time (nominal), sec	250
Throttling range, %	100 to 70
Gimbal angle, typical, degrees	3.5

[a]Liquid oxygen/Russian kerosene (similar to RP-1).
From NPO Energomash, Khimki, Russia

6.7. ROCKET ENGINES FOR MANEUVERING, ORBIT ADJUSTMENTS, OR ATTITUDE CONTROL

These engines usually include a set of small thrusters that are installed at various places in a vehicle and a common pressurized feed system, similar to Figs. 1–3, 4–14 and 6–14. They are called *reaction control systems, auxiliary rocket propulsion systems*, or *attitude control systems* in contrast to higher-thrust *primary or boost propulsion systems*. Most have multiple small thrusters, produce low thrust, use storable liquid propellants, and require accurate repeatable pulsing, a long life in space and/or long-term storage with loaded propellants in the flight tanks. Typical thrust levels in small thruster are between 0.1 and 1000 lbf (0.445 and 4,448.2 N). Figure 4–14 indicates that 12 thrusters are required for the application of pure torques about three vehicle axes. If three-degree-of-freedom rotations are not needed, or if torques can be combined with some translation maneuvers, fewer thrusters will be required. Such *auxiliary rocket engines* are commonly used in spacecraft and missiles for accurate *control of flight trajectories, orbit adjustments*, or *attitude control* of the vehicle. References 6–1 and 6–17 give information on several of these maneuvers and on small-thruster history. Figure 6–14 shows a simplified flow diagram for a postboost control rocket engine, with one larger rocket thrust chamber for changing the velocity vector and eight small thrusters for attitude control.

Section 4.5 describes various space trajectory correction maneuvers and satellite station–keeping maneuvers that are typically performed by these small auxiliary liquid propellant rocket engines with multiple thrusters. Table 6–8 lists typical applications for rocket engines with small thrusters.

Attitude control can be provided during two occasions, while a primary propulsion system (of a vehicle or of a stage) is operating and while its small thruster rocket system operates by itself. For instance, this is done to point satellite's telescope into a specific orientation or to rotate a spacecraft's main thrust chamber into the desired direction for a vehicle turning maneuver.

A common method for achieving accurate velocity corrections or precise angular positions is to operate (fire) some of the thrusters in a *pulsing mode* (e.g., fire repeatedly for 0.010 to 0.020 sec, each time followed by a pause of perhaps 0.020 to 0.150 sec). The guidance system determines the maneuver to be undertaken and the vehicle control system sends command signals to specific thrusters for the number of pulses needed to accomplish such maneuver. Small liquid propellant engine systems are uniquely capable of such pulsing operations. Some small thrusters have been tested with more than 1 million pulses. For very short pulse durations the specific impulse degrades by 5 to 25% because pressure and performance during the period of thrust buildup and thrust decay is lower as transient times become a major portion of the total pulse time.

Ballistic missile defense vehicles usually have highly maneuverable upper stages. These require substantial side forces, also called *divert forces* (typically 200 to 6000 N), during the final closing maneuvers just prior to target interception. In concept, the system is similar to that of Fig. 6–14, except that a larger thrust chamber

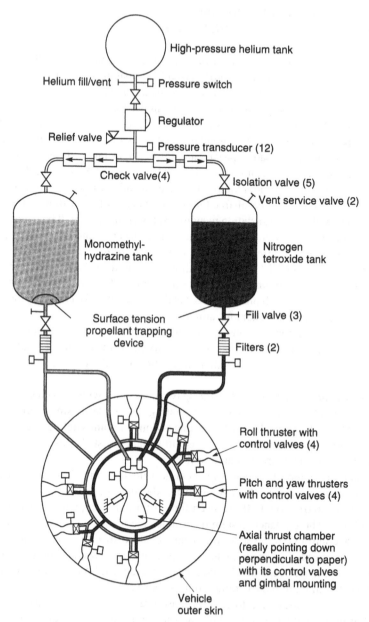

High-pressure helium tank

Helium fill/vent — Pressure switch

Regulator

Relief valve

Pressure transducer (12)

Check valve(4)

Isolation valve (5)

Vent service valve (2)

Monomethyl-hydrazine tank

Nitrogen tetroxide tank

Surface tension propellant trapping device

Fill valve (3)

Filters (2)

Roll thruster with control valves (4)

Pitch and yaw thrusters with control valves (4)

Axial thrust chamber (really pointing down perpendicular to paper) with its control valves and gimbal mounting

Vehicle outer skin

FIGURE 6–14. Schematic flow diagram of a helium-pressurized, bipropellant rocket engine system for the fourth stage of the Peacekeeper ballistic missile, which provides the terminal velocity (in direction and magnitude) to each of several warheads. It has one large gimbaled thrust chamber for trajectory translation maneuvers and eight small thrusters (with scarfed nozzles) for attitude control in pitch, yaw, and roll. For clarity, the tanks and feed system are shown outside the vehicle skin though they are located within. Courtesy of USAF

TABLE 6–8. Typical Applications for Small Thrusters

Flight path (or orbit) corrections or changes:
Minor flight velocity adjustments
Orbit station keeping (correcting for deviations from orbit), or orbit maintenance
Orbit injection for small satellites
Deorbit maneuver for satellites
Divert and other maneuvers of terminal interceptor stages
Flight path control of some tactical missiles
Attitude control for:
Satellites, stages of space launch vehicles, space stations, missiles
Roll control for a single gimbaled larger rocket engine
Pointing/orienting antennas, solar cells, mirrors, telescopes, etc.
Correct the misalignment of principal, larger thrust chamber
Velocity tuning of warheads (postboost control system) for accurate targeting
Settling of liquid propellants in tanks prior to gravity-free start of main engine
Docking/rendezvous of 2 space vehicles with one another
Flywheel desaturation

Source: Mostly from Ref. 6–17.

would be at right angles to the vehicle axis. A similar system for terminal maneuvers, but using solid propellants, is shown in Figs. 12–27 and 12–28.

The Space Shuttle performed its reaction control with 38 different thrusters assembled in four pods, as shown schematically in Fig. 1–14; this setup included several duplicate (spare or redundant) thrusters. Selected thrusters were used for different maneuvers, such as space orbit corrections, station keeping, or positioning the orbiting vehicle for reentry or visual observations. These small restartable rocket engines were also used for space *rendezvous* or *docking maneuvers*, where one spacecraft slowly approaches another and locks itself to the other, without causing excessive impact forces during this docking maneuver. Docking operations require rotational and translational maneuvers from different rocket engines.

The application of pure torque to spacecraft can be divided into two classes, *mass expulsion* types (rockets) and *nonmass* types. Nonmass types include momentum storage (fly wheels), gravity gradient, solar radiation, and magnetic systems. Some space satellites are equipped with both mass expulsion and nonmass types. *Reaction wheels* or flywheels, momentum storage devices, are particularly well suited to obtaining vehicle angular position control with high accuracies (less than $0.01°$ deviation) and low vehicle angular rates (less than 10^{-5} degrees/sec) with relatively little expenditure of energy. A vehicle's angular momentum is changed by accelerating (or decelerating) the wheel. Of course, when the wheel reaches its maximum (or minimum) permissible speed, no further electrical motor torquing is possible; the wheel must be decelerated (or accelerated) to have its momentum removed (or augmented), a function usually accomplished through the simultaneous use of two small attitude control rocket thrusters, which apply a torque to the vehicle in opposite directions. This procedure has been called "*desaturation* of the flywheel."

Propellants for *auxiliary rockets* fall into three categories: *cold gas jets* (also called inert gas jets), *warm or heated gas jets*, and *chemical combustion gases*, such as fuel-rich liquid bipropellants. The specific impulse is typically 50 to 120 sec for cold gas systems, and 105 to 250 sec for warm gas systems. Warm gas systems can use inert gases from an electric heater or a monopropellant, which is catalytically and/or thermally decomposed. Bipropellant attitude control thrust chambers reach an I_s of 220 to 325 sec and can vary from 5 to 4000 N thrust; the highest thrusts being associated with large spacecraft. All basically use pressurized feed systems with multiple thrusters or thrust chambers equipped with fast-acting, positive-closing precision valves. Many systems use small, uncooled, metallic thrusters with supersonic exhaust nozzles which are strategically located on the periphery of the spacecraft pointing in different directions. Gas jets are typically used for low thrusts (up to 10 N) and low total impulses (up to 4000 N-sec). They have been used on small satellites and often only for roll control. See Ref. 6–17 and Section 7.6.

Small liquid monopropellant and liquid bipropellant rocket units are commonly used as auxiliary rocket systems for thrust levels typically above 1.0 N and total impulse values above 3000 N-sec. Hydrazine is the most common monopropellant used in auxiliary control rockets. The MESSENGER's probe propulsion system discussed in Section 6.3 has monopropellant thrusters as seen in Fig. 6–7. Nitrogen tetroxide and monomethylhydrazine is a common bipropellant combination. Chapter 7 contains data on all three categories of liquid propellants, and Chapter 8 shows small thrusters.

Each specific mission requirement needs to be carefully analyzed to determine which type or thruster combination is most advantageous for a particular application.

6.8. ENGINE FAMILIES

An engine family is made up of a series of related rocket engines which have evolved over a period of several years. These originate from the same rocket engine organization, and each engine has been tailored to a specific application. Family engines strongly resemble each other, use the same engine concept, usually the same propellants, and some identical or somewhat modified components of the same type. When an existing proven liquid propellant rocket engine can be modified and/or up-rated (or down-rated) to fit a new application, the newer modified engine can share much proven hardware, test data, qualified vendors, technical and fabrication personnel, and software from earlier engines.

An example is the RL 10 family of upper-stage rocket engines. It was developed by Aerojet Rocketdyne over a period of more than 55 years and is shown in Table 6–9. These data are from Ref. 6–16. Each engine has a specific application and is a modification and/or uprating of an earlier model. Some principal changes from one engine model to the next include increases in thrust, increases in performance (a somewhat higher specific impulse) by using higher chamber pressure, improved injector designs, and increases in nozzle exit area ratios. They all use LOX/LH$_2$ propellants, the same basic engine concept with an expander engine cycle, the same tubular cooling jacket approach for the chamber and nozzle throat region, the same generic geared turbopump arrangement, often the same or similar valves, and power level control by a bypass of hydrogen gas around the turbine. In the turbopump, the fuel pump and turbine are on the same high-speed shaft and the LOX pump is driven efficiently through a gear train at a lower speed. All engines are gimbal mounted (most at 4° maximum deflection), and most have space restart capability. Figure 8–17 shows the extendable nozzle of the RL 10B-2. In the RL 10A-3–3A thrust chamber, a high-conductivity ring is silver brazed into the nozzle throat, thus enabling higher chamber pressures. In some versions, the cooling jacket tubes were brazed with silver and were compatible. Figure 6–10 shows a flow sheet of an RL 10 expander engine cycle.

It is noteworthy to observe how thrust or specific impulse changed with time and with the model. To date, the specific impulse listed in Table 6–9 for the RL 10B-2 is the highest of any flying liquid propellant rocket engine and the extendable nozzle exit segment of the RL 10A-4 was a first for liquid propellant rocket engines.

In summary, when compared to a brand new engine, the principal benefits of adopting a modified engine based on an earlier family of proven engines, are savings in costs (costs of design, development, less new fabrication, less testing, qualification, and operation), attaining a high engine reliability more quickly, and often also a shorter schedule. Heritage of earlier proven similar engines allows the use of older engine or component data, having trained experienced personnel, proven subcontractors, a higher confidence level of reliability, and often, but not always, the use of the same debugged materials, fabrication, and test facilities. An intangible benefit is that the vehicle developer or prime contractor, or whoever plans to use one of these engines, will have more confidence. A brand new engine may reach a better performance and/or a somewhat lower inert engine mass and other improvements but it will be more costly, take longer to develop, and take more time to reach an equivalent high level of reliability.

6.9. VALVES AND PIPELINES

Valves control the flow of liquids and gases, and pipes conduct these fluids to their intended components. They are essential parts of a rocket engine. There are many different types of valves and all those chosen must be reliable, lightweight, leakproof, and must withstand intensive vibrations and very loud noises. Table 6–10 gives several key classification categories for rocket engine valves. Any one engine will use only some of the valves listed here.

Designing and making valves is an art largely based on experience. A single book section describing valve design and operation cannot do justice to this field. References 6–1 and 6–2 describe the design of specific valves, lines, and joints. Often design details, such as clearance, seat materials, or opening time delays present some development difficulties. With many of these valves, any significant internal or

TABLE 6–9. The RL 10 Engine Family[a]

Engine Model	Year Qualified	Vehicle	p_1 psia	Thrust, lbf	Mixture Ratio	Weight, lbf	I_s (vac) sec	Nozzle Area Ratio	No. Engines per Stage	Comments
RL 10A-1	1961	Atlas Centaur	300	15,000	5.0:1	300	424	40.0:1	2	Never fired in space. Two were on the first Centaur vehicle launch attempt that had a booster failure.
RL 10A-3C	1962	Atlas Centaur	292	15,000	5.0:1	292	427	40.0:1	2	Early Centaur missions. Improved injector.
RL 10A-3S	1962	Saturn IV	292	15,000	5.0:1	296	427	40.0:1	6	Saturn IV upper stage. Solenoid added to separate LOX cooldown from fuel cooldown.
RL 10A-3-1	1964	Atlas Centaur	292	15,000	5.0:1	291	433	40.0:1	2	Improved injector incorporated. Early Surveyor missions.
RL 10A-3-3	1966	Atlas & Titan Centaur	396	15,000	5.0:1	282	442.4	57.0:1	2	New chamber/nozzle and turbopump. Used on Centaur for both Atlas and Titan. Used for Surveyor, Mariner, Pioneer, Helios, Viking, and Voyager missions.
RL 10A-3-3A	1981	Atlas & Titan & Shuttle Centaur	475	16,500	5.0:1	310	444.4	61.0:1	2	Engine modified to operate with reduced propellant inlet pressures because of boost pump removal from vehicle. Silver throat insert in chamber.

Engine	Year	Vehicle								Comments
RL 10A-3-3B	1986	Shuttle Centaur	425	15,000	6.0:1	310	436	61.0:1	2	Modified to handle long space stay time for AF version of Shuttle/Centaur. Mixture ratio increased to 6.0 changed. Never flew.
RL 10A-4	1991	Atlas Centaur	570	20,800	5.5:1	370	448.9	84.0:1	2	New thrust chamber with no silver throat. Turbopump modified for increased F and p_c. Extendible radiation cooled columbium nozzle.
RL 10A-4-1	1994	Atlas Centaur	610	22,300	5.5:1	370	450.5	84.0:1	1 or 2	Improved injector.
RL 10A-4-1A	1999	Titan Centaur	580	20,500	5.0:1	321	444.4	61.0:1	2	Derivative of RL 10A-4-1 derated to 20.5K thrust and 5.0 O/F. Only 2 engines flown. No nozzle extension.
RL 10A-4-2	2001	Atlas Centaur	610	22,300	5.5:1	370	450.5	84.0:1	1 or 2	Dual spark plugs and igniter flow path modifications. Can use fixed or translating nozzle extension.
RL 10A-5	1993	DC-X	470	13,400	6.0:1	316	365.1	13.0:1	4	Chamber modified for sea-level operation and controls modified to enable 3 to 1 throttling. Delta Clipper Experimental vehicle.
RL 10B-2	1998	DELTA III and IV	633	24,750	5.88:1	664	465.5	385.0:1	1	New chamber/nozzle and high area ratio extendable composite nozzle.

[a]0As of July 2013, 385 RL10 engines have flown in space with 826 engine firings.

Courtesy of Aerojet Rocketdyne.

TABLE 6–10. Classification of Valves Used in Liquid Propellant Rocket Engines[a]

1. *Fluid*: fuel; oxidizer; cold or heated pressurized gas; hot turbine gas
2. *Application or use*: main propellant control; thrust chamber valve (dual or single); bleed; vent; drain; fill; bypass; preliminary stage flow; pilot valve; safety valve; overboard dump; regulator; gas generator or preburner control; sequence control; prevent back flow; isolation of part or all of feed system; latch valve
3. *Mode of actuation*: automatically operated (by solenoid, pilot valve, trip mechanism, pyrotechnic, etc.); manually operated; pressure-operated by high-pressure air, gas, propellant, or hydraulic fluid (e.g., check valve, tank vent valve, pressure regulator, relief valve), with or without position feedback, rotary or linear actuator
4. The *flow* magnitude and allowable pressure drop determine the size of the valve
5. *Duty cycle*: single operation or multiple operation during the same flight, short duration, pulsed operation; reusable for other flights; long or short life
6. *Valve type*: normally open; normally closed; normally partly open; two-way; three-way, with/without valve position feedback; ball valve, gate valve, butterfly type, spring loaded, low pressure drops, latch valve
7. *Temperature* and *pressure* allow classification by high, low, or cryogenic temperature fluids, or high or low pressure or vacuum capability
8. *Accessible or not accessible* to inspection, servicing, or replacement of valve or its seal

[a]This list is neither comprehensive nor complete.

external leakage or valve failure (stuck open or stuck close) may cause failure of the engine itself. All valves must be tested for two qualities prior to installation; they are tested for leaks—through the seat and also through the glands—and for functional soundness or performance.

Propellant valves in high-thrust units handle relatively large flows at high service pressures. Therefore, the forces needed to actuate these valves can be substantial. Hydraulic or pneumatic pressure, controlled by pilot valves, is used to operate the larger valves; these pilot valves are in turn actuated by a solenoid or a mechanical linkage. Essentially this is a means of "power boost."

Two valves commonly used in pressurized feed systems are *isolation valves* (when shut, they isolate or shut off a portion of the propulsion system) and *latch valves*, which require power for brief periods during movements, such as to open or shut, but need no power when latched or fastened into either an open or a closed position.

A very simple and very light valve concept is a *burst diaphragm*. It is essentially a circular disk of material blocking a pipeline and is designed with grooves so that it will fail and burst at a predetermined pressure differential. Burst diaphragms provide positive seals and prevent leakage, but they can be used only once and cannot stop the flow. The German *Wasserfall* anti-aircraft WWI missile used four burst disks; two were in high-pressure air lines and two were in the propellant lines.

Figure 6–15 shows a large main liquid oxygen valve. This is a normally closed, rotary-actuated, cryogenic, high-pressure, high-flow, reusable ball valve with low-pressure losses in the open position, which allows continuous throttling, a controlled

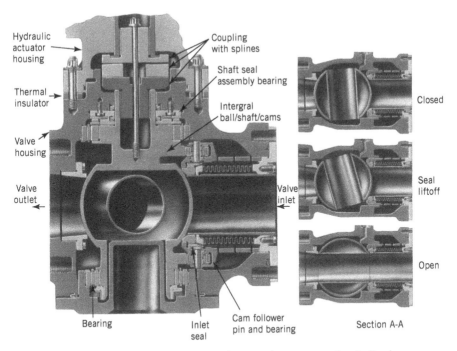

Hydraulic actuator housing

Coupling with splines

Shaft seal assembly bearing

Thermal insulator

Intergral ball/shaft/cams

Valve housing

Valve outlet

Valve inlet

Bearing

Inlet seal

Cam follower pin and bearing

Closed

Seal liftoff

Open

Section A-A

FIGURE 6–15. The SSME main oxidizer valve was a low-pressure drop ball valve representative of high-pressure large valves used in rocket engines. The ball and its integral shaft rotate in two bearings. The seal is a machined plastic ring spring-loaded by a bellows against the inlet side of the ball. Two cams on the shaft lift the seal a short distance off the ball within the first few degrees of ball rotation. The ball is rotated by a precision hydraulic actuator (not shown) through an insulating coupling. Courtesy of Aerojet Rocketdyne

rate of opening through a crank and hydraulic piston (not shown), with position feedback and anti-icing controls.

Pressure regulators are special valves that are frequently used to control gas pressures. Usually the discharge pressure is regulated to a predetermined standard pressure value by continuously throttling the flow, using a piston, flexible diaphragm, or electromagnet as the actuating mechanism. Regulators can be seen in the flow paths of Figs. 1–3 and 6–14.

The various fluids in a rocket propulsion engine are transported by *pipes*, *ducts* or *lines*, usually made of metal and joined by fittings or welds. Their design must provide for thermal expansion and provide support to minimize vibration effects. For gimballed thrust chambers it is necessary to provide flexibility in the piping to allow the thrust axis of the chamber to be rotated through small angles, typically ±3 to ±10°. This flexibility is provided by flexible pipe joints and/or by allowing pipes to deflect slightly when using two or more right-angle turns in the lines. High-pressure propellant feed lines in many liquid rocket engines provide both flexible joints and right-angle bends, as shown in Figs. 6–1 and 6–16. Such joints had flexible bellows

Bearing   Bellows seal

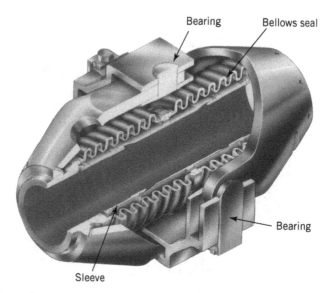

Bearing

Sleeve

FIGURE 6–16. Flexible high-pressure joint with external gimbal rings for a high-pressure hot turbine exhaust gas. Courtesy of Aerojet Rocketdyne

as seals and a universal joint-type mechanical linkage with two sets of bearings for carrying the separating loads imposed by the high pressures.

Sudden closing of valves may cause *water hammer* in pipelines, leading to unexpected pressure rises that can be destructive to propulsion system components. Analysis of this water hammer phenomenon allows determination of the approximate maximum pressure (Ref. 6–19). Friction in pipes and the branching of pipelines reduce this maximum pressure. Water hammer can also occur when admitting an initial flow of high-pressure propellant into evacuated pipes (surge flow). Pipes are normally under vacuum to remove moisture and prevent the formation of gas bubbles in the propellant flow, which can cause combustion problems.

All liquid rocket engines have one or more *filters* in their lines. These are necessary to prevent dirt, particles, or debris, such as small pieces from burst diaphragms, from entering precision valves or regulators (where debris can cause a malfunction) or from plugging small injection holes, which may cause hot streaks in the combustion gases, in turn causing thrust chamber failures.

Occasionally, a convergent–divergent *venturi section*, with a sonic velocity at its throat, is placed into one or both of the liquid propellant lines. It has also been called a *cavitating venturi* when the local throat pressure goes below the vapor pressure. Its merits are that it maintains constant flow and prevents pressure disturbances from traveling upstream. This can prevent the propagation of chamber pressure oscillations or the coupling with thrust chamber combustion instabilities. Venturi sections can also help in minimizing some water hammer effects in systems with multiple banks of thrust chambers.

6.10. ENGINE SUPPORT STRUCTURE

Most larger rocket engines have their own mounting or support structure where all major components are mounted. This structure also usually transmits the thrust force to the vehicle. Welded tube structures or metal plate/sheet metal assemblies have often been used for support structures. In some large engines the thrust chamber is used as a structure and the turbopump, control boxes, or gimbal actuators are attached to it.

In addition to the thrust load, an engine structure has to withstand forces imposed by vehicle maneuvers (in some cases a side acceleration of 10 g_0), vibration forces, actuator forces for thrust vector control motions, and loads from transportation over rough roads.

In low-thrust engines with multiple thrusters there often is no separate engine mounting structure; the major components are in different locations of the vehicle, connected by tubing, wiring, or piping, and each is usually mounted directly to the vehicle or spacecraft structure.

SYMBOLS

c	effective exhaust velocity, m/sec (ft/sec)
c_v, c_p	specific heats constant volume or pressure, J/kg-K (Btu/lbm-°R)
C_F	thrust coefficient
F	thrust force, N (lbf)
g_0	acceleration of gravity at sea level, 9.8066 m/sec^2
I_s	specific impulse, sec
m	propellant mass, kg (lbm)
k	specific heat ratio
p	pressure, N/m^2 (psi)
Δp	pressure difference, N/m^2 (psi)
$\dot{m}$	mass flow rate, kg/sec (lb/sec)
r	mixture ratio (oxidizer to fuel mass flow rates)
R	gas constant per unit mass, J/kg-K (ft-lbf/lbm-°R)
T	absolute temperature, K
$\dot{V}$	volume flow rate, m^3/sec (ft^3/sec)
V	volume, m^3 (ft^3)
w	total propellant weight, N (lbf)
$\dot{w}$	weight flow rate, N/sec (lbf/sec)

Subscripts

f	Fuel
0	initial condition or stagnation condition
g	gas tank
o	oxidizer
p	propellant tank or power cutoff

PROBLEMS

1. In an engine with a gas generator engine cycle, the turbopump has to do more work in pumping, if the thrust chamber operating pressure is raised. This requires an increase in turbine gas flow, which, when exhausted, adds little to the engine specific impulse. If the chamber pressure is raised too much, the decrease in performance due to an excessive portion of the total propellant flow being sent through the turbine and the increased mass of the turbopump will outweigh the gain in specific impulse that can be attained by increased chamber pressure and also by increased thrust chamber nozzle exit area ratio. Outline in detail a method for determining the optimum chamber pressure where the sea-level performance will be a maximum for a rocket engine that operates in principle like the one shown in Fig. 1–4.

2. The engine performance data for a turbopump rocket engine system are as follows:

Propellants	Liquid oxygen/kerosene
Engine system specific impulse (steady state)	272 sec
Engine system mixture ratio	2.52
Rated engine system thrust	40,000 N
Oxidizer vapor flow to pressurize oxidizer tank	0.003% of total oxidizer flow
Propellant flow through turbine at rated thrust	2.1% of total propellant flow
Gas generator mixture ratio	0.23
Specific impulse of turbine exhaust	85 sec

Determine performance of the thrust chamber I_s, r, F (see Section 11.2).

3. For a pulsing rocket engine, assume a simplified parabolic pressure rise of 0.005 sec, a steady-state short period of full chamber pressure, and a parabolic decay of 0.007 sec approximately as shown in the sketch. Plot curves of the following ratios as a function of operating time t from $t = 0.013$ to $t = 0.200$ sec: (a) average pressure to ideal steady-state pressure (with zero rise or decay time); (b) average I_s to ideal steady-state I_s; (c) average F to ideal steady-state F.

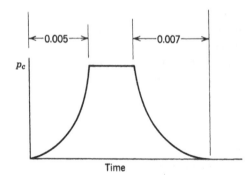

4. For a total impulse of 100 lbf-sec compare the volume and approximate system weights of a pulsed propulsion system using different gaseous propellants, each with a single spherical gas storage tank (at 3500 psi and 0 °C). A package of small thrust nozzles with piping, valves, and controls is provided that weighs 3.2 lbf. The gaseous propellants are hydrogen, nitrogen, or argon (see Table 7–3).

5. Compare several systems for a potential roll control application which requires four thrusters of 1 lbf each to operate for a cumulative duration of 2 min each over a period of several days, which allows a constant gas temperature. Include the following:

Pressurized helium	70 °F temperature
Pressurized nitrogen	70 °F ambient temperature
Pressurized krypton	70 °F ambient temperature
Pressurized helium	300 °F (electrically heated)

The pressurized gas is stored at 5000 psi in a single spherical fiber-reinforced plastic tank; use a tensile strength of 150,000 psi and a density of 0.050 lbm/in.^{3} with a 0.012-in.-thick aluminum inner liner as a seal against leaks. Neglect the gas volume in the pipes, valves, and thrusters, but assume the total hardware mass of these to be about 1.3 lbm. Use Table 7–3. Make estimates of the tank volume and total system weight. Discuss the relative merits of these systems.

6. A sealed propellant tank contains hydrazine. It is stored for long periods of time, and therefore the propellant and the tank will reach thermal equilibrium with the environment. At an ambient temperature of 20 °C and an internal pressure of 1.2 atm the liquid occupies 87% of the tank volume and the helium pressurizing gas occupies 13%. Assume no evaporation of the propellant, no dissolving of the gas in the liquid, and no movement of the tank. Use the hydrazine properties from Figs. 7–1 and 7–2 and Table 7–1. What will be the approximate volume percentages and the gas pressure at the extreme storage temperatures of 4 and 40 °C?

7. A liquid hydrogen/liquid oxygen thrust chamber has a constant bipropellant flow rate of 347 kg/sec at a mixture ratio of 6.0. It operates at full thrust for exactly 2 min. The propellants in the vehicle's tanks are initially vented to the atmosphere at the propellant boiling points and are assumed to be of uniform initial temperature at start. Use data from Table 7–1 for the propellant specific gravities. Assuming no losses, find the masses of (a) fuel and (b) of oxidizer used to produce the thrust for the nominal duration. (c) What was the volume of the liquid hydrogen actually used? (d) Assuming 4.0% extra fuel mass (for unusable propellant residual, evaporation, hardware cooling, or venting just prior to start, or propellant consumed inefficiently during start-up and shutdown) and a 10% ullage volume (the void space above the liquid in the tank), what will be the volume of the fuel tank? Assume other losses can be neglected.

Answer: (a) 5941 kg, (b) 35,691 kg, (c) 83.83 m^3 (d) 95.6 m^3.

8. Prepare dimensioned rough sketches of the four propellant tanks needed for operating a single gimbal-mounted RD 253 engine (Table 11–3) for 80 sec at full thrust and an auxiliary rocket system with a separate pressurized feed system using the same propellants,

with two gimbal-mounted small thrust chambers, each of 150 kg thrust, a duty cycle of 12% (fires only 12% of the time), but for a total flight time of 1.00 hr. For propellant properties see Table 7–1. Describe any assumptions that were made with the propellant budget, the engines, or the vehicle design, as they affect the amount of propellant.

9. Table 11–3 shows that the RD 120 rocket engine can operate at thrusts as low as 85% of full thrust and with a mixture ratio variation of ±10.0%. Assume a 1.0% unavailable residual propellant. The allowance for operational factors, loading uncertainties, off-nominal rocket performance, and a contingency is 1.27% for the fuel and 1.15% for the oxidizer.

a. In a particular flight the average main thrust was 98.0% of nominal and the mixture ratio was off by +2.00% (oxidizer rich). What percent of the total fuel and oxidizer loaded into the vehicle will remain unused at thrust termination?

b. If we want to run at a fuel-rich mixture in the last 20% of the flight duration (in order to use up all the intended flight propellant), what would the mixture ratio have to be for this last period?

c. In the worst possible scenario with maximum throttling and extreme mixture ratio excursion (±3.00%, but operating for the nominal duration), what is the largest possible amount of unused oxidizer or unused fuel in the tanks?

REFERENCES

6–1. D. K. Huzel and D. H. Huang, *Design of Liquid Propellant Rocket Engines*, rev. ed., AIAA, Reston, VA, 1992.

6–2. Personal communication with J. S. Kincaid of Aerojet-Rocketdyne, 2013–2015.

6–3. "NASA Conducts First Test Fire of Shuttle-Era Engine for SLS," *Space News*, Vol. 26, No. 2, 2015, pp. 14.

6–4. J. J. Pocha, "Propellant Slosh in Spacecraft and How to Live with It," *Aerospace Dynamics*, Vol. 20, Autumn 1986, pp. 26–31; B. Morton, M. Elgersma, and R. Playter, "Analysis of Booster Vehicle Slosh Stability During Ascent to Orbit," AIAA Paper 90–1876, July 1990.

6–5. J. R. Rollins, R. K. Grove, and D. R. Walling, Jr. "Design and Qualification of a Surface Tension Propellant Tank for an Advanced Spacecraft," AIAA Paper 88–2848, 24th Joint Propulsion Conference, 1988.

6–6. *Design Guide for Pressurized Gas Systems*, Vols. I and II, prepared by IIT Research Institute, NASA Contract NAS7-388, March 1966.

6–7. H. C. Hearn, "Evaluation of Bipropellant Pressurization Concepts for Spacecraft," *Journal of Spacecraft and Rockets*, Vol. 19, July 1982, pp. 320–325.

6–8. H. C. Hearn, "Design and Development of a Large Bipropellant Blowdown Propulsion System," *Journal of Propulsion and Power*, Vol. 11, No. 5, September–October 1995, pp. 986–991; G. F. Pasley, "Prediction of Tank Pressure History in a Blowdown Propellant Feed System," *Journal of Spacecraft and Rockets*, Vol. 9, No. 6 (1972), pp. 473-475, doi: 10.2514/3.61718.

6–9. Personal communication with O. Morgan of Aerojet-Rocketdyne, Redmond, WA, and indirectly with C. Engelbrecht of Johns Hopkins University's Applied Physics Laboratory, Laurel, MD; K. Dommer of Aerojet Rocketdyne, Sacramento, CA, 2014, 2015.

6–10. K. Dommer and S. Wiley, "System Engineering in the Development of the MESSEN-GER Propulsion System"; AIAA 2006-5216; 42nd AIAA/ASME/SAE/ASEE Joint Propulsion Conference & Exhibit, Sacramento, CA, July 9–12, 2006.

6–11. Personal communications with J. H. Morehart, The Aerospace Corp., 2000–2015.

6–12. D. Manski, C. Goertz, H. D. Sassnick, J. R. Hulka, B. D. Goracke, and D. J. H. Levack, "Cycles for Earth to Orbit Propulsion," *Journal of Propulsion and Power, AIAA,* Vol. 14, No. 5, Sept.–Oct. 1998, pp. 588–604.

6–13. Personal communications with J. S. Kincaid (2012 to 2015), D. Adamski (2014), and R. Berenson (2015), RS-68 and other liquid propellant rocket engine facts.

6–14. B. K. Wood, "Propulsion for the 21st Century", AIAA Paper 2002-4324, July 2002.

6–15. D. Conley, N. Y. Lee, P. L. Portanova and B. K. Wood, "Evolved Expendable Launch Vehicle System: RS-68 Main Engine Development," Paper IAC-02.S.1.01, 53rd International Astronautical Congress, Oct. 10–19, 2002, Houston, Texas.

6–16. Personal communication with C. Cooley and P. Mills, Aerojet Rocketdyne, 2002–2015.

6–17. G. P. Sutton, "History of Small Liquid Propellant Thrusters," presented at 52nd JAN-NAF Propulsion Meeting, Las Vegas, NV, May 12, 2004; published by Chemical Propulsion Information Analysis Center, Columbia, MD.

6–18. "RD-191 Engine Scheme," data sheet published by NPO Energomash (undated).

6–19. R. P. Prickett, E. Mayer, and J. Hermel, "Waterhammer in Spacecraft Propellant Feed Systems," *Journal of Propulsion and Power*, Vol. 8, No. 3, May–June 1992, pp. 592–597, doi: 10.2514/3.23519; G. P. Sutton, Section 4.6, "Small Attitude Control and Trajectory Corrections," *History of Liquid Propellant Rocket Engines,* AIAA, 2006.

CHAPTER 7

LIQUID PROPELLANTS

The classification of liquid propellants was first introduced in Section 6.1. In this chapter, we discuss properties, performance, hazards, and other characteristics of commonly used propellants that are stored as liquids (and a few as gases). These characteristics influence engine and vehicle design, test facilities, and propellant storage and handling. At the present time, we ordinarily use three liquid bipropellant combinations: (1) the cryogenic *oxygen–hydrogen propellant system*, used in upper stages and sometimes booster stages of space launch vehicles, giving the highest specific impulse nontoxic propellant combination and one that is best for high vehicle velocity missions; (2) the *liquid oxygen–hydrocarbon propellant combination*, used for booster stages (and a few second stages) of space launch vehicles —having a higher average density allows more compact booster stages with less inert mass when compared to the previous combination (historically, it was developed first and was originally used with ballistic missiles); (3) not a single bipropellant combination but several ambient temperature *storable propellant combinations* used in large rocket engines for first and second stages of ballistic missiles and in almost all bipropellant low-thrust, auxiliary or reaction control rocket engines (this term is defined below); these allow for long-term storage and almost instant readiness (starting without the delays and the precautions that come with cryogenic propellants). Each of these propellant systems is further described in this chapter. Presently, Russia and China favor nitrogen tetroxide as the oxidizer and unsymmetrical dimethylhydrazine, or UDMH, as the fuel for ballistic missiles and for auxiliary engines. The U.S. has used nitrogen tetroxide and a fuel mixture of 50% UDMH with 50% hydrazine in the Titan II and III missiles' large engines. For auxiliary engines in many satellites and upper stages, the United States uses a nitrogen-tetroxide/monomethylhydrazine bipropellant. A subcategory of item (3) above is the *storable monopropellant* such as hydrogen

244

peroxide or hydrazine. The International Space Station and many U.S. satellites use monopropellant hydrazine for low-thrust auxiliary engines.

No truly new liquid propellant has been adopted for operational rocket flight vehicles in the past 30 years. Some new propellants (such as hydroxyl ammonium nitrate) were synthesized, manufactured, and ground tested in thrust chambers and flown in experimental vehicles in the past two decades, but they have not found their way into operational rocket engine applications. Between 1942 and 1975, a number of other propellants were successfully flown; these included ammonia (X-15 Research test aircraft), ethyl alcohol (German V-2 or U.S. Redstone missile), and aniline (WAC Corporal). They each had some disadvantages and are no longer used in operational flights today. Liquid fluorine and fluorine containing chemicals (such as chlorine pentafluoride and oxygen difluorine) give excellent performance and have been investigated and experimentally evaluated but, because of their extreme toxicity, they are no longer being considered.

A comparative listing of various performance quantities for a number of propellant combinations is given in Table 5–5 and in Ref. 7–1. Some important physical properties of selected common liquid propellants are shown in Table 7–1 (water is also listed for comparison). Specific gravities and vapor pressures are shown in Figs. 7–1 and 7–2. The *specific gravity* is defined to represent the ratio of the density of any given liquid to that of water at standard conditions (273 K and 1.0 atm) and thus carries no dimensions.

Green propellants (Ref. 7–2), a recently minted term, represent those liquid propellants and their exhaust gases that are "environmentally friendly" and can be used without causing damage to people, equipment, or the surroundings. An excellent example is the liquid oxygen–liquid hydrogen propellant combination—they are not toxic, not corrosive, not hypergolic and will not decompose or explode. Some authors use a more restricted interpretation for the green propellant category, namely, one that can replace a toxic and/or potentially explosive propellant with a chemical substance that is harmless.

7.1. PROPELLANT PROPERTIES

It is important to distinguish between characteristics and properties of *liquid propellants* (i.e., fuel and oxidizer liquids in their unreacted condition) and those of the *hot gas mixture* resulting from their reaction in the combustion chamber. The chemical nature of liquid propellants and their mixture ratio determine the properties and characteristics of both storage and reaction products. Because none of the known practical propellants encompass all properties deemed desirable, the selection of propellant combinations is usually a compromise between various economic factors, such as those listed below.

Economic Factors

Availability in quantity and a *low cost* are very important considerations in propellant selection. In military applications, consideration has to be given to the *logistics* of

TABLE 7–1. Physical Properties of Liquid Propellants

Propellant	Liquid Oxygen	Nitrous oxide	Nitrogen Tetroxide	Nitric Acid[a] (99% pure)	Rocket Fuel RP-1, RP-2
Chemical formula	O_2	N_2O	N_2O_4	HNO_3	Hydrocarbon $CH_{1.97}$
Molecular mass	31.988	44.013	92.016	63.016	~175
Melting or freezing point (K)	54.8	182.29	261.95	231.6	225
Boiling point (K)	90.2	184.67	294.3	355.7	460–540
Heat of vaporization (kJ/kg)	213	374.3 (at 1 atm)	413[2]	480	246[b]
Specific heat (kcal/kg-K)	0.4 (65 K)	0.209	0.374 (290 K) 0.447 (360 K)	0.042 (311 K) 0.163 (373 K)	0.48+ (298 K)
Specific gravity[c]	1.14 (90.4 K) 1.23 (77.6 K)	1.23[b]	1.38 (293 K) 1.447 (322 K)	1.549 (273.15 K) 1.476 (313.15 K)	0.58 (422 K) 0.807 (289 K)
Viscosity (centipoises)	0.87 (53.7 K) 0.19 (90.4 K)	0.0146 (gas at 300 K)	0.47 (293 K) 0.33 (315 K)	1.45 (273 K)	0.75 (289 K) 0.21 (366 K)
Vapor pressure (MPa)	0.0052 (88.7 K)	5.025 (293 K)	0.1014 (293 K) 0.2013 (328 K)	0.0027 (273.15 K) 0.605 (343 K)	0.002 (344 K) 0.023 (422 K)

[a]Red fuming nitric acid (RFNA) has 5 to 20% dissolved NO_2 with an average molecular mass of about 60, and a density and vapor pressure somewhat higher than those of pure nitric acid.
[b]At boiling point.
[c]Reference for specific gravity ratio: 10^3 kg/m^3 or 62.42 lbm/ft^3.

production, supply, storage, along with other factors. The production process should require only ordinarily available chemical equipment and available raw materials. It is more expensive to use toxic or cryogenic propellants than storable, nontoxic ones, because the former require additional steps in their operation, more safety provisions, additional design features, longer check-out procedures prior to launch, and often better trained personnel.

Performance of Propellants

The performance rocket engines may be compared on the basis of their *specific impulse, exhaust velocity, characteristic velocity,* and/or other engine parameters. These were introduced in Chapters 3, 5, and 6. The specific impulse and exhaust

Liquid Hydrogen	Liquid Methane	Monomethyl-hydrazine	Hydrazine	Unsymmetrical Dimethyl-hydrazine	Water
para–H_2	CH_4	$CH_3 NHNH_2$	$N_2 H_4$	$(CH_3)_2 NNH_2$	H_2O
2.016	16.04	46.072	32.045	60.099	18.02
14.0	90.67	220.7	275.16	216	273.15
20.27	111.7	360.8	387.46	335.5	373.15
446	510[b]	808	1219[b]	543	2253[2]
2.34[b]	0.835[b]	0.700	0.736	0.704	1.008
(20.27 K)		(298 K)	(293 K)	(298 K)	(273.15 K)
—	—	0.735	0.758	0.715	
		(393 K)	(338 K)	(340 K)	
0.071	0.424	0.8702	1.0037	0.7861	1.002
(20.4 K)	(111.5 K)	(298 K)	(298 K)	(298 K)	(373.15 K)
0.076	—	0.857	0.952	0.784	1.00
(14 K)		(311 K)	(350 K)	(244 K)	(293.4 K)
0.024	0.12	0.775	0.97	0.492	0.284
(14.3 K)	(111.6 K)	(298 K)	(298 K)	(298 K)	(373.15 K)
0.013	0.22	0.40	0.913	0.48	1.000
(20.4 K)	(90.5 K)	(344 K)	(330 K)	(300 K)	(277 K)
0.2026	0.033	0.0066	0.0019	0.0223	0.00689
(23 K)	(100 K)	(298 K)	(298 K)	(298 K)	(312 K)
0.87	0.101	0.638	0.016	0.1093	0.03447
(30 K)	(111.7 K)	(428 K)	(340 K)	(339 K)	(345 K)

velocity are functions of pressure ratio, specific heat ratio, combustion temperature, mixture ratio, and molecular mass. Equilibrium values of performance parameters for various propellant combinations can be calculated with a high degree of accuracy and several are listed in Table 5–5. Very often, performance is also expressed in terms of *flight performance parameters* for specific rocket applications, as explained in Chapter 4. Here, average propellant density, total impulse, and engine mass ratio usually enter into the various flight relation descriptions.

For high performance, *high chemical energy content* per unit of propellant mixture is desirable because this yields high chamber temperatures. A *low molecular mass* for the combustion products gases is also desirable. This can be accomplished by using fuels rich in hydrogen obtained when a significant portion of the hydrogen gas injected or produced remains uncombined. In general, therefore, the best mixture ratio for many bipropellants is not stoichiometric (which results in complete oxidation and yields the highest flame temperature) but fuel-rich, containing a large amounts of low-molecular-mass reaction products as shown in Chapter 5.

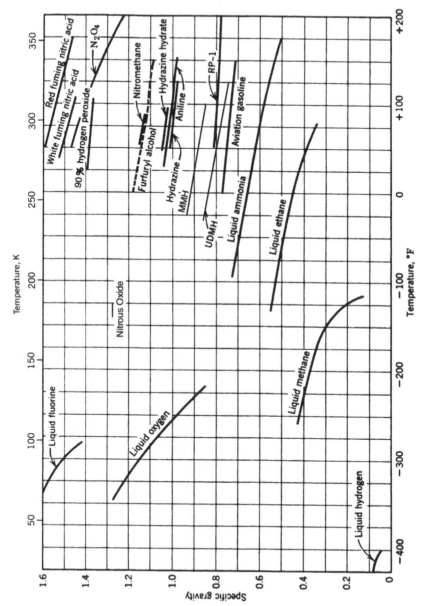

FIGURE 7–1. Specific gravities of several liquid propellants as a function of temperature.

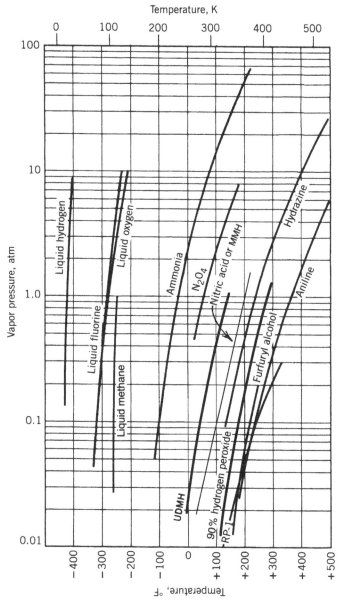

FIGURE 7–2. Vapor pressures of several liquid propellants as a function of temperature.

When relatively small metallic fuel particles (such as beryllium or aluminum) are suspended in the liquid fuel, it is theoretically possible to increase the specific impulse by between 9 and 18%. A particular chemical propellant combination having the highest ideal specific impulse known (approximately 480 sec at 1000 psia chamber pressure and expansion to sea-level atmosphere, and 565 sec in a vacuum with a nozzle area ratio of 50) uses a toxic liquid fluorine oxidizer with hydrogen fuel plus suspended toxic solid particles of beryllium; as yet, acceptably safe and practical means for storing such seeded propellants and their practical rocket engine use remain undeveloped.

Gelled propellants are materials that have additives that make them thixotropic. They have the consistency of thick paint or jelly when at rest, but do liquefy and flow through pipes, valves, pumps and/or injectors when an adequate pressure or shear stress is applied. In spite of extensive research and development work and demonstrations of better safety and certain "green" qualities, to date they have not been adopted for any production rocket engine. Gelled propellants have been described in the Sixth, Seventh, and Eighth editions of this book.

Common Physical Hazards

Although several hazard categories are described below, they do not all apply to each propellant or to every bipropellant combination. Hazards can be different for each specific propellant and must be carefully understood before working with it. The consequences of unsafe operation or unsafe design are usually also unique to each propellant.

Corrosion. Several propellants, such as nitrogen tetroxide, nitric acid, nitric oxide and/or hydrogen peroxide, can only be handled in containers and pipelines made from special materials. When any propellant gets contaminated with corrosion products, its physical and chemical properties may sufficiently change to make it unsuitable for its intended operation. Corrosion caused by expelled gaseous reaction products is most critical in applications where the reaction products are likely to damage launch or ground test structures and parts of the vehicle, and/or affect communities and housing near a test facility or launch site.

Explosion Hazard. Over time some propellants (e.g., hydrogen peroxide or nitromethane) can become unstable in their storage tanks and may even detonate under certain conditions, depending on local impurities, temperatures, and shock magnitudes. When liquid oxidizers (e.g., liquid oxygen) and fuels get unintentionally mixed together, detonation may sometimes result. Unusual flight vehicle launch mishaps or transport accidents have caused such mixing and subsequent explosions to occur (see Refs. 7–3 and 7–4).

Fire Hazard. Many oxidizers will react with a large variety of organic compounds. Nitric acid, nitrogen tetroxide, fluorine, and/or hydrogen peroxide react spontaneously when in contact with many organic substances resulting in fires. Most of rocket fuels exposed to air are readily ignitable when heated. Also some household dusts, certain paints, or smoke particles can oxidize. Oxygen by itself will not usually start a fire with organic materials but it will greatly enhance an existing fire.

Accidental Spills. Unforeseen mishaps during engine operation or traffic accidents on highways or railroads while transporting hazardous materials, including many propellants, have on occasion caused spills which expose people to intense fires and/or potential health hazards. The U.S. Department of Transportation has strict rules for marking and containing hazardous materials during transport and also guidelines for emergency actions (see Ref. 7–5).

Health Hazards. Exposure to many commonly used propellants represents a health hazard. Toxic unburned chemicals or poisonous exhaust species affect the human body in a variety of ways and the resulting health disorders are propellant specific. Nitric acid causes severe skin burns and tissue disintegration. Skin contact with aniline or hydrazine may cause nausea and other adverse health effects. Hydrazine, monomethylhydrazine, unsymmetrical dimethylhydrazine, or hydrazine hydrate are known animal and suspected human carcinogens. Many propellant vapors cause eye irritation, even in small concentrations. Inadvertent ingestion of many propellants may result severe health degradation.

Inhalation of toxic exhaust gases or gaseous or vaporized liquid propellants is perhaps the most common health hazard. It can cause severe damage if the exposure is long or in concentrations that exceed established maximum threshold values. In the United States, the Occupational Safety and Health Administration (OSHA) has established limits or thresholds on the allowable exposure and concentration for most propellant chemicals. Several of these threshold limits are mentioned later in this chapter. References 7–3 and 7–6 give more information on toxic effects.

Toxic Propellants. These require special safety provisions, strict rules, and specific procedures for handling, transferring to other containers, road transport, inspection, and for working on rocket engines that contain them (such as the removal of residuals after testing or after return from a space mission). Instruments are available to detect toxic vapors or toxic contents in liquids or water. For personnel protection, face shields and gas masks (some with an oxygen supply), special gloves and boots, sealed communications equipment, medical supplies and periodic medical examinations need to be available. For accidental toxic spills or leaks there should be equipment for diluting with water or other suitable chemical, and for safe disposal of the contaminants from a test stand or launch platform. Chemicals for neutralization and detoxification should always be on hand. In case of mishaps instructions and procedures for notifying management, certain government offices and relevant others should be available and not overlooked. When compared to nontoxics, operations with toxic-propellants require two to four times as many trained workers. Only subsets of the precautions listed above apply to any one operation. Toxic propellant safety is further discussed in Chapters 20 and 21.

Materials Compatibility. For several liquid propellants there are only a limited number of truly compatible materials, both metals and nonmetals, particularly for making gaskets or O-rings. There have been many failures (causing fires, leakage, corrosion, or other malfunctions) when improper or incompatible hardware materials have been used in rocket engines. Depending on the specific component and loading conditions, structural materials have to withstand high stresses, stress corrosion, and in some applications high temperatures and/or abrasion. Several specific material limitations are mentioned in the next section. Certain storage materials may act to catalyze the self-decomposition of hydrogen peroxide into water and oxygen making long-term storage difficult and causing any closed containers to explode. Many structural materials, when exposed to cold (cryogenic) propellants become unacceptably brittle.

Desirable Physical Properties

Low Freezing Point. This permits operation of rockets in cold environments. The addition of small amounts of freezing point depressants has been found to help lower the freezing point in some liquid propellants that might otherwise solidify at environmental storage conditions.

High Specific Gravity. Denser propellants provide a larger propellant mass for a given vehicle tank volume. Alternatively, for a given mass, they permit smaller tank volumes and, consequently, lower structural vehicle mass and lower aerodynamic drag. The specific gravity of a propellant, therefore, has an important effect on the maximum flight velocity and range of any rocket-powered vehicle or missile flying within the Earth's atmosphere as explained in Chapter 4. Specific gravities for various propellants are plotted in Fig. 7–1. Variations of ambient temperature in stored propellants cause changes of the liquid level in their storage tanks.

For a given bipropellant mixture ratio r, the *average specific gravity* of any propellant combination δ_{av} can be determined from the specific gravities of the fuel δ_f and of the oxidizer δ_o. This average specific gravity δ_{av} is defined below. Here r is the mixture ratio; it represents the oxidizer mass flow rate divided by the fuel mass flow rate (see Eq. 6–1):

$$\delta_{av} = \frac{\delta_o \delta_f (1 + r)}{r\delta_f + \delta_o} \qquad (7-1)$$

Values of δ_{av} for various propellant combinations are listed in Table 5–5. The value of δ_{av} can be increased by adding high density materials to the propellants, either by solution or colloidal suspension. An identical equation can be written for the average density ρ_{av} in terms of the fuel and oxidizer densities:

$$\rho_{av} = \frac{\rho_o \rho_f (1 + r)}{\rho_f r + \rho_o} \qquad (7-2)$$

Though the specific gravity is unitless, in the SI system it has the same numerical value as the density expressed in units of grams per cubic centimeter or kg/liter.

In some performance comparisons the parameter *density specific impulse I_d* is used. It is defined as the product of the average specific gravity δ_{av} and the specific impulse I_s:

$$I_d = \delta_{av} I_s \qquad (7-3)$$

A *propellant density increase* will allow increases in mass flow and total propellant mass when all other system parameters remain the same. For example, Fig. 7–1 shows that lowering the temperature of liquid oxygen from –250 to –280 °F raises its specific gravity by approximately 8%. Therefore, the system's mass flow rate and the total mass will increase by approximately the same amount. These changes will enable increases in chamber pressure, total impulse, and thrust; changes of less than 1% in specific impulse may also be noticed. So there is some benefit to vehicle performance in operating with the liquid propellants at their lowest practical temperature. In ground-based systems, cooling of liquid oxygen may be achieved with a (colder) liquid nitrogen heat exchanger just before launch. The percent increase in performance with propellants other than oxygen is usually much smaller.

Stability. Proper *chemical stability* means no decomposition of the liquid propellant during operation or storage, even at elevated temperatures. With many propellants, insignificant amounts of *deterioration* and/or *decomposition* during long-term (over 15 years) storage and minimal *reaction with the atmosphere* have been attained. A desirable liquid propellant should also experience no *chemical deterioration* when in contact with tubing, pipes, tank walls, valve seats, and gasket materials even at relatively high ambient temperatures. No appreciable *absorption of moisture* and no adverse effects of small amounts of *impurities* are also desirable properties. There should also be no appreciable chemical deterioration when liquids flow through hot cooling jacket passages of a regeneratively cooled thrust chamber. When carbon-containing coolants decompose and form (carbonaceous) deposits on hot inside surfaces of cooling passages, such deposits may harden, reducing the heat flow, increasing metal temperatures locally, and thus may cause the unit to weaken and eventually fail. Unavoidably, even in well-insulated tanks, between 1 and 20% of a cryogenic propellant may evaporate daily when stored in the flight vehicle.

Heat Transfer Properties. *High specific heat, high thermal conductivity, low freezing temperature*, and *high boiling or decomposition temperature* are all desirable properties for propellants used for thrust chamber cooling (see Section 8.5).

Pumping Properties. Low *vapor pressures* permit not only easier handling of propellants, but also more effective pump designs in applications where propellants must be pumped. Low vapor pressures also reduce the potential for pump cavitation, as explained in Chapter 10. When the propellant *viscosity* is too high, then pumping and engine-system calibrations become difficult. Propellants with high vapor pressures (such as LOX, liquid hydrogen, and other liquefied gases) require special design provisions, unusual handling techniques, and special low-temperature materials.

Temperature Variation of Physical Properties. *Temperature variations* in the physical properties of any liquid propellant should be small and should be very similar for the fuel and oxidizer. For example, wide temperature variations in vapor pressure and density or unduly high changes in viscosity with temperature make it difficult to accurately calibrate rocket engine flow systems and/or to predict their performance over any reasonable range of operating temperatures. While stored in a flight vehicle, if one of the propellants experiences a larger temperature change than the other, this may cause a noticeable change in mixture ratio and in specific impulse, and possibly a significant increase in unusable or undesirable propellant residue.

Ignition, Combustion, and Flame Properties

When a propellant combination is *spontaneously ignitable* burning is initiated as soon as the oxidizer and the fuel come in contact with each other. Spontaneously or self-ignitable propellant combinations are called *hypergolic* propellants. Although ignition systems are not necessarily objectionable, their elimination simplifies the propulsion system. In order to reduce potential explosion hazards during starting, all rocket propellants should be readily ignitable and exhibit acceptably short ignition time delays in order to eliminate potential explosion hazards. Starting and ignition problems are discussed further in Section 8.6.

Nonspontaneously ignitable propellants must be energized by external means for ignition to begin. Igniters are devices that accomplish a localized initial chamber pressurization and initial heating of the propellant mixture to a state where steady flow combustion can be self-sustained. The amount of energy needed from the igniter to activate the propellants should be small so that low-power, light-weight ignition systems may be used. The energy required for satisfactory ignition usually diminishes with increasing propellant storage temperature. At low ambient temperatures ignition can become relatively slow (0.05 to 0.02 sec).

Certain propellant combinations burn very smoothly, without vibration (i.e., gas pressure oscillations). Other propellant combinations do not exhibit such *combustion stability* and, therefore, become less desirable. Combustion stability is treated in Chapter 9.

Smoke formation is objectionable in many applications because it may deposit on the surrounding equipment and parts. *Smoke* and brilliantly *luminous exhaust flames* are objectionable for certain military applications since they can be easily detected. In some applications, the condensed species from gaseous exhausts can cause *surface contamination* on spacecraft windows or optical lenses, and the presence of free electrons in a flame may cause undesirable *interference* or *attenuation of communications radio signals*. See Chapter 20 for information on exhaust plumes.

Property Variations and Specifications

Propellant properties and quality must not vary from delivered batch to batch because this can affect engine performance and combustion from changes in physical and/or chemical properties. The same propellant must have consistent composition and storage properties; rocket operating characteristics should not change even when propellants are manufactured at different times or made by different manufacturers. For these reasons propellants are purchased to conform to strict specifications which define ingredients, maximum allowable impurities, packaging methods and/or compatible materials, allowable tolerances on physical properties (such as density, boiling point, freezing point, viscosity, or vapor pressure), quality control requirements, container cleaning procedures, documents of inspections, laboratory analyses, and/or test results. A careful chemical analysis of composition and impurities is always necessary. Reference 7–7 describes some of these methods of analysis.

Additives

Altering and tailoring propellant properties may be achieved with additives. For example, a reactive ingredient is added to make a nonhypergolic fuel become hypergolic (readily ignitable). To desensitize concentrated hydrogen peroxide and reduce self-decomposition, it is diluted with 3 to 15% of pure water. To increase the density or to alleviate certain combustion instabilities, a fine powder of a heavy solid material (such as aluminum) is suspended in the fuel. The use of additives to lower the freezing-point temperature of nitrogen tetroxide is treated later.

7.2. LIQUID OXIDIZERS

The most energetic known oxidizer producing the highest specific impulse and having the highest density is liquid fluorine. It has been tested in several complete experimental rocket engines but abandoned because of its extreme hazards. Many different types of new storable and cryogenic liquid oxidizer propellants have been synthesized, tested in small thrust chambers, or proposed; these included mixtures of liquid oxygen and liquid fluorine, oxygen difluoride (OF_2), chlorine trifluoride (ClF_3), and chlorine pentafluoride (ClF_5). None of these are used today because they are highly toxic and very corrosive. Several of the most commonly used oxidizers are listed below.

Liquid Oxygen (O_2) (LOX)

Liquid oxygen is listed in Table 7–1. It is widely used as an oxidizer and burns with a bright white-yellow flame with most hydrocarbon fuels. It has been used in combination with alcohols, jet fuels (kerosene-type), gasoline, and hydrogen. As shown in Table 5–5, attainable performances are relatively high, and LOX is therefore a desirable and commonly used propellant in large rocket engines. The following missiles and space launch vehicles burn oxygen: (1) with jet fuel or RP-1 (kerosene)—Atlas V and Soyuz (Russia); (2) with hydrogen—Ariane-V (France), Delta IV, and Centaur upper stage; (3) with ethanol—the historic V-2 (German) and the Redstone missile. Figs. 1–12, 1–13, and 6–1 show units that use LOX; Figs. 5–1 to 5–6 present theoretical performance information for LOX with one kerosene-type fuel.

Although it usually does not burn spontaneously with organic materials at ambient pressures, combustion and/or explosions have occurred when confined mixtures of oxygen and organic matter suddenly come under pressure. Impact tests show that mixtures of LOX with many commercial oils or organic materials will detonate.

Liquid oxygen also supports and accelerates the combustion of other materials. Handling and storage can be safe only when all contact materials are clean. Liquid oxygen is a noncorrosive and nontoxic liquid that will not cause the deterioration of clean container walls. Because of its low temperature, this propellant causes severe frostbite during any prolonged contact with human skin. Because LOX evaporates rapidly at ambient conditions, it cannot be stored readily for any extended length of time. When LOX is used in large quantities, it is often produced very close to its geographical point of application. Liquid oxygen can be obtained in several ways, but most commonly from the fractionated distillation of liquid nitrogen out of liquid air.

It is necessary to insulate all lines, tanks, valves, and parts that contain liquid oxygen in order to reduce evaporation losses. Rocket propulsion systems which have to remain filled with liquid oxygen for several hours and liquid oxygen storage systems need to be well insulated against heat absorption from the surroundings. External drainage provisions have to be made on all LOX tanks and lines to allow the water that condenses on the cold outside walls to drain from the rocket during launch preparations. Provisions for minimizing and removing ice formations are also needed.

Example 7–1.   Estimate the approximate temperature and volume change of LOX in an oxygen tank that has been under 8.0 atm pressure for long times before engine start. Assume the tank was 60% full at one atmosphere and that the evaporated oxygen is caught, refrigerated, and recondensed (i.e., a constant mass).

SOLUTION. Using Table 7–1 and Figs. 7–1 and 7–2, the vapor pressure is changed from 1.0 atm (0.1 MPa) to 8 atm (about 0.8 MPa) so that the equilibrium temperature changes from the boiling point of 90 K at 1.0 atm to about 133 K at 8 atm. The corresponding specific gravities are 1.14 and 0.88, respectively. This is a change of 1.14/0.88 = 1.29 (a 29% increase of liquid volume). At 133 K the tank would now be 60x1.29 = 77.4% full.

In tanks with turbopump feed systems the actual tank pressures are lower (typically 2 to 4 atm) and the evaporated oxygen is vented, causing a cooling effect on the liquid surface. So the numbers calculated above are too large (8 atm was selected to clearly show the effect). Warming of a cryogenic liquid occurs during long launch hold delays on pressurized propellant tanks and is most noticeable when the final portions of the liquid are being emptied. The higher temperature, higher vapor pressure, and lower density may cause changes in mixture ratio, required tank volumes, and pump suction conditions (see Section 10.5). Therefore, tanks with cryogenic propellants are heavily insulated (to minimize heat transfer) and are pressurized only shortly before engine start, so as to keep the propellant at its lowest possible temperature and highest density.

Hydrogen Peroxide (H_2O_2)

This propellant is not only a powerful liquid oxidizer when burning with an organic fuel but also clean burning. It produces a nontoxic exhaust when used as a monopropellant. In rocket applications, hydrogen peroxide needs to be in a highly concentrated form of 70 to 98% (known as high-test peroxide or HTP), the remainder being mostly water. Hydrogen peroxide was used in gas generators and rocket applications between 1938 and 1965 (X-1 and X-15 research aircraft). Since that time, 90% hydrogen peroxide has been the most common concentration for rocket engine use and gas generator applications.

As a monopropellant, it decomposes according to the following chemical reaction, forming superheated steam and hot gaseous oxygen:

$$H_2O_2 \rightarrow H_2O + \frac{1}{2}O_2 + \text{heat}$$

This decomposition is brought about by the action of catalysts such as silver screens, various liquid permanganates, solid manganese dioxide, platinum, or iron oxide but, in fact, many materials can catalyze hydrogen peroxide. Theoretical specific impulses for 90% hydrogen peroxide can be around 154 sec, when used as a monopropellant with a solid catalyst bed. As bipropellant, H_2O_2 is hypergolic with hydrazine and will burn well with kerosene.

Concentrated peroxide causes severe burns when in contact with human skin and may ignite when in contact with wood, oils, and many other organic materials. In the past, rocket engines with hydrogen peroxide as oxidizer have been used for aircraft boosting (German Me 163, and U.S. F 104) and for satellite launching (Britain: Black Arrow). As a monopropellant it has been used for attitude control thrusters in the Soyuz space capsule and for steam generators driving the main propellant pumps in the Soyuz launch vehicle. It was not used in the United States for some time, partly because it was hard to predict its long-term storage stability. In the past, concentrated hydrogen peroxide would self-decompose during storage at about 1% per year. This progressively diluted this propellant until no longer useful. As oxygen gas would bubble out of the stored liquid, the decomposition rate would accelerate. Small amounts of impurities in the liquid or in wall materials would also accelerate decomposition and cause a rise in liquid temperature. Before reaching 448 K (175 °C), warm stored H_2O_2 or contaminated H_2O_2 must be diluted and discarded because an explosion may occur.

Recent progress in the manufacture of storage tanks and piping materials (and stricter cleaning methods) has much reduced the amount of impurities and lengthened storage life of HTP from about 3 to 4 years to 12 to 16 years, and there may further improvements. This has much renewed interest in this dense oxidizer and several different investigations and development programs have been underway to date (Ref. 7–8). To date, the authors have not become aware of any new production applications for 90% H_2O_2.

Nitric Acid (HNO₃)

Several types of nitric acid mixtures were used as oxidizers between 1940 and 1965; these are not used extensively today in the U.S. The most common type, *red fuming nitric acid* (RFNA), consists of concentrated nitric acid (HNO_3) that contains between 5 and 27% dissolved nitrogen dioxide (NO_2). Compared to concentrated nitric acid (also called *white fuming nitric acid*), RFNA is more energetic, more stable in storage, and slightly less corrosive to many tank materials.

Only certain types of stainless steel, gold, and a few other materials are satisfactory as storage containers and/or tubing materials for concentrated nitric acid. Small additions of fluoride ion (less than 1% of hydrofluoric acid, or HF) inhibit nitric acid corrosion, forming a protective fluoride layer on walls and reducing the corrosion with many metals. This combination is called *inhibited red fuming nitric acid* (IRFNA). Even with an inhibitor, nitric acid reacts with many wall materials and forms dissolved nitrates and sometime insoluble nitrates. This changes the properties of this oxidizer and may cause blocking of valve and injector orifices. Nitric acid reacts with gasoline, various amines, hydrazine, dimethylhydrazine, and alcohols. It ignites spontaneously with hydrazine, furfuryl alcohol, aniline, and other amines. The specific gravity of nitric acid varies from 1.5 to 1.6, depending on the percentages of nitrogen dioxide, water, and impurities. This high density permits compact vehicle construction.

When RFNA is exposed to air, the evaporating red-brown fumes are exceedingly annoying and poisonous. In case of accidental spilling, the acid must be quickly diluted with water or chemically deactivated. Lime and alkali metal hydroxides and carbonates are common neutralizing agents. However, any nitrates formed by the neutralization also act as oxidizing agents and must be handled accordingly. Vapors from nitric acid or red fuming nitric acid have an OSHA eight-hour personnel exposure limit or a threshold work allowance of 2 ppm (parts per million, or about 5 mg/m^3) and a short-term exposure limit of 4 ppm. Droplets on the skin cause burns producing sores that do not heal readily.

Nitrogen Tetroxide (N_2O_4) (NTO)

The proper chemical name for N_2O_4 is 'dinitrogen tetroxide' as widely used by chemists and the chemical industry, but the rocket industry has always called it nitrogen tetroxide or NTO and this is the designation used in this book. It is a high-density yellow-brown liquid (specific gravity of 1.44). NTO is hypergolic with hydrazine, monomethylhydrazine (MMH), and unsymmetrical dimethylhydrazine (UDMH); it is the most common storable oxidizer employed today. Its liquid temperature range is narrow and thus easily accidentally frozen or vaporized. It is only mildly corrosive when pure but forms strong acids when moist or allowed to mix with water, readily absorbing moisture from the air. It can be stored indefinitely in sealed containers made of compatible materials. NTO is hypergolic with many fuels but can also cause spontaneous ignition with many common materials, such as paper, grease, leather, or wood. When it decomposes, the resulting NO_2 fumes are reddish brown and extremely toxic. Because of its high vapor pressure, it must be kept sealed in relatively heavy tanks. The freezing point of N_2O_4 may be lowered (for example, by adding a small amount of nitric oxide or NO) but at the expense of a higher vapor pressure and slightly reduced performance. Such mixtures of NO and N_2O_4 are called mixed oxides of nitrogen (MON) and different grades are available between 2 and 30% by weight NO content. Several MON oxidizers have flown.

Nitrogen tetroxide is used as the oxidizer with UDMH in many Russian engines and in almost all their small thrusters. It was also used with MMH fuel in the Space Shuttle orbital maneuver system and multithruster reaction control system, and is still used in many U.S. spacecraft propulsion systems. For many applications, care has to be taken to avoid freezing the nitrogen tetroxide. The OSHA eight-hour personnel exposure limit is 5 ppm NO_2 or 9 mg/m^3.

Nitrous Oxide (N_2O)

This oxidizer is also known as dinitrogen monoxide and as "laughing gas" (a medical anesthetic), see Refs. 7–9 and 7–10. It is a less potent oxidizer than the other four listed above. At ambient temperature it is not flammable but supports combustion at elevated temperatures. Its handling and safety are discussed in Refs. 7–10 and 7–11. Compared to nitrogen tetroxide (N_2O_4) it is much less toxic, its eight-hour exposure limit being about 20 times less severe. Various catalysts will cause the decomposition of N_2O into O_2 and N_2, and many materials and some impurities are incompatible with N_2O. Its cryogenic liquid temperature range is very narrow, see Table 7–1 for boiling and freezing points; liquid N_2O is not as cold as LOX. Although somewhat difficult to control, this oxidizer can self-pressurize itself, avoiding the need for separate helium pressurizing systems.

Nitrous oxide has flown in space for the last 14 years. One principal recent flight application has been that of liquid nitrous oxide with HTPB (hydroxyl-terminated polybutadiene) as a solid fuel in a hybrid rocket propulsion system (Ref. 7–12) for a suborbital manned space plane, which is discussed in Chapter 16. In a bipropellant combination, it has also been tested as the oxidizer with organic liquid fuels and gaseous monopropellants using a catalyst.

Oxidizer Cleaning Process

The internal surfaces of all *newly manufactured* components for rocket engine liquid oxidizer systems (pipes, tanks, valves, *pumps*, *seals*, injectors, etc.) usually undergo a special cleaning process for removing all traces of organic materials or other impurities that can react with the oxidized and cause bubbles or overpressure. This cleaning includes removal of even minute amounts of grease, oils, machine cuttings, fluids, paints, and small deposits of carbon, plastics, and common dust. The cleaning process may differ for each particular oxidizer and consists of scraping-off *visible deposits*, successive multiple flushings and rinsing (with appropriate liquids), followed by drying with clean hot air. After the components are *"oxygen clean"* they are usually sealed to avoid any contamination prior to assembly or operation.

7.3. LIQUID FUELS

As with oxidizers, many different chemicals have been proposed, investigated, and tested as fuels. Liquid fuels *not* listed in this chapter that have been used in experimental rocket engines, in older experimental designs, and in older production engines include aniline, furfuryl alcohol, xylidine, gasoline, hydrazine hydrate, borohydrides, methyl and/or ethyl alcohol, ammonia, and some mixtures of these.

Hydrocarbon Fuels

Petroleum derivatives encompass a large variety of different hydrocarbon chemicals that can be used as rocket fuels. The most common are types already in use with other engines applications, such as gasoline, kerosene, diesel oil, and turbojet fuel. Their physical properties and chemical composition vary widely with the type of crude oil from which they were refined, with the chemical process used in their production, and with the accuracy of control exercised in their manufacture. Typical values are listed in Table 7–2.

In general, these petroleum fuels form yellow-white, brilliantly radiating flames and give good performance. They are relatively easy to handle and abundant at relatively low cost. Specifically refined petroleum products particularly suitable as rocket propellants carry the designation of RP-1 or -2 (rocket propellant number 1 or 2). These are basically a kerosene-type fuel mixture of hydrocarbons with a somewhat narrow range of densities and vapor pressures. Under some conditions, hydrocarbon fuels may form carbon deposits on the inside of cooling passages, impeding heat transfer and raising wall temperatures. Ref. 7–13 indicates that such carbon formations depend on fuel temperature in the cooling jacket, the particular fuel, the rate of heat transfer, and the chamber wall material. Fortunately, RP-1 is low in olefins and aromatics (the ones likely to cause such solid carbonaceous deposits inside fuel cooling passages). RP-1 has been used with liquid oxygen in many early rocket engines

TABLE 7–2. Properties of Some Typical Hydrocarbon Fuels Made from Petroleum

	Jet Fuel	Kerosene	Aviation Gasoline 100/130	RP-1	RP-2
Specific gravity at 289 K	0.78	0.81	0.73	0.80–0.815	0.80–0.815
Freezing point (K)	213 (max.)	230	213	225	225
Viscosity at 289 K (centipoise)	1.4	1.6	0.5	0.75 (at 289 K)	0.75 (at 289 K)
Flash point[a] (K)	269	331	244	333	333
ASTM distillation (K)					
10% evaporated	347	—	337	458–483	458–483
50% evaporated	444	—	363	—	—
90% evaporated	511	—	391	—	—
Reid vapor pressure (psia)	2 to 3	Below 1	7	—	—
Specific heat (kcal/kg-K)	0.50	0.49	0.53	0.50	0.50
Average molecular mass (g/mol)	130	175	90	—	—
Sulfur, total mg/kg	—	—	—	30 (max)	1 (max)

[a]Tag closed cup method.

(see Figs. 5–1 to 5–6 and Ref. 7–14). A very similar kerosene-type fuel is used in Russia today. In about 2003, RP-2 was substituted for RP-1 in U.S. rocket applications; the principal difference is a reduced sulphur content (see Table 7–2) because this impurity was believed to have caused corrosion in cooling jackets of certain thrust chambers. See Ref. 7–14.

Methane (CH_4) is a cryogenic liquid hydrocarbon and the main constituent of ordinary "natural gas." Liquid methane is abundant and relatively low in cost. Compared to petroleum-refined hydrocarbons, it has highly reproducible properties. A bipropellant consisting of LOX and methane gives a lower characteristic velocity c^* and/or specific impulse I_s than LOX-hydrogen but higher than LOX-kerosene or LOX-RP-1, see Table 5–5. Several rocket propulsion organizations in different countries have been developing and ground testing experimental liquid propellant thrust chambers and/or small reaction control thrusters with LOX-CH_4. Since the density of liquid methane is approximately six times higher than liquid hydrogen, methane fuel tanks can be much smaller and less costly. Methane is being considered as a bipropellant fuel (with LOX) for future rocket engines in large, multistage space launch vehicles for both manned and unmanned missions to Mars. In anticipation of three Mars flights, a new larger oxygen-methane engine is being developed in the United States; while these investigations are presently ongoing, there have not been any flights (as of 2015, personal communication with J. H. Morehart of The Aerospace Corporation).

Liquid Hydrogen

As shown in Table 5–5, hydrogen when burned with oxygen gives a high performance colorless flame (shock waves may however be visible in the rocket plume). Liquid hydrogen is an excellent regenerative coolant. Of all known fuels, liquid hydrogen is by far the lightest and the coldest, having a specific gravity of 0.07 and a boiling point of about 20 K. This extremely low fuel density requires bulky and large fuel tanks, which necessitate rather large vehicle volumes. The extremely low storage temperatures limit available materials for pumps, cooling jackets, tanks, and piping because many metals become brittle at such temperatures.

Because of their very low temperatures, liquid hydrogen tanks and supply lines have to be well insulated to minimize hydrogen evaporation or the condensation of moisture or air on the outside, with subsequent formations of liquid or solid air and ice. A vacuum jacket has often been used in addition to the insulating materials. All common liquids and gases solidify in liquid hydrogen. Such solid particles in turn plug orifices and valves. Therefore, care must be taken to purge all lines and tanks of air and moisture (flushing with helium and/or creating vacuums) before introducing any liquid hydrogen propellant. Mixtures of liquid hydrogen and solid oxygen or solidified air can actually explode.

Liquid hydrogen is manufactured from gaseous hydrogen by successive compression, cooling, and expansion processes. Hydrogen has two isomers, namely, orthohydrogen and parahydrogen, which differ in the orientation of their nuclear spin state. As hydrogen is liquefied and cooled to lower and lower temperatures, the relative equilibrium composition of ortho- and parahydrogen changes. The transformation from orthohydrogen to parahydrogen is exothermic and results in excessive boil-off, unless complete conversion to parahydrogen is achieved during liquefaction. Parahydrogen is the only stable form of liquid hydrogen.

Hydrogen gas, when mixed with air, is highly flammable and explosive over a wide range of mixture ratios. To avoid this danger, escaping excess hydrogen gas (at tank vent lines) is often intentionally ignited or "flared" in air. Liquid hydrogen is used with liquid oxygen in the Delta IV, Ariane V and -HII launch vehicles, Centaur upper stage, and upper stage space engines developed in Japan, Russia, Europe, India, and China.

Hydrogen burning with oxygen forms a nontoxic essentially invisible exhaust. This propellant combination has been applied successfully to space launch vehicles because of its high specific impulse (payload capability usually increases greatly for relatively small increases in specific impulse) even if the low density of liquid hydrogen makes for a very bulky fuel tank, a large vehicle, and relatively higher drag. Studies have shown that, when burned with liquid oxygen, hydrocarbons (such as methane or RP-1) give a small advantage in space launch vehicle first stages. Here, the higher average propellant density allows a smaller vehicle with less mass and lower drag, which compensates for the lower specific impulse of the hydrocarbon when compared to hydrogen. Therefore, some concepts exist for operating the same booster-stage rocket engines initially with a hydrocarbon fuel and then switching during flight to hydrogen. Engines using LOX with these two fuels, namely, hydrocarbon and hydrogen, are called *tripropellant* engines. They have not yet been fully developed or flown. Some work on experimental tripropellant engines was done in Russia, but there is no current effort known to the authors.

Hydrazine (N_2H_4)

Reference 7–15 gives a good discussion of this propellant, which is widely used as a bipropellant fuel as well as a monopropellant. Hydrazine, monomethylhydrazine (MMH), and unsymmetrical dimethylhydrazine (UDMH) have similar physical and thermochemical properties. Hydrazine is a toxic, colorless liquid with a high freezing point (275.16 K or 35.6 °F). Hydrazine tanks, pipes, injectors, catalysts, and valves are usually electrically heated to prevent freezing in cool ground weather or in outer space. Hydrazine has a short ignition delay and is spontaneously ignitable with nitric acid, nitrogen tetroxide, and concentrated hydrogen peroxide.

Pure anhydrous hydrazine is a stable liquid; it has been safely heated to near 416 K. It has been stored in sealed tanks for over 15 years. Space probes with monopropellant hydrazine have been operating for about 40 years while traveling beyond the boundary of the solar system. With impurities or at higher temperatures it decomposes releasing energy. Under pressure shock (blast wave or adiabatic compression) hydrazine vapor or hydrazine mist can decompose at temperatures as low as 367 K. Under some conditions this decomposition can be a violent detonation, and this has caused problems in cooling passages of experimental injectors and thrust chambers. Harmful effects to personnel may result from ingestion, inhalation of its toxic vapors, or prolonged contact with skin. The American Conference of Government Industrial Hygienists (ACGIH) recommended eight-hour personnel exposure limit is 0.01 ppm or $0.013 mg/m^3$. Hydrazine is a known animal carcinogen and a suspected carcinogen for people.

Hydrazine vapors may form explosive mixtures with air. If hydrazine is spilled on a porous surface or a cloth, spontaneous ignition with air may occur. It reacts with many materials, and care must be exercised to avoid contact with storage materials that cause decomposition (see Refs. 7–15 and 7–16). Tanks, pipes, injectors, catalysts, or valves must be cleaned and free of all traces of impurities. Compatible materials include certain stainless steels (303, 304, 321, or 347), nickel, and the 1100 and 3003 series of aluminum. But iron, copper, and its alloys (such as brass or bronze), monel, magnesium, zinc, and some types of aluminum alloy must be avoided.

Hydrazine cannot be safely used in cooling jackets of bipropellant thrust chambers because it will explode when a certain detonation temperature is exceeded. Such explosions occur when heat transfer is unusually high or immediately after thrust termination when the cooling fluid is stagnant; heat soaking back from hot inner walls can overheat any trapped hydrazine and this may cause a violent decomposition. Presently, bipropellant chambers with hydrazine are radiation cooled (using molybdenum or rhenium metals) and/or are ablatively cooled since internal thrust chamber insulation layers can erode and oxidize. Hydrazine injectors have to be designed without inside stagnant pockets or manifolds where liquid hydrazine can remain just after shutdown. Helium purges may be used to solve this situation. Radiation cooling and ablation cooling are discussed in Chapter 8. Reference 7–2 describes recent work on green-propellant hydrazine replacements.

Unsymmetrical Dimethylhydrazine [$(CH_3)_2NNH_2$]

This derivative of hydrazine, abbreviated as UDMH, is often used instead of or in mixtures with hydrazine because it forms a thermally more stable liquid. Furthermore, it has a lower freezing point (215.9 K) and a lower boiling point (335.5 K) than hydrazine itself. When UDMH is burned with an oxidizer it gives only slightly lower values of I_s than pure hydrazine. UDMH has been used when mixed with 30 to 50% hydrazine or with 25% hydrazine hydrate. UDMH is used in many Russian and Chinese small thrusters and some main rocket engines. The ACGIH (Ref. 7–6) recommended eight-hour personnel exposure limit for vapor is 0.01 ppm, and UDMH is a known animal carcinogen.

Monomethylhydrazine (CH_3NHNH_2)

Monomethylhydrazine, abbreviated as MMH, is being used extensively in U.S. spacecraft rocket engines, particularly in small attitude control engines, usually with N_2O_4 as the oxidizer. It has a better shock resistance to blast waves and a better liquid temperature range than pure hydrazine. Like hydrazine, its vapors are easily ignited in air; flammability limits are from 2.5 to 98% by volume at atmospheric sea-level pressure and ambient temperature. Any materials compatible with hydrazine are also compatible with MMH. The specific impulse with storable oxidizers usually is 1 or 2% lower with MMH than with N_2H_4.

Both MMH and UDMH are soluble in many hydrocarbons, though pure hydrazine is not. Monomethylhydrazine, when added in quantities of 3 to 15% by volume to hydrazine, has a substantial quenching effect on the explosive decomposition of hydrazine. MMH decomposes at 491 K, whereas pure hydrazine can explode at 369 K when subjected to certain pressure shocks. MMH is a known animal carcinogen and the ACGIH recommended personnel 8-hour exposure limit is 0.01 ppm. Of all hydrazines, MMH vapor is the most toxic when inhaled.

7.4. LIQUID MONOPROPELLANTS

The simplicity associated with monopropellant feed and control systems makes this kind of propellant very attractive for certain applications. Hydrazine is used extensively as monopropellant in small attitude and trajectory control rockets for the control of satellites and other spacecraft and also in hot gas generators (this is discussed in a preceding section). Other monopropellants (ethylene oxide or nitromethane) were tried experimentally, but are no longer used today. Concentrated hydrogen peroxide (usually 90%) was used for small thrusters between 1945 and 1965 and for monopropellant gas generator in the United States, United Kingdom or Britain, and Germany and is still being used in Russia and elsewhere (see Section 7.2. and Example 5–1). Presently, green monopropellants based on hydroxylammonium nitrate or ammonium dinitrate aqueous solutions are under development as alternatives to hydrazine.

Decomposition of monopropellants can be achieved thermally (electrically or flame heated) or by a catalytic material. A monopropellant must be chemically and thermally sufficiently stable to ensure proper liquid storage properties, and yet it must be easily decomposed and reactive to quickly give complete decomposition.

Hydrazine as a Monopropellant

Hydrazine stores well and is used as monopropellant when decomposed by a suitable solid catalyst; such catalyst often needs to be preheated for fast startups and/or for extending the useful catalyst life. Iridium on a porous alumina base is an effective catalyst at room temperature. At elevated temperature (about 450 K) many other materials decompose hydrazine, including iron, nickel, and cobalt. See Ref. 7–15. Different catalysts and different configurations produce different decomposition products, resulting in gases of varying composition and temperature. Monopropellant hydrazine units are used in gas generators and/or attitude control rockets, and as propellant in electrothermal and arcjet thrusters. A typical hydrazine monopropellant thrust chamber, its injection patterns, and its decomposition reaction are described in Fig. 8–14. Typical operating parameters are shown in Fig. 7–3 and Table 7–1.

The catalytic decomposition of hydrazine can be described ideally as a two-step process; such a simplified scheme ignores other steps and intermediate products. First, hydrazine (N_2H_4) decomposes into gaseous ammonia (NH_3) and nitrogen (N_2); this reaction is highly exothermic (e.g., it releases heat). Second, hot

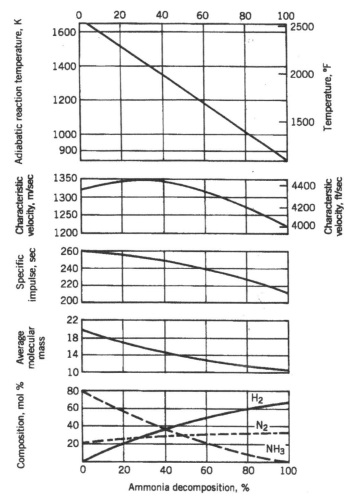

FIGURE 7–3. Operating parameters for decomposed hydrazine at the exit of a catalytic reactor as a function of the ammonia dissociation fraction. Adapted with permission from Ref. 7–15

ammonia decomposes further into nitrogen and hydrogen gases, but this reaction is endothermic and absorbs heat. These simplified reaction steps may be written as

$$3N_2H_4 \rightarrow 4(1-x)NH_3 + (1+2x)N_2 + 6xH_2 \qquad (7-4)$$

Here, x is the degree of ammonia dissociation; it is a function of catalyst type, size, and geometry; chamber pressure; and dwell time within the catalyst bed. Figure 7–3 shows several calculated rocket engine parameters for hydrazine monopropellant as a function of x. The shown values are for an ideal thruster at 1000 psia chamber pressure with a nozzle area ratio of 50 expanding at high altitudes. The best specific impulse is attained when little ammonia is allowed to dissociate.

For more than 25 years now a number of investigations have been conducted to find a substitute for hydrazine monopropellant because it is toxic, its freezing point is relatively high (275 K or 2 °C), and it requires heating of all parts it comes in contact with. To date, none of the propellants that have been evaluated so far are fully satisfactory. One possible candidate is a monopropellant (identified as LMP-1035) made from ammonium dinitramine (ADN, a solid) dissolved in small amounts of water; it was first flown in the European satellite PRISMA, which was launched by the Russian DNEPR launch vehicle on June 15, 2010. This relatively new experimental monopropellant has a higher density than hydrazine, is not toxic, and has an acceptable liquid temperature range but a slightly lower performance. However, LMP-1035 will decompose only when its catalyst has been preheated to 533 K or 500 °F. This higher decomposition temperature requires extra thermal insulation and this heating may shorten the catalyst life. Another substitute candidate is based on hydroxyl ammonium nitrate (HAN) and this monopropellant has also been under investigation. Both HAN- and ADN-based monopropellants are scheduled for tests in U.S. satellites, but to date neither has been fully proven. Investigations continue on other potential hydrazine monopropellant replacements.

7.5. GASEOUS PROPELLANTS

Propellants that store as a gas at ambient temperatures or *cold gas propellants* have been used successfully in reaction control systems (RCSs) for more than 70 years. The phrase "cold gas" distinguishes them from "warm gas," those expanded after being heated. The applicable engine system components are relatively simple, consisting of one or more high-pressure gas tanks, multiple simple nozzles (often aluminum or plastic) each with an electrical control valve, a pressure regulator, and provisions for filling and venting the gas. Tank sizes are smaller when the storage pressures are high. Because these pressures can typically be between 300 and 1000 MPa (about 300 to 10,000 psi), strong (often thick-walled and massive) gas tanks are needed.

Typical cold gas propellants and some relevant properties and characteristics are listed in Table 7–3. Nitrogen, argon, dry air, and helium have been employed for spacecraft RCS. With high-pressure hydrogen or helium as the cold gas, the specific impulse is much higher, but because their densities are lower they require much larger gas storage volumes and/or more massive high-pressure tanks; in most applications any extra inert mass outweighs the advantages of better performance. In a few applications the gas (and sometimes also its storage tank) may be heated electrically or chemically. This improves the specific impulse and allows for smaller tanks, but it also introduces complexity (see Chapter 17).

Selection of propellant gas, storage tanks, and RCS design depend on many factors, such as volume and mass of the storage tanks, maximum thrust and total impulse, gas density, required maneuvers, duty cycle, and flight duration. Cold gas systems have been used for producing total impulses of up to 22,200 N-sec or 5000 lbf-sec; higher values usually require liquid mono- or bipropellants.

TABLE 7–3. Properties of Gaseous Propellants Used for Auxiliary Propulsion

Propellant	Molecular Mass	Density[a] (lbm/ft^3)	Specific Heat Ratio k	Theoretical Specific Impulse[b] (sec)
Hydrogen	2.0	1.77	1.40	284
Helium	4.0	3.54	1.67	179
Methane	16.0	14.1	1.30	114
Nitrogen	28.0	24.7	1.40	76
Air	28.9	25.5	1.40	74
Argon	39.9	35.3	1.67	57
Krypton	83.8	74.1	1.63	50

[a] At 5000 psia and 20 °C.
[b] In vacuum with nozzle area ratio of 50:1 and initial temperature of 20 °C.

During short operations (most of the gas utilized in a few minutes while the main engine is running), gas expansions will be close to adiabatic (no heat absorption by gas) which are often analyzed as isentropic expansions. The gas temperature in high-pressure storage tanks, the (unregulated) pressures, and the specific impulse will drop as the gas is being utilized. For intermittent low-duty-cycle operations (months or years in space) heat from the spacecraft may transfer to the gas and tank temperatures stay essentially constant; then expansions will be nearly isothermal.

Advantages and disadvantages of cold gas thrusters and systems are further described in Section 8.3 in the discussion of low thrust.

7.6. SAFETY AND ENVIRONMENTAL CONCERNS

To minimize any hazards and potential damages inherent in reactive propellant materials, it is necessary to be very conscientious of all likely risks and hazards (see Refs. 7–5, 7–17, and 7–18). These relate to toxicity, explosiveness, fires and/or spill dangers, and others mentioned in Section 7.1. Before an operator, assembler, maintenance mechanic, supervisor, or engineer is allowed to transfer or use any particular propellant, he or she should receive safety training in that particular propellant, its characteristics, its safe handling or transfer, potential damage to equipment or the environment, and in countermeasures for limiting the consequences in case of accidents. Staff must also be aware of potential hazards to personnel health, first aid remedies in case of contact exposure of the skin, ingestion, or inhaling, and of how to use safety equipment. Examples of safety equipment are protective clothing, face shields, detectors for toxic vapors, remote controls, warning signals, and/or emergency water deluges. Personnel working with or close to highly toxic materials usually must undergo periodic health monitoring. Also, rocket engines need to be designed for safety to minimize the occurrence of leaks, accidental spills, unexpected fires, and/or other potentially unsafe conditions. Most organizations have one or more safety specialists who review the safety of test plans, manufacturing operations, designs, procedures, and/or safety equipment. With proper training, equipment, precautions, and design safety features, all propellants can be handled safely.

When safety violation occur or if an operation, design, procedure, or practice is found to be (or appears to be) unsafe, then a thorough investigation of the particular item or issue should be undertaken, the cause of the lack of safety should be investigated and identified, and an appropriate remedial action should be selected and initiated as soon as possible.

The discharge of toxic exhaust gases to the environment and their wind dispersion may cause exposure to operating personnel as well as the general public in nearby areas, and result in damage to plants and animals. This is discussed in Section 20.2. The dumping or spilling of toxic liquids contaminates subterranean aquifers and surface waters, and their vapors pollute the air. Today the type and amount of gaseous and liquid discharges are regulated and monitored by government authorities. These discharge quantities must be controlled or penalties will be assessed against violators. Obtaining a permit to discharge can be a lengthy and involved procedure.

One way to enhance safety against accidents (explosions, fires, spills, bullet impacts, etc.) is to use *gelled propellants*. They have additives to make them thixotropic materials, with the consistency of very thick paint when at rest; they readily flow through valves or injectors when under a pressure gradient or shear stress. Different gelling agents have been extensively investigated with several common propellants and the reader should consult the literature for details (e.g., Ref. 7–19 or Section 7.5 of the Eighth edition of this book). As far as the authors know, there is as yet no production engine using gelled propellants.

SYMBOLS

I_d	density specific impulse, sec
I_s	specific impulse, sec
k	ratio of specific heats for cold-gas propellants
m_f/m_0	ratio of final to initial mass
r	mixture ratio (mass flow rate of oxidizer to mass flow rate of fuel)

Greek Letters

δ_{av}	average specific gravity of mixture
δ_f	specific gravity of fuel
δ_o	specific gravity of oxidizer
$\rho_{av}, \rho_f, \rho_o$	densities average, fuel, and oxidizer, kg/m^3 (lbm/ft^3)

PROBLEMS

1. Plot the variation of the *density specific impulse* (product of average specific gravity and specific impulse) against mixture ratio and explain the meaning of the curve, using the theoretical shifting specific impulse values in Fig. 5–1 and the specific gravities from Fig. 7–1 or Table 7–1 for the liquid oxygen/RP-1 propellant combination.

   Answer: Check point at $r = 2.0$; $I_s = 290$; $I_d = 303$; $\delta_{av} = 1.01$.

2. Prepare a list comparing the relative merits of liquid oxygen and of nitrogen tetroxide as rocket engine oxidizers.

3. Derive Eq. 7–1 for the average specific gravity.

4. A rocket engine uses liquid oxygen and RP-1 as propellants at a design mass mixture ratio of 2.40. The pumps used in the feed system are basically constant-volume flow devices. The RP-1 hydrocarbon fuel has a nominal temperature of 298 K which can vary by about ±25 °C. The liquid oxygen is nominally at its boiling point (90 K), but, after the tank is pressurized, this temperature may increase by up to 30 K during long time storage. What are the extreme mixture ratios that result under unfavorable temperature conditions? If this engine has a nominal mass flow rate of 100 kg/sec and duration of 100 sec, what is the maximum residual propellant mass when the other propellant is fully consumed? Use the curve slopes of Fig. 7–1 to estimate changes in density. Assume that the specific impulse is constant for relatively small changes in mixture ratio, that small vapor pressure changes have no influence on the pump flow, that there is no evaporation of the oxygen in the tank, and that the engine has no automatic control for mixture ratio. Assume also zero residual propellant at the design condition.

5. The vehicle stage propelled by the rocket engine in Problem 7–4 has a design mass ratio m_f/m_0 of 0.50 (see Eq. 4–6). For the specific impulse use a value halfway between the shifting and the frozen equilibrium curves of Fig. 5–1. How much will the worst combined changes in propellant temperatures affect the mass ratio and the ideal gravity-free vacuum velocity?

6. **a.** What should be the approximate percent ullage volume for nitrogen tetroxide tank when the vehicle is exposed to ambient temperatures between about 50 °F and about 150 °F?

 b. What is maximum tank pressure at 150 °F.

 c. What factors should be considered in part (**b**)?

 Answer: (**a**) 15 to 17%; the variation is due to the nonuniform temperature distribution in the tank; (**b**) 6 to 7 atm; (**c**) vapor pressure, nitrogen monoxide content in the oxidizer, chemical reactions with wall materials, or impurities that result in largely insoluble gas products.

7. An insulated, long, vertical, vented liquid oxygen tank has been sitting on the sea-level launch stand for a period of time. The surface of the liquid is at atmospheric pressure and is 10.2 m above the closed outlet at the bottom of the tank. What will be the temperature, pressure, and density of the oxygen at the tank outlet if (**a**) liquid oxygen is allowed to circulate within the tank and (**b**) if it is assumed that there is no heat transfer throughout the tank wall to the liquid oxygen?

REFERENCES

7–1. S. F. Sarner, *Propellant Chemistry*, Reinhold, New York, 1966.

7–2. C. H. McLean et al., "Green Propellant Infusion Mission Program Overview, Status, and Flight Operations," AIAA Paper 2015-3751, Orlando, FL, 2015; R. L. Sackheim and R. K. Masse, "Green Propulsion Advancement—Challenging the Propulsion Maturity of Monopropellant Hydrazine," AIAA Paper 2013-3988, July 2013; K. Anflo and R. Moellerberg, "Flight Demonstration of a New Thruster and Green Propellant Technology on the PRISMA Satellite," *Acta Astronautica,* Vol. 65, 1238–1249, Nov. 2009.

7–3. Chemical Rocket Propellant Hazards, Vol. 1, *General Safety Engineering Design Criteria*, Chemical Propulsion Information Agency (CPIA) Publication 194, Oct. 1971; http://handle.dtic.mil/100.2/AD0889763

7–4. L. C. Sutherland, "Scaling Law for Estimating Liquid Propellant Explosive Yields, *Journal of Spacecraft and Rockets,* Mar.–Apr. 1978, pp. 124–125; doi: 10.2514/3.28002.

7–5. *NIOSH Pocket Guide to Chemical Hazards*, DHHS (NIOSH) Publication No. 2005-149, Department of Health and Human Services, Washington DC, 454 pages, Sept. 2005; http://www.cdc.gov/niosh/docs/2005-149/pdfs/2005-149.pdf

7–6. 2014 *Threshold Limit Values for Chemical Substances and Physical Agents and Biological Exposure Indices*, American Conference of Government Industrial Hygienists, Cincinnati, OH, 2013, ACGIH Publication #0113, ISBN: 978-1-607260-59-2 (revised periodically).

7–7. H. E. Malone, *The Analysis of Rocket Propellants*, Academic Press, New York, 1976.

7–8. M. C. Ventura, "Long Term Storability of Hydrogen Peroxide," AIAA Paper 2005-4551, Jul. 2005; A. Pasini et al., "Testing and Characterization of a Hydrogen Peroxide Monopropellant Thruster," *Journal of Propulsion and Power*, Vol. 24, No. 3, May–June 2008.

7–9. "Nitrous Oxide Summary Properties," M4 Liquid Propellant Manual, Chemical Propulsion Information Analysis Center, Columbia, MD, November 2009.

7–10. F. Mackline, C. Grainger, M. Veno, and S. Benson, "New Applications for Hybrid Propulsion," AIAA Paper 2003-5202, 20-23 July 2003.

7–11. R. Haudy, "Nitrous Oxide/Hydrocarbon Fuel Advanced Chemical Propulsion: DARPA Contract Overview," Qualis Corporation, Huntsville AL, 2001.

7–12. Yen-Sen Chen, A. Lau, T. H. Ch, and S.S. Wu, "N_2O-HTPB Hybrid Rocket Combustion Modeling with Mixing Enhancement Designs," AIAA Paper 2013-3645, 14-17 July, 2013.

7–13. K. Liang, B. Yang, and Z. Zhang, "Investigation of Heat Transfer and Coking Characteristics of Hydrocarbon Fuels," *Journal of Propulsion and Power*, Vol. 14, No. 5, Sept.–Oct. 1998; doi: 10.2514/2.5342.

7–14. *Detail Specification—Propellant Rocket Grade Kerosene*, Department of Defense, Washington DC, MIL-DTL-25576D, 20 May 2005.

7–15. E. W. Schmidt, *Hydrazine and Its Derivatives, Preparation, Properties, Applications*, 2nd ed., John Wiley & Sons, New York, 2001.

7–16. O. M. Morgan and D. S. Meinhardt, "Monopropellant Selection Criteria—Hydrazine and other Options," AIAA Paper 99-2595, June 1999; doi: 10.2514/6.1999-2595.

7–17. J. A. Hannum, Hazards of Chemical Rockets and Propellants, Vol. I, *Safety, Health and the Environment*, AD-A160951, and Vol. III, *Liquid Propellants*, Chemical Propulsion Information Analysis Center, AD-A158115, CPIA Publication 394, 1984.

7–18. Emergency Response Guidebook, U.S. Dept. of Transportation, Pipeline and Hazardous Materials Safety Administration, Emergency Response Guidebook 2012, 392 pages. http://phmsa.dot.gov/staticfiles/PHMSA/DownloadableFiles/Files/Hazmat/ERG2012.pdf

7–19. K. Madlener, et al., "Characterization of Various Properties of Gel Fuels with Regard to Propulsion Application," AIAA Paper 2008-4870, Jul. 2008; J. von Kampen, F. Alberio, and H. K. Ciezki, "Spray and Combustion Characteristics of Aluminized Gelled Fuels with an Impinging Jet Injector," *Aerospace Science and Technology*, Vol. 11, No. 1, Jan. 2007, pp. 77–83.

CHAPTER 8

THRUST CHAMBERS

Thrust chambers are an essential subassembly of liquid propellant rocket engines. This chapter describes chemical rocket *thrust chambers* and their components, including the topics of ignition and heat transfer. In the thrust chamber liquid propellants are metered, injected, atomized, vaporized, mixed, and burned to form hot reaction gaseous products, which are subsequently accelerated and ejected at supersonic velocities (see Refs. 6–1, 6–2, and 8–1). Chamber assemblies (e.g., Figs. 8–1 and 8–2) comprise one or more *injectors, a combustion chamber*, a *supersonic nozzle*, and various *mounting provisions*. All these parts have to withstand the extreme combustion environments and various forces, including those that transmit thrust to the vehicle. An ignition system must be also present when nonspontaneously ignitable propellants are utilized. Some thrust chamber assemblies also include integrally mounted propellant valves and occasionally thrust vector control devices, as described in Chapter 18. Table 8–1 presents data on five thrust chambers each having different kinds of propellants, cooling methods, injectors, feed systems, thrust levels, and/or nozzle expansions. Some engine parameters are also listed. Several terms used in this table are explained later in this chapter.

Basic formulations for a thrust chamber's specific impulse and combustion temperature are given in Chapters 3 and 5, other basic design parameters (thrust, flow, chamber pressure, or throat area) are introduced in Chapter 3, and unsteady combustion phenomena are treated in Chapter 9.

Although in this book we use the phrase *thrust chamber* (for rocket engines larger than about 1000 lbf thrust), some technical publications use other terms such as *thrust cylinder, thrust cell*, or *rocket combustor*. We use the term *thruster* when referring to small thrust units, such as attitude control thrusters, and for electrical propulsion systems.

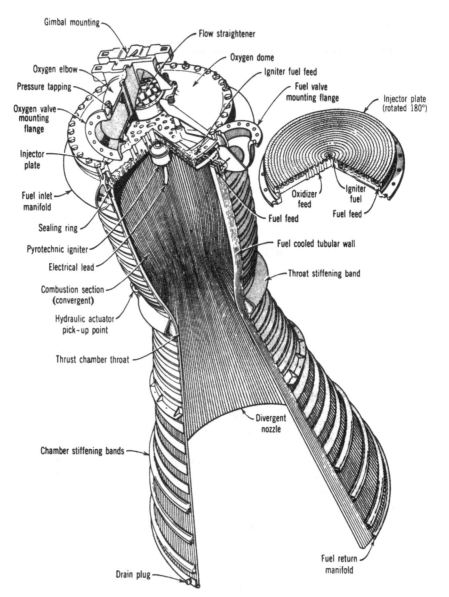

FIGURE 8–1. Construction of an early regeneratively cooled tubular thrust chamber using a kerosene-type fuel with liquid oxygen, as originally used in the Thor missile. The nozzle throat inside diameter is about 15 in. The sea-level thrust was originally 120,000 lbf, but was uprated to 135,000, then 150,000, and finally to 165,000 lbf by increasing the flow and chamber pressure and strengthening and modifying the hardware. The cone-shaped nozzle exit was replaced by a bell-shaped nozzle exit. Later in this chapter, Fig. 8–9 shows how the fuel flows down through every other cooling jacket tube and returns through the adjacent tube before flowing into the injector. Figure 8–5 shows a similar injector. Developed by Rocketdyne and licensed to Rolls Royce, England.

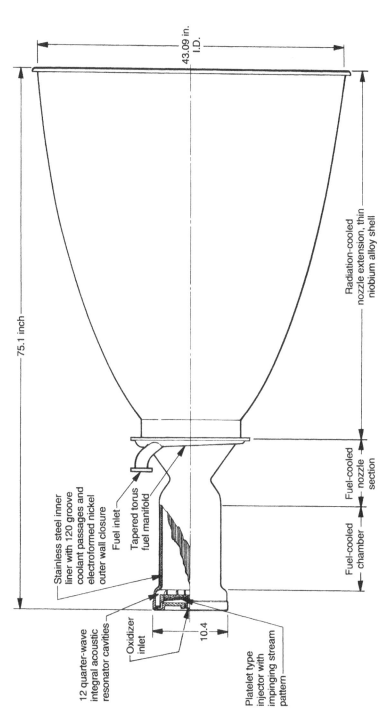

FIGURE 8–2. Simplified half-section of one of the two thrust chambers of the orbital maneuvering engines used on the Space Shuttle Orbiter. Each developed a vacuum thrust of 6000 lbf (26,689 N) and delivered a minimum vacuum specific impulse of 310 sec, using nitrogen tetroxide and monomethyl hydrazine at a nominal mixture ratio of 1.65 and a nominal chamber pressure of 128 psia. It was designed for 100 flight missions, a service life of 10 years, and a minimum of 500 starts. These engines provided the thrust for final orbit attainment, orbit circularization, orbit transfer, rendezvous, and deorbit maneuvers. The nozzle area ratio was 55:1. Courtesy of Aerojet Rocketdyne.

TABLE 8–1. Thrust Chamber Characteristics

	Engine Designation				
	RL 10B-2	LE-7 (Japan)	RCS	RS-27	AJ-10-1181
Application	Delta-III and IV upper stage	Booster stage for H-II launcher	Attitude control	Delta II Space Launch booster	Delta II Second stage
Manufacturer	Aerojet Rocketdyne	Mitsubishi Heavy Industries	Aerojet Rocketdyne	Aerojet Rocketdyne	Aerojet Rocketdyne
Thrust Chamber					
Fuel	Liquid H_2	Liquid H_2	MMH	RP-1 (kerosene)	50% N_2H_4/50% UDMH
Oxidizer	Liquid O_2	Liquid O_2	N_2O_4	Liquid oxygen	N_2O_4
Thrust chamber thrust	No sea-level firing	190,400	12	164,700	NA
at sea level (lbf)					
at sea level in vacuum (lbf)	24,750	242,500	18	207,000	9850
Thrust chamber mixture ratio	5.88	6.0	2.0	2.35	1.90
Thrust chamber specific impulse					
at sea level (sec)	NA	349.9	200	257	
in vacuum (sec)	465.5	445.6	290	294	320
Characteristic exhaust velocity, c^* (ft/sec)	7578	5594.8	5180	5540	5606
Thrust chamber propellant flow (lbm/sec)	53.2	346.9	0.062	640	30.63
Injector end chamber pressure (psia)	640	—	70	576	125
Nozzle end stagnation pressure (psia)	NA	1917	68	534	
Thrust chamber sea-level weight (lbf)	<150	1560	7	730	137
Gimbal mount sea-level dry weight (lbf)	<10	57.3	NA	70	23
Chamber diameter (in.)	—	15.75	1.09	21	11.7
Nozzle throat diameter (in.)	5.2	9.25	0.427	16.2	7.5
Nozzle exit diameter (in.)	88	68.28	3.018	45.8	60
Nozzle exit area ratio	285	54:1	50:1	8:1	65:1
Chamber contraction area ratio	—	2.87	6:1	1.67:1	2.54:1

Characteristic chamber length L^* (in.)	—	30.7	18	38.7	30.5
Thrust chamber overall length (in.)	90	14.8	11.0	86.15	18.7
Fuel jacket and manifold volume (ft³)	—	3.5	—	2.5	—
Nozzle extension	Carbon–carbon	None	None	None	None
Cumulative firing duration (sec)	>360[a]	a	a	a	>150
Restart capability	Yes	No	Yes	No	Yes
Cooling system	Stainless steel tubes, 1 ½ passes regenerative cooled	Regenerative (fuel) cooled, stainless steel tubes	Radiation cooled, niobium	Stainless steel tubes, single pass, regenerative cooling	Ablative layer is partly consumed
Tube diameter/channel width (in.)	NA	0.05 (channel)	NA	0.45	Ablative material: Silica phenolic
Number of tubes	NA	288	0	292	0
Jacket pressure drop (psi)	253	540	NA	100	NA
Injector type and its combustion vibration control feature	Concentric annular swirl element resonator cavities	Hollow post/sleeve elements; baffle and acoustic cavities	Drilled holes	Flat plate, drilled rings and baffles	Outer row: shower head; triplets and doublets, with dual tuned resonator
Injector pressure drop—oxidizer (psi)	100	704	50	156	40
Injector pressure drop—fuel (psi)	54	154	50	140	40
Number of oxidizer injector orifices	216	452 (coaxial)	1	1145	1050
Number of fuel injector orifices	216	452 (coaxial)	1	1530	1230

Engine Characteristics

Feed system and engine cycle	Turbopump with expander cycle	Turbopump; staged combustion	Pressure fed tanks	Turbopump with gas generator cycle	Pressure fed tanks
Engine thrust (at sea level) (lbf)	NA	190,400	12	165,000	NA
Engine thrust (altitude) (lbf)	24,750	242,500	18	207,700	9850
Engine specific impulse at sea level	NA	349.9	200	253	320
Engine specific impulse at altitude	465.5	445.6	290	288	320
Engine mixture ratio (oxidizer/fuel)	5.88	6.0	2.0	2.27	1.90

[a]Limited only by available propellant.
Sources: Companies listed and NASA

8.1. INJECTORS

The several functions of fuel injectors are to introduce and meter liquid propellant flows into the combustion chamber, to break up liquid jets into small droplets (a process called atomization), and to distribute and mix the propellants so that the desired fuel and oxidizer mixture ratio will result, with uniform propellant mass flows and composition over the chamber cross section.

There are two common design approaches for admitting propellants into the combustion chamber. Older types used a set of propellant jets going through a multitude of holes on the injector face. Many rocket injectors developed in the United States used this type for both large and small thrust chambers. Different hole arrangements are shown in Fig. 8–3. The second type has individual cylindrical *injection elements*,

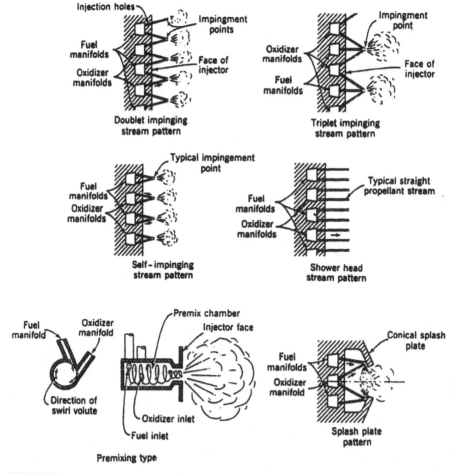

FIGURE 8–3. The upper four sketches show common injection designs using holes. The premix chamber-type (its igniter is not shown) is used today on a few large LOX/LH$_2$ thrust chambers. Used with permission from Ref. 8–1.

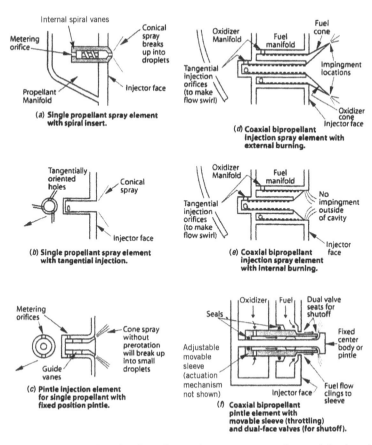

FIGURE 8-4. Cross-section sketches of several common types of spray injection elements. These injector elements can be designed to protrude into the combustion chamber forming barriers that prevent some combustion instabilities. Used with permission from Ref. 8-1.

which are inserted and fastened (welded, brazed, or soldered) into the injector face and each element delivers a conically shaped spray of propellants into the combustion chamber. Figure 8-4 shows several such spray injection elements with conical sheets of propellant issuing either from slots or from the internal edges of a hollow cylinder in the injection element. This type has been used with liquid oxygen (LOX)/liquid hydrogen (LH_2) thrust chamber worldwide, including the Space Shuttle thrust chamber. It has also been the preferred approach in Russia, being used with most of their propellants and thrust chamber sizes. There are also other injection element designs and combined types that use jets and sprays together. Typical injector hole patterns can be seen in Figs. 8-1 and 8-5.

Injection patterns on the injector face are closely related to its internal manifolds or feed passages. These distribute propellant from inlets to injection holes or spray elements. A large complex manifold volume allows for low passage velocities and proper flow distributions over the chamber cross section. Small manifold volumes permit lighter injectors, faster starts, and reduce "dribble" (flow after the main valves

FIGURE 8–5. Early 10 in. diameter injector with 90° self-impinging (fuel-against-fuel and oxidizer-against-oxidizer)-type countersunk doublet injection pattern (a concept originally developed at General Electric). Large holes are inlets to fuel manifolds. Predrilled rings are brazed alternately over an annular fuel manifold or groove and a similar adjacent oxidizer manifold or groove. A section through a similar but larger injector is shown in Fig. 8–1.

are shut). Higher passage velocities usually cause uneven flows through identical injection holes and thus yield poorer distributions and wider local composition variations. Upon thrust termination, any propellant dribbling results in after-burning—an inefficient irregular combustion effect that produces some "cutoff" thrust. For applications with very accurate terminal vehicle velocity requirements, cutoff impulses must be very small and reproducible; often valves are built into an injector that act to minimize propellant passage volumes.

Doublet *impinging-stream-type, multiple-hole injectors* are commonly used with oxygen–hydrocarbon and also with storable propellants; these are shown in Fig. 8–3. For *unlike doublet* patterns, propellants are injected through a number of separate holes in such a manner that fuel and oxidizer streams impinge upon each other. Impingement forms thin liquid fans and aids in atomization and droplet distribution. Discharge coefficients of specific injector orifices are given in Table 8–2. Impinging hole injectors are also used for *like-on-like* or *self-impinging patterns* (e.g., fuel-on-fuel and oxidizer-on-oxidizer). Here, two liquid streams form a fan which then breaks up into droplets. Unlike doublets, they work best when the fuel hole size (more exactly, volumetric flow) is about equal to that of the oxidizer and when ignition delays are long enough to allow the formation of fans. For uneven volume flows, triplet patterns often seem to be more effective.

TABLE 8–2. Injector Discharge Coefficients

Orifice Type	Diagram	Diameter (mm)	Discharge Coefficient
Sharp-edged orifice		Above 2.5	0.61
		Below 2.5	0.65 approx.
Short tube with rounded entrance $L/D > 3.0$		1.00	0.88
		1.57	0.90
		1.00	
		(with $L/D \sim 1.0$)	0.70
Short tube with conical entrance		0.50	0.7
		1.00	0.82
		1.57	0.76
		2.54	0.84–0.80
		3.18	0.84–0.78
Short tube with spiral effect		1.0–6.4	0.2–0.55
Sharp-edged cone		1.00	0.70–0.69
		1.57	0.72

Nonimpinging or *shower head* injectors employ streams of propellant usually emerging normal to the face of the injector. They rely on turbulence and diffusion to achieve mixing. The German World War II V-2 rocket used this type of injector. They are no longer used because they require relatively large chamber volumes for efficient combustion.

Sheet or *spray-type injectors* produce cylindrical, conical, and/or other types of spray sheets, which generally intersect and thereby promote mixing and atomization. See Fig. 8–4. The droplets thus formed subsequently vaporize. Droplet size distributions from spray injection elements are usually more uniform than those with streams impinging from holes. By changing some spray element internal dimensions (change size or number of tangential feed holes, the length/protrusion of an internal cylinder, or the angle of an internal spiral rib), it is possible to change the conical sheet angle, the impingement location of fuel and oxidizer spray sheets, and to affect mixture ratios, combustion efficiency, or stability. By varying sheet widths (through an axially movable sleeve), it is possible to throttle propellant flows over a wide range without excessive reductions in injector pressure drop. This type of variable area concentric tube injector was used on the descent engine of the Apollo lunar excursion module where it was throttled over a 10:1 flow range with only small effects on mixture ratio or performance.

Coaxial hollow post or spray injectors have been used for liquid oxygen and gaseous hydrogen injectors in most rocket designs. These are shown in sketches *d* and *e* of Fig. 8–4. They only work well when the liquid hydrogen has been gasified. This gasified hydrogen flows at high speed (typically, 330 m/sec or 1000 ft/sec); the liquid oxygen flows far more slowly (usually at less than 33 m/sec or 100 ft/sec), and this differential velocity causes a shearing which helps to break up the oxygen stream into small droplets. This injector has a multiplicity of coaxial elements on its face. In Russia and Germany spray injector elements have also been used with storable propellants.

The original method for making injection holes was to carefully drill them and then round out or chamfer their inlets. This is still being done today, although drilling with intense ultrafast lasers can now produce much higher precision injector holes (see Ref. 8–2). Because it has been difficult to align holes accurately (for good impingement) and avoid burrs and surface irregularities with conventional techniques, one method developed to overcome these problems and allow the production of a large number of small accurate injection orifices is to use multiple etched, very thin plates (often called platelets) that are then stacked and diffusion bonded together to form a monolithic structure as shown in Fig. 8–6. The photo-etched pattern on each individual plate or metal sheet then provides not only for many small injection orifices at the injector face but also for the internal distribution of flow passages in the injector and sometimes also for a fine-mesh filter inside the injector body. The platelets can be stacked parallel to or normal to the injector face. Finished injectors are called platelet injectors and have been patented by Aerojet Rocketdyne.

Injector Flow Characteristics

Differences between the various injector element configurations shown in Figs. 8–3 and 8–4 become apparent as different hydraulic flow–pressure relationships, starting characteristics, atomization patterns, resistance to self-induced vibrations, and as combustion efficiencies.

Hydraulic injector characteristics may be accurately evaluated and designed for orifices with any desired injection pressures, injection velocities, flows, and mixture ratios. For a given thrust F and a given effective exhaust velocity c, the total propellant mass flow $\dot{m}$ is given by $\dot{m} = F/c$ from Eq. 2–6. Equations 6–1 to 6–4 give relations between the mixture ratio and the oxidizer and fuel flow rates. For the *flow* of an incompressible fluid through hydraulic orifices, the volumetric flow rate Q and the mass flow rate are given by

$$Q = C_d A \sqrt{2\Delta p/\rho} \tag{8-1}$$

$$\dot{m} = Q\rho = C_d A \sqrt{2\rho\Delta p} \tag{8-2}$$

where C_d is a dimensionless discharge coefficient, ρ the propellant mass density, A the cross-sectional area of the orifice, and Δp the pressure drop across the injector elements. These relationships are general and can be applied to any one propellant feed system section, to the injector, or to the overall liquid flow system.

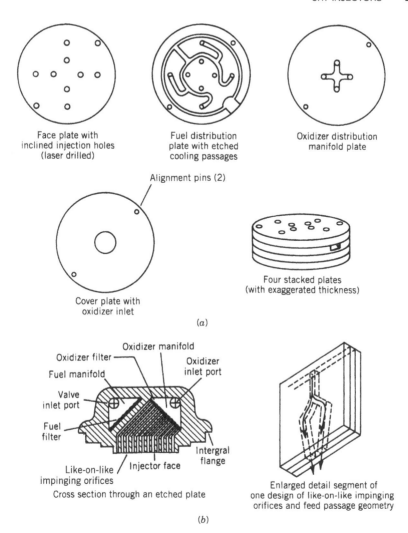

FIGURE 8–6. Simplified diagrams of two types of injector using a bonded platelet construction technique: (a) injector for low thrust with four impinging unlike doublet liquid streams; the individual plates are parallel to the injector face; (b) like-on-like impinging stream injector with 144 orifices; plates are perpendicular to the injector face. The thrust chamber in Fig. 8–2 used a platelet type injector. Courtesy of Aerojet Rocketdyne.

For any given pressure drop, injection orifices usually establish *mixture ratio* and *propellant flows* in the rocket propulsion unit. Using Eqs. 6–1 and 8–2 the mixture ratio *r* becomes

$$r = \dot{m}_o/\dot{m}_f = [(C_d)_o/(C_d)_f](A_o/A_f)\sqrt{(\rho_o/\rho_f)(\Delta p_o/\Delta p_f)} \qquad (8\text{–}3)$$

Values in the preceding equations have to be chosen so that the desired design mixture ratio is attained. Orifices whose discharge coefficients are constant over a large range of Reynolds numbers and whose ratio $(C_d)_o/(C_d)_f$ remains essentially invariant should preferably be selected. For a given injector, it is usually difficult to maintain the mixture ratio constant at low flows, such as during starting.

Injector quality can be checked by performing cold tests with inert simulant liquids instead of reactive propellant liquids. Water is often used to confirm calculated pressure drops through the fuel or oxidizer side at different flows, and this allows determination of the pressure drops with actual propellants and of the discharge coefficients. Nonmixable inert liquids are used with a special apparatus to determine local cold flow mixture ratio distributions over the chamber cross section. The simulant liquid should be of approximately the same density and viscosity as the actual propellant. Because these cold flow tests usually do not simulate the vapor pressure, new injectors are often also hot fired and tested with actual propellants.

Actual mixture ratios can be estimated from cold flow test data, measured hole areas, and discharge coefficients by correcting with the square root of the density ratio of the simulant liquid and the propellant. When water at the same pressure is fed alternately into both the fuel and the oxidizer sides, $\Delta p_f = \Delta p_o$ and $\rho_f = \rho_o$ and the mixture ratio with water flows will be

$$r = [(C_d)_o/(C_d)_f]A_o/A_f \tag{8-4}$$

Therefore, any mixture ratio measured in water tests can be converted into the actual propellant mixture ratio by using Eq. 8–3. Mechanisms for propellant atomization with simultaneous vaporization, partial combustion, and mixing are difficult to analyze and performance of injectors has to be experimentally evaluated within the burning rocket thrust chamber. The *injection velocity* is given by

$$v = Q/A = C_d\sqrt{2\Delta p/\rho} \tag{8-5}$$

Values of discharge coefficients for various types of injection orifices are shown in Table 8–2. The velocity is a maximum for a given injection pressure drop when the discharge coefficient approaches one. Smooth and well-rounded entrances to injection holes and clean bores give high values of the discharge coefficient, and such entry hole designs are most common. Small differences in chamfers, hole-entry radii, or hole-edge burrs may cause significant variations in the discharge coefficient and jet flow patterns, and these in turn can alter the quality and distribution of the atomized small droplets, the local mixture ratio, and the local heat transfer rates. An improperly manufactured hole may cause local chamber or injector burnout.

When oxidizer and fuel jets impinge, the *resultant momentum* can be calculated from the following relation, based on the principle of conservation of momentum. Figure 8–7 illustrates a pair of impinging jets and defines γ_o as the angle between the chamber axis and the oxidizer stream, γ_f as the angle between the chamber axis and the fuel stream, and δ as the angle between the chamber axis and the average

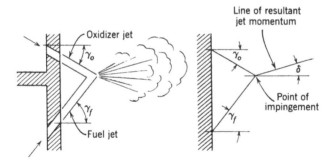

FIGURE 8–7. Angular relation for a unlike doublet impinging-stream injection pattern.

resultant stream. Equating the total axial momentum of the two jets before and after impingement results in the following:

$$\tan \delta = \frac{\dot{m}_o v_o \sin \gamma_o - \dot{m}_f v_f \sin \gamma_f}{\dot{m}_o v_o \cos \gamma_o + \dot{m}_f v_f \cos \gamma_f} \tag{8-6}$$

Adequate performance is often obtained when the resultant momentum of impinging streams is approximately axial. In such a case, $\delta = 0$ and $\tan \delta = 0$, and the angular relation for an axially directed momentum jet becomes

$$\dot{m}_o v_o \sin \gamma_o = \dot{m}_f v_f \sin \gamma_f \tag{8-7}$$

From these equations the relation between γ_f, γ_0, and δ may be determined. A sample injector analysis is shown in Section 8.8.

Factors Influencing Injector Behavior

A completely satisfactory theory relating injector design parameters to both rocket engine performance and combustion phenomena has yet to be devised and therefore approaches to design and development of liquid propellant rocket injectors have been largely empirical. Some analyses (see Ref. 8–3) have been useful for understanding the phenomenology and in indicating directions for injector development. Available data reveal several important factors that affect performance and operating characteristics of injectors; some of these factors are briefly enumerated here. They differ for injectors where both propellants are liquid (such as nitrogen tetroxide and hydrazine) from where one propellant is liquid and the other gaseous, as with LOX/gasified liquid hydrogenLH$_2$. Monopropellant injectors feed propellant into a catalyst bed and they are yet different, as described in Section 8.3.

Propellant Combination. Particular combinations of fuel and oxidizer affect such features as relative chemical reactivity, ease and speed of vaporization, droplet formation, ignition temperature, diffusion of hot gases, volatility, and/or surface tension. In general, hypergolic (self-igniting) propellants require injector designs somewhat different from those needed by propellants that must be ignited. Injector designs that perform efficiently with one combination do not necessarily work well with different propellant combinations.

Injection Element Pattern and Orifice Size. With individual hole elements or sprays at the injector plate, there appears to be an optimum performance and/or heat transfer condition for each of the following parameters: orifice size, angle of impingement, angle of resultant momentum, distance of the impingement locus from the injector face, number of injection orifices per unit of injector face surface, flow per unit of injection element, and distribution of orifices over the injector face. These parameters are largely determined experimentally or derived from similar earlier successful injectors.

Transient Conditions. Starting and stopping may require special provisions (temporary plugging of holes, accurate valve timing, insertion of disposable cups over holes to prevent entry of one propellant into the manifold of the other as was done on the German A-4 or V-2 thrust chamber, inert gas purges, check valves) to permit satisfactory transient operation.

Hydraulic Characteristics. Orifice type and pressure drop across the injection orifice determine injection velocities. A low-pressure drop is desirable to minimize feed system mass and pumping power. High-pressure drops are often used to increase a rocket engine's resistance to combustion instabilities and to enhance atomization of the liquids thereby improving performance. Reference 8–4 discusses atomization and combustion modeling.

Heat Transfer. Injectors affect heat transfer rates in rocket thrust chambers. Low heat transfer rates have been observed when injection patterns result in intentionally rich mixtures near the chamber walls and nozzle throat region or when chamber pressures are low. In general, higher performance injectors have larger heat transfer rates to the walls of the combustion chamber, the nozzle, and the injector face. (See Section 8.5.)

Structural Design. Injectors become highly loaded by pressure forces from the combustion chamber and from its propellant distribution manifolds. During transition (starting or stopping) these pressure conditions can cause internal transient structural stresses that sometimes exceed steady-state operating conditions. The face in most modern injectors is flat and must be reinforced by suitable structures that should not obstruct existing hydraulic manifold passages; these structures must also be sufficiently flexible to allow for thermal deformations (caused by heating at the injector face from the combustion gases or cooling of certain flow passages by cryogenic propellants). Injector designs must also provide for positive seals between fuel and oxidizer injector manifolds (internal leaks may result in manifold explosions or internal fires). No seals between chamber and propellant feed pipes or between chamber and injector can be allowed to leak, even under thermal deformations. In large gimbal-mounted thrust chambers, injectors also often carry the main thrust load, and gimbal mounts are often directly attached to the injector, as shown in Figs. 6–1 and 8–1.

Combustion Stability. Injection holes with their resulting spray pattern, impingement patterns, hole or spray element distributions, and pressure drops all have a strong influence on combustion stability; some types can be more resistant to pressure disturbances than others. As explained in Section 9.3, resistance to vibration is determined experimentally, and often special antivibration devices, such as baffles or resonance cavities, are designed directly into the injector.

8.2. COMBUSTION CHAMBER AND NOZZLE

The *combustion chamber* is that portion of the thrust chamber where nearly all propellant burning takes place. Because combustion temperatures are much higher than the melting point of ordinary chamber wall materials, it is necessary to either cool these walls (as described in a later section of this chapter) or to stop rocket operation before critical wall areas overheat. A thrust chamber will fail when at any location the heating is so high that wall temperatures exceed their operating limit. Heat transfer to thrust chambers is described later in this chapter. Section 8.9 gives a sample analysis of a thrust chamber and Ref. 8–3 describes analyses for their design and development.

Volume and Shape

Spherical chambers expose the least internal surface area and have the lowest inert mass per unit chamber volume—several have been tried even though they are expensive to build. Today, cylindrical chambers (or slightly tapered cone frustums) are preferred with flat injector faces and, at the other end, converging–diverging nozzles. Chamber volume is defined as the volume from injector face up to nozzle throat section so as to include the cylindrical chamber and the converging cone frustum of the nozzle. Neglecting the effect of any corner radii, the chamber volume V_c for a cylindrical chamber is given by

$$V_c = A_1 L_1 + A_1 L_c \left(1 + \sqrt{A_t/A_1} + A_t/A_1 \right) \tag{8–8}$$

Here, L_1 is the cylinder length, A_t/A_1 is the reciprocal of the chamber contraction area ratio, and L_c is the length of the converging conical frustum. Chamber surfaces exposed to heat transfer from hot gases include the injector face, the inner surface of the cylinder chamber, and the inner surface of the converging cone frustum. *Volume and shape* are selected after evaluating the following parameters:

1. The volume needs to be large enough for adequate *atomization, mixing, evaporation*, and for nearly *complete combustion* of propellants. For different

propellants, chamber volumes vary with the time delay necessary to vaporize and activate the propellants and with the speed of the combustion reaction. When chamber volumes are too small, combustion is incomplete and the performance becomes poor. With higher chamber pressures or with highly reactive propellants and with injectors that yield improved mixing, smaller chamber volumes become permissible.

2. Chamber diameter and volume influence the *cooling requirements*. With larger chamber volumes and chamber diameters, gas velocities are lower and wall rates of heat transfer reduced, areas exposed to heat larger, and walls become somewhat thicker. Conversely, if volumes and cross sections are small, inner wall surface areas and their inert mass will be smaller, but chamber gas velocities and heat transfer rates will increase. Therefore, for the same thrust chamber requirements, there is an optimum chamber volume and diameter where the total heat absorbed by the walls will be a minimum. This condition is important when the available capacity of the coolant is limited (e.g., oxygen–hydrocarbon at high mixture ratios) or when the maximum permissive coolant temperature must be limited (for safety reasons as with hydrazine cooling). Total heat transfer may be further reduced by going to fuel rich mixture ratios or by adding film cooling (as discussed below).

3. All inert components should have *minimum mass*. In addition to thrust chamber wall composition, mass depends on chamber dimensions, chamber pressures, and nozzle area ratios, and on cooling methods.

4. Manufacturing considerations favor simple chamber geometries (such as a cylinder with a double cone bow-tie-shaped nozzle), low-cost materials, and standard fabrication processes.

5. In some applications the *length* of both the chamber and nozzle directly affect the overall length of the vehicle. Large-diameter but short chambers and/or short nozzles may allow shorter vehicles with a lower structural inert vehicle mass.

6. *Gas pressure drops* needed to flow the combustion products within the chamber should be a minimum; any pressure losses ahead of the nozzle inlet reduce the exhaust velocity and thus vehicle performance. These losses may become appreciable when the chamber area is less than three times the throat area.

7. For the same thrust, combustor volume and nozzle throat area become smaller as the operating chamber pressure is increased. This means that chamber length and nozzle length (for the same nozzle area ratio) also become shorter with increasing chamber pressure. But while performance may slightly increase, heat transfer rates do go up substantially with chamber pressure.

Some of the preceding chamber considerations conflict with each other. It is, for instance, impossible to have a large chamber that gives complete combustion but has a low inert mass. Depending on application, a compromise solution that will satisfy most of these considerations is therefore usually selected and then tested.

Some designers use a converging chamber geometry instead of a straight cylindrical shape; here, the injector face is somewhat larger than the nozzle inlet face dimension.

A *characteristic chamber length* is defined as the length that a chamber of the same volume would have if it were a straight tube whose diameter is the nozzle throat diameter, that is, if it had no converging nozzle section:

$$L^* \equiv V_c/A_t \qquad (8\text{--}9)$$

Here, L^* (pronounced el star) is this characteristic chamber length, A_t is the nozzle throat area, and V_c is the chamber volume. Such chamber includes all the volume up to the throat area. Typical values for L^* are between 0.8 and 3.0 m (2.6 to 10 ft) for several bipropellants and higher for some monopropellants. Because this parameter does not consider any variables except the throat area, it is useful only for some particular propellant combinations and narrow ranges of mixture ratio and chamber pressures. Today, chamber volume and shape are chosen using data from successful thrust chambers of prior similar designs and identical propellants.

The *stay time t_s* of propellant gases is the average value of time spent by each flow element within the chamber volume. It is defined by

$$t_s = V_c/(\dot{m}V_1) \qquad (8\text{--}10)$$

Above, $\dot{m}$ is the propellant mass flow, V_1 is the average specific volume (or volume per unit mass of propellant gases in the chamber), and V_c is the chamber volume. A minimum stay time for good performance defines a chamber volume that gives nearly complete combustion. Stay time represents the time necessary for vaporization, activation, and nearly complete burning of the propellant. Stay times vary for different propellants and with their storage temperatures; they need to be experimentally determined. Stay times vary between 0.001 to 0.040 sec for different types and sizes of thrust chambers and for different propellants.

Most *nozzle* configurations and dimensions can be determined from the analyses presented in Chapter 3. The converging section of a nozzle experiences much higher internal gas pressures than its diverging section and therefore wall designs for converging walls are similar to those for the cylindrical chamber; exact contours for the converging nozzle section are not critical because most have effectively no separation losses. Many thrust chambers use shortened bell shapes in the diverging nozzle section. Nozzles with exit area ratios over 400 have been developed and flown.

In Chapter 3 it was stated that very large nozzle exit area ratios allow significant improvements in specific impulse, particularly at higher altitudes; however, the extra length, extra nozzle mass, and extra vehicle mass necessary to house such large nozzles often makes them unattractive. This disadvantage can be mitigated with multipiece diverging nozzle sections that are stored as annular pieces around the engine during launch and vehicle ascent and self-assembled in space after launch vehicle separation but before upper stage firing. This concept, known as the *extendible nozzle*, has been successfully employed with solid propellant rocket motors for space applications for about 35 years. The first flight with an extendible nozzle of a liquid propellant engine took place in 1998 with a modified version of an Aerojet Rocketdyne RL 10 upper stage engine. Its flight performance is listed in Table 8–1. The engine is shown later in Fig. 8–17 and its carbon–carbon extendible nozzle cone is described in the section on materials and fabrication of this chapter.

Special thrust chamber configurations with a center body, such as the aerospike nozzle or the expansion-deflection nozzle shown in Fig. 3–11 have been thoroughly investigated and ground tested but have not been adopted in production flight vehicles. Discussion of these has therefore been deleted from this edition; those interested in such advanced throat chambers, which offer an optimum nozzle expansion at all altitudes, should consult the Seventh and Eighth Editions of this book, where the aerospike thrust chamber is described in some detail.

Heat Transfer Distribution

Heat is transmitted to all internal hardware surfaces exposed to the hot gases, namely the injector face, internal chamber and nozzle walls. *Heat transfer rates* or *heat transfer intensities* (i.e., heat transfer per unit local wall area) vary within the thrust chamber. A typical heat transfer rate distribution is shown in Fig. 8–8. In well-designed chambers, only between 0.5 and 5% of the total energy generated in the gas is lost to the walls. For a typical rocket of 44,820 N or 10,000 lbf thrust the heat rejection rate to the wall may be between 0.75 and 3.5 MW, depending on exact operating conditions and configurations. See Section 8.5.

The amount of heat transferred by *conduction* from chamber gases to walls in rocket thrust chambers may be neglected. By far the largest portion of the heat is transferred by means of *convection* along with some attributable to *radiation* (usually 5 to 35%).

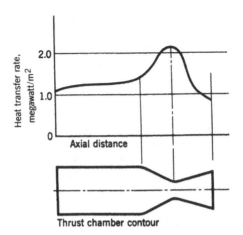

FIGURE 8–8. Typical heat transfer rate axial distribution for both liquid propellant thrust chambers and solid propellant rocket motors. The peak value is always at the nozzle throat or slightly upstream of it and the lowest value is at or near the nozzle exit.

As thrust levels rise, for constant chamber pressures, a chamber wall's surface increases less rapidly than its volume. Thus, cooling of chambers is generally easier in the large thrust sizes, and the capacity of wall materials or coolants to absorb heat is generally more critical in smaller sizes because of such volume–surface relationship.

Higher chamber pressures lead to higher vehicle performance (higher I_s). However, resulting *increases of heat transfer with chamber pressure* often impose design or material limits on maximum practical chamber pressures for both liquid and solid propellant rocket propulsion systems.

Heat transfer intensities in chemical rocket propulsion can vary from less than 50 W/cm^2 or 0.3 Btu/in.^{2}-sec to over 16 kW/cm^2 or 100 Btu/in.^{2}-sec. The higher values reflect the nozzle throat region of large bipropellant high-chamber-pressure thrust chambers and high-pressure solid rocket motors. The lower values represent gas generators, nozzle exit sections, and/or small thrust chambers at low chamber pressures.

Cooling of Thrust Chambers

The primary objective of cooling is to prevent chambers and nozzle walls from failing, a condition where they no longer withstand the imposed loads or stresses (these are discussed in a following section). Most wall materials lose strength as their temperature increases. With rising heat rates, most wall materials ultimately fail and eventually melt. Cooling must therefore be implemented to reduce wall temperatures to acceptable levels.

Basically, there are two cooling methods presently in common use. The first is the *steady-state method* where heat transfer rates and chamber temperatures reach *thermal equilibrium*. It includes *regenerative cooling* and *radiation cooling*. Their duration is limited only by the available supply of the cooling propellant.

Regenerative cooling is done with cooling jackets built around the thrust chamber where one liquid propellant (usually the fuel) circulates through before it is fed to the injector. This cooling technique is used primarily in bipropellant chambers of medium to large thrust capacity. It has been very effective in applications with high chamber pressures and high heat transfer rates. Also, most injectors use regenerative cooling along their hot faces.

In *radiation cooling*, the chamber and/or nozzle have a single wall made of a high-temperature material, such as niobium, carbon-carbon, or rhenium. When it reaches thermal equilibrium, this wall may glow red or white as it radiates heat away to the surrounding medium or to empty space. Radiation cooling is used with bipropellant and monopropellant thrusters, bipropellant and monopropellant gas generators, and in diverging nozzle exhaust sections beyond an area ratio of about 6 to 10 in larger thrust chambers (see Fig. 8–2). A few small bipropellant thrusters are also radiation cooled. This cooling scheme has worked well with lower chamber pressures (less than 250 psi) and with moderate heat transfer rates.

The second cooling method relies on *transient* or *unsteady heat transfer*. It is also called *heat sink cooling*. Here, thrust chambers do not reach thermal equilibrium, and temperatures continue to increase with operating duration. The heat-absorbing capacity of the enclosure determines its maximum duration, which can be relatively short (a few seconds for an all-metal construction). Rocket combustion operation has to be stopped just before any exposed walls reach the critical temperature at which it would fail. This method has mostly been used with low chamber pressures and low heat transfer rates. Heat-sink cooling of thrust chambers may also be done by absorbing heat in an inner liner made of *ablative materials*, such as fiber-reinforced plastics. Ablative materials burn, evaporate or erode slowly and their cumulative operating durations can amount to minutes. Ablative materials are used extensively in solid propellant rocket motors and are discussed further in Chapter 15. An analysis of both of these cooling methods is given in the next section of this chapter.

Film cooling and *special insulation* materials are supplementary techniques occasionally used with both methods to locally augment their cooling capabilities. They are described further in this chapter.

Cooling also helps to reduce wall material oxidation and/or the rate at which walls would wear away. Rates of chemical oxidizing reactions between hot gases and wall materials do increase dramatically with wall temperature. This oxidation problem can be minimized not only by limiting the wall temperature but also by burning liquid propellants at mixture ratios where the percentage of hot aggressive gases (such as oxygen or hydroxyl) is relatively small, and by coating specific wall materials with an oxidation-resistant layer; for example, iridium has been coated on the inside of rhenium walls.

Cooling with Steady-State Heat Transfer. *Cooled thrust chambers* must include provisions for cooling all metal parts that come into contact with hot gases, such as chamber walls, nozzle walls, and injector faces. Internal cooling passages, cooling jackets, or cooling coils circulate the fluid *coolant*. Jackets may consist of separate inner and outer walls or of assemblies of contoured, adjacent tubes (see Figs. 8–1 and 8–9 where the inner walls confine the hot gases, and the spaces between the walls serve as cooling passages). Because the *nozzle throat region*

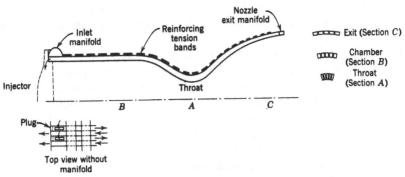

FIGURE 8–9. Diagram of a tubular cooling jacket. The tubes are bent to the chamber and nozzle contours; they are formed to give a variable cross section and to permit the same number of tubes at the throat and exit diameters. Coolant enters through the inlet manifold into every other tube and proceeds axially to the nozzle exit manifold, where it then enters the alternate tubes and returns axially to go directly to the injector.

experiences the highest heat transfer rates it is therefore the most difficult to cool. For this reason its cooling jacket is usually designed by restricting the coolant passage cross section so that the coolant velocity is highest at the nozzle throat, and so that fresh cold coolant enters the jacket at or near the nozzle throat. While selection of coolant velocities and their variation along the wall for any given thrust chamber design only depends on heat transfer considerations, the design of coolant passages depends additionally on pressure losses, stresses, and on life and manufacturing considerations. *Axial flow cooling jackets*, or *tubular walls*, have low hydraulic friction losses but are practical only with large coolant flows (above approximately 9 kg/sec); for small coolant flows and small thrust units, design tolerances of the cooling jacket width between the inner and outer walls or the diameters of the tubes become too small, or dimensional tolerances become prohibitive. Hence, most small thrust chambers use radiation cooling or ablative materials.

In *regenerative cooling* the heat absorbed by the coolant is not discarded; it augments the energy content of the propellant prior to injection increasing the exhaust velocity (or specific impulse) slightly (0.1 to 1.5%). This method is called regenerative cooling because of its similarity to steam regenerators. Tubular chamber and nozzle designs combine the advantages of thin walls (good for reducing thermal stresses and high wall temperatures) with cool, lightweight structures. Tubes are formed into special shapes and contours (see Figs. 8–1 and 8–9), usually by hydraulic means, and then brazed, welded, or soldered together (see Ref. 8–5). In order to take gas pressure loads in hoop tension, they are reinforced on the outside by high-strength bands or wires. While Fig. 8–9 shows alternate tubes for up and down flows, some chambers have the fuel inlet manifold downstream of the nozzle throat area so coolant flow is up and down in the nozzle exit region, but only unidirectionally up in the throat and chamber regions.

Radiation cooling is the other steady-state method of cooling. It is rugged and simple and is used extensively in low-heat transfer applications previously listed. Further discussion of radiation cooling is given in Section 8.4. In order for heat to be mostly radiated out, it is necessary for the nozzle and chamber outer surfaces to stick out of the vehicle. Since a glowing *radiation-cooled* chamber and/or nozzle surface can be a potent radiator, it may cause undesirable heating of adjacent vehicle or engine components. Therefore, many include insulation (see Fig. 8–13) or simple external radiation shields to minimize these thermal effects; however, in these cases the actual chamber or nozzle wall temperatures are higher than they would be without the outside insulation or shielding.

Cooling with Transient Heat Transfer. *Thrust chambers with unsteady heat transfer* are basically of two types. One consists of a *simple all-metal chamber* (steel, copper, stainless steel, etc.) made with walls sufficiently thick to absorb the required heat energy, often used for the short-duration testing of new injectors or new propellants, or for combustion stability ratings, or for very short duration rocket-propelled missiles such as an antitank weapon. The other type uses certain organic materials on all inner walls commonly labeled as *ablative cooling* or *heat sink cooling* method. The inner surfaces of these organic layers recede due to a combinations of endothermic reactions (breakdown or distillation of matrix materials into smaller compounds and gases), pyrolysis of organic materials, and localized melting. An ablative material typically consists of a series of strong, oriented fibers (such as glass, Kevlar, or carbon fibers) engulfed by a matrix of organic binder materials (such as plastics, epoxy resins or phenolic resins). As depicted in Fig. 15–11, heat causes the binder to decompose and form gases that seep out of the matrix and form a protective cooling layer film on inner wall surfaces. The fibers and the residues of these matrixes form a hard black char or porous coke-like material that helps to preserve the wall contour shapes.

Orientation, number, and type of fiber determine the ability of a composite ablative material to withstand significant stresses in its preferred directions. For example, internal pressure produces longitudinal as well as hoop stresses in thrust chamber walls and thermal stresses produce compression on the inside walls and tensile stresses on the outside. There are techniques to place fibers with fiber orientations in two or three directions, which make them anisotropic. These techniques can produce two-directional (2-D) and/or (3-D) fiber orientations.

An array of strong carbon fibers in a matrix of amorphous carbon is a special, but favorite, type of material. It is often abbreviated as C–C or carbon–carbon. Carbon materials only lose their ability to carry loads at temperatures of about 3700 K or 6200 °F, but carbon oxidizes readily to form CO or CO_2. Its best applications are with fuel-rich propellant mixtures that have little or no free oxygen or hydroxyl in their exhaust. They have also been used in nozzle throat inserts. Properties for one type of C–C are given in Table 15–5. A nozzle extension made of C–C is shown later in Fig. 8–17. See Ref. 8–6.

Ablative cooling came first and is still extensively used with solid propellant rocket motors. It has since been successfully applied to liquid propellant thrust chambers, particularly of low chamber pressures, where static gas temperatures are relatively low; it is still used today in nozzle extension materials, such as the RS-68 in Fig. 6–9a, where it can operate for several minutes. It is also used as a chamber and nozzle liner at low chamber pressures. An example is the axial gimbaled thruster of the Peacekeeper fourth stage, which is seen in Figure 6–14. Ablatively lined small thrusters (100 lbf thrust or less) were flown extensively in the 1950s and 1960s in the Apollo missions and in other applications for attitude control and minor maneuvers. They are no longer used today because they are relatively heavy and because eroded particles, droplets, and/or certain exhaust plume gases tend to deposit or condense on optical surfaces of spacecraft (mirrors, solar cells, or windows).

It is often advantageous to use a different cooling method for the downstream part of the diverging nozzle section, because its heat transfer rate per unit area is much lower than in the chamber or the converging nozzle section, particularly with nozzles of large area ratio. There can be small savings in inert engine mass, small increases in performance, and cost savings, if the chamber and the converging nozzle section and the throat region (up to an area ratio of perhaps 5 to 10) use regenerative cooling while the remainder of the nozzle exit section is radiation cooled (or sometimes ablative cooled). See Fig. 8–2 and Ref. 8–6.

Film Cooling. This is an auxiliary method applied to chambers and/or nozzles for augmenting either a marginal steady state or a transient cooling method. It can be applied to a complete thrust chamber or just to the nozzle throat region, where heat transfer is the highest. Film cooling is a method of cooling whereby a relatively cool fluid film covers and protects internally exposed wall surfaces from excessive heat transfer. Figure 8–10 shows several film-cooled chambers. Films can be introduced by injecting small quantities of extra fuel or an inert fluid at very low velocities through a number of orifices along the exposed surfaces in such a manner that a protective relatively cool gas film (or cold boundary layer) forms. More uniform protective boundary layers can be obtained by using slots for coolant film injection instead of multiple holes. In liquid propellant rocket engines fuel can also be admitted through extra injection holes at the outer layers of the injector (or alternatively, at low mixture ratios, through special injector spray elements at the injector's face periphery); thus, a propellant mixture is achieved (at the periphery of the chamber), which has a lower combustion temperature. This differs from film cooling or transpiration cooling which enters at the inner walls of the combustion chamber or the nozzles and not through the injector.

The RS-191 thrust chamber shown in Fig. 8–11 has slots for film injection. The major portion of fuel coolant flow is provided to the throat and converging nozzle regions of the cooling jacket where heating is highest. Another portion of fuel coolant, estimated at 10 to 15%, cools the nozzle divergent section. The nozzle

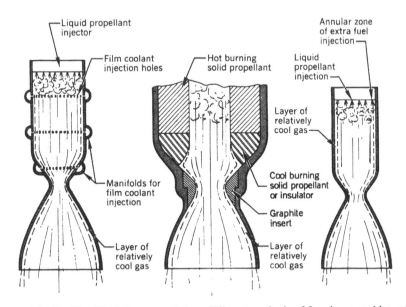

FIGURE 8–10. Simplified diagrams of three different methods of forming a cool boundary layer in the nozzle.

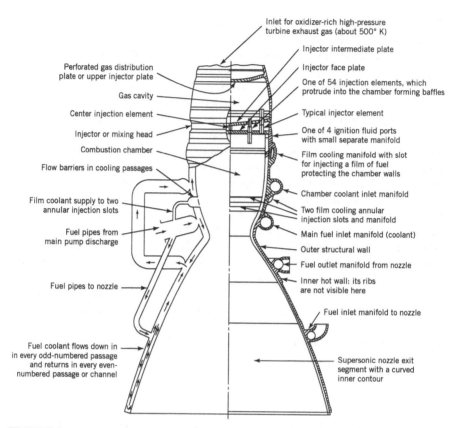

Inlet for oxidizer-rich high-pressure
turbine exhaust gas (about 500° K)

Injector intermediate plate

Injector face plate

One of 54 injection elements, which
protrude into the chamber forming baffles

Typical injector element

One of 4 ignition fluid ports
with small separate manifold

Film cooling manifold with slot
for injecting a film of fuel
protecting the chamber walls

Chamber coolant inlet manifold

Two film cooling annular
injection slots and manifold

Main fuel inlet manifold (coolant)

Outer structural wall

Fuel outlet manifold from nozzle

Inner hot wall: its ribs
are not visible here

Fuel inlet manifold to nozzle

Supersonic nozzle exit
segment with a curved
inner contour

Perforated gas distribution
plate or upper injector plate

Gas cavity

Center injection element

Injector or mixing head

Combustion chamber

Flow barriers in cooling passages

Film coolant supply to two
annular injection slots

Fuel pipes from
main pump discharge

Fuel pipes to nozzle

Fuel coolant flows down in
in every odd-numbered passage
and returns in every even-
numbered passage or channel

FIGURE 8–11. Sectioned view of the RD-191 thrust chamber showing a coolant fuel flow diagram on the left side and on the right a structural depiction of walls, manifolds, and three film cooling slots. Most of the fresh fuel (at its initial storage temperature) first flows into the part of the cooling jacket that surrounds the throat nozzle region (where the heat transfer is highest), and subsequently it flows through other parts of the cooling jacket and then to the injector. This injector is shown in Figure 9–6 (see Ref. 9–22). From NPO Energomash, Khimki, Russia.

return flow together with the main flow from the throat region then go from the cooling jacket through external pipes to the chamber portion of the cooling jacket and then into the injector. A very small portion of gas goes directly to two of the three film cooling slots; the third film slot is supplied through metering holes and through a separate small manifold. The hypergolic start propellant slug (a mixture of tri-ethyl aluminum/borane) enters the chamber in four jets from separate small manifolds near the injector.

Film cooling by itself (without other cooling methods) has been effective in keeping chamber and nozzle materials from overheating. The very first thrust chambers developed by Robert H. Goddard in the 1920s were film cooled. However, his film coolant did not burn effectively and there was a 5 to 17% reduction in specific impulse. Today, film cooling is used in small quantities (1 to 6% of fuel) to locally supplement other cooling methods and performance losses are only 0.5 to 2%. In solid propellant rocket engines film cooling can be accomplished by inserting a ring of cool-burning propellant upstream of the nozzle, as shown in the center figure of Fig. 8–10 (not in production) or by wall insulation materials, whose ablation and charring will release relatively cool gases into the boundary layer.

Turbine discharge gases (400 to 800 °C or 752 to 1972 °F) have also been successfully used as film coolants for uncooled nozzle exit sections in large liquid propellant rocket engines. Of course, any gas layer injection at the inner wall chambers and nozzle, at a temperature lower than the maximum possible value, causes a small decrease in specific impulse. Therefore, it is always desirable to reduce both the thickness and the total mass flow of cooler gas layers, relative to the total flow, to any practical minimum value at which cooling will still be effective.

In a special type of film cooling, termed *sweat cooling* or *transpiration cooling*, a porous wall material that admits a coolant through pores uniformly spread over the surface is used. This technique was used successfully for cooling the injector faces in the 5 upper stage engines (J-2) of the moon Saturn V launch vehicle with hydrogen fuel. Sweat cooling has apparently not been used since in chambers or nozzles, because materials with changing porosity are difficult to fabricate.

Thermal Insulation. Hypothetically, appropriate thermal insulation layers on the gas side of the chamber wall should be very effective in reducing chamber wall heat transfer and wall temperatures. However, efforts with known insulation materials such as refractory oxides or ceramic carbides have not been successful. They cannot withstand the differential thermal expansions between the wall and the coating materials without cracking or spalling. Any sharp edges at the surface (from cracks or flaked-off pieces of insulator) will cause sudden local temperature rises (up to the stagnation temperature) and most likely lead to local wall failures. Asbestos is a good insulator and was used several decades ago but it is no longer used because it is cancer causing. Development efforts on rhenium coatings and coatings with other materials are continuing. Insulation layers and/or heat shields have been successfully applied on the exterior of radiation-cooled thrust chambers to reduce heat transfer rates from thrust chamber walls to adjacent sensitive equipment or structures.

With hydrocarbon fuels small carbon particles (soot) or other solid carbon forms may develop in the combustion region resulting in thin carbon deposits on the gas side of the chamber and/or nozzle walls. Thin, mildly adhesive soot deposits insulating layers are difficult to control; more often, soot forms hard, caked deposits that spall off as localized flakes and form sharp edges. Any sharp edges will cause the local gas temperature to rise to near stagnation, which leads to losses of strength in hot retaining metal walls. Most designers prefer to avoid such deposits by using film cooling or extra high coolant velocities in the cooling jackets (particularly in the nozzle throat region) and by using injector patterns that minimize the formation of adhesive carbon deposits.

Hydraulic Losses in the Cooling Passage

Cooling coils or jackets should be designed so that the cooling fluid absorbs all the heat transferred across the inner thrust chamber walls with acceptably small coolant pressure drops.

While higher pressure drops may allow higher coolant velocities in cooling jackets that cool better, they require heavier feed systems slightly increasing engine mass and thus total inert vehicle mass. For many liquid propellant rockets, coolant velocities in the chamber vary from approximately 3 to 10 m/sec or 10 to 33 ft/sec and at the nozzle throat from 6 to 24 m/sec or 20 to 80 ft/sec.

A cooling passage may be considered to be a hydraulic pipe, and its *friction loss* can be calculated accordingly. For straight pipes,

$$\Delta p/\rho = \frac{1}{2} f v^2 (L/D) \tag{8-11}$$

where Δp is the friction pressure loss, ρ the coolant mass density, L the length of coolant passage, D the equivalent diameter, v the average velocity in the cooling passage, and f a friction loss coefficient. In English Engineering units the right side of this equation has to be divided by g_0, a factor proportional to the sea-level acceleration of gravity (32.174 ft/sec^2). The friction loss coefficient is a function of Reynolds number and has values between 0.02 and 0.05. This coefficient can be found in tables on hydraulic pipes. *Typical pressure losses* in cooling jackets correspond to between 5 and 25% of chamber pressure.

Large pressure drops in cooling jackets usually occur in locations where the flow direction or the flow-passage cross-sectional changes. Such sudden expansions or contractions cause a loss, sometimes larger than the velocity head $v^2/2$. These hydraulic situations exist at inlet and outlet chamber manifolds, injector passages, valves, and expansion joints.

Pressure losses in cooling passages of thrust chambers may be calculated, but more often they are measured. They are usually determined from cold flow tests (with an inert fluid, such as water, instead of the propellant and without combustion), and then the measured values are corrected for the actual propellant's different physical properties and for the hot chamber conditions; higher temperatures change propellant densities and viscosities, and in some designs also affect the needed cooling flow passage cross sections.

Thrust Chamber Wall Loads and Stresses

Analyses of loads and stresses are performed on all propulsion system components during design. Their purpose is to assure the propulsion designer and the flight vehicle user that (1) all components are strong enough to carry all imposed loads under all operating conditions so as to fulfill their intended function; (2) known potential failures have been identified and remedied; and (3) all components have been reduced to their practical minimum mass. In this section, we focus on describing loads and stresses at thrust chambers walls, where high heat fluxes and large thermal stresses complicate stress analyses. Some of the given information on safety factors and stress analysis also applies to other propulsion systems, including solid propellant motors and electric thrusters.

Safety factors (really "margins for ignorance") need to be relatively small in rocket propulsion systems when compared to commercial machinery, where these factors can be two to six times larger. Several *load conditions* have to be considered for each rocket component and these are:

1. *Maximum expected working load* is the largest operating load under all likely operating conditions or transients. Examples include operating at slightly higher chamber pressures than nominal as set by tolerances in design or fabrication (e.g., the tolerance in setting the tank pressure regulator) or any likely transient overpressures from ignition shocks.

2. *Design limit load,* typically set at 1.20 times the maximum expected working load, provides a safety margin. When there are significant variations in material composition or properties and with the uncertainties in methods of stress analysis or predicted loads, larger factors should be selected.

3. *Damaging loads* can be based either on yield, or ultimate, or endurance limit loads, whichever have the lowest value. A yield load causes permanent changes or deformations and is typically set as 1.10 times the design limit load. Endurance limits may be given by fatigue or creep considerations (such as during pulsing). Damaging loads induce stresses equal to the ultimate strength of the material, where significant elongations and area reductions can lead to failure. Typically, damaging loads are set at 1.50 times the design limit load.

4. *The proof test load* is applied to engines or their components during development and manufacturing inspection. It is often equal to the design limit load, provided this load condition can be simulated in the laboratory. For thrust chambers and other components whose high thermal stresses are difficult to simulate, so actual hot firing tests are used to obtain this proof, often with loads that approach the design limit load (e.g., with higher than nominal chamber pressures or mixture ratios that result in hotter combustion products).

During rocket operation, all thrust chamber walls experience radial and axial *loads from chamber pressures, flight accelerations* (axial and transverse), *vibrations,* and *thermal stresses.* These walls also have to withstand a momentary *ignition pressure surge or shock,* often due to excessive propellant accumulation in the chamber—such surge may exceed the nominal chamber pressure. In addition, chamber walls have to *transmit thrust loads* as well as other forces and, in some applications, also moments imposed by *thrust vector control* devices (described in Chapter 18). Walls also have to survive any "thermal shocks," namely, initial thermal stresses at starting. Because walls start at ambient temperatures, initially they experience higher heating rates than after reaching operating temperatures. Because loads differ in almost every design, each unit has to be considered individually in determining wall strengths.

As stated, heat transfer analyses are usually done only at the most critical wall regions, such as at and near the nozzle throat and at crucial locations in the chamber, and sometimes at the nozzle exit. Thermal stresses induced by temperature differences across a wall often result in the most severe stresses and any change in heat transfer or wall temperature distribution will affect these stresses. Specific failure criteria (wall temperature limits, yield stresses, and/or maximum coolant temperatures, etc.) need to be established prior to analysis.

Temperature differentials across wall chambers introduce compressive stresses on the inside and tensile stresses on the outside; this stress, s, can be readily calculated for simple cylindrical chamber walls that are thin in relation to their radius as

$$s = 2\lambda E\Delta T/(1 - v) \qquad (8-12)$$

where λ is the coefficient of thermal expansion of the wall material, E its modulus of elasticity, ΔT the temperature drop across the wall, and v the Poisson ratio of the wall material. Equation 8–12 only applies to elastic deformations. Temperature stresses can frequently exceed a material's yield point and values of E, v, and λ change with temperature. Effects of yielding in relatively thick-walled thrust chambers and nozzles appear as small and gradual contractions of the throat diameter after each operation (perhaps a 0.05% reduction after each firing) and as progressive crack formations of on inside chamber wall surfaces and on the throat inner surfaces after successive runs. Such phenomena limit the useful life and the number of starts and/or temperature cycles of thrust chambers (see Section 8–7 and Refs. 8–7 and 8–8).

In selecting the working stress for thrust chamber materials, variations of wall strength with temperature and temperature stresses present over the wall thickness have to be considered. Temperature drops across inner walls are typically between 50 and 550 K, and an average temperature is sometimes used for estimating material properties. The most severe thermal stresses may occur during starting as the hot combustion gases thermally shock the hardware, initially at ambient temperature. These transient thermal gradients may result in severe strains and local yielding.

Figure 8–12 depicts a typical steady-state stress distribution resulting from pressure loads and thermal gradients in a relatively thick inner wall. Here the inner wall surface is subjected to a compressive pressure differential from a high liquid pressure in the cooling jacket and a relatively large temperature gradient. In large rocket chambers, such as those used in the Redstone missile or the German V-2, the wall thickness of their steel nozzle may be up to 7 mm and any temperature differential across it may readily exceed several hundred degrees. Such large temperature gradients cause the hot inner wall side to expand more than the coolant side and impose high compressive thermal stresses on the inside and high tensile thermal stresses on the coolant side. For thick walls, pressure load induced stresses are usually small compared to thermal stresses. The resultant stress distribution in thick inner walls (shown shaded in the sample stress diagram of Fig. 8–12) indicates that the stress in the third of the wall thickness adjacent to the hot gases has exceeded the material's yield point. Because the modulus of elasticity and the yield point diminish with temperature, stress distributions are not linear over any yielded portion of the wall. In effect, this inner portion acts as a heat shield for the outer portion, which carries the load.

Because of the differential expansion between the hot inner shell and the relatively cold outer shell, it is necessary to provide for axial *expansion joints* to prevent severe temperature stresses. This is particularly critical in larger double-walled thrust chambers. The German V-2 thrust chamber expanded over 5 mm in an axial and 4 mm in a radial direction during firing.

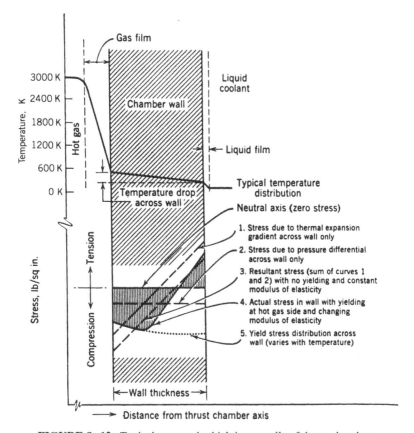

FIGURE 8-12. Typical stresses in thick inner walls of thrust chambers.

Cooling tubes at the walls of cylindrical thrust chambers are subjected to several different stress conditions. Only that portion of an individual cooling tube exposed to hot chamber gases experiences high thermal stresses and deformations (as shown later in Fig. 8-15). Cooling tubes have to withstand the internal coolant pressure, absorb the one-sided thermal stresses, and contain the chamber's gas pressure. The hottest temperature occurs at the center of the outer surface of that portion of tube exposed to the hot gases. Here, thermal stresses are relatively low, since temperature gradients are small. Typical copper alloy tube materials have a high conductivity and their walls are relatively thin (0.5 to 2 mm). Coolant pressure-induced loads on cooling tubes are relatively high, particularly when the thrust chamber operates at high pressures. The internal coolant pressure tends to separate the tubes. Gas pressure loads in the chamber are usually accommodated by reinforcing bands that are placed over the outside of the tubular jacket assembly (see Figs. 8-1 and 8-9). Joints between tubes must be gas tight and this can be accomplished by soldering, welding, or brazing.

When high-area-ratio nozzles are operated at sea level or at low altitudes, outer portions of the nozzle structure experience compression because the pressure in the nozzle near the exit is actually below atmospheric pressure. This can cause nozzle exit wall deformations and oscillations (between any existing circular and slightly elliptical nozzle exit shapes), which may lead to nozzle failure. Therefore, high-area-ratio nozzles usually have stiffening rings on the outside of the nozzle near the exit to maintain their circular shape. During development testing at low altitudes (or sea level) of thrust chambers that feed high area ratio nozzles it is common practice to substitute another thrust chamber with a "stub" nozzle, which has a much lower area ratio to prevent flow separation inside the nozzle (see Section 3.3) and exit wall oscillations or nozzle flutter. Test results with stub nozzles then have to be corrected to reflect actual flight conditions with higher area ratios.

8.3. LOW-THRUST ROCKET THRUST CHAMBERS OR THRUSTERS

Many spacecraft, certain tactical missiles, missile defense vehicles, and upper stages of ballistic missiles use special multiple thrusters in their small, liquid propellant rocket secondary engines. They generally have thrust levels between about 0.5 and 10,000 N or 0.1 to 2200 lbf, depending on vehicle size and mission. As mentioned in Sections 4.5 and 6.7, they are used for trajectory corrections, attitude control, docking, terminal velocity control in spacecraft or ballistic missiles, divert or side movements, propellant settling, and other necessary functions. Many operate in a pulsing mode with multiple restarts of relatively short duration during the major part of their duty cycle. As mentioned before, they can be classified as *hot gas thrusters* (high-performance bipropellant with combustion temperatures above 2600 K and vacuum I_s of 230 to 325 sec), *warm gas thrusters* such as monopropellant hydrazine (temperatures between 500 and 1600 K and I_s of 180 to 245 sec), and *cold gas thrusters* such as high-pressure stored nitrogen (200 to 320 K) with low specific impulse (40 to 120 sec).

A typical small thruster using bipropellants is shown in Fig. 8–13 and one using hydrazine as a monopropellant in Fig. 8–14. For attitude control angular motions these thrust chambers are usually arranged in pairs as explained in Section 4.5 and shown in Fig. 4–14; the same control signal activates valves on both paired units. For translation maneuvers a single thruster can be fired (often in a pulsing mode) and its thrust axis usually goes through the center of gravity of the vehicle. The smaller space rocket systems use pressurized feed systems, some with positive expulsion provisions, as described in Section 6.2. Vehicle missions and automatic control systems in the vehicle often require frequent pulses to be applied by pairs of attitude control thrust chambers for vehicle rotational control, usually operating for short periods (as low as 0.01 to 0.02 sec per pulse). This type of frequent and short-duration thrust application is also known as *pulsed thrust operation*; typical pulsing frequencies are between 150 and 500 pulses per minute. Resulting accelerations will depend on thrust magnitude and thruster location on the vehicle; these accelerations can be axial or at an angle to the flight velocity vector.

For any given thruster, throttling may be achieved by varying (1) the time between pulses (less total cumulative impulse per unit time), or (2) by limiting the total number of pulses for a given maneuver, and/or (3) by reducing the pulse duration. There is

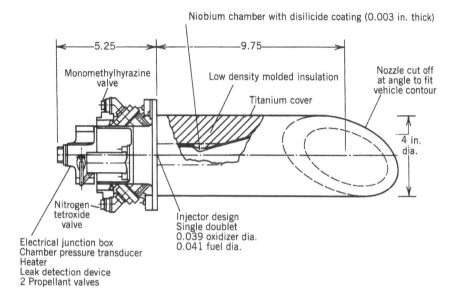

Niobium chamber with disilicide coating (0.003 in. thick)

FIGURE 8–13. This radiation-cooled, insulated noncryogenic bipropellant vernier thruster was one of several used on the reaction control system of the Space Shuttle vehicle for orbit stabilization and orientation, rendezvous or docking maneuvers, station keeping, deorbit, or entry. The nozzle was cut off at an angle to fit the contour of the vehicle. Operation could be pulsed (firing durations between 0.08 and 0.32 sec with minimum off time of 0.08 sec) or steady (0.32 to 125 sec). Demonstrated life was 23 hours of cumulative operation and more than 300,000 starts. Courtesy of Aerojet Rocketdyne.

a performance degradation with decreasing pulse duration for supersonic-nozzle types of thrusters because propellants are used inefficiently during thrust buildup and the decay, when they operate below full chamber pressure and nozzle expansion characteristics are not optimum. Specific impulse greatly suffers when pulse durations become very short. In Section 3.5 the actual specific impulse of a rocket operating at a steady state was estimated as no higher than 92% of its value. With very short pulses (0.01 sec) this estimate can be lower than 50% but with pulses of 0.10 sec it may reach around 75 to 88%. Also, the reproducibility of the total impulse delivered from short pulses decreases after prolonged use. Preheating monopropellant catalyst beds results in longer lifetimes with little performance degradation (i.e., in pressure rise during pulse width). Heat transfer is also reduced by pulsing.

One way to minimize impulse variations in short pulses and to maximize the effective actual specific impulse is to minimize liquid propellant passage volumes between the control valve and the combustion chamber. Propellant flow control valves for pulsing attitude control thrust chambers are therefore often designed as an integral part of the thrust chamber–injector assembly, as shown in Fig. 8–13 (and later in Fig. 8–16). Special electrically actuated, leakproof, fast-acting valves with response times ranging from 2 to 25 msec for both opening and closing operations are used. *Life of small thrusters* is limited by the same criteria mentioned for large cooled thrust chambers;

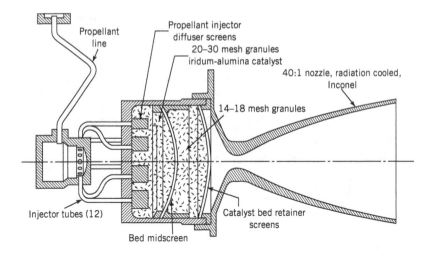

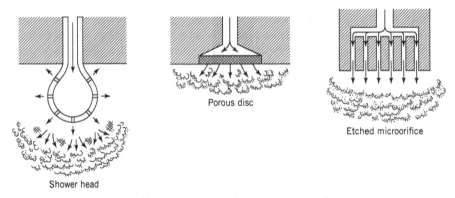

FIGURE 8–14. Typical hydrazine monopropellant small thrust chamber with catalyst bed, showing different methods of injection.

however, such life can also be affected by the failure of other components in propulsion systems utilizing multiple thrusters for reaction control. The life of each system component (such as propellant valves, pressurizing gas valves, and pressure switches or measuring instruments) has to be longer than that of the assembled unit. Overall life in any rocket propulsion system can be enhanced by redundancies (e.g., by a spare thruster that can be activated when the original on fails), or by extra qualification testing at 5 to 10 times the number of cycles or starts over mission requirements. With pulsing units the number of starts may become too large and proof tests are usually only taken as 5 to 10 times the number of starts. Valves must operate reliably with predictable characteristics for perhaps up to 40,000 to 80,000 starts and this turn often requires an equal number of endurance proof test cycles.

Liquid storable bipropellants such as N_2O_4-monomethylhydrazine are used when high performance is mandatory. Some units have utilized ablative materials for thrust chamber construction, as in the Gemini command module. The Space Shuttle small thrusters used radiation cooling with refractory metals, as shown in Fig. 8–13 (another radiation-cooled unit is shown later in Fig. 8–16). Rhenium and carbon-based materials made of woven strong carbon fibers in a carbon matrix have also been utilized in radiation-cooled bipropellant thrusters.

Hydrazine monopropellant thrusters are used when system simplicity is important and moderate performance is acceptable. They have nontoxic, clear, clean exhaust plumes. For a catalyst, virtually all attitude control rockets use finely dispersed iridium or cobalt deposited on porous-ceramic (aluminum oxide) substrate pellets, 1.5 to 3 mm in diameter. Figure 8–14 shows a typical design for the catalyst pellet bed of an attitude control thruster designed for both pulse and steady-state operations. Each injection passage is covered with a cylindrical screen section that extends into a part of the catalyst bed and distributes the hydrazine propellant. Figure 8–14 also depicts other successful types of hydrazine injectors. Several arrangements of catalyst beds have been employed; some have a spring loading to keep the pellets firmly packed. Hydrazine monopropellant units range in size from 0.2 to 2500 N of thrust but the vast majority of catalyzed hydrazine monopropellant thrusters are small, with thrust levels below 22 N (5 lbf). Scaling procedures are empirical and each size and design requires extensive testing. The amount of ammonia decomposition, as shown in Fig. 7–3, may be controlled by the design of the catalyst bed and its decomposition chamber.

Several mechanical, thermal, and chemical problems need to be addressed in designing a catalyst bed for decomposing hydrazine, the more important of which are catalytic attrition and catalyst poisoning. *Catalytic attrition* or physical loss of catalyst material stems from pellet motion and abrasion resulting in the production and loss of very fine particles. Crushing of pellets can occur from thermal expansions and from momentary overpressure spikes. As mentioned in Chapter 7, catalytic activity can also decline because of *poisoning* by trace quantities of contaminants present in commercial hydrazine, such as aniline, MMH, UDMH, sulfur, zinc, sodium, and/or iron. Some of these contaminants come with the hydrazine and some are added by the materials used in the tanks, and propellant plumbing. High-purity grade hydrazine (with less than 0.003% aniline and less than 0.005% carbonaceous material) does not contaminate catalysts. Catalyst degradation, regardless of cause, results in ignition delays, overpressures and pressure spikes, thereby decreasing specific impulse and the impulse bit per pulse in attitude control engines.

Thruster modules that contain subassemblies of two or more small thruster are commonly used. These can be part of a multithrust control system for small or medium sized space vehicles or part of stages in a multistaged vehicle. Figure 4–14 shows a schematic of 4 modules with 4 thrusters each. Thruster modules can save considerable assembly time.

Figures 17–4 and 17–6 show combinations of chemical and electrical propulsion with a monopropellant. Electrical postheating of reaction gases issuing from catalysis allows an increase of the vacuum specific impulse from 240 sec to about

290 or 300 sec. A number of these combination auxiliary thrusters have successfully flown on a variety of satellite applications and are particularly suitable for space-craft where ample electrical power is available and extensive short-duration pulsing is needed.

Cold gas thrusters use stored high-pressure inert gases as the propellant; they and their performance are mentioned in Section 7.5 and relevant propellants and specific impulses are listed in Table 7–3. These may be used with pressurized feed systems for pulsing and for low thrust and low total impulse operations. Thrusters, valves, and piping can be made from aluminum or plastics. Early versions of the Pegasus air-launched launch vehicle used cold gas thrusters for roll control. Some advantages of cold gas systems are: (a) they are very reliable and have been proven in space flights lasting more than 15 years; (b) these systems are simple and relatively inexpensive; (c) ingredients are nontoxic; (d) no deposits or contamination occurs on sensitive spacecraft surfaces, such as mirrors; (e) they are relatively safe; and (f) capable of random pulsing. The disadvantages are: (a) the units are relatively heavy with poor propellant mass fractions (0.02 to 0.19); (b) specific impulses and vehicle velocity increments are low, when compared to mono- or bipropellant systems; and (c) they occupy relatively large volumes.

8.4. MATERIALS AND FABRICATION

The choice of materials for inner chamber walls in combustion chamber and nozzle throat regions (i.e., the most critical locations) is influenced by hot gas composition, maximum allowable wall temperatures, heat transfer rates, and duty cycles. A large variety of materials have been utilized. Table 8–3 lists some typical materials selected for several applications, thrust sizes, and propellants. For high-performance, high-heat-transfer, regeneratively cooled thrust chambers, any high-thermal-conductivity material with a thin-wall design will reduce thermal stresses. Copper and some of its alloys are excellent conductors and they will not noticeably oxidize in fuel-rich noncorrosive gas mixtures—it is often used with oxygen and hydrogen below a mixture ratio of 6.0. Inner walls are therefore usually made of copper alloys (with small additions of zirconium, silver, or silicon), which have a conductivity not quite as good as pure (oxygen-free) copper but display improved high-temperature strength.

Figure 8–15 shows partial sections through five cooling jacket configurations. All involve curved or doubly curved components and precision fits for good fastening or joining. The top configuration has an intermediate thin corrugated sheet of metal, which has been used extensively in Russia; it is soldered together under external pressure in a furnace. It is used in thrust chamber locations where the heat transfer is mild or of intermediate intensity. The milled slot design usually can accept the highest heat transfer intensities expected. One fabrication technique in design (b) is to machine (usually on a milling machine) nearly rectangular grooves or cooling channels of varying width and depth into the inner surface of a relatively thick contoured inner wall using a high conductivity material such as a copper alloy. The grooves are then filled with wax and with an electrolytic plating technique an outer wall is built of a suitable metal (such as nickel) so as to enclose the coolant channels; the wax

TABLE 8–3. Typical Materials Used in Several Common Liquid Propellant Thrust Chambers

Application	Propellant	Components	Cooling Method	Typical Materials
Bipropellant TC, cooled, high pressure (booster or upper stage)	Oxygen—hydrogen	C, N, E	F	Copper alloy
		I	F	Transpiration cooled porous stainless steel face. Structure is stainless steel
		Alternate E	R	Carbon fiber in a carbon matrix, or niobium
		Alternate E	T	Steel shell with ablative inner liner
Bipropellant TC cooled, high pressure (booster or upper stage)	Oxygen–hydrocarbon or storable propellant[a]	C, N, E, I	F	Stainless steel with tubes or milled slots
		Alternate E	R	Carbon fiber in a carbon matrix, or niobium
		Alternate E	T	Steel shell with ablative inner liner
Experimental TC (very limited duration—only a few seconds)	All types	C, N, E	U	Low-carbon steel
Small bipropellant TC	All types	C, N, E	R	Carbon fiber in carbon matrix, rhenium, niobium
			T	Steel shell with ablative inner liner
		I	F	Stainless steels, titanium
Small monopropellant TC	Hydrazine	C, N, E,	R	Inconel, alloy steels
		I	F	Stainless steel
Cold gas TC	Compressed air, nitrogen	C, N, E, I	U	Aluminum, steel, or plastic

[a]HNO_3 or N_2O_4 oxidizer with N_2H_4, MMH, or UDMH as fuels (see Chapter 7). Abbreviation: TC, thrust chamber; C, chamber wall; N, nozzle converging section walls, diverging section walls, and throat region walls; E, walls at exit region of diverging section of nozzle; I, injector face; F, fuel cooled (regenerative); R, radiation cooled; U, uncooled; T, transient heat transfer or heat sink method (ablative material).

is then melted out. A unique alternate here is to use HIP (Hot Isostatic Pressing) to bond the inner slotted/curved liner throat support components with an outer cylinder. The third, design (c), shows a tubular construction; it has been used extensively in the U.S. for the larger thrust chambers with intermediate to high heat transfer. Although it is not clearly discernable, the cooling jackets of the thrust chamber in Figs. 6–1 and 8–1 have the coolant flow through formed tubes to cool the walls. Figure 8–9

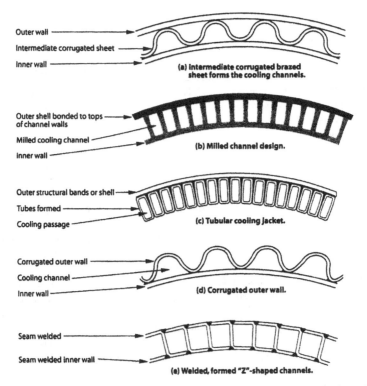

FIGURE 8–15. Sketches of sections through a portion of the cooling jacket of several different cooling schemes in regeneratively cooled thrust chambers. Used with permission from Ref. 8–1.

shows individual tubes that are shaped to the contour of the thrust chamber including its nozzle; round tube shapes are altered into nearly rectangular cross sections. These formed tubes are then brazed or soldered together in a furnace and the outer structural shell or outer bands are also brazed to the assembly. The corrugated outer wall concept (d) is perhaps the simplest and often the lightest cooling jacket configuration. It has been used in locations where heat transfer is modest. The bottom shell, design (e), represents a welded construction made of stainless steel. It has a higher allowable maximum wall temperature than a brazed construction. It has been used in nozzles designed and built in Europe.

Cooling channel depths and widths vary with chamber profile location and with local thrust chamber and nozzle diameters. The nozzle throat region, where heat transfer is the highest, is therefore the location of the highest coolant velocity and where total cooling passage area is the smallest. Often, two or sometimes three different cooling jacket schemes are designed into the same thrust chamber. Typically, the milled groove configuration is used for the nozzle throat region and other configurations for the chamber and the nozzle exit regions.

Failure modes of regeneratingly cooled thrust chambers often are seen as hot wall bulgings on the gas side and/or the appearance of cracks. During hot firing, strains at the hot surface may exceed the local yield point thus giving it local permanent compressive deformations. With the cooldown after operation and with each successive firing, some additional yielding and further plastic deformation will occur until cracks form. As successive firings take place, these cracks can become deep enough for leaks of cooling propellant into the chamber or nozzle to take place, and then the thrust chamber will usually fail. The *useful life* of a thrust chamber is the maximum number of firings (and sometimes also the cumulative firing duration) without any failures. The prediction of wall failures is not simple and Refs. 8–7 and 8–8 explain this in more detail. The life of a regeneratively cooled, large thrust chamber is typically 6 to 10 starts. For small thrusters without cooling jackets (e.g., radiation cooled) their life can be limited by ablation or erosion after thousands of pulses or starts. Useful life can also be limited by the storage life of soft components (O-rings, gaskets, valve stem lubricant) and, for small thrusters with many pulses, also the fatigue of valve seats. There always is, therefore, an upper limit on the number of firings that any thrust chamber can withstand safely (see Section 8.7).

For *radiation cooling*, several different carbon-based materials work well in reducing or fuel-rich hot-gas environments; they can be used up to wall temperatures of perhaps 3300 K or 6000 °R. At leaner gas mixtures carbon oxidizes at elevated temperatures (when its surfaces glow red or white). Carbon materials and ablative materials are used extensively in solid propellant rocket motors and are discussed further in Chapter 15.

For some small radiation-cooled bipropellant thrusters with storable propellants, such as those used as reaction control thrusters on the Space Shuttle orbiter (see Fig. 8–13), hot walls are made of niobium coated with disilicide (which are good up to 1120 K or 2050 °R). To prevent damage, a fuel-rich mixture or film cooling is often used. In small thrusters, rhenium walls protected by iridium coatings (oxidation resistant) have more recently been used up to about 2300 K or 4100 °R (see Refs. 8–1 and 8–9). Other high-temperature materials, such as tungsten, molybdenum, alumina, and tantalum, have been tried but have had problems in manufacture, cracking, hydrogen embrittlement, and/or with excessive oxidation.

A small radiation-cooled bipropellant thruster is shown in Fig. 8–16. It uses three different nozzle and chamber materials. This thruster's injector has extra fuel injection holes (not shown in Fig. 8–16) that provide film cooling to keep wall temperatures below their failure limits. High-temperature copper-nickel alloys or stainless steels are used for the radiation cooled nozzle-and chamber walls of thrusters operating with hydrazine monopropellant. See Fig. 8–14.

Until recently, it had not been possible to manufacture large pieces with carbon–carbon materials. That was one reason why large nozzle sections and integral nozzle exit cone pieces in solid motors were made from carbon phenolic cloth lay-ups. Progress in manufacturing equipment and technology has now made it possible to build and fly larger pieces made of carbon fiber in a carbon matrix. A three-piece extendible carbon nozzle exit cone of 2.3 m (84 in.) diameter and 2.3 to 3 mm thickness has flown on an upper-stage engine. This thrust chamber with its movable nozzle

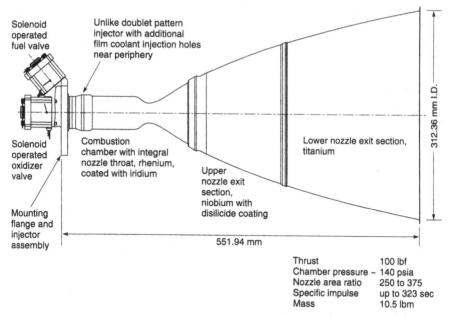

Solenoid operated fuel valve

Unlike doublet pattern injector with additional film coolant injection holes near periphery

Solenoid operated oxidizer valve

Combustion chamber with integral nozzle throat, rhenium, coated with iridium

Mounting flange and injector assembly

Upper nozzle exit section, niobium with disilicide coating

Lower nozzle exit section, titanium

312.36 mm I.D.

551.94 mm

Thrust	100 lbf
Chamber pressure ~	140 psia
Nozzle area ratio	250 to 375
Specific impulse	up to 323 sec
Mass	10.5 lbm

FIGURE 8–16. Radiation-cooled reaction control thruster R-4D-15 that uses nitrogen tetroxide and monomethylhydrazine propellants. The large nozzle area ratio allows good vacuum performance. It has three different nozzle materials, each with a lower allowable temperature (Re 4000 °F; Nb 3500 °F; Ti 1300 °F). Courtesy of Aerojet Rocketdyne.

extension is shown in Fig. 8–17; its parameters are listed in Table 8–1, its testing is reported in Ref. 8–6. The RL-10B-2 has flown successfully many times.

Material properties must be evaluated under all likely operating conditions, loads, starting conditions, temperature changes, pressure variations, and the like before selection for any specific thrust chamber application. This evaluation must include physical properties, such as tensile and compressive strengths, yield strength, fracture toughness, modulus of elasticity (for determining deflections under load), thermal conductivity (a high value is best for steady-state heat transfer), coefficient of thermal expansion (some large thrust chambers can grow by 3 to 10 mm when hot, causing problems with their piping connections and/or structural supports), specific heat (capacity to absorb thermal energy), reflectivity (for radiation heat transfer), or density (ablatives require more volume than steel). These properties change with temperature and sometimes they change noticeably with small changes in materials composition. The temperature at which a material loses perhaps 60 to 75% of its ambient temperature strength is often selected as the maximum allowable wall temperature, a value well below its melting point. Since listing of all the key properties of any single material requires many entries, it is not possible to give them here, but they are usually available from suppliers and other sources. Other important material properties are erosion resistance, acceptably low chemical activity with the propellants or the hot gases, reproducible decomposition

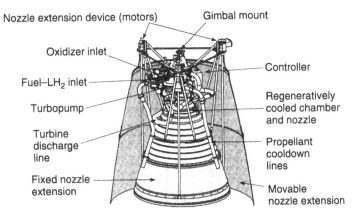

Nozzle extension device (motors)

Gimbal mount

Oxidizer inlet

Controller

Fuel–LH$_2$ inlet

Turbopump

Regeneratively cooled chamber and nozzle

Turbine discharge line

Propellant cooldown lines

Fixed nozzle extension

Movable nozzle extension

(*a*) Half section of nozzle extension in stowed position

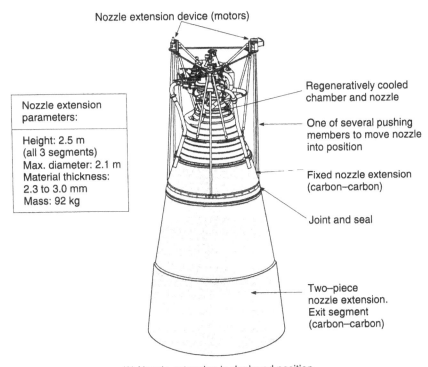

Nozzle extension device (motors)

Nozzle extension parameters:

Height: 2.5 m (all 3 segments)
Max. diameter: 2.1 m
Material thickness: 2.3 to 3.0 mm
Mass: 92 kg

Regeneratively cooled chamber and nozzle

One of several pushing members to move nozzle into position

Fixed nozzle extension (carbon–carbon)

Joint and seal

Two–piece nozzle extension. Exit segment (carbon–carbon)

(*b*) Nozzle extension in deployed position

FIGURE 8–17. The RL-10B-2 rocket engine has an extendible nozzle cone or skirt, which is placed around the engine during the ascent of the Delta III and IV launch vehicles. This extension is lowered into position by electromechanical devices after the launch vehicle has been separated from the upper stage at high altitude and before firing. Courtesy of Aerojet Rocketdyne.

or vaporization for ablative materials, ease and cost of fabrication (welding, cutting, forming, etc.), consistency of batch composition (including identified impurities) for the different types of material needed (metals, organics, seals, insulators, lubricants, cleaning fluids), and ready availability and competitive cost for the raw materials.

A relatively new fabrication technology known as *Additive Manufacturing* (AM) is being adapted to make certain key components in a large variety of manufactured products. For rocket propulsion systems, a process called Selective Laser Melting (SLM) has been under development at Aerojet Rocketdyne. Here, the first layer of a selected powdered metal is deposited with a predetermined 2-D pattern on a horizontal flat bed, which is made of a high temperature material. The powder is then melted by precision laser heating in an argon atmosphere. This process may be repeated numerous times, each with a new metal powder layer on top of the now solidified layer below, until the desired object or part has been created. The SLM process uses sophisticated numerical-control programming for the three-dimensional movement of the bed, and for the position, timing, and laser power level, and for the feeding of metal powder at the desired locations. In rocket propulsion systems, only certain SLM parts offer significant technical or economic advantages over present fabrication methods; these items depend on part geometry, selected metal powder, bed material, and in particular on the number of identical parts needed. The amortization of special fabrication machinery and equipment also plays a role. Lockheed Martin employs electron beam welders for additively manufactured large rocket propellant tanks using strands of titanium wire.

An experimental liquid propellant rocket engine injector, fabricated by AM, was successfully hot fired in a thrust chamber by NASA in 2013. Also, small thrusters and parts for rocket engines currently in production have been fabricated using AM (as of 2015, no AM fabricated parts are known to have flown). Merits claimed for AM parts include desirable quality high-strength products, reductions in mass and/or lower manufacturing costs (depending on part geometry, material cost, and number of identical parts). This technique is being developed for a variety of industries worldwide. See Refs. 8–10 and 8–11.

8.5. HEAT TRANSFER ANALYSIS

For actual rocket engine development not only is heat transfer analyzed, but rocket units are almost always tested to assure that heat transfer is handled satisfactorily under all operating and emergency conditions. Heat transfer calculations provide a most useful guide in design, testing, and failure investigations. Rocket combustion devices that are regeneratively cooled or radiation cooled can reach thermal equilibrium and steady-state heat transfer relationships may be applied. Transient heat transfer conditions exist not only during thrust buildup (starting) and shutdown in all rocket propulsion systems, but also for cooling techniques that never reach equilibrium, such as those with heat sinks such as ablative materials.

Sophisticated *finite element analysis* (FEA) programs for heat transfer analysis have been available for over 30 years, and several different FEA computer programs are used for thrust chamber steady-state and transient calculations, with a variety of chamber geometries and different materials, and with temperature-variant properties. Any detailed descriptions of such powerful and sophisticated techniques are beyond the scope of this book, but can be found in Refs. 8–12, 8–13, and 8–14. Major rocket propulsion organizations develop their own computer programs. NASA builds and tests scale-down models with advanced diagnostic tools to ascertain heat transfer rates in new designs. In this section, we give some basic relationships that are a foundation for FEA programs; they are intended only to provide an understanding of phenomena and underlying principles.

General Steady-State Heat Transfer Relations

For heat *conduction* the following general relation applies:

$$\frac{Q}{A} = -\kappa\frac{dT}{dL} \approx -\kappa\frac{\Delta T}{t_w} \tag{8–13}$$

where Q is the heat transferred across a surface area A, dT/dL the temperature gradient, t_w the wall thickness, and k the thermal conductivity (expressed as the amount of heat transferred per unit time through a unit area of surface for 1° temperature difference over a unit wall thickness). The negative sign indicates that temperature always decreases in the direction of heat transfer.

The steady-state heat transfer through a chamber wall of a liquid-cooled rocket chamber can be treated as a series-resistance-type, steady-state heat transfer problem with a large temperature gradient across the gaseous film on the inside of the chamber wall, a temperature drop across the wall, and, in cases of cooled chambers, a third temperature drop across the film of the moving cooling fluid. This model requires a combination of *convection* at the flow boundaries and *conduction* through the chamber walls as shown schematically in Fig. 8–18.

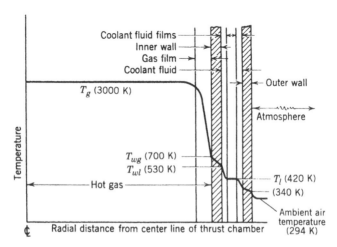

FIGURE 8–18. Applicable temperature gradients in cooled rocket thrust chamber. The temperatures shown are typical.

The typical steady-state *convection* heat transfer relation is shown in Eq. 8–14. For regeneratively cooled thrust chambers, it may be variously expressed as follows (refer to Fig. 8–18):

$$q = h(T_g - T_l) = Q/A \tag{8–14}$$

$$= \frac{T_g - T_l}{1/h_g + t_w/\kappa + 1/h_l} \tag{8–15}$$

$$= h_g(T_g - T_{wg}) \tag{8–16}$$

$$= (\kappa/t_w)(T_{wg} - T_{wl}) \tag{8–17}$$

$$= h_l(T_{wl} - T_l) \tag{8–18}$$

where q is heat transferred per unit area per unit time, T_g the absolute chamber gas temperature, T_l the absolute coolant liquid temperature, T_{wl} the absolute wall temperature on the liquid side of the wall, T_{wg} the absolute wall temperature on the gas side of the wall, h an overall film coefficient, h_g the gas film coefficient, h_l the coolant liquid film coefficient, t_w the thickness of the chamber wall, and k the conductivity of the wall material. Any consistent set of units may be used in these equations. These simple relations assume that the heat flow is only radial; often, quasi-one-dimensional theories also assume that the thermal conductivity and the film coefficients are at average values and not functions of temperature or pressure. A two- or three-dimensional finite element model would be needed to analyze the heat transfer in any axial directions (usually occurring at nozzle throat wall regions where non-negligible heat is transferred to regions upstream and downstream of it).

Because film coefficients, gas and liquid coolant temperatures, wall thicknesses, and surface areas usually vary with the axial distance L within a combustion chamber (assuming axial heat transfer symmetry), the *total heat transfer per unit time Q* can be found by integrating the local heat transfer over the entire internal (circular) surface area A of the chamber and the nozzle:

$$Q = \int q\,dA = \pi \int D q\,dL \tag{8–19}$$

Because both the heat transfer rate per unit area q and the diameter D can often be complicated functions of the thrust chamber length L, the integration is usually carried out by dividing the rocket chamber into finite lengths. Here, L is taken as zero at the injector face. Assuming that q is given by Eqs. 8–13 to 8–19 and remains constant over the length of each element can only give approximate solutions.

The fluid film boundaries established by the combustion products on one side of the wall and by the coolant flow on the other are important quantities controlling the heat transfer across rocket chamber walls. Gas film coefficients establish the amount of the heat transfer rate, and liquid films largely determine the value of the wall temperatures. Film coefficient determinations for use in Eqs. 8–16 and 8–18 may be difficult because of the inherent complex geometries, nonuniform velocity profiles, surface roughness, boundary layer behavior, and combustion oscillations usually present in such systems.

Conventional heat transfer theory is usually presented in terms of several dimensionless parameters (Refs. 8–12, 8–13 and 8–14). Formulations of convection heat transfer in turbulent flows are largely empirical, and one preferred relation for the heating inside circular tubes has been

$$\frac{h_g D}{\kappa} = 0.023 \left(\frac{D \upsilon \rho}{\mu} \right)^{0.8} \left(\frac{\mu c_p}{\kappa} \right)^{0.4} \tag{8–20}$$

where h_g is the film coefficient, D the diameter of the chamber or of the nozzle, υ the calculated average local gas velocity, k the conductivity of the gas, μ the absolute gas viscosity, c_p the specific heat of the gas at constant pressure, and ρ the gas density.

In Eq. 8–20 the quantity $h_g D/k$ is known as the Nusselt number, the quantity $D\upsilon\rho/\mu$ as the Reynolds number, and the quantity $c_p\mu/k$ as the Prandtl number Pr (often tabulated as a property of the fluid). The gas film coefficient h_g may also be determined from Eq. 8–21:

$$h_g = 0.023 \frac{(\rho \upsilon)^{0.8}}{D^{0.2}} \mathrm{Pr}^{0.4} \kappa / \mu^{0.8} \tag{8–21}$$

where $\rho\upsilon$ is the local mass velocity, and the constant 0.023 is dimensionless when compatible units are used. Boundary layer temperature gradients affect the various gas properties in propellant combustion and combustion phenomena are propellant specific. Conventional theoretical approaches, using Eq. 8–20 or 8–21, describe steady-state flows in relatively long heated circular tubes where "fully developed velocity profiles" can be attained. Heat typically flows into the tube from all sides (360°); however, in thrust chambers the heat flow to the coolant passage comes only from one side of the passage. When chamber lengths are relatively short, equilibrium flow profiles may not form. Actual flows in combustion chamber are highly turbulent containing liquid droplets which are evaporating and often there is no equilibrium. For all these reasons Eqs. 8–20 and 8–21 must be viewed only as good approximations.

Equations where the coefficients have been validated by actual experimental data tend to be more reliable and they are used in design. Bartz (Ref. 8–14) has surveyed the agreement between theory and experiment and developed semiempirical correction factors incorporated the equation below:

$$h_g = \frac{0.026}{D^{0.2}} \left(\frac{c_p \mu^{0.2}}{\mathrm{Pr}^{0.6}} \right) (\rho \upsilon)^{0.8} \left(\frac{\rho_{am}}{\rho'} \right) \left(\frac{\mu_{am}}{\mu_0} \right)^{0.2} \tag{8–22}$$

The subscript 0 refers to properties evaluated at the stagnation or chamber combustion temperature; the subscript *am* refers to properties at the arithmetic mean temperature of the local free-stream static temperature and the wall temperatures; ρ' is the free-stream value of the local gas density. Again, the empirical constant 0.026 is dimensionless when compatible dimensions are used for the other terms. The gas velocity v is the local free-stream velocity corresponding to the density ρ'. Since the density to the 0.8th power is roughly proportional to the pressure and the gas film coefficient is roughly proportional to the heat flux, so it follows that the heat transfer rate increases approximately linearly with the chamber pressure. These semiempirical heat transfer equations have been further modified and validated for common propellants, limited chamber pressure ranges, and specific injectors and such analysis is often proprietary to specific design organizations.

Temperature drops across inner walls along with the maximum temperature are reduced when the wall is thin and made of a high-thermal-conductivity material. Wall thickness is determined from strength considerations and thermal stresses, and some designs have as little as 0.025 in. thickness. Effects of changing the film coefficients are shown in Example 8–1.

Surface roughness has a large outcome on the value of film coefficients and thus on the heat flux. Measurements have shown that the heat flow can be increased by a factor of up to 2 by surface roughness and to higher factors when designing turbulence-creating obstructions in the cooling channels. Major surface roughness on the gas side will cause the gas to come close to its stagnation temperature locally. However, surface roughness on the liquid coolant side of the wall will enhance turbulence and the absorption of heat by the coolant and reduce wall temperatures.

Example 8–1. The effects of varying film coefficients on the heat transfer and wall temperatures of liquid cooled chambers are to be explored. The following data are given:

Wall thickness	0.445 mm
Wall material	Low-carbon steel
Average conductivity, k	43.24 W/m²-K/m
Average gas temperature, T_g	3033 K or 2760 °C
Average liquid bulk temperature, T_l	311.1 K or 37.8 °C
Gas-film coefficient, h_g	147 W/m²-°C
Liquid-film coefficient, h_l	205,900 W/m²-°C

Vary h_g (at constant h_l), then vary h_l (at constant h_g), and then determine the resulting changes in heat transfer rate and wall temperatures on the liquid and the gas side of the wall.

SOLUTION. Use Eqs. 8–13 to 8–18 and solve for q, T_{wg}, and T_{wl}. The answers shown in Table 8–4 indicate that variations in the gas-film coefficient have a profound influence on the heat transfer rate but relatively little effect on the wall temperature. The exact opposite is true for variations in the liquid-film coefficient; here, changes in h_l produce little change in q but a fairly substantial change in the wall temperature.

Figure 8–19 shows heat flow directions, temperature distributions, and locations for maximum wall temperatures in a milled cooling jacket design. This design is represented in (c), the third sketch of Fig. 8–15. The inner wall should be thin, so that the temperature difference across this wall is low and therefore thermal stresses are also low.

TABLE 8–4. Change in Film Coefficient for Example 8–1

Change in Film Coefficient (%)		Change in Heat Transfer (%)	Wall Temperature (K)	
Gas Film	Liquid Film		Gas Side, T_{wg}	Liquid Side, T_{wl}
50	100	50	324.4	321.1
100	100	100	337.2	330.5
200	100	198	362.8	349.4
400	100	389	415.6	386.1
100	50	99	356.1	349.4
100	25	98	393.3	386.7
100	12.5	95	460.0	397.8
100	6.25	91	596.7	590.5

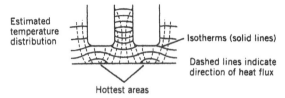

Estimated temperature distribution — Isotherms (solid lines)

Dashed lines indicate direction of heat flux

Hottest areas

FIGURE 8–19. Results from a two-dimensional analysis of the heat transfer in two cooling channels of a milled slot-cooling jacket. The outer wall and the upper parts of the channels are not shown.

Transient Heat Transfer Analysis

An uncooled (high melting point) metal thrust chamber is the simplest type to analyze because there are no chemical or phase changes. Here, thermal equilibrium is not reached. Uncooled walls act essentially as heat sinks continuously absorbing energy from the combustion gases. With the aid of experimental data to determine some typical coefficients, it is possible in some cases to predict the transient heating of uncooled walls.

During propellant combustion, a changing temperature gradient exists across the walls because varying heat is being transferred from the hot gases. The heat being transferred from hot walls to the surrounding atmosphere and to the structure, by conduction in metal parts, can be negligibly small during transient heating. Each wall location experiences a rising temperature as the burning process progresses in time. After the completion of the rocket engine's operation, wall temperatures will tend to equalize.

When axial heat transfer within a metal wall can be neglected, then the heat balance across any wall cross section may be expressed with the one-dimensional unsteady heat conduction equation (e.g., Ref. 8–12):

$$\frac{\partial}{\partial x}\left(\kappa\frac{\partial T}{\partial x}\right) = \rho\bar{c}\frac{\partial T}{\partial t} \qquad (8\text{--}23)$$

Above, T is a function of both the thickness coordinate x and time t. The heat conductivity k may depend on the wall material and its temperature; ρ is the wall material density and $\bar{c}$ its average specific heat. A typical temperature–time–location history is given in Fig. 8–20. Here, the lower horizontal line at $T = 21°C$ denotes the initial wall condition before the rocket engine operates; the various curves show temperature profiles across the wall at successive time intervals after initiation of

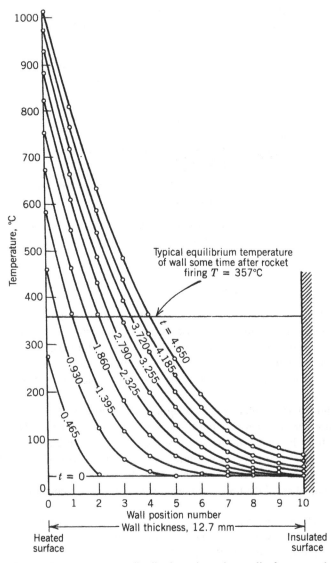

FIGURE 8–20. Typical temperature distributions through a wall of an uncooled relatively thick metal thrust chamber as a function of heating time t (in seconds).

combustion. The line at $T = 357°C$ shows an end temperature of the wall a finite time after cutoff.

The heat transferred into any hot wall surface (and distributed within the wall) must not exceed any critical temperature related to the heat-absorbing capacity of the wall. Equation 8–15 shows that Q/A depends on the hot gas temperature, the wall temperature, and the film coefficients. For calculations, the chamber and nozzle walls may be divided into cylindrical or conical segments as appropriate and each wall segment in turn divided into an arbitrary number of axisymmetric concentric layers, each of a finite thickness. At any given time, the heat conducted by any one wall layer exceeds the heat conducted into the next layer by the amount of heat absorbed (raising the temperature of that particular layer). This iterative approach readily lends itself to two- or three-dimensional computer analyses, resulting in data similar to Fig. 8–20. It is usually sufficient to only determine the heat transfer at the most critical locations. Three-dimensional analyses must be undertaken when the wall geometry is more complex than purely cylindrical, heat is conducted also in directions other than normal to the axis, temperature variable properties are involved, boundary layers vary with time and location, and more than one material layer is present in the wall.

A number of mathematical simulations of transient heat transfer in *ablative materials* have been devised, many with limited success. Any such approach must include simulation for the pyrolysis, chemical decomposition, char depth, and outgassing effects on the film coefficient, and requires proper material property data. Most simulations for transient heat transfer require some experimental data.

Steady-State Transfer to Liquids in Cooling Jacket

The term *regenerative cooling* applies to rocket engines when one of the propellants is circulated through cooling passages around the thrust chamber prior to being injected and subsequently burned in the chamber. This is represented by a forced convection heat transfer situation. The term *regenerative* is perhaps not altogether appropriate here as it bears little relation to the meaning given to it in steam-turbine practice. It is intended to convey the fact that heat absorbed by the coolant propellant is not wasted but augments its initial temperature, raising its internal energy level before injection. This increase in internal energy can be calculated as a correction to the enthalpy of the propellant (see Chapter 5). However, regenerative cooling's overall effect on rocket engine performance is usually very slight. With some propellants the specific impulse can be 1% larger when the propellants are preheated through a temperature differential of 100 to 200°C. In hydrogen-cooled thrust chambers and in small combustion chambers, where the wall-surface-to-chamber volume ratio is relatively large, the temperature rise in the regenerative coolant will be high, and the resulting increase in specific impulse may exceed 1%.

The *liquid film* behavior is critical for controlling the wall temperatures in forced convection cooling of rocket devices at high heat fluxes (see Table 8–4 and Refs. 8–12 and 8–16). At least four different film types appear to exist, as can be seen in Fig. 8–21. Here, the dependence of q, the heat transfer rate per unit wall

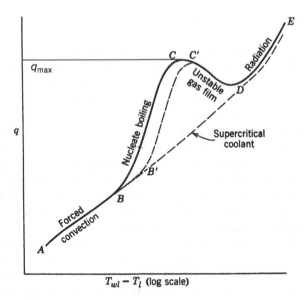

FIGURE 8–21. Regimes in transferring heat from a hot wall to a flowing boiling liquid.

surface, is shown as a function of the difference between the wall temperature on the liquid side T_{wl} and the bulk temperature of the liquid T_l.

1. At low heat fluxes, normal *forced convection* regions appear to form liquid boundary layers of predictable characteristics. This is indicated by region A–B in Fig. 8–21. Here, the wall temperature is usually below the boiling point of the liquid at the prevailing cooling jacket pressure. In steady-state heat transfer analysis, the liquid-film coefficient can be approximated by the following equation (see Refs. 8–11 to 8–14):

$$h_l = 0.023 \, \bar{c} \, \frac{\dot{m}}{A} \left(\frac{D \upsilon \rho}{\mu} \right)^{-0.2} \left(\frac{\mu \bar{c}}{\kappa} \right)^{-2/3} \tag{8–24}$$

where $\dot{m}$ is the liquid fluid mass flow rate, $\bar{c}$ its average specific heat, A the cross-sectional cooling jacket flow area, D the equivalent diameter of the coolant passage cross section,[*] υ the fluid velocity, ρ the coolant density, μ its absolute viscosity, and k its conductivity. Many liquid-cooled rocket devices operate in this heat transfer regime. Values of the physical properties of several propellants are given in Tables 8–5 and 7–1.

---

[*]The grooves, tubes, or coolant passages in liquid propellant rocket chambers are often of complex cross section. The equivalent diameter, needed for fluid-film heat transfer calculations, is usually defined as four times the hydraulic radius of the coolant passage; the hydraulic radius is the cross-sectional flow area divided by the wetted perimeter.

2. When the wall temperature T_{wl} exceeds the boiling point of the liquid by perhaps 10 to 50 K, small vapor bubbles form at the hot wall surface. These small, nuclei-like bubbles cause local turbulence, break away from the wall, and collapse in the cooler liquid. This phenomenon is known as *nucleate boiling*. The turbulence induced by the bubbles changes the character of the liquid film and, augmented by the vaporization of some of the propellant, the heat transfer rate is increased without much proportional increase in the temperature drop across the film, as can be seen by the steep slope $B-C$ of the curve in Fig. 8–21. If the pressure of the fluid is raised, then the boiling point is also raised and the nucleate boiling region shifts to the right, to $B'-C'$. Such boiling permits a substantial increase in the heat transfer beyond that predicted by Eq. 8–24. This phenomenon often occurs locally at the nozzle throat area cooling regions. The maximum feasible heat transfer rate (point C) is indicated as q_{max} in Fig. 8–21 and Table 8–5 and appears to be a strong function of cooling-fluid properties, the presence of dissolved gases, pressure, and flow velocity. In Table 8–5 it can be seen that hydrazine is a good heat absorber, but kerosene is poor (low q_{max} and low critical pressure).

3. As the temperature difference is increased further, the rate of bubble formation and the bubble size become so great that the bubbles are unable to escape from the wall rapidly enough. This region (shown as $C–D$ in Fig. 8–21) is characterized by an unstable *gas film* where it is difficult to obtain reproducible results. When a gas film forms along the hot wall surface, it acts as an insulation layer, causing a decrease in heat flux and, usually, a rapid increase in wall temperature (often resulting in a burnout or melting of the wall material). Cooling flow systems must be designed to avoid this unstable gas film regime.

4. As the temperature difference across the film continues to increase, the wall temperatures reach values in which heat transfer by *radiation* becomes important. Region $D–E$, however, is not of interest to cooling jacket designers.

Cooling has also been accomplished with fluids above their critical point with coolants such as hydrogen because in this case there is no nucleate boiling and the heat transfer increases with the temperature difference, as shown by the supercritical (dashed) line in Fig. 8–21. Liquid hydrogen is an excellent coolant, has a high specific heat, and leaves no residues. (See Ref. 8–16.)

Chemical changes in liquid coolants can seriously influence the heat transfer. Cracking of hydrocarbon fuels, with an attendant formation of insoluble gases, tends to reduce the maximum heat flux and thus promote failure. Hydrocarbon fuel coolants (methane, jet fuel) can also break down to form solid, sticky carbon deposits inside cooling channels, impeding the heat transfer. Other factors that influence steady-state coolant heat transfer rates are gas radiation to container walls, bends in the coolant passage, improper welds or manufacture, and flow oscillations caused by turbulence or by combustion unsteadiness. Some propellants, such as hydrazine, can decompose spontaneously and explode inside the cooling passage when they become hot enough.

To achieve a good *heat-absorbing capacity in the coolant*, the pressure and the coolant flow velocity are selected so that boiling takes place locally at the hot surface,

TABLE 8–5. Heat Transfer Characteristics of Several Liquid Propellants

Liquid Coolant	Boiling Characteristics		Critical Temp. (K)	Critical Pressure (MPa)	Nucleate Boiling Characteristics			
	Pressure (MPa)	Boiling Temp. (K)			Temp. (K)	Pressure (MPa)	Velocity (m/sec)	q_{max} (MW/m²)
Hydrazine	0.101	387	652	14.7	322.2	4.13	10	22.1
	0.689	455					20	29.4
	3.45	540			405.6	4.13	10	14.2
	6.89	588					20	21.2
Kerosene	0.101	490	678	2.0	297.2	0.689	1	2.4
	0.689	603					8.5	6.4
	1.38	651			297.2	1.38	1	2.3
	1.38	651					8.5	6.2
Nitrogen tetroxide	0.101	294	431	10.1	288.9	4.13	20	11.4
	0.689	342			322.2			9.3
	4.13	394			366.7			6.2
Unsymmetrical dimethyl hydrazine	0.101	336	522	6.06	300	2.07	10	4.9
	1.01	400					20	7.2
	3.45	489			300	5.52	10	4.7

but the bulk of the coolant does not reach this boiling condition. The total heat rejected by the hot gases to the surface of the hot walls, as given by Eq. 8–14, becomes

$$qA = Q = \dot{m}\bar{c}(T_1 - T_2) \tag{8–25}$$

where $\dot{m}$ is the coolant mass flow rate, $\bar{c}$ the average specific heat of the liquid, T_1 the initial temperature of the coolant as it enters the cooling jacket, and T_2 its final temperature; Q is the rate of heat absorption per unit time and A the heat transfer area. The total heat absorbed must be less than the permitted temperature rise in the coolant, namely, value of T_2 should remain below the boiling point prevailing at the cooling jacket pressure.

Radiation

Radiation cooling is the electromagnetic energy exchanged by a substance (gas, liquid, or solid) with its environment by the virtue of its higher temperature and at the expense of its internal energy. It is mostly in the infrared but may cover the wavelength range from 10,000 to 0.0001 µm, which includes the visible range of 0.39 to 0.78 µm at the higher temperatures. Radiation heat transfer occurs most efficiently in a vacuum (because there is no absorption by intervening media) and when the surroundings are relatively much cooler.

The amount of heat radiated depends primarily on the temperature of the radiating bodies and on their surface properties. The energy E radiated depends on the fourth power of the absolute temperature T as follows:

$$E = f\varepsilon\sigma A T^4 \tag{8–26}$$

The energy E radiated by a body is defined as a function of the emissivity ε, which is a dimensionless surface-condition factor less or equal to one, the Stefan–Boltzmann constant σ (5.67×10^{-8} W/m^2 – K^4), the radiating surface area A, the absolute surface temperature T, and a geometric factor f which depends on the geometrical arrangement of adjacent surfaces and their shapes. At low wall temperatures (below 800 K) radiation can usually be neglected.

In rocket propulsion here are primary heat transfer concerns:

1. Emission from hot gases to the internal walls of a combustion chamber and its nozzle converging section, or to a solid propellant grain, or to a hybrid propellant grain.
2. Emission to the surroundings or to space from external hot hardware surfaces (radiation-cooled chambers, nozzles, or electrodes in electric thrusters).
3. Radiation from hot plume gases (downstream of the nozzle exit). This effect is described in Chapter 20.

In rocket combustion devices at gas temperatures between 1900 and 3900 K or about 3000 to 6600 °F radiation may represent a significant portion of the total heat transfer, between 3 to 40% of the heat transfer to the chamber walls depending on the reaction gas composition, chamber size, geometry, and chamber temperatures. In solid propellant motors radiation heating of grain surfaces may critically affect the burning rate, as discussed in Chapter 12. Radiation from surrounding walls follows essentially the same law as Eq. 8–26. Metal surfaces and formed tubes reflect much radiant energy (low emissivity), whereas ablative materials and solid propellants seem to absorb most of the incident radiation (absorptivity $\approx$ emissivity). Thus highly reflective surfaces on the inside wall of a combustor tend to reduce radiant absorption and to minimize wall temperature increases. The relative amounts of convective and radiative heat transfer are discussed in Ref. 8–17.

Only few hot reaction gases in rocket combustion chambers are potent radiation sources. Gases with symmetrical molecules, such as hydrogen, oxygen, and nitrogen, have few strong emission bands in wavelength regions of importance in radiant heat transfer. Also, they do not really absorb much radiation and do not contribute significant energies to radiant heat transfer. However, heteropolar gases, such as water vapor, carbon monoxide, carbon dioxide, hydrogen chloride, hydrocarbons, ammonia, oxides of nitrogen, and the alcohols do have strong infrared emission bands (radiation from these molecules is associated with broadened quantum changes in their energy levels of rotation and vibration). In general, for all gases, radiation intensity increases with volume, partial pressure, and fourth power of their absolute temperature. For small thrust chambers at low chamber pressures, gas radiation may usually be neglected.

When the hot flowing gases contain small solid particles or liquid droplets, the heat radiated may increase dramatically, by a factor of 2 to 10. Particulates greatly increase the amounts of radiated energy as explained in Section 20.1. As examples we cite reaction gases from certain organic liquid fuels that contain small solid carbon particles and those of many solid propellants containing fine aluminum powder. When burned to form aluminum oxide, the propellant's heat of combustion and combustion temperature are increased and the specific impulse is raised somewhat. These oxides may exist in liquid droplet form (in the chamber) or solid particles (in the nozzle diverging section), depending on the local gas temperature. Furthermore, with wall impacts these particulates cause additional increases in heat transfer, mainly at the nozzle throat walls and immediately upstream of the nozzle throat region. Wall impacts may also cause unwanted erosion and/or abrasion at the walls.

8.6. STARTING AND IGNITION

Thrust chamber starting has to be controlled to achieve timely and even propellant ignition so that the flow and the thrust built up smoothly and quickly to their rated value (see Refs. 6–1, 8–18 and 8–19). Note that any *initial propellant flow* is always less than *full flow* and that any *starting mixture ratio* is usually different from the *operating mixture ratio*. A low initial flow prevents strong water hammer, gives a lower initial heat release, and for nonhypergolic propellants prevents unwanted or excessive accumulations of unignited liquid propellants in the chamber.

Because during starting injection velocities are low, the initial vaporization, atomization, and mixing of propellants in a cold combustion chamber is often incomplete and there can be localized regions of lean and rich mixtures. With cryogenic propellants the initial chamber temperature can be below ambient. The optimum starting mixture is therefore only one of a range of mixture ratios, all of which must readily ignite. Mixture ratios near stoichiometric have the highest heat release per unit propellant mass and therefore bring the chamber and its gases up to equilibrium faster than with other mixtures. Operating mixture ratios, however, are nearly always fuel rich usually selected for optimum specific impulse. One method for modeling cryogenic propellant ignition of is given in Ref. 8–18.

Ideally, *time delays for starting* any thrust chamber should consist of the following time increments:

1. Time needed to fully open propellant valves (typically between 0.002 to over 1.00 sec), depending on valve type and size, and upstream pressure).

2. Time needed to fill liquid passage volumes (piping, manifolds, internal injector feed holes, and cavities) between the valve seat and the injector face.

3. Time for forming discrete streams or sprays of liquid propellant, for initial atomization into small droplets, and for mixing these droplets. With cryogenic propellants initial flows are usually gaseous so this time would differ.

4. With current hypergolic propellant combinations at ambient initial temperatures, combustion starts within a few milliseconds after the fuel (droplet or vapor) comes into contact with the oxidizer (droplet or vapor). This delay lengthens at lower ambient temperatures and at off-design mixtures ratios.

5. For nonhypergolic propellant combinations, an igniter system must bring the mixed propellant flow to its ignition temperature before combustion can start. This igniter is usually started before any propellant is admitted to the chamber, before propellant valves are opened. The igniter operating time before propellant comes in contact with it can be one or more seconds in larger thrust chambers (if properly sensed, when an igniter fails to operate engine controls can be made to block the opening of propellant valves).

6. Time needed for droplets to vaporize and ignite (laboratory tests show this can be as short as 0.02 to 0.05 sec, depending on the propellants and the existing heat transfer).

7. Once ignition is achieved at one particular location in the chamber, there is a finite flame spread time plus a time to heat all the mixed propellant that is entering into the chamber, vaporizing it, and raising it to ignition temperature.

8. Time needed to raise the chamber pressure and temperature to the condition where combustion is self-sustaining. Only then has the system been raised to its full operating state.

There are overlaps in these delays because several of them may occur simultaneously. Large injectors and/or large-diameter chambers cause longer delays. Small thrusters can usually be started relatively quickly, in a few milliseconds, while larger units require one second and sometimes as much as 5 sec to reach full thrust.

In the starting of thrust chambers, it is difficult to exactly synchronize the fuel and oxidizer feed systems so that propellants reach the chamber simultaneously from all injection holes or spray elements. Frequently, more reliable ignitions are assured when one of the propellants is intentionally made to reach the chamber first. For example, for fuel-rich starting mixtures the fuel is admitted first. Reference 8–19 describes the control of such propellant lead.

Other factors influencing flow starting, propellant lead or lag, and some of the delays mentioned above relate to the pressures supplied to the liquid injectors (e.g., regulated pressure), the temperature of the propellant (some can be stored close to their vapor point), and to the amount of insoluble gas (air bubbles) mixed with the initial quantity of propellants.

Propellant valves (and flow passages between these valves and the injector face) are often so designed and controlled that they operate in a definite sequence, thereby assuring an intentional lead of one of the propellants and a controlled buildup of flow and mixture ratio. Often these valves are only partially opened because it is easier to ignite a small flow thus avoiding accumulation of hazardous unburned propellant mixtures in the chamber. Once combustion is established, valves are fully opened and full flows reach the thrust chamber assembly. This initial reduced flow burning period is called the *preliminary stage*. Section 11.4 describes the starting controls.

Full flows in larger thrust chambers are not initiated with non-self-igniting propellants until the controller received a signal of successful ignition. The verification of ignition or initial burning is often built into engine controls using image detectors (photocells), heat detectors (pyrometers), fusible wire links, or by sensing pressure rises. If starting controls are not designed properly, unburnt propellant may accumulate in the chamber, which may then explode upon ignition, possibly causing severe damage to the rocket engine. Starting controls and engine calibrations are discussed in Sections 11.4 and 11.5.

Nonspontaneously ignitable propellants are activated by the absorption of energy prior to combustion initiation. This energy is supplied by an *ignition system* as described below. Once ignition has begun, the flame becomes self-supporting. The igniter has to be located near the injector in such a manner that a satisfactory starting mixture at low initial flows is present at the time of igniter activation, yet it should not hinder or obstruct the steady-state combustion process. At least five different types of successful propellant ignition systems have been used:

Spark plug ignition has been used successfully on liquid oxygen–gasoline and on oxygen–hydrogen thrust chambers, particularly for multiple starts during flight. The spark plug is often built into the injector. The RS-25 (Aerojet Rocketdyne's designator for the SSME) uses a redundant augmented spark igniter in the main combustion chamber and preburners.

Ignition by electrically heated wires has been accomplished but at times has proven to be less reliable than spark ignition with liquid propellants.

Pyrotechnic ignition uses solid propellant squibs or grains of a few seconds' burning duration. This solid propellant "charge" is electrically ignited and burns with a hot flame within the combustion chamber. Almost all solid propellant rockets and many liquid rocket chambers are ignited in this fashion. The igniter container may be designed to fit directly onto the injector or the chamber (as in Fig. 8–1), or may be held in the chamber from outside through the nozzle. The latter ignition method can only be used once; thereafter, its "charge" has to be replaced.

Ignition has been achieved in a *precombustion chamber* also called *premix chamber;* see Fig. 8–3; it is a small chamber built next to the main combustion chamber and connected through an orifice; this is similar to the precombustion chamber used in some automobile engines. A small amount of fuel and oxidizer is injected into the precombustion chamber and ignited. The burning mixture enters the main combustion chamber in a torchlike fashion and ignites the larger main-propellant flow as it is being been injected into the main chamber. This ignition procedure permits the repeated starting of thrust chambers and has proven successful with the liquid oxygen–gasoline and oxygen–hydrogen thrust chambers.

Auxiliary fluid ignition is a method whereby some hypergolic liquid or gas, in addition to the regular fuel and oxidizer, is injected into the combustion chamber for a short period during the starting operation. This fluid produces spontaneous combustion with either the fuel or the oxidizer. The combustion of nitric acid and some organic fuels can, for instance, be initiated by the introduction of a small quantity of hydrazine or aniline at the beginning of rocket operation. Liquids that ignite with air (e.g., zinc diethyl or triethyl aluminum), when preloaded in the fuel piping, can also accomplish hypergolic ignition. A mixture of triethyl aluminum and triethyl borane has been successfully used for liquid oxygen and RP-1 (kerosene) ignition in several U.S. and Russian rocket engines. Flow diagrams for two Russian rocket engines, the RD 191 in Fig. 6–13 and the RD 170 in Fig. 11–2, show cylindrical containers (start fuel tanks or ampoules) with a hypergolic liquid for each of the high-pressure fuel supply lines; this liquid is pushed out (by the initial fuel) into the thrust chambers and into the preburners to start ignition.

Vehicles with multiple engines or thrust chambers are required to start two or more together, although it is often difficult to get precisely simultaneous starts. Usually, the passage or manifold volumes of each thrust chamber and their respective values are designed to be the same. Careful control of temperature of the initial propellant fed to each thrust chamber and of lead time of the first quantity of propellants entering into the chambers is needed. This is required, for example, when two thrusters are used to apply roll torques to a vehicle. It is also one reason why large space launch vehicles are not released from their launch facility until there is assurance that all thrust chambers are started and operating.

During storage or between ground tests, *nozzle throat plugs* or alternatively round nozzle exit covers are routinely used. These prevent dust, moisture, and small creatures (e.g., ants, mice) from entering the chamber and/or injectors. These units are usually removed manually just before thrust chamber operation or may be ejected automatically by the rising chamber pressure during starts.

8.7. USEFUL LIFE OF THRUST CHAMBERS

The useable life of a thrust chamber is given by the number of starts or full-thrust operations that can be performed safely, without undue risk of failure. There are two relevant categories: First, for flight missions where the thrust chamber can be recovered between flights, an opportunity exists to inspect and if needed maintain and repair before the next flight. A typical of large recoverable thrust chamber was the Space Shuttle Main Engine (SSME, Aerojet Rocketdyne) and, presently, the Merlin engine for the Falcon 9 vehicle (SpaceX). NASA wanted the SSMEs designed for 100 flights but they only flew between 5 and 20 times because of factors not related to thrust chamber lifetimes.

There are several life-limiting sources in recoverable units. At their inner walls, thrust chambers deteriorate by gradual oxidation from the combustion gases (O, O_2, H_2O, etc.), though these species may only be present in relatively small amounts, and from crack formations usually near the throat. Moreover, as inner walls of cooling jackets are heated, they expand and grow slightly all directions. However, because of their construction, these inner walls are restricted from freely expanding by the cooler outer walls (see Figs. 8–9 and 8–15). Having a lower yield stress at higher temperatures, an inner wall will yield in compression causing slight shrinking on the hotter side. After shutdown, an inner wall will also (slightly) shrink as it cools but will not get back to exactly the original size, thus having to yield under tension. These effects cause the formation of small surface cracks on inside inner walls. With each new start and shutdown, such cracks deepen and become more numerous. Ultimately, the damaged inner walls can no longer tolerate the chamber pressure loads and the thrust chamber will fail. Most thrust chambers have relatively thin inner walls of high conductivity materials to decrease thermal stresses and thus achieve longer lives.

The second category relates to the small-pulse multiple thrusters typically found in the reaction control system of nonrecoverable but long-duration space missions, such as long-term orbit maintenance or deep space flights. Here, the cumulative number of starts or pulses could be between 1000 and 50,000 for multiyear mission durations. During development, one or more thrusters are usually endurance tested for several hundred thousand cycles in a vacuum chamber. The useful life of small thrusters is limited by the fatigue failure of valves and other structures in the reaction control system, and also by oxidation and/or erosion of walls exposed to propellant gases. Many types of small thrusters have been life-tested, including units with radiant cooling and with ablative inner liners in both their monopropellant and bipropellant versions (Refs. 8–8 and 8–20). With monopropellant small thrusters, thruster life is often limited by the attrition and/or poisoning of the catalyst (see Section 8.3).

8.8. RANDOM VARIABLE THRUST

Only some applications require randomly variable engine thrust, for example, rocket descents to another planet or moon or of the upper stage of anti-ballistic missiles. One advantage of liquid propellant rocket engines is that they can be designed to be throttled (i.e., to randomly vary their thrust) over wide ranges during flight. Several approaches exist to achieve throttling (all involve the thrust chamber and therefore this section has been placed here). One early scheme was to use multiple engines or multiple thrust chambers on the same engine and then stopping operation of one or more of them. The now historic Reaction Motors 6,000C-4 engine had four thrust chambers and they could be turned on or off individually producing step changes in thrust.

Today we distinguish between two modes of operation, though both depend on reducing propellant flow. The first is *moderate throttling*, typically over a thrust range of factors of two or three, which can usually be accomplished with modified propellant valves without significant design changes to the engine. In one application, the thrust is throttled during the ascent of a booster vehicle in order to prevent excessive aerodynamic pressure or heating of the vehicle. The other mode, often called *deep throttling*, varies the thrust by factors between 6 and 30 for specific engines. It applies, for example, to a planetary landing rocket engine with controlled deceleration. To achieve this deep throttling, engines require some special features, some of which are mentioned below.

Thrust is mostly proportional to propellant mass flow (Eq. 2–13) and therefore reducing propellant flow will reduce thrust. Moderate throttling can be achieved by partially closing the main fuel and oxidizer valves (matching flow characteristics) simultaneously or by slowing down the rotary speed of the turbopumps (by reducing the gas flow to their turbine) with hydraulically matched fuel and oxidizer pumps. At the lower flows, the pressure drop across the injector and the injection velocity will diminish, the atomization and mixing of the propellants will be somewhat less efficient and the combustion will be less complete, and chamber pressures will be reduced. These decrease the specific impulse at the lower thrust levels of perhaps 1.5 to 9%, depending on specific engine design. Such method for varying thrust is used in a number of booster engines for space launch vehicles.

With the deeper throttling, oxidizer and fuel flows tend to oscillate and will usually no longer remain at the original mixture ratio, the propellant injection streams or sprays and their impingement locations will wander, and their flow is likely to become erratic. To prevent these, engines need to have some special features like those found in the engine used in the first lunar landing rocket—this engine, developed by the predecessor of Northrop Grumman, flew in the late 1960s and 1970s and was throttled by a factor of up to 10:1; each propellant flow control valve included a cavitating venturi with a movable tapered pintle that allowed flow area variations in the venturi throat of the thrust chamber valve; this assured predetermined steady reduced flows of propellants at all thrust levels, and maintained a constant mixture ratio; furthermore, propellant injection was accomplished through a pintle injector with two annular slots and the width of these slots could be reduced by an actuator built into the injector. A lower right sketch in Fig. 8–3 shows this feature. It allows for high injection velocities of liquid propellants, giving good atomization and adequate combustion with relatively small losses in performance at low thrusts. The highest thrust variation (327 to 1.0) has been obtained with the sustainer engines of the Lance surface-to-surface missile. It had two concentric thrust chambers—a small one inside a larger one. Its engine, developed by Rocketdyne (today Aerojet Rocketdyne), first flew in the late 1960s and was deployed in the 1970s until 1990. The specific impulse at low thrust was poor (more than 15% loss). See Ref. 8–1.

Another approach to achieve variable thrust is to vary the nozzle throat area (see Eq. 3–31). This scheme requires a movable tapered pintle in the main nozzle throat area. The pintle has to be made of heat-resistant materials or has to be regeneratively cooled, and its position is usually hydraulically controlled. This will allow the unit to maintain essentially constant chamber pressure at all thrust levels. Several experimental liquid propellant engines and solid propellant motors with random variable nozzle throat areas have been built and tested. To the best of the authors' knowledge, none have flown.

For pulsed, small reaction control thrusters the average thrust can be noticeably reduced by changing the pulsing mode. This is accomplished by controlling the number of cycles or pulses (each having one short fixed-duration thrust pulse plus a short fixed-duration zero-thrust pause), by modulating the duration of individual pulses (with short pauses between pulses), or alternatively by lengthening the pause between pulses.

8.9. SAMPLE THRUST CHAMBER DESIGN ANALYSIS

In this example we show how the preliminary design of a thrust chamber is strongly influenced by overall vehicle system requirements and mission parameters; it illustrates vehicle design procedures and factors that are considered in the selection of some key thrust chamber parameters., In order to arrive at key design parameters, each engine goes through a series of rationalizations and requirements. This is outlined in the design section of Chapter 11 and in the discussion of the selection of propulsion systems in Chapter 19. In this example, we describe one of several ways of deriving thrust chamber parameters from both vehicle and engine requirements. Overall system requirements relate to the mission, its purpose, environment, trajectories, reusability, reliability, and to restraints such as allowable engine mass, or maximum dimensional envelope. Even though we list only some relevant requirements, this example shows how theory needs to be blended with experience to arrive at initial design choices; these will differ when done by different design teams.

Example 8–2.   We examine here the application of a new upper stage to an existing multistage space launch vehicle for propelling a payload into deep space. This means continuous firing (no restart or reuse), operating in the vacuum of space (high nozzle area ratios), modest acceleration (not to exceed $5g_0$), reasonably low costs, moderately high performance (specific impulse), and a thrust magnitude which depends not only the payload but also on the flight path and acceleration limits. The desired mission velocity increase of the new stage is 3400 m/sec. The engine is attached to its own stage, which is subsequently disconnected and dropped from the payload stage. The payload stage (3500 kg) consists of a payload of 1500 kg (for scientific instruments, power supply, or communications and flight control equipment) with its own propulsion systems (including propellant) of 2000 kg (for trajectory changes, station keeping, attitude control, and/or emergency maneuvers). There are two geometric restraints: The vehicle has an outside diameter of 2.0 m, but when the structure, conduits, certain equipment, thermal insulation, fittings, and assembly are considered, it really is only about 1.90 m. The restraint on the stage length of 4.50 m maximum will affect the length of the thrust chamber. Below, we summarize the key requirements:

Application	Uppermost stage to an existing multistage launch vehicle
Payload	3500 kg
Desired velocity increase Δu	3400 m/sec in gravity-free vacuum
Maximum stage diameter	1.90 m
Maximum stage length	4.50 m
Maximum acceleration	$5g_0$

DECISIONS ON BASIC PARAMETERS. The following engine design decisions or parameter selections should be made early in the design process:

> Propellant combination
> Chamber pressure
> Nozzle area ratio
> Feed system, using pumps or pressurized tanks
> Thrust level

From a performance point of view, the best *propellant combination* would be liquid oxygen with liquid hydrogen. Because this bipropellant would have a low average specific gravity (0.36), there is not enough available volume in this upper stage to store sufficient propellant mass for attaining the desired vehicle velocity increment Δu (the new stage is limited in volume and cross section). The lower stages of the existing launch vehicle are known use liquid oxygen with RP-1 fuel with an average specific gravity of about 1.014, and the launch pad is already equipped for supplying these. Because of these factors the propellant combination of liquid oxygen and RP-1 (a type of kerosene) is selected. From Fig. 5–1 we see that the theoretical specific impulse can be between 280 and 300 sec, depending on the mixture ratio and whether frozen or shifting chemical equilibrium applies to the flow expansion. Figure 5–1 also shows that the maximum value of the characteristic velocity c^* is reached at a mixture ratio of about 2.30, which is a fuel-rich mixture; we select this mixture ratio. The resulting combustion temperature is lower than with higher mixture ratios and this should make the cooling of the thrust chamber easier. We will see later that cooling may be problematic. Based on common experience, we select a value of I_s part way (about 40%) between the values for frozen and shifting equilibrium, namely 292 sec at the standard chamber pressure of 1000 psi or 6.895 MPa, and a nozzle only big enough for expansion to sea level. From Fig. 5–1 and Table 5–5 we find the molecular mass to be 23 kg/kg-mol and the specific heat ratio k to be about 1.24. Later, we will correct this value of I_s from this standard reference condition to the actual vacuum specific impulse of the thrust chamber with its actual nozzle exit area ratio.

Next, we select a chamber pressure, a nozzle area ratio, and a feed-system concept. Historically, there has been favorable experience with this propellant combination at chamber pressures between 100 and 3400 psia with nozzle area ratios up to about 60 with both gas generator cycles and staged combustion cycles, indicating that these choices are feasible. The following considerations enter into this selection:

1. Higher chamber pressures allow smaller thrust chambers and (for the same nozzle exit pressure) shorter nozzle divergent sections with smaller nozzle exit diameters. The thrust chamber can be small enough for a toroidal tank to be built around it, and this will conserve stage length. This arrangement not only saves vehicle space but also some inert mass in the vehicle and the engine. Figure 8–22 shows three thrust chamber relative sizes with different pressures and with two nozzle area ratios (having ε values of 100 and 300). The nozzle length and exit diameter cannot exceed the values given in the requirements, which, as can be seen, rules out the 100 psia chamber pressure. Calculations for the dimensions pictured in Fig. 8–22 are shown later in this analysis.

2. The heat transfer rate, which varies with the gas-film coefficient, is nearly proportional to the gas density, which itself is nearly proportional to the chamber pressure, as implied from Eq. 8–20 or 8–22. In some thrust chambers using this propellant mixture there have been problems with formations of solid carbon layers or deposits either inside the cooling jacket (increasing wall temperatures) or on the inner walls of the combustion chamber (solids can flake off and cause burnout). These effects favor the lower chamber pressures.

3. Concern over leak-free seals for both static and dynamic seals increases with chamber pressure, which in turn causes all feed pressures also to increase.

4. A feed system using pressurized gas is feasible, but its inert tank and engine masses are favorable only when chamber pressures are very low, perhaps around 100 psia or less.

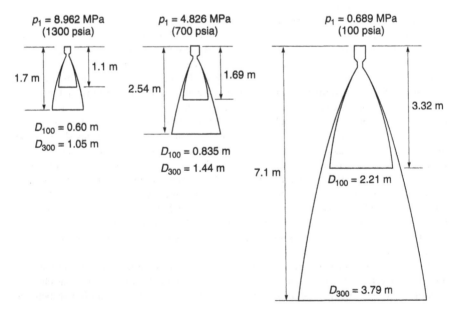

FIGURE 8–22. Comparison of thrust chamber sizes for three arbitrary chamber pressures and two nozzle area ratios (100 and 300).

Tanks for propellants and pressurizing gases will become heavy and the thrust chamber will be very large and exceed the dimensional restraints given. For this application, we therefore cannot use this type of feed system or its very low chamber pressures.

5. If we use a pump feed system, the power needed to drive the pumps increases directly with chamber pressure p_1. In a gas generator engine cycle, this means a slightly reduced performance as the value of p_1 goes up. For a staged combustion cycle it means high pressures, particularly high-pressure hot gas flexible piping, and a more complex, heavier, and expensive engine. We therefore select a gas generator cycle (see Fig. 1–4) at a low enough chamber pressure so that the thrust chamber (and the other inert hardware) will just fit the geometrical constraints, and the engine inert mass and the heat transfer will be reasonable.

For the above reasons we pick a *chamber pressure* of 700 psia or 4.825 MPa and an *area ratio* of 100. With further scrutiny we could have chosen p_1 more precisely at a slightly lower value. Next, we correct the *specific impulse* to the operating conditions using a ratio of thrust coefficients. We can use Eq. 3–30 or interpolate between Figs. 3–6 and 3–7 for a value of $k = 1.24$. The reference or standard condition (see Fig. 3–5) is for a pressure ratio p_1/p_3 of $1000/14.7 = 68$, which corresponds to an area ratio of about 8, then $(C_F)_{\text{standard}} = 1.58$. For the actual high-altitude operation the pressure ratio is close to infinite. The nozzle has an area ratio of 100; we determine the thrust coefficient by interpolating with pressure at $k = 1.24$ and the result is $(C_F)_{\text{vacuum}} \approx 1.90$. The new ideal specific impulse value for a chamber threshold of 700 psia and a nozzle area ratio of 100 is therefore $292 \times (1.90/1.58) = 351.1$ sec. In order to correct for losses (divergence, boundary layers, incomplete combustion, film cooling, etc.), we use an exhaust velocity correction factor of 0.96 giving a thrust chamber specific impulse of 337.1 sec. The engine uses a gas generator, and this will reduce the engine specific impulse further by a factor of approximately 0.98 or $(I_s)_{\text{engine}} = 330.3$ sec resulting in an effective exhaust velocity of approximately 3237 m/sec.

STAGE MASSES AND THRUST LEVEL. An estimate of the stage masses will be made next. We assume that the inert hardware (tanks, gas, generator, turbopumps, etc.) amounts to about 7% of the propellant mass m_p, which is conservative when compared to existing engines. In a full-fledged engine design this number would be verified or corrected once an estimated but more detailed mass budget becomes available. From Eq. 4–7

$$e^{\Delta u/v} = \frac{m_o}{m_f} \quad \text{or} \quad e^{3400/3237} = \frac{m_p + 0.07m_p + 3500}{0.07m_p + 3500}$$

Solve for $m_p = 7483$ kg. The final and initial masses m_f and m_0 of the stage are then 4023 kg and 11,510 kg, respectively. Residual propellants are to be neglected here.

The maximum *thrust* is limited by the maximum allowed acceleration of $5g_0$. It has the value $F_{\text{max}} = m_0 a = 11,510 \times 5 \times 9.81 = 564,400\,\text{N}$. This would become a relatively large and heavy thrust chamber; considerable saving in inert mass can be obtained if a smaller thrust magnitude (but longer firing duration) is chosen. Since it is known that this same thrust chamber would be used for another mission where an acceleration of somewhat less than 1.0 g_0 is wanted, a thrust level of 50,000 N or 11,240 lbf is chosen. The maximum acceleration of the stage occurs just before cutoff, $a = F/m_f = 50,000/4023 = 12.4\,\text{m/sec}^2$ or about $1.26g_0$. This value fits the thrust and acceleration requirements.

The following have now been determined:

Propellant	Liquid oxygen and liquid kerosene (RP-1)
Mixture ratio (O/F)	2.30 (engine)
Thrust	50,000 N or 11,240 lbf
Chamber pressure	700 psia or 4.826 MPa
Nozzle area ratio	100
Specific impulse (engine)	330.3 sec
Specific impulse (thrust chamber)	337.1 sec
Engine cycle	Gas generator
Usable propellant mass	7483 kg
Estimated nozzle exit exhaust velocity	3237 m/sec or 10,613 ft/sec

PROPELLANT FLOWS AND DIMENSIONS OF THRUST CHAMBER. From Eq. 2–6 we obtain the propellant mass flow rate:

$$\dot{m} = F/c = 50{,}000/3237 = 15.446 \text{ kg/sec}$$

When this total flow and the overall mixture ratio are known, then the fuel flow $\dot{m}_f$ and oxidizer flow $\dot{m}_o$ for the engine, its gas generator, and its thrust chamber can be determined from Eqs. 6–3 and 6–4 as shown below:

$$\dot{m}_f = \dot{m}/(r+1) = 15.446/(2.3+1) = 4.680 \text{ kg/sec}$$
$$\dot{m}_o = \dot{m}r/(r+1) = (15.446 \times 2.30)/3.30 = 10.765 \text{ kg/sec}$$

The gas generator flow $\dot{m}_{gg}$ consumes about 2.0% of the total flow and operates at a fuel-rich mixture ratio of 0.055; this results in a gas temperature of about 890 K. Moreover,

$$(\dot{m}_f)_{gg} = 0.2928 \text{ kg/sec} \qquad (\dot{m}_o)_{gg} = 0.0161 \text{ kg/sec}$$

The flows through the thrust chamber are equal to the total flow diminished by the gas generator flow, which is roughly 98.0% of the total flow or 15.137 kg/sec:

$$(\dot{m}_f)_{tc} = 4.387 \text{ kg/sec} \qquad (\dot{m}_o)_{tc} = 10.749 \text{ kg/sec}$$

The *duration* is the total effective propellant mass divided by the mass flow rate:

$$t_b = m_p/\dot{m}_p = 7483/15.446 = 484.5 \text{ sec. (or a little longer than 8 min)}$$

The *nozzle throat area* is determined from Eq. 3–31. Note this calculation does not represent any transient conditions:

$$A_t = F/(p_1 C_F) = 50{,}000/(4.826 \times 10^6 \times 1.90) = 0.005453 \text{ m}^2 \text{ or } 54.53 \text{ cm}^2$$

The nozzle throat diameter is $D_t = 8.326$ cm. The internal diameter of the nozzle at exit A_2 is determined from the area ratio of 100 to be $D_2 = \sqrt{100} \times D_t$ or 83.26 cm. A shortened or *truncated bell nozzle* (as discussed in Section 3.4) will be used equivalent to 80% of the length of a 15° conical nozzle, but with the same performance as the 15° cone. The nozzle length (from the throat to the exit) can be determined by an accurate layout or by an equivalent 15° conical nozzle exit $L = (D_2 - D_t)/(2 \tan 15)$ as 139.8 cm. For an 80% shortened bell nozzle this length would be about 111.8 cm. The contour or shape of a shortened bell nozzle can be approximated by a parabola (its equation is $y^2 = 2ax$). Using an analysis (similar to that in Fig. 3–13), the maximum angle of the diverging section at the inflection point would be about $\theta_i = 34°$ and the nozzle exit angle $\theta_e = 7°$. The approximate contour consists of a short segment of radius $0.4r_t$ of a 34° included angle (between points T and I in Fig. 3–13) and a parabola with two known points at I and E. Knowing the tangent angles (34 and 7°) and the y coordinates $[y_e = r_2$ and $y_i = r_t + 0.382 r_t(1 - \cos\theta_i)]$ allows the determination of the parabola by geometric analysis. Before any detailed design is undertaken, a more accurate contour, using the method of characteristics, is suggested.

The *chamber diameter* should be about twice the nozzle throat diameter to avoid pressure losses in the combustion chamber ($D_c = 16.64$ cm). Using the approximate length of prior successful smaller chambers and a characteristic length L^* of about 1.1 m, the chamber length (together with the converging nozzle section) becomes about 11.8 in. or 29.9 cm. The overall length of the thrust chamber (169 cm) is the sum of the nozzle length (111.8 cm), chamber (29.9 cm), injector thickness (estimated at 8 cm), mounted valves (estimated at 10 cm), a support structure, and possibly also a gimbal joint. The middle sketch of the three thrust chambers in Fig. 8–22 roughly corresponds to these numbers.

We have now the stage masses, propellant flows, nozzle, and chamber configuration. Since this example is aimed at a thrust chamber, data on other engine components or parameters are needed only when they relate directly to the thrust chamber or its parameters.

Next, we check if there is enough available vehicle volume (1.90 m diameter and 4.50 m long) to allow making a larger nozzle area ratio and thus gain a little more performance. First, we determine how much of this volume is occupied by propellant tanks and how much might be left over or be available for the thrust chamber. This analysis would normally be done by tank design specialists. The average density of the propellant mixture can be determined from Eq. 7–1 to be 1014 kg/m^3 and the total usable propellant of 7483 kg. Using densities from Table 7–1 the fuel volume and the oxidizer volume are calculated to be about 2.8 and 4.6 m^3, respectively. For a diameter of 1.90 m, a nearly spherical fuel tank, a separate oxidizer cylindrical tank with elliptical ends, 6% ullage, and 2% residual propellant, a layout would show an overall tank length of about 3.6 m in a space that is limited to 4.50 m. This would leave only 0.9 m for the length of the thrust chamber, which not long enough. Therefore, we would need to resort to a more compact tank arrangement, such as using a common bulkhead between the two tanks (not recommended with cryogenic propellants) or building a toroidal tank around the engine. It is not the aim to design the tanks in this example, but the above conclusion affects the thrust chamber. Since the available volume of the vehicle is limited, it is not possible to make the thrust chamber any bigger.

This diversion into the tank design shows how a single vehicle parameter affects thrust chamber design. For example, if the tank design would turn out to be difficult or the tanks would become too heavy, then one of the following thrust chamber options can be considered: (1) go to a higher chamber pressure (makes the thrust chamber and nozzle smaller, but heavier), (2) go to a lower thrust engine (will be smaller and lighter), (3) store the nozzle of the upper-stage thrust chamber in two pieces and assemble them during the flight once the lower stages have been used and discarded (see the extendible nozzle in Fig. 8–17, it is more complex and somewhat heavier), or (4) use more than one thrust chamber in the engine (will be heavier but shorter and can provide roll control). We will not pursue these or other options here.

HEAT TRANSFER. The choice of computer software for estimating heat transfer and cooling parameters in thrust chambers depends on the experience of the analyst and the rocket organization. Typical computer programs divide the internal wall surface of the chamber and nozzle into incremental axial steps. Usually in preliminary analyses heat transfer calculations are done only at critical locations such as for the throat and perhaps the chamber.

From Fig. 5–1 and Eq. 3–12 or 3–22 we determine the following gas temperatures for the chamber, nozzle throat region, and at a location in the diverging exit section. These are: $T_1 = 3600\,K$, $T_t = 3214\,K$, and at the exit $T_e = 1430\,K$ for an area ratio of 6.0 in the diverging nozzle section. The chamber and nozzle down to this exit area ratio of 6 will need to be cooled by the fuel. Stainless steel has been successfully used for inner wall materials with this propellant combination and for these elevated wall temperatures.

Note that beyond an area ratio of 6, the nozzle free-stream gas temperatures are sufficiently low so that uncooled high-temperature metals can be used in this region. Radiation cooling, using high-temperature materials such as niobium (coated to prevent excessive oxidation) or carbon fibers in a nonporous carbon matrix can used between area ratios of from 6 to about 25. For the final nozzle section, where the temperatures are even lower, less expensive materials such as stainless steel or titanium should be suitable. Ablative materials have been ruled out because of the long mission duration and of the aggressive ingredients in the exhaust gas—the gas compositions shown in Figs. 5–2 and 5–3 indicate that some free oxygen and hydroxyl would be present.

We have now identified some likely materials for key chamber and nozzle components. The radiation-cooled exit segment of the nozzle (beyond area ratio of 6) is best cooled by having it protrude from the vehicle structure so the heat can radiate directly to space. One way to accomplish this is to discard the vehicle structure around the nozzle exit after stage separation and before second-stage start.

As depicted in Fig. 8–8, the maximum heat transfer rate will be at the nozzle throat region. Various software programs are normally available for estimating this heat transfer, but when a suitable computer program is not accessible an approximate steady-state analysis can be made using Eqs. 8–14 to 8–18. Relevant physical properties (specific heat, thermal conductivity, and density) of RP-1 at elevated temperatures together with film coefficients from Eqs. 8–22 or 8–24 will be needed. This is not done in this example, in part because data tables for physical properties are not always reliable (and would take up a lot of space). Data from prior thrust chambers with the same propellants indicate that heat transfer rates at the nozzle throat region may exceed 10 Btu/in.^{2}-sec or $1.63 \times 10^7\,W/m^2$.

RP-1 fuel is an unusual coolant since it does not have a distinct boiling point. Its composition is not consistent because it depends on the oil stock from which it was refined and the refining process (it is distilled or evaporated gradually over a range of temperatures). Sufficiently hot walls in a cooling passage may cause RP-1 to locally break down into carbon-rich material and to partially evaporate or gasify. As long as the vapor bubbles are small and are recondensed when mixed with the cooler portions of the coolant flow, steady heat transfer will occur. When the heat transfer is high enough, then these bubbles will not disappear and may contain noncondensable gases, and the flow will contain substantial gas bubbles and become unsteady, causing local overheating. Recondensing is aided by high cooling passage velocities (more than 10 m/sec at the throat region) and by turbulence in these passages. A coolant flow velocity, found adequate to prevent large gas bubbles, of 15 m/sec is selected for the nozzle throat region, 7 m/sec for the chamber region and somewhat less, 3 m/sec for the cooled nozzle exit segment.

The material for the cooling jacket will be stainless steel to resist the oxidation and erosion of the fast-moving, aggressive hot gases, which contain a small amount of free oxygen and hydroxyl species. Forced cooling by fuel must assure that the stainless steel temperatures are well below its softening temperature of about 1050 K (1430 °F).

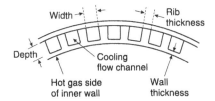

FIGURE 8–23. Segment of cooling jacket with milled channels and an electrodeposited outer wall.

The construction of the cooling jacket can be tubular, as shown in Figs. 8–1 and 8–9, or it can consist of milled channels as shown in Figs. 8–15 and 8–23. The cross section of each tube or cooling channel should be a minimum at the throat region, gradually become larger, and be about two or more times as large at the chamber and diverging nozzle regions. The wall thickness (on the hot gas side) should be as small as possible to reduce the temperature drop across the wall (this reduces thermal stresses and allows lower wall temperatures) and to minimize any yielding of materials that result from thermal deformation and pressure loads. Figure 8–12 shows this behavior, but for a thick wall. Practical considerations such as manufacturability, number of test firings before flight, deformation under pressure loads, temperature gradient and dimensional tolerances also enter into wall thickness selection. An inner wall thickness of 0.5 mm and a cooling velocity of 15 m/sec have been chosen for the throat region of the cooling jacket. Milled slots (rather than tubes) are also selected for this thrust chamber.

Selection of the number of milled slots, their cross sections, and their wall thickness depends on the coolant mass flow, its pressure, wall stresses, wall material, and on the shape of the channel. Figure 8–23 and Table 8–6 describe channel widths and depths for different channels at different locations. The fuel coolant flow (which is diminished by the gas generator fuel flow of 0.293 kg/sec) is about 4.387 kg/sec. For this flow and a cooling velocity of 15 m/sec in the

TABLE 8–6. Alternative Milled Channel Configurations for Fuel (Cooling) Flow of 4.387 kg/sec.

Throat Section			Chamber Section		
Wall thickness	0.05 cm		Wall thickness	0.06 cm	
Rib thickness	0.08 cm		Rib thickness	0.08	
Total flow area	3.653 cm^2		Total flow area	7.827 cm^2	
Flow velocity	15 m/sec		Flow velocity	7.0 m/sec	
Number of Channels	Channel Width, cm	Channel Depth, cm	Number of Channels	Channels Width, cm	Channel Depth, cm
80	0.257	0.177			
100	0.193	0.189	100	0.456	0.171
120	0.145	0.210	120	0.367	0.179
140	0.113	0.231	140	0.303	0.184
150	0.100	0.243	150	0.277	0.188
160	0.092	0.247	160	0.255	0.192
180	0.070	0.289	180	0.218	0.196

throat region the cumulative cross-sectional area for all the channels is only about 3.62 cm². The cooling velocity is lower in the chamber and other nozzle regions and the cumulative channel flow area will be larger there. The variables consist of the number of channels, the thickness of the hot wall, the rib thickness between channels, the cooling velocity, the gas temperature, and the location along the thrust chamber profile. The number of channels or tubes will determine the shape of the coolant flow cross section, ranging from deep and thin to almost square. Figure 8–9 shows the shape of the formed tubes for three different locations of a thrust chamber. Effects of varying the number of channels or channel dimensions and shape are shown in Table 8–6. For a minimum inert mass in the cooling jacket and low friction losses occur, the shape (which varies axially throughout the jacket) should be, on an average, close to a square. On the basis of such criteria, as shown in the table, a 150-channel design has been selected, which provides a favorable cross section, reasonable dimensions for ease of fabrication, sufficient cooling, and often low thermal wall stresses.

Reinforcing bands will be needed on the outside of the tubes or milled channels to hold the internal gas pressure during operation, to contain the coolant pressures and any surge pressures during start transients or arising from water hammer in the lines. We assume here a surge pressure of 50% above chamber pressure, a safety factor of 1.3, and the strength for steel as 120,000 psi. In the chamber the inside diameter is 16.7 cm (6.57 in.), the walls and channels each are 0.3 cm deep, and the pressure is 700 psia or 4.826 MPa. If one band reinforces a chamber section of length 3.0 in., then the cross-sectional area of that reinforcing band will be

$$A = pDL/(2\sigma) = [700 \times 1.5 \times 1.3 \times (6.57 + 0.3) \times 3]/(2 \times 120,000) = 0.117 \text{ in.}^2$$

If the bands were 1.0 in. wide, their thickness would be 0.11 in., and if it were 3 in. wide, they would be 0.037 in. thick. Because large nozzle exit sections have been observed to experience flutter or cyclic deformation, some external stiffening rings may be needed near the nozzle exit.

The capacity of the fuel to absorb heat (Eq. 8–25) is approximately $c_p \dot{m}_f \Delta T = 0.5 \times 4.81 \times 200 = 278,000 \text{ J/sec}$. The maximum ΔT is established by keeping the fuel well below its chemical decomposition point. This calculated heat absorption turns out to be less than the heat transfer from the hot gases. It is therefore necessary to reduce the gas temperature near the chamber and nozzle walls. This can be accomplished by (1) introducing film cooling by injection into the chamber just ahead of the nozzle, by (2) modifying the injection patterns, so that cooler, fuel-rich thick internal boundary layers form, or (3) by allowing some nucleate boiling in the throat region. The analysis of these three methods is not given here. Item (2), supplementary cooling, is selected because it is easiest to design and build and can be based on extensive data of prior favorable experience. However, it causes a small loss of performance (up to about 1% in specific impulse). The vacuum specific impulse of the engine then becomes 327 sec.

INJECTOR DESIGN. The injector pattern can be any one of the several types shown in Figs. 8–3 and 8–4. For this propellant combination, we have used both doublets (like and unlike) and triplets in the United States, and the Russians have used multiple hollow double-walled posts with swirling or rotation of the flow in the outer annulus. Based on favorable experience and demonstrated combustion stability with similar designs, we select a doublet self-impinging-type stream pattern (Fig. 8–3) and an injector structure design similar to Fig 8–5. The impinging streams form fans of liquid propellant, which break up into droplets. Oxidizer and fuel fans alternate radially. (We could also use a platelet design, as in Fig. 8–6.)

The pressure drop across the injector is usually set at values between 15 and 25% of the chamber pressure, in part to obtain high injection velocities, which aid in atomization and droplet breakup. This in turn leads to more complete combustion (and thus better performance) and to stable combustion. We will use 20% or 140 psi or 0.965 MPa for the injector pressure drop. There is also a small pressure loss in the injector passages. Injection velocities are found from Eqs. 8–1 and 8–5. Such equations are solved for the area A, which is the cumulative cross-section area of all the injection holes of one of the propellants in the injector face.

With rounded and clean injection hole entrances the discharge coefficient will be about 0.80 for a short hole as shown in Table 8–2. Solving with the fuel and the oxidizer flows for the cumulative injection hole area gives 1.98 cm^2 for the fuel and 4.098 cm^2 for the oxidizer. A typical hole diameter in this size of injector would be between 0.5 to 2.5 mm. We will use a hole size of 1.5 mm for the fuel holes (with 90% of the fuel flow) and 2.00 mm for the oxidizer hole size, resulting in 65 doublets of oxidizer holes and 50 doublets of fuel. By using a slightly smaller fuel injection hole diameter, we could match the number of 65 doublets used with the oxidizer holes. These injection doublets will be arranged on the injector face in concentric patterns similar to Fig. 8–5. We may obtain a slightly higher performance by going to smaller hole sizes and a large number of fuel and oxidizer holes. In addition there could be extra fuel holes on the periphery of the injector face to help provide the cooler boundary layer, which is needed to reduce heat transfer. These will use approximately 10% of the fuel flow and, for a 0.5-mm hole diameter, the number of holes will be about 100. To make a proper set of liquid fans, equal inclination angles of about 25° are used with the doublet impingements. (See Fig. 8–7.)

IGNITER DIMENSIONS. A pyrotechnic (solid propellant) igniter will be used. It needs to supply sufficient energy and run long enough to provide the pressure and temperature in the thrust chamber for good ignition. Its largest diameter has to be small enough to be inserted through the throat, namely 8.0 cm maximum diameter and it can be 10 to 15 cm long. The actual igniter will, most likely, be smaller than this.

LAYOUT DRAWINGS, MASSES, FLOWS, AND PRESSURE DROPS. We now have found enough key design parameters for the selected thrust chamber, so a preliminary layout drawing or an initial CAD (computer-aided design) can be made. Before this can be properly done, we need to analyze or estimate the thrust chamber manifolds for the fuel and oxidizer, valve mounting provisions and their locations, any nozzle closure part used during storage, a thrust structure, and possibly an actuator and gimbal mount, if gimbaling is required by the mission. A detailed layout or CAD image (not included here) would allow a more accurate picture and a better determination of the mass of the thrust chamber and its center of gravity both with and without propellants.

Estimates of gas pressures, liquid pressures (or pressure drops) in the flow passages, injector, cooling jacket, and the valves are needed for stress analysis, so that various wall thicknesses and component masses can be determined. Material properties will need to be gathered from references or laboratory tests. Some of these estimates and designs may actually change portions of the data we selected or estimated early in this sample analysis, and some of the calculated parameters may have to be re-analyzed and revised. Further need for changes in the thrust chamber design may become evident in the design of the engine, the tanks, or the interface with the vehicle. Methods, processes, tooling and fixtures for manufacturing and testing (and the number and types of tests) will have to be selected and evaluated; the number of thrust chambers to be built has to be decided before any reasonable manufacturing plans, schedules, and cost estimates can be started.

SYMBOLS

A	area, m^2 (ft^2)
c_p	gas specific heat at constant pressure, J/kg-K (Btu/lbm°R)
$\bar{c}$	average liquid or solid specific heat, J/kg-K (Btu/lbm°R)
C_d	discharge coefficient
C_F	thrust coefficient
D	diameter, m (ft)
E	modulus of elasticity, N/m^2 ($lbf/in.^2$), or radiation energy, W
f	friction loss coefficient, or geometric factor in radiation
g_0	sea-level acceleration of gravity, 9.806 m/sec^2 (32.17 ft/sec^2)
h	convective film coefficient, $W/(m^2\text{-}K)$; (Btu/(ft^2-°R))
I_s	specific impulse, sec
κ	specific heat ratio
L	length, m (ft)
L^*	characteristic chamber length, m (ft)
m	mass, kg
$\dot{m}$	mass flow rate, kg/sec (lb/sec)
p	pressure, N/m^2 or Pa ($lbf/in.^2$)
Pr	Prandtl number ($c_p \mu/k$)
q	heat transfer rate or heat flow per unit area, J/m^2-sec (Btu/ft^2-sec)
Q	volume flow rate, m^3/sec (ft^3/sec), or heat flow rate, J/sec
R	Reynolds number $Dv\rho/\mu$
r	flow mixture ratio (oxidizer to fuel); or radius, m (ft)
s	stress N/m^2 ($lbf/in.^2$)
t	time, sec, or thickness, m (ft)
t_s	stay time, sec
t_w	wall thickness, m (in.)
T	absolute temperature, K (°R)
Δu	flight velocity increment m/sec (ft/sec)
v	velocity, m/sec (ft/sec)
V_1	specific volume, m^3/kg (ft^3/lb)
V_c	combustion chamber volume (volume up to throat), m^3(ft^3)
x, y	coordinates of a parabola with constant a

Greek Letters

γ_o	angle between chamber axis and oxidizer stream
γ_f	angle between chamber axis and fuel stream
δ	angle between chamber axis and the resultant stream
ϵ	nozzle exit area ratio ($\epsilon = A_2/A_t$)

ε	emissivity of radiating surface, dimensionless
θ	angle
k	thermal conductivity, J/(m^2-sec-K)/m (Btu/in.^{2}-sec^2-°R/in.)
λ	coefficient of thermal expansion, m/m-K (in./in.-°R)
μ	absolute gas viscosity, kg/(m-sec) or lbf/ft^2-sec
v	Poisson ratio
ρ	density, kg/m^3(lbf/ft^3)
σ	Stefan–Boltzmann constant (5.67×10^{-8} W/m^2-K^4); also stress N/m^2(lbf/in^2) N/m^2(lbf/in^2)

Subscripts

am	arithmetic mean
c	Chamber
f	fuel or final condition
g	Gas
gg	gas generator
inj	Injector
l	Liquid
o	Oxidizer
t	Throat
tc	thrust chamber
w	Wall
wg	wall on side of gas
wl	wall on side of liquid
0	initial condition
1	inlet or chamber condition
2	nozzle exit condition
3	atmosphere or ambient condition

PROBLEMS

1. How much total heat per second can be absorbed in a thrust chamber with an inside wall surface area of 0.200 m^2 if the coolant is liquid hydrogen and the coolant temperature does not exceed 145 K in the jacket? Coolant flows at 2 kg/sec. What is the average heat transfer rate per second per unit area? Use the data from Table 7–1 and the following:

Heat of vaporization near boiling point	446 kJ/kg
Thermal conductivity (gas at 21 K)	0.013 W/m-K
(gas at 194.75 K)	0.128 W/m-K
(gas at 273.15 K)	0.165 W/m-K

2. During a static test a certain steel thrust chamber is cooled by water in its cooling jacket. The following data are given for the temperature range and pressure of the coolant:

Average water temperature	100 °F
Thermal conductivity of water	1.07×10^{-4} Btu/ sec -ft^2-°F/ft
Gas temperature	4500 °F
Specific gravity of water	1.00
Viscosity of water	2.5×10^{-5} lbf- sec /ft^2
Specific heat of water	1.3 Btu/lb-°F
Cooling passage dimensions	$\frac{1}{4} \times \frac{1}{2}$ in.
Water flow through passage	0.585 lb/sec
Thickness of inner wall	$\frac{1}{8}$ in.
Heat absorbed	1.3 Btu/in.^{2}-sec
Thermal conductivity of wall material	26 Btu/hr-ft^2-°F/ft

Determine (a) the film coefficient of the coolant; (b) the wall temperature on the coolant side; (c) the wall temperature on the gas side.

3. In Problem 2 (above) determine the water flow required to decrease the wall temperature on the gas side by 100 °F. What is the percentage increase in coolant velocity? Assume that the various properties of the water and the average water temperature do not change.

4. Determine the absolute and relative reduction in wall temperatures and heat transfer caused by applying insulation in a liquid-cooled rocket chamber with the following data:

Tube wall thickness	0.381 mm
Gas temperature	2760 K
Gas-side wall temperature	1260 K
Heat transfer rate	15 MW/m^2-sec
Liquid-film coefficient	23 kW/m^2-K
Wall material	Stainless steel AISI type 302

A 0.2-mm-thick layer of insulating paint is applied on the gas side; the paint consists mostly of magnesia particles. The average conductivity of this magnesia paint is 2.59W/m^2-K/m over the temperature range. The stainless steel has an average thermal conductivity of 140 Btu/hr.–ft^2-°F/in and a specific gravity of 7.98

5. A small thruster has the following characteristics:

Propellants	Nitrogen tetroxide and monomethyl hydrazine
Injection individual hole size	0.063 in. for oxidizer and 0.030 in. for fuel
Injection hole pattern	Unlike impinging doublet
Thrust chamber type	Ablative liner with a carbon–carbon nozzle throat insert
Specific gravities	1.446 for oxidizer and 0.876 for fuel
Impingement point	0.25 in. from injector face
Direction of jet momentum	*Parallel to chamber axis after impingement*
$r = 1.65$ (fuel rich)	$(I_s)_{\text{actual}} = 251$ sec
$F = 300$ lbf	$t_b = 25$ sec (burning time)
$p_1 = 250$ psi	$A_1/A_t = 3.0$
$(\Delta p)_{\text{inj}} = 50.0$ psi	$(C_d)_o = (C_d)_f = 0.86$

Determine the number of oxidizer and fuel injection holes and their angles. Make a sketch to show the symmetric hole pattern and the feed passages in the injector. To protect the wall, the outermost holes should all be fuel holes.

6. A large, uncooled, uninsulated, low-carbon-steel thrust chamber burned out at the throat region during testing. The wall (0.375 in. thick) had melted and there were several holes. The test engineer said that he estimated the heat transfer to have been about 15 Btu/in.^{2}. The thrust chamber was repaired, and assume that you are responsible for the next test. Someone suggested that a series of water hoses be hooked up to spray plenty of water on the outside of the nozzle wall at the throat region during the next test to prolong the firing duration. The steel's melting point is estimated to be 2550 °F. Because of the likely local variation in mixture ratio and possibly imperfect impingement, you anticipate some local gas regions that are oxidizer rich and could start the rapid oxidation of the steel. You therefore decide that 2150 °F should be the maximum allowable inner wall temperature. Besides knowing the steel weight density (0.284 lbf/in.^{3}), you have the following data for steel for the temperature range from ambient to 2150 °F: the specific heat is 0.143 Btu/lbm-°F and the thermal conductivity is 260 Btu/hr-ft^2-°F/in. Determine the approximate time for running the next test (without burnout) both with and without the water sprays. Justify any assumptions you make. If the water spray seems to be adequate (getting at least 10% more burning time), make sketches with notes on how the mechanic should arrange for this water flow during testing so it will be most effective.

7. The following conditions are given for a double-walled cooling jacket of a rocket thrust chamber assembly:

Rated chamber pressure	210 psi
Rated jacket pressure	290 psi
Chamber diameter	16.5 in.
Nozzle throat diameter	5.0 in.
Nozzle throat gas pressure	112 psi
Average inner wall temperature at throat region	1100 °F
Average inner wall temperature at chamber region	800 °F
Cooling passage height at chamber and nozzle exit	0.375 in.
Cooling passage height at nozzle throat	0.250 in.
Nozzle exit gas pressure	14.7 psi.
Nozzle exit diameter	10 in.
Wall material	1020 carbon steel
Inner wall thickness	0.08 in.
Safety factor on yield strength	2.5
Cooling fluid	RP-1
Average thermal conductivity of steel	250 Btu/hr-ft^2-°F/in.

Assume other parameters, if needed. Compute the outside diameters and the thickness of the inner and outer walls at the chamber, at the throat, and at the nozzle exit.

8. Determine the hole sizes and the angle setting for a multiple-hole, doublet impinging stream injector that uses alcohol and liquid oxygen as propellants. The resultant

momentum should be axial, and the angle between the oxygen and fuel jets $(\gamma_o + \gamma_f)$ should be 60°. Assume the following:

$(C_d)_o$	0.87	Chamber pressure	300 psi
$(C_d)_f$	0.91	Fuel pressure	400 psi
ρ_o	71 lb/ft^3	Oxygen pressure	380 psi
ρ_f	51 lb/ft^3	Number of jet pairs	4
r	1.20	Thrust	250 lbf
	Actual specific impulse 218 sec		

Answer: 0.0197 in.; 0.0214 in.; 32.3°; 27.7°.

9. Table 11–3 shows that the RD-120 rocket engine can operate down to 85% of full thrust and at a mixture ratio variation of $\pm$ 10.0%. In a particular static test the average thrust was held at 96% of nominal and the average mixture ratio was 2.0% fuel rich. Assume a 1.0% residual propellant, but neglect other propellant budget allowances. What percentage of the fuel and oxidizer that have been loaded will remain unused at thrust termination? If we want to correct the mixture ratio in the last 20.0% of the test duration and use up all the available propellant, what would be the mixture ratio and propellant flows for this last period?

10. Make a simple cross-section sketch approximately to scale of the thrust chamber that was analyzed in Section 8.9. The various dimensions should be close, but need not be accurate. Include or make separate detailed-section sketches of the cooling jacket and the injector. Also compile a table of all the key characteristics, similar to Table 8–1, but include gas generator flows and key materials. Make estimates or assumptions for any key data that is not mentioned in Section 8.9.

REFERENCES

8–1. G. P. Sutton, *History of Liquid Propellant Rocket Engines*, AIAA, Reston, VA, 2006.

8–2. M. M. Mielke et al., "Applications of Ultrafast Lasers in Microfabrication," *Journal of Laser Micro/Nanoengineering*, Vol. 28, No.2, Aug. 2013, pp. 115–123. doi: 102961/jlmn2013.02.0001.

8–3. V. Yang, M. Habiballah, J. Hulka, and M. Popp, (Eds.), *Liquid Rocket Thrust Chambers: Aspects of Modeling, Analysis, and Design*, Progress in Astronautics and Aeronautics (Series), Vol. 200, AIAA, Reston, VA, 2004.

8–4. K. W. Brinckman et al., "Impinging Fuel Injector Atomization and Combustion Modelling," AIAA Paper 2015-3763, Orlando, FL, 2015.

8–5. R. D. McKown, "Brazing the SSME," *Threshold, an Engineering Journal of Power Technology*, No. 1, Rocketdyne Division of Rockwell International (now Aerojet-Rocketdyne.), Canoga Park, CA, Mar. 1987, pp. 8–13.

8–6. R. A. Ellis, J. C. Lee, F. M. Payne, M. Lacoste, A. Lacombe, and P. Joyes, "Testing of the RL 10B-2 Carbon-Carbon Nozzle Extension," AIAA Conference Paper 98-3363, Jul. 1998.

8–7. M. Niino, A. Kumakawa, T. Hirano, K. Sumiyashi, and R. Watanabe, "Life Prediction of CIP Formed Thrust Chambers," *Acta Astronautica*, Vol. 13, Nos. 6–7, 1986, pp. 363–369 (fatigue life prediction).

8–8. J. S. Porowski, W. J. O'Donnell, M. L. Badlani, B. Kasraie, and H. J. Kasper, "Simplified Design and Life Predictions of Rocket Thrust Chambers," *Journal of Spacecraft and Rockets*, Vol. 22, No. 2, Mar.–Apr. 1985, pp. 181–187.

8–9. A. J. Fortini and R. H. Tuffias, "Advanced Materials for Chemical Propulsion: Oxide-Iridium/Rhenium Combustion Chambers," AIAA Paper 99-2894, Jun. 1999.

8–10. U. Gotzig et al., "Development and Test of a 3D Printed Hydrogen Peroxide Fight Control Thruster," AIAA Paper 2015-4161, Orlando FL, 2015; H. Kenjon, "TRUST: 3-D Manufacturing's Holy Grail," *Aerospace America*, Vol. 53, No. 7, Jul.–Aug. 2015, pp. 42–45.

8–11. Personal information from J. D. Haynes, Additive Manufacturing Program Manager, Aerojet Rocketdyne, 2015.

8–12. F. P. Incropera, D. P. DeWitt, T. L. Bergman, and A. S. Lavine, *Introduction to Heat Transfer*, 5th ed., John Wiley & Sons, Hoboken, NJ, 2006.

8–13. A. A. Samarskii and P. N. Vabishchevich, *Computational Heat Transfer,* Vol. 1. *Mathematical Modeling,* and Vol. 2. *The Finite Difference Methodology.* John Wiley & Sons, New York, 1995, 1996.

8–14. R. W. Lewis, Perumal Nithiarasu, and Kankanhalli Seetharamu, *Fundamentals of the Finite Element Method for Heat and Fluid Flow.* John Wiley & Sons, Hoboken, NJ, 2004.

8–15. D. R. Bartz, "Survey of Relationships between Theory and Experiment for Convective Heat Transfer in Rocket Combustion Gases," in *Advances in Rocket Propulsion,* S. S. Penner (Ed.), AGARD, Technivision Services, Manchester, England, 1968.

8–16. J. M. Fowler and C. F. Warner, "Measurements of the Heat-Transfer Coefficients for Hydrogen Flowing in a Heated Tube," *American Rocket Society Journal*, Vol. 30, No. 3, March 1960, pp. 266–267.

8–17. B. Betti et al., "Convective and Radiative Contributions to Wall Heat Transfer in Liquid Rocket Engine Thrust Chambers," AIAA Paper 2015-3757, Orlando, FL, 2015.

8–18. P. A. Baudart, V. Duthoit, T. Delaporte, and E. Znaty, "Numerical Modeling of the HM 7 B Main Chamber Ignition," AIAA Paper 89-2397, 1989.

8–19. A. R. Casillas, J. Eninger, G. Josephs, J. Kenney, and M. Trinidad, "Control of Propellant Lead/Lag to the LEA in the AXAF Propulsion System," AIAA Paper 98-3204 Jul. 1998.

8–20. C. Gafto and B. Nakazro, "Life Test Results of a MONARCH5 1 lbf Monopropellant Thruster with the Haraeus Catalyst," AIAA Paper 2014-3795, 2014.

CHAPTER 9

LIQUID PROPELLANT COMBUSTION AND ITS STABILITY

In this chapter, the complexities of combustion chamber phenomena found in liquid bipropellant thrust chambers are treated. We describe combustion behavior and its analysis in general terms, together with several types of combustion instability with their resulting undesirable effects, and discuss semiempirical remedies on how to avoid these effects. An objective here is to operate at very high combustion efficiencies while preventing any occurrence of disruptive or destructive combustion instabilities. Thrust chambers need to operate with stable combustion over a wide range of operating conditions. Further treatment of these subjects can be found in Refs. 9–1 to 9–7.

Liquid propellant combustion is very efficient in well-designed thrust chambers. Combustion efficiencies of between 95 to 99.5% are typical, in contrast to turbojets or ordinary furnaces which may range from 50 to 97%. Such high efficiencies stem from the very fast reaction rates at high combustion temperatures resulting from the thorough mixing of fuel and oxidizer (from proper injection flow distributions, uniform atomization, and turbulent gas diffusion). Losses arise from nonuniform mixture ratios resulting in incomplete burning or inadequate mixing. Combustion efficiencies can be well below 95% in very small bipropellant thrust chambers or small gas generators because their injectors lack sufficient injection orifices or elements for proper mixing.

9.1. COMBUSTION PROCESS

For understanding combustion processes, it is helpful to divide the combustion chamber into a series of discrete zones as shown in Fig. 9–1. The configuration depicted has a flat injector face with many small injection orifices that introduce both fuel

344

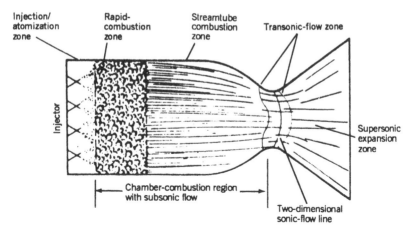

FIGURE 9–1. Division of combustion chamber into zones for analysis. Modified from Y. M. Timnat, *Advanced Chemical Rocket Propulsion*, Academic Press, New York, 1987.

and oxidizer liquids into the combustion chamber and this results in many discrete liquid propellant streams, or thin sprays or sheets. The relative thickness, behavior, and transition of each discrete zone in the axial direction is strongly influenced by specific propellant combination, operating conditions (pressure, mixture ratio, etc.), and by injector design and chamber geometry. The boundaries between the zones shown in Fig. 9–1 are really neither flat nor steady surfaces. They can be undulating or movable irregular boundaries with local changes in velocity, temporary bulges, locally intense radiation emissions, and/or variable temperatures. Table 9–1 shows all major interacting physical and chemical processes that may occur in the chamber. This table is a modification of tables and data in Refs. 9–1, 9–2, and 9–7.

Combustion behavior is highly propellant injection dependent. With a hypergolic propellant combination there is an initial chemical reaction in the liquid phase as droplets of fuel impinge on droplets of oxidizer. Experiments show that such contacts can create local mini-explosions with enough energy release to suddenly vaporize thin layers of fuel and oxidizer locally at the droplet's contact face—there immediately follows a vapor chemical reaction and droplets blow-apart and breakup. Small local explosions have been documented (Refs. 9–8 and 9–9). In contrast, when a cryogenic fuel like hydrogen comes from cooling the thrust chamber, the injected hydrogen is gaseous and relatively warm (60 to 240 K); here, there are no hydrogen droplets to evaporate.

Combustion processes are typically analyzed as steady flow, but in reality this is not truly so. When observing any one location within the chamber, one finds turbulent flows with local fluctuations in pressure, flow, temperature, density, mixture ratio, and/or radiant emission. Some of the processes listed in the table occur sequentially, while others may occur simultaneously. Not all listed processes happen with each propellant. Three-dimensional turbulence consistently forms in all parts of the combustion chamber and in the nozzle. Most combustion also generates a great deal of noise.

TABLE 9–1. Physical and Chemical Processes in the Combustion of Liquid Propellants

Injection	Atomization	Vaporization
Liquid jets or sprays enter chamber at relatively low velocities	Impingement and breakup of jets or sheets	Droplet gasification and gas diffusion
Sometimes gaseous propellant is injected (e.g., hydrogen)	Formation of liquid fans or spray cones	Further heat release from local chemical reactions
Partial evaporation of liquids	Formation of droplets	Relatively low gas velocities and some cross flow
Interaction of jets or sprays with high-pressure hot gas	Secondary breakup of droplets	Heat absorbed in droplets by radiation, convection, and conduction; some blowback of turbulent gases from the hot reaction zone into injector zone
	Liquid mixing and some liquid–liquid chemical reactions	
	Oscillations of jets or fans or spray liquid sheets as they become unstable during breakup	Acceleration to higher velocities
	Vaporization begins and some vapor reactions occur	Vaporization rate influenced by turbulence, pressure or temperature oscillations and acoustic waves

Mixing and Reaction	Expansion in Chamber
Turbulent mixing (three-dimensional)	Chemical kinetics causes attainment of final combustion temperature and final equilibrium reaction gas composition
Multiple chemical reactions and major heat releases	Acceleration to high chamber velocities
Interactions of turbulence with droplets and chemical reactions	Gas dynamics displays turbulence and increasing axial gas velocities
Temperature rise reduces densities and increases volume flow	Formation of a boundary layer
Local mixture ratios, reaction rates, or velocities are not uniform across chamber and vary rapidly with time	Streamlined high-velocity axial flow with very little cross flow
Some tangential and radial flows, much of it near the injector	Flow toward nozzle
	Heat transfer to walls

Injection/Atomization Zone

Typically, two different liquids are being injected, either storable bipropellants or liquid oxygen–hydrocarbon combinations. These are injected through orifices or slots at speeds usually between 7 and 60 m/sec or about 20 to 200 ft/sec. Injector design has a profound influence on combustion behavior and some seemingly minor design changes may have major effects on instabilities. Orifice pattern, sizes, number, distribution, and types influence combustion behavior, as do pressure drops, manifold geometries, and/or surface roughness at the injection orifice walls. Individual jets, streams, and/or liquid sheets break up into droplets by mutual impingement (or with a surface), by impingement of conical spray sheets, by the intrinsic instability of liquid jets or sprays, and/or by their interaction with gases at different velocities and temperatures. In the first zone of Fig. 9–1, injected liquids atomize, yielding large numbers of small droplets (see Refs. 9–2, 9–6 to 9–9). Heat is transferred to these droplets by radiation from the very hot rapid-combustion zone and by convection from moderately hot gases within the first zone. As the droplets evaporate they create local regions rich in either fuel or oxidizer vapor.

This first zone is highly heterogeneous; it contains liquids and vaporized propellant as well as some burning hot gases. With liquids being located at discrete sites, there are large gradients in all directions with respect to fuel and oxidizer mass fluxes, mixture ratios, size and dispersion of droplets, and/or gas properties. While chemical reactions begin to occur in this zone, the rate of heat generation is relatively low, partly because the liquids and the gases are still relatively cold and partly because droplet vaporization causes fuel-rich and fuel-lean regions, which do not burn as quickly. Some hot gases recirculating back from the rapid combustion zone may create local gas flows across the injector face. These hot gases, which typically flow in unsteady vortexes or turbulent patterns, play an essential role in initial evaporation rates.

All injection, atomization, and vaporization processes change when one of the propellants is gaseous as, for example, with liquid oxygen and hydrogen in thrust chambers or precombustion chambers where liquid hydrogen has been previously gasified. Such hydrogen gas contains no droplets and ordinarily enters at much higher injection velocities (above 120 m/sec) than the liquid propellant. The shear forces thus created on liquid jets produce more rapid droplet formation and gasification. The preferred injector designs for gaseous hydrogen and liquid oxygen are quite different from the individual liquid jet streams used often with storable propellants, as shown in Chapter 8.

Rapid Combustion Zone

In this zone, intense and rapid chemical reactions occur at increasingly higher temperatures; here, any remaining liquid droplets are quickly vaporized by convective heating and pockets of fuel-rich and fuel-lean gases mix robustly. This mixing is enabled by local three-dimensional turbulent diffusion within the gas flows.

Further breakdown of propellant chemicals into intermediate fractions and smaller simpler chemicals together with oxidation of fuel fractions occur rapidly in this zone. The rate of heat release greatly increases, and this causes the "specific volume" of the gas mixture together with the local axial velocity to increase by factors of 100 or more. This rapid expansion of the heated gases also creates a series of small local transverse gas flows and even some localized temporary back flows from hot high-burning-rate sites to colder low-burning-rate sites. Any liquid droplets that may still persist in the upstream portion of this zone cannot follow the gas flow quickly, having a large inertia to transverse-direction movements; therefore, local zones of fuel-rich or oxidizer-rich gases may persist reflecting orifice spray patterns from the upstream injection zone. Hence, mixtures may approach but never quite become completely uniform, even though gas composition and mixture ratio across chamber sections are becoming more uniform as gases move through this zone. As reaction-product gases accelerate, they are subject to further heat releases, and their lateral velocity becomes relatively small compared to their increasing axial velocities. Our understanding of such process has been aided by high-speed photography. See Ref. 9–8.

Streamtube Combustion Zone

In this zone oxidation reactions continue, but at a lower rate, and some additional heat is released. Chemical reactions continue as the mixture is driven toward its equilibrium composition. Axial velocities are high (200 to 600 m/sec) whereas transverse flow velocities are relatively small. Streamlines form and there is relatively little turbulent mixing across their boundaries. Locally, flow velocity and pressure may fluctuate somewhat. A streamline type of inviscid flow and some further chemical activity may persist, not only throughout the remainder of the combustion chamber but also extending into the nozzle.

Residence times in this zone decrease compared to residence times in the previous two zones. Overall, propellant residence times in the combustion chamber are relatively short, usually less than 10 msec. Combustion rates inside liquid rocket engines are very fast (with the volumetric heat release being approximately 370 MJ/m^3–sec) compared to those in turbojets—the higher temperatures in rocket chambers accelerate chemical reaction rates, which increase exponentially with temperature.

9.2. ANALYSIS AND SIMULATION

In analyzing the combustion process and its instabilities it has been convenient to divide the associated acoustical characteristics into their linear and nonlinear parts. A number of computer simulations with linear analyses have been developed over the past 55 years to understand combustion processes in liquid propellant combustion devices and to predict combustion oscillation frequencies. Nonlinear behavior (i.e., why does a disturbance suddenly change an apparently stable combustion to unstable?) is not well understood and has not been properly simulated. Mathematical descriptions require a number of simplifying assumptions to permit solutions (see Refs. 9–2 to 9–5, 9–7, 9–11, and 9–12). Good models exist for isolated phenomena such as propellant droplet vaporization (see Refs. 9–1 and 9–10), burning in gaseous environments, and/or for steady-state gas flows under heat release; some programs also consider turbulence and film cooling effects. The thermochemical equilibrium principles mentioned in Chapter 5 also apply here.

In analytical models, the following phenomena are usually *not* treated or are greatly simplified: cross flows; nonsymmetrical gradients; flow unsteadiness (such as time variations in local temperatures, local velocities, or local gas composition); local thermochemical reactions at off-design mixture ratios and at different kinetic rates; vaporization enhancement by acoustic fields (see Ref. 9–9); uncertainties in sprayed-droplets spatial and size distribution; and/or drag forces on the droplets.

While computer program results give valuable information regarding any particular design and are useful guides in interpreting actual test results, by themselves they are not sufficient to define such designs, select specific injector patterns, or predict combustion instabilities occurrences.

Most computer software known to the authors is suitable only for steady-state flow descriptions, usually at a predetermined average mixture ratio and chamber pressure. However, during engine starts, thrust changes (such as throttling), and stopping transients, local and overall mixture ratios and pressures may change sufficiently to make routine analyses of these transient conditions relatively inadequate.

As previously stated, combustion is strongly influenced by injector design. The following are some injection parameters which influence combustion behavior: injector spray or jet patterns and their impingement locations, injector hole sizes and/or hole distribution, chamber/injector geometry, angle of jets or injection sheets, and liquid injection pressure drops. Also of influence are: pressurant gas saturation, droplet evaporation, mixture ratio distributions, pressure and/or temperature gradients near the injector, and initial propellant temperature. Presently, attempts to analyze all these effects together require advanced computational methods.

Computational fluid dynamics (CFD) techniques allow for the most comprehensive description of interrelated fluid dynamic and thermodynamic phenomena. They may model the time history of all parameters and may include many nonlinear effects. For complex geometries the information is tracked with a multitude of discrete locations to account for changes in gas composition; thermodynamic conditions; equilibrium reactions; phase changes; viscous or nonviscous flow; one-, two-, or three-dimensional flows; and steady-state or transient conditions. CFD has been applied to resonant cavities in injectors and/or chambers and to the flow of burning gases through turbines.

9.3. COMBUSTION INSTABILITY

When rocket combustion processes are not well controlled, combustion instabilities may grow and very quickly cause excessive pressure-induced vibrational forces (that may break engine parts) or excessive heat transfer (that may melt thrust chamber parts). The engineering aim here must be to prevent all occurrences of such damaging instabilities by proper design and to maintain reliable operation (see Refs. 9–3, 9–5, and 9–11 to 9–13). Much progress has been made in understanding and avoiding combustion instabilities; since about 1975, hardly any thrust chambers have experienced destructive instabilities.

References 9–2 to 9–5 and 9–7 describe analyses of combustion instabilities in thrust chambers as practiced in the United States. References 9–4 and 9–13 present the same information but as practiced in Russia; there are some differences in the assumptions, analysis, and testing, but their basic approach is essentially similar. Any new or modified thrust chamber and/or rocket engine is tested to demonstrate that combustion instabilities will not occur. Combustion stability for hybrid propulsion systems is discussed in Chapter 16 and for solid propellant motors in Chapter 14.

TABLE 9–2. Principal Types of Combustion Instability

Type and Word Description	Frequency Range (Hz)	Cause Relationship
Low frequency, called chugging or feed system instability	10–400	Linked with pressure interactions between propellant feed system, if not the entire vehicle, and combustion chamber
Intermediate frequency, called acoustic[a] instability, buzzing, or entropy waves	400–1000	Linked with mechanical vibrations of propulsion structure, injector manifold, flow eddies, fuel/oxidizer ratio fluctuations, and propellant feed system resonances
High frequency, called screaming, screeching, or squealing	Above 1000	Linked with combustion process forces (pressure waves) and chamber acoustical resonance properties

[a]Use of the word *acoustic* stems from the fact the frequency of the oscillations is related to combustion chamber dimensions and velocity of sound in the combustion gas.

Table 9–2 lists the principal types of combustion generated vibrations encountered in liquid rocket thrust chambers (see Refs. 9–2 and 9–9). Admittedly, liquid propellant rocket thrust chamber combustion is never perfectly smooth because minor fluctuations of pressure, temperature, and velocity are always present. When these fluctuations interact with the natural frequencies of the propellant feed system (with or without vehicle structure) or with the chamber acoustics, large periodic superimposed oscillations develop. In normal rocket propulsion practice, *smooth combustion* is said to occur when pressure fluctuations during operation do not exceed ± 5% of the mean chamber pressure. Combustion that generates larger pressure fluctuations, as measured at chamber wall locations and occurring at completely random intervals is called *rough combustion*. Unstable combustion, or *combustion instability*, displays organized oscillations occurring at well-defined intervals with pressure peaks that may be maintained, may increase, or may die out. These periodic peaks represent fairly large concentrations of vibratory energy and can be easily recognized against the random-noise background using high-frequency pressure measurements (see Fig. 9–2).

Chugging, the first type of combustion instability listed in Table 9–2, stems mostly from the elastic nature of the feed systems and vehicle's structures and/or from the imposition of propulsive forces upon the vehicle. Chugging in an engine or thrust chamber assembly may occur at a test facility or during flight, especially with low-chamber-pressure engines (100 to 500 psia). It may originate from propellant pump cavitation, gas entrapment in propellant flows, tank pressurization control fluctuations, and/or vibration of engine supports and propellant lines. It may also be caused by resonances in the engine feed system (such as oscillating bellows

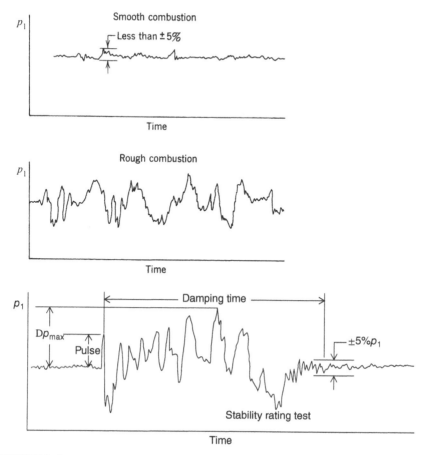

FIGURE 9–2. Typical simplified oscillograph traces of chamber pressure p_1 with time for different combustion events.

inducing periodic flow fluctuations) or by the coupling of structural and feed system frequencies.

When both the vehicle structure and the propellant liquid in the feed system have similar natural frequencies, then a coupling of forces occurs that may not only maintain but also strongly amplify existing oscillations. Propellant flow rate disturbances, usually at 10 to 50 Hz, give rise to such low-frequency longitudinal combustion instabilities, producing longitudinal vibration modes in the vehicle. *Pogo instability* refers to a vehicle's long-feed pipe instability since it is similar to a pogo jumping stick motion. Pogo instabilities can occur in the large, long propellant feed lines of large vehicles such as space launch vehicles or in ballistic missiles. See Refs. 9–14 and 9–15.

Avoiding objectionable engine–vehicle coupled oscillations is best accomplished during initial vehicle design, in contrast to applying "fixes" later, as had been the case with rocket engines of older large flight vehicles. Analytical methods exist for understanding most vibration modes and damping tendencies of major vehicle components, including propellant tanks, tank pressurization systems, propellant flow lines, engines, and basic vehicle structures. Figure 9–3, a simplified spring–mass model of a typical two-stage vehicle, illustrates the complexity of the analytical problem. Fortunately, most vibrational characteristics of the rocket assembly can be substantially controlled by introducing damping into major components or subassemblies. Techniques for damping pogo instabilities include the use of energy absorption devices in fluid flow lines, perforated tank liners, special tank supports, and properly designed engine, interstage, and payload support structures.

A partially gas-filled, spherical, small *pogo accumulator* attached to the main oxidizer feed line has been used as an effective damping device. It was used in the oxidizer feed line of the Space Shuttle main engine (SSME) between the oxidizer booster pump and the main oxidizer pump (see Figs. 6–1 and 6–11).

Along with the bending or flexing of pipes, joints, bellows, or long tanks, the dynamic characteristics of propellant pumps can further influence pogo-type vibrations, as examined in Ref. 9–16. Pogo frequencies will also change as propellant is consumed and as remaining mass of propellant in the vehicle changes.

Buzzing, an intermediate frequency type of instability, seldom derives from pressure perturbations greater than 5% of the mean in the combustion chamber and is usually not accompanied by large vibratory energies. It is more often an annoying noise than a damaging effect, although the occurrence of buzzing may sometimes initiate high-frequency instabilities. It is characteristic of a coupling between the combustion process and the flow in a portion of the propellant feed system. Buzzing initiation is thought to originate from the combustion process itself. Acoustic resonances of the combustion chamber with some critical portion of the propellant flow system, sometimes originating in a pump, promote continuation of these buzzing effects. This type of instability seems to be more prevalent in medium-size engines (2000 to 250,000 N thrust or about 500 to 60,000 lbf) than in larger engines.

The third type of instability, *screeching* or *screaming*, produces high frequencies (4 to 20 kHz) and is the most perplexing and common feature in new engine development. Many liquid rocket engines and solid propellant motors experience some high-frequency instability during their developmental phase. Since energy content increases with frequency, this type can be the most damaging, capable of destroying an engine in less than 1 sec. Once encountered, it is often difficult to prove that any newly incorporated "fixes" or improvements will render the engine "stable" under all launch and flight conditions. Screeching may be treated as a phenomenon solely related to the combustion chamber and not generally influenced by feed systems or structures.

High-frequency instabilities occur in two basic modes, *longitudinal* and *transverse*. The *longitudinal mode* (sometimes called *organ pipe mode*) propagates along axial planes of the combustion chamber, and its pressure waves are reflected at the injector face and at the converging nozzle entrance section. The *transverse modes* propagate along planes perpendicular to the chamber axis and can be broken down into *tangential* and *radial* modes. Transverse mode instabilities predominate in large

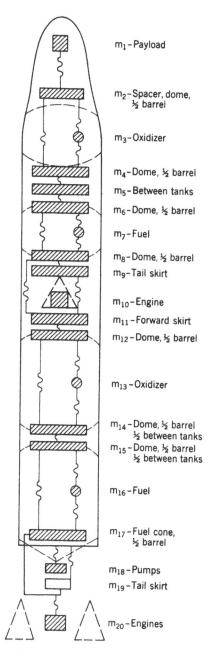

m_1 – Payload

m_2 – Spacer, dome, ½ barrel

m_3 – Oxidizer

m_4 – Dome, ½ barrel
m_5 – Between tanks
m_6 – Dome, ½ barrel

m_7 – Fuel

m_8 – Dome, ½ barrel
m_9 – Tail skirt

m_{10} – Engine
m_{11} – Forward skirt
m_{12} – Dome, ½ barrel

m_{13} – Oxidizer

m_{14} – Dome, ½ barrel
       ½ between tanks
m_{15} – Dome, ½ barrel
       ½ between tanks

m_{16} – Fuel

m_{17} – Fuel cone,
       ½ barrel

m_{18} – Pumps
m_{19} – Tail skirt

m_{20} – Engines

FIGURE 9–3. Diagram of a typical two-stage vehicle spring–mass model used in analysis of pogo vibration in the vertical direction. The cross-hatched areas represent masses and the wiggly lines indicate springs.

liquid rockets, particularly in the vicinity of the injector. Figure 9–4 shows pressure distributions at various time intervals in a combustion chamber of cylindrical cross section that is encountering transverse mode instabilities. Two kinds of wave form have been observed in tangential vibrations; one can be considered a *standing* wave because it remains in a fixed position while its pressure amplitude fluctuates; the other

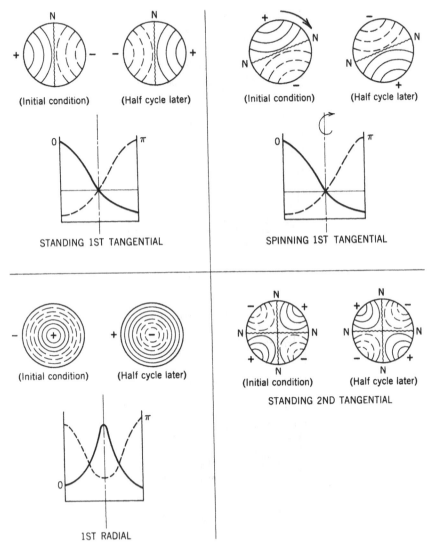

FIGURE 9–4. Simplified representation of transverse pressure oscillation modes at two time intervals in a cylindrical combustion chamber. The solid-line curves indicate pressures greater than the normal or mean operating pressure and the dashed lines indicate lower pressures. The N–N lines show node locations for these wave modes.

is a *spinning* or *traveling* tangential wave with an associated rotation of the whole vibratory system—this waveform can be visualized as maintaining constant amplitude while the wave rotates. Combinations of transverse and longitudinal modes may also occur and their frequency can also be estimated. See Refs. 9–2 to 9–6.

Screeching is believed to be predominantly driven by acoustical-energy-stimulated variations in droplet vaporization and/or mixing, by local detonations, and by acoustic changes in combustion rates. Thus, with any assisting acoustic properties, once triggered, high-frequency combustion instabilities can rapidly induce destructive modes. Invariably, any distinct boundary layer may disappear and heat transfer rates may increase by an order of magnitude causing metal melting and wall burn-throughs, sometimes within less than 1 sec. Here tangential modes appear to be the most damaging, with heat transfer rates during an instability often increasing 2 to 10 times and instantaneous pressure peaks about twice as high as with stable operation.

One source that possibly triggers high-frequency pressure-wave instabilities is a rocket combustion phenomenon called *popping*. Popping is an undesirable random high-amplitude pressure disturbance that arises during the steady-state operation of rocket engines that use hypergolic propellants. These "pops" exhibit some of the characteristics of a detonation wave. The pressure rise times are a few microseconds and the pressure ratios across the wave can be as high as 7:1.

Some combustion instabilities are induced by *pulsations in the liquid flow* originating at the turbopumps. Unsteady liquid flows may result from irregular cavitation at the leading edge of inducer impellers or main pump impellers (Ref. 9–16). Also, when an impeller's trailing edge passes a rib or stationary vane in the volute, a small pressure perturbation always results in the liquid as it travels downstream to the injector. These two types of pressure fluctuation can be greatly amplified if they coincide with the natural frequencies of combustion induced vibrations in the chamber. Also, liquid propellant feed systems may generate oscillations under certain conditions. See Ref. 9–17. Ground stability studies with liquid-oxygen/liquid-methane thrust chambers are reported in Refs. 9–18 and 9–19.

An estimate for the natural combustion gas frequencies can be determined from the wavelength l and the distance traveled per cycle together with the acoustic velocity a (see Eq. 3–10). The frequency (number of cycles per second) is

$$\text{frequency} = a/l = (1/l)\sqrt{kTR'/\mathfrak{M}} \tag{9-1}$$

where k is the specific heat ratio, R' the universal gas constant, $\mathfrak{M}$ the effective average molecular mass of the hot chamber gases, and T the local average gas absolute temperature. The wavelength l depends on the type of vibrational mode, as shown in Fig. 9–4. Smaller chambers give higher frequencies.

Table 9–3 lists estimated vibrational frequencies for one version of the French Vulcain HM 60 rocket thrust chamber; it operates with liquid hydrogen and liquid oxygen propellants and has a vacuum thrust of 1008 kN, a nominal chamber pressure of 10 MPa, and a nominal mixture ratio of 5.6 (see Ref. 9–20). Values in the table are based on acoustic measurements at ambient conditions (with corrections for an appropriate sonic velocity correlation); since the chamber has a shallow conical shape

TABLE 9–3. Estimated Acoustic Hot Gas Frequencies for Nominal Chamber Operating Conditions for the Vulcain HM-60 Thrust Chamber

Mode[a]	(L, T, R)	Frequency (Hz)	Mode[a]	(L, T, R)	Frequency (Hz)
T1	(0, 1, 0)	2424	L1T3	(1, 3, 0)	6303
L1T1	(1, 1, 0)	3579	T4	(0, 4, 0)	6719
T2	(0, 2, 0)	3856	L2R1	(2, 0, 1)	7088
R1	(0, 0, 1)	4849	T5	(0, 5, 0)	8035
L1T2	(1, 2, 0)	4987	TR21	(0, 2, 1)	8335
T3	(0, 3, 0)	5264	R2	(0, 0, 2)	8774
L1R1	(1, 0, 1)	5934			

[a]Modes are classified as L (longitudinal), T (tangential), or R (radial), and the number refers to the first, second, or third natural frequency.
Source: Reprinted with AIAA permission from Ref. 9–20

with no discrete converging nozzle section, any purely longitudinal vibration modes would be weak; in fact, no pure longitudinal modes have been detected.

Figure 9–5 shows time-sequenced diagrams of frequency–pressure–amplitude measurements taken with the Vulcain HM 60 rocket engine during the first 8 sec of a static thrust chamber test while operating at off-nominal design conditions. Chugging can be seen at low frequencies (up to 500 Hz) during the first few seconds; the action around 1500 Hz is attributed to the natural resonance frequency of the oxygen injector dome structure where a high-frequency pressure transducer was mounted. The continued oscillations observed at about 500 and 600 Hz are likely resonances associated with the feed system. The frequency–pressure–amplitude diagram shown in Fig. 9–5 for a liquid propellant engine is very similar to that shown in Fig. 14–6 for a solid propellant rocket motor.

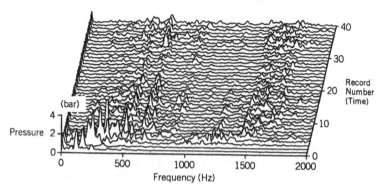

FIGURE 9–5. Graphical representation of a series of 40 superimposed frequency–amplitude diagrams taken 0.200 sec apart during the start phase (for the first 8 sec) of the Vulcain HM 60 thrust chamber. In this static hot-firing test, the thrust chamber was operating at 109 bar chamber pressure and an oxidizer-to-fuel mass flow mixture ratio of 6.6. Used with permission from Ref. 9–20

An investigation of combustion stability in liquid-oxygen/liquid-methane thrust chamber (5,000 lbf thrust) is reported in Ref. 9–17 and a transverse combustion instability analysis is found in Ref. 9–18. Reference 9–19 uses a particular analytical model to simulate combustion instabilities in liquid rocket engines.

Rating Techniques

The objective of any "rating technique" is to measure and demonstrate the ability of an engine system to quickly return to its normal or rated operation and stable combustion after the combustion process has intentionally been perturbed. Semiempirical techniques have been developed for artificially disturbing the combustion in rocket thrust chambers during testing in order to evaluate their resistance to instabilities (see Refs. 9–6 and 9–21). These include: (1) nondirectional "bombs" placed within the combustion chamber; (2) oriented explosive pulses (radial or tangential) from a "pulse gun" directed through special fittings in the chamber sidewall; and (3) directed flows of inert gas (bubbles) through a sidewall into the chamber. For these tests, heavy, thick-walled prototype thrust chambers are preferred because they are less expensive and more resistant to damage than flight-weight engines. The first two methods for inducing instability are for medium and large thrust chambers and the third method is used with small ones. Other important techniques used less widely, especially for small engines, include: (1) momentary operation at "off-mixture ratios," (2) introduction of "slugs" of inert gas into a propellant line, and (3) purposeful "hard starts," achieved by introducing some unreacted propellant at the beginning of the operation.

Assuming that stable combustion resumes, all testing techniques that intentionally introduce shock waves into the combustion chamber or otherwise perturb the combustion process afford opportunities for measuring recovery times from predetermined overpressure disturbances. Important to the magnitude and mode of the instability are the type of explosive charge selected, the size of the charge, the location and direction of the charge, and the duration of the exciting pulse. The bottom curve in Fig. 9–2 characterizes the recovery of stable operation after the combustion chamber was "bombed." The time interval to recover and the magnitude of explosive or perturbation pressure are then used to rate the resistance of the engine to such an instability. Reference 9–21 gives guidelines on how to determine and experimentally verify avoidance to combustion instabilities.

As stated above, the nondirectional bomb method and the explosive pulse-gun method are the two techniques commonly in use in larger thrust chambers. Although the latter requires modifications of the combustion chamber, this technique affords directional control, which is important in the formation of tangential modes of high-frequency instabilities and allows several data points to be observed during a single test run by installing multiple pulse guns on one combustion chamber. These pulse guns may be fired in sequence, introducing successive pressure perturbations (approximately 150 msec apart), each of increasing intensity, into the combustion chamber. For small thrusters the introduction of gas bubbles through sidewalls or propellant feed lines has been effective in triggering combustion vibration.

Control of Instabilities

Control (or really complete elimination) of instabilities is an important task during the design and development phases of a rocket engine. Designers rely on both prior experience with similar engines and on tests of new experimental engines; analytical tools are also utilized. The design selection has to be proven to be free of instabilities in actual experiments over a wide range of transient and steady-state operating conditions. Some of these experiments may be accomplished with subscale rocket thrust chambers that have similar injectors, but most tests have to be done on the full-scale engine or thrust chamber.

Design features needed to control instabilities are different for each of the three types of vibrations described in Table 9–2. Chugging is usually avoided by eliminating resonances in the propellant feed system and/or its couplings with elastic vehicle structures or any one of their components. Increased injection pressure drops and the addition of artificial damping devices in the propellant feed lines have been successfully used. Chugging and other acoustical instabilities can sometimes be traced to the natural frequency of a particular feed system component that is free to oscillate, such as a loop of piping that vibrates, an injector dome, or a bellows whose oscillations cause a pumping effect.

With the choice of the propellant combination usually fixed early in the planning of a new engine, the designer may alter combustion feedbacks (depressing the driving mechanism) by altering injector details (such as changing the injector hole pattern, hole sizes, or increasing the injection pressure drop), or alternatively by increasing acoustical damping within the combustion chamber. Of the two methods, the second has been favored in recent years because of its greater effectiveness—it is also better understood. This leads to the application of *injector face baffles, discrete acoustic energy absorption cavities*, and *combustion-chamber liners*, and/or to other changes in injector design from trial-and-error approaches.

Injector face baffles (see Fig. 9–6) have been a widely accepted anti-vibration remedy since the 1960s for overcoming or preventing high-frequency instabilities mostly in medium and larger size rocket engines. Baffle design is predicated on the assumption that, along with its driving source, the most severe instability oscillations are found in or near the atomization zone at the injector-end of the combustion chamber. Baffles minimize any influential coupling and amplification of gas dynamic forces within the chamber particularly with transverse oscillations. Obviously, these baffles must be strong, have excellent resistance to the high combustion temperatures (they are usually cooled internally by the propellants), and must protrude enough into the chamber to be effective, yet not so far as to act as a separate combustion chamber with its own acoustical characteristics. The identical injector of Fig. 9–6 has been used in several Russian engines. It has nine circular sets of injector elements operating at chamber pressures above 3500 psi and with relatively high area ratios for a booster engine (producing an I_s above 330 sec. in a vacuum). In addition to the baffle pattern, Fig. 9–6 shows other hardware remedies for combustion instabilities such as a large oxygen-rich gas cavity which reduces feed system instability and a fuel manifold pattern which results in different mixture ratios at different burn locations thus

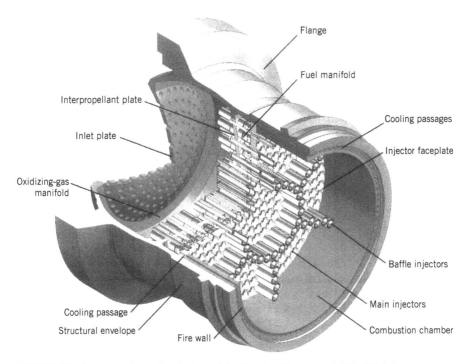

Flange

Fuel manifold

Interpropellant plate

Cooling passages

Inlet plate

Injector faceplate

Oxidizing-gas manifold

Baffle injectors

Main injectors

Cooling passage

Structural envelope

Fire wall

Combustion chamber

FIGURE 9–6. Isometric sectional view of the Russian Energomash RD-170 injector assembly (it is essentially the same as the RD-191 injector assembly). It has 271 injector elements, and 54 of these protrude into the combustion chamber to form seven baffle compartments. It is a good example of baffles used to control radial and tangential combustion instabilities. Copied with NASA's permission from Ref. 9–22

minimizing any high frequency instabilities. Chapter 8.4 of Ref. 9–6 shows additional information on these injector elements (in Fig. 8.4–20) and the injector face (in Fig. 8.4–19).

As stated, all combustion processes generate abundant *noise* and large thrust chambers and large plumes produce acoustically painful levels. Strong noise may also induce combustion instabilities the exact mechanisms of which are still controversial (see Ref, 9–1).

Various *energy absorption* or *vibration damping* mechanisms already exist in thrust chambers (although damping by wall friction in the combustion chambers is considered to be relatively insignificant). The nozzle acts as a main damper for longitudinal mode oscillations; wave reflections at a convergent nozzle entrance depart significantly from those predicted for an ideal closed end. The principal damping mechanism affecting wave propagation in a transverse plane is the combustion itself. The great volumetric change in going from liquid to burned gases and the momentum imparted by the fast moving gas to the particles (solid and/or liquid) both constitute damping mechanisms in that they absorb energy fromthe combustion. Unfortunately, the combustion process itself may generate a great deal more pressure-driven oscillatory energy than is absorbed by its inherent damping mechanisms.

Acoustical absorbers are applied usually as discrete *cavities* along or inside the combustion chamber walls near the injector end. See Refs. 9–2 and 9–23. These cavities act as Helmholtz resonator arrays (i.e., as enclosed cavities with small passage entries) that remove vibratory energy which would otherwise act to maintain the pressure oscillations. Figure 9–7 shows the application of discrete cavities (interrupted slots) at the "corner" or periphery of an injector face. A corner location usually minimizes fabrication problems, and it is the only location in a combustion chamber where pressure antinodes exist for all resonant modes of vibration (including longitudinal, tangential, radial, and combinations of these). Velocity oscillations are also minimal at this location, which favors absorber effectiveness; hence transverse modes of instability are best damped by locating absorbers at corners. Figure 9–7 also shows a Helmholtz resonator cavity and its working principles in simplified form. For each resonator element, the mass of gas in the orifice together with the volume of gas behind it form an oscillatory system analogous to the spring–mass system shown in the figure (see Refs. 9–18 and 9–23).

Absorption cavities designed as Helmholtz resonators placed in or near the injector face offer relatively high absorption bandwidths and energies per cycle. Such resonators dissipate energy twice each cycle because jets are formed upon inflow and outflow. Modern design practices favor acoustic absorbers over baffles. The rocket engine shown in Fig. 8–2 has acoustic absorption cavities in the chamber wall at a location next to the injector.

The resonance frequency f of a simple Helmholtz cavity can be estimated as

$$ f = \frac{a}{2\pi} \sqrt{\frac{A}{V(L + \Delta L)}} \qquad (9-2) $$

Here, a is the local acoustic velocity, A is the restrictor area $[A = (\pi/4)d^2]$, and other symbols are as shown in Fig. 9–7. ΔL is an empirical factor between 0.05 and $0.9L$ to allow for any additional oscillating gas mass. It varies with the L/d ratio and the edge condition of the restricted orifice (sharp edge, rounded, chamfered). Resonators in thrust chambers are tuned or designed to perform their maximum damping at predicted frequencies.

Small changes in injector geometry may cause unstable combustion to become stable and vice versa. New injectors, therefore, use geometry designs of proven, stable prior units with the same propellants. For example, the individual pattern of concentric tube injector elements used with gaseous hydrogen and liquid oxygen (shown in Fig. 8–4d) are likely to be more stable when the hydrogen gas is relatively warm and its injection velocity is at least 10 times larger than that of the liquid oxygen.

There are other combustion-driven vibration remedies besides baffles and resonant cavities. They include higher injector pressure drops or higher injection velocities, avoidance of resonances from certain key engine component structures, changes

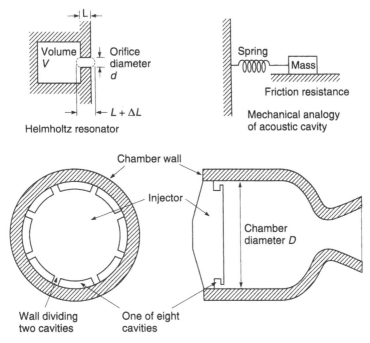

FIGURE 9–7. Simplified diagram of acoustic energy absorber cavities at the periphery of an injector. In the thrust chamber the cavity restriction is a slot (in the shape of sections of a circular arc) and not a hole. Details of the chamber cooling channels, injector holes, or internal feed passages are not shown.

of critical injector dimensions, and changing or modifying the propellants. References 9–2 and 9–6 describe several other methods, such as cooled flame holders, critical dimensions of injector elements, varying the number of injector elements and their distribution over the chamber cross section, making the mixture ratio nonuniform near the face of the injector, and/or adding temporary consumable baffles, which are effective only during starting. Small solid particulates in the reaction gas may cause some damping in nozzle oscillating flows.

In summary, the designer needs to (1) be familiar with and use data from prior successful engines and simulation programs to establish key design features in order to estimate all likely resonances of the new engine and its major components, (2) design the feed system and structure to avoid these identified resonances, (3) use an injector design robust enough to provide appropriate mixing and propellant dispersion and be resistant to disturbances, and if needed, (4) include tuned damping devices (such as cavities) to overcome acoustic oscillations. To validate that a particular thrust chamber is stable, it is necessary to test it over the entire range of likely operating conditions without encountering instabilities. Analyses are needed (e.g., Ref. 9–24) to determine the maximum and minimum likely propellant temperatures, the maximum and minimum probable chamber pressures, and the highest and lowest mixture ratios, using a propellant budget such as shown in Section 11.1. Determining these ranges then establishes the variations of test conditions for the experiments. Presently, because of an improved understanding, the amount of testing needed to validate stability has been greatly reduced compared to that required 30 to 50 years ago.

PROBLEMS

1. For a certain liquid propellant thrust chamber the following data are given:

Chamber pressure	68 MPa
Chamber shape	Cylindrical
Internal chamber diameter	0.270 m
Length of cylindrical section	0.500 m
Nozzle convergent section angle	45°
Throat diameter and radius of wall curvature	0.050 m
Injector face	Flat
Average chamber gas temperature	2800 K
Average chamber gas molecular weight	20 kg/kg-mol
Specific heat ratio	1.20

Assume the gas composition and temperature to be uniform in the cylindrical chamber section. Stating any other assumptions that may be needed, determine the approximate resonance frequencies in the first longitudinal, radial, and tangential modes.

2. Discuss how the three frequencies from Problem 1 will change with changes of combustion temperature, chamber pressure, chamber length, chamber diameter, and throat diameter.

3. Discuss why heat transfer increases during combustion instabilities?

4. Prepare a list of steps for undertaking a series of tests to validate the stability of a new pressure-fed medium sized liquid bipropellant rocket engine. State your assumptions.

5. Estimate the resonant frequency for a set of each of nine cavities similar to Fig. 9–7. Here, the chamber diameter $D = 0.20$ m, the slot width is 1.0 mm, and the width and height of the cavity are each 20.0 mm. The walls separating the individual cavities are 10.0 mm thick. Assume $L = 4.0$ mm, $\Delta L = 2.0$ mm, and $a = 1050$ m/ sec.

REFERENCES

9–1. K. K. Kuo, *Principles of Combustion*, 2nd ed., John Wiley & Sons, Hoboken, NJ, 2005.

9–2. V. Yang and W. Anderson (Eds.), *Liquid Rocket Engine Combustion Instability*, Vol. 169 of *Progress in Astronautics and Aeronautics*, AIAA, 1995, in particular Chapter 1, F.E.C. Culick and V. Yang, "Overview of Combustion Instabilities in Liquid Propellant Rocket Engines."

9–3. M. S. Natanzon and F. E. C. Culick, *Combustion Instability, Progress in Astronautics and Aeronautics*, Vol. 222, AIAA, Reston, VA, 2008.

9–4. W. Sirigano, A. Merzhanov, and L. De Luca, *Advances in Combustion Science: In Honor of Ya. B. Zel'dovich, Progress in Astronautics and Aeronautics Series*, V-173, American Institute of Aeronautics and Astronautics, 1997.

9–5. D. T. Harrje (Ed.), *"Liquid Propellant Rocket Combustion Instability,"* NASA SP-194, U.S. Government Printing Office, No. 3300–0450, 1972; NASA NTRS Doc. ID 19720026079; http://hdl.handle.net/2060/19720026079

9–6. G. P. Sutton, *History of Liquid Propellant Rocket Engines*, AIAA, 2006, Chapter 4.10, "Combustion and Vibrations," and Chapter 8.4, "NPO Energomash."

9–7. V. Yang, M. Habiballah, J. Hulka and M. Popp (Eds.), *Liquid Rocket Thrust Chambers: Aspects of Modeling, Analysis and Design, Progress in Astronautics and Aeronautics (server)*, Vol. 200, American Institute of Aeronautics and Astronautics, Reston, VA, 2004; K. Kobayashi et al., "Studies on Injection-Coupled Instability for Liquid Propellant Rocket Engines," AIAA Paper 2015-3843, Orlando, FL, 2015.

9–8. B. R. Lawver, "Photographic Observations of Reactive Stream Impingement," *Journal of Spacecraft and Rockets*, Vol. 17, No. 2, Mar.—Apr. 1980, pp. 134–139.

9–9. M. Tanaka W. Daimon and I. Kimura, "Explosion Phenomena from Contact of Hypergolic Liquids," *Journal of Propulsion and Power*, Vol. 1, No. 4, 1985, pp. 314–316; doi: 10.2514/3.22801; B. R. Lawver, "Photographic Observations of Reactive Stream Impingement," *Journal of Spacecraft and Rockets*, Vol. 17, No. 2, Mar.–Apr. 1980, pp. 134–139; doi: 10.2514/3.57719.

9–10. R. I. Sujith, G. A. Waldherr, J. I. Jagoda, and B. T. Zinn, "Experimental Investigation of the Evaporation of Droplets in Axial Acoustic Fields," *Journal of Propulsion and Power*, Vol. 16, No. 2, Mar.–Apr. 2000, pp. 278–285; doi: 10.2514/2.5566.

9–11. P. Y. Liang, R. J. Jensen, and Y. M. Chang, "Numerical Analysis of the SSME Preburner Injector Atomization and Combustion Process," *Journal of Propulsion and Power*, Vol. 3, No. 6, Nov.–Dec. 1987, pp. 508–513; doi: 10.2514/3.23018.

9–12. M. Habiballah, D. Lourme, and F. Pit, "PHEDRE—Numerical Model for Combustion Stability Studies Applied to the Ariane Viking Engine," *Journal of Propulsion and Power*, Vol. 7, No. 3, May–Jun. 1991, pp. 322–329; doi: 10.2514/3.23330.

9–13. M. L. Dranovsky (author), V. Yang, F. E. C. Culick, and D. G. Talley (Eds.), *Combustion Instability in Liquid Rocket Engines, Testing and Development; Practices in Russia, Progress in Astronautics and Aeronautics*, Vol. 221, AIAA, Reston, VA, 2007.

9–14. B. W. Oppenheim and S. Rubin, "Advanced Pogo Analysis for Liquid Rockets," *Journal of Spacecraft and Rockets*, Vol. 30, No. 3, May–Jun. 1993, pp. 360–373; doi: 10.2514/3.25524.

9–15. K. W. Dotson, S. Rubin, and B. M. Sako, "Mission Specific Pogo Stability Analysis with Correlated Pump Parameters," *Journal of Propulsion and Power*, Vol. 21, No. 4. Jul.–Aug. 2005, pp. 619–626; doi: 10.2514/1.9440.

9–16. T. Shimura and K. Kamijo, "Dynamic Response of the LE-5 Rocket Engine Liquid Oxygen Pump," *Journal of Spacecraft and Rockets*, Vol. 22, No. 7, Mar.–Apr. 1985, pp. 195–200; doi: 10.2514/3.25731.

9–17. J. C. Melcher and R. L. Morehead, "Combustion Stability Characteristics of the Project Morpheus Liquid-Oxygen/Liquid-Methane Main Engine," AIAA Paper 2014-3681, July 2014.

9–18. K. Shipley and W. Anderson, "A Computational Study of Transverse Combustion Instability Mechanism," AIAA Paper 2014-3679, Cleveland, OH, Jul. 2014.

9–19. J. Pieringer, T. Settelmayer, and J. Fassel, "Simulation of Combustion Instabilities in Liquid Rocket Engines with Acoustic Perturbation Equations," *Journal of Propulsion and Power*, Vol. 25, No. 5, 2009, pp. 1020–1031.

9–20. E. Kirner, W. Oechslein, D. Thelemann, and D. Wolf, "Development Status of the Vulcain (HM 60) Thrust Chamber," AIAA Paper 90-2255, Jul. 1990.

9–21. *Guidelines for Combustion Stability Specifications and Verification Procedures for Liquid Propellant Rocket Engines,*" CPIA Publication 655, Chemical Propulsion Information Agency, John Hopkins University, Jan. 1997.

9–22. V. Yang, D. D. Ku, M. L. R. Walker, and L. T. Williams, Daniel Guggenheim School of Aerospace Engineering, Georgia Institute of Technology, "Liquid Oxygen (LOX)/Kerosene Rocket Engine with Oxidizer –Rich Preburner, Staged Combustion Cycles", for NASA Marshall Apace Flight Center/Jacobs Technology, December 10, 2013.

9–23. T. L. Acker and C. E. Mitchell, "Combustion Zone–Acoustic Cavity Interactions in Rocket Combustors," *Journal of Propulsion and Power*, Vol. 10, No 2, Mar.–Apr. 1994, pp. 235–243; doi: 10.2514/3.23734.

9–24. J. Garford, B. Barrewitz, S. Rari, and R. Frederic, "An Analytical Investigation Characterizing the Application of Single Frequency/Acoustic Modulation for High Frequency Instability Suppression," AIAA Paper 2014–3679, Cleveland, OH, July 2014.

CHAPTER 10

TURBOPUMPS AND THEIR GAS SUPPLIES

10.1. INTRODUCTION

A turbopump (TP) is a high-precision, high-speed rotating machine usually consisting of a gas turbine driving one or two centrifugal pumps. Its purpose is to take propellants from the vehicle's tanks, raise their pressure, and deliver them into suitable piping systems. These pressurized propellants are then fed to one or more thrust chambers, where they are burned forming hot gases. A TP is a unique key component of all larger liquid propellant rocket engines with pumped feed systems. This chapter discusses various common types of TPs, describes major TP components, mentions some of their principal design issues, and outlines one design approach.

A well-designed TP must deliver the intended propellant flows at the intended pump discharge pressures and mixture ratio, must have an acceptable reliability (e.g., it will not malfunction or fail during the flight duration), and will run at the highest practical energy efficiencies. Furthermore, TPs should not cause any significant vibration (in the engine or entire vehicle) nor should be adversely affected by externally caused vibrations, should function well under all operating conditions (such as different initial propellant temperatures, a range of ambient temperatures, start and stop transients, accelerations in the flight direction and other directions), and have the minimum practical inert TP mass. It is important for pumps not cavitate during operation because cavitation produced bubbles reduce the steady nominal propellant mass flow and cause combustion instabilities. Cavitation is further discussed in Section 10.5. The TP's bearings and seals need to be adequately cooled (by small secondary flows of propellant within the TP assembly) to prevent their overheating and malfunctioning. Since unexpected leaks in the seals and secondary flow passages within a TP cannot be tolerated, there can be no leakages between bipropellants inside the TP, between turbine stages (gas side) or pump stages (liquid side), between a propellant and hot gas, or between propellant or hot gas and the outside atmosphere.

The *selection of a specific TP configuration*, such as seen later in this chapter, will depend on the propellant combination, the desired flow and pump discharge pressures, the engine cycle, the available pump suction pressures, and the number of units to be built and delivered, among other factors. Available heat-resistant materials and their maximum working temperatures for the turbine blades or the maximum load capacity of their bearings can limit any design selection. The best TP placement location within a rocket engine represents a compromise between several design considerations. For example, its inlet and outlet flanges should be placed to minimize piping and propellant trapping, holes should be available to allow visual inspection of bearings and turbine blades on reusable engines, mountings should allow for thermal expansion without high stresses, and each design should have a minimum number of TP components—all these will influence the choice of the design configuration. References 10–1 to 10–5 give further information on TP selection and design.

10.2. DESCRIPTIONS OF SEVERAL TURBOPUMPS

Typical TPs can be seen in Figs. 10–1 to 10–3 and 6–9a, and many key components can be identified by their callouts. Most named parts in these figures will be discussed in subsequent paragraphs of this chapter. TPs are also seen as a component of an engine in Figs. 6–1, 6–10, 6–11, and 8–17. A number of different arrangements of key TP components for several common assemblies are shown later in Fig. 10–4. This figure includes different numbers of pumps and turbines, different component arrangements on TP shafts, with or without gear cases, and with or without booster pumps. These will be discussed later in this chapter. Figures 10–1 and 10–2 depict experimental TPs shown here because they clearly identify key components. See Ref. 10–6. Figure 10–1 shows a single-stage propellant pump; it has a screw-type axial-flow inducer impeller ahead of the main impeller and is driven by a single-stage axial-flow turbine. The hot combustion gases that drive this turbine are burned in a separate gas generator (or a preburner) at a fuel-rich mixture ratio that produces gases between 900 and 1200 K; this is sufficiently cool, so that any heated turbine hardware (blades, nozzles, manifolds, or disks) will have sufficient strength without needing forced cooling. Gases are expanded (accelerated) in an annular set of converging–diverging supersonic turbine inlet nozzles, usually cast into the turbine inlet housing. Accelerated gases then enter a set of rotating blades mounted on a wheel or turbine disk. These blades essentially remove the tangential energy of the gas flow. The exhaust gas velocity from the blades is relatively low and its direction is essentially parallel to the shaft. The pump is driven by the turbine through a shaft which is supported by two (experimental) hydrostatic bearings. The propellant enters the pump through an *inducer,* a special spiral flow impeller where the pressure of the propellant is raised only slightly (perhaps 5 to 10% of the total pressure rise). This should be just enough pressure so that there will be no cavitation as the flow enters the main pump impeller. Most of the kinetic energy given to the flow by the pump

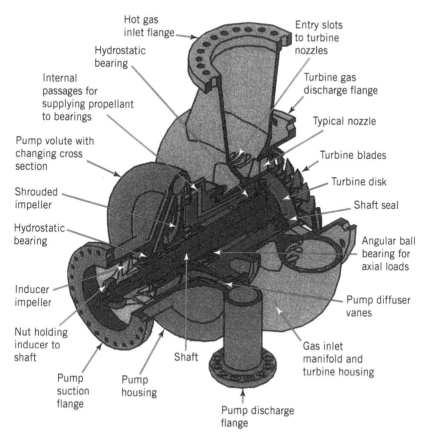

FIGURE 10–1. Cut-away view of an experimental turbopump demonstrator with a single-stage liquid oxygen pump impeller, an inducer impeller, and a single-stage turbine (one row of blades) on the same shaft. Courtesy of Aerojet Rocketdyne.

impeller is converted into hydrostatic pressure in the diffusers (the diffuser vanes are not clearly visible, since they are inclined) and/or volutes of the pump. Two hydrostatic bearings support the shaft radially. All bearings and shaft seals generate heat as they run so they are cooled and lubricated with small propellant flows, which are supplied from the pump discharge through drilled passages. One bearing (near the pump) is cold and the other is hot, since it is close to the hot turbine. The angular ball bearing accepts axial net loads from the unbalanced hydrodynamic pressures around the shrouded impeller, the inducer, and also from the turbine blades or the turbine disk.

The (experimental) high-speed, compact, and light weight liquid hydrogen TP shown in Fig. 10–2 and discussed in Refs. 10–6 and 10–7 was intended to be used with an upper-stage hydrogen–oxygen rocket engine that would deliver about 50,000 lbf (22.4 kN) thrust. The unique single-piece titanium rotor (turning nominally at 166,700 rpm) has two machined sets of pump vanes, a machined inducer impeller, a set of machined radial inflow turbine blades, and radial as well as

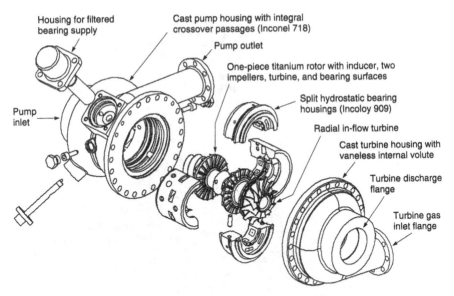

Housing for filtered bearing supply

Cast pump housing with integral crossover passages (Inconel 718)

Pump outlet

One-piece titanium rotor with inducer, two impellers, turbine, and bearing surfaces

Split hydrostatic bearing housings (Incoloy 909)

Radial in-flow turbine

Cast turbine housing with vaneless internal volute

Turbine discharge flange

Turbine gas inlet flange

Pump inlet

FIGURE 10–2. Exploded view of an advanced high-speed, two-stage experimental liquid hydrogen fuel pump driven by a radial flow turbine. Used with permission of Aerojet Rocketdyne. Adapted from Ref. 10–7.

axial hydrostatic bearing surfaces. A small filtered flow of hydrogen lubricates the hydrostatic bearing surfaces. The cast pump housing has internal crossover passages between stages. Its unique radial inflow turbine (3.2 in. dia.) produces about 5900 hp at an efficiency of 78%. The hydrogen pump impellers are only 3.0 in. diameter and produce a pump discharge pressure of about 4500 psi at a fuel flow of 16 lbm/sec and efficiencies of 67%. A pump inlet pressure of about 100 psi is needed to assure cavitation-free operation. The TP can operate down to about 50% flow (at 36% discharge pressure and 58% of rated speed). In this design the number of pieces to be assembled has been greatly reduced, compared to more conventional TPs, thus enhancing its inherent reliability.

The geared TP in Fig. 10–3 has high turbine and pump efficiencies, the rotary speed of the two-stage turbine is higher than the pump shaft speeds, and the turbine is smaller than a comparable single-shaft TP. An auxiliary power package (e.g., a hydraulic pump) was used in early applications. The precision ball bearings and seals on the turbine shaft can be seen, but the pump bearings and seals are not visible in this figure.

Table 10–1 gives selected data on relatively large TPs for two large liquid-oxygen /liquid-hydrogen rocket engines. It shows that the main LOX pumps have single-stage impellers, while the main liquid hydrogen pumps are multistage with two or three impellers in series. The Space Shuttle Main Engine (SSME) had axial-flow inducer impellers on its main pumps and also on its two booster pumps, which added together to raise the pressure of the flows going to the inlets of the respective main

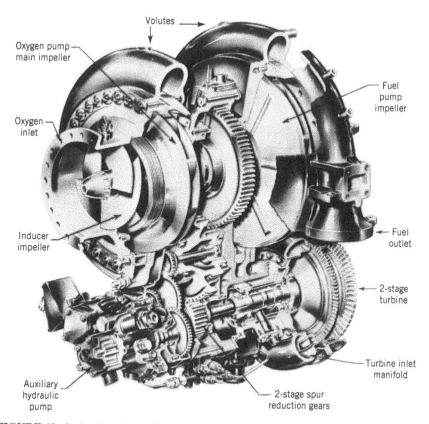

Volutes →

Oxygen pump main impeller

Fuel pump impeller

Oxygen inlet

Oxygen inlet

Inducer impeller

Fuel outlet

2-stage turbine

Turbine inlet manifold

Auxiliary hydraulic pump

2-stage spur reduction gears

FIGURE 10–3. Section view of early, typical geared turbopump assembly similar to the one used on the RS-27 engine (Delta I and II launch vehicles) with liquid oxygen and RP-1 propellants. Courtesy of Aerojet Rocketdyne.

centrifugal pumps. Booster pump data are also listed. The Japanese LE-7 engine feed system does not use booster pumps and it features pump inducers ahead of the main impellers. The turbine blade shapes for both TPs are a combination of impulse turbines with some reaction turbines contours. These terms are explained later.

One relatively small company in New Zealand has recently developed a clever way to drive propellant pumps with a direct current (DC) electric motor. This scheme simplifies pump drives and avoids the use of hot gas turbine gas generators or preburners; since it includes batteries it is probably limited to low pump power levels. Their engine uses liquid RP-1 (kerosene) and liquid oxygen propellants and the engine is self-pressurizing (no tank pressurization system). As of 2015, engines using electric pump drives have not been completely developed. Nine of these engines of 600 lbf thrust each and with a vacuum specific impulse of 327 sec are intended for the lower stage of a two-stage satellite lander (from *The New Zealand Herald,* October 9, 2015).

TABLE 10–1. Turbopump Characteristics

Engine:	Space Shuttle Main Engine[a]					LE-7[b]		
Feed System Cycle:	Modified Staged Combustion Cycle					Modified Staged Combustion Cycle		
Propellants:	Liquid Oxygen and Liquid Hydrogen					Liquid Oxygen and Liquid Hydrogen		

Pumps

Designation[c]	LPOTP[c]	LPFTP	HPOTP[c]		HPFTP[c]	HPFTP	HPOTP	
Type	Axial flow	Axial flow	Dual inlet		Radial flow	Radial flow	Radial flow	
No. of impeller stages	1	1	1 + 1[d]		3	2	1 + 1[d]	
No. of aux. or inducers	—	1	—		1	1	1	
Flow rate (kg/sec)	425	70.4	509	50.9	70.4	35.7	211.5	46.7[d]
Inlet pressure (MPa)	0.6	0.9	2.70	NA	1.63	0.343	0.736	18.2[d]
Discharge pressure (MPa)	2.89	2.09	27.8[d]	47.8[d]	41.0	26.5	18.2	26.7[d]
Pump efficiency (%)	68	75	72	75[d]	75	69.9	76.5	78.4[d]

Turbines

	LPOTP	LPFTP	HPOTP	HPFTP	HPFTP	HPOTP
No. of stages	1	2	3	2	2	
Type	Hydraulic LOX driven	Reaction-impulse	Reaction-impulse	Reaction-impulse	Reaction-impulse	Reaction-impulse
Flow rate (kg/sec)			27.7	66.8	33.1	15.4
Inlet temperature (K)	105.5	264	756	1000	871	863
Inlet pressure (MPa)	26.2	29.0	32.9	32.9	20.5	19.6
Pressure ratio	NA	1.29	1.54	1.50	1.43	1.37
Turbine efficiency (%)	69	60	74	79	73.2	48.1
Turbine speed (rpm)	5020	15,670	22,300	34,270	41,600	18,300
Turbine power (kW)	1120	2290	15,650	40,300	25,350	7012
Mixture ratio, O/F	LOX only	H₂ only	~0.62	~0.88	~0.7	~0.7

[a] Data courtesy of Aerojet Rocketdyne, at flight power level of 104.5% of design thrust.
[b] Data courtesy of Mitsubishi Heavy Industries, Ltd.
[c] LPOTP, low-pressure oxidizer turbopump; HPFTP, high-pressure fuel turbopump.
[d] Boost impeller stage for oxygen.

10.3. SELECTION OF TURBOPUMP CONFIGURATION

The selection of features, performance, and configuration of any specific new TP arrangement can be complicated. First, it depends on engine requirements identified for the particular flight application (thrust, propellant combination, mixture ratio, chamber pressure, duration, low cost, schedule, etc.). These identified engine criteria have then to be evaluated, analyzed, and prioritized. Furthermore, selection depends on the engine cycle (see Section 6.6 and Fig. 6–8), propellant physical properties (such as vapor pressure, density, viscosity), nominal flows and pump discharge pressures (sum of chamber pressure, hydraulic losses in valves, cooling jacket, injector, and pipes), minimum practical inert TP mass, component reliability (with no flight failures allowed), available suction pressure, maximum and minimum propellant and hardware initial temperatures, and on the component arrangements of the assembly. It is also influenced by the size of the TP, the location of its inlet and outlet flanges relative to the thrust chamber inlets and the suction pipes outlets, maximum allowable turbine gas inlet temperature (which depends on the turbine material), number of hot starts during its lifetime, critical speed of the rotating assembly, and the simplicity of the design (e.g., fewest number of parts). If reuse is an engine requirement, then additional factors have to be considered, such as ease of access to bearings, seals, or turbine blades for inspection of wear or cracks. With engine throttling, the TP has to operate efficiently over a range of shaft speeds. There can be other TP selection criteria; some preliminary selection and design criteria are explained in Refs. 10–1 to 10–3 and 10–5 and basic texts on pumps are found in Refs. 10–8 and 10–9.

Shaft speed is related to *impeller diameter or TP size* and thus to *TP mass*. The tip speed of pump impeller blades or the speed of turbine blades at their mean diameter can actually be the same at different diameters and rotational speeds. It is well known that the highest practical shaft speed gives the smallest diameter, and thus usually the lowest inert TP mass. This is often in conflict with other design criteria, such as cavitation avoidance, which is likely to be more difficult as impeller speed increases. These topics are further discussed later in this section.

The *arrangement of the various key TP components* into an assembly can be important design criterion. Figure 10–4 shows several common arrangements. The two most common have one turbine with two propellant pumps on the same shaft (Figs. 10–4a and 10–4b), and another arrangement with two smaller separate TPs, one turbine drives an oxidizer pump and the other turbine drives a fuel pump (Fig. 10–4e). In Fig. 10–4a the fuel pump inlet has a shaft going through it and that will affect the suction pressure. In Fig. 10–4b the turbine requires additional seals. Although not discussed here, the placement and selection of bearings and seals also influences TP selection.

Centrifugal pumps are used in all known production TPs, and these are basically constant-volume flow devices (Refs. 10–3, 10–4, 10–5, 10–8 and 10–9). If the two propellants have similar densities (say within about 40%), such as nitrogen tetroxide (NTO) and unsymmetrical dimethylhydrazine (UDMH) or LOX and kerosene, and the volume flow of oxidizer and of fuel are similar then the same type of impeller (running at the same speed on a single shaft) can be used for both of them.

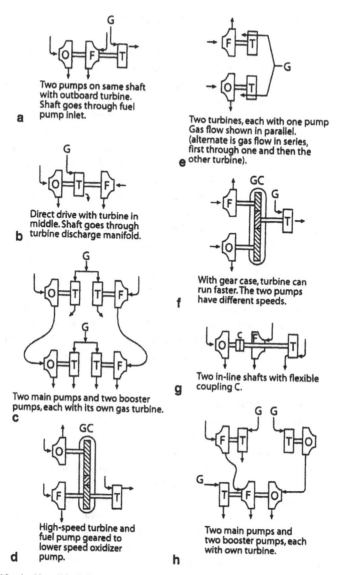

a Two pumps on same shaft with outboard turbine. Shaft goes through fuel pump inlet.

e Two turbines, each with one pump Gas flow shown in parallel. (alternate is gas flow in series, first through one and then the other turbine).

b Direct drive with turbine in middle. Shaft goes through turbine discharge manifold.

f With gear case, turbine can run faster. The two pumps have different speeds.

c Two main pumps and two booster pumps, each with its own gas turbine.

g Two in-line shafts with flexible coupling C.

d High-speed turbine and fuel pump geared to lower speed oxidizer pump.

h Two main pumps and two booster pumps, each with own turbine.

FIGURE 10–4. Simplified diagrams of several different component arrangements of turbopumps. F, fuel pump; O, oxidizer pump; T, turbine; G, hot gas; C, shaft coupling; and GC, gear case. Used with permission from Ref. 10–1.

When the densities are quite different (e.g., LOX and liquid hydrogen), a single shaft would not be practical or efficient. Instead two separate TPs are used—rotating at the higher speed for the fuel pump and the lower speed for the oxidizer pump—thus giving good pump performance. This corresponds to Fig. 10–4e.

It is at times difficult to package a relatively large TP inside an engine. A two-TP configuration is often easier to integrate into an engine assembly; here both pumps will be able to run at a relatively high efficiency, each TP will have fewer seals, and since the shafts are shorter, they are lighter for the same shaft stiffness.

Figure 10–4g shows *two shafts, in line with each other, with a mechanical coupling* between them (Ref. 10–10). In this example the fuel pump and the turbine are on one shaft and the oxidizer pump on the other. The two shafts are connected by a flexible coupling, and this allows considerable shaft misalignment and, as a result, more generous fabrication and assembly tolerances for the TP. The first two-shaft inline TP assembly was flown in the German V-2 rocket engine (1938) with a six pin/sleeve type of flexible coupling. The Russians developed a curvic serrated sleeve as a coupling, which was smaller and lighter. References 10–1 and 10–4 show a section view of these two TPs. The United States has not produced TPs with two inline coupled shafts. When compared to a larger single shaft, smaller spans between bearings on each of the two shafts result in substantially smaller shaft diameters, which in turn can reduce the mass in the shafts, the diameter and mass of pumps and pump housings, along with the inertia of the rotary assembly, and can allow some increases of the shaft speed. Key TP design objectives include the selection of high enough shaft speeds (which result in the smallest and lightest TP but not susceptible to excessive cavitation), low enough pump suction pressures (which give the lightest vehicle tank mass while not promoting pump cavitation), and an impeller configuration that will be strong and efficient and prevent dangerous cavitation levels under all operating conditions.

Figures 10–4d and 10–4f show TPs with a *gear case* to transmit the power from the turbine to the pumps (Ref. 10–11). The purpose here is to allow the turbine to run at higher speeds than the pumps, attaining higher efficiencies and thus minimizing the amount of gas generator propellant needed to drive the TP. Gear cases for TPs were common in the United States beginning in the early 1950s and one is still in use. In Fig. 10–4e, the turbine runs at a higher speed and the two pumps each run at lower but somewhat different speeds. In Fig. 10–4f, the fuel is liquid hydrogen; the fuel pump and the turbine run at the higher speed on the same shaft and the oxygen pump is geared down to the lower speed.

Figures 10–4c and 10–4h represent two arrangements for using *booster TPs* in addition to the two main TPs. See Ref. 10–12. Their purpose is to further reduce the vehicle's tank pressure and inert mass, thus improving vehicle performance. A typical booster TP provides about 10% of the propellant pressure rise and the respective main pump the remaining 90%. Booster pumps typically use axial-flow impellers, and a few can operate with considerable cavitation at the leading edges of the impellers. With a booster pump, the pressure at the inlet to the main pump is raised enough so cavitation does not occur. Low rotating speeds causes booster pumps to have a relatively larger diameter; in some liquid propellant rocket engines, they can be larger than the main pump. Since pressures in booster pumps are low, the walls are thin, and the booster pump inert mass can be low. Most turbines in booster TPs are driven by gas expansion, but some are powered by high-pressure liquid propellant taken from the discharge of the main pump. The total worldwide number of rocket engines with booster pumps is small, perhaps a dozen. Booster pumps provide low suction pressure and good cavitation resistance to the main pumps of the engines, but these engines are more complex and more costly.

A representative booster TP used in a staged combustion cycle liquid propellant rocket engine is shown in Fig. 10–5. It supplies the main oxygen pump of the RD-180 Russian engine with a high enough inlet pressure to prevent cavitation in the main oxygen pump impeller. The RD-180 has a nominal thrust of 933 klbf (vacuum) or 861 klbf (sea level), gives a specific impulse of 337.8 sec (vacuum) or 311.3 sec (sea level) at a relatively high chamber pressure of 3,868 psi and a nominal mixture ratio of 2.72. It is used in the first stage of the Atlas V Space launch vehicle.

Bearings and seals will overload and may fail when shafts deflect at or near a bearing (even by amounts as small as 0.001 in. in some TPs). See Refs. 10–13 and 10–14. In order to minimize *shaft deflection*, the shaft has to be stiff enough (less deflection for a given side load) and that translates to large diameter and mass. Rotating assemblies (turbine, pumps, seals, bearings) have to be carefully balanced because

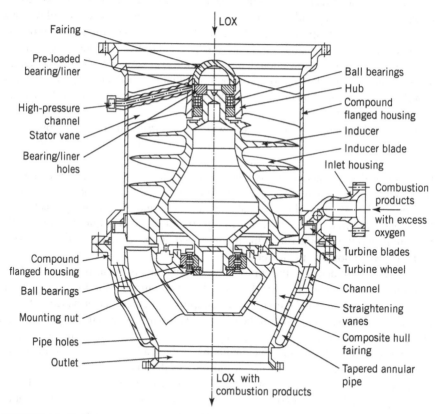

FIGURE 10–5. Cross section of the LOX booster turbopump of a Russian RD-180 engine with an axial flow spiral-type pump impeller and a single row of short turbine blades. An oxidizer rich gas mixture (at 558 K) from the engine's preburner drives the turbine and its exhaust is mixed with the pressurized LOX flow within the turbopump assembly. This turbopump has a LOX flow of 917 kg/sec, delivers the mixed flow at a pressure of 12 bar and mixture ratio of about 52. From NPO Energomash, Khimki, Russia.

unbalances cause side loads (which can be substantial at high rotational speed) that produce shaft deflections. Accurate balancing, both static and dynamic, is usually required during assembly to avoid bearing overloading. The operating speeds of rotating assemblies are never chosen to be the same as their *resonant or critical shaft speeds*, because here deflections become very high and bearings, seals, and even the shaft can fail. Usually, the *synchronous or critical speed* is well above the operating speed; this is discussed in Section 10–7. In some TPs, the operating speed is actually above the critical speed; here, the shaft must go momentarily through the critical speed during startup and shutdown.

Frequently, there are several *secondary flow passages* built into TPs. Bearings, rotating shaft seals, or impeller wear rings are all usually cooled and/or lubricated with small propellant flows, which become heated in the process and can then be fed back to the suction side of a pump or dumped overboard. To route these small cooling and lubricating flows requires a set of small external tubes or internal passages in the TP assembly. Since there are many seals and bearings, the distribution system for the cooling propellant can be intricate; some routes have to be throttled to the desired small flows and mixing of unlike propellants has to be avoided. The RS-68 flow diagram in Fig, 6–9b shows tubes that dump overboard the flow of coolant/lubricant used. When bearings or seals are close to a turbine, they will warm further (needing more coolant) and when cryogenic propellants are being pumped the enclosing hardware will become very cold. Clearances for these bearings or seals have to be so designed that the intended gaps are correct at their operating conditions.

Bubbles within propellant flows are indeed unwanted. They may originate from extensive flow *cavitation* at the pumps (see Refs. 10–4 and 10–9), leaks in low-pressure suction piping or from improper priming or filling of propellant lines before or during start. Bubbles reduce the effective density of the propellant, which in turn reduces the mass flow and the mixture ratio and may reduce the thrust level somewhat below a nominal value. If the intended operating mixture ratio was to be fuel rich and bubbles were to reduce the fuel flow, the new mixture ratio will approach stochiometric; this creates hotter gases and could cause rapid failure of cooling jackets. Another major concern with bubbles is that they can readily trigger combustion instabilities (see Chapter 9), which often lead to thrust chamber failure.

A *balance of axial pressure forces* within TP rotating assemblies and their housings is necessary to prevent excessive axial hydraulic pressure forces (on surfaces normal to the shaft) from overloading certain parts of the assembly (Refs. 10–1 and 10–15). Unbalances cause small axial movements of the rotating assembly (typically just a few thousands of an inch) often leading to intense rubbing between stationary and rotating parts. Such pressure forces are created by fluids moving through pumps (increasing their pressure) and by turbine gases flowing through moving and stationary sets of turbine blades (decreasing their pressure) as well as by the radial placement of wear rings and/or seals. Some TPs have built-in balancing pistons and some use the fluid pressures on the back side of an impeller or turbine disk to make an effective balancing piston. Pressures on these pistons are controlled by slight axial movements that change clearances at the edge of the piston. During transient operations (start, stop, change of thrust), there are some periods of unbalance caused by changes in pressure distributions of the internal flows, but they are typically of short duration. Nevertheless, there can be noticeable rubbing or axial contact between a rotating and a stationery surface normal to the shaft during the brief periods of unbalance. If such rubbing contact persists, there will often be some cumulative damage to the TP.

10.4. FLOW, SHAFT SPEEDS, POWER, AND PRESSURE BALANCES

The design of a TP requires careful balancing of propellant flows, shaft speeds, and power among the key pumps, turbines and pressure distributions of the flowing propellants along their paths. The relationship between shaft speeds, torques, and power may readily be seen by reviewing a TP flow diagram such as Fig. 1–4. For TPs with two pumps and one gear transmission, the relation between shaft speeds may be written as

$$N_t = a_o N_o = a_f N_f \qquad (10\text{--}1)$$

Here, N is the shaft speed and the subscripts t, f, and o stand for turbine, fuel pump, and oxidizer pump, respectively. The a_o and a_f are the gear ratios for the oxidizer and fuel pumps (when there are no gears, a_o and a_f become 1.0). For TPs similar to Fig. 10–4a and 10–4b:

$$N_t = N_f = N_o \qquad \text{and} \qquad L_t = L_o + L_f + L_b \qquad (10\text{--}2)$$

Here, L_b is the torque to overcome friction in bearings, seals, and/or auxiliaries. For all TPs, as in Figs. 10–4a and 10–4b, the power of the turbine P_t has to equal the sum of the powers of the pumps plus any losses. This power balance can also be expressed as the product of the torque L and the speed N:

$$P_t = \sum P_p + P_b \qquad (10\text{--}3)$$

$$P_t = L_t N_t = L_f N_f + L_o N_o + P_b \qquad (10\text{--}4)$$

Here, P_b is the power for overcoming the friction in all the bearings and seals, and in some TP configurations also the power loss in gears, and auxiliaries (such as an oil pump for lubricating gears). The power P_b used by bearing, seals, and auxiliaries is usually small.

For the separated two-TP scheme in Fig. 10–4e, each TP would produce its own internal losses. In this configuration, the speeds N and the power P and even the torques L must individually balance:

$$(N_t)_o = (N_p)_o \qquad \text{and} \qquad (N_t)_f = (N_p)_f \qquad (10\text{--}5)$$

$$(P_t)_o = (P_p)_o + (P_b)_o \qquad \text{and} \qquad (P_t)_f = (P_p)_f + (P_b)_f \qquad (10\text{--}6)$$

The subscripts t and p stand for turbine and pump. The pressure balance equation for the fuel line (the oxidizer line equation would be similar) at a point just past the fuel pump discharge flange may be written as

$$(p_f)_d = (p_f)_s + (\Delta p_f)_{\text{pump}} \qquad (10-7)$$

$$= (\Delta p)_{\text{main fuel system}} + p_1$$

$$= (\Delta p)_{\text{gas generator fuel system}} + p_{\text{gg}}$$

Here, the fuel pump discharge pressure $(p_f)_d$ equals the fuel pump suction pressure $(\Delta p_p)_s$ plus the pressure rise across the pump $(\Delta p)_{\text{pump}}$. This discharge pressure in turn has to equal the chamber pressure p_1 modified by all the pressure drops in the high-pressure main fuel flow system downstream of the pump and upstream of the injector face. This usually includes the pressure losses in the cooling jacket, injector, piping, and in the open fuel valve. This pump discharge pressure furthermore has to equal the gas generator combustion pressure p_{gg} modified by all the pressure losses between the fuel pump discharge and the gas generator combustion chamber. A similar pressure balance is needed for the oxidizer systems.

A fourth type of balance equations differs from the other three (i.e., flow, pressure, and power balance); it is aimed at preventing certain high-pressure *axial hydraulic loads* and/or *axial gas pressure* loads from causing damage inside TP assemblies. See Refs. 10–1, 10–4, and 10–15. These axial internal loads (side forces on impellers or turbine disks) can be relatively high, particularly in large sized TPs. Though certain ball bearings are capable of withstanding considerable axial loads, in addition to their usual radial loads, axial movement of rotating assemblies is limited by design to be insignificant. The method of controlling axial forces with ball bearings is used in many TPs, but the maximum axial load of ball bearings is usually limited; therefore, ball bearings are mostly used with small-diameter TPs or in low-pressure TPs. As mentioned earlier, the balancing of axial fluid pressure forces within a large TP rotating assembly and its large stationary housing assembly can be accomplished by balance pistons.

A phenomenon characteristic of rotating machinery is the relative displacement and very small movement of the center of the rotating shaft within the confinement of small clearances of a fixed sleeve bearing; this is known as *whirl*. Reference 10–16 describes a condition where relative small movements of the shaft's center are synchronous (the same revolutions per unit time) as the rotation movement of the shaft itself.

The above equations relate to steady-state operations. *Transients and other dynamic conditions* are also important and many have been analyzed using computers. Many conditions have been tested, as reported in Ref. 10–17; they include start, stop, and thrust changes. For example, starting conditions include tank pressurization, the sequencing and opening controls of appropriate valves, filling of pipes, pumps, or manifolds with liquid propellants, filling of turbines and their manifolds with heated gas, ignition of thrust chambers and gas generators, and/or thrust buildup.

However, no detailed discussion of these transient conditions will be given here because dynamic conditions can be quite complicated, are related to combustion behavior, and are thus difficult to analyze. Each major rocket propulsion organization has developed some method for analyzing these transients, often specific to certain engines. Similar yet simpler transient analyses of flow and pressure drops must be performed for engines with pressurized gas feed systems.

10.5. PUMPS

Classification and Description

Centrifugal pumps are generally considered to be the most suitable for propellant pumping in medium and large-sized rocket engines. They are efficient as well as economical in terms of mass and of space requirements for the large flows and high pressures involved.

Figure 10–6 is a schematic drawing of a centrifugal pump. Fluid entering the *impeller*, essentially a wheel with spiral curved vanes rotating within a *casing*, is accelerated within the impeller channels and leaves its periphery with a higher velocity, enters the *volute* or collector, and thereafter enters the *diffuser* (not shown) where conversion from kinetic energy (velocity) to potential energy (pressure) takes place. In some pumps curved diffuser vanes appear upstream of the collector or volute (see Fig. 10–1). Three-dimensional hydraulic designs of impeller vanes, diffuser vanes, and volute passages are accomplished with computers programmed to account for efficiency and strength. Internal leakages, or circulation between the high-pressure (discharge) side and the low-pressure (suction) side of an impeller, are held to a minimum by maintaining close clearances between the rotating and stationary parts

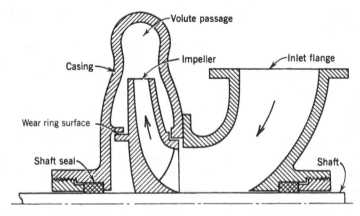

FIGURE 10–6. Simplified schematic half cross section of a typical centrifugal pump.

at the *seals* or *wear ring surfaces*. External leakages along a shaft are minimized or prevented by the use of *shaft seals*. Single-stage pumps (i.e., one impeller only) are stress limited in the pressure rise they can impart to the liquid, and multiple-stage pumps are therefore needed for high *pump head*,[*] such as with liquid hydrogen. References 10–2, 10–5, 10–8, 10–9, and 10–17 give information on different pumps. There is a free flow passage through these pumps at all times and no positive means for shutoff are provided. Pump characteristics, that is, the pressure rise, flow, and efficiency, are functions of pump speed and of the impeller, vane shape, and casing configuration. Figure 10–7 shows a typical set of curves for centrifugal pumps. The negative slope on the head versus flow curves indicate stable pump behavior. Reference 10–7 describes the development of a small TP and the testing of a spiral high-speed first-stage impeller, called an inducer.

Shrouded impellers have a shroud or cover (in the shape of a curved surface of revolution) on top of the vanes as depicted in Figs. 10–1, 10–3, and 10–6. This type usually has higher stresses and lower leakage around the impeller. In an unshrouded impeller or turbine the vanes are not covered as seen in the turbine vanes in Fig. 10–2.

Pump Parameters

This section outlines some important parameters and features that need to be considered in the design of rocket propellant centrifugal pumps under steady-flow conditions.

The *required pump flows* are established for the rocket engine design given the thrust, effective exhaust velocity, propellant densities, and the mixture ratio. In addition to flow requirements in the thrust chamber, propellant consumption in gas generators and, in some designs, also bypass flows around the turbine and its auxiliaries have to be considered in determining pump flows. Required *pump discharge pressures* are determined from the chamber pressure and the hydraulic losses in valves, lines, cooling jacket, and injectors (see Eq. 10–7). To obtain rated flows at rated pressures, an additional adjustable pressure drop for a control valve or orifice is usually included; this permits calibration adjustments or changes in required feed pressures. Regulation of pump speeds can also change the required adjustable pressure drops. As described in Section 11.5, such adjustment of head and flow is necessary to allow for hydraulic and performance tolerances on pumps, valves, injectors, propellant density, and the like.

It is possible to predict *pump performance at various speeds* when such performance is known at any given speed. Because the fluid velocity in a given pump is proportional to the pump speed N, the flow quantity or discharge Q is also proportional to that speed while the head H is proportional to the square of the speed.

---

[*] *Pump head* means the difference between pump discharge and pump suction head. Its units are meters or feet. Head represents the height of a column of liquid with equivalent pressure at its bottom. The conversion from pounds per square inch into feet of head is: (X) psi $= 144$ (X)/density (lb/ft^3). To convert Pascals (N/m^2) of pressure into column height (m), divide by the density (kg/m^3) and g_0 (9.806 m/sec^2).

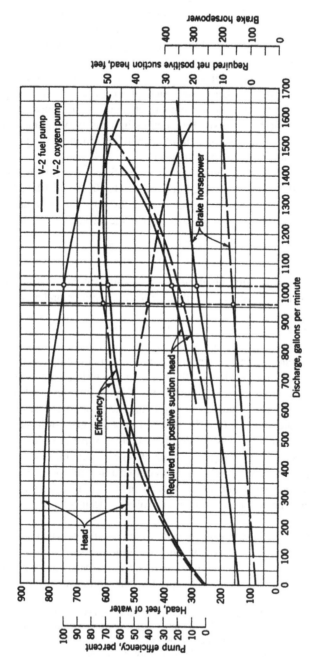

FIGURE 10-7. Water test performance curves of the centrifugal pumps of the German V-2 rocket engine. The propellants are 75% ethyl alcohol and liquid oxygen.

This gives the following relations:

$$Q(\text{flow}) \sim N(\text{rpm or rad/sec})$$

$$H(\text{pump head}) \sim N^2 \qquad (10\text{--}8)$$

$$P(\text{pump power}) \sim N^3$$

From these relations it is possible to derive a parameter called the *specific speed* N_s. It is a dimensionless number derived from analysis of pump parameters as shown in Ref. 10–8.

$$N_s = N\sqrt{Q_e}/(g_0\,\Delta H_e)^{3/4} \qquad (10\text{--}9)$$

Any set of consistent units will satisfy the equation: for example, N in radians per second, Q in m^3/sec, $g_0 = 9.806$ m/sec^2, and H in meters. The subscript e refers to a *maximum efficiency condition*. In U.S. pump practice, it has become customary to delete g_0, express N in rpm, and Q in gallons per minute or ft^3/sec. Much of the existing U.S. pump data is given in these units. This results a different value for N and leads to a modified form of Eq. 10–9, where N_s is no longer dimensionless, namely,

$$N_s = 21.2N\sqrt{Q_e}/(\Delta H_e)^{3/4} \qquad (10\text{--}10)$$

The factor 21.2 applies when N is in rpm, Q is in gallons/min, and H is in ft. For each range of specific speed, only certain shapes and impeller geometries proved most efficient, as shown in Table 10–2. Because of its low density, hydrogen can be pumped effectively by axial-flow devices.

The *impeller tip speed* in centrifugal pumps is limited by design and material strength considerations and ranges between about 200 to 450 m/sec or roughly 655 to 1475 ft/sec. Using titanium (lower in density than steel) and with machined

TABLE 10–2. Pump Types—Simple sketches and range of N_s and efficiency

	Impeller type				
	Radial	Francis	Mixed Flow	Near Axial	Axial
Basic shape (half section)					
Specific speed N_s					
U.S. nomenclature	500–1000	1000–2000	2000–3000	3000–6000	Above 8000
SI consistent units	0.2–0.3	0.4	0.6–0.8	1.0–2.0	Above 2.5
Efficiency %	50–80	60–90	70–92	76–88	75–82

Adapted from Ref. 10–8.

unshrouded impellers tips speeds of over 2150 ft/sec are now possible and have been used on the pump shown in Fig. 10–2. For cast impellers this limiting value is lower than for machined impellers. This maximum impeller tip speed determines the maximum head that can be obtained from a single stage. The impeller vane tip speed u is the product of the shaft speed, expressed in radians per second, and the impeller radius and is related to the pump head ΔH by

$$u = \psi \sqrt{2g_0 \, \Delta H} \qquad (10\text{–}11)$$

where ψ, a velocity correction factor, has values between 0.90 and 1.10 for different designs, though for many pumps, $\psi = 1.0$.

The volumetric flow rate Q defines the impeller inlet and outlet areas according to the equation of continuity. Diameters obtained from this equation should be in the proportion indicated by the diagrams for a given specific speed in Table 10–2. The continuity equation for an incompressible liquid is

$$Q = A_1 v_1 = A_2 v_2 \qquad (10\text{–}12)$$

where the subscripts 1 and 2 may refer to impeller inlet and outlet sections, all areas A being measured normal to their respective flow velocity v. The inlet velocity v_1 ranges usually between 2 and 6 m/sec or 6.5 to 20 ft/sec and the outlet velocity v_2 between 3 and 15 m/sec or 10 to 47 ft/sec. For a "somewhat compressible liquid," such as liquid hydrogen, the density ρ will change with pressure. Here, the continuity equation is given in terms of mass flow rate $\dot{m}$

$$\dot{m} = A_1 v_1 \rho_1 = A_2 v_2 \rho_2 \qquad (10\text{–}13)$$

The head developed by the pump will then also depend on the change in density. The *liquid power output* of the pump may be found from the volumetric flow rate Q (m³/sec) multiplied by the pressure rise of the pump (discharge pressure minus suction pressure, in N/m²).

Pump performance is limited by *cavitation*, a phenomenon that occurs when the static pressure at any point in a fluid flow passage becomes less than the fluid's local vapor pressure. Cavitation is discussed in Refs. 10–1 to 10–5, 10–8, 10–9, and 10–17. When the local vapor pressure of the liquid exceeds the local absolute pressure, then vapor bubbles will form at that location and such vapor bubbles formation in the low-pressure regions cause cavitation. These bubbles collapse when they reach regions of higher pressure, that is, when the static pressure in the fluid is above the vapor pressure. In centrifugal pumps cavitation is most likely to occur just behind the leading edge of the pump impeller vane at the impeller inlet because here is where the lowest absolute pressure is encountered. Any excessive formations of vapor cause the pump discharge mass flow to diminish and fluctuate and reduces thrust and makes combustion erratic.

Bubbles that travel along the pump impeller surface from the low-pressure region (at the leading edge of the vane where they are formed) downstream to the higher-pressure region collapse. These sudden collapses create local high-pressure pulses that may cause excessive stresses and erosion at the impeller surface. In most rocket applications, this cavitation erosion is not as serious as in common water or chemical pumps usage because their cumulative duration is relatively short and the erosion of the impeller is not usually extensive. However, it has been a concern with some test facility transfer pumps. It is also believed that excessive bubbles may be a cause for combustion vibrations (see Chapter 9).

The *required suction head* $(H_s)_R$ is the limit value of the head at the pump inlet (above the local vapor pressure); below this value cavitation in the impeller may occur. It is a function of the pump and impeller design and its value increases with flow as can be seen in Fig. 10–7. To avoid cavitation, the *suction head* above the vapor pressure *required* by the pump $(H_s)_R$ must always be *less* than the *available* or *net positive suction head* furnished by the suction pipe line up to the pump inlet $(H_s)_A$, that is, $(H_s)_R \leq (H_s)_A$. The required suction head above vapor pressure can be determined from the suction-specific speed S and the volume flow rate Q_e at maximum efficiency:

$$S = 21.2N\sqrt{Q_e}/(H_s)_R^{3/4} \qquad (10\text{--}14)$$

The suction-specific speed S depends on the quality of design and the specific speed N_s, as shown in Table 10–2. The suction-specific speed S has values between 5000 and 60,000 when using ft-lbf units. For pumps with poor suction characteristics, it has values nearer 5000; for the best pump designs without cavitation, it has values near 10,000 and 25,000, and for pumps with limited and controllable local cavitation, it has values above 40,000. In Eq. 10–14 the required suction head $(H_s)_R$ is usually defined as the critical suction head at which the developed pump discharge head has been diminished arbitrarily by 2% in pump tests with increasing throttling in the suction side. Turbopump development has, over the last several decades, led to impeller designs that can operate successfully with considerably more cavitation than this commonly accepted 2% head loss limit. Inducers are now designed to run stably with extensive vapor bubbles near the leading edge of their vanes, but these bubbles collapse at the vane's trailing end. Inducers can now operate at S-values above 80,000. A discussion of one design method for impeller blades can be found in Ref. 10–8.

The head that is available at the pump suction flange, or *net positive suction head* or *available suction head above vapor pressure*, $(H_s)_A$ is an absolute head value determined from the tank pressure (the absolute gas pressure in the tank above the liquid level), the elevation of the propellant level above the pump inlet, diminished by the friction losses in the line between tank and pump and by vapor pressure of the fluid. When a flying vehicle is undergoing accelerations, the head due to elevation must be corrected accordingly. These various heads are defined in Fig. 10–8. The term $(H_s)_A$ is often abbreviated as NPSH and is the maximum head available for suppressing cavitation at the inlet to the pumps:

$$(H_s)_A = H_{\text{tank}} + H_{\text{elevation}} - H_{\text{friction}} - H_{\text{vapor}} \qquad (10\text{--}15)$$

If additional head is required by the pump, the propellant may have to be pressurized by external means, such as by the addition of another pump in series (called a booster

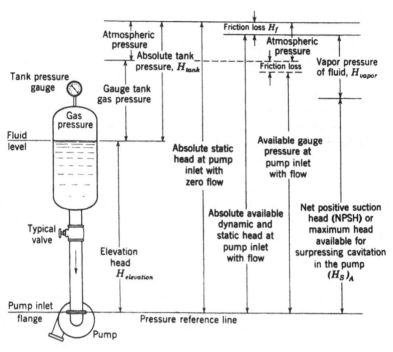

FIGURE 10–8. Definition of pump suction head.

pump) or by gas pressurization of the propellant tanks. This latter method requires thicker tank walls and, therefore, heavier tanks and a bigger gas-pressurizing system. For example, the oxygen tank of the German V-2 was pressurized to 2.3 atm, partly to avoid pump cavitation. For a given value of $(H_s)_A$, propellants with high vapor pressure require correspondingly higher tank pressures and heavier inert tank masses; here, pumps with low required suction pressures usually permit designs with high shaft speeds, small diameters, and low pump inert mass. A small value of $(H_s)_R$ is desirable because it may permit a reduction of the requirements for tank pressurization and, therefore, a lower inert tank mass. The value of $(H_s)_R$ will be small when the impeller and fluid passages are well designed and when shaft speeds N are low. Very low shaft speeds, however, require large-diameter pumps, which become excessively heavy. The trend in selecting centrifugal pumps for rocket applications has been to select the highest shaft speed that gives a pump with a low enough value of $(H_s)_R$, that does not require excessive tank pressurization or other design complications, thereby permitting relatively lightweight pump designs. This places a premium on pumps with good suction characteristics.

There have been some low-thrust, low-flow, experimental engines that have used positive displacement pumps, such as diaphragm pumps, piston pumps, or rotary displacement pumps (gear and vane pumps). For low values of specific speed N_s, these pumps have much better efficiencies but their discharge pressure fluctuates with each stroke and they are noisy.

One method to provide a lightweight TP with low vehicle tank pressure is to use an *inducer*, which is a special pump impeller usually mounted on the same shaft and rotating at the same speed as the main impeller. It has a low head rise and therefore a relatively high specific speed. Inducers are located immediately upstream of the main impeller. They are basically axial-flow pumps with a spiral impeller, and many will operate under slightly cavitating conditions. Here, the inducer stage's head rise (typically, 2 to 10% of the total pump head) has to be just large enough to suppress cavitation in the main pump impeller; this allows a smaller, lighter, higher-speed main pump. Figures 10–2 and 10–9 and Reference 10–18 show such inducers. Reference 10–19 describes pump testing with an inducer. Most TPs today have inducers ahead of the pump impellers.

Several TPs use impellers with double inlets. Not shown in Figs. 10–1, 10–3 and schematically in 10-4 are pumps, impellers and housings with two inlets (one on each side of the volute). Such design doubles the pump inlet area, reduces the inlet velocity to about half and raises the pump inlet pressure, all of which improves the pump's cavitation resistance.

Influence of Propellants

For the same power and mass flow rate, pump head is inversely proportional to propellant density. Since liquid pumps are basically constant-volume flow devices, the propellant with the highest density requires less head, less power, and thus allows smaller pump assemblies.

FIGURE 10–9. Fuel pump inducer impeller of the Space Shuttle main engine low-pressure fuel turbopump. It had a diameter about 10 in., a nominal hydrogen flow of 148.6 lbm/sec, a suction pressure of 30 psi, a discharge pressure of 280 psi at 15,765 rpm, an efficiency of 77%, and a suction specific speed of 39,000 when tested with water. Courtesy of Aerojet Rocketdyne.

Because many propellants are dangerous to handle, special provisions have to be made to prevent any leakage through the shaft seals. With spontaneously ignitable propellants, leakages can lead to fires in the pump compartment and may also cause explosions. Multiple seals are often used with drainage provisions that safely remove or dispose any propellants that flow past the first seal. Inert-gas purges of seals have also been used to remove hazardous propellant vapors. The sealing of corrosive propellants puts very severe requirements on sealing materials and design. With cryogenic propellants pump bearings are usually lubricated by the propellant itself since ordinary lubricating oils would freeze at such low pump hardware temperatures.

Centrifugal pumps need to operate at their highest possible *pump efficiency*. This efficiency increases with volume flow rate and reaches a maximum value of about 90% for very large flows (above 0.05 m³/sec) and specific speeds above about 2500 (see Refs. 10–1 and 10–8). Most propellant pump efficiencies are between 30 and 70%. Pump efficiencies are reduced by surface roughness of casing and impellers, the power consumed by seals, bearings, and stuffing boxes, and by excessive wear-ring leakage and poor hydraulic designs. The pump efficiency η_P is defined as the fluid power divided by the pump shaft power P_P:

$$\eta_P = \rho Q \Delta H / P_P \qquad (10\text{–}16)$$

A correction factor of 550 ft-lbf/hp has to be added when P_P is given in horsepower, the head H in feet, and the volume flow Q in ft³/sec.

Example 10–1. Determine shaft speed and overall impeller dimensions for a liquid oxygen pump which delivers 500 lbm/sec of propellant at a discharge pressure of 1000 psia and a suction pressure of 14.7 psia. The oxygen tank is pressurized to 35 psia. Neglect friction in the suction pipe and any suction head changes due to acceleration and propellant consumption. The initial tank level is 15 ft above the pump suction inlet. See Fig. 10–8.

SOLUTION. The density of liquid oxygen is 71.2 lbm/ft³ at its boiling point (from Table 7–1). Hence the volume flow will be $500/71.2 = 7.022$ ft³/sec. The vapor pressure of the oxygen is given as 1 atm = 14.7 psi = 29.8 ft. The tank head is $35 \times 144/71.2 = 70.8$ ft. The pressure at the pump inlet is $(H_s)_A = 70.8 + 29.8 = 100.6$ ft. The available suction head above vapor pressure (Eq. 10–15) is $(H_s)_A = 70.8 + 15.0 - 0 - 29.8 = 56.0$ ft. The discharge head is $1000 \times 144/71.2 = 2022$ ft. The head delivered by the pump is then the discharge head minus the suction head $2022 - 100.6 = 1921$ ft.

The required suction head will be taken as 80% of the available suction head in order to provide a margin of safety for cavitation $(H_s)_R = 0.80 \times 100.6 = 80.48$ ft. Assume a suction specific speed of 15,000, a reasonable value if no test data are available. From Eq. 10–14 solve for the shaft speed N:

$$S = 21.2 N \sqrt{Q}/(H_s)_R^{3/4} = 21.2 N \sqrt{7.022}/80.48^{0.75} = 15,000$$

Solve for $N = 7174$ rpm or 751.7 rad/sec.
The specific speed, from Eq. 10–14, becomes

$$N_s = 21.2 N \sqrt{Q}/H^{3/4} = 21.2 \times 7174 \sqrt{7.022}/1921^{0.75} = 1388$$

According to Table 10–2, the impeller shape for this value of N_s will be a Francis type. The impeller discharge diameter D_2 can be evaluated from the tip speed by Eqs. 10–11 and 10–12:

$$u = \psi \sqrt{2g_0\ \Delta H} = 1.0\sqrt{2 \times 32.2 \times 1921} = 352 \text{ ft/sec}$$

Assume the pump efficiency is $\eta = 88\%$, typical of a Francis impeller,

$$D_2 = 2u/N\eta = 2 \times 352/751.7 \times 0.88 = 1.064\text{ft} = 12.77 \text{ in.}$$

The impeller inlet diameter D_1 can be found from Eq. 10–12 by assuming a typical inlet velocity of 15 ft/sec and a shaft cross section 5.10 in.^{2} (2.549 in. diameter).

$$A = Q/v_1 = 7.022/15 = 0.468 \text{ ft}^2 = 67.41 \text{ in.}^2$$

The pump inlet flow passage area of 67.41 in.^{2} needs to be increased by the shaft cross section area of 5.10 in.^{2} to obtain the pump inlet diameter,

$$A_1 = \frac{1}{4}\pi D_1^2 = 67.41 + 5.10 = 72.51 \text{ in.}^2$$
$$D_1 = 9.61 \text{ in.(internal flow passage diameter)}$$

This is enough information to draw a preliminary sketch of the impeller.

10.6. TURBINES

The turbine must provide adequate shaft power for driving propellant pumps (and sometimes also auxiliaries) at the desired shaft speeds and torques. Turbines derive their energy from a gaseous working fluid expansion through fixed nozzles and rotating turbine blades. Such blades are mounted on disks which are fastened to the shaft. The working gas expands to a high, nearly tangential, velocity through inclined nozzles and then flows through specially shaped *blades*, where the flow is turned as the gas kinetic energy is converted into tangential forces on each blade. These forces cause the turbine wheel to rotate (see Refs. 10–1 to 10–5, 10–20 and 10–21).

Classification and Description

The majority of turbines have blades at their disk periphery and the gas flow is axial, similar to the axial-flow pattern shown for pumps in Table 10–2 and the single-stage turbine of Fig. 10–1. However, there are some turbines with radial flow (particularly those operating at high shaft speeds), as the one shown in Fig. 10–2. Generally, there are two types of axial-flow turbines of interest to rocket pump drives: impulse turbines and reaction turbines, as shown in Fig. 10–10. In *impulse turbines*, the enthalpy of the working fluid is converted into kinetic energy within the first set of stationary turbine nozzles and not in the rotating blade elements. High-velocity gases are delivered

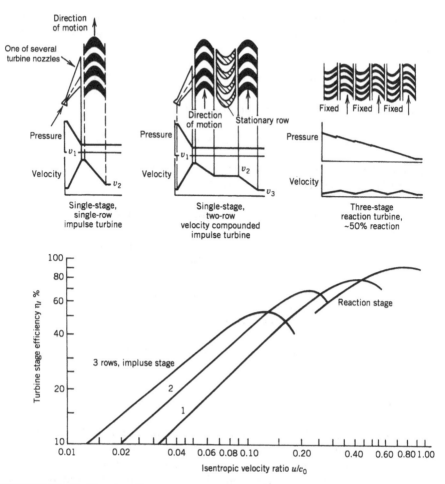

FIGURE 10–10. Top view diagram, pressure and velocity profiles, and efficiency curves for impulse and reaction type turbines. The velocity ratio is the pitch line velocity of the rotating blades u divided by the theoretical gas spouting velocity c_0 derived from the enthalpy drop. Adapted with permission from Refs. 10–3 and 10–19.

(at a small angle to a tangential direction) to the rotating blades, and blade rotation takes place as a result of the impulse imparted by the momentum of the fluid stream of high kinetic energy to the rotating blades that are mounted on the turbine disk. The *velocity-staged impulse turbine* has a stationary set of blades that change the flow direction after the gas leaves the first set of rotating blades and directs the gas to enter a second set of rotating blades in which the working fluid gives up further energy to the turbine wheel. In *pressure-staged impulse turbines,* the expansion of the gas takes place in all the stationary rows of blades. In *reaction turbines* the expansion of the gas is roughly evenly split between rotating and stationary blade elements. The high-pressure drop available for turbine working-fluid expansion in gas generator cycles favors simple, lightweight one- or two-stage impulse turbines for high thrust engines. Many rocket propulsion turbines are neither pure impulse nor reaction turbines, but often operate fairly close to impulse turbines with a small reaction in the rotating vanes. In some turbines, the rotating blades are mechanically fastened to the rotating turbine disk. Other turbines use a single piece of high-strength alloy metal where the blades have been machined out—this type is called "blisk." The RS-68 rocket engine has such blisks.

With some gas generator engine cycles the turbine exhaust gases pass through a *supersonic nozzle* at the exit of the turbine exhaust duct flow (see Fig. 1–14). The relatively high turbine outlet pressure gives critical flow conditions at the nozzle throat (particularly at high altitudes) and thereby assures constant turbine outlet pressure and constant turbine power which will not vary with altitude. Furthermore, this nozzle provides a small additional thrust to the engine (see Chapter 3).

Turbine Performance and Design Considerations

The power supplied by the turbine is given by a combined version of Eqs. 3–1 and 3–7:

$$P_t = \eta_t \dot{m}_t \Delta h \tag{10–17}$$

$$P_t = \eta_t \dot{m}_t c_p T_1 [1 - (p_2/p_1)^{(k-1)/k}] \tag{10–18}$$

The power P_t delivered by the turbine is proportional to the turbine efficiency η_t, the mass flow through the turbine $\dot{m}$, and the available enthalpy drop per unit of flow Δh. The units in this equation must be consistent (1 Btu = 778 ft − lbf = 1055 J). As seen, this enthalpy decrease is a function of the specific heat c_p, the nozzle inlet temperature T_1, the pressure ratio across the turbine, and the ratio of the specific heats k of the turbine gases. For gas generator cycles the pressure drop between the turbine inlet and outlet is relatively high, but the turbine flow is small (typically 2 to 5% of full propellant flow). For staged combustion cycles this pressure drop is very much lower, but the turbine flow is much larger.

In very large liquid propellant engines with high chamber pressure, turbine power can reach over 250,000 hp, and for small engines it can perhaps be around 35 kW or 50 hp.

According to Eq. 10–3, the power delivered by the turbine P_t equals the sum of the power required by the propellant pumps, the auxiliaries mounted on the TP (such as hydraulic pumps, electric generators, tachometers, etc.), and power losses in hydraulic friction, bearings, gears, seals, and wear rings. Usually, these losses are small enough to be neglected. Effects of turbine gas flows on the specific impulse of rocket engines are discussed in Section 6.6. For gas generator engine cycles, the rocket engine designer is interested in obtaining high turbine efficiencies and high turbine inlet temperatures T_1 in order to reduce the flow of turbine working fluids, and for gas generator cycles also in raising the overall effective specific impulse and, therefore, reducing the propellant mass required for driving the turbine. Computer analyses of gas flow behavior and turbine blade geometry have yielded efficient blade designs.

Presently, better turbine blade materials (such as unidirectionally solidified single crystals) and specialty alloys allow turbine inlet temperatures up to 1400 K (or about 2050 °F) and perhaps to 1600 K (or 2420 °F); at these higher temperatures or higher gas enthalpies the required turbine flow is reduced. When using special steel alloys for blade and disk materials, reliability, gas temperature variations (nonuniformities), and cost considerations have reduced actual turbine inlet temperatures to more conservative values, such as 1150 to 1250 °F or about 900 to 950 K. *Turbine stage efficiencies* for certain types of rocket TPs are shown in Fig. 10–10. Maximum *blade speeds* with good design and strong high-temperature materials are typically 400 to 700 m/sec or about 1300 to 2300 ft/sec; higher blade speeds generally get improved efficiencies. For the efficiency to be high turbine blade and nozzle profiles need to have smooth surfaces. Small clearances at the turbine blade tips are also needed to minimize leakage around the blades.

Different organizations define *turbine efficiency* differently. One way to define turbine efficiency η_t is to divide the turbine power output LN (torque L and shaft speed N) by the ideal isentropic enthalpy drop (note that actual pressure drops and actual Δhs are usually higher than ideal):

$$\eta_t = L_t N_t / \dot{m} \ \Delta h \qquad (10-19)$$

Here, Δh is the isentropic gas expansion enthalpy drop across the turbine nozzles and the turbine buckets or blades (moving and stationary) per unit gas mass flow $\dot{m}$. It is typically based on a uniform gas with uniform properties flowing across a section, the perfect gas equation of state, no leakage or alternatively only nominal leakage around the blades, and precise blade contours. Even small clearances at the tip of turbine blades may cause substantial losses, particularly at small turbine diameters and small blade heights. Some organizations also include a few unavoidable losses in defining the ideal power.

A proper turbine efficiency may be realized only when the TP design has efficient turbine blade contours and allows high blade speeds. Power input to the turbine can be regulated by controlling the gas flow and gas temperatures to the turbine inlet through valves or orifices.

10.7. APPROACH TO TURBOPUMP PRELIMINARY DESIGN

This section presents one approach to TP analysis and selection of key TP features, and gives estimates of some key parameters for preliminary design. Principal criteria (high performance or efficiency, minimum mass, high reliability, and low cost) need to be evaluated and prioritized for each vehicle mission including all major rocket engine components. For example, high efficiency and low mass usually lead to low design margins, and thus lower reliability. A higher shaft speed will allow lower mass TPs, but ones that cavitate more readily and require higher tank pressures and thus heavier vehicle tanks (these usually outweigh any mass savings in the TP).

Engine requirements provide the basic goals for TP preliminary designs, namely, propellant flows, pump outlet or discharge pressures (which must equal the chamber pressure plus pressure drops in the piping, valves, cooling jacket, and injector), the best engine cycle (shown in Fig. 6–8), as well as any start delays and needs for restart or throttling. Also, propellant properties (density, vapor pressure, viscosity, or boiling point) must be known. Some design criteria are explained in Refs. 10–2 and 10–3, and basic texts on turbines and pumps are listed as Refs. 10–4 to 10–8.

Usually, pump preliminary analyses are done first. Avoiding excessive cavitation in the pump sets a key pump parameter, namely the maximum pump shaft speed. This is the highest possible shaft speed, which in turn allows the lightest TP mass. When excessive cavitation occurs at the leading edge of the main impeller, then the flow becomes unsteady, leading to lower thrust and possible combustion instabilities. The amount of total pressure in the vehicle (gas pressure in propellant tanks plus the static elevation pressure) that can be made available to the rocket engine (at the pump inlet) for suppressing cavitation has to be larger than the impeller vanes' own pressure limit to cavitate. This allows determination of shaft speed, which in turn establishes approximate pump efficiencies, impeller tip speeds (usually also limited by the material strength of the impeller), number of pump stages, and/or key dimensions of the impeller.

There are several design variations or geometrical arrangements for transmitting turbine power to one or more propellant pumps and an initial selection has to be made; some are shown schematically in Fig. 10–4. If there is a mismatch between optimum pump speed and optimum turbine speed (which is usually higher), it may save inert mass and turbine drive gas mass to interpose a gear reduction between their shafts. See Fig. 6–10 and Figs. 10–4d and 10.4f. For the past three decades designers have preferred to use a direct drive, which avoids the complication of a gear case but at a penalty both in efficiency and in the amount of turbine drive propellant gas required. See Figs. 6–11, 10–1, 10–2 and 10–4e.

Key turbine parameters are estimated by equating the power output of the turbine to the power demand of the pumps. If the pump is driven directly, that is, without a gear case, then pump speed and turbine speed must be equal. From the properties of the turbine drive gas (temperature, specific heat, etc.), the strength limits of turbine materials, and likely pressure drops, it is possible to determine the basic blade dimensions (pitch line velocity, turbine nozzle outlet velocity, number of rows [stages] of blades, turbine type, or turbine efficiency). Any particular arrangement or geometry of major TP components results from their selection process. Most propellant pumps have a single-stage main impeller but for liquid hydrogen, because of the very low density, a two- or three-stage pump is normally needed. Usually, some design limit is reached which requires one or more iterations. The arrangement of the major TP components (Fig. 10–4) is also influenced by the position of the bearings on the shaft. For example, the placement of a bearing in front of an impeller inlet is not desired because this will cause turbulence, distort the flow distribution, raise the suction pressure requirement, and make cavitation more likely to occur. Also, bearings positioned close to a turbine will experience high temperatures, and these influence lubrication needs from the propellant and may demand more cooling of the bearings.

The use of *booster pumps* allows lower tank pressures and thus a lower inert vehicle mass while providing adequate suction pressures to the main pump inlet.

See Ref. 10–12. Booster pumps have been used in the Space Shuttle main engine and the Russian RD-170, as seen in Figs. 6–11 and 11–2. Some booster pumps have been driven by a booster turbine using small flows of a high-pressure liquid propellant tapped off the discharge side of the main pump. The discharged turbine liquid then gets mixed with the main propellant flow at the discharge of the booster pump.

Mass is at a premium in all flying vehicles and feed systems are selected to have a minimum combined mass of turbines, pumps, gas generator, valves, tanks, and gas generator propellants. Design considerations of TPs include thermal stresses, warpage due to thermal expansion or contraction, axial loads, adequate clearances to prevent internal rubbing yet minimizing leakage, alignment of bearings, provisions for dynamic balancing of rotating subassemblies, mountings on elastic vehicle frames without inducing external forces, and avoiding undue pressure loads in pipes flowing liquids or gases.

The critical speed of any rotating assembly equals its natural vibration frequency when not rotating but supported on knife edges at the same two locations as the center of its bearings. At the critical speed and under load, shaft deflections will amplify and any small shaft-slope changes (e.g., 0.001 to 0.003 in.) at any bearing may cause bearing failure. Some types of shaft seals will also likely fail. The rotating assembly of a TP consists of the shaft, turbine drive and blades, pump impeller, and parts of bearings and seals. A TP's operating shaft speed must never reach any critical value at which it would fail. Therefore, TP operating speeds are usually well below the critical speed. There are some exceptions when the operating speed can exceed critical and these occur during start and stop transients.

Designers can tailor the critical speed by using large-diameter stiff shafts, rigid bearings with stiff bearing support housings, short distances between the two bearings, and little or no shaft overhangs beyond the bearings—all these will increase critical shaft speeds. A higher critical speed will allow higher pump operation speeds and result in smaller, lighter TPs. Also, the natural resonant frequency of key components (such as piping subassemblies, injector domes, or thrust structures) must not coincide with the critical shaft speed because this can lead to higher stresses and flow disturbances in these components. Critical speeds and resonance frequencies of key components can usually be determined analytically and designs may be altered through changes in mass and stiffness (spring forces).

Solving methods for various internal vibration problems, such as whirl in bearings and blade vibrations, are reported in Ref. 10–16, and the dynamics of propellant flow in pumps can be found in Ref. 10–15. Whenever a pump blade tip of passes the stationary tongue of a pump volute, a pressure wave is generated in the liquid; the frequency of this wave is the product of the number of pump vanes and the speed of the vane tip. Vibrations external to TPs, such as those generated by the thrust chamber (see Chapter 9) should not influence TP operation.

The *bearings* in most existing TPs are high precision, special metal alloy ball bearings (Ref. 10–13). A few are roller bearing types which have a higher radial load capacity. Some ball bearings can take both radial and axial loads. Early ball and roller bearings were limited in the loads and speeds at which they could operate reliably. In some TP designs, limits on bearing loads and speeds (and thus minimum TP sizes) were determined by shaft speeds rather than by cavitation limits of the pumps. Various types of bearings and many types of seals have been explored and tested in laboratory fixtures or in experimental TPs. They include hydrostatic (precision sleeve type) bearings, foil bearings, and magnetic bearings (Ref. 10–13). As far as the authors know, none of these have as yet found their way into a production TP. The variety of available static and dynamic seals is relatively large and seal selection is strongly influenced by the preferences of the design organization (Ref. 10–14).

When the TP is part of a reusable rocket engine, it becomes a more complex system. For example, such system may include provisions to allow for inspection and automatic condition evaluation after each mission or flight. This may encompass access holes for boroscope instruments for bearing inspections, checking for cracks in highly stressed parts (turbine blade roots or hot-gas, high-pressure manifolds), and measurement of shaft torques (to detect possible binding or warpage).

The number of different *materials for TP construction* has increased. For example, for high-speed, high-load, ball bearings a new ball material (silicon nitride) has been successfully introduced. Relatively common materials, such as stainless steels, have in part been replaced by superalloys, such as Inconel. Powder metallurgy has found its way into pump impellers and turbine parts, see Ref. 10–22; although their strength is not really better than forged or cast materials, they have smoother surfaces (lower friction), uniform physical properties, and can be fabricated into more complex shapes (Ref. 10–22).

There are *no allowable warm-up times* available for rocket turbines. The sudden admission of hot gases at full flow causes severe thermal shock and thermal distortion, increasing chances for rubbing between moving and non-moving metal parts. The most severe stresses in turbine blades are often thermal stresses; they arise during engine start-ups when the blade leading edge becomes very hot while its other parts are still cold.

For low-thrust engines shaft speeds can be very high, such as over 100,000 rpm. Also, turbine blade height can be very short and friction as well as other losses may become prohibitive. In order to obtain reasonable blade heights, partial admission turbine designs have been used; here a portion of the turbine nozzles are effectively plugged or eliminated.

10.8. GAS GENERATORS AND PREBURNERS

The purpose of gas generators or preburners is to supply the "warm" gases (usually between 600 and 2000 °F or 315 to 1200 °C) that drive the turbine of a TP. *Gas generators* are used exclusively with liquid propellant rocket engines that operate on a *gas generator engine cycle,* and *preburners* are used exclusively with rocket engines that operate on a *staged combustion cycle* and usually at higher combustion pressures. Table 10–3 details some differences between these two and Table 6–6 and Fig. 6–8 show the engine cycles. See Refs. 10–1 to 10–5. The gas temperatures mentioned above allow for an uncooled turbine with uncooled inlet and outlet ducts, and uncooled blades and turbine nozzles.

TABLE 10–3. Comparison of Key Characteristics of Gas Generators and Preburners

Parameter	Gas Generator	Preburner
Engine cycle	Gas generator cycle	Staged combustion cycle
Chamber pressure	Usually equal or lower than thrust chamber pressure	30–90% higher than its thrust chamber pressure
Mass flow as % of total propellant flow	1–7	45–90
Cooling	Usually uncooled	Usually uncooled, but may be partially cooled
Inert mass	Relatively light	Heavy
Size	Relatively small	Can be large
Flow of warm turbine exhaust gas	Out of vehicle into ambient atmosphere	Into injector of main thrust chamber

Gas generators and combustion devices each comprise a combustion chamber, an injector, and pipes or ducts leading to the turbine. See Ref. 10–6. They all have separate dedicated propellant control valves. Warm gases flow subsonically from the gas generator or preburner through pipes or manifolds into multiple nozzles in the turbine and then flow supersonically through row(s) of turbine blades that extract the driving energy for the propellant pumps. Requirements for gas generators or preburners include the delivery of warm gases at the intended mass flows, pressures, and design temperatures together with essentially uniform gas temperatures across the flow path to the turbine and without high gas temperature spikes. Combustion instability problems are extremely rare.

The typical gas generator has relatively small flows, typically 2 to 3% of the total engine propellant flow. It makes little difference in pump power or TP design if the gas in the gas generator is oxidizer rich or fuel rich. The authors have found that a majority of gas generators run fuel rich.

For staged combustion rocket engines with oxidizer-rich gas preburners, oxidizer flows with a small portion fuel flows supply the preburner. The gas pressure in the preburner is substantially higher than the thrust engine's chamber pressure. Preburner gases flow through the turbine that drives the propellant pumps, and then the turbine exhaust flows to the thrust chamber injector. For an oxidizer-rich preburner with a liquid oxygen/RP-1 propellant combination a typical mixture ratio would be 54 parts liquid oxygen to one part fuel. Oxidizer pump powers need to be substantially higher than those for engines without a preburner resulting in longer and heavier (but perhaps more efficient) pumps. The much smaller fuel flows to the preburner can utilize a separate "kick" pump or an extra low flow stage on the fuel pump. The majority of operational engines with a staged combustion cycle operate with an oxidizer-rich preburner gas.

For fuel rich preburners, all the fuel goes to the preburner and only a small oxidizer flow is needed. The preburner operates at higher propellant pressures compared to engines without preburners and this also results in heavier thrust chambers. For a liquid oxygen/RP-1rocket engine, the mixture ratio in the preburner would typically be about three parts of fuel to one part of oxidizer (or 0.33).

In ground-based test facilities, any unexpected leak or spill of hot fuel-rich gases will most likely ignite with ambient air resulting in a fire whereas a hot oxidizer-rich leak will not. However, the latter gases are very corrosive and can react with all surfaces they come in contact with (oxidizing metallic surfaces and oxidizing and/or causing fires with organic materials such as rubber gaskets).

Gas generators are shown as a component of the engine in Fig. 1–4 and Ref. 10–4, and preburners are shown in Figs. 6–1, 6–11, and 11–2. Propellants supplied to the gas generator or the preburner are usually tapped off from discharges of the engine main pumps. When starting an engine, the turbomachinery needs to be brought up to rated speed before propellants can be supplied to the thrust chambers at their full operating pressure. Such required gas generator starting has also been done with a solid propellant starting cartridge (running only a few seconds), or an auxiliary set of small propellant tanks pressurized by a cold gas (also running only for short duration), or by letting the engine "bootstrap" itself into starting using the tank pressure augmented by the existing liquid column head in the vehicle tanks and feed system pipe lines—usually called "tank head" start (which requires more time to start). A discussion of engine starts and tank pressurization can be found in Section 6.5 and of thrust chamber starts in Section 8.6.

In the past, monopropellant gas generators were common using either 80 or 90% hydrogen peroxide or pure hydrazine to provide the warm gases, usually through catalytic decomposition in a bed of solid catalysts. Here, a simpler gas generator system is required (only one tank and one set of valves, instead of two), having no mixture ratio adjustments, and with predictable, fully reproducible uniform warm gas temperatures without potential temperature spikes. The key disadvantages were the complications for providing a third propellant, potential propellant hazards, lower performances, and/or the higher mass of propellant for making the warm gases necessary for providing the required power.

SYMBOLS

a	gear ratio
A	area, m^2 (ft^2)
c_p	specific heat at constant pressure, J/kg K (Btu/lb°R)
c_o	ideal nozzle exit gas velocity
D	diameter, m (ft)
g_0	sea-level acceleration of gravity, 9.806 m/sec^2 (32.17 ft/sec^2)
H	head, m (ft)
$(H_s)_A$	available pump suction head above vapor pressure, often called net positive suction head, m (ft)
$(H_s)_R$	required pump suction head above vapor pressure, m (ft)
k	ratio of specific heats
L	torque, Nm (ft-lbf)

$\dot{m}$ mass flow rate, kg/sec (lbm/sec)

N shaft speed, rpm (rad/sec)

N_s specific speed of pump

p pressure, N/m^2 (lbf/in.^{2})

P power, W (hp)

P_b power of auxiliaries, bearings, rubbing seals, friction

Q volumetric flow rate, m^3/sec (ft^3/sec)

S suction specific speed of pump

T absolute temperature, K ($^\circ$R)

u impeller tip speed or mean blade speed, m/sec (ft/sec)

v liquid flow velocity, m/sec (ft/sec)

Greek Letters

η Efficiency

ρ density, kg/m^3 (lb/ft^3)

ψ velocity correction factor

Subscripts

e maximum efficiency condition

f Fuel

o Oxidizer

p Pump

t Turbine

0 initial condition

1 pump or turbine inlet, or chamber condition

2 pump or turbine outlet, or chamber condition

PROBLEMS

1. A rocket propulsion system with two TPs delivers the fuel, namely unsymmetrical dimethylhydrazine (UDMH), at a pump discharge pressure of 555 psia, a suction pressure of 25 psi, a flow of 10.2 lbm/sec, at 3860 rpm, and a fuel temperature of 68 °F. Using UDMH properties from Table 7–1 and Fig. 7–1 and its efficiency from Table 10–2, determine the following:

   a. The fuel pump power for these nominal conditions.

   b. If the fuel flow is reduced to 70% of nominal, what will be the approximate power level, discharge pressure, and shaft speed? Assume that the oxidizer pump is also reduced by 70% and also the gas flow to the turbines, but that the gas temperature is unchanged.

   c. If the anticipated temperature variation of the propellants is between − 40°F and + 120°F, describe qualitatively how this variation will affect the power level, shaft speed, and the discharge pressure of the fuel pump.

2. What are the specific speeds of the four SSME pumps? (Use the data given in Table 10–1.)

3. Compute the turbine power output for a gas consisting of 64% by weight of H_2O and 36 % by weight of O_2, if the turbine inlet is at 30 atm and 658 K with the outlet at 1.4 atm and with 1.23 kg flowing each second. The turbine efficiency is 37%.

4. Compare the pump discharge gauge pressures and the required pump powers for five different pumps using water, gasoline, alcohol, liquid oxygen, and diluted nitric acid. The respective specific gravities are 1.00, 0.720, 0.810, 1.14, and 1.37. Each pump delivers 100 gal/min, a head of 1000 ft, and has arbitrarily a pump efficiency of 84%.

Answers: 433, 312, 350, 494, and 594 psi; 30.0, 21.6, 24.3, 34.2, and 41.1 hp.

5. The following data are given for a liquid propellant rocket engine:

Thrust	40,200 lbf
Thrust chamber specific impulse	210.2 sec
Fuel	Gasoline (specific gravity 0.74)
Oxidizer	Red fuming nitric acid (sp. gr. 1.57)
Thrust chamber mixture ratio	3.25
Turbine efficiency	58%
Required pump power	580 hp
Power to auxiliaries mounted on turbopump gear case	50 hp
Gas generator mixture ratio	0.39
Turbine exhaust pressure	37 psia
Turbine exhaust nozzle area ratio	1.4
Enthalpy available for conversion in turbine per unit of gas	180 Btu/lbm
Specific heat ratio of turbine exhaust gas	1.3

Determine the engine system mixture ratio and the system specific impulse.

Answer: 3.07 and 208 sec.

REFERENCES

10–1. G. P. Sutton, "Turbopumps, a Historical Perspective," AIAA Paper 2006-7531, Jul. 2006.

10–2. D. K. Huzel and D. H. Huang, "Design of Turbopump Feed Systems," Chapter 6 in *Design of Liquid Propellant Rocket Engines*, rev. ed., Vol. 147, *Progress in Astronautics and Aeronautics*, AIAA, Reston, VA, 1992.

10–3. M. L. Strangeland, "Turbopumps for Liquid Rocket Engines," *Threshold, an Engineering Journal for Power Technology*, No. 3, Rocketdyne Propulsion and Power, Summer 1988, pp. 34–42.

10–4. G. P. Sutton, "Turbopumps," Chapter 4.4, and "Gas Generators, Preburners and Tank Pressurization," Chapter 4.5, in *History of Liquid Propellant Rocket Engines*, AIAA, Reston, VA, 2006.

10–5. "Turbopump Systems for Liquid Rocket Engines," NASA Space Vehicle Design Monograph, NASA SP-8107, Aug. 1974.

10–6. Personal communications with personnel from Pratt & Whitney Rocketdyne, Northrop Grumman, and The Aerospace Corporation, 2006 to 2008.

10–7. A. Minick and S. Peery, "Design and Development of an Advanced Liquid Hydrogen Turbopump," AIAA Paper 98-3681, Jul. 1998, and G. Crease, R. Lyda, J. Park, and A. Minick, "Design and Test Results of an Advanced Liquid Hydrogen Pump," AIAA Paper 99–2190, 1999.

10–8. I. J. Karassik, W. C. Krutzsch, W. H. Frazer, and J. P. Messina (Eds.), *Pump Handbook*, McGraw-Hill, New York, 1976 (water hammer and pumps); I. J. Karassik, Chapter 14-2 in *Marks' Standard Handbook for Mechanical Engineers,* 10th Ed., McGraw-Hill, New York, 1978.

10–9. C. E. Brennan, *Hydrodynamics of Pumps*, Concepts ETI, Inc. and Oxford University Press, Oxford, England, 1994.

10–10. "Liquid Rocket Engines Turbopump Shafts and Couplings," NASA Space Vehicle Design Monograph, NASA SP-8101, Sept. 1972.

10–11. "Liquid Rocket Engines Turbopumps Gears," NASA Space Vehicle Design Monograph, NASA SP-8100, Mar. 1974.

10–12. Y. V. Demyanenko, A. I. Dimitrenko, and I. I. Kalatin, "Experience of Developing Propulsion Rocket Engine Feed Systems Using Boost Turbopump Units," AIAA Paper 2003-5072, 2003.

10–13. "Liquid Rocket Engine Turbopump Bearings," NASA Space Vehicle Design Monograph, NASA SP-8048, Mar. 1971.

10–14. Liquid Rocket Engine Turbopump Rotating Shaft Seals. NASA SP-8121, Feb. 1978.

10–15. J. Kurokawa, K. Kamijo, and T. Shimura, "Axial Thrust Analysis on LOX-Pump," AIAA Paper 91-2410, June 1991.

10–16. M. C. Ek, "Solving Synchronous Whirl in High Pressure Turbine Machinery of the Space Shuttle Main Engine," *Journal of Spacecraft and Rockets*, Vol. 17, No. 3, May–Jun. 1980, pp. 208–218.

10–17. R. S. Ruggeri and R. D. Moore, "Method for Prediction of Pump Cavitation Performance for Various Liquids, Liquid Temperatures, and Rotating Speeds," NASA TN D5292, Jun. 1969.

10–18. "Liquid Rocket Engine Turbopump Inducers," NASA Space Vehicle Design Monograph, NASA SP-8052, May 1971.

10–19. T. Shimura and K. Kamijo, "Dynamic Response of the LE-5 Rocket Engine Oxygen Pump," *Journal of Spacecraft and Rockets*, Vol. 22, No. 2, Mar.–Apr., 1985.

10–20. "Liquid Rocket Engine Turbines," NASA Space Vehicle Design Criteria Monograph, NASA SP-8110, Jan. 1974.

10–21. S. Andersson and S. Trollheden, "Aerodynamic Design and Development of a Two-Stage Supersonic Turbine for Rocket Engines," AIAA Paper 99–2192, 1999.

10–22. D. Guichard and A. DuTetre, "Powder Metallurgy Applied to Impellers of Vinci Turbopump," International Symposium for Space Transportation of the XXI Century, in CD ROM of the symposium, May 2003.

CHAPTER 11

ENGINE SYSTEMS, CONTROLS, AND INTEGRATION

In this chapter, we examine propellant budget, performance of multiple or complete rocket propulsion systems, designs of liquid propellant rocket engines with pressurized or with turbopump (TP) feed systems, engine controls, engine calibration, system integration, and system optimization. Some content also applies to solid propellant motors and hybrid propulsion systems.

11.1. PROPELLANT BUDGET

In all liquid propellant rocket engines the amount of propellant inserted into the vehicle tanks is always somewhat greater than the nominal amount of propellant needed to accomplish the intended mission. The extra propellant is needed for purposes other than providing thrust (e.g., for auxiliary functions like valve actuation), to compensate for changes from engine to engine (such as dimensional tolerances causing slight changes in flow), for uncertainties of engine construction and for minor variations in the flight plan. A propellant budget represents the sum of all the propellant utilization categories and losses in an engine—11 are identified below. Such a budget defines how much propellant has to be loaded and is aimed at minimizing the amounts of extra propellant.

1. Enough propellant must be available for achieving the *required vehicle velocity increase* and any *nominal set of attitude control maneuvers* for the specified application of the entire flight vehicle or stage. The nominal velocity increment is defined from a system analysis and mission optimization calculations based on Eqs. 4–19 or 4–20 and 4–35. When there are alternative flight

paths or missions for the same vehicle, the mission with the most propellant consumption, such as one with higher drag or different orbit or the highest total impulse, is selected. Mission-required propellants constitute by far the largest portion of the total propellant loaded into the vehicle tanks.

2. In TP systems using a gas generator cycle, a small portion of the overall propellant is used by the *gas generator*. It operates at lower flame temperatures and different mixture ratios than the thrust chamber; this causes a slight change to the overall mixture ratio on the propellants flowing from the tanks, as shown later in Eqs. (11–3) and (11–5).

3. In a rocket propulsion system with *thrust vector control* (TVC), such as a swiveling thrust chamber or nozzle, the thrust vector may rotate by a few degrees; this causes a slight decrease in the axial thrust that reduces the vehicle velocity increment in item 1. The extra propellant needed to compensate for this small velocity reduction is determined from mission requirements and TVC duty cycles. It could amount to between 0.1 and 4% of the total propellant depending on the average angle position of the thrust vector. Thrust vector control systems are described in Chapter 18.

4. In some engines with cryogenic propellants, a small portion of propellant is vaporized and then used to *pressurize its own tank*. As shown schematically in Fig. 1–4, a heat exchanger is used to heat liquid oxygen (LOX) from the pump discharge and pressurize the oxygen tank. This method is used in the hydrogen and oxygen tanks of several space vehicles.

5. Auxiliary rocket engines that provide for *trajectory corrections, station keeping, maneuvers*, or *attitude control* normally have a series of small restartable thrusters (see Chapter 4). Propellants for these auxiliary thrusters have to be included in the propellant budget when they are supplied from the same feed system and tanks as the larger rocket engine. Depending on the mission, the duty cycle, and the propulsion system concept, auxiliary propulsion systems may consume a significant portion of the budgeted propellants.

6. Any *residual propellant* that clings to tank walls or remains trapped in valves, pipes, injector passages, or cooling passages is unavailable for producing thrust. It can typically amount to between 0.5 and 2% of the total propellant load. All unused residual propellants increase the final vehicle mass at thrust termination and slightly reduce the final vehicle velocity.

7. A *loading uncertainty* always exists due to variations in tank volume or changes in propellant density or liquid level in the tank. This typically amounts to 0.25 to 0.75% of the total propellant. It depends, in part, on the accuracy of the method of measuring the propellant mass during loading (weighing the vehicle, flow meters, level gauges, etc.).

8. *Off-nominal rocket performance* refers to manufacturing variations on the of hardware from one engine to another (such as slightly different pressure losses in cooling jackets, in injectors and valves, or somewhat different pump characteristics); these cause slight changes in combustion behavior, mixture ratio, and/or specific impulse. When there are such variations in *mixture ratio*, one of the two liquid propellants will be fully consumed and a residue will remain

in the other propellant's tank. If a minimum total impulse requirement has to be met, extra propellant has to be tanked to allow for these mixture ratio variations. This can amount to up to 2.0% for each of the propellants.

9. *Operational factors*, such as filling more propellant than needed into a tank or incorrectly adjusting regulators or control valves and can also include changes in flight acceleration from the nominal value, may result in additional propellant requirements. For an engine that has been carefully calibrated and tested, this factor can remain small, usually between 0.1 and 1.0%.

10. When using cryogenic propellants, an *allowance for evaporation and for engine cooling down* has to be included. This represents the extra propellant mass that is allowed to evaporate (and vented overboard while the vehicle is waiting to be launched) and that is fed through the engine to lower its temperature just before start. It is common practice to replace the evaporated propellant by feeding fresh cryogenic propellant into the tank; this process is called "topping off" and it occurs just before launch. If there are significant time delays between tapping off and launch, additional cryogenic propellant will evaporate, so there is often some uncertainty about the precise amount of propellant in the tanks at launch.

11. Finally, an *overall contingency* or ignorance factor needs to be included to allow for unforeseen propellant needs or inadequate or uncertain estimates of any of the items above. This can also include allowances for such factors as atmospheric drag uncertainties, variations in the guidance and control systems, or propellant or gas leaks.

Only some items above provide axial thrust (items 1, 2, and sometimes also 3 and 5), but all items above need to be considered in determining the total propellant mass and tank volume.

Table 11–1 shows a propellant budget example for a spacecraft pressure-fed engine system, where the majority of the monopropellant is consumed in a larger axial thrust chamber, and the second largest amount of propellant is fed to a set of small thrusters for extensive attitude control maneuvers. For flights where the mission may be more demanding or the engine performance may be lower, extra propellant will be needed to accomplish their missions. Conversely, when engine performance is actually better than nominal or where the mission can be accomplished with less total impulse (operate with fewer or lower orbits), then the engine will consume less than the nominal amount of propellant and the residual propellant can be larger than budgeted.

11.2. PERFORMANCE OF COMPLETE OR MULTIPLE ROCKET PROPULSION SYSTEMS

The simplified relations given here complement Eqs. 2–23 to 2–25. They represent a basic method for determining overall specific impulse, total propellant flow, and overall mixture ratio as a function of the corresponding component performance terms for complete rocket engine systems. They apply to engine systems consisting

TABLE 11–1. Example of a Propellant Budget for a Spacecraft Propulsion System with a Pressurized Monopropellant Feed System

Budget Element	Typical Value
1. Main thrust chamber (increasing the velocity of stage or vehicle)	85–96% (determined from mission analysis and system engineering)
2. Flight control function (for reaction control thrusters and flight stability)	2–10% (determined by control requirements)
3. Residual propellant (trapped in valves, lines, tanks, on walls, etc.)	0.5–2% of total load
4. Loading uncertainty	Up to 0.5% of total load
5. Allowance for off-nominal performance	0.1–1.0% of total load
6. Allowance for off-nominal operations	0.1–1.0% of total load
7. Mission margin (reserve for first two items above)	1–5% of items 1 and 2
8. Contingency	1–5% of total load

Source: Engineering estimates.

of one or more thrust chambers, exhaust nozzles, gas generators, turbines, and venting of evaporative propellant pressurization systems, all these operating at the same time.

Refer to Eqs. 2–5 and 6–1 for relations involving the specific impulse I_s, propellant flow rate $\dot{w}$ or $\dot{m}$ and mixture ratio r. The overall thrust F_{oa} is the sum of all the thrusts from thrust chambers firing in parallel together with any turbine exhausts, and the overall flow $\dot{m}$ is the sum of their flows as shown in in Section 2.5. The subscripts oa, o, and f designate the overall engine system, the oxidizer, and the fuel, respectively.

$$(I_s)_{oa} = \frac{\sum F_{oa}}{\sum \dot{w}} = \frac{\sum F_{oa}}{g_0 \sum \dot{m}} \tag{11–1}$$

$$\dot{w}_{oa} = \sum \dot{w} \quad \text{or} \quad \dot{m}_{oa} = \sum \dot{m} \tag{11–2}$$

$$r_{oa} \approx \frac{\sum \dot{w}_o}{\sum \dot{w}_f} = \frac{\sum \dot{m}_o}{\sum \dot{m}_f} \tag{11–3}$$

These same equations represent the overall performance when more than one rocket engine is contained in a vehicle propulsion system (all operating simultaneously). They also apply to multiple solid propellant rocket motors and combinations of liquid propellant rocket engines and solid propellant rocket booster motors, as shown in Figs. 1–12 to 1–14. All nozzles and exhaust jets must be pointed in the same direction in Eqs. 2–23, 11–1, and 11–4 (since we are dealing with vectors).

Example 11–1. For an engine system (LOX/kerosene) with a gas generator similar to the one shown in Fig. 1–4, determine a set of equations that will express (1) the overall engine performance and (2) the overall mixture ratio of the propellant flows from the tanks. Let the following subscripts be used: c, thrust chamber; gg, gas generator; and tp, tank pressurization. For a nominal burning time t, a 1% residual propellant, and a 6% overall reserve factor, give a formula for the amount of fuel and oxidizer propellant required with constant propellant flow. Ignore stop and start transients, thrust vector control, and evaporation losses.

SOLUTION. Only the oxidizer tank is pressurized by vaporized propellant. Although this pressurizing propellant must be considered in determining the overall mixture ratio, it should not be considered in determining the overall specific impulse since it stays with the vehicle and is not usually exhausted overboard.

$$(I_s)_{oa} \approx \frac{F_c + F_{gg}}{(\dot{m}_c + \dot{m}_{gg})g_0} \tag{11-4}$$

$$r_{oa} \approx \frac{(\dot{m}_o)_c + (\dot{m}_o)_{gg} + (\dot{m}_o)_{tp}}{(\dot{m}_f)_c + (\dot{m}_f)_{gg}} \tag{11-5}$$

$$m_f = [(\dot{m}_f)_c + (\dot{m}_f)_{gg}] \, t \, (1.00 + 0.01 + 0.06)$$

$$m_o = [(\dot{m}_o)_c + (\dot{m}_o)_{gg} + (\dot{m}_o)_{tp}] \, t \, (1.00 + 0.01 + 0.06)$$

For the gas generator cycle the engine mixture ratio or r_{oa} is different from the thrust chamber mixture ratio $r_c = (m_o)_c / (m_f)_c$. Similarly, the overall engine specific impulse is slightly lower than the thrust chamber specific impulse $I_c = F_c / \dot{m}_c g_0$. However, for an expander cycle or a staged combustion cycle, these two mixture ratios and two specific impulses are the same, provided that there is no gasified propellant used for tank pressurization. Engine cycles are explained in Section 6.6.

Overall engine specific impulse is influenced by the propellants, the nozzle area ratio, the chamber pressure, and to a lesser extent by the engine cycle together with its mixture ratio. Table 11–2 describes 10 rocket engines, designed by different companies in different countries, that use liquid oxygen and liquid hydrogen propellants and shows the sensitivity of the specific impulse to these parameters. References 11–1 to 11–3 give additional data on several of these engines and later versions.

11.3. ENGINE DESIGN

Approaches, methods, and resources utilized for rocket engine preliminary and final design usually differ for each design organization and for each major type of engine. They also differ by the degree of novelty.

1. A *totally new engine* with *new major components* and *novel design* and *manufacturing concepts* might result in an optimum engine design for a given application, but this is by far the most expensive and longest development approach. One major development cost comes from the necessary testing of engine components and from additional testing of several engines under various environmental and performance limit conditions that must be done to establish reliability data and enough credibility and confidence to allow initial flights

TABLE 11–2. Comparison of Rocket Engines Using Liquid Oxygen and Liquid Hydrogen Propellants[a]

Engine Designation, Engine Cycle, Manuf. or Country (Year Qualified)	Vehicle	Thrust in Vacuum, kN (lbf)	Specific Impulse in Vacuum (sec)	Chamber Pressure, bar (psia)	Mixture Ratio	Nozzle Area Ratio	Engine Mass (dry), kg
SSME, RS-25 formerly, staged combustion, Aerojet Rocketdyne (1998/2010)	Space launch systems, Space Shuttle (3 required)	2183 (490,850)	452.5	196 (2747)	6.0	68.8	3400
RS-68, gas generator, Aerojet Rocketdyne (2000)	Delta	3313 (745,000)	410	97.2 (1410)	6.0	21.5	6800
LE-5A, Expander bleed, MHI, Japan, (1991)	HII	121.5 (27,320)	452	37.2 (540)	5.0	130	255
LE-7, staged combustion, MHI, Japan (1992)	HII	1080 (242,800)	445.6	122 (1769)	6.0	52	1720
Vulcain, gas generator, SEP (circa 1996)	Ariane 5 2nd stage	1120 (251,840)	433	112 (1624)	5.35	45	1585
HM-7, gas generator, SEP, France (1986)	Ariane 1,2,3,4 3rd stage	62.7 (14,100)	444.2	36.2 (525)	5.1	45	155
RL 10-A3, Aerojet Rocketdyne (1965)	Various upper stages	73.4 (16,500)	444.4	32.75 (475)	5.0	61	132
RL 10-B2, same as above (1998)	Same as above	110 (24,750)	465.5	43.6 (633)	5.88	385	275
YF 73, China (circa 1981)	Long March 3rd stage	44.147 (10,000)	420	26.28 (381)	5.0	40	236
YF 75 (2 required), China (circa 1991)	Same	78.45 (17,600)	440	36.7 (532)	5.0	80	550

[a]Additional information on some of these engines is given in this book; use the index to find it.

and initial production. Since the state of the art is relatively mature today, design and development of a truly novel engine does not happen very often.

2. *Engine designs using some new major components* or *somewhat modified key components from proven existing engines* represent the most common approach today. The design of such an engine requires working within the capability and limits of existing or slightly modified components. This approach often needs the least amount of testing for proving reliability.

3. *Uprated, improved, or modified versions of an existing, proven engine.* This approach is quite similar to the second. It is needed when an installed engine for a given mission requires more payload (i.e., higher thrust) and/or longer burning duration (more total impulse). Uprating often means more propellant (larger tanks), higher propellant flows and higher chamber and feed pressures, and more feed system power. Usually, an uprated engine also has an increased inert engine mass (thicker walls).

In a simplified way, we proceed to illustrate a typical process for designing an engine. Chapter 19 and Refs. 11–4 and 11–5 also describe such a process and the selection of propulsion systems but from a different point of view. The basic function together with all requirements for the new engine must be first established. These engine requirements are derived from the vehicle mission and vehicle constraints, usually determined by the customer and/or the vehicle designers, often in cooperation with one or more engine designers. Engine requirements may include key parameters such as thrust level, desired thrust–time variation, restart or pulsing, altitude flight profile, duty cycle, maximum accelerations, engine placements within the vehicle, and limitations or restraints on cost, engine envelope, test locations, or schedules. If an existing engine can be adapted to meet these requirements, any subsequent design process will be simpler and quite different than for a truly new engine.

Usually, some tentative decisions about the engine are first made, such as selection of propellants, their mixture ratio, or the cooling approach for the hot components. These must be based on mission requirements, customer preferences, past experiences, relevant analyses, and the experience of key decision makers. After some study, additional selection decisions may be made, such as having one or more thrust chambers fed from the same feed system, redundancy of auxiliary thrusters, and/or type of ignition system.

A systematic approach using systems engineering or other proper set of analyses, together with good coordination with the customers, key vendors and vehicle designers are all needed during preliminary and final design efforts. Before a meaningful proposal for an engine can be prepared, all preliminary designs have to be completed. See Refs. 11–5 to 11–7. One early design decision is the choice of feed system: pressurized gas feed or pump feed; the next two paragraphs give some guidelines.

A *pressurized feed system* (see Fig. 1–3) gives better vehicle performance for low values of total impulse (thrust less than about 4.5 kN or 1000 lbf with up to perhaps 2 min duration). A pump feed system gives better vehicle performance for high thrust (say above 50,000 lbf or approximately 222 kN) and a long cumulative duration—more than a few minutes. For intermediate values of total impulse the choice can go either way, and there is no simple criterion based only on total impulse. If the chamber pressure of a pressurized feed system is relatively high (say about 2.4 to 3.5 MPa or about 350 to 500 psia and occasionally more), then the inert weight of the thrust chamber will also be high, but the thrust chamber will be small and can usually fit into most available engine compartments. Because the propellant tanks and pressurizing gas tank will have to be at relatively high pressures, they will be heavy. For relatively low chamber pressures (0.689 to 1.379 MPa or 100 to 200 psia), the vehicle tank pressures will be lower and their tank walls thinner, but the thrust chamber size will be relatively large and often will exceed the limits of the engine compartment, unless it has a low nozzle exit area ratio, which implies lower performance. Pressurized feed systems and relatively low chamber pressures are preferred for reaction control systems, also known as attitude control systems with multiple small thrusters. Pressurized feed systems are relatively simple, very reliable, and they allow fast starts and fast restarts. Because of their unmatched proven reliability, NASA has at times conservatively selected pressurized feed systems for certain space applications, such as the Apollo service module engine (21,900 lbf thrust), even though there was a major weight penalty and a somewhat inferior vehicle performance compared to a pump-fed system of equal total impulse. Also, a decision needs to be made on using either a pressurized system with a gas pressure regulator or alternatively a blow-down system. See Table 6–3 and Section 6.4.

For a *turbopump-fed liquid propellant rocket engine* (see Fig. 1–4) the overall inert weight of propellant tanks and engine will be considerably lighter and usually vehicle performance will be somewhat better for the longer total impulse applications. These engines commonly operate at high chamber pressures (3.5 to 24.1 MPa or about 500 to 3500 psia), and the thrust chamber is not normally protruding from the vehicle. Smaller and shorter thrust chambers often allow a shortening of the vehicle with a savings in vehicle structural mass. This modestly improves the vehicle's performance and the higher specific impulse will slightly reduce the amount of propellant needed for the mission. Compared to an engine with a pressurized feed system, the savings in inert mass (thin vehicle tank walls) and less propellant will allow a smaller, lighter, and probably lower cost vehicle with a somewhat superior performance. Engines with TPs are more complex with more parts and the engine itself will generally be heavier and cost more; however, the vehicles' propellant and gas tanks will be much lighter and that will often more than compensate for the heavier engine. Also, it takes more tests and more effort to prove high reliability in an engine with a pump feed system. At the higher chamber pressures, the heat transfer will be higher and cooling is more challenging, but many high heat transfer cases have been solved successfully in earlier high-pressure rocket engines and high reliability has been achieved in many TP-fed large rocket engines. Restartable engines will have more complexity and here one of several engine cycles will have to be selected.

At this step, trade-off studies between several available options are appropriate. With a modified existing engine these parameters may be well established and require fewer such trade-off studies. Initial analyses of the pressure balances, power distribution between pumps and turbines, gas generator flow, propellant flows and reserves, or the maximum cooling capacity are appropriate. Sketches and preliminary estimates of inert mass of key components need to be made, such as tanks, thrust chambers, TPs, feed and pressurization systems, thrust vector control, or support structure. Alternate arrangements of components (layouts) are usually now examined, often to get the most compact configuration and/or control of the travel of the center of gravity. Initial evaluations of combustion stability, stress analysis of critical components, water hammer, engine performance at some off-design conditions, safety features, testing requirements, cost, and schedule are often performed here. The participation of appropriate experts from manufacturing, field service, materials, stress analysis, and/or safety, can be critical for selecting the proper engine and the key design features. A design review is usually conducted on the selected engine design and the rationale for new or key features.

Test results from subscale or full-scale components, or from related or experimental engines, can have a strong influence on this design process. Any key engine selection decisions need to be later validated in the development process by testing all new components and new engines.

The *inert mass of the engine and other mass properties (center of gravity or moment of inertia)* are key parameters of interest to vehicle designers. They are needed during preliminary design and again, in more detail, in the final design. Engine mass is usually determined by summing up all component or subsystem masses, each of which is either actually weighted or estimated by calculating their volumes and knowing their densities and locations. Sometimes early estimates are based on known similar parts or subassemblies.

Preliminary engine performance estimates are often based on data from prior similar engines. If these are not available, then theoretical performance values can be used (see Chapters 2, 3, and 5) for F, I_s, or others, using appropriately established correction factors (see Chapter 3). Well-measured static test data are, of course, better than estimates. Final performance values are obtained from flight tests, or simulated altitude tests where airflow and altitude effects can interact with the vehicle or the plume.

If the preliminary design does not meet the engine requirements, then initial engine decisions need to be changed and, if that is not sufficient, sometimes also mission requirements. Components, pressure balances, and other items will be reanalyzed, resulting in a modified version of the engine configuration, its inert mass, and performance. This process is iterated until all requirements are met and a suitable engine design has been found. The initial efforts culminate in preliminary layouts of the engine, preliminary inert mass estimates, estimated engine performance, a cost estimate, and a tentative schedule. These preliminary design data may now form the basis for a written proposal to the customer for undertaking the final or detail design, development, testing, and for delivering engines.

Optimization studies help to select the best engine parameters for meeting all requirements; some are done before a suitable engine has been identified, some afterwards. They are described further in Section 11.6. We optimize parameters such as chamber pressure, nozzle area ratio, thrust, mixture ratio, and/or number of large thrust chambers supplied by the sameTP. Results of optimization studies identify the

TABLE 11–3. Data on Three Russian Large Liquid Propellant Rocket Engines Using a Staged Combustion Cycle

Engine Designation	RD-120	RD-170	RD-253
Application (number of engines)	Zenit second stage (1)	Energia launch vehicle booster (4), Zenit first stage (1), and Atlas V (1)	Proton vehicle booster (1)
Oxidizer	Liquid oxygen	Liquid oxygen	N_2O_4
Fuel	Kerosene	Kerosene	UDMH
Number and types of turbopumps (TPs)	One main TP and Two boost TPs	One main TP and Two boost TPs	Single TP
Thrust control, %	Yes	Yes	±5
Mixture ratio control, %	±10	±7	±12
Throttling (full flow is 100%), %	85	40	None
Engine thrust (vacuum), kg	85,000	806,000	167,000
Engine thrust (SL), kg	—	740,000	150,000
Specific impulse (vacuum), sec	350	337	316
Specific impulse (SL), sec	—	309	285
Propellant flow, kg/sec	242.9	2393	528
Mixture ratio, O/F	2.6	2.63	2.67
Length, mm	3872	4000	2720
Diameter, mm	1954	3780	1500
Dry engine mass, kg	1125	9500	1080
Wet engine mass, kg	1285	10,500	1260
Thrust Chamber Characteristics			
Chamber diameter, mm	320	380	430
Characteristic chamber length, mm	1274	1079.6	999.7
Chamber area contraction ratio	1.74	1.61	1.54
Nozzle throat diameter, mm	183.5	235.5	279.7
Nozzle exit diameter, mm	1895	1430	1431
Nozzle area ratio	106.7	36.9	26.2
Thrust chamber length, mm	2992	2261	2235
Nominal combustion temperature, K	3670	3676	3010
Rated chamber pressure, kg/cm^2	166	250	150
Nozzle exit pressure, kg/cm^2	0.13	0.73	0.7
Thrust coefficient, vacuum	1.95	1.86	1.83
Thrust coefficient, SL	—	1.71	1.65
Gimbal angle, degree	Fixed	8	Fixed
Injector type	Hot, oxidizer-rich precombustor gas plus fuel		

With a staged combustion cycle the thrust, propellant flow, and mixture ratio for the thrust chamber have the same values as for the entire engine.

TABLE 11–3. (*Continued*)

Engine Designation	RD-120		RD-170		RD-253	
	Turbopump Characteristics[b]					
Pumped liquid	*Oxidizer*	*Fuel*	*Oxidizer*	*Fuel*	*Oxidizer*	*Fuel*
Pump discharge pressure, kg/cm^2	347	358	614	516	282	251
Flow rate, kg/sec	173	73	1792	732	384	144
Impeller diameter, mm	216	235	409	405	229	288
Number of stages	1	1	1	1 + 1[a]	1	1 + 1[a]
Pump efficiency, %	66	65	74	74	68	69
Pump shaft power, hp	11,210	6145	175,600	77,760	16,150	8850
Required pump NPSH, m	37	23	260	118	45	38
Shaft speed, rpm	19,230		13,850		13,855	
Pump impeller type	Radial flow		Radial flow		Radial flow	
Turbine power, hp	17,588		257,360		25,490	
Turbine inlet pressure, main turbine, kg/cm^2	324		519		239	
Pressure ratio	1.76		1.94		1.42	
Turbine inlet temperature, K	735		772		783	
Turbine efficiency, %	72		79		74	
Number of turbine stages	1		1		1	
	Preburner Characteristics					
Flow rate, kg/sec	177		836		403.5	
Mixture ratio, O/F	53.8		54.3		21.5	
Chamber pressure, kg/cm^2	325		546		243	
Number of preburners	1		2		1	

[a]Fuel flow to precombustor goes through a small second-stage pump.
[b]Includes booster pump performance where applicable.
(From NPO Energomash, Khimki, Russia.)

best parameters that will give some further (usually small) improvement in vehicle performance, propellant fraction, engine volume, or cost.

Once the engine proposal has been favorably evaluated by the vehicle designers and after the customer has provided authorization and funding to proceed, then the final design can begin. Some of analyses, layouts, and estimates will be repeated in more detail, specifications and manufacturing documents will be written, vendors will be selected, and tooling will be built. Any selection of key parameters (particularly those associated with technical risk) will need to be validated. After other design review, key components and prototype engines are built and ground tested as part of a planned development effort. If proven reliable, one or two sets of engines will be installed in vehicles and operated during flight. In those programs where a fair number of vehicles are to be built, the engine will then be produced in the required quantity.

Table 11–3 shows detailed parameters for three different Russian staged-combustion-cycle engines designs (from Ref. 11–6). It shows primary engine

FIGURE 11–1. The RD-170 rocket engine, shown here on a transfer cart, can be used as an expendable or reusable engine (up to 10 flights). It has been used on Energiya launch vehicles. The tubular engine structure supports the four hinged thrust chambers and its control actuators. It has the highest known thrust of liquid rocket engines. From NPO Energomash, Khimki, Russia, Ref. 11–6.

parameters (chamber pressure, thrust, specific impulse, mass, propellant combination, nozzle area ratio, dimensions, etc.) that influence vehicle performance and alternate configurations. It also shows secondary parameters, those internal to the engine but important in component design and engine studies, health monitoring systems, and optimization. Figure 11–1 shows the RD-170 engine with four thrust chambers (and their thrust vector actuators) supplied by a centrally located single large TP (257,000 hp; not visible in the photo) and one of the two oxidizer-rich preburners. Figure 11–2 shows a flow diagram schematic for this RD-170 rocket engine—it identifies key components of the large main TP, the two oxidizer-rich

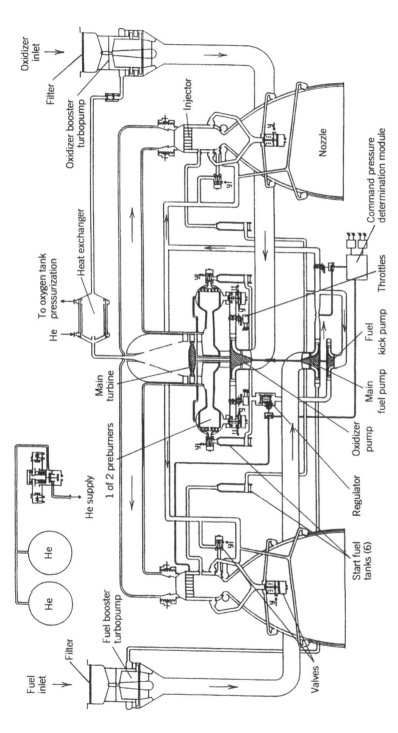

FIGURE 11–2. Simplified flow diagram of the RD-170 high-pressure rocket engine. The single-stage reaction turbine, two fuel pumps, and a single-stage oxygen pump with an inducer impeller. All of the oxygen and a small portion of the fuel flow supply two preburners. The oxidizer-rich gas drives the turbine, then entering the four thrust chamber injectors (only two are shown). The two booster pumps prevent cavitation in the main pumps. The pressurized helium subsystem (only partially shown) supplies various actuators and control valves; it is indicated by the symbol y. Ignition is accomplished by injecting a hypergolic fuel into the two preburners and the four thrust chambers. From NPO Energomash, Khimki, Russia, from Ref. 11–6.

411

preburners, and the two booster TPs; one is driven by a turbine using oxygen-rich gas bled from the turbine exhaust (this gas is condensed when it mixes with the liquid oxygen flow) and the other by a liquid turbine driven by a high-pressure liquid fuel. A version of this RD-170 engine, identified as the RD-180 rocket engine with two thrust chambers is used in the first stage of the U.S. Atlas V space launch vehicle. A single chamber derivative is the current RD-191 discussed in Chapters 6, 8, and 9.

Much of today's engine preliminary design and design optimization can be performed with computers. These involve codes for calculating stress/strain and heat transfer, mass properties, water hammer, engine performance, feed system analyses (for balance of flow, pressures, and power), gas pressurization, combustion vibrations, and various exhaust plume effects (Refs. 11–4 and 11–5 are examples of such analyses and design). Some customers require certain analyses results (e.g., safety, static test performance) to be delivered to them prior to engine deliveries.

Many computer programs are specific to a certain category of applications (e.g., interplanetary flight, air-to-air combat, long-range ballistic missile, or ascent to earth orbit), and many are specific to a particular engine cycle. For example, an *engine balance program* balances the pressure drops in the fuel, oxidizer, and pressurizing gas flow systems; similar programs balance the pump and turbine power, speeds, and torques (see Section 10.4), compare different TP configurations (see Section 10.3); some balance programs also calculate approximate masses for engine, tanks, or turbine drive fluids. Such programs allow iterations with various pressures and pressure drops, mixture ratios, thrust levels, number of thrust chambers, distribution of total velocity increment between different vehicle stages, trades between constant thrust (or propellant flow) and decreasing thrust (throttling) or pulsed (intermittent) thrust.

11.4. ENGINE CONTROLS

All liquid propellant rocket engines must have controls that accomplish some or all of these tasks (see Refs. 11–7, 11–8, and 11–9):

1. Start rocket operation.
2. Shut down rocket operation.
3. Restart, if desired. With small thrusters there may be thousands of restarts.
4. Maintain programmed operation (e.g., predetermined thrust profile, presets of propellant mixture ratio and flow). A constant flow of propellants can be achieved with *sonic flow venturis* in the feed lines as discussed in the last paragraph of Section 6.9.
5. Use an engine health monitoring system to prevent certain engine failures or performance losses (as explained later in this chapter).
6. Fill propellants into their vehicle tanks.
7. Drain excess propellant after operation of a reusable or test engines.

8. Cool (with cryogenic propellants) pipes, pumps, cooling jackets, injectors, and valves that must be at low temperatures prior to start, by bleeding cryogenic propellant through them. This cooling propellant is not used to produce thrust; its flow has to be periodically controlled.

9. Check out proper functioning of critical components or a group of components without actual hot operation before and/or after flight or ground tests.

10. Provide features to perform checks and recycle engines to a ready condition, as needed for recoverable or reusable rocket engines and for engines used in ground/development tests.

The *complexity* of these control elements together with the complexity of engine systems strongly depend on vehicle mission. In general, rockets that are used only once (single-shot devices), that are filled with storable propellants at the factory, that operate at nearly constant propellant flow, and that operate over a narrow range of environmental conditions tend to be simpler than rocket systems intended for repeated use or for applications where satisfactory operation must be demonstrated prior to use, and for manned vehicles. Because of the nature of liquid propellants, most control actuation functions are achieved remotely by valves, regulators, pressure switches, valve position indicators, or calibrated orifices. The use of dedicated computers for automatic control in large engines is now common. Flow control devices, namely valves, are discussed in Section 6.9 and other controls are discussed in this section.

Safety controls are intended to protect personnel and equipment in case of malfunctions. This applies mostly to development engines during ground tests but can also apply to certain flight test engines and to certain operating engines. For example, any control system is usually so designed that failure of an electrical power supply to the rocket causes a nonhazardous shutdown (all electrical valves automatically returning to their normal position), and no mixing or explosion of unreacted propellant can occur. Another example is the use of an electrical interlock device which prevents the opening of the main propellant valves until the igniter has functioned properly.

Check-out controls permit a simulation of the operation of critical control components without actual hot operation of the rocket unit. For example, many rockets have provisions for permitting actuation of the principal valves without having propellant or pressure in the system.

Control of Engine Starting and Thrust Buildup

During a rocket engine's *starting* and *stopping* processes, the mixture ratio should be expected to vary considerably from the rated design mixture ratio (because of a lead in of one of the propellants and because the hydraulic resistances to propellant flow are usually not the same for the fuel and the oxidizer passages). During this transition period, it is possible for the rocket engine to pass through regions of chamber pressure and mixture ratio which produce combustion instabilities. The starting and stopping of a rocket engine thus requires very critical timing, valve sequencing, and transient characteristics. A good control system must be designed to avoid undesirable transient effects. Close *controls* of propellant *flow,* of *pressure,* and of *mixture ratio* are necessary to obtain reliable, repeatable and safe rocket performance. The starting and ignition of thrust chambers is discussed in Section 8.6.

Fortunately, most rocket units operate with a nearly constant propellant consumption and constant mixture ratio, which simplifies the operating control problem. Stable operation of liquid propellant flows can be accomplished without automatic control devices because liquid flow systems in general tend to be inherently stable. This means that the propellant feed system reacts to any disturbance in the flow of propellant (a sudden flow increase or decrease) in such a manner as to reduce the effect of the disturbance. The system, therefore, usually has a natural tendency to control itself. However, in some cases natural resonances of the system and/or its components can exist at frequency values that tend to destabilize the system.

Start delay times for a pressure feed system are always present but usually small. Prior to start, the pressurization system has to be activated and any ullage volume pressurized. This start delay is also the time to purge the system (if needed), open valves, initiate combustion, and raise the flow and chamber pressure to rated values. Turbopump systems usually require more time to start; in addition to the above starting steps for pressurized systems, TPs have to allow a period for starting gas generators or preburners and for bringing the unit up to full speed and to a discharge pressure at which combustion can be self-sustained. If the propellant is nonhypergolic, additional time has to be allowed for an igniter to operate and for feedback to confirm that it is working properly. All these events need to be controlled. Table 11–4 describes many of these typical steps, but not all of them belong with every engine.

Starting small thrusters with pressurized feed systems can be relatively fast, as short as 3 to 15 msec, enough time for a small valve to open, propellant to flow into the chamber and ignite, and for the small chamber volume to be filled with high-pressure combustion gases. In an engine with a pressurized feed system the initial flow of each propellant is often considerably higher than the rated flow at full thrust because the pressure differential ($p_{tank} - p_1$) is much higher, with the chamber pressure initially being at its lowest value. These higher flows can lead to propellant accumulation in the chamber and may lead what is called a "hard start," with an initial surge of chamber pressure. In some cases, this surge has damaged the chamber. One solution has been to slowly open the main propellant valves or to build a throttling mechanism into them.

For TP-fed systems and larger thrust engines, the time from start signal to full chamber pressure is about 1 to 5 sec in part because pump rotors have inertia, the igniter flame has to heat the relatively large initial mass of propellants, the propellant line volumes to be filled are large, and the number of events or steps that need to take place are more numerous.

Large TP-fed rocket engines have been started in at least the following four ways:

1. *A solid propellant start grain* or *start cartridge* is used to pressurize the gas generator or preburner, and this starts turbine operations. This method was used on Titan III hypergolic propellant (first and second stages) and on the H-1 (nonhypergolic) rocket engines—where the start grain flame also ignited the liquid

TABLE 11–4. Major Steps in the Starting and Stopping of a Typical Large Liquid Bipropellant Rocket Engine with a Turbopump Feed System

---

1. *Prior to Start*

Check out functioning of certain components (without propellant flow), such as the thrust vector control or some valve actuators (optional).

Make sure that tanks and pipes are clean. Fill tanks with propellants.

Bleed liquid propellants to eliminate pockets of air or gas in all pipes up to the propellant valves.

When using propellants that can interact with air (e.g., hydrogen can freeze air, small solid air crystals can plug injection holes, and solid air crystals with liquid hydrogen can form an explosive mixture; some hypergolic start propellant will burn in air), it is necessary to purge the piping system (including injector, valves, and cooling jacket) with an inert, dry gas (e.g., helium) to remove air and moisture. In many cases, several successive purges are undertaken.

With cryogenic propellants the piping system needs to be cooled to cryogenic temperatures to prevent vapor pockets. This is done by repeated bleeding of cold propellant through the engine system (valves, pumps, pipes, injectors, etc.) just prior to start. The vented cold gas condenses moisture droplets in the air and this looks like heavy billowing clouds escaping from the engine.

Refill or "top off" tank to replace cryogenic propellant that has evaporated or been used for cooling the engine.

Pressurize vehicle's propellant tanks just before start.

2. *Start: Preliminary Operation*

Provide start electric signal, usually from vehicle control unit or test operator.

With nonhypergolic propellants, start the ignition systems in gas generator or preburner and main thrust chambers; for nonhypergolic propellants a signal has to be received that the igniter is burning before propellants are allowed to flow into the chambers.

Initial operation: opening of valves (in some cases only partial opening or a bypass) to admit fuel and oxidizer at low initial flows to the high-pressure piping, cooling jacket, injector manifold, and combustion chamber(s). Valve opening rate and sequencing may be critical to achieve proper propellant lead. Propellants start to burn and turbine shaft begins to rotate.

Using an automated engine control, make checks (e.g., shaft speed, igniter function, feed pressures) to assure proper operation before initiating next step.

In systems with gearboxes the gear lubricant and coolant fluid start to flow.

For safety reasons, one of the propellants must reach the chamber first.

3. *Start: Transition to Full Flow/Full Thrust*

Turbopump power and shaft speed increase.

Propellant flows and thrust levels increase until they reach full-rated values. May use controls to prevent exceeding limits of mixture ratio or rates of increase during transient.

Principal valves are fully opened. Attain full chamber pressure and thrust.

In systems where vaporized propellant is fed into the propellant tanks for tank pressurization, the flow of this heated propellant gas is initiated.

Systems for controlling thrust or mixture ratio or other parameter are activated.

4. *Stop*

Signal to stop deactivates the critical valve(s).

Key valves close in a predetermined sequence. For example, the valve controlling the gas generator or preburner will be closed first. Pressurization of propellant tanks is stopped.

As soon as turbine drive gas supply diminishes, the pumps will slow down. Pressure and flow of each propellant will diminish quickly until it stops. The main valves are closed, often by spring forces, as the fluid pressures diminish. In some engines the remaining propellant trapped in the lines or cooling jacket may be blown out by propellant vapor or inert gas purge.

propellants in the gas generator. This is usually the fastest starting method, but it does not provide for restarts.

2. The *tank head start* (used on the SSME) method is slower, does not require a start cartridge, and permits engine restarts. The "liquid head" from the vehicle tanks in vertically launched large vehicles (the term *head* is defined in Section 10.5), plus the tank pressure cause a small initial flow of propellants; then slowly more pressure is built up as the turbine begins to operate, and in a few seconds the engine "bootstraps"; its flows and pressures then rise to their rated values.

3. A small *auxiliary pressurized propellant feed system* with its own propellant tanks is used to feed an initial quantity of fuel and oxidizer (at essentially full pressure) to the thrust chamber and gas generator. This method was used on one version of the RS-27 engine in the first stage of a Delta II space launch vehicle.

4. The *spinner start* method uses stored high-pressure gas from a separate tank to spin the turbine (usually at less than full speed) until the engine provides enough hot gas to drive the turbine. High-pressure tanks are heavy and their connections add complexity; in booster engines, the gas tank can be part of the ground equipment. This method is used on the RS-68 engine where its high-pressure helium tank is part of the ground equipment during launch.

Sample Start and Stop Sequences. This is an example of the transient start and stop sequence now used in the RS-25 rocket engine and previously in the SSME (Space Shuttle Main Engine now retired). Both are complex staged combustion cycle engines designed for a "tank head start." The flow diagram in Fig. 6–11 and the engine view of Fig. 6–1 identify the location within the engine of the key components mentioned below, and Fig. 11–3 shows the sequence and events of these transients. This section and figure are based on information provided by Aerojet-Rocketdyne some years ago.

As stated earlier, for tank head starts, the energy to start the turbines spinning is all derived from initial propellant tank pressures (fuel and oxidizer) and from gravity (the head of a liquid column). Combining the tank head start with a staged combustion cycle consisting of four TPs, two preburners, and a main combustion chamber (MCC) results in a complicated and sophisticated start sequence, but one which proved to be robust and reliable. Prior to the start, the TPs and ducting (down to the main propellant valves) are chilled with liquid hydrogen and liquid oxygen (LOX) to cryogenic temperatures to ensure liquid propellants temperatures for proper pump operation. At engine start command, the main fuel valve (MFV) is opened first, providing chilling below the MFV and a fuel lead to the engine. Three oxidizer valves sequence the main events during the crucial first 2 sec of start—the fuel preburner oxidizer valve (FPOV) is ramped to 56% to provide LOX for ignition at the fuel preburner (FPB) in order to provide initial turbine torque for the high-pressure fuel turbopump (HPFTP). Fuel system oscillations (FSOs), occurring due to heat transfer downstream of the initially chilled system, could result in flow rate dips that can lead to damaging

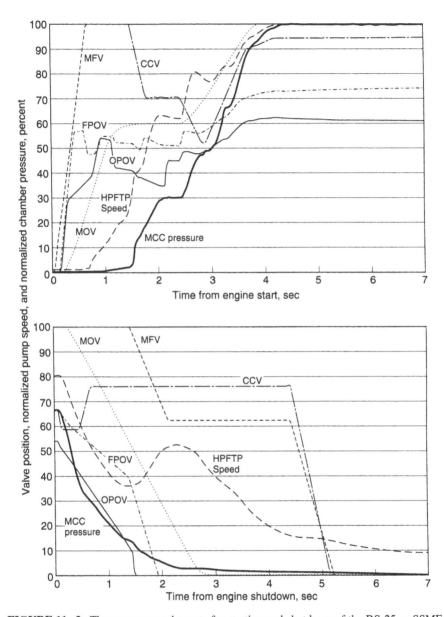

FIGURE 11–3. The sequence and events for starting and shutdown of the RS-25 or SSME (Space Shuttle Main Engine). This particular start sequence leads to a chamber pressure of 2760 psia (normalized here to 100%), a high-pressure fuel turbopump speed of 33,160 rpm (100%), at a sea-level thrust of 380,000 lbf (shown as 100%). This shutdown occurs at altitude when the engine has been throttled to 67% of its power level or a vacuum thrust of 312,559 lbf, which is shown as 67% of the chamber pressure of the main combustion chamber. Courtesy of Aerojet Rocketdyne.

temperature spikes in the FPB as well as in the oxidizer preburner (OPB) at ignition and at 2 Hz cycles thereafter, until the hydrogen is above critical pressure. The oxidizer preburner oxidizer valve (OPOV) and the main oxidizer valve (MOV) were next ramped-open to provide LOX for OPB and MCC ignition.

The next key event was FPB prime which consists of filling the LOX system upstream of the injectors with liquid propellant. This results in increased combustion and higher power. This event occurred around 1.4 sec into start. The HPFTP speed was automatically checked at 1.24 sec into start to ensure it would be at a high enough level before the next key event, MCC prime, which was controlled by the MOV. Priming and valve timing were critical. We mention next some of the events that could go wrong: At MCC prime, an abrupt rise in backpressure on the fuel pump/turbine occurs. If flow rate through the fuel pump at this time is not high enough (as indicated by the shaft speed), then the heat imparted to the fluid as it is being pumped can vaporize it, leading to unsatisfactory flow in the engine, and subsequent high mixture ratio with high gas temperatures and possible system burnouts in hot gas system. This occurs when the MCC primes too early or HPFTP speed is abnormally low. If the MCC primes too late, the HPFTP may accelerate too fast due to low backpressures after FPB prime and exceed its safe speed. The MCC prime normally occurs at 1.5 sec. The OPB was primed last since it controls LOX flow; here a strong fuel lead and healthy fuel pump flow are desirable to prevent engine burnouts (due to high mixture ratios). The OPOV provided minimal flow rates during the early part of the start to force the oxidizer prime to last for 1.6 sec into start. Again, the FSO influences temperature spikes in the OPB and was sequenced around and prior to the MCC prime which raises the fuel pressure above critical in the fuel system. At 2 sec into start, the propellant valves were sequenced to provide 25% of rated power level (RPL). During the first 2.4 sec of start, the engine was in an open-loop mode, but here proportional control of the OPOV was used, based on MCC pressure. At this point, additional checks were carried out to ensure engine health, and a subsequent ramp to mainstage at 2.4 sec was done using closed-loop MCC-chamber-pressure/OPOV control. At 3.6 sec, closed-loop mixture ratio/FPOV control was activated.

The chamber cooling valve (CCV) was opened at engine start and sequenced to provide optimum coolant fuel flow to the nozzle cooling jacket, the chamber and preburners during the ignition and main stage operation. It diverted flow to the cooling passages in the nozzle after MCC prime causes the heat load to increase. The description above is simplified and does not mention several other automatic checks, such as verifying ignition in the MCC or FPB or the fuel or chamber pressure buildup, which were sensed and acted upon at various times during the start sequence. Spark-activated igniters were built into the three injectors (MCC, FPB, OPB) using the same propellants. They are not mentioned above or shown in the flow sheet.

The shutdown sequence was initiated by closing the OPOV, which powered down the engine (reducing oxygen flow, chamber pressure, and thrust); this was followed quickly by closing the FPOV, so the burning would shut down fuel rich. Shortly thereafter the MOV was closed. The MFV stayed open for a brief time and then was moved into an intermediate level to balance with the oxygen flow (from trapped oxygen downstream of the valves). The MPV and the CCV were closed after the main oxygen mass had been evaporated or expelled.

Automatic Controls

Automatically monitored controls are frequently used in liquid propellant rockets to accomplish thrust control or mixture ratio control. Automatic control of thrust vectors is discussed in Chapter 18.

Before electronic controls became common for large engines, pneumatic controls with helium gas were used. Helium is still used to actuate large valves, but no longer for logic control. A pressure ladder sequence control has been used where pressures (and a few other quantities) were sensed and, if satisfactory, the next step of the start sequence was pneumatically initiated. This arrangement was used on the U.S. H-1 engine.

Most automatic controls utilize servomechanisms. In general, they consist of three basic elements: a *sensing mechanism,* which measures or senses the variable quantity to be controlled; a *computing or controlling mechanism,* which compares the output of the sensing mechanism with a reference value and gives a control signal to the third component, the *actuating device,* which manipulates the variable to be controlled. Additional discussion of computer control with automatic data recording and analysis is given in Chapter 21.

Figure 11–4 shows a typical simple thrust control system for a gas-generator-cycle engine aimed at regulating chamber pressure (and therefore thrust) during flight to a predetermined value. A pressure-measuring device with an electric output is used as the sensing element, and an automatic control device compares this gauge output signal with a signal from the reference gauge or a computer referenced voltage value and thus computes an error signal. This error signal is amplified, modulated, and fed to the actuator of the throttle valve. By controlling the propellant flow to the gas generator, the generator pressure is regulated and, therefore, also the pump speed and the main propellant flow; indirectly, the chamber pressure in the thrust chamber is also regulated and, therefore, also the thrust. These quantities are varied until such time as the error signal approaches zero. Such system has been vastly simplified here, for the sake of illustration; in actual practice the system may have to be integrated with other automatic controls. In Figure 11–4, the mixture ratio in the gas generator is controlled by the pintle shapes of the fuel and oxidizer valves of the gas generator and by yoking these two valves together and having them moved in unison with a single actuator.

In the expander cycle shown schematically in Fig. 6–10, the thrust is regulated by maintaining a desired chamber pressure and controlling the amount of hydrogen gas flowing to the turbine by means of a variable bypass. The flow through this bypass is small (typically 5% of gas flow) and is controlled by the movement of a control valve.

In *propellant utilization* systems, the mixture ratio is varied to ensure that both fuel and oxidizer propellant tanks are simultaneously and completely emptied; no undue useable propellant residue should remain because it increases the empty mass

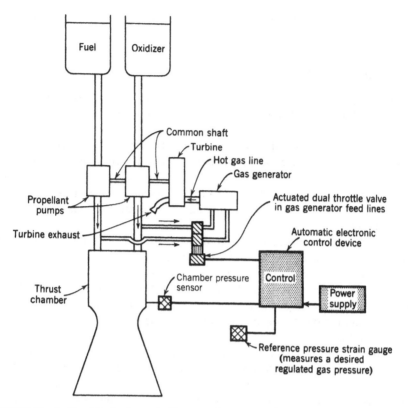

FIGURE 11–4. Simplified schematic diagram of an automatic servomechanism-type chamber pressure control of a liquid propellant rocket engine with a turbopump feed system, a gas generator, and a tank head, boot strap (self-pumping) starting system.

of the vehicle, which in turn detrimentally decreases the vehicle mass ratio and the vehicle's flight performance (see Chapter 4). For example, the flow rate of oxidizer may be somewhat larger than normal due to its being slightly denser than normal or due to a lower than normal injector pressure drop; if uncontrolled, some fuel residue would remain at the time of oxidizer exhaustion; however, the control system could cause the engine to operate for a period at a propellant mixture ratio slightly more fuel-rich than normal, to compensate and assure almost simultaneous emptying of both propellant tanks. Such control system requires accurate measurements of the amount of propellant remaining in the two propellant tanks during the flight.

Any one of the three principal components of an automatic control system may have several different forms. Typical sensing devices include those that measure chamber pressures, propellant pressures, pump rotational speeds, tank levels, and/or propellant flows. An actuating device can throttle propellant flow or control a bypass device or the gas generator discharge. There are many operating mechanisms for the controller, such as direct electrical devices, electronic analog or digital computers, hydraulic or pneumatic devices, and mechanical devices. Actuators can be driven by electrical motors or hydraulic, pneumatic, or mechanical power. Hydraulic actuators provide very high forces and quick responses. The exact type of component, the nature of the power supply, the control logic, the system type, and the operating mechanisms for the specific control will depend on details of the application and the requirements. Controls are discussed further in Refs. 11–4, 11–8 and 11–9.

In applications where the final vehicle velocity must be accurately determined, the amount of impulse that is imparted to the vehicle during any cutoff transient may be sufficiently variable to exceed any desired velocity tolerance. Therefore, for these applications close control over the thrust decay curve is necessary, and this can be accomplished by automatic control over the sequencing and closing rates of the main propellant valves and the location of the valves in relation to the injector.

Control by Computer

Early rocket engines used simple timers and, later, a pressure ladder sequence to send commands to the engine for actuating valves and other steps in the operation. Pneumatic controllers were also used in some engines for starting and stopping. For the last 35 years *digital computers* have been used in large liquid propellant rocket engines for controlling their operation. In addition to controlling engine start and stop, they can do much more and have contributed to making engines more reliable. Table 11–5 gives a list of typical functions undertaken by modern engine control computers. This list covers primarily one or more large TP-fed engines but does not include consideration of multiple small thruster attitude control rocket engines.

Actual designs of control computers are not presented in this text. In general, designers have to carefully consider all possible engine requirements, all functions that need be monitored, all likely potential failure modes and their compensating or ameliorating steps, all sensed parameters and their scales, methods of control (such as open, closed, or multiple loops, adaptive or self-learning/expert systems), system architecture, software approaches and their interrelation and division of tasks with other computers on board the vehicle or on the ground, and methods for validating events and operations. It is also convenient to have reprogrammable software that will allow changes (which may become necessary because of engine developments or failures) and allow the control of several parameters simultaneously. While the number of functions performed by control computers has increased in the past 35 years, their size and mass has decreased considerably.

The control computer is usually packaged in a waterproof, shockproof metal box, which is mounted on the engine. Fire-resistant and waterproof cable harnesses lead from this box to all the instrument sensors, valve position indicators, tachometers, accelerometers, actuators, and other engine components, to the power supply and the vehicle's controller; an umbilical, severable multiwire harness then leads to ground support equipment. Reference 11–9 describes the controller used for the Space Shuttle Main Engine.

TABLE 11–5. Typical Functions Performed by Digital Computers in Monitoring and Controlling the Operation of a Large Liquid Propellant Rocket Engine

1. *Sample the signals from significant sensors* (e.g., chamber pressure, gas and hardware temperatures, tank pressure, valve position, etc.) at frequent intervals, say once, 10, 100, or 1000 times per second. For parameters that change slowly (e.g., the temperature of the control box), sampling every second or every 5 sec may be adequate, but chamber pressure would be sampled at a high frequency.
2. *Keep a record of all the significant signals* received and all the signals generated by the computer and sent out as commands or information. Old records have at times been very important.
3. *Control and verify the steps and sequence of the engine start.* Figure 11–3, and Table 11–4 list typical steps that have to be taken, but do not list the measured parameters that will confirm that the commanded step was implemented. For example, if the igniter is activated, a signal change from a properly located temperature sensor or a radiation sensor could verify that the ignition had indeed happened.
4. *Control the shutdown of the engine.* For each of the steps listed at the bottom of Table 11–4 or in Fig. 11–3 there often has to be a sensing of a pressure change or other parameter change to verify that the commanded shutdown step was taken. An *emergency shutdown* may be commanded by the controller during development testing, when it senses certain kinds of malfunctions that allow the engine to be shut down safely before a dramatic failure occurs. This emergency shutdown procedure must be done quickly and safely and may be different from a normal shutdown, and must avoid creating a new hazardous condition.
5. *Limit the duration of full thrust operation.* For example, cutoff is to be initiated just before the vehicle attains the desired mission flight velocity.
6. *Safety monitoring and control.* Detect combustion instability, overtemperatures in precombustors, gas generators, or TP bearings, violent TP vibration, TP overspeed, or other parameter known to cause rapid and drastic component malfunction that can quickly lead to engine failure. Usually, more than one sensor signal will show such a malfunction. If detected by several sensors, the computer may identify it as a possible failure whose in-flight remedy is well known (and preprogrammed into the computer); then a corrective action or a safe shutdown may be automatically commanded by the control computer. This applies mostly to development engines during ground tests.
7. *Analyze key sensor signals for deviation from nominal performance* before, during, and after engine operation. Determine whether sensed quantities are outside of predicted limits. If appropriate and feasible, if more than one sensor indicates a possible out-of-limit value, and if the cause and remedy can be predicted (preprogrammed), then the computer can automatically initiate a compensating action. Parts of or combinations of items 6 and 7 have been called *engine health monitoring systems.* They are discussed in Section 11.5.
8. *Control propellant tank pressurization.* The tank pressure value has to be within an allowable range during engine operation and also during a coasting flight period prior to a restart. Sensing the activation of relief valves on the tank confirms overpressure. Automatically, the computer can then command stopping or reducing the flow of pressurant.
9. *Perform automatic closed-loop control of thrust and propellant utilization* (described before).
10. *Transmit signals to a flying vehicle's telemetering system,* which in turn can send them to a ground station, thus providing information on the engine status, particularly during experimental or initial flights.
11. *Self-test the computer and software.*

11.5. ENGINE SYSTEM CALIBRATION

Although engines are designed to deliver a specific performance (F, I_s, $\dot{m}$, r), a newly manufactured engine will not usually perform precisely at their nominal parameters. The calibration process provides necessary corrections to the engine system, so it will perform at the rated/intended operating conditions. When deviations from nominal performance are more than a few percent, the vehicle will probably not complete its intended mission. There are several sources for these deviations. Because of unavoidable dimensional tolerances on the hardware, the flow–pressure time profile or injector jet impingements (related to combustion efficiency) will deviate slightly from nominal design values. Even a small change in mixture ratio can cause a significant increase of residual propellant. Also, minor changes in propellant composition or storage temperatures (which affect their density and viscosity) may cause significant deviations. Other factors involve regulator setting tolerances or changes in flight acceleration (affecting the static head). Engine calibration is the process of adjusting some of its internal parameters so that it will deliver the intended performance within the allowed tolerance bands. See Refs. 11–4 and 11–5.

Hydraulic and pneumatic components (valves, pipes, expansion joints) can readily be tested on water-flow benches to determine their pressure drop at rated flow (corrected for the propellant density and viscosity). Components that operate at elevated temperatures (thrust chambers, turbines, preburners, etc.) have to be hot fired and cryogenic components (pumps, some valves) often have to be tested at the cryogenic propellant temperatures. Engine characteristics may be estimated by adding together the corrected values of pressure drops at the desired mass flow. Furthermore, the ratio of rated flows $\dot{m}_o/\dot{m}_f$ has to equal the desired mixture ratio r. This is shown in the example below. Adjustments include adding pressure drops with judiciously placed orifices or changing valve positions or regulator settings.

In most pressurized feed systems, the gas is supplied from its high-pressure tank through a regulator to pressurize both the fuel and the oxidizer in their respective tanks. The two pressure drop equations for the oxidizer and the fuel (subscripts o and f) are given below for a pressurized feed system at nominal flows:

$$p_{\text{gas}} - (\Delta p_{\text{gas}})_f = p_1 + \Delta p_f + (\Delta p_{\text{inj}})_f + (\Delta p_j)_f + \frac{1}{2}\rho_f v_f^2 + La\rho_f \quad (11\text{–}6)$$

$$p_{\text{gas}} - (\Delta p_{\text{gas}})_o = p_1 + \Delta p_o + (\Delta p_{\text{inj}})_o + \frac{1}{2}\rho_o v_o^2 + La\rho_o \quad (11\text{–}7)$$

The gas pressure available in both of the propellant tanks is the regulated pressure p_{gas}, diminished by the pressure losses in the gas line Δp_{gas}, which includes the pressure drop across a pressure regulator. The static head of a liquid, $La\rho$ (L is liquid level distance above the thrust chamber, a is flight acceleration, and ρ is propellant density), acts to augment the gas pressure. It has to equal the chamber pressure p_1 plus all other pressure drops in the liquid piping or valves, namely, Δp (Δp_f or Δp_o), the injector's Δp_{inj}, the cooling jacket's Δp_j, and the dynamic flow head $\frac{1}{2}\rho v^2$. When the required liquid pressures (right-hand side of Eq. 11–6 and 11–7.) do not equal the gas pressure in the propellant tank at the nominal propellant flow (left hand side of equations), then additional pressure drops (from calibration orifices) have to be inserted. A good design should provide extra pressure drop margins for this purpose.

Two methods are available for precise control of engine performance parameters. One uses an automatic system with feedback, throttling valves and a digital computer to control any deviations in real time. The other relies on the initial static calibration of the engine system. The latter approach is simpler and is preferred, and can be quite accurate.

Pressure balancing is the process of balancing the available pressure supplied to the engine (by pumps, the static head, and/or pressurized tanks) against the liquid pressure drops plus the chamber pressure. This balancing is done in order to calibrate the engine so it will operate at the desired flows and mixture ratios. Figure 11–5 shows pressure balances for one propellant branch of a bipropellant engine with a pressurized feed system. It displays the pressure drops (for injector, cooling jacket passages, pressurizing gas passages, valves, propellant feed lines, etc.) and chamber pressure against propellant flow, using actual component pressure drop measurements

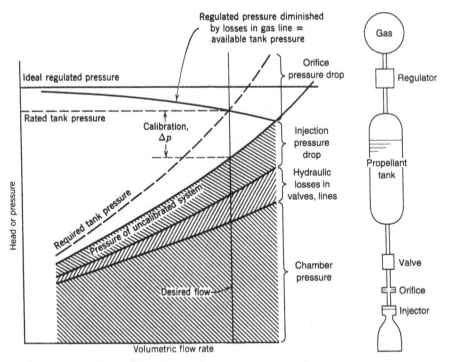

FIGURE 11–5. Simplified flow diagram and balance curves for the fuel or the oxidizer of a typical gas-pressurized bipropellant feed system. This diagram is also the same for a monopropellant feed system, except that here there may not be a calibration orifice; calibration is done by setting the proper regulated pressure. Hydraulic losses include friction in both the liquid piping and the cooling jacket.

(or estimated values) corrected for the different flows. These curves are generally plotted in terms of a "head loss" and of the "volumetric flow" to eliminate the fluid density as an explicit variable for a particular regulated pressure. These regulated pressures are usually the same in the fuel and oxidizer pressure balance and also can be adjusted. This balance of head (the term *head* is defined in a footnote in Section 10.5) and flow must be made for both the fuel and oxidizer systems because the ratio of their flows establishes the actual mixture ratio and the sum of their flows establishes the thrust. The pressure balance between available and required tank pressure, both at the desired flow, is achieved by adding a calibration orifice into one of the lines, as can be seen in Fig. 11–5. Not shown in the figure is the static head provided by the elevation of the liquid level, since it is relatively small for many space launch systems. However, with high acceleration and dense propellants, it can be a significant addition to the available head.

For a pumped feed system of a bipropellant engine, Fig. 11–6 shows a balance diagram for one branch of the two propellant systems. Here, pump speed is an additional variable. Calibration procedures are usually more complex for TP systems because pump calibration curves (flow–head–power relation) cannot readily be estimated without good test data and cannot be easily approximated analytically. Propellant flows to a gas generator or preburner also need to be calibrated. In this case, the turbine shaft torque has to equal the torque required by the pumps plus the losses in bearings, seals, and/or windage. Thus, a power balance must be achieved in addition to the matching of pressures and the individual propellant flows. Since these parameters are interdependent, the determination of calibration adjustments may not always be simple. Many rocket organizations have developed computer programs to analyze some or all this required balancing.

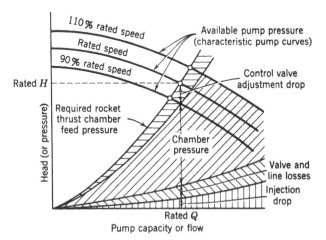

FIGURE 11–6. Simplified diagram of the balance of available and required feed pressures versus flow for one of the propellants in a bipropellant rocket engine with a turbopump feed system. Chamber pressure includes the liquid column.

Example 11–2.   This example shows one way for converting component test data into nominal engine data. The following component information and design requirements are given for a pressurized liquid propellant rocket system similar to that in Figs. 1-3 and 11-5: fuel, 75% ethyl alcohol; oxidizer, liquid oxygen; desired mixture ratio, 1.30; desired thrust, 500 lbf at sea level. For this propellant combination, the combustion gases $k = 1.22$. The tank temperatures are 70°F for the alcohol fuel and boiling point (90 K or 162°R) for the liquid oxygen. Assume they do not vary during operation or from run to run. The nozzle throat to exit area ratio is 4.0. Assume a nominal chamber pressure of 300 psia, but correct later as needed.

Component test data: Pressure losses in gas systems, including the pressure regulator, were found to be 8.0 psi average for the operating duration at rated propellant flow. Fuel valve and fuel line losses were 9.15 psi at a water flow of 0.963 lbm/sec. Oxidizer valve and line losses were 14.2 psi at a flow of 1.28 lbm/sec of liquid oxygen. Fuel cooling jacket pressure loss was 35.0 psi at a water flow of 0.961 lbm/sec. Oxidizer injector pressure drop was 35.0 psi at 1.40 lbm/sec of oxygen flow under thrust chamber operating conditions but without fuel or combustion. Fuel injector pressure drop was 40.0 psi at 1.02 lb/sec of fuel flow under thrust chamber operating conditions, but without combustion. Average measured results of several sea-level actual hot thrust chamber tests were: thrust = 541 lbf; mixture ratio = 1.29; specific impulse = 222 sec; and the chamber *pressure* = 328 psia. Determine the sea-level regulator setting and size and location of calibration orifices to achieve 500 lbf of thrust.

SOLUTION. First, the corrections necessary to obtain the desired thrust chamber conditions have to be determined. The experimental thrust chamber data must be adjusted for deviations in mixture ratio, thrust, and specific impulse. The variation of specific impulse with *mixture ratio* is determined from experimental data or (on a relative basis) from theoretical calculations similar to those that are the basis of Fig. 5–1. Because the value of I_s at the desired mixture ratio of 1.30 is within 0.08% of the value of I_s under the actual test conditions (where $r = 1.29$), any mixture ratio correction of I_s is small enough to be neglected here.

One has to include the *head of the liquid column* (between the liquid level in the propellant tank and the inlet of the thrust chamber). It is a maximum at the beginning of flight and decreases as propellant is consumed. For small engines this head can often be neglected when the propellant tanks are close to the thrust chamber. We will assume an average head of about 2.5 ft or 0.92 psi of fuel and about 4.0 ft or approximately 2.0 psi of liquid oxygen.

The correction of *specific impulse* for the actual chamber pressure is made next. The specific impulse is essentially proportional to the thrust coefficients as determined from Eq. 3–30. For $k = 1.22$, the actual measured pressure ratios $p_1/p_3 = 328/14.7 = 22.2$, and the nominal value $300/14.7 = 20.4$, the values of C_F can be calculated as 1.420 and 1.405, respectively. In this calculation p_2 has been determined for isentropic conditions, such as those in Fig. 3–6 or 3-7 for the given nozzle area ratio. The sea-level-specific impulse is therefore corrected to $I_s = 222(1.425/1.420) = 223$ sec. The *chamber pressure* has to be reduced from 328 psi to a lower value in order to bring the thrust from its test value of 541.0 lbf to the design value of 500 lbf. In accordance with Eq. 3–31, $F = C_F A_t p_1$. The throat area A_t is the same for the experimental and flight thrust chambers. The chamber pressure is inversely proportional to the thrust coefficient C_F and proportional to the thrust, and therefore

$$p_1/p_1' = (F_1/F_1')(C_F'/C_F) \tag{11–8}$$

The primes refer to the component test condition. The desired value of p_1 is

$$p_1 = 328(500/541)(1.420/1.425) = 304 \text{ psi}$$

This is 1.3% off the value assumed originally (300 psi) and close enough for preliminary design. The desired total *propellant weight flow rate* at sea level is, from Eq. 2–5,

$$\dot{w} = F/I_s = 500/220 = 2.27 \text{ lbf/sec}$$

For a mixture ratio of 1.30, the desired *fuel and oxidizer flows* are obtained from Eqs. 6–3 and 6–4 as $\dot{w}_f = 0.975$ lbf/sec and $\dot{w}_o = 1.267$ lbf/sec. Next, various measured component pressure drops are corrected to the desired flow values and to the corrected propellant densities in accordance with Eq. 8–2, which applies to all hydraulic devices. By neglecting variations in discharge coefficients, which are taken to be very small, this equation can be rewritten in the following convenient form:

$$\dot{w}/\dot{w}' = \sqrt{\rho/\rho'}\sqrt{\Delta p/\Delta p'} \qquad (11\text{–}9)$$

Solve above for the estimated pressure drop during operating conditions. Again, the prime is for the actual experimental data from component and engine tests

$$\Delta p = \Delta p' \left(\frac{\dot{w}}{\dot{w}'}\right)^2 \left(\frac{\rho'}{\rho}\right)$$

and for the fuel injector

$$(\Delta p_f)_{\text{inj}} = 40.0 \text{ psi} \left(\frac{0.975}{1.02}\right)^2 \left(\frac{1.00}{0.85}\right) = 42.5 \text{ psi}$$

With this result and the known specific gravity values (from Fig.7-1) of 1.14 for cryogenic oxygen, 0.85 for diluted ethyl alcohol, and 1.0 for water, the new pressure drops for the corrected flow conditions can be found, and these are tabulated below with flow values given in pound-mass per second and pressure values in pounds per square inch.

Component	Component Test Data			Design Conditions		
	Fluid	$\dot{m}$	Δp	Fluid	$\dot{m}$	Δp
Fuel injector	Water	1.02	40.3	Fuel	0.975	42.9
Oxidizer injector	Oxygen	1.30	35.0	Oxygen	1.267	30.9
Fuel cooling jacket	Water	0.961	35.0	Fuel	0.975	42.4
Fuel valve and line	Water	0.963	9.15	Fuel	0.975	11.0
Oxidizer valve and line	Oxygen	1.28	14.2	Oxygen	1.267	13.7

Adding the liquid line pressure drop, the chamber pressure and the tank column pressures required, one obtains the tank pressures:

$$(p_o)_{\text{tank}} = 30.9 + 304 - 2 + 13.7 = 346.6 \text{ psi}$$

$$(p_f)_{\text{tank}} = 42.8 + 304 - 0.92 + 11 + 42.4 = 399.9 \text{ psi}$$

To equalize tank pressures so that a single gas pressure regulator can be used and so that the flow system will balance, an *additional pressure loss must be introduced into the oxygen system*. Correction to this simple pressurized liquid propellant system is accomplished by means of an orifice, which must be placed in the propellant piping between the oxidizer tank and the thrust chamber. Allowing 10 psi for regulator functioning, the pressure drop available for the calibration orifice will be $\Delta p = 399.9 - 346.6 = 52.7$ psi. The *regulator setting* should be adjusted to give a regulated pressure downstream of the regulator of 399.9 psi under flow conditions. The orifice area (assume a discharge coefficient $C_d = 0.60$ for a sharp-edged orifice) can be obtained from Eq. 8–2, but corrected with g_0 to work with English units:

$$A = \frac{\dot{m}}{C_d\sqrt{2g_0\rho\,\Delta p}} = \frac{1.274}{0.60\sqrt{2 \times 32.2 \times 71.1 \times 52.7}}$$
$$= 0.00406 \text{ in.}^2(\text{or } 0.072 \text{ in. diameter})$$

The set of balancing equations may be programed into a computer to assist in the calibration of engines. It can also include some of the system's dynamic analogies that enable proper calibration and adjustment for transient engine performance, as during start. There is a trend to require tighter tolerances on rocket engine parameters (such as thrust, mixture ratio, or specific impulse), and therefore all measurements, calibrations, and adjustments are also being performed to much tighter tolerances than were customary 50 years ago.

Engine Health Monitoring System

A *health monitoring system* (HMS) for rocket engines also called *condition monitoring system* represents a sophisticated engine control system. HMSs evolved from conventional sets of measuring instruments about 30 years ago, when their first rudimentary forms were used. Today there are several variations or types of HMSs. References 11–4, 11–10 and 11–11 show the different aspects of a health monitoring system.

HMSs are used to monitor the performance and behavior of an operating liquid propellant engine by measuring and recording in real time key engine parameters of such as chamber pressure, pump speed, or turbine gas inlet temperature. They can also be used for engine calibration; here, a computer would compare the actually measured but corrected data with the intended or desired nominal data of an engine operating at the intended design condition (obtained through analysis or existing data from earlier engines which performed satisfactorily). The computer would analyze results and provide output indicating any required actions, such as changing calibration orifices, trimming an impeller, or valve timing adjustments. Such remedial actions identified by the HMS would then be done by test technicians or at the factory. This improved calibration can be done during ground tests of either research and development (R&D) engines or production engines.

HMSs are also used extensively during ground development testing where similar parameters are monitored and measured, but where many impending potential or incipient failures can also be detected; here, the HMS computer can quickly take remedial action before an actual failure occurs. This has saved much R&D hardware. Also, the first function and the test stand-failure-remedy function may be combined. A description of this HMS application can be found in Section 21.3.

In a third application, the HMS is used during lift-off of a launch vehicle and it is discussed below. The HMS monitors the booster engines during starting (by checking key engine parameters in real time against their intended design values) and determines if the engine is healthy or is likely to experience an impending failure after it has been fully started while on the launch stand but just before the vehicle is released and allowed to fly. This provides a safety feature not only for the booster engine(s), but also for the launch vehicle. Measurements are performed for a few seconds only during the start period and can include readings from instruments for various valve positions, TP shaft speed, pump suction pressure, gas temperature of the gas generator or preburner, or chamber pressure. If the HMS determines that the liquid propellant rocket engine(s) are healthy, it gives a signal to the vehicle computer and/or the launch facility computer and the vehicle will be released, allowing the launch to proceed. If the HMS detects a major potential or impending failure in one of the booster engines (and this may be sensed in many engines before the full thrust is attained), it sends a signal to the vehicle not to launch. It also initiates the safe shutdown of all the booster engines installed in the vehicle held on the launch stand, before any impending failure and major damage can take place. The RS-25/SSME start sequence in Fig. 11–3 shows that starting required about 4.4 sec to reach full thrust and about 3 sec to reach 50% of full thrust, so there was ample time for an HMS application.

The nominal intended value of each measured parameter is usually based on analysis of transient behavior of the engine during start and validated and modified by actual data from measurements of prior similar rocket engines. With each nominal intended value there is associated an upper and a lower limit line, known colloquially as "red line limits." If the actual measured value fall between the two limit lines, then the measured parameter is satisfactory. If it goes over one of the limit lines, then it is an indication of improper engine behavior. By itself a single off-limit measurement is not necessarily an indication of failure. If there should be a potential impending failure, the HMS must verify this fact; it cannot rely on just a single malfunction indication because the measuring instrument or its signal processing may be flawed. Usually when there is a real likelihood of failure, more than one instrument will show measurements which exceed a red limit line. For example, if the fuel pump discharge line is well below its intended pressure or flow values at any particular time during the start sequence, it may indicate an insufficient supply of fuel to the pump, or fuel that may be warmer than should be, a low pump speed, or a possible propellant leak. These would be validated by measurements of the suction pressure to the pump, temperature measurements at the pump, shaft rpm, or a sensor for fuel in the engine compartment. With three such out-of-tolerance readings and with an assessment of the severity of the potential failure, the HMS will automatically register a real impending failure. It will immediately send a signal to both the vehicle computer and the ground control computer of an impending engine failure (which should result in stopping vehicle release or launch). This same signal will initiate a safe shutdown of all booster engines.

Once the vehicle is launched, the HMS continues to check and record the performance and various engine parameters but it is not normally programmed to initiate remedies during flight (such as changing gas temperature or changing thrust level). However, with multiple parallel engines, one engine can be shut down and the mission can then be completed by firing the remaining functioning engines for a longer duration to provide the required total impulse. The flight controller in the vehicle has to be programmed to allow for a lower thrust of longer duration; the flight path might be different but the mission can be completed. (This shutdown of one engine during flight actually happened in a five-engine cluster and the flight mission was satisfactorily completed on four engines).

11.6. SYSTEM INTEGRATION AND ENGINE OPTIMIZATION

Rocket engines as part of an overall vehicle must interact and be well integrated with other vehicle subsystems. There must exist *interfaces* (connections, wires, or pipelines) between the engine and the vehicle's structure, electric power system, flight control system (commands for start or thrust vector control), and ground support system (for checkout or propellant supply). The rocket engine also affects other vehicle components by its heat emissions, noise, and vibrations.

Integration means that the engine and the vehicle are compatible with each other, all interfaces are properly designed, and there is no interference or unnecessary duplication of functions with other subsystems. The engine must work properly with other subsystems to enhance the vehicle's performance and reliability, and reduce the cost. Some organizations use "system engineering" techniques to achieve this integration. See Ref. 11–7. In Chapter 19, we describe the process of selecting rocket propulsion systems and include a discussion of interfaces and of vehicle integration; the discussion in Chapter 19 is supplementary and applies to several different rocket propulsion systems. The present section only concerns liquid propellant rocket engines.

Since the propulsion system is usually the major mass component of the vehicle, its structure (which usually includes the tanks) often becomes a key structural element for the vehicle and has to withstand not only its thrust force but also various vehicle loads, such as aerodynamic forces, inertia forces, or vibration. In the design stages, several alternate tank geometries and locations (fuel, oxidizer, and pressurizing gas tanks), different tank pressures, and different structural connections need to be evaluated to determine the best arrangement.

The thermal behavior of the vehicle is strongly affected by the heat generation (hot plume, hot engine components, and/or aerodynamic heating) and the heat absorption (the liquid propellants are usually heat sinks), and by the heat rejection to its surroundings. Many vehicle components must operate within narrow temperature limits, and their thermal designs must be critically evaluated in terms of an overall heat balance before, during, and after rocket engine operation.

Optimization studies are conducted to select best values or to optimize various vehicle parameters such as vehicle performance (see below), thrust, number of restarts, or engine compartment geometry. These studies are usually performed by vehicle designers with help from propulsion system designers. Rocket engine designers conduct optimization studies (together with vehicle engineers) on engine parameters such as chamber pressure (or thrust), mixture ratio (which affects average propellant density and specific impulse), number of thrust chambers, nozzle area ratio, or engine volume. By modifying one or more of these parameters, it is usually possible to make some improvement to the vehicle performance (0.1 to 5.0%), its reliability, or to reduce costs. Depending on the mission or application, optimization studies are aimed at maximizing one or more vehicle parameter such as range, vehicle velocity increment, payload, circular orbit altitude, propellant mass fraction, or minimizing costs. For example, the mixture ratio of hydrogen–oxygen engines for maximum specific impulse is about 3.6, but most launch engines operate at mixture ratios between 5 and 6 because the total propellant volume is less, and this allows a reduced mass for the propellant tanks (resulting in a higher vehicle velocity increment) and a reduced vehicle drag (more net thrust). Selection criteria for the best nozzle area ratio were introduced in Chapter 3; they depend on things like the flight path's altitude–time history and on whether an increase in specific impulse is offset by extra nozzle weight and length. The best thrust–time profile can also usually be optimized, for a given application, by using trajectory analyses.

SYMBOLS

a	acceleration, m/sec^2 (ft/sec^2)
A	area, m^2 (ft^2)
C_F	thrust coefficient (see Eq. 3–30)
C_d	orifice discharge coefficient
F	thrust, N (lbf)
g_0	sea-level or standard acceleration of gravity, 9.806 m/sec^2 (32.174 ft/sec^2)
H	head, m (ft)
I_s	specific impulse, sec
k	specific heat ratio
L	length, m (ft)
$\dot{m}$	mass flow rate, kg/sec (lbm/sec)
p	pressure, N/m^2(lbf/in.^{2})
Q	volume flow rate, m^3/sec (ft^3/sec)
r	mixture ratio (oxidizer to fuel flow)
t	time, sec
T	absolute temperature, K ($^\circ$R)
v	fluid or liquid velocity, m/sec^3 (ft/sec)
$\dot{w}$	weight flow rate kg $-$ m/sec^3 (lbf/sec)

Greek Letters

ζ_d	discharge correction factor
ζ_v	velocity correction factor
ρ	density, kg/m^3(lb/ft^3)

Subscripts

c	chamber
f	fuel
gas	related to propellant tank pressure
gg	gas generator
inj	Injector
o	oxidizer
oa	overall engine system
p	pump
T	turbine
t_p	tank pressurization
0	initial condition
1	inlet or chamber condition
2	outlet or nozzle exit condition

PROBLEMS

1. Estimate the mass and volume of nitrogen gas required to pressurize an N_2O_4 – MMH feed system for a 4500 N thrust chamber of 25 sec duration ($\zeta_v = 0.92$, the ideal $I_s = 285$ sec at 1000 psi or 6894 N/m² and expansion to 1 atm). The chamber pressure is 20 atm (absolute) and the mixture ratio is 1.65. The propellant tank pressure is 30 atm, and the initial gas tank pressure is 150 atm. Allow for 3% excess propellant and 50% excess gas to allow some nitrogen to dissolve in the propellant. The nitrogen regulator requires that the gas tank pressure does not fall below 29 atm.

2. A rocket engine operating on a gas generator engine cycle has the following test data:

Engine thrust	100,100 N
Engine specific impulse	250.0 sec
Gas generator flow	3.00% of total propellant flow
Specific impulse of turbine exhaust flowing through a low area ratio nozzle	100.2 sec

Determine the specific impulse and thrust for the single thrust chamber.

Answer: 254.6 sec and 98,899 N.

3. This problem concerns various potential propellant loss/utilization categories; the first section of this chapter identifies most of them. This engine has a TP feed system with a single TP, a single fixed-thrust chamber (no thrust vector control) with no auxiliary small thrusters, storable propellants, good priming of the pumps prior to start, and a gas generator engine cycle.
   Prepare a list of propellant utilization/loss categories for which propellant have to be provided. Which of these categories will have a little more propellant or a little less propellant, if the engine is operated with either warm propellant (perhaps 30 to 35 °C) and alternatively in a cold space environment with propellants at −25 °C? Give brief reasons, such as "higher vapor pressure will be more likely to cause pump cavitation."

4. What happens to the thrust and the total propellant flow if an engine (calibrated at 20 °C) is supplied with propellants at higher or at lower storage temperatures?

REFERENCES

11–1. P. Brossel, S. Eury, P. Signol, H. Laporte, and J. B. Micewicz, "Development Status of the Vulcain Engine," AIAA Paper 95-2539, 1995.

11–2. R. Iffly, "Performance Model of the Vulcain Ariane 5 Main Engine," AIAA Paper 1996–2609, 1996; J.-F. Delange et al., "VINCI®, the European Reference for Ariane 6 Upper Stage Cryogenic Propulsive System," AIAA Paper 2015-4063, Orlando FL, 2015.

11–3. G. Mingchu and L. Guoqui, "The Oxygen/Hydrogen Engine for Long March Vehicle," AIAA Paper 95-2838, 1995.

11–4. D. K. Huzel, and D. H. Huang, Chapter 7, "Design of Rocket Engine Controls and Condition Monitoring Systems," and Chapter 9, "Engine System Design Integration," of *Modern Engineering Design of Liquid Propellant Rocket Engines*, rev. ed., Vol. 147 of *Progress in Astronautics and Aeronautics* (Series), AIAA, Reston, VA, 1992.

11–5. R. W. Humble, G. N. Henry, and W. J. Larson, Chapter 5, "Liquid Rocket Propulsion Systems," in *Space Propulsion Analysis and Design*, McGraw-Hill, New York, 1995.

11–6. Copied from Eighth Edition of this book.

11–7. P. Fortescue, J. Stark, and G. Swinerd, *Spacecraft System Engineering*, 3rd ed., John Wiley & Sons, Chichester, England, 2003, reprinted 2005.

11–8. A. D'Souza, *Design of Control Systems*, Prentice Hall, New York, 1988.

11–9. R. M. Mattox and J. B. White, "Space Shuttle Main Engine Controller," NASA Technical Paper 1932, 1981.

11–10. H. Zhang, J. Wu, M. Huang, H. Zhu, and Q. Chen, "Liquid Rocket Engine Health Monitoring Techniques," *Journal of Propulsion and Power*, Vol. 14, No. 5, Sept.–Oct. 1988, pp. 657–663.

11–11. A. Ray, X. Dai, M-K. Wu, M. Carpino, and C. F. Lorenzo, "Damage-Mitigating Control of a Reusable Rocket Engine," *Journal of Propulsion and Power*, Vol. 10, No. 2, Mar.–Apr. 1994, pp. 225–234.

CHAPTER 12

SOLID PROPELLANT ROCKET MOTOR FUNDAMENTALS

This is the first of four chapters dealing exclusively with solid propellant rocket motors—the word *motor* is as common to solid propellants as the word *engine* is to liquid propellants. In this chapter, we cover burning rates, grain configurations, rocket motor performance, and structural issues. In solid propellant rocket motors the propellant is contained and stored directly within the combustion chamber, sometimes hermetically sealed for long-time storage (5 to 20 years). Motors come in many different types and sizes, varying in thrust from about 2 N to over 12 million N (0.4 to over 3 million lbf). Historically, solid propellant rocket motors have been credited with having no moving parts. This is still true of many, but some rocket motor designs include movable nozzles and actuators for vectoring (rotating the line of thrust relative to the motor axis). But in comparison to liquid rockets, solid rockets are typically much simpler, are easy to attach (often constituting most of the vehicle structure), do not leak, are ready to ignite, and require little servicing; however, they cannot be fully checked out prior to use and for most applications thrust cannot be randomly varied in flight.

The subjects of thrust vector control, exhaust plumes, and testing are omitted from these four chapters but are treated for both liquid and solid propellant units in Chapters 18, 20, and 21, respectively. Chapter 19 provides a comparison of the advantages and disadvantages of solid and liquid propellant rocket units. Chapters 2 to 5 are needed as background for these four chapters on solid propellants.

Figures 1–5 and 12–1 depict the principal components and features of two relatively simple solid propellant rocket motors. The *grain* is the solid body of the hardened *propellant* and typically accounts for 82 to 94% of the total rocket motor mass. Designs and stress profiles for these grains are described later in this chapter and propellants and their properties are described in the next chapter. An electrically

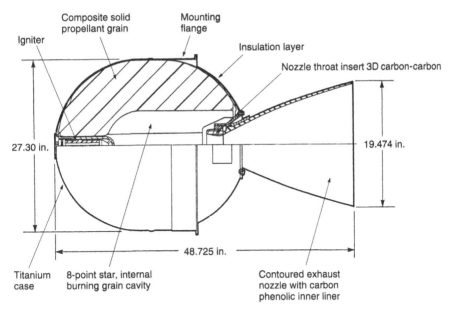

FIGURE 12–1. Cross section of the STAR™ 27 rocket motor, which has been used for orbit and satellite maneuvers. It has an altitude thrust of 6000 lbf, nominally burns for 34.4 sec and has an initial mass of 796 lbm. This motor is no longer used but is a good example of a simple, high-performance, high-mass-ratio unit. For relevant data see Table 12–3. (Courtesy of Orbital ATK.)

activated *igniter* is needed to start combustion. The propellant grain starts to burn on all its exposed inner surfaces. Combustion and ignition of solid propellants are discussed in Chapter 14. In Fig. 12–1 the grain configuration has a central cylindrical cavity with eight tapered slots, forming an eight-pointed star cross section (a typical star cross section is shown later in Fig. 12–16). Many grains have slots, grooves, holes, or other geometric features that alter the initial burning surface and thus determine the initial mass flow rate and the initial thrust. Hot combustion gases flow along the *perforation* or *port inside cavity* toward the nozzle. The motor case body is either made of metal (such as steel, aluminum or titanium) or a composite fiber-reinforced plastic material. Any inner surfaces of the case, which are exposed directly to hot gas, must have thermal protection or *insulation layers* to keep the case from overheating a condition when it would no longer being able carry its pressure and other loads.

The motor's *nozzle* has the role of efficiently accelerating the hot gases issuing from the combustion chamber through its purposely shaped convergent–divergent passage (see Chapter 3). Nozzles are made of high-temperature materials (usually a graphite and/or an ablative heat absorbing material to absorb the heat) to withstand the high-temperatures and erosive environment. The majority of all solid rocket motors have a simple fixed nozzle, as shown, but some nozzles have provisions to slightly rotate them so as to allow vehicle steering. Variable throat area moveable plug nozzles (or "pintle nozzles") have been developed but are as yet too complex for adoption.

Chapter 15 describes nozzles, cases, insulators, liners, and the overall design of solid propellant rocket motors.

Each motor must be fastened to its vehicle by a thrust-carrying *structure*. In Fig. 12–1 there is a skirt (with a flange) integral with the case; this mounting flange fastens to the vehicle. As indicated above, since there is no active cooling the rocket motor must be designed to withstand the transient heat loads without failure for the flight duration.

Applications for solid propellant rockets are listed in Tables 1–3 to 1–5, and 12–1, each having its own mission requirements and associated propulsion requirements. Figures 12–1 to 12–4 illustrate representative designs for some of the major categories of rocket motors listed in Table 12–1, namely, large booster or second stage, space flight motor, and tactical missile motor. Reference 12–1 is useful for component and design information. The Atlas V solid rocket strap-on booster (SRB) shown in Fig. 12–2 is representative of a modern, large solid booster designed by Aerojet Rocketdyne. Depending on the payload, there can be up to five such boosters mounted in the manner shown in the sketch of Fig. 4–15 labeled *parallel staging*. Several SRBs can be seen in Fig. 1–13 boosting the Atlas V. In this motor the grain design is tubular with slots at the aft end; this style of grain corresponds to the third sketch from the top shown later on Fig. 12–16. The Atlas-V propellant consists of aluminized hydroxyl-terminated polybutadiene (HTPB)/ammonium perchlorate (AP) (see Chapter 13). Its case structure is a large, lightweight graphite-fiber-composite unit that, unlike the Shuttle's Solid Rocket Motors (SRM) (see Fig. 15–2), has no segments or joints. This motor has been designed with mechanical release device, which separate it from the space launch vehicle after SRB burnout. An erosion-resistant internal insulation based on EPDM (ethylene propylene diene monomer, see Section 13.6.) has been used—EPDM is a low-density rubber-like material. In common with other SRMs, some inert material (i.e., nonpropellant) is ejected during the burn, such as small abraded particles of insulation, nozzle liner and throat insert, which have been decomposed, eroded, charred, and sometimes gasified during propellant burning; these can amount up to 1% of the propellant mass. In the Shuttle SRM the *propellant mass fraction* (see Eq. 4–4) was 88.2%.

There are several ways of classifying solid propellant rocket motors. Some are listed in Table 12–2 together with their definitions. Table 12–3 gives characteristics for three specific rocket motors, and these data exemplify the magnitudes of key parameters. As is common in this field, Table 12–3 lists more than one value for thrust and chamber pressure—the maximum value is used to determine loads on structures and the average value is used in performance analyses. These motors are shown in Figs. 12–1, 18–5, and 18–9.

Nearly all rocket motors are used only once. Any hardware that remains after all the propellant has been burned and the mission has been completed—namely, the nozzle, case, and/or thrust vector control device—is not made to be reusable. In very rare applications, such as NASA's Space Shuttle solid booster, was the hardware recovered, cleaned, refurbished, and reloaded with new propellant; reusability

TABLE 12–1. Major Application Categories for Solid Propellant Rocket Motors

Category	Application	Typical Characteristics
Large booster and second-stage motors	Space launch vehicles; long-range ballistic missiles (see Figs. 12–2 and 15–2)	Booster diameter (above 48 in.); L/D of *case* = 2–7; burn time t = 60–120 sec; low-altitude operations with low nozzle area ratios of 6 to 22
High-altitude motors	Upper stages of multistage ballistic missiles; space launch vehicles; space maneuvers. See Figs. 12–1, 12–3	High-performance propellant; large nozzle area ratio (20–300); L/D of *case* = 1–2; burn time t = 40–120 sec
Tactical missiles	1. High acceleration: short-range bombardment, antitank missile	Tube launched, L/D = 4–13; very short burn time (0.25–1 sec); small diameter (2.75–18 in.); some are spin stabilized
	2. Modest acceleration, guided or unguided: air-to-surface, surface-to-air, short-range guided surface-to-surface, and air-to-air missiles. Infantry support weapons, shoulder fired antitank or anti-aircraft missiles, mortars	Small diameter (2.5–18 in.); L/D of *case* = 5–10; usually has fins and/or wings; thrust is high at launch and then is reduced (boost-sustain); many have blast tubes (see Fig. 12–4); wide ambient temperature limits: −65°F to +160°F; usually high acceleration; often low-smoke or smokeless propellant
Ballistic missile defense	Defense against long- and medium-range ballistic missiles. Two or three stages	Booster rocket motors; upper maneuverable stage with attitude control nozzles and a larger divert nozzle. Throttling liquid or solid propellant
Gas generator	Pilot emergency escape; push missiles from submarine launch tubes or land mobile canisters; actuators for TVC; short-term power supply; jet engine starter; munition dispersion; rocket turbine drive starter; automotive air bags	Usually low gas temperature (<1300°C); many different configurations, designs, and propellants; purpose is to create high-pressure, energetic gas rather than thrust
Solid propellant augmented artillery projectiles	Increase range of gun-fired projectiles	Withstand very high acceleration in gun barrel transit (<20,000 g_0)

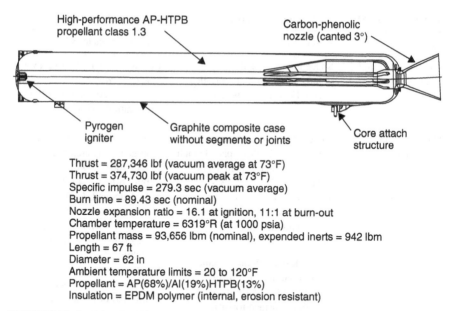

Thrust = 287,346 lbf (vacuum average at 73°F)
Thrust = 374,730 lbf (vacuum peak at 73°F)
Specific impulse = 279.3 sec (vacuum average)
Burn time = 89.43 sec (nominal)
Nozzle expansion ratio = 16.1 at ignition, 11:1 at burn-out
Chamber temperature = 6319°R (at 1000 psia)
Propellant mass = 93,656 lbm (nominal), expended inerts = 942 lbm
Length = 67 ft
Diameter = 62 in
Ambient temperature limits = 20 to 120°F
Propellant = AP(68%)/Al(19%)HTPB(13%)
Insulation = EPDM polymer (internal, erosion resistant)

FIGURE 12–2. Atlas V solid rocket booster (SRB) cross-sectional view. This motor has a large monolithic carbon-composite case and other novel features representative of modern solid boosters. The grain has tapered aft slots. (Courtesy of Aerojet Rocketdyne.)

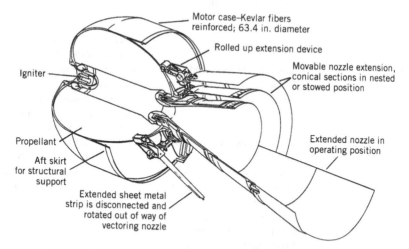

FIGURE 12–3. Inertial upper-stage (IUS) Orbus rocket motor with an extendible exit cone (EEC). This motor was used for propelling upper launch vehicle stages or spacecraft. The grain was simple (internal tube perforation). With the EEC and a thrust vector control, the motor had a propellant mass fraction of 0.916. When launched, and while the two lower vehicle stages were operating, the two conical movable nozzle segments were stowed around the smaller inner nozzle segment. Each of the movable segments was then deployed in space and moved into its operating position by three rotary actuators. The nozzle area ratio increased from 49.3 to 181; overall this improved the specific impulse by about 14 sec. This motor (without the EEC) is described in Table 12–3 and a similar motor is shown in Fig. 18–5. (Courtesy of United Technologies Corp.)

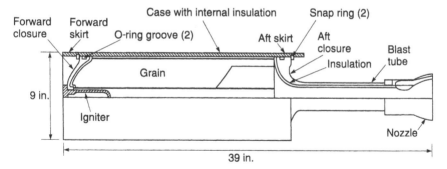

FIGURE 12–4. Simplified cross section through a typical tactical motor. The blast tube allows the grain to be close to the center of gravity of the vehicle so it moves very little during motor operation. The nozzle is at the missile's aft end. The annular space around the blast tube is usually filled with guidance, control, and other nonpropulsive equipment. A freestanding grain with bonded external insulation (see Fig. 12–14) is loaded before the aft closure is assembled.

makes the design more complex, but if the hardware is to be reused often enough a major cost saving will result. Unlike some liquid propellant rocket engines, a solid propellant rocket motor and its key components cannot be operationally pretested. As a result, individual motor reliability must be inferred by assuring its structural integrity and by verifying the manufactured quality on entire populations of motors.

Solid propellant rocket motor technologies have remained relatively mature in the last 10 years since many U.S. operational systems use designs developed in the 1970s (Ref. 12–2). Structural materials for motor cases and nozzles continue to be improved (Ref. 12–3) but high-energy propellants appear to be near their practical limits. There are many more solid propellant rocket motors than liquid propellant engines (presently, over one million worldwide mostly in tactical military applications compared to a few thousand liquid propellant rocket engines worldwide).

12.1. BASIC RELATIONS AND PROPELLANT BURNING RATE

A rocket motor's operation and its design depend on the propellant's combustion characteristics such as burning rate, burning surface, and grain geometry. The branch of applied science describing these is known as *internal ballistics*; a treatment of the first two follows and effects of grain geometry are treated in Section 12.3.

The burning surface of a propellant grain recedes in a direction essentially perpendicular to it. The rate of regression, usually expressed in cm/sec, mm/sec, or in./sec, is the *burning rate r*. In Fig. 12–5, we show changes of the grain geometry by drawing successive burning surfaces at a constant time interval between adjacent surface contours. Figure 12–5 depicts this for a two-dimensional grain with five slots residing in a central cylindrical cavity. Success in rocket motor design and development depends significantly on knowledge of burning rate behavior of the selected propellant under all motor operating conditions and design limits. Burning rate is a function

TABLE 12–2. Classification of Solid Rocket Motors

Basis of Classification	Examples of Classification
Application	See Table 12–1.
Diameter/length	0.005–6.6 m or 0.2–260 in./ 0.025–45 m or 1–1800 in.
Propellant. See Chapter 13	*Composite:* Heterogeneous (physical) mixture of powdered metal (fuel), small crystal oxidizer and polymer binder
	Double-base: Homogeneous mixture (colloidal) of two explosives (nitroglycerin in nitrocellulose)
	Composite-modified double-base: Combines composite and double-base ingredients
	Gas generator and others
Case design	*Steel monolithic:* One-piece steel case
	Fiber monolithic: Filament wound (high-strength fibers) with a plastic matrix
	Segmented: Case (usually alloy steel) and large grain are in segments which are transported separately and fastened together at launch site
Grain configuration	*Cylindrical:* Cylindrically shaped, usually hollow
	End-burning: Solid cylinder propellant grain
	Other configurations: See Figs. 12–16 and 12–17
Grain installation	*Case-bonded:* Adhesion exists between grain and case or between grain and insulation or liner and case; propellant is usually cast into the case
	Cartridge-loaded: Grain is formed separately from the motor case and then assembled into case
Explosive hazard	*Class 1.3:* Catastrophic failure shows evidence of burning and explosion, not detonation
	Class 1.1: Catastrophic failure shows evidence of detonation
Thrust action	*Neutral grain:* Thrust remains essentially constant during the burn period. See Fig. 12–15
	Progressive grain: Thrust increases with time. See Fig. 12–15
	Regressive grain: Thrust decreases with time. See Fig. 12–15
	Pulse rocket: Two independent thrust pulses or burning periods
	Step-thrust rocket: Usually, two distinct levels of thrust
Toxicity	Toxic and nontoxic propellant and exhaust gases

of the propellant composition. For composite propellants (see Chapter 13) it can be increased by changing the propellant characteristics as follows:

1. Add a burning rate *catalyst*, often called burning rate *modifier* (0.1 to 3.0% of propellant) or increase percentage of existing catalyst.
2. *Decrease* the *oxidizer particle size.*
3. *Increase oxidizer percentage.*
4. Increase the *heat of combustion* of the binder and/or the plasticizer.
5. Imbed *wires* or *metal staples* in the propellant.

FIGURE 12–5. Diagram of successive burning surface contours, each a fixed small time interval apart. It shows how the internal cavity grows. When the lengths of these contour lines are roughly the same (within ±15%), the burning area is considered to be constant. The burning surface area A_b diminishes greatly near the end causing reduced thrust and chamber pressure. When the pressure falls below the conflagration value, combustion will stop leaving some unburnt propellant (slivers).

For any given propellant formulation the burning rate can be increased or otherwise modified by the following:

1. Higher temperatures of solid propellant prior to start
2. Higher combustion gas chamber pressures
3. Higher combustion gas temperatures
4. Higher gas flow velocities parallel to its burning surface
5. Rocket motor motions (acceleration and spin-induced grain stress)

All these influencing factors are separately discussed in this chapter. An explanation for the behavior of the burning rate with various parameters is largely based on the combustion mechanisms of solid propellants, which are treated in Chapter 14. Analytical models for the burning rate and the combustion process in general exist and are useful for preliminary design and for extending existing test data; for detailed designs and for evaluation of new or modified propellants, some actual test data is necessary. *Burning rate data* are usually obtained in three ways—namely, from testing by:

1. Standard *strand burners*, often called *Crawford burners*
2. Small-scale *ballistic evaluation motors*
3. *Full-scale rocket motors*, properly instrumented

A strand burner is a small pressure vessel (usually with windows) in which a thin strand or bar of propellant is ignited at one end and burned to completion. The strand can be inhibited with an external nonflammable coating so that it will burn only on the

TABLE 12-3. Characteristics of a Missile Motor and Two Space Motors

Characteristic	First-Stage Minuteman I Missile Motor[a]	Orbus-6 Inertial Upper-Stage Motor[b]	STAR™ 27 Apogee Motor[a]
Motor Performance (70°F, sea level)			
Maximum thrust (lbf)	201,500	23,800	6,404 (vacuum)
Burn time[c] average thrust (lbf)	194,600	17,175	6,010 (vacuum)
Action time[c] average thrust (lbf)	176,600	17,180	5,177 (vacuum)
Maximum chamber pressure (psia)	850	839	569
Burn time average chamber pressure (psia)[c]	780	611	552
Action time average chamber pressure (psia)[c]	720	604	502
Burn time/action time (sec)[c]	52.6/61.3	101.0/103.5	34.35/36.93
Ignition delay time (sec)	0.130		0.076
Total impulse (lbf-sec)	10,830,000	1,738,000	213,894
Burn time impulse (lbf-sec)	10,240,000	1,737,000	
Altitude specific impulse (sec)	254	289.6 (vacuum)	290.8 (vacuum)
Temperature limits (°F)	60–80	45–82	20–100
Propellant			
Composition:			
NH_4ClO_4 (%)	70	68	72
Aluminum (%)	16	18	16
Binder and additives (%)	14	14	12
Density (lbm/in.^{3})	0.0636	0.0635	0.0641
Burning rate at 1000 psia (in./sec)	0.349	0.276	0.280
Burning rate exponent	0.21	0.3–0.45	0.28
Temperature coefficient of pressure (% °F)	0.102	0.09	0.10
Adiabatic flame temperature (°F)	5790	6150	5,909
Characteristic velocity (ft/sec)	5180	5200	5,180
Propellant Grain			
Type	Six-point star	Central perforation	Eight-point star
Propellant volume (in.^{3})	709,400	94,490	11,480
Web (in.)	17.36	24.2	8.17
Web fraction (%)	53.3	77.7	60
Sliver fraction (%)	5.9	0	2.6
Average burning area (in.^{2})	38,500	3905	1,378
Volumetric loading (%)	88.7	92.4	92.6
Igniter			
Type	Pyrogen	Pyrogen	Pyrogen
Number of squibs	2	2 through-the-bulkhead initiators	2
Minimum firing current (A)	4.9	NA	5.0

(*continued*)

TABLE 12–3. (*Continued*)

Characteristic	First-Stage Minuteman I Missile Motor[a]	Orbus-6 Inertial Upper-Stage Motor[b]	STAR™ 27 Apogee Motor[a]
	Weights (lbf)		
Total	50,550	6515	796.3
Total inert	4719	513	60.6
Burnout	4264	478	53.4
Propellant	45,831	6000	735.7
Internal insulation	634	141	12.6
External insulation	309	0	0
Liner	150	Incl. with insulation	0.4
Igniter	26	21	2.9 (empty)
Nozzle	887	143	20.4
Overall length (in.)	294.87	72.4	48.725
Outside diameter (in.)	65.69	63.3	27.30
	Case		
Material	Ladish D6AC steel	Kevlar fibers/epoxy	6 Al-4V titanium
Nominal thickness (in.)	0.148	0.35	0.035
Minimum ultimate strength (psi)	225,000	—	165,000
Minimum yield strength (psi)	195,000	—	155,000
Hydrostatic test pressure (psi)	940	<1030	725
Hydrostatic yield pressure (psi)	985	NA	767
Minimum burst pressure (psi)	—	1225	—
Typical burst pressure (psi)	—	>1350	—
	Liner		
Material	Polymeric	HTPB system	TL-H-304
	Insulation		
Type	Hydrocarbon–asbestos	Silica-filled EPDM	Polyisoprene
Density (lbm/in.^{3})	0.0394	0.044	0.044
	Nozzle		
Number and type	4, movable	Single, flexible	Fixed, contoured
Expansion area ratio	10:1	47.3:1	48.8/45.94
Throat area (in.^{2})	164.2	4.207:1	5.900
Nozzle exit cone half angle (deg)	11.4	Initial 27.4, Final 17.2	Initial 18.9, Exit 15.5
Throat insert material	Forged tungsten	3D carbon–carbon[d]	3D carbon–carbon[d]
Shell body material	AISI 4130 steel	NA	NA
Exit cone material	NA	2D carbon–carbon[d]	Carbon phenolic

[a]Courtesy of Orbital ATK.
[b]Courtesy United Technologies Corp., there is also a version Orbus-6 E (see Fig. 12–3) with an extendible, exit nozzle; it has a specific impulse of 303.8 sec, a total weight of 6604 lbf and a burnout weight of 567 lbf.
[c]Burn time and action time are defined in Fig. 12–13.
NA: not applicable or not available.
[d]2D and 3D carbon-carbon refer to "two-directional and three-directional" reinforcements.

exposed cross-sectional surface; chamber pressure is simulated by pressurizing the container with an inert gas. Burning rates can be measured by electric signals from embedded wires, by ultrasonic waves, or by optical means. The burning rate measured on strand burners is usually lower than that obtained from full-scale motor firing (by 4 to 12%) because it does not completely simulate the hot chamber environment. Small ballistic evaluation motors usually also have a slightly lower burning rate than full-scale larger motors because of scaling factors. The relationship between these three types of measured burning rates must be determined empirically for each propellant category and grain configuration. Strand burner data are useful in screening propellant formulations and in quality control operations. Because of their cost advantage, strand burners and other substitutes for full-scale motor tests are used to explore as many needed variables as practicable but data from full-scale rocket motors tested under a variety of conditions constitute the best and final proof of burning rate behavior.

During development, a new or modified solid propellant is extensively examined or *characterized*. This includes burn rate testing (in several different ways) under different temperatures, pressures, and impurity conditions. Characterization also includes measurements of physical and chemical properties, as well as manufacturing properties, ignitability, aging, sensitivity to various energy inputs or stimuli (e.g., shock, friction, and fires), moisture absorption, and compatibility with other materials (liners, insulators, cases). Characterization is a lengthy, expensive, and often hazardous process involving many tests, samples, and studies.

The burning rate of propellants in a motor is a function of several parameters; at any instant of time the *mass flow rate* $\dot{m}$ of the hot gases generated and flowing from the motor is given by:

$$\dot{m} = A_b r \rho_b \tag{12-1}$$

Here, A_b is the *propellant grain burning area*, r the *burning rate*, and ρ_b the *solid propellant density* prior to motor ignition. The total effective mass m of propellant burned is determined by integrating Eq. 12–1:

$$m = \int \dot{m}\,dt = \rho_b \int A_b r\,dt \tag{12-2}$$

Here, A_b and r may vary with time (and pressure), but not ρ_b. Grains can also be designed for A_b to remain essentially constant (within $\pm 15\%$).

Mass Flow Relations

A first basic performance relation comes from the principle of conservation of matter. The gaseous propellant mass evolving from a burning surface per unit time must equal the sum of the change in gaseous mass storage per unit time in the combustion chamber (due to increases in volume of the grain cavity) and the mass flowing out through the exhaust nozzle per unit time,

$$A_b r \rho_b = d(\rho_1 V_1)/dt + A_t p_1/c^* \tag{12-3}$$

The left side of the equation represents the rate of gas generation from Eq. 12–1. The first term on the right represents the rate of change in storage of the hot gas in the combustion chamber volume, and the last term represents the propellant flow rate through the nozzle according to Eqs. 3–24 and 3–32. Note that ρ_b is the solid propellant density whereas ρ_1 is the chamber hot-gas density and that here in derivative form V_1 represents the rate of increase chamber gas cavity volume; A_t is the nozzle throat area; p_1 the chamber pressure; c^* is the characteristic velocity (which is proportional to T_1, the absolute chamber temperature found from the thermochemistry for a given propellant); and k is the specific heat ratio of the combustion gases (see Eq. 3–32). Equation 12–3 is most useful in numerical solutions of transient conditions, such as during start-up or shutdown. Though the rate of change of hot gas in the grain cavity is always important during start-up, it is seldom included in preliminary designs.

The magnitude of the burning surface A_b may or may not change significantly with time and this is a function of grain design as described in Section 12.3. For preliminary performance calculations, the nozzle throat area A_t is usually taken as constant (Eqs. 3–24, 3–32, 12–3, or 12–4) for the entire burning duration but, for accurate performance predictions during hot firings, it is necessary to account for oxidation, erosion and abrasion in the nozzle material which increases the nozzle throat area as the propellant burns; such nozzle enlargements are usually small (0.05 to 5%) as described in Section 15.2. With time, as the throat area enlarges a noticeable decrease of chamber pressure, burning rate, and thrust ensues.

The chamber gas volume V_1 will measurably increase with burn time. But to fill this void requires relatively small amounts of gaseous propellant mass compared to what flows through the nozzle (the volume change of a unit of mass is about 1000 to 1 as the propellant gasifies during the combustion). As a result, the term $d(\rho_1 V_1)/dt$ can usually be neglected except for very short operating durations. This then yields the commonly used relation for the pressure in steady burning:

$$p_1 = K\rho_b rc^* \quad \text{where} \quad K \equiv A_b/A_t \tag{12–4}$$

Here, K is an important new dimensionless motor parameter, the ratio of burning area to nozzle throat area, which has typical values much larger than one. For steady flow and steady burning the value of K (or A_b) must remain essentially constant. When noticeable abrasion/erosion takes place during burning, values of K listed must be interpreted as "initial area ratios." Equation 12–4 by itself does not properly represent most of the observed dependence of r on T_b and p_1 so an additional equation for the burning rate is introduced; this equation is empirical and is discussed next.

Burning Rate Relation with Pressure

Classical relations describing the burning rate are empirical generalizations helpful in data extrapolation and in understanding the phenomena involved. Though useful in preliminary design, they can only deal with the influence of some of the important parameters. However, strict analytical modeling and supportive research have yet to adequately predict the burning rate of a new propellant in a new rocket motor. Unless otherwise stated, the burning rates specified here are based on an ambient temperature of 70°F or 294 K for the propellant grain (prior to ignition) and a reference chamber operating pressure of 1000 psia or 6.895 MPa.

With many propellants it is possible to approximate the *burning rate* as a function of *chamber pressure*, at least over a limited range of chamber pressures. A log–log set of plots is shown in Fig. 12–6. For a majority of production-type propellants the most commonly used empirical equation is

$$r = ap_1^n \tag{12-5}$$

where r, the burning rate, is in mm/sec or in./sec and the chamber operating pressure p_1 is in MPa or psia. Known as the *temperature coefficient*, a is an empirical constant influenced by the ambient grain temperature (T_b)—the dimensions of a are defined by those of the other terms in Eq. 12–5. The *burning rate exponent* n (a pure number), sometimes also called the *pressure exponent* or the *combustion index*, is taken to be independent of the propellant initial temperature but influences the chamber operating pressure and the burning rate. For combustion stability $n < 1.0$ (see Ref. 12–4), otherwise, when $n > 1.0$, any pressure disturbances present will be amplified in the chamber. Equation 12–5 applies to all commonly used double-base, composite, or composite double-based propellants, several of which are described in

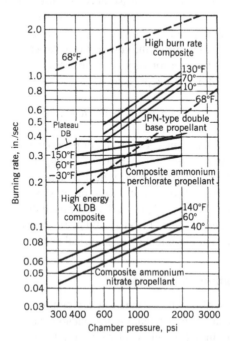

FIGURE 12–6. Calculated burning rates versus chamber pressure for several typical solid rocket propellants, some at three different ambient grain temperatures. A particular *double-base* (DB) *plateau propellant* shows constant burning rate over a wide pressure range.

the next chapter. While changes in ambient grain temperature (T_b) do not alter the propellant's chemical energy available for release during combustion, they do change the rate of reaction at which the energy is released and have a slight effect on c^* through changes in T_1.

The curves shown in Fig. 12–6 are calculated over limited pressure ranges of interest and appear as straight lines on a log–log plot; however, most actual burning rates deviate somewhat from such linearity and the actual data display slight bending in parts of the curve as the pressure range increases, as seen in the samples of Fig. 12–7. While analyses for production propellants are based on data such as shown in Fig.12–7, preliminary design and comparative analyses use the linear versions shown in Fig.12–6. For any particular propellant and for wide temperature and pressure limits, the burning rate can vary by factors of 3 or 4. For all propellants this translates to a range from about 0.05 to 75 mm/sec or 0.02 to 3 in./sec; the higher values being presently difficult to obtain. To achieve such rates, combinations of very small sized ammonium perchlorate (AP), burning rate catalysts, additives, or embedded metal wire are needed.

Inserting Eq. 12–5 into Eqs. 12–4 and 12–1, we may now write K and the generated mass flow rate as

$$K = p_1^{(1-n)}/(\rho_b a c^*) \tag{12-6}$$

$$\dot{m} = A_b \rho_b a p_1^n \tag{12-7}$$

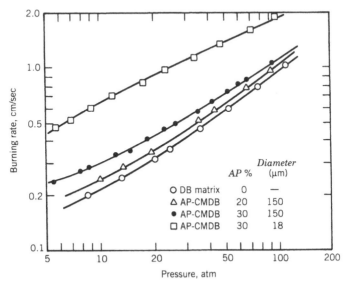

FIGURE 12–7. Measured burning rate characteristics of a double-base (DB) propellant and three composite-modified double-base (CMDB) propellants that contain an increasing percentage of small-diameter (159-μm) particles of ammonium perchlorate (AP). When the size of the AP particles is reduced or the percentage of AP is increased, an increase in burning rate is observed. (Reproduced with permission of the AIAA from Chapter 1 of Ref. 12–4.)

Example 12–1.   Show that n > 1.0 is unrealistic for propellant operation under steady-state conditions. Use Eq. 12–3 (neglecting the cavity gas storage term) and Eq. 12–5 to explore what happens to the mass flow rate when p_1 fluctuates by ±0.1 MPa from its nominal, steady-state value. Take one propellant at $n = 0.5$ and another (hypothetical) at $n = 1.5$, both operating at a nominal pressure $p_1 = 7.00$ MPa. Assume that the (small) pressure changes act as the only independent variable.

SOLUTION.   Propellant flow through the nozzle is proportional to p_1 (see Eq. 3–32), whereas according to Eq. 12–5 the hot-gas mass generation is proportional to p_1^n so that these flows will change unequally as the pressure changes. Defining the symbol 'Δ' as the difference between the mass flowing through the nozzle and the mass generated by the propellant combustion,

$$\Delta \equiv p_1 A_t / c^* - A_b \rho_b a p_1^n = (A_t / c^*)(p_1 - K c^* \rho_b a p_1^n)$$

or

$$(c^*/A_t)\Delta = (p_1 - b_1 p_1^n) \quad \text{where} \quad b_1 \equiv Kc * \rho_b a$$

The new constant b_1 above is found from the nominal pressure where $\Delta = 0$ because both flow rates are designed to be equal at that pressure. For this problem (at 7.0 MPa), $b_1 = 2.646$ for $n = 0.5$ and $b_1 = 0.378$ for $n = 1.5$. The table below shows values of pressure together with values of $(c^*/A_t)\Delta$ for the two stated propellants. The units are not shown but are consistent:

	$(c^*/A_t)\Delta$	
Pressure p_1 (MPa)	$n = 0.5$	$n = 1.5$
6.90	−0.05	+0.05
7.00 (nominal)	0.00	0.00
7.10	+0.05	−0.05

These results indicate that with $n < 1.0$ the Δ-values are negative when the pressure drops below nominal—a situation which will drive the pressure up; when the pressure rises above nominal, Δ-values are positive and this will drive the pressure down because more flow can pass through the nozzle than is generated. The precise opposite is true with $n > 1.0$. Here, as the pressure decreases the nozzle can accommodate more flow than is being generated further dropping the pressure. From stability considerations, for the $n < 1.0$ case the flow is able to physically adjust, whereas in the $n > 1$ it cannot. Hence, n must be less than one for solid propellant motor (with constant throat area) steady operation. It will be seen that $n = 1.0$ becomes problematic from other considerations. Propellants with $n < 1.0$ can operate in a 'quasi-steady-state' fashion because the chamber pressure fluctuations are small enough for the trends presented to be valid, but for $n > 1.0$ the steady state condition cannot be realistic.

From inspection of the results in the Example 12–1 or from Eq. 12–7, it can be seen that the hot gas flow rate is quite sensitive to the exponent n. High values on n give rapid changes of burning rate with pressure. This implies that even an ordinarily small variation in chamber pressure can induce substantial changes in the amount of hot gas produced. Most commercially available propellants have a pressure exponent n ranging between 0.2 and 0.6. As n approaches 1.0, the burning rate and chamber pressure become very sensitive to one another and a disastrous rise in chamber pressure may occur in only a few milliseconds. On the other hand, a propellant having a pressure exponent of zero displays essentially zero change in burning rate over a wide pressure range. *Plateau propellants* is the name given to those that exhibit nearly constant burning rate over a limited pressure range, and they are desirable for minimizing effects of changes in initial temperature on motor operation as described in the next section. One plateau propellant is shown as a horizontal dashed line in Fig. 12–6. These propellants are only insensitive to changes in chamber pressure over a limited range. Several double-based propellants and a few composite propellants are known have this desirable plateau characteristic. Table 13–1 lists nominal burning rates r and pressure exponents for several operational (production) propellants.

Burning Rate Relation with Ambient Temperature (T_b)

Because temperature influences chemical reaction rates, the initial (i.e., prior to combustion) or *ambient temperature* of a propellant (T_b) noticeably changes the burning rate as shown in Figs. 12–6 and 12–8. Common practice in developing and testing larger rocket motors is to "condition" the motor for many hours at a particular temperature before ground-test firing it to ensure that the propellant grain is uniformly at the desired temperature since rocket motor performance characteristics must stay within specified acceptable limits. For air-launched missile motors the extremes are usually 219 K (−65°F) and 344 K (160°F) and for silo or submarine launched motors 266 K (20°F) to 300 K (80°F). Motors using typical composite propellant mixtures may experience a 20 to 35% variation in chamber pressure and a 20 to 30% variation in operating time over such a range of propellant temperatures (see Fig. 12–8). In large rocket motors, an uneven heating of the grain (e.g., by the sun heating on one side) may cause a sufficiently large difference in burning rate so that noticeable thrust misalignments can result (see Ref. 12–5). Thus, effects of ambient temperature on motor

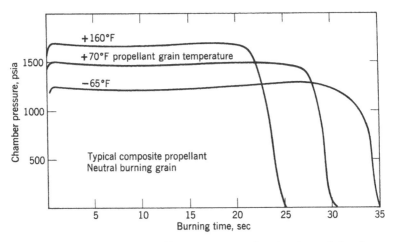

FIGURE 12–8. Effect of propellant initial temperature on burning time and chamber pressure in a particular motor. The integrated areas under the curves are proportional to the total impulse, which is the same for the three curves.

performance are of some importance and a discussion of this topic is necessary to understand solid propellant behavior.

The sensitivity of the burning rate to propellant temperature can be expressed in the form of temperature coefficients, the two most common being

$$\sigma_p = \left(\frac{\partial \ln r}{\partial T_b}\right)_{p_1} = \frac{1}{r}\left(\frac{\partial r}{\partial T_b}\right)_{p_1} \tag{12-8}$$

$$\pi_K = \left(\frac{\partial \ln p_1}{\partial T_b}\right)_{K} = \frac{1}{p_1}\left(\frac{\partial p_1}{\partial T_b}\right)_{K} \tag{12-9}$$

with σ_p known as the *temperature sensitivity of burning rate*, expressed as the change of burning rate per degree change in propellant ambient temperature at a fixed value of chamber pressure, and π_K known as the *temperature sensitivity of pressure* expressed as the change of chamber pressure per degree change of propellant ambient temperature at a particular value of K. Here, K is the geometric factor introduced earlier, namely, the ratio of the burning surface A_b to the nozzle throat area A_t.

The coefficient σ_p for a new propellant is usually found from *strand burner test data*, and π_K is usually found from *small-scale or full-scale motors*. Values of σ_p typically range between 0.001 and 0.009 per °C, or 0.002 and 0.04 per °F, and for π_K they range between 0.00067 and 0.0027 per °C, or 0.0012 and 0.005 per °F. Since these sensitivity coefficients are small numbers, Eqs. 12–8 and 12–9 are sometimes written in terms of differences for more convenient manipulation. Furthermore, when π_K remains sufficiently constant over the temperature interval of interest, we may integrate Eq. 12–9 at constant K for the pressure and obtain a useful equation for predicting chamber pressure excursions with ambient temperature changes from any given or defined reference condition, subscripted as 0 (i.e., p_{01} and T_{0b}):

$$\ln p_1 = \ln p_{01} + \pi_K(T_b - T_{0b}) \tag{12-10}$$

Values of σ_p and π_K depend primarily on the nature of the propellant burning rate, the composition, and the combustion mechanisms of the propellant. Because of variations in manufacturing tolerances act in addition to changes of ambient temperature, it is not simple to predict motor performance. Both sensitivity coefficients above are used for setting temperature limits and maximum pressures for a given rocket motor. Reference 12–5 reports on an analysis for predicting burning time.

In order to arrive at a relationship between the two sensitivity coefficients introduced above, Eq. 12–5 is rewritten in log-form and then derivatives are taken under the assumption that the coefficient a depends only on T_b and that n and K are constant:

$$\ln r = \ln a + n \ln p_1$$

$$(\partial \ln r / \partial T_b)_{p_1} = d(\ln a)/dT_b \tag{12-11}$$

hence

$$\sigma_p = d(\ln a)/dT_b \tag{12-12}$$

Next Eq. 12–6 is solved for p_1 and written in log-form. Its ambient temperature derivative is then taken, keeping the second term in the right-hand-side brackets as constant:

$$\ln p_1 = \frac{1}{1-n}[\ln a + \ln(K\rho_b c^*)] \qquad (12\text{--}13)$$

hence

$$\pi_K = \frac{1}{1-n}\frac{d\ln a}{dT_b} = \frac{\sigma_p}{1-n} \qquad (12\text{--}14)$$

Note that this results from taking the entire product $K\rho_b c^*$ as constant in T_b which involves more than the geometry factor K (i.e., chamber temperature, propellant density, and composition do not change); according to Eq. 12–4, a good indicator for the constancy of $K\rho_b c^*$ in any data is how constant the ratio p_1/r remains during measurements. Since σ_p is an often tabulated material/propellant property, Eq. 12–14 shows how π_K may be obtained from σ_p and this highlights a strong dependence of π_K on burning rate exponent n values near one. Both these conclusions hinge on the assumption that the two sensitivities do remain unchanged over limited temperature and pressure ranges.

Example 12–2. For a given propellant with a neutrally burning grain (see Fig. 12–15) the value of the temperature sensitivity at constant burning area is $\pi_K = 0.005/°F$ and the value of the pressure exponent n is 0.6. The burning rate r is 0.32 in./sec at 70°F at a chamber pressure of $p_1 = 1500$ psia for an effective nominal burning time (t_b) of 55 sec. Determine the variation in p_1 and in t_b for a change of $\pm 40°F$ (from $+30°$ to $+110°F$) assuming that the variation is linear.

SOLUTION. We may use Eq. 12–10, taking exponents on both sides to calculate the pressure excursions, i.e., $p_1 = p_{01}e^{\pi_K(T_b-T_{0b})}$:

$$(p_1)_1 = 1500e^{[0.005\times(-40)]} = 1230 \text{ psia}$$
$$(p_1)_2 = 1500e^{[0.005\times(+40)]} = 1830 \text{ psia}$$

This translates into a total excursion of 600 psi or 40% of the nominal chamber pressure. This large change in p_1 is typical of many solid rocket motors.

It is proper to assume that the total impulse or the chemical energy released remains constant as grain ambient temperature varies; only the rate at which the energy is released changes. When pressure thrust is negligible, thrust is essentially proportional to the chamber pressure and it will change accordingly (i.e., when A_t and C_F may be taken as constant in Eq. 3–31, namely, $F = C_F p_1 A_t$). For a constant total impulse, then

$$I_t = Ft_b = \text{constant} \quad \text{or} \quad (p_1)_1 t_{b1} = (p_1)_2 t_{b2} = (1500) \times (55) = 82{,}500 \text{ psia–sec}$$

Thus,

$$t_{b1} = 67.1 \text{ sec} \quad \text{and} \quad t_{b2} = 45.1 \text{ sec}$$

A change of 22 sec between the two values of t_b represents 40% of the nominal burning time. These trends are similar to what is shown in Fig. 12–8. The inclusion of actual C_F changes would affect these results by only a few percent.

In the example above, the variation of chamber pressure is shown to affect both the thrust and burning time of the rocket motor. Between warm or cold grain operations, the thrust can easily vary by a factor of 2 and this may cause significant changes of flight performance during atmospheric flight (because of differences in drag and in the vehicle's flight path). Thrust and chamber pressure increases become more pronounced as n approaches one; the least variations in thrust or chamber pressures occur at small n's (0.2 or less) and with relatively low temperature sensitivities.

Variable Burning Rate Exponent n

A close look at burning rate data (e.g., Fig. 12–7) indicates that n in Eq. 12–5 may not be really constant but a function of p_1 (which indirectly makes it also a function of T_b). Here, it may no longer be possible to accurately predict the pressure excursions with ambient temperature changes using Eq. 12–10; this can be seen from Eq. 12–14 by noting that in the constant-n case π_K depends only on a material/propellant property (σ_p) and on n. An upward concavity in the propellant burning rate data may result in larger excursions of chamber pressure than constant-n cases would predict. The opposite would hold true for a downward concavity. Such variations may be accommodated by piecewise curve fitting the data with constant n values in preselected ranges of interest (see Ref. 12–5), provided the pressure increments in these curve fits are small enough.

When the piecewise constant-burning-rate-exponent n approach is not satisfactory, a useful approach would be to assume that n depends only on p_1, keeping all previous dependences intact (retaining a in Eq. 12–5 as the only function of T_b). This then can be shown to imply that π_K (a rocket motor parameter) becomes now a function of p_1 and that the differential relation shown below needs to be solved:

$$\ln p_1 \frac{dn}{d \ln p_1} = (1 - n) - \frac{\sigma_p}{\pi_K} \tag{12–15}$$

The above reverts to the form of Eq. 12–14 when n is constant. Results from Eq. 12–15 depend on having suitable (empirical) information on n as a function of p_1. In order to obtain an explicit relation for p_1 that reflects ambient temperature changes with the influence of a variable n, we introduce Eq. 12–9 and solve (keeping K, ρ_b and c^* constant):

$$(1 - n) \ln p_1 - (1 - n_0) \ln p_{01} = \sigma_p \int \frac{d \ln p_1}{\pi_K} = \sigma_p (T_b - T_{0b})$$

or

$$\left(\frac{1 - n}{1 - n_0} \right) \ln p_1 = \ln p_{01} + \frac{\sigma_p}{1 - n_0} (T_b - T_{0b}) \tag{12–16}$$

which appropriately reverts back to the form of Eq. 12–10 when n is unchanging. Note that because n has been assumed only a function of p_1, values of n may be obtained at any T_b from available burning rate data as a function of chamber pressure for the propellant in question (provided that the variation of n with p_1 is monotonic over the interval). The nonexplicit nature of Eq. 12–16 is such that it will require several trials to solve it, for example, first, let $n = n_0$ and find p_1 (which is equivalent to just getting the constant-n constant-K solution), then update n at this new pressure from the relevant empirical information and repeat solving Eq. 12–16 until the solution converges for the new ambient temperature; Problem 11 provides an application for this procedure; another approach would be to introduce a polynomial fit for n as a function of p_1 into Eq. 12–6. As to results, with $\Delta T_b > 0$ and for given values of σ_p and p_{01}, when $n > n_0$ the final chamber pressure will be greater than that resulting from Eq. 12–10 and when $n < n_0$ the final chamber pressure will be less. While these trends are consistent with intuition, any quantitative inferences must be tempered by the nature of the above assumptions. As stated earlier, Eq. 12–5 is only approximately met for many solid propellants.

Burning Enhancement by Erosion

Erosive burning refers to any increase in the propellant burning rate caused by the high-velocity flow of combustion gases across the burning propellant surface. It can seriously affect the performance of solid propellant rocket motors. It occurs primarily in port passages or in grain perforations as the combustion gases flow toward the nozzle; it is more likely to occur when the port passage cross-sectional area A is not large relative to the throat area A_t, with a port-to-throat area ratio of 4 or less. High-velocity gases near the burning surface and their turbulent mixing in the boundary layers increase the heat transfer to the solid propellant and thus increase the burning rate. Chapter 10 of Ref. 12–6 surveys some 29 different theoretical analytical treatments and a variety of experimental techniques aimed at a better understanding of erosive burning.

Erosive burning raises chamber pressure and thrust during the early portion of the burning, as shown in Fig. 12–9. As soon as burning enlarges the flow passage (without a major increase in burning area), the port area flow velocity is reduced and erosive burning diminishes until normal burning will again occur. Since propellant has been consumed more rapidly during the early erosive burning, there is also a reduction of burning time, and of flow and thrust at the end of burning. Erosive burning (defined in Section 12.3 and Ref. 12–7) also causes an early burnout of the web; for certain grain configurations, early web burnouts cause it to lose structural integrity resulting in thin web breakups which produce pieces of unburned propellant that are ejected thru the nozzle. Erosion can also occur at or near the end of the burning period with grains designed for progressive burning (see Figs. 12–15 and 16) where the burning area is intentionally designed to increase in order to raise thrust and chamber pressure shortly before thrust termination. In general, only mild erosion can be tolerated and, in designing rocket motors, erosive burning is either avoided or controlled to be reproducible from one motor to the next. Correlations of erosive burning data are discussed in Ref. 12–8.

A relatively simple model for erosive burning, based on heat transfer (see Section 8.5), was first developed in 1956 by Lenoir and Robillard and has since been

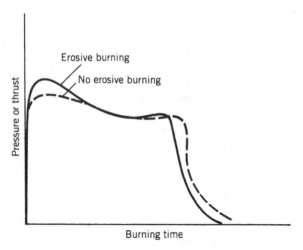

FIGURE 12–9. Typical simplified pressure–time curve with and without erosive burning.

improved and used widely in motor performance calculations. It is based on adding together two burn rates: r_0, which is primarily a function of pressure and ambient grain temperature (basically Eq. 12–5) without erosion, and r_e, the increase in burn rate due to gas erosion:

$$r = r_0 + r_e$$
$$= a p_1^n + \alpha G^{0.8} D^{-0.2} \exp(-\beta r \rho_b / G) \qquad (12\text{–}17)$$

Here, G is the mass flow velocity per unit area in kg/m²-sec, D is a characteristic dimension of the port passage (usually, $D = 4A_p/S$, where A_p is the port area and S is its perimeter), ρ_b is the density of the unburned propellant (kg/m³), and α and β are empirically constants. Apparently, β is independent of propellant formulation and has a value of about 53 when r is in m/sec, p_1 is in pascals, and G is in kg/m²-sec. An expression of α has been determined from convective heat transfer considerations to be

$$\alpha = \frac{0.0288 c_p \mu^{0.2} \mathrm{Pr}^{-2/3}}{\rho_b c_s} \frac{T_1 - T_s}{T_2 - T_b} \qquad (12\text{–}18)$$

Here c_p is the average specific heat at constant pressure of the combustion gases in kcal/kg-K, μ the gas viscosity in kg/m-sec, Pr the dimensionless Prandtl number ($\mu c_p/k$) based on the molecular properties of the gases, k the thermal conductivity of the gas, c_s the heat capacity of the solid propellant in kcal/kg-K, T_1 the combustion gas reaction absolute temperature, T_s the solid propellant surface temperature, and T_b the initial ambient temperature within the solid propellant grain.

Figure 12–10 shows the augmentation ratio r/r_0, or the ratio of the burning rate with and without erosive burning, as a function of gas velocity for two similar propellants, one of which has an iron oxide burn rate catalyst. Augmentation ratios up

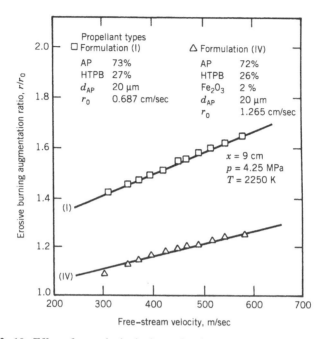

FIGURE 12–10. Effect of gas velocity in the perforation or grain cavity on the erosive burning augmentation factor, which is the burning rate with erosion r divided by the burning rate without erosion r_0 for two composite propellants. (Reproduced with permission of the AIAA from Chapter 10 of Ref. 12–6.)

to 3 can be found in some motor designs. There is a pressure drop from the forward end to the aft end of the port passage, because static pressure is being converted to kinetic gas energy as the flow is accelerated. This pressure differential during erosive burning causes an extra axial load and deformation on the grain, which must be considered in the stress analysis. Erosion rates or burn rate augmentations are not the same throughout the length of the port passage. Erosion is increased locally by flow turbulence resulting from discontinuities such as protrusions, edges of inhibitors, structural supports, or gaps between segmented grains.

Other Burning Rate Enhancements

Enhancement of the burning rate can be expected in vehicles that *spin* the rocket motor about its longitudinal axis (necessary for spin-stabilized flight) or that have high *lateral* or *longitudinal accelerations*, as typical in antimissile rockets. These effects have been experienced with a variety of propellants, with and without aluminum fuel, where propellant formulation is one of the controlling variables (see Fig. 12–11). Whether acceleration comes from spin or longitudinal forces, burning surfaces that form an angle of 60 to 90° with the acceleration vector are most prone to burning rate enhancement. For example, spinning cylindrical internal

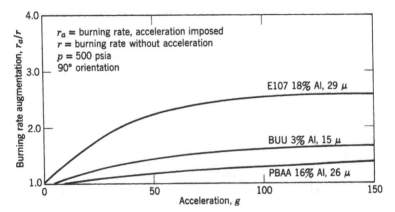

FIGURE 12–11. Acceleration effect on burning rate for three different propellants. See Ref. 12–8 for symbols. (Adapted with permission from Ref. 12–8.)

burning grains are heavily affected. The effect of spin on a motor with an operational composite propellant internal burning grain is shown in Fig. 12–12. Accelerated burning behavior of candidate propellants for new motor designs is often determined in small-scale motors, or with a test apparatus that subjects burning propellant to acceleration (Refs. 12–9 and 12–10). Any stresses induced by rapid acceleration or rapid chamber pressure rises may cause cracks to develop (see Ref. 12–11), which then expose additional burning surfaces.

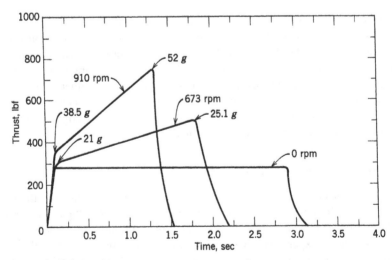

FIGURE 12–12. Effect of axial spin on the thrust–time behavior of a rocket motor with composite propellant using aluminum and PBAN (polybutadiene acrylonitrile) as fuels. (Adapted with permission from Ref. 12–8.)

The *embedding of wires* or other metallic (good-heat-conductor) shapes in the propellant grain increases the burning rate. One technique used has several silver wires arranged longitudinally in an end-burning grain (see Ref. 12–12). Depending on wire size and the number of wires per grain cross-sectional area, the burning rate can easily be doubled. Aluminum wires are about half as effective as silver wires. Other forms of heat conductors have been wire staples (short bent wires) mixed with the propellant prior to the casting operation. The so-called Sprint propellants achieved average burning rates of 3.6 in./sec at 2000 psi using staples.

Intense *radiation emissions* from the hot gases inside the grain cavity also transfer heat to the burning propellant surfaces. The more energetic radiation causes an increase in burning rate. Radiation from exhaust plumes (outside of the nozzle) and effects of particles in the gas are discussed in Chapter 20.

Combustion instability, also called oscillatory combustion, can affect the burning rate of the propellant because of increased heat transfer rates from the changing gas velocities and higher pressures. Combustion instabilities induced by structural vibrations at frequencies where the rocket motor has its highest response function have been reported and investigated in Ref. 12–13. Combustion instability is discussed in Chapter 14.

12.2. OTHER PERFORMANCE ISSUES

Parameters that govern the burning rate and mass discharge rate of motors are called *internal ballistic properties*; these include r, K, σ_p, π_K, along with their dependence on pressure, propellant ingredients, gas velocity, and/or acceleration. The remaining solid propellant rocket parameters are *performance parameters*; these include thrust, ideal exhaust velocity, specific impulse, propellant mass fraction, flame temperature, equipment temperature limits, and duration.

The *ideal nozzle exhaust velocity* of a solid propellant rocket is described by the thermodynamics as given by Eqs. 3–15 and 3–16. As explained in Chapter 5, Eq. 3–16 holds only for frozen equilibrium conditions; for shifting equilibrium the exhaust velocity is best found in terms of the enthalpy drop $(h_1 - h_2)$, and computed from $v_2 = \sqrt{2(h_1 - h_2)}$. Here, it is assumed that the approach velocity of gases upstream of the nozzle is small and can be neglected. This is true if the port area A_p (the flow area of gases between and around the propellant grains) is sufficiently large compared to the nozzle throat area A_t, i.e., when the port-to-throat-area ratio A_p/A_t is greater than about 4.

Internal material erosion always causes a small decrease in performance. This erosion is usually highest at the nozzle throat where the diameter may grow by 0.01 to 0.15 mm/sec. during rocket operation. This enlargement depends on the propellant exhaust gases, any solid particles embedded in the flow, and on the nozzle material. Any nozzle-expansion area ratio reduction decreases rocket motor performance. In rocket motor specifications, delivered specific impulse is labeled the *effective specific impulse* and it is somewhat lower than the initial or theoretical specific impulse (the *total-impulse-to-loaded-weight ratio* also differs from the effective specific impulse because it accounts for the inert mass expended. In many motors insulation and liner are partially consumed during burning).

Thrust for solid propellant rocket motors is given by the identical definitions developed in Chapters 2 and 3, namely, Eqs. 2–13 and 3–29. The *flame* or *combustion temperature* is a thermochemical property of the propellant formulation and the chamber pressure. It not only affects the exhaust velocity, but also heat transfer to the grain, hardware design, materials selection, and external flame radiation emissions. In Chapter 5, methods for combustion temperature calculations are presented. The determination of the nozzle throat area, nozzle expansion area ratio, and nozzle dimensions is discussed in Chapter 3.

The *specific impulse* I_s and the *effective exhaust velocity c* are defined by Eqs. 2–3, 2–4, and 2–6. It is experimentally difficult to measure the instantaneous propellant flow rate or the effective exhaust velocity. However, total impulse and total propellant mass consumed during the test can be measured. The motor's total effective propellant mass and an approximate mass flow rate can be determined by weighing the rocket before and after a ground test. The effective propellant mass is often slightly less than the total propellant mass because some grain designs lead to small portions of the propellant (called slivers) to remain unburned during combustion, as explained later in this chapter. Also, portions of the hot nozzle surface and insulation material erode and/or vaporize during the burning, and this reduces the final inert mass of the rocket motor and also slightly increases the nozzle mass flow. This explains the difference between the total inert mass and the burnout mass in Table 12–3. It has been found that the *total impulse* can be accurately determined in testing by integrating the area under a thrust time curve. For this reason the average specific impulse is usually calculated from total measured impulse and effective propellant mass. The total impulse I_t is defined by Eq. 2–1 as the integration of thrust F over the operating duration t_b:

$$I_t = \int_0^{t_b} F dt = \overline{F} t_b \qquad (12\text{–}19)$$

where $\overline{F}$ is an average value of thrust over the burning duration t_b. For real rocket motors, two time intervals are used, burning time and action time as shown in Fig. 12–13 and further described in Section 12.3.

The definitions for the *burning time, action time*, and pressure rise or *ignition rise time* are defined in Fig. 12–13. Time zero is given as the instance when a starting voltage is applied to the ignition squib and primer charge; a common method is the aft tangent bisector method, as shown in Fig. 12–13 (Ref. 12–14), but this time can also be determined by computer analysis (Ref. 12–15). Exhaust gases can actually be seen from the rocket nozzle for periods longer than the action time, but the effluent mass flow ahead of and behind the action time is actually very small; the end of burning has been determined by several methods. The above definitions are somewhat arbitrary but are commonly in use and documented by standards such as Ref. 2–2.

For flight tests it is possible to arrive at the instantaneous thrust from the measured flight path acceleration (reduced by any estimated drag) and from the estimated

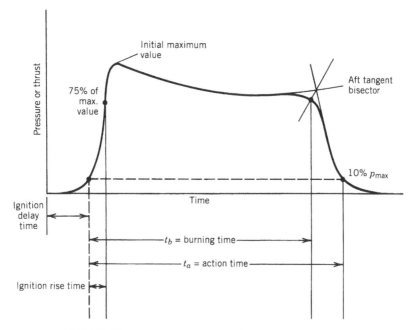

FIGURE 12–13. Definitions of burning time and action time.

instantaneous motor mass from chamber pressure measurements, which is essentially proportional to the rocket nozzle mass flow; this gives another way to calculate specific impulse and total impulse.

There are at least four values of *specific impulse*: (1) *theoretical specific impulse*, (2) *delivered* or *actual* values as measured from flight tests, static tests, or demonstrations (see Ref. 12–16), (3) delivered specific impulse *at standard or reference conditions,* and (4) the *minimum guaranteed value*. Merely quoting a number for specific impulse without further qualification leaves many questions unanswered. This notion is similar to the "four performance parameters" listed in Section 3.5. Specific impulse as diminished by several losses can be predicted as shown in Ref. 12–17.

Losses include nozzle inefficiencies due to friction in the viscous boundary layers, nonaxial nozzle exit flows as described in Chapter 3, thrust vector deflection as described in Chapter 18, residual propellants, heat losses to the walls or insulators, incomplete combustion, and to the presence of solid particles and/or small droplets in the gas which need to be accelerated. There are also some performance *gains*—other gases (created by ablation of the ablative nozzle and insulators or the igniter propellants) contribute to an increased mass flow, in many cases also to a somewhat lower effective average molecular mass, and to a slight reduction of the final inert mass after rocket motor operation.

When particles are present, *two-phase flow* equations for calculating specific impulse can be solved when the size distribution, shape, and percentage of solid particles in the exhaust gas are known. The assumption of a uniform average spherical particle diameter simplifies the analysis (Ref. 12–17), and this diameter can be estimated from specific impulse measurements on rocket motor tests (Ref. 12–18). Section 3.5 gives a simple theory for two-phase flow of solid particles within a gas flow.

Propellants burn to varying degrees of completion depending on the fuel/oxidizer types and ratios, energy losses, and on the environment within the rocket motor. Propellants with nonmetal fuels usually require the use a velocity correction factor (see Section 3.5) of 97 or 98%, as contrasted to 90 to 96% for propellants with aluminum powder as the fuel. Any solid or liquid particles in the exhaust do not contribute to the gas expansion or require energy to be accelerated, and hence two-phase flow is less efficient. However, the addition of the aluminum increases the heat of combustion, i.e., the chamber gas temperature and thus the exhaust velocity or specific impulse. This increase can be made to outweigh any losses incurred.

The *propellant mass fraction* ζ is defined in Eq. 2–8 as $\zeta = m_p/m_0$, and it is directly related to the *motor mass ratio* and therefore also to the flight performance of the vehicle. The initial rocket motor mass m_0 is taken as the sum of the useful solid propellant mass m_p and the nonburning, inert hardware mass of the rocket motor. For a vehicle's total propellant mass fraction, the payload mass and the nonpropulsion inert mass (vehicle structure, guidance and control, communications equipment, and power supply) have to be included. A high value of ζ indicates a low inert motor mass and an efficient hardware design, but often high stresses. This parameter (ζ) has been used to make approximate preliminary design estimates. It depends on motor size or mass, thrust level, nozzle area ratio, and combustion case materials. For very small motors (less than 100 lbm) the value of the propellant fraction is between 0.3 and 0.75. Medium-sized motors have ζ values between 0.8 and 0.91. For larger motors ($1000 < m_0 < 50,000$ lbm) ζ is between 0.88 and 0.945. A range of values is given for each category because of the influence of the following other variables: Medium- and large-sized motors with steel cases generally have lower ζ values than those with titanium cases, and their values are lower than for cases made of Kevlar fibers in an epoxy matrix. The highest values are for cases made of graphite or carbon fibers in an epoxy matrix. The ζ values are lower for larger area ratio nozzles and motors with thrust vector control. The STAR^{TM} 27 rocket motor, shown in Fig. 12–1 and described in Table 12–3, has a propellant mass fraction of 0.924. This is a high value for a medium-sized motor with a titanium metal case and a relatively large nozzle exit section.

A number of performance parameters are used to evaluate solid propellant rocket motors and to compare the quality of design of one rocket motor with another. The first is the *total-impulse-to-loaded-weight ratio* (I_t/w_G), also called the "effective specific impulse." The loaded gross weight w_G is the sea-level initial gross weight of propellant and the rocket propulsion system hardware. Typical values for I_t/w_G are between 100 and 230 sec, with the higher values representative of high-performance rocket propellants and highly stressed hardware, which means a lower inert mass. The total-impulse-to-loaded-weight ratio ideally approaches the value of the specific impulse as seen from Eq. 2–11, when the weight of hardware, metal parts, inhibitors, and so on becomes very small in relation to the propellant weight w_p, i.e., the ratio I_t/w_G approaches I_t/w, which is the definition of the specific impulse

(Eq. 2–4). The higher the value of I_t/w_G, the better the design of a rocket motor. Another parameter used for comparing propellants is the *volume impulse*; it is defined as the total impulse per unit volume of propellant grain, or I_t/V_b.

The *thrust-to-weight ratio* F/w_G is a dimensionless parameter that represents the acceleration of the rocket propulsion system (expressed in multiples of g_0) if it could fly by itself in a gravity-free vacuum; it excludes other vehicle component weights (see Section 4.3). It is peculiar to each application and can vary from very low values of less than one g_0 to over $1000 g_0$ for high acceleration applications of solid *propellant* rocket motors. Some rocket-assisted gun munitions have accelerations of 20,000 g_0.

Two ambient *temperature limits* for the grain are commonly listed in rocket motor specifications or motor parameter lists. The first comprises the minimum and maximum allowable temperatures for starting rocket motor operation. The second involves the minimum and maximum allowable storage temperatures. Also listed is a *storage time* which includes any transportation time together with the time in a warehouse and the time when loaded in a vehicle. These limits are intended to minimize excessive internal grain stresses.

It is difficult to measure actual burning rates or specific impulse on full-scale solid propellant rocket motors (SPRMs) because it is essentially impossible to measure rates of propellant flow or burning area changes directly without changes to the hardware. Some direct measurements of these have been made in experimental rocket motors, such as measuring changes of burning areas or burn rates by means of X-rays during ground tests. The following measurements are typical of SPRM ground tests (but only some are used during any one test): thrust-time profile, pressures (including chamber pressure), action time, burning time, ignition delay, total propellant consumed (with small amounts of eroded insulation/ nozzle materials) by weighing the SPRM before and after testing, initial and final throat and exit diameters, temperature of various components, local stresses and strains, and vibrations. Production SPRMs have fewer direct measurements. Additional data are obtained from analyses, laboratory tests, and design information such as initial volumes of the chamber or cavity, initial burning area, expected combustion temperature, and physical properties and composition of the propellant (such as laboratory-measured specific heats, strand burning rate, propellant density, etc.).

Example 12–3. The following requirements are given for making preliminary estimates of performance for solid of a propellant rocket motor:

Sea-level thrust	2000 lbf average
Thrust duration	10 sec
Chamber pressure	1000 psia
Ambient temperature	70 °F
Propellant	Ammonium nitrate-hydrocarbon

The properties of this propellant are: $k = 1.26$; $R = 76.82$ ft-lbf/lbm-°R; $T_1 = 2790°R$; $r = 0.10$ in./sec at 1000 psia and 70°F; $\rho_b = 0.0546$ lbm/in³. A given *actual value* of $c^* = 3967$ ft./sec can be shown to imply a correction factor $\zeta_{c^*} = 0.997$. Assume moreover a thrust-coefficient correction factor of $\zeta_{C_F} = 0.98$, and optimum operation at sea level.

Determine the specific impulse, the throat and exit areas, the propellant flow rate, the total propellant weight at sea level, the total impulse, the burning area, and an estimated sea-level mass assuming that the total impulse-to-weight ratio, I_t/w_G, is 143 sec (a moderately efficient design).

SOLUTION. From Figs. 3–4 and 3–6, $C_F = 1.57$ and $\varepsilon = 7.8$ (using a pressure ratio of $1000/14.7 = 68.03$ and $k = 1.26$ and an optimum nozzle expansion). The actual thrust coefficient becomes $C_F = 1.57 \times 0.98 = 1.54$. The *specific impulse* is (Eq 3–32)

$$I_s = c^* C_F / g_0 = (3967 \times 1.54)/32.2 = 190 \text{ sec}$$

The require *throat area* is obtained from Eq. 3–31:

$$A_t = F/(p_1 C_F) = 2000/(1000 \times 1.54) = 1.30 \text{ in.}^2$$

The exit area is $7.8 \times 1.30 = 10.1 \text{ in.}^2$. The average propellant weight flow rate at sea level is obtained from Eq. 2–5, namely, $\dot{w} = F/I_s = 2000/190 = 10.5 \text{ lbf/ sec}$. The effective propellant weight for a duration of 10 sec is about 105 lbf. Allowing for residual propellant and for inefficiencies in combustion and lower pressure during thrust buildup, the total loaded propellant weight may be assumed 2% larger or 107 lbf.

The total impulse is from Eq. 2–2: $I_t = 2000 \times 10 = 20,000 \text{ lbf} - \text{sec}$. The propellant burning surface is found by using Eq. 12–4, $p_1 = K \rho_b r c^*$ together with the definition of K:

$$A_b = A_t p_1 / \rho_b r c^* = [1.30 \times 1000/(0.0546 \times 0.10 \times 3967)] \times 32.2 = 1933 \text{ in.}^2$$

The factor of 32.2 is needed to convert the units of the density. Now, the loaded gross weight of the rocket motor only can be estimated. Using the value

$$I_t/w_G = 143 \text{ sec}$$
$$w_G = I_t/(I_t/w_G) = 20,000/143 = 140 \text{ lbf}$$

Because the propellant itself accounts for 107 lbf, the estimate for all the rocket motor hardware parts then becomes 33 lbf. This includes the igniter, case, nozzle, instruments and insulation.

12.3. PROPELLANT GRAIN AND GRAIN CONFIGURATION

The *grain* is the shaped mass of processed solid propellant inside the rocket motor. The material and geometrical configuration of the grain govern motor performance characteristics. Propellant grains are cast, molded, or extruded bodies and their appearance and feel is similar to that of hard rubber or plastic. Once ignited, the grain will burn on all its exposed surfaces forming hot gases that are then exhausted through a nozzle. Most rocket motors have a single grain. A few have more than one grain inside a single case or chamber, and very few grains have segments made of different propellant composition (e.g., to allow different burning rates).

There are two methods of holding the grain in its case, as seen in Fig. 12–14. *Cartridge-loaded or freestanding grains* are manufactured separately from the case

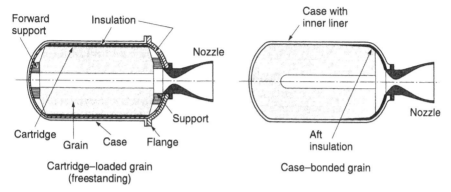

FIGURE 12–14. Schematic diagrams of a freestanding (or cartridge-loaded) and a case-bonded grain.

(by extrusion or by casting into a cylindrical mold or cartridge) and then loaded into or assembled into the case. In *case-bonded grains* the case is used as a mold and the propellant is cast directly into the case and is bonded to the case, its liner or case insulation. Freestanding grains can more easily be replaced when the propellant grain has aged excessively. Cartridge-loaded grains are used in some small tactical missiles and a few medium-sized rocket motors. They often have a lower cost and are easier to inspect. Case-bonded grains give a somewhat better performance, a little less inert mass (no holding device, support pads, and less insulation), and a better volumetric loading fraction, but are more highly stressed and often somewhat more difficult and expensive to manufacture. Today almost all larger motors and many tactical missile motors use case bonding. Stresses in grains are briefly discussed under structural design in the next section.

Definitions and terminology relevant to *grains* include:

Configuration: The shape or geometry of the initial burning surfaces of a grain as it is intended to operate inside a motor.

Cylindrical grain: A grain in which the internal cross section is constant along the axis regardless of perforation shape (see Fig. 12–3).

Neutral burning: A grain for which thrust, pressure, and burning surface area remain approximately constant during burning (see Fig. 12–15), typically within about ±15%. Many grains are neutral burning.

Progressive burning: A grain in which thrust, pressure, and burning surface area increase with burn time (see Fig. 12–15).

Regressive burning: A grain in which thrust, pressure, and burning surface area decrease with burn time (see Fig. 12–15).

Perforation: The central cavity port or flow passage of a propellant grain; its cross section may be a cylinder, a star shape, and the like (see Fig. 12–16).

Sliver: Unburned propellant residue or propellant lost—i.e., expelled through the nozzle at the time of web burnout.

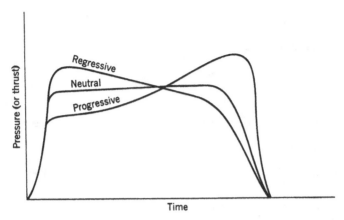

FIGURE 12–15. Classification of grains according to their pressure–time characteristics.

Burning time, or effective burning time, t_b: Usually, the interval from 10% maximum initial pressure (or thrust) to web burnout, with web burnout usually taken as the aft tangent-bisector point on the pressure–time trace (see Fig. 12–13).

Action time, t_a: The burning time or typically, the interval between the initial and final 10% pressure (or thrust) points on the pressure–time trace (see Fig. 12–13).

Deflagration limit: The minimum pressure at which combustion can still be barely self-sustained and maintained without adding energy. Below this pressure the combustion ceases altogether or may be erratic and unsteady with the plume appearing and disappearing periodically.

Inhibitor: A layer or coating of slow or nonburning material (usually, a polymeric rubber type with filler materials) applied (glued, painted, dipped, or sprayed) to a part of the grain's propellant surface to prevent burning on that surface. By preventing burning on inhibited surfaces the initial burning area can be reduced and controlled. Also called *restrictor*.

Liner: A sticky, inert non-self-burning thin layer, flexible and relatively low density, rubber-like polymeric material that is applied to the cases prior to casting the propellant in order to promote good bonding between the propellant and the case or between the propellant and the insulator. It also allows some axial motion between the grain periphery and the case.

Internal insulator: An internal layer between the case and the propellant grain made of an adhesive, thermally insulating material that will not burn readily. Its purpose is to limit the heat transfer to and the temperature rise of the case during rocket operation. Liners and insulators can be seen in Figs. 12–1, 12–4, and 12–14, and are described in Section 13–6.

Web thickness, b: The minimum thickness of the grain from the initial internal burning surface to the insulated case wall or to the intersection of another

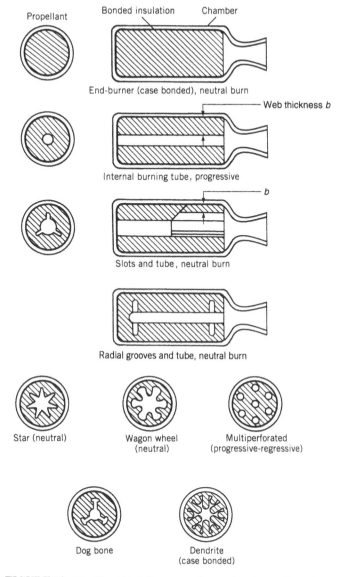

FIGURE 12–16. Simplified diagrams of several grain configurations.

burning surface; for an end-burning grain, b equals the length of the grain (see Fig. 12–16).

Web fraction, b_f: For a case-bonded internal burning grain, the ratio of the web thickness b to the outer radius of the grain:

$$b_f = b/\text{radius} = 2b/\text{diameter} \qquad (12\text{–}20)$$

Volumetric loading fraction, V_f: The ratio of propellant volume V_b to the chamber volume V_c (excluding nozzle) available for propellant, insulation, and restrictors. Using Eq. 2–4 and $V_b = m/\rho$:

$$V_f = V_b/V_c = I_t/(I_s\rho_b g_0 V_c) \qquad (12–21)$$

where I_t is the total impulse, I_s the specific impulse, and ρ_b the propellant density.

Grain designs have to satisfy several interrelated *requirements*:

1. *Rocket motor requirements* can be determined from the *flight mission*. These have to be known before the grain can be designed. They are usually established by the vehicle designers. Motor requirements can include total impulse, desired thrust–time curve and tolerance thereon, motor mass, ambient temperature limits during storage and operation, available vehicle volume or envelope, desired location or movement of rocket motor's center of gravity, and accelerations caused by vehicle forces (vibration, bending, aerodynamic loads, etc.).

2. The *grain geometry* is selected to fit motor requirements; it should be compact efficiently using the available volume, have an appropriate burn surface versus time profile to match the desired thrust–time curve, and avoid or predictably control possible erosive burning (many motors with progressive burning can tolerate short periods of erosive burning). Any remaining unburned propellant slivers, and often also the shift of the center of gravity during burning, should be minimized. Selection of the geometry can be complex, as discussed in Refs. 12–1 and 12–8 as well as below in this section.

3. The *propellant* is usually selected on the basis of its performance capability (e.g., characteristic velocity c^*), mechanical properties (e.g., strength), ballistic properties (e.g., burning rate r), manufacturing characteristics, exhaust plume characteristics, and aging properties. If necessary, the propellant formulation may be slightly altered or "tailored" to fit more exactly the required burning time or grain geometry. Propellant selection is discussed in Chapter 13 and in Ref. 12–8.

4. Grain *structural integrity*, including its liner and/or insulator, must be analyzed to assure that the grain will not fail from stress or strain under all conditions of loading, acceleration, or thermal stress. Grain geometry can be changed to reduce excessive stresses. This is discussed in the next section of this chapter.

5. Any *internal cavity volume* made of perforations, slots, ports, and fins increases with burning time. These cavities need to be evaluated for resonance, vibration damping, and *combustion stability*. This is discussed in Chapter 14.

6. The *processing* of the grain and the *fabrication* of the propellant should be repeatable, simple, low cost (see Chapter 13), and only cause acceptable thermal stresses.

Though grain configuration designs aim to satisfy most of these requirements, there are cases when some of the six above categories can only be partially met. Grain geometry is crucial in its design. For a neutral burning grain (nearly constant thrust), for example, the burning surface A_b has to remain sufficiently constant, and for a regressive burning grain the burning area has to diminish during the burning time. The trade-off between burning rate and the burning surface area is evident from Eqs. 12–5 and 12–6 where it can be seen that changes of burning surface with time have a strong influence on chamber pressure and thrust. Because the density of most modern propellants falls within a narrow range (about 0.066 lbm/in.^{3} or 1830 kg/m^3, +2 to −15%), propellant density has only a small influence on grain design.

As a result of rocket motor developments of the past five decades, many *grain configurations* are available to motor designers. As new methods evolved for increasing the propellant burning rate, the number of configurations needed decreased. Current designs concentrate on relatively few configurations, since the needs of a wide variety of solid rocket motor applications can now be fulfilled by combining known configurations or by slightly altering a classical configuration. The trend has been to discontinue configurations that give weak grains or which form cracks more readily, or produce high sliver residues, or have a low volumetric loading fraction, or are more expensive to manufacture.

The results of propellant burning on internal surface areas are readily apparent for simple geometric shapes such as rods, tubes, wedges, and slots, as shown in the top four configurations of Fig. 12–16. Other basic surface shapes burn as follows: external burning rod—regressive; external burning wedge—regressive. Most propellant grains combine two or more of these basic surfaces to obtain the desired burning characteristic. The star perforation, for example, combines the wedge and the internal burning tube. Figure 12–17 indicates typical single grains with combinations of two basic shapes. The term *conocyl* is a contraction of the words *cone* and *cylinder*. A *finocyl* has *fins* on a *cylinder*.

Configurations that combine both radial and longitudinal burning, as does the 'internal–external burning tube without restricted ends', are frequently referred to as "three-dimensional grains" even though all grains are geometrically three-dimensional. Correspondingly, grains that burn only longitudinally or only radially are called "two-dimensional grains." Grain configurations can be classified according to their web fraction b_f, their length-to-diameter ratio L/D, and their volumetric loading fraction V_f. These three interdependent variables are often used in selecting a grain configuration in the preliminary design of a motor for a specific application. Obvious overlap of characteristics may exist with some configurations, as given in Table 12–4 and shown by simplified sketches in Fig. 12–16. The configurations listed above the horizontal line inside the table are common in recent designs. The bottom three were used in earlier designs and usually are more difficult to manufacture or support in a motor case. An end-burning grain has the highest volumetric loading fraction, the lowest grain cavity volume for a given total impulse, and a relatively low burning area or thrust with a long duration. The internal burning tube is relatively easy to manufacture and can be neutral burning with unrestricted

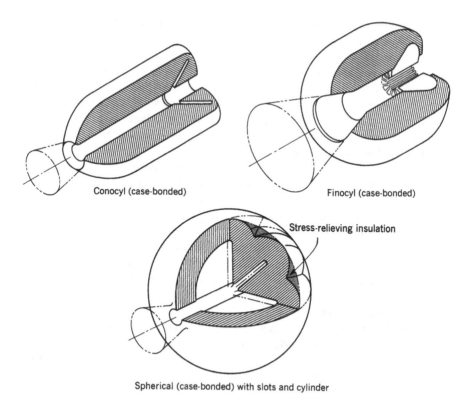

Conocyl (case-bonded)   Finocyl (case-bonded)

Spherical (case-bonded) with slots and cylinder

FIGURE 12–17. Typical common grain configurations using combinations of two basic shapes for the grain cavity.

ends of $L/D \approx 2$. By adding fins or cones (see Fig. 12–17) this configuration works for $2 < L/D < 4$. The star configuration is ideal for web fractions of 0.3 to 0.4; it is progressive above 0.4 but can burn neutral with fins or slots. The wagon wheel is structurally superior to the star shape around 0.3 and is a necessary shape for a web fraction of 0.2 (high thrust and short burn time). Dendrites are used in the lowest web fraction when a relatively large burning area is needed (high thrust and short duration), but stresses may be high. Although the limited number of configurations given in this table may not encompass all the practical possibilities for fulfilling a nearly constant thrust–time performance requirement, combinations of these features may be considered for achieving neutral pressure–time traces and high volumetric loadings before any relatively unproven configuration is accepted. The capabilities of basic configurations listed in these tables can be extended by alterations. Movements of the center of gravity (CG) influence flight stability of the vehicle. Relative values of this CG shift are also shown in Table 12–4. Most solid propellant manufacturers have specific approaches and sophisticated computer programs for analyzing and optimizing grain geometry alternatives for burn surface and cavity volume analyses. See Refs. 12–19 and 12–20 and Chapters 8 and 9 of Ref. 12–1.

TABLE 12–4. Characteristics of Several Grain Configurations

Configuration	Web Fraction	L/D Ratio	Volumetric Fraction	Pressure–Time Burning Characteristics	CG Shift
End burner	>1.0	NA	0.90–0.98	Neutral	Large
Internal burning tube (including slotted tube, trumpet, conocyl, finocyl)	0.5–0.9	1–4	0.80–0.95	Neutral[a]	Small to moderate
Segmented tube (large grains)	0.5–0.9	>2	0.80–0.95	Neutral	Small
Internal star[b]	0.3–0.6	NA	0.75–0.85	Neutral	Small
Wagon Wheel[b]	0.2–0.3	NA	0.55–0.70	Neutral	Small
Dendrite[b]	0.1–0.2	1–2	0.55–0.70	Neutral	Small
Internal–external burning tube	0.3–0.5	NA	0.75–0.85	Neutral	Small
Rod and tube	0.3–0.5	NA	0.60–0.85	Neutral	Small
Dog bone[b]	0.2–0.3	NA	0.70–0.80	Neutral	Small

[a]Neutral if ends are unrestricted, otherwise progressive.
[b]Has up to 4 or sometimes 8% sliver mass and thus a gradual thrust termination.
NA: not applicable or not available.

The *end-burning grain* (i.e., burning like a cigarette) is unique; it burns solely in the axial direction and maximizes the amount of propellant that can be placed in a given cylindrical motor case. In larger motors (over 0.6 m diameter) these end burners eventually show a progressive thrust curve. Figure 12–18 indicates how the burning surface forms a conical shape, causing the rise in pressure and thrust. Although this behavior is not fully understood, two factors may contribute to higher burning rate near the bondline: chemical migration of the burning rate catalyst into and toward the

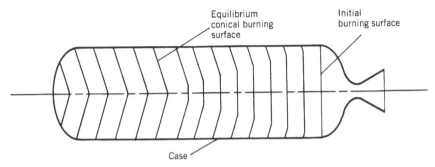

FIGURE 12–18. Schematic diagram of end-burning grain coning effect. In larger sizes (above approximately 0.5 m diameter) the burning surface does not remain flat and perpendicular to the motor axis, but gradually assumes a conical shape. The lines in the grain indicate successively larger-area burning surface contours leading from neutral to progressive burning.

bondline, and local high propellant stresses and strains at the bond surface, creating local cracks (Ref. 12–21).

Rocket motors used in air-launched or in certain surface-launched missile applications, weather rockets, certain anti-aircraft or antimissile rockets, and other tactical applications actually benefit by a reducing thrust with burn time. A high thrust is desired to apply the initial acceleration to attain flight-speed quickly, but, as propellant is consumed and the vehicle mass is reduced, a decrease in thrust is desirable; this limits the maximum acceleration on the rocket-propelled vehicle and on any sensitive payload, often reduces the drag losses, and usually permits a more effective flight path. Therefore, there is a benefit to vehicle mass, flight performance, and cost in having a higher initial thrust during the *boost phase* of the flight, followed by a lower thrust (often 10 to 30% of boost thrust) during the *sustaining phase* of the powered flight. Figure 12–19 shows several grains which give two discrete thrust periods in a single burn operation. The configurations are actually combinations of the configurations listed in Table 12–4.

In a single-propellant dual-thrust level solid rocket motor, factors relating to the sustain flight portion usually dominate in the selection of the propellant type and

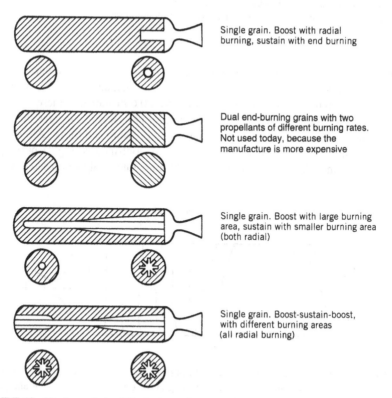

FIGURE 12–19. Several simplified schematic diagrams of grain configurations for an initial period of high thrust followed by a lower-thrust period.

grain configuration when most of the propellant volume is used during the longer sustain portion.

A *restartable rocket motor* has advantages in a number of tactical rocket propulsion systems used for aircraft and missile defense applications. Here two (or sometimes three) grains are contained inside the same case, each with its own igniter. The grains are physically separated typically by a structural bulkhead or by an insulation layer. One method for accomplishing this is shown in Fig. 12–20. The timing between thrust periods (sometimes called *thrust pulses*) can be controlled and commanded by the missile guidance system, so as to change the trajectory in a nearly optimum fashion and to minimize the flight time to target. The separation mechanism has to prevent the burning-hot pressurized gas of the first grain from reaching the other grain and causing its inadvertent ignition. When the second grain is ignited, the separation devices are automatically removed, fractured, or burned, but in such a manner that the fragments of hardware pieces will not plug the nozzle or damage the insulation (see Ref. 12–22).

Slivers

Any remaining unburnt propellant is known as *slivers*. Figure 12–5 and the figure in Problem 12–6 show small slivers or pieces of unburnt propellant remaining at the periphery of the grain, because the pressure there was below the deflagration limit (see Ref. 12–23). About 45 years ago grain designs routinely produced 2 to 7% propellant slivers; any such residue material causes a reduction in propellant mass fraction and vehicle mass ratio. The technology of grain design has advanced so that today there are almost no slivers (usually less than 1%). If slivers were to occur in a new grain design, the designer would try to replace the sliver volume with a lower-density

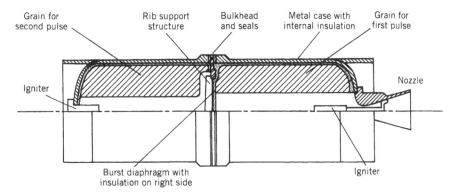

FIGURE 12–20. Simplified diagram of one concept for a two-pulse experimental rocket motor with two grains separated by a bulkhead. During the first pulse operation the metal diaphragm is supported by a spider-web-like structure made of high-temperature material. Upon ignition of the second stage, the scored burst diaphragm is loaded in the other direction; it breaks and its leaves peel back. The bulkhead opening has a much larger area than the nozzle throat.

insulator, which gives less of a mass ratio penalty than the higher-density propellant residue.

12.4. PROPELLANT GRAIN STRESS AND STRAIN

The objective of stress analysis in rocket motors is to design the configuration of the grain, the liners, and/or the grain support in such a way that stresses or strains that may lead to failure will not occur. Static and dynamic loads and stresses are always imposed on propellant grains during manufacture, transportation, storage, and operation. Structurally, a rocket motor is a thin shell of revolution (the case) almost completely filled with a viscoelastic material, the propellant, which normally accounts for 80 to 94% of the motor mass. Propellants have some mechanical properties that are not found in ordinary structural materials and to date these have received relatively little study. The viscoelastic nature of solid propellants is time–history dependent and these materials accumulate damage from repeated stresses; this is known as the *cumulative-damage phenomenon*.

The most common *failure modes* are:

1. *Surface cracks* that are formed when surface strains are excessive. They open up additional burning surfaces, and this in turn causes the chamber pressure as well as the thrust to increase. A higher, shorter duration thrust may cause the vehicle to fly a different trajectory, and this may result in a missed mission objective. With many cracks or a few deep cracks, the case becomes over-pressurized and will fail. The limiting strain depends on the stress level, grain geometry, temperature, propellant age, load history, and the sizes of flaws or voids. At the higher strain rates, deeper, more highly branched cracks readily form (see Ref. 12–11).

2. Small locally detached or broken *unbonded area* may also exist at the grain periphery—a very thin gap may form between the propellant and the liner or the case or the insulation. It can be seen with X-rays or proper heat conduction instruments. As any part of such unbonded areas become exposed to the hot, high-pressure combustion gases in the grain cavity the design burning area increases suddenly. Such enlarged burning area results in a higher chamber pressure (and larger thrust). Having even a few large unbonded areas often results in rocket motor failure.

Other failure modes, such as having a high enough ambient grain temperature to cause large reductions in the physical strength properties, ultimately result in grain cracks and/or debonding. Air bubbles, porosity, and/or an uneven density in the grain may locally reduce the propellant strength to cause failure, again by cracks or debonds. X-ray inspections are routinely used by manufacturers, which can detect all but the smallest debonded areas and surface cracks. When the X-ray inspection uncovers debonding or cracks that cannot be tolerated, the solid propellant unit is rejected and scrapped. Other failure modes are excessive deformations of the grain (e.g., *slump of large grains* can restrict the port area) and involuntary ignitions due to heat generated in viscoelastic propellants from excessive mechanical vibrations (e.g., prolonged bouncing during transport).

Material Characterization

Before any structural analysis can be performed it is necessary to specify the materials and obtain data on their properties. Grain materials (propellant, insulator, and liner) are all nearly incompressible rubber-like substances. They typically have a *bulk modulus in compression* of at least 1400 MPa or about 200,000 psi in their original state (undamaged). Since there should be very few voids in properly made propellants (much less than 1%), their compression strain is low. However, propellants are easily damaged by applied tension and shear loads. When the strength of any propellant in tension and shear (typically between 50 and 1000 psi) is exceeded, the grain will be damaged or fail locally. Since grains are three-dimensional, all stresses are combined and not purely compression stresses, and thus grains are easily damaged. This *damage* may take the form of a "dewetting" of the adhesion between individual solid particles and the binder in the propellant and appears initially as many small voids or porosity. Those very small holes or debonded areas next to or around the solid particles may initially be under vacuum, but they become larger with strain growth.

Propellants, liners, and insulators with solid fillers behave as viscoelastic materials which display nonlinear viscoelastic behaviors. This means that the maximum stress and maximum elongation or strain will diminish each time a significant load is applied. The material becomes weaker and may suffer cumulative damage with each loading cycle or thermal stress application. Other physical properties also change with the time rate of applied loads; for example, very fast pressurizations actually give stronger materials. Because certain binders, such as HTPB (see Chapter 13), give good elongation and are stronger propellants than other polymers used with the same percentage of binder, today HTPB is a preferred binder. Physical properties are also affected by the manufacturing process. For example, tensile specimens cut from the same conventionally cast grain of composite propellant may show a 20 to 40% variation in strength between samples of different orientation which may derive from the local casting slurry flow direction. Though viscoelastic material properties change as a function of prior loading and damage history, they have some capability to "reheal and partially recover" following damage. In time, chemical deterioration will also degrade the properties of many propellants. All these phenomena make it difficult to characterize such materials and predict their behavior or physical properties in engineering terms.

Several kinds of laboratory tests on small samples are routinely performed to determine the physical properties of these materials. (see Refs. 12–24 and 12–25). Simple tests, however, cannot properly represent any complex nonlinear behavior. Laboratory tests are conducted under ideal conditions—mostly uniaxial stresses instead of complex three-dimensional stresses—with a uniform temperature instead of a thermal gradient and usually with no prior damage to the material. The application of laboratory test results to real structural analysis therefore must involve several assumptions and empirical correction factors. Test data are transformed into certain derived parameters for determining safety margins and useful life, as described in Chapter 9 of Ref. 12–1. No complete agreement exists yet on how best to characterize these materials. Nevertheless, laboratory tests do provide useful information and several are described below.

The most common test is a simple *uniaxial tensile test* at constant strain rate. One set of results is shown in Fig. 12–21. This test is commonly used for manufacturing quality control, propellant development, and for determining failure criteria. Once the sample has been loaded, unloaded, and restressed several times, the damage to the material changes its response and properties as shown by the dashed curve in Fig. 12–21.

The *dewetting strain* is, by definition, the strain (and corresponding maximum stress) where incipient failure of the interface bonds between small solid oxidizer crystals and the rubbery binder occurs. The dewetting stress is analogous to the yield point in elastic materials, because it represents when internal material damage begins

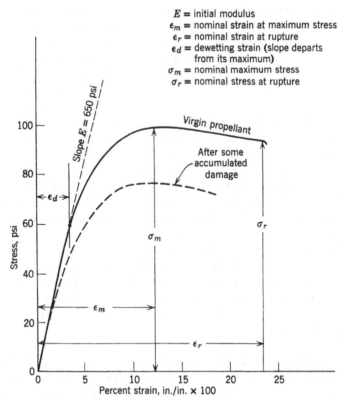

FIGURE 12–21. Stress–strain curves for a typical composite-type solid propellant showing the effect of cumulative damage. The maximum stress σ_m is higher than the rupture stress σ_r, of the tensile test sample.

to happen. The slope E, the modulus at low strain, is not ordinarily used in design, but is often used as a quality control parameter. Data from several such uniaxial tests at different temperatures can be then manipulated to arrive at allowable stresses, permissible safe strains, and a derived artificial modulus, as described later. Once a case-bonded grain has been cooled down from its casting temperature, it will have shrunk and be under multidirectional strain. Samples cut from different parts of a temperature-cycled grain will usually give different tensile test results.

Biaxial strength tests are also performed frequently in the laboratory. One type is described in Ref. 12–24. Meaningful three-dimensional stress tests are difficult to perform in the laboratory and are usually not done. Other sample tests give information about propellant behavior, such as strain endurance tests to obtain the strain levels at which the propellant has long endurance and does not suffer significant damage, tests at constant stress levels, fracture tests of samples with known cracks or defects, tensile tests under simulated chamber pressures, and/or tests to measure thermal coefficients of expansion. Peel tests for the adhesive bonds of propellants to liners or insulators are very common and their failures are discussed in Ref. 12–25. In addition, strain or stress measurements are made occasionally on full-scale, experimental, flight-weight motors using special embedded sensors. Care must be taken that the implanting of these sensors into the grain does not disturb the local stress–strain distribution, which would lead to erroneous measurements. The application and interpretation of all these depend on the test conditions in the grain and on experimenter preferences.

The maximum failure stress for most solid propellants is relatively low compared to that of plastic materials. Typical values range from about 0.25 to 8 MPa or about 40 to about 1200 psi, with average values between 50 and 300 psi, and elongations range from 4 to 250%, depending on the specific propellant, its temperature, and its stress history. Table 12–5 shows such properties for a relatively strong propellant. A few double-base propellants and binder-rich composite propellants can withstand higher stresses (up to about 32 MPa or 4600 psi). Because pressure and strain rate have a major influence on the physical properties, tensile tests performed at actual chamber pressures give a higher strength than those done at atmospheric pressure,

TABLE 12–5. Range of Tensile Properties of a Reduced Smoke Composite Propellant for a Tactical Missile[a]

	Temperature (°F)		
	158	77	−40
Maximum stress (psi)	137–152	198–224	555–633
Modulus (psi)	262–320	420–483	5120–6170
Strain at maximum stress–strain at ultimate stress (%)	54/55–65/66	56/57–64/66	46/55–59/63

[a]Polybutadiene binder with reduced aluminum and ammonium perchlorate; data are from four different 5-gallon mixes.

Source: Data taken with permission of the AIAA from Ref. 12–26.

in some cases by a factor of 2 or more. High strain rates (sudden-start pressurization) can also temporarily improve propellant properties.

Commonly, strength properties of grain materials are determined over a range of propellant temperatures. For air-launched missiles these limits are wide, with −65 and +160°F or 219 and 344 K often representing the lower and upper extremes expected during motor exposure. Propellant grains must be strong enough and have sufficient elongation capability to meet the high stress concentrations present during shrinkage at low temperatures and also under the dynamic load conditions of ignition and motor operation. Mechanical properties (strength, elongation) can be increased by increasing the percent of binder material in the propellant, but at a reduction in performance.

Structural Design

Structural analyses of any typical case-bonded grain have to consider not only the grain itself but also the liner, insulator, and case, all of which interact structurally with the propellant grain under various loading conditions (see Chapter 9 of Ref. 12–1). The need to obtain strong bonds between propellant and liner, liner and insulator, or insulator and case is usually satisfied by using selected materials and manufacturing procedures that assure a proper set of bonds. Liners are usually flexible and can accept relatively large strains without failure as they transmit vehicle loads from the case (which is usually part of the vehicle structure) into the propellant.

When a propellant is oven-cured, it is assumed to have a uniform internal temperature and be free of thermal stresses. As the grain cools and shrinks after curing it reaches equilibrium at the uniform ambient temperature (say, from −40 to +75°F), the propellant thus experiences internal stresses and strains which can be relatively large even at these temperatures. Stresses are further increased because the case material usually has a thermal coefficient of expansion that is smaller than that of the propellant by an order of magnitude. The stress-free temperature range of a propellant can be changed by curing the motor under pressure. Since this usually reduces the stresses at ambient temperature extremes, such pressure cure is now being used more commonly.

Structural analyses may proceed only after all loads have been identified and quantified. Table 12–6 lists typical loads that are experienced by solid propellant motors during their life cycle and some of the failures induced. Since some loads are unique to specific applications, individual loads and their timing during the life cycle of a solid propellant rocket motor have to be analyzed for each application and each motor. They uniquely depend on the motor design and use. Although ignition and high accelerations (e.g., impact on a rocket motor that falls off a truck) usually cause high stresses and strains, these commonly do not produce the critical loads. Stresses induced by ambient environmental temperature cycling or gravity slumps are often relatively small; however, they are additive to other stresses and thus can be critical. A motor used in space that is to be fired within a few months after manufacture presents a different problem than a tactical motor that is to be transported, temperature cycled, and vibrated for longer times, and this is yet different from large-diameter ballistic missile motors that may sit in a temperature-conditioned silo for over 10 years.

TABLE 12–6. Summary of Loads and Likely Failure Modes in Case-Bonded Rocket Motors

Load Source	Description of Load and Critical Stress Area
1. Cool-down during manufacture after hot cure	Temperature differential across case and grain; tension and compression stresses on grain surfaces; hot grain, cool case
2. Thermal cycling during storage or transport (cold night, warm day)	Alternating hot and cold environment; critical condition is with cold grain, hot case; two critical areas: bond-line tensile stress (tearing), inner-bore surface cracking
3. Improper handling and transport vibrations	Shock and vibration, 5–$30g_0$ forces during road transport at 5–300 Hz (5–2500 Hz for external aircraft carry) for hours or days; critical failure: grain fracture or grain debonding
4. Ignition shock/initial pressure loading	Case expands and grain compresses; axial pressure differential more severe with end-burning grains; critical areas; fracture and debonding at grain periphery
5. Friction of internal gas flow in cavity	Axially rearward force on grain
6. Launch and axial flight acceleration	Inertial load mostly axial; shear stress at bond line; slump deformation in large motors can reduce port area
7. Flight maneuvers (e.g., antimissile rocket)	High side accelerations cause unsymmetrical stress distribution; can result in debonding or cracks
8. Centrifugal forces in spin-stabilized projectiles/missiles	High strain at inner burning surfaces; cracks can form
9. Gravity slump of grain during storage; only in large motors (such as 10 ft in diameter)	Stresses and deformation in perforations minimized by rotating the motor periodically; port area cross section can be reduced by slump
10. External air friction when case is also the vehicle's skin—at low altitudes only	Aerodynamic heating of case lowers strength of internal insulation in some propellants causing premature failure. Induces thermal stresses. External insulation is a remedy.

Any structural analysis also requires knowledge of material characteristics and of *failure criteria:* namely, the maximum stress and strains that can safely be accepted by the propellant under various conditions. Failure criteria are derived from cumulative damage tests, classical failure theories, actual motor failures, and from fracture mechanics. Such analyses may need to be iterative because materials and grain geometries must be changed when analysis shows that desired margins of safety are exceeded.

Analyses based on nonlinear viscoelastic stress theory have not been reliable (see Ref. 12–1). Viscoelastic material behavior modelling, though feasible, is relatively complex requiring material-property data that are difficult to obtain and uncertain in value. Most structural analyses have been based on elastic material models; they are relatively simpler and two- and three-dimensional finite element analysis computer programs of such an approach have been developed by rocket motor manufacturing companies. Admittedly, the theory does not fit all the facts, but with appropriate empirical corrections some such analyses have given satisfactory answers to structural grain design problems. An example of a two-dimensional finite element grid from a computer output is shown in Fig. 12–22 for a segment of a grain using an elastic model (see Ref. 12–27).

With elastic materials, stress is essentially proportional strain and independent of time, and when the load is removed they return to their original condition. Neither of these propositions is valid for grains or their propellant materials; in viscoelastic material a time-related dependence exists between stresses and strains and this relationship is not linear, being influenced by the rate of strain. Stresses are not one-dimensional as many laboratory tests are, but three-dimensional and more difficult to visualize and when the load is removed the grain does not return to its exact original position. References 12–28 and 12–29 and Chapters 9 and 10 of Ref. 12–1 discuss three-dimensional analysis techniques in viscoelastic design. A satisfactory analytical description that predicts the influence of cumulative damage has yet to be developed.

Various means have been used to compensate for nonelastic behavior by using allowable stress values degraded for nonlinear effects and/or an effective modulus that uses complicated approximations based on laboratory strain test data. Many use a modified modulus (maximum stress–strain at maximum stress or σ_m/ε_m in Fig. 12–21) called the *stress relaxation modulus* E_R in a master curve against temperature-compensated time to failure, as shown in Fig. 12–23 (see Ref. 12–29 for details). It is constructed from data collected from a series of uniaxial tests at constant strain rate (typically 3 to 5%) performed at different temperatures (typically

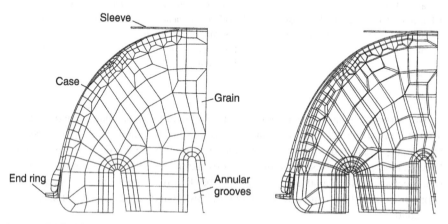

FIGURE 12–22. Partial view of a finite element analysis grid of the forward end of a cast grain in a filament-wound plastic case. The grain has an internal tube (not fully visible) and annular grooves. The left diagram shows the model grid elements and the right shows one calculated strain or deformation condition. (Reprinted with permission from A. Turchot, Chapter 10 of Ref. 12–1.)

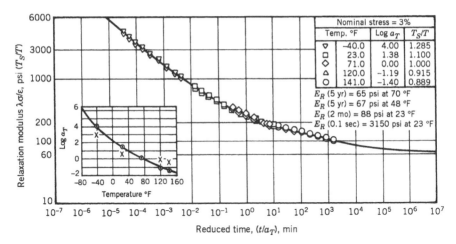

FIGURE 12–23. Stress–relaxation modulus master curve for a particular composite solid propellant constructed with data taken from a number of uniaxial tensile tests at constant strain rate but at different temperatures. (Reproduced with permission of United Technologies Corp., from Ref. 12–29.)

−55 to +43 °C). A shifted temperature ratio T_s/T is shown in the curve on the upper right for 3% strain rate and sample tests taken at different temperatures. The factor λ in the ordinate corrects for the necking down of the tension sample during test. The small inset in this figure explains the correction for temperature that is applied to the reduced time to failure. The empirical time–temperature shift factor a_T is set to zero at ambient temperatures (25 °C or 77 °F) and graphically shifted for higher and lower temperatures. Such a master curve then provides time-dependent stress–strain data to calculate the response of the propellant for structural analysis (see Ref. 12–24 and Chapter 9 of Ref. 12–1).

Usually, several different grain loading and operating conditions need to be analyzed. Such structural analyses are useful for identifying locations of maximum stress or strain and for finding any structural members or grain sectors that are too weak or too heavy. The choice of the best analytical tool and the best pseudo-viscoelastic compensation factors depends on the experience of the stress analyst, the specific motor design conditions, the complexity of the motor, the geometry, and on suitable, available and valid propellant property data.

For case-bonded rocket motors, special provisions are required to reduce stress concentrations at the grain ends where the case and grain interface, especially in motors expected to operate satisfactorily over a wide range of temperatures. Basically, high stresses arise from two primary sources: The first one comprises the physical properties, including the coefficient of thermal expansion of the case material and the propellant, when they are grossly dissimilar. The coefficient of expansion of a typical solid propellant is 1.0×10^{-4} m/m-K, which is five times as great as that of a typical steel motor case. The second one arises from the aft-end and head-end geometries at the grain–case juncture that often present a discontinuity, where the grain stress theoretically would approach infinity. Actually, finite stresses arise because viscoplastic deformations occur in the propellant, the liner, and the case insulation. Calculating the stress in a given case–grain termination arrangement is usually impractical, and designers rely on approximations supported by empirical data.

For simple cylindrical grains, the highest stresses usually occur at the outer and inner surfaces, at discontinuities such as the bond surface termination point, or at stress concentration locations such as sharp radii at the roots or tips of star or wagonwheel perforations, as shown in Fig. 12–16. Figure 12–24 shows a *stress relief flap*, sometimes called a *boot*, a device to reduce local grain stresses at the cylindrical walls. It is usually an area on the outside of the grain near its aft end (and sometimes also its forward end), where the liner material is not sticky but has a nonadhesive coating that permits the grain to shrink away from the wall. It allows for a reduction of the grain at the bond termination point moving the location of highest stress into the liner or the insulation at the flap termination or hinge. Normally, the liner and insulation are much stronger and tougher than the propellant.

Parametric studies of propellant and case-bond stresses of a typical grain–case termination design (Fig. 12–24) reveal the following:

1. Flap length is less significant than the thickness of the insulation of the separate flap boot, if one is used, in controlling the local level of stresses at the grain–case termination.

2. The distribution of stresses at the grain–case termination is sensitive to the local geometry; the level of stress at the case bond increases with web fraction and length-to-diameter ratio under loading by internal pressure and thermal shrinkage.

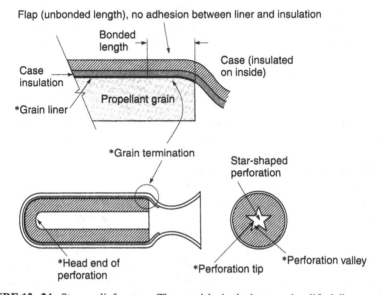

FIGURE 12–24. Stress relief system. The asterisks in the bottom simplified diagram denote potentially critical failure areas. The top sketch is an enlargement of the aft termination region of the grain and shows a boot or stress relief flap.

3. As the L/D and web fraction increase, the inner-bore hoop stress and the radial stress at the grain–case bond increase more rapidly than does the grain–case termination stress under internal pressure and thermal shrinkage loads.

4. The radial case-bond stress level at the grain–case termination is much larger than the case-bond shear stress under axial acceleration loading as well as under internal pressure and thermal shrinkage loading.

Aging of propellants in rocket motors refers to deterioration in their physical properties with time. It is caused by any *cumulative damage* done to the grain (such as by thermal cycling, and/or repetitive load applications) during storage, handling, or transport. It can also be caused by chemical changes with time, such as the gradual depletion (evaporation) of certain liquid plasticizers and/or moisture absorption. The ability to carry stress or to allow for elongation in propellants diminishes with cumulative damage. The *aging limit* is the estimated time when the motor is no longer able to perform its operation reliably or safely (see Refs. 12–30 and 12–31). Depending on the propellant and the grain design, this age limit or motor life can be between 8 and 25 years. Before this limit is reached, the motor may be deactivated and have its propellant removed and replaced; this refurbishing of propellant is routinely done on larger and more expensive rocket motors in military inventories.

With small tactical rocket motors the aging limit is usually determined by full-scale motor-firing tests at various time intervals after manufacture, say two or three years, together with extrapolation to longer time periods. Accelerated temperature aging (more severe thermal cycles) and accelerated mechanical pulse loads and overstressing are often used to reduce the time needed for these tests. For the larger rocket motors, which are more expensive, the number of full-scale tests has to be relatively small and aging criteria are then developed from structural analysis, laboratory tests, and subscale rocket motor tests.

Many of the early grains were *cartridge loaded* and kept the grain isolated from the motor case to minimize the interrelation of the case and any grain stresses and strains resulting from thermal expansion or contraction. Also, upon pressurization, the case expands but the grain shrinks. The *case-bonded* grain presents a far more complex problem in stress analysis. With the propellant grain bonded firmly to the case, being a semirubbery and relatively weak material, it is forced to respond to case strains. As a result, several critically stressed areas exist in every case-bonded motor design; some are shown with an asterisk in Fig. 12–24.

The physical character of propellants brings about the varying nature of the stress analysis problem. In summary, solid propellants are relatively weak in tension and shear, are semielastic, grow softer and weaker at elevated temperatures, become hard and brittle at low temperatures, readily absorb and store energy upon being vibrated, degrade physically during long-term storage because of decomposition and chemical or crystalline changes, and accumulate structural damage under load, including cyclic load. This last phenomenon is shown graphically in Fig. 12–25 and is particularly important in the analysis of rocket motors that are to have a long shelf life (more than 10 years).

No a priori reason is known for materials to exhibit *cumulative damage*, but propellants and their bond-to-case materials do exhibit this trait even under constant load,

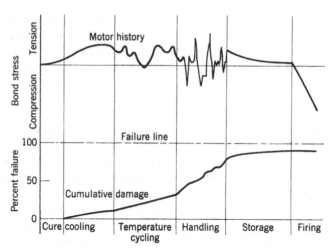

FIGURE 12–25. Representation of the progress in cumulative damage to the bond between the grain and the case in a case-bonded rocket motor experiencing a hypothetical stress history. (Adapted from Ref. 12–32.)

as shown in Fig. 12–26. Valid theories and analytical methods applicable to cumulative damage include a consideration of both the stress–strain history and the loading path (of the material affected). The most important environmental variables affecting the shelf life of a motor are time, temperature cycles, propellant mass, stress (gravity forces for large motors), and shock and vibration. Failure due to cumulative damage usually appears as cracks in the face of the perforation or as local "unbonds" in case-bonded motors.

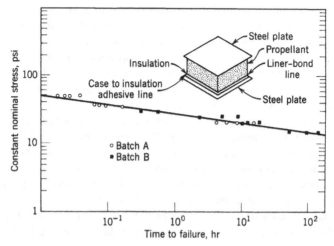

FIGURE 12–26. Time-dependent reduction of the propellant–liner–insulator bond strength when subjected to constant load at 77 °F. (From Ref. 12–33.)

The strength of most propellants is sensitive to the rate of strain; in effect they appear to become more brittle at any given temperature as the strain rate is increased, a physical trait that becomes important during the ignition process.

12.5. ATTITUDE CONTROL AND SIDE MANEUVERS WITH SOLID PROPELLANT ROCKET MOTORS

An ingenious attitude control (also called reaction control) system with solid propellants used with some ballistic missiles is shown schematically in Fig. 12–27. Its hot reaction gases have a low enough temperature so that uncooled hardware can be used for long operating durations. Ammonium nitrate composite propellant (mentioned as gas generator propellants in Tables 13–1 and 13–2) or any propellant consisting of a nitramine (RDX or HMX, described in Chapter 13) with a polymer binding is suitable. The version shown in Fig. 12–27 provides pitch and yaw control, see Chapter 18; hot gas flows continuously through insulated manifolds, open hot gas valves, and all four nozzles. When one of these valves is closed, it causes an imbalance of gas flow producing a side force. The chamber pressure rises when any valve is closed. Four roll-control thrusters have been deleted from this figure for simplicity.

With this type of attitude control system it is possible to achieve variable duration thrust pulsing operations and random pitch, yaw, and roll maneuvers. It is competitive with multithruster liquid propellant attitude control systems. The solid propellant versions are usually heavier because they have heavy insulated hardware and require more propellant (for continuous gas flow), whereas the liquid version is operated only when attitude control vehicle motions are required.

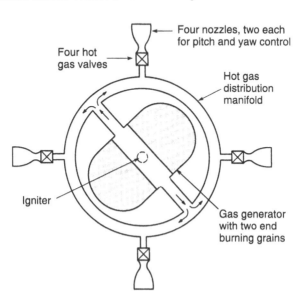

FIGURE 12–27. Simplified diagram of a rocket attitude control system using solid propellant. All four valves are normally open and gas flows equally through all nozzles.

A similar approach with hot gas valves applies to upper stages of interceptor vehicles and is used for missile defense; there is little time available for maneuvers of the upper stage to reach the incoming missile or aircraft and therefore the burning durations are usually short. The solid propellant gas temperatures are higher than with gas generators (typically 1260 °C or 2300 °F), but lower than with typical composite propellants (3050 K or 5500 °F), but this requires the valves and manifolds to be made of high-temperature material (such as rhenium or carbon). In addition to attitude control, the system provides a substantial side force or divert thrust, which displaces the flight path laterally. Figure 12–28 shows such a system. Since all hot gas valves are

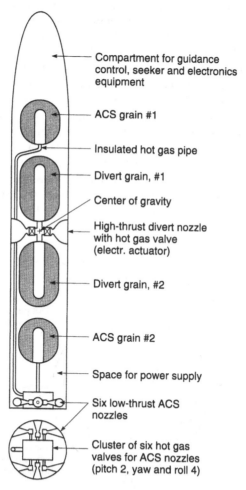

Compartment for guidance control, seeker and electronics equipment

ACS grain #1

Insulated hot gas pipe

Divert grain, #1

Center of gravity

High-thrust divert nozzle with hot gas valve (electr. actuator)

Divert grain, #2

ACS grain #2

Space for power supply

Six low-thrust ACS nozzles

Cluster of six hot gas valves for ACS nozzles (pitch 2, yaw and roll 4)

FIGURE 12–28. Simplified diagram of two propulsion systems for one type of maneuverable upper stage of an interceptor missile. The side or divert forces are relatively large and go essentially through the center of gravity (CG) of the upper-stage vehicle. To minimize the CG travel two grains are above and two grains are below the CG. Each nozzle has its own hot gas valve, which is normally open and can be pulsed. The attitude control system (ACS) is fed from the reaction gas of two grains and has six small nozzles.

normally open, a valve has to be closed to obtain thrust as explained in the previous figure. The attitude control system provides pitch, yaw, and roll control to stabilize the vehicle during its flight, to orient the divert nozzle into the desired direction, and sometimes to orient the seeker (at the front of the vehicle) toward the target. Another example is NASA's Orion launch abort control system, which has eight proportionally controlled valves designed to provide up to 7000 lbf of net steering force in any direction. Thrust vector control is treated in Chapter 18.

SYMBOLS

a	burning rate constant, also called temperature coefficient
A_b	solid propellant burning area, m^2 (ft^2)
A_p	port area (flow area of gases inside grain cavity or between and around propellant grains), m^2 (ft^2).
A_t	nozzle throat cross-sectional area, m^2 (ft^2)
b	web thickness, m (in.)
b_f	web fraction, or web thickness-to-radius ratio
c	effective exhaust velocity, m/sec (ft/sec)
c^*	characteristic exhaust velocity, m/sec (ft/sec)
c_p	specific heat of gas, kcal/kg-K
c_s	specific heat of solid, kcal/kg-K
C_F	thrust coefficient
D	diameter or other dimension, m (ft.)
E_R	relaxation modulus, MPa (psi)
F	thrust, N (lbf)
$\overline{F}$	average thrust, N (lbf)
g_0	acceleration due to gravity at sea level, 9.81 m/sec^2 (32.2 ft/sec^2)
G	mass flow velocity per unit area kg/m^2 sec
h	enthalpy per unit mass, J/kg (Btu/lbm)
I_s	specific impulse, sec
I_t	total impulse, N-sec (lbf-sec)
k	specific heat ratio
K	ratio of burning surface to throat area, A_b/A_t
L	length, m (ft)
m	mass, kg (lbm)
$\dot{m}$	mass flow rate, kg/sec
n	burning rate exponent
p	pressure, MPa ($lbf/in.^2$)
p_1	chamber pressure, MPa ($lbf/in.^2$)
Pr	Prandtl number, $\mu c_p/k$
r	propellant burning rate (velocity of consumption), m/sec or mm/sec or in./sec
R	gas constant, J/kg-K (Btu/lbm-°R)
S	perimeter, m
t	time, sec
t_a	action time, sec

t_b burn time, sec
T absolute temperature, K (°R)
T_1 chamber temperature, K (°R)
T_b propellant ambient temperature, °F(°C)
T_s Propellant initial temperature, °F(°C)
v_2 theoretical exhaust velocity, m/sec (ft/sec)
V_b propellant volume, m³ (ft³)
V_c chamber volume, m³ (ft³)
V_f volumetric loading fraction, %
w total effective propellant weight, N (lbf)
w_G total loaded rocket weight, or gross weight, N (lbf)
$\dot{w}$ weight rate of flow, N/sec (lbf/sec)

Greek Letters

α heat transfer factor
β Constant
ϵ elongation or strain
κ Conductivity
μ Viscosity
π_K temperature sensitivity coefficient of pressure, K^{-1} ($°R^{-1}$)
ρ density, kg/m³ (lbm/ft³)
σ stress, N/cm² (psi)
σ_p temperature sensitivity coefficient of burning rate, K^{-1} ($°R^{-1}$)
ζ propellant mass fraction

Subscripts

a action time
b solid propellant burning conditions
e erosive index
m Maximum
p pressure or propellant or port cavity
t throat conditions
0 initial or reference condition
1 chamber condition
2 nozzle exit condition

PROBLEMS

1. What is the ratio of burning area to nozzle throat area for a solid propellant motor with the characteristics shown below? Also, calculate the temperature coefficient (a) and the temperature sensitivity of pressure (π_K).

Propellant specific gravity	1.71
Chamber pressure	14 MPa

Burning rate	38 mm/sec
Temperature sensitivity σ_p	$0.007 \ (K)^{-1}$
Specific heat ratio	1.27
Chamber gas temperature	2220 K
Molecular mass	23 kg/kg-mol
Burning rate exponent n	0.3

2. Plot the burning rate against chamber pressure for the motor in Problem 1 using Eq. 12–5 between chamber pressures of 11 and 20 MPa.

3. What would the area ratio A_b/A_t in Problem 1 be if the pressure were increased by 10%? (Use curve from Problem 2 or Eq. 12–5.)

4. Design a simple rocket motor for the conditions given in Problems 1 and 2 for a thrust of 5000 N and for a duration of 15 sec. Determine principal dimensions and approximate weight.

5. For the Orbus-6 rocket motor described in Table 12–3 determine the total impulse-to-weight ratio, the thrust-to-weight ratio, and the acceleration at start and burnout if the vehicle inert mass and the payload come to about 6000 lbm. Use burn time from Table 12–3 and assume g ≈ 32.2 ft./sec².

6. For a cylindrical grain with two slots the burning progresses in finite time intervals approximately as shown by the successive burn surface contours in the following drawing. Draw a similar set of progressive burning surfaces for any one configuration shown in Figure 12–16, and draw an approximate thrust–time curve from these plots, indicating the locations where slivers might remain. Assume the propellant has a low value of n and thus the motor experiences little change in burning rate with chamber pressure.

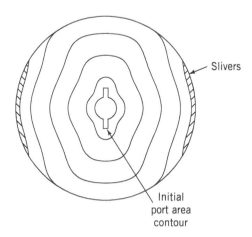

7. Discuss the significance of the web fraction, the volumetric loading ratio, and the L/D ratio in terms of vehicle performance and design influence.

8. Equations 12–8 and 12–9 express the influence of temperature on the burning of a solid propellant. Describe how a set of tests should be set up and what should be measured in order to determine these coefficients over a given range of operating conditions.

9. What would be the likely changes in burning rate (at 1000 psi), action time, average thrust, action-time-average chamber pressure, burn time, and action-time total impulse for the three rocket motors described in Table 12–3 if each were operated with the rocket motor at a storage temperature lower than that given in the table? Describe what results qualitatively.

10. A newly designed case-bonded rocket motor with a simple end-burning grain failed and exploded on its first test. The motor worked well for about 20% of its burn time, when the record showed a rapid rise in chamber pressure. It was well conditioned at room temperature before firing, and the inspection records did not show any flaws or voids in the grain. Make a list of possible causes for this failure and give suggestions on what to do in each case to avoid a repetition of the failure.

11. For the AP-CMDB (30%, 150 μm) propellant shown in Fig. 12–7 as the solid dots, find the chamber pressure that would result from an increase of $\sigma_p \Delta T_b = 0.3$ (this would correspond to a 30°F change with $\sigma_p = 0.01/°F$). Take the reference values as 28 atm and 70°F. Values of $n(p_1)$ at various pressure ranges from Fig. 12–7 may be taken as:

$$n(5 - 10 \text{ atm}) = 0.38, n(10 - 30 \text{ atm}) = 0.54, n(30 - 100 \text{ atm}) = 0.58$$

Answer: $p_1 = 78.6$ atm.

12. What will be the percent change in nominal values of A_t, r, I_s, T_0, t_b, A_b/A_t, and the nozzle throat heat transfer rate, if the Orbus-6 rocket motor listed in Table 12–3 is to be downgraded in thrust for a particular flight by 15% by substituting a new nozzle with a larger nozzle throat area but the same nozzle exit area? The propellants, grain, insulation, and igniter will be the same.

13. What would be the new values of I_t, I_s, p_1, F, t_b, and r for the first stage of the Minuteman rocket motor described in Table 12–3, if the motor were fired at sea level with the grain temperature 20 °F hotter than the data shown. Use only data from this table.
Answers:

$$I_t = 10,240,000 \text{ lbf} - \text{sec}, I_s = 254 \text{ sec}, p_1 = 799 \text{ psia}, F = 1.99 \times 10^5 \text{ lbf},$$

$$t_b = 51.4 \text{ sec}, r = 0.339 \text{ in.} / \text{sec}.$$

14. Calculate K-ratio of the burning area to the nozzle throat area in two separate ways for the STAR™ 27 motor using *only* the data found in Table 12–3. Refer to Figure 12–13 for several of the definitions. Compare the two values of K and comment.

15. We may take K as constant when the chamber contour lengths remain within ±15% during propellant burning (see Fig. 12–5). Calculate the variation in thrust that a ±8% change would represent in a rocket motor with a fixed nozzle geometry and $n = 0.5$. What would be the corresponding changes in thrust and mass-flow rate. State your assumptions noting that in the real world some nozzle erosion and other effects may occur.

REFERENCES

12–1. P. R. Evans, Chapter 4A, "Composite Motor Case Design"; H. Badham and G. P. Thorp, Chapter 6, "Considerations for Designers of Cases for Small Solid Propellant Rocket Motors"; B. Zeller, Chapter 8, "Solid Propellant Grain Design"; D. I. Thrasher, Chapter 9, "State of the Art of Solid Propellant Rocket Motor Grain Design in the United States"; and A. Truchot, Chapter 10, "Design and Analysis of Rocket

Motor Internal Insulation"; all in *Design Methods in Solid Propellant Rocket Motors*, AGARD Lecture Series 150, Revised Version, NATO, Brussels, 1988.

12–2. L. H. Caveney, R.L. Geisler, R, A. Ellis, and T. L. Moore, "Solid Rocket Enabling Technologies and Milestones in the United States," *Journal of Propulsion and Power*, Vol. 19, No. 6, Nov.—Dec. 2003, pp. 1038–1066.

12–3. E. Gautronneau, M. Darant, and E. Vari, "Vega Program the P80 FW SRM Nozzle," Report ESA 42-113-1, 2008.

12–4. N. Kubota, Chapter 1, "Survey of Rocket Propellants and Their Combustion Characteristics," in *Fundamentals of Solid Propellant Combustion*; K. K. Kuo and M. Summerfield (Eds.), *Progress in Astronautics and Aeronautics,* Vol. 90, AIAA, New York, 1984. See also A. Davenas, Development of Modern Solid Propellants, *Journal of Propulsion and Power*, Vol. 19, No. 6, May–Jun. 2003, pp. 1108–1128; V. Yang, T. B. Brill, and W.-Z. Ren (Eds.), *Solid Propellant Chemistry, Combustion, and Motor Interior Ballistics, Progress in Aeronautics and Astronautics*, Vol. 185, AIAA, Reston, VA, 2000.

12–5. S. D. Heister and J. Davis, Predicting Burning Time-Variations in Solid Rocket Motors, *Journal of Propulsion and Power*, Vol. 8, No. 3, May–June 1992; J. R. Osborn and S. D. Heister, Solid Rocket Motor Temperature Sensitivity, *Journal of Propulsion and Power*, Vol. 10, No. 6, November–December 1994, pp. 908–910.

12–6. M. K. Razdan and K. Kuo, Chapter 10, "Erosive Burning of Solid Propellants," in *Fundamentals of Solid Propellant Combustion*, K. K. Kuo and M. Summerfield (Eds.), *Progress in Astronautics and Aeronautics,* Vol. 90, AIAA, New York, 1984.

12–7. E. M. Landsbaum, "Erosive Burning Revisited," AIAA Paper 2003-4805, Jul. 2003.

12–8. "Solid Propellant Selection and Characterization," *NASA SP-8064*, Jun. 1971 (N72-13737).

12–9. M. S. Fuchs, A. Peretz, and Y. M. Timnat, "Parametric Study of Acceleration Effects on Burning Rates of Metallized Solid Propellants," *Journal of Spacecraft and Rockets*, Vol. 19, No. 6, Nov.–Dec. 1982, pp. 539–544.

12–10. P. Yang, Z. Huo, and Z. Tang, Chapter 3, "Combustion Characteristics of Aluminized HTPB/AP Propellants in Acceleration Fields," in V. Yang, T. B. Brill and W.-Z. Ren (Eds.), *Solid Propellant Chemistry, Combustion, and Motor Interior Ballistics, Progress in Aeronautics and Astronautics*, Vol. 185, AIAA, Reston VA, 2000.

12–11. K. K. Kuo, J. Moreci, and J. Mantzaras, "Modes of Crack Formation in Burning Solid Propellant," *Journal of Propulsion and Power*, Vol. 3, No. 1, Jan.–Feb. 1987, pp. 19–25.

12–12. M. K. King, "Analytical Modeling of Effects of Wires on Solid Motor Ballistics," *Journal of Propulsion and Power*, Vol. 7, No. 3, May–Jun. 1991, pp. 312–320.

12–13. D. R. Greatrix, "Correlation of Pressure Rise with Radial Vibration Level in Solid Rocket Motors," AIAA Paper 95-2880, July 1995; "Simulation of Axial Combustion Instability Development and Suppression in Rocket Motors," *International Journal of Spray and Combustion Dynamics*, Vol. 1, No, 1, 2009, pp. 143–168.

12–14. R. Mannepalli, "Automatic Computation of Burning Time of Solid Rocket Motors Using the Aft-Tangent Method," Report SRO-SHAR-TR-07-96-92, ISRO (Indian Space Rocket Organization, Bangalore), 1992.

12–15. R. Mannepalli, "Automatic Computation of Burning Time of Solid Rocket Motors Using the Chord-Midpoint Method", Report SRO-SHAR-TR-07-95-90, ISRO (Indian Space Rocket Organization, Bangalore), 1990.

12–16. "Solid Rocket Motor Performance Analysis and Prediction," *NASA SP-8039*, May 1971 (N72-18785).

12–17. E. M. Landsbaum, M. P. Salinas, and J. P. Leavy, "Specific Impulse Predictions of Solid Propellant Motors," *Journal of Spacecraft and Rockets*, Vol. 17, 1980, pp. 400–406.

12–18. R. Akiba and M. Kohno, "Experiments with Solid Rocket Technology in the Development of M-3SII," *Acta Astronautica*, Vol. 13, No. 6–7, 1986, pp. 349–361.

12–19. "SPP'04TM," Computer Program from Software & Engineering Associates, Inc., http://www.seainc.com

12–20. R. J. Hejl and S. D. Heister, "Solid Rocket Motor Grain Burnback Analysis Using Adaptive Grids," *Journal of Propulsion and Power*, Vol. 11, No. 5, Sept.–Oct. 1995.

12–21. W. H. Jolley, J. F. Hooper, P. R. Holton, and W. A. Bradfield, "Studies on Coning in End-Burning Rocket Motors," *Journal of Propulsion and Power*, Vol. 2, No. 2, May–Jun. 1986, pp. 223–227.

12–22. L. C. Carrier, T. Constantinou, P. G. Harris, and D. L. Smith, "Dual Interrupted Thrust Pulse Motor," *Journal of Propulsion and Power*, Vol. 3, No. 4, Jul.–Aug. 1987, pp. 308–312.

12–23. C. Bruno et al., "Experimental and Theoretical Burning of Rocket Propellant near the Pressure Deflagration Limit," *Acta Astronautica*, Vol. 12, No. 5, 1985, pp. 351–360.

12–24. F. N. Kelley, Chapter 8, "Solid Propellant Mechanical Property Testing, Failure Criteria and Aging," in C. Boyars and K. Klager (Eds.), *Propellant Manufacture Hazards and Testing*, Advances in Chemistry Series 88, American Chemical Society, Washington, DC, 1969.

12–25. T. L. Kuhlmann, R. L. Peeters, K. W. Bills, and D. D. Scheer, Modified Maximum Principal Stress Criterion for Propellant Liner Bond Failures, *Journal of Propulsion and Power,* Vol. 3, No. 3, May–Jun. 1987.

12–26. R. W. Magness and J. W. Gassaway, "Development of a High Performance Rocket Motor for the Tactical VT-1 Missile," AIAA Paper 88-3325, July 1988.

12–27. I-Shih Chang and M. J. Adams, "Three-Dimensional, Adaptive, Unstructured, Mesh Generation for Solid-Propellant Stress Analysis," AIAA Paper 96-3256, July 1996.

12–28. G. Meili, G. Dubroca, M. Pasquier, and J. Thenpenier, "Nonlinear Viscoelastic Design of Case-Bonded Composite Modified Double Base Grains," AIAA Paper 80-1177R, July 1980; and S. Y. Ho and G. Care, "Modified Fracture Mechanics Approach in Structural Analysis of Solid-Rocket Motors," *Journal of Propulsion and Power*, Vol. 14, No. 4, Jul.–Aug. 1998.

12–29. P. G. Butts and R. N. Hammond, "IUS Propellant Development and Qualification," Paper presented at the 1983 JANNAF Propulsion Meeting, Monterey, CA, Feb. 1983.

12–30. A. G. Christianson et al., "HTPB Propellant Aging," *Journal of Spacecraft and Rockets*, Vol. 18, No. 3, May–Jun. 1983; D. Zhou et al., "Accelerated Aging and Structural Integrity Analysis Approach to Predict Service Life of Solid Rocket Motors," AIAA Paper 2015-4240, Orlando, FL, 2015.

12–31. D. I. Thrasher and J. H. Hildreth, "Structural Service Life Estimates for a Reduced Smoke Rocket Motor," *Journal of Spacecraft and Rockets*, Vol. 19, No. 6, Nov. 1982, pp. 564–570.

12–32. S. W. Tsa, Ed., *Introduction to Viscoelasticity*, Technomic, Stanford, CT, 1968.

12–33. J. D. Ferry, *Viscoelastic Properties of Polymers*, John Wiley & Sons, New York, 1970.

CHAPTER 13

SOLID PROPELLANTS

This is the second of four chapters dealing with solid propellant rocket motors. Here we describe several common solid rocket propellants, their principal categories, ingredients, hazards, manufacturing processes, and quality control. We also discuss liners and insulators, propellants for igniters, propellant tailoring, and propellants for gas generators.

Thermochemical analyses are needed to characterize the performance of any given propellant and specific methods are described in Chapter 5. Such analyses provide values for the effective average molecular mass, combustion temperature, average specific heat ratio, and characteristic velocity—these are all functions of propellant composition and chamber pressure. Specific impulses can also be computed for given nozzle configurations and exhaust conditions.

The term *solid propellant* has several connotations, including: (1) the rubbery or plastic-like mixture of oxidizer, fuel, and other ingredients that have been processed (including curing) and constitute the finished grain; (2) the processed but uncured product; (3) a single ingredient, such as the fuel or the oxidizer. In this field, acronyms and chemical symbols are used indiscriminately as abbreviations for propellant and ingredient names; only some of these will be shown in this chapter.

13.1. CLASSIFICATION

Historically, the early rocket motor propellants used to be grouped into two classes: *double-base* (DB)* propellants were the first production propellants and subsequently the development of polymers as binders made the *composite* propellants feasible.

---

*Acronyms, symbols, abbreviations, and chemical names of propellant ingredients are explained in Tables 13–6 and 13–7 in Section 13.4.

Processed modern propellants are more finely classified as described below. Such classifications are helpful but they are neither rigorous nor complete. Sometimes the same propellant will fit into two or more classifications.

1. Propellants are often tailored to and classified by *specific applications*, such as space launch booster propellants or tactical missile propellants, each having specific chemical ingredients, different burning rates, different physical properties, and different performance. Table 12–1 shows four *rocket motor applications* (each with somewhat different propellants), plus several *gas generator applications* and an *artillery shell* application. Propellants for rocket motors produce hot (over 2400 K) gases and are used for thrust, but gas generator propellants operate with lower-temperature combustion gases (800 to 1200 K in order to use uncooled hardware) and are used to produce power, not thrust.

2. *Double-base* (DB)[*] propellants form a *homogeneous* propellant grain, usually a nitrocellulose (NC)[*]—a solid ingredient that absorbs liquid nitroglycerine (NG), plus minor percentages of additives. The major ingredients are highly energetic materials and they contain both fuel and oxidizer. Both *extruded double-base* (EDB) and *cast double-base* (CDB) propellants have found extensive applications, mostly in small tactical missiles of older design. By adding crystalline nitramines (HMX or RDX)[*] both performance and density can be improved; these are sometimes called *cast-modified double-base* propellants. Adding an elastomeric binder (rubber-like, such as crosslinked polybutadiene) further improves the physical properties and allows more nitramine and thus increasing performance slightly. The resulting propellant is called *elastomeric-modified cast double-base* (EMCDB). These four classes of double-base propellants have nearly smokeless exhausts. Adding some solid ammonium perchlorate (AP) and aluminum (Al) increases the density and the specific impulse slightly, but exhaust gases becomes smoky—such propellant is called *composite-modified double-base propellant* or CMDB.

   Two operational systems produced by ATK that use double-based propellants are the AGM-114 Hellfire (which uses XLDB, a minimum smoke crosslinked propellant) and the Hydra 70 rocket (with a plateau burning propellant, see Fig. 12–6).

3. *Composite propellants* form a *heterogeneous* propellant grain between oxidizer crystals and powdered fuel (usually aluminum) held together in a matrix of synthetic rubber (or plastic) binder, such as polybutadiene (HTPB).[*] Composite propellants are cast from a mix of solid (AP crystals, Al powder) and liquid (HTPB, PPG)[*] ingredients. The propellant is hardened by crosslinking or curing the liquid binder polymer with a small amount of curing agent, and curing it in an oven, where it becomes solid. In the past four decades composites have been the most commonly used class of propellant. Composites can be further subdivided:

   a. Conventional *composite propellants,* which usually contain between 60 and 72% AP as crystalline oxidizer, up to 22% Al powder as a metal fuel, and 8 to 16% of elastomeric binder (organic polymer) including its plasticizer.

b. Modified composite propellant where an *energetic nitramine* (HMX or RDX) is added for obtaining some added performance and also a somewhat higher density.

c. Modified composite propellant where an *energetic plasticizer* such as nitroglycerine (used in double-base propellants) is added to give increased performance. Sometimes HMX is also added.

d. *High-energy composite solid propellant* (with added aluminum), where the organic elastomeric binder and the plasticizer are largely replaced by highly energetic materials and where some of the AP is replaced by HMX and RDX. Hexanitrohexaazaiso-wurtzitane or CL-20 is a recent propellant ingredient being used; it is produced outside of the United States (see Section 13.4 and Ref. 13–1). Some propellants are called elastomer-modified cast double-base propellants (EMCDB). Most are experimental propellants. Their theoretical specific impulse can be between 250 and 275 sec at standard conditions as explained below.

e. *Lower-energy composite propellant*, where *ammonium nitrate* (AN) is the crystalline oxidizer (not AP). These are used for gas generator propellants. When large amounts of HMX are added, they become minimum smoke propellants with fair performance.

4. Propellants may also be classified by the smoke density in the exhaust plume as *smoky, reduced smoke*, or *minimum smoke* (essentially smokeless). Aluminum powder, a desirable fuel ingredient for performance, is oxidized to aluminum oxide during burning, which yields visible, small, solid smoky particles in the exhaust gas. Most composite propellants (e.g., AP) are also smoky. By replacing AP with HMX and RDX and by using energetic binders and plasticizers to compensate for eliminating aluminum, the amount of smoke may be considerably reduced in composite propellants. Carbon (soot) particles and metal oxides, such as zirconium oxide or iron oxide, are also visible in high enough concentrations. This is further discussed in Chapter 20.

5. *Safety ratings* for detonation can distinguish propellants as a potentially *detonable* material (class 1.1) or as a *nondetonable* material (class 1.3), as described in Section 13.3. Examples of class 1.1 propellant are double-base propellants and composite propellants containing a significant portion of a solid explosive (e.g., HMX or RDX), together with certain other ingredients.

6. Propellants can be classified by some of the principal manufacturing processes used. A *cast propellant* is made by mechanically mixing solid and liquid ingredients, followed by casting and curing; it is the most common process for composite propellants. *Curing* of many cast propellants takes place through a chemical reaction between binder and curing agent at above ambient temperatures (45 to 150°C); however, there are some that can be cured at ambient temperatures (20 to 25°C) or hardened by nonchemical processes such as crystallization. Propellants can also be made by a *solvation* process (dissolving a plasticizer in a solid pelletized matrix, whose volume is expanded). *Extruded propellants* are made by mechanical mixing (rolling into sheets) followed by

extrusion (pushing through a die at high pressure). Solvation and extrusion processes are applied primarily to double-base propellants.

7. Propellants have also been classified by their *principal ingredient,* such as the *principal oxidizer (ammonium perchlorate propellants, ammonium nitrate propellants,* or *azide-type propellants)* or their *principal binder* or *fuel ingredient,* such as *polybutadiene propellants* or *aluminized propellants.* This classification of propellants by ingredients is further described later in Section 13.4 and Table 13–8.

8. Propellants with *toxic* and *nontoxic* exhaust gases are discussed in more detail in Section 13.3.

9. *Experimental* and/or *production propellants.* Such propellants are selected after extensive testing (preflight test, qualification tests) and demonstrated safety, life, and other essential properties. The culmination of any successful research and development (R&D) program is to have a propellant selected for production in a flight vehicle application.

Figures 13–1 and 13–2 show general regions for the specific impulse, burning rate, and density for the more common classes of propellants. The ordinate in these

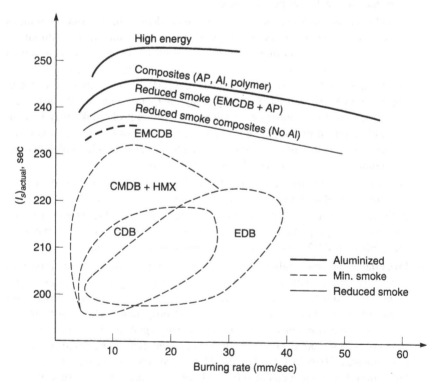

FIGURE 13–1. Typical delivered specific impulse and burning rate for several solid propellant categories. Adapted and reproduced from Ref. 13–2 with permission of the AIAA.

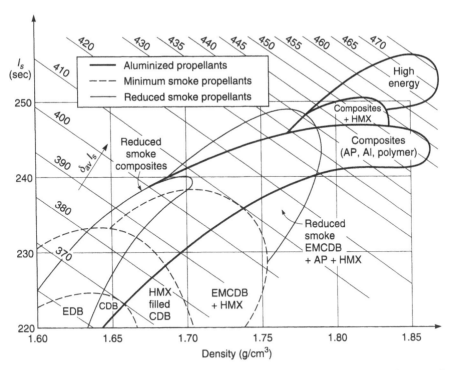

FIGURE 13-2. Typical delivered specific impulse and density-specific impulse for several solid propellant categories with an expansion of 7 to 0.1 MPa. Adapted and reproduced from Ref. 13-2 with permission of the AIAA.

figures is an actual or estimated specific impulse at standard conditions (1000 psi chamber pressure and expansion to sea-level pressure). These results do not reflect pressure drops in the chamber, nozzle erosion, or combustion losses and scaling assumptions. Composite propellants are shown to have a wide range of burning rates and densities; most of them have specific gravities between 1.75 and 1.81 and burning rates between 7 and 20 mm/sec. Composite propellants give higher densities, specific impulse, and a wider range of burning rates than others. Table 13-1 lists performance characteristics for several propellants. DB propellants and AN propellants have lower performance and density. Most composite propellants display similar performance and density but with a wider range of burning rates. The highest performance indicated is for a CMDB propellant whose ingredients are identified as DB/AP-HMX/Al, though it is only 4% higher than the next.

Several of the above listed classifications tend to be confusing. The term composite-modified double-base propellant (CMDB) has been used for a DB propellant, where some AP, Al, and binder are added; alternatively, the same propellant could be classified as a composite propellant to which some double-base ingredients have been added.

TABLE 13–1. Characteristics of Some Operational Solid Propellants

Propellant Type[a]	I_S Range (sec)[b]	Flame Temperature[e] (°F)	(K)	Density or Spec. Gravity[e] (lbm/in³)	(SG)	Metal Content (mass %)	Burning Rate[c,e] (in./sec)	Pressure Exponent[e] n	Hazard Classification[d]	Stress (psi)/Strain (%) −60°F	+150°F	Typical Processing Method
DB	220–230	4100	2550	0.058	1.61	0	0.05–1.2	0.30	1.1	4600/2	490/60	Extruded
DB/AP/Al	260–265	6500	3880	0.065	1.80	20–21	0.2–1.0	0.40	1.3	2750/5	120/50	Extruded
DB/AP–HMX/Al	265–270	6700	4000	0.065	1.80	20	0.2–1.2	0.49	1.1	2375/3	50/33	Solvent cast
PVC/AP/Al	260–265	5600	3380	0.064	1.78	21	0.3–0.9	0.35	1.3	369/150	38/220	Cast or extruded
PU/AP/Al	260–265	5700	3440	0.064	1.78	16–20	0.2–0.9	0.15	1.3	1170/6	75/33	Cast
PBAN/AP/Al	260–263	5800	3500	0.064	1.78	16	0.25–1.0	0.33	1.3	520/16 (at −10°F)	71/28	Cast
CTPB/AP/Al	260–265	5700	3440	0.064	1.78	15–17	0.25–2.0	0.40	1.3	325/26	88/75	Cast
HTPB/AP/Al	260–265	5700	3440	0.067	1.86	4–17	0.25–3.0	0.40	1.3	910/50	90/33	Cast
HTPE[7]/AP/Al	248–269	5909	3538	0.07	1.70		0.4–0.7	0.50	1.3	174/44	(77°F)	Cast
PBAA/AP/Al	260–265	5700	3440	0.064	1.78	14	0.25–1.3	0.35	1.3	500/13	41/31	Cast
AN/Polymer	180–190	2300	1550	0.053	1.47	0	0.06–0.5	0.60	1.3	200/5	NA	Cast

[a]Al, aluminum; AN, ammonium nitrate; AP, ammonium perchlorate; CTPB, carboxy-terminated polybutadiene; DB, double-base; HMX, cyclotetramethylene tetranitramine; HTPB, hydroxyl-terminated polybutadiene; HTPE hydroxyl terminated polyether; PBAA, polybutadiene–acrylic acid polymer; PBAN, polybutadiene–acrylic acid–acrylonitrile terpolymer; PU, polyurethane; PVC, polyvinyl chloride.

[b]At 1000 psia expanding to 14.7 psia, ideal or theoretical value at reference conditions.

[c]At 1000 psia.

[d]See hazard classification section.

[e]I_s, flame temperature, density, burn rate, and pressure exponent will vary slightly with specific composition.

Data from Ref. 13–3, CPIAC, and Orbital ATK.

A large variety of *chemical ingredients* and propellant formulations have been synthesized, analyzed, and tested in experimental rocket motors. A typical solid propellant has between 4 and 12 different ingredients and this chapter discusses about a dozen basic propellant types, but many other types are still being investigated. Table 13–2 presents some advantages and disadvantages of selected propellant classes. Representative formulations for three types of propellant are given in Table 13–3. In actual practice, each propellant manufacturer uses its own formulation and processing procedures. The exact percentages of ingredients, even for a given propellant such as PBAN, can not only vary among manufacturers but often vary from one rocket motor application to another. The practice of adjusting mass percentages together with adding or deleting one or more of the minor ingredients (additives) is known as *propellant tailoring*. Tailoring is the practice of taking an established propellant and changing it slightly to fit a new application, different processing equipment, altered motor ballistics, storage life, temperature limits, and/or even a change in ingredient source.

New propellant formulations are typically developed using laboratory-size mixers, curing ovens, and related equipment with the propellant mixers (1 to 5 liters), and operated by remote control for safety reasons. Process studies usually accompany the development of new propellant formulations to evaluate the "processability" and to guide the design of any special production equipment needed in preparing ingredients, mixing, casting, or curing it.

Historically, black or gun powder (a pressed mixture of potassium nitrate, sulfur, and an organic fuel such as ground peach stones) was the very first propellant to be used. Many other types of propellants ingredients have been used in experimental motors, including fluorine compounds, propellants containing powdered beryllium, boron, hydrides of boron, lithium, or beryllium, or new synthetic organic plasticizer and binder materials with azide or nitrate groups. Most of these have not yet been considered satisfactory or practical for production in rocket motors.

13.2. PROPELLANT CHARACTERISTICS

Propellant selection is critical to rocket motor design. *Desirable propellant characteristics* are listed below and further discussed in other parts of this book. Many requirements for particular solid propellant rocket motors will influence priorities for choosing these characteristics:

1. High performance or *high specific impulse*; this implies a high gas temperature and/or low exhaust gas molecular mass.
2. Predictable, reproducible, and initially adjustable *burning rate* to fit grain-design needs and thrust-time requirements.
3. For minimum variations in thrust or chamber pressure during burning, both the *pressure* or *burning rate exponent* and the *temperature coefficient* should be small.

TABLE 13–2. Characteristics of Selected Propellants

Propellant Type	Advantages	Disadvantages
Double base (extruded)	Modest cost; nontoxic clean exhaust, smokeless; good burn rate control; wide range of burn rates; simple well-known process; good mechanical properties; low-temperature coefficient; very low pressure exponent; plateau burning is possible	Freestanding grain requires structural support; low performance, low density; high to intermediate hazard in manufacture; can have storage problems with NG bleeding out; diameter limited by available extrusion presses; class 1.1[a]
Double base (castable)	Wide range of burn rates; nontoxic smokeless exhaust; relatively safe to handle; simple, well-known process; modest cost; good mechanical properties; good burn rate control; low-temperature coefficient; plateau burning can be achieved	NG may bleed out or migrate; high to intermediate manufacture hazard; low performance; low density; higher cost than extruded DB; class 1.1[a]
Composite modified double base or CMDB with some AP and Al	Higher performance; good mechanical properties; high density; less likely to have combustion stability problems; intermediate cost; good background experience	Complex facilities; some smoke in exhaust; high flame temperature; moisture sensitive; moderately toxic exhaust; hazards in manufacture; modest ambient temperature range; the value of n is high (0.8–0.9); moderately high temperature coefficient
Composite AP, Al, and PBAN or PU or CTPB binder	Reliable; high density; long experience background; modest cost; good aging; long cure time; good performance; usually stable combustion; low to medium cost; wide temperature range; low to moderate temperature sensitivity; good burn rate control; usually good physical properties; class 1.3	Modest ambient temperature range; high viscosity limits at maximum solid loading; high flame temperature; toxic, smoky exhaust; some are moisture sensitive; some burn-rate modifiers (e.g., aziridines) are carcinogens
Composite AP, Al, and HTPB binder; most common composite propellant	Slightly better solids loading % and performance than PBAN or CTPB; wide ambient temperature limits; good burn-rate control; usually stable combustion; medium cost; good storage stability; wide range of burn rates; good physical properties; good experience; class 1.3	Complex facilities; moisture sensitive; fairly high flame temperature; toxic, smoky exhaust

TABLE 13–2. (*Continued*)

Propellant Type	Advantages	Disadvantages
Modified composite AP, Al, PB binder plus some HMX or RDX	Higher performance; good burn-rate control; usually stable combustion; high density; moderate temperature sensitivity; can have good mechanical properties	Expensive, complex facilities; hazardous processing; harder-to-control burn rate; high flame temperature; toxic, smoky exhaust; can be impact sensitive; can be class 1.1,[a] high cost; pressure exponent 0.5–0.7
Composite with energetic binder and plasticizer such as NG, and with AP, HMX	Highest performance; high density; narrow range of burn rates	Expensive; limited experience; impact sensitive; high-pressure exponent; class 1.1[a]
Modified double-base with HMX	Higher performance; high density; stable combustion; narrow range of burn rates	Same as CMDB above; limited experience; most are class 1.1[a]; high cost
Modified AN propellant with HMX or RDX added	Fair performance; relatively clean; smokeless; nontoxic exhaust	Relatively little experience; can be hazardous to manufacture; need to stabilize AN to limit grain growth; low burn rates; impact sensitive; medium density; class 1.1 or 1.3[a]
Ammonium nitrate plus polymer binder (gas generator)	Clean exhaust; little smoke; essentially nontoxic exhaust; low-temperature gas; usually stable combustion; modest cost; low-pressure exponent	Low performance; low density; need to stabilize AN to limitgrain growth and avoid phase transformations; moisture sensitive; low burn rates

[a]Class 1.1 and 1.3—see Section 13.3 on Hazard Classification.

4. Adequate *physical properties* (including bond strengths) over the intended operating temperature range with allowance for some degradation due to cumulative damage.

5. High *density* (resulting in a small-volume rocket motor).

6. Predictable, reproducible ignition qualities (such as acceptable ignition overpressures).

7. Desirable *aging characteristics* and *long life*. Aging and life predictions depend on the propellant's chemical and physical properties, cumulative damage criteria with load cycling (see Section 12.4) and thermal cycling, and from actual tests on propellant samples and test data from failed motors.

8. Low *moisture* absorption, because moisture often causes chemical deterioration.

9. Simple, reproducible, safe, low-cost, controllable, and low-hazard *manufacturing*.

TABLE 13–3. Representative Propellant Formulations

Double Base (JPN Propellant)		Composite (PBAN Propellant)		Composite Double Base (CMDB Propellant)	
Ingredient	Mass %	Ingredient	Mass %	Ingredient	Mass %
Nitrocellulose	51.5	Ammonium perchlorate	70.0	Ammonium perchlorate	20.4
Nitroglycerine	43.0	Aluminum powder	16.0	Aluminum powder	21.1
Diethyl phthalate	3.2	Polybutadiene–acrylic acid–acrylonitrile	11.78	Nitrocellulose	21.9
Ethyl centralite	1.0	Epoxy curative	2.22	Nitroglycerine	29.0
Potassium sulfate	1.2			Triacetin	5.1
Carbon black	<1%			Stabilizers	2.5
Candelilla wax	<1%				

Source: Courtesy of Air Force Phillips Laboratory, Edwards, California.

10. Guaranteed availability of all *raw materials* and *purchased components* over the production and operating life of the propellant, and acceptable control over undesirable impurities.

11. *Low technical risk*, such as a favorable history of prior applications.

12. Relative *insensitivity* to certain external energy stimuli as described Section 13.3, the hazards section.

13. *Nontoxic* and *noncorrosive exhaust* gases, also called *green exhausts*.

14. Not prone to *combustion instability* (see Chapter 14).

15. Equivalent composition, performance and properties with every new propellant batch.

16. No slow or long-term chemical reactions or migrations between propellant ingredients or between propellant and insulator/liner.

Some of these desirable characteristics will also apply to all materials and purchased components used in solid rocket motors, such as the igniter, insulator, case, or safe-and-arm device but several of these characteristics can sometimes be in conflict with each other. For example, increasing the physical strength (more binder and or more crosslinker) will reduce propellant performance and density. So a modification of the propellant for one of these characteristics may cause changes in a few others.

Several illustrations will now be given on how characteristics of a propellant change when the concentration of one of its major ingredients is changed. Figure 13–3 shows calculated variations in combustion or flame temperature, average product gas molecular mass, and specific impulse as a function of oxidizer concentration for composite propellants that use a polymer binder [hydroxyl-terminated

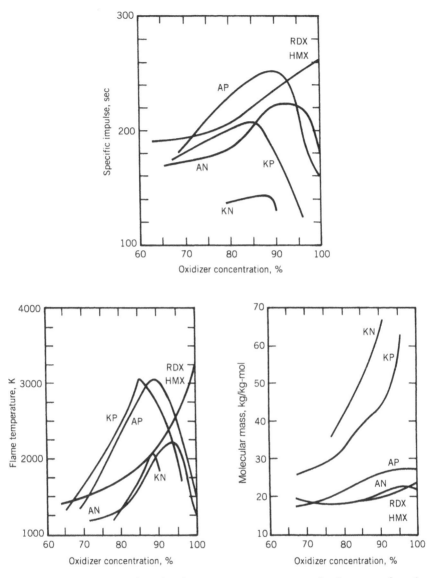

FIGURE 13–3. Variation of combustion temperature, average molecular mass of combustion gases, and theoretical specific impulse (at frozen equilibrium) as a function of oxidizer concentration for HTPB-based composite propellants. Data are for a chamber pressure of 68 atm and nozzle exit pressure of 1.0 atm. Reproduced from Ref. 13–4 with permission of the AIAA.

polybutadiene (HTPB)] and various crystalline oxidizers; these results are taken from Ref. 13–4, based on a thermochemical analysis as explained in Chapter 5. The maximum values of I_s and T_1 occur at approximately the same concentration of oxidizer. For practical reasons, this optimum percentage for AP (about 90 to 93%) and AN (about 93%) cannot be implemented because concentrations greater than about 90% total solids (including the aluminum and solid catalysts) cannot be processed in a mixer—a castable slurry that will flow into a mold requires more than 10 to 15% liquid content.

A typical composition diagram for a composite propellant is depicted in Fig. 13–4. It shows how the specific impulse varies with changes in the composition of the three principal ingredients: the solid AP, solid Al, and viscoelastic polymer binder.

For DB propellants variations of I_s and T_1 are shown in Fig. 13–5 as a function of nitroglycerine (NG) concentration. The theoretical maximum specific impulse occurs at about 80% NG. In practice, NG, which is a liquid, is seldom found in concentrations over 60% because its physical properties are poor at the higher concentrations. Other major solid or soluble ingredients are also needed to make a usable DB propellant.

For CMDB propellants the addition of either AP or a reactive nitramine such as RDX allows for higher I_s than with ordinary DB (where AP or RDX percent is zero), as shown in Fig. 13–6. Both AP and RDX greatly increase the flame temperature and

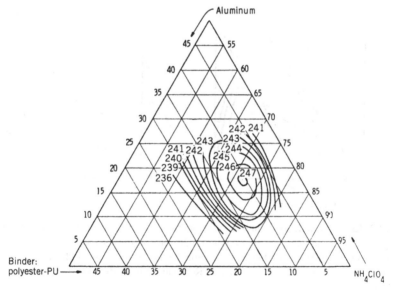

FIGURE 13–4. Composition diagram of calculated specific impulse for an ammonium perchlorate–aluminum–polyurethane propellant (PU is a polyester binder) at standard conditions (1000 psi and expansion to 14.7 psi). The maximum value of specific impulse occurs at about 11% PU, 72% AP, and 17% Al. Reproduced from Ref. 13–5 with permission of the American Chemical Society.

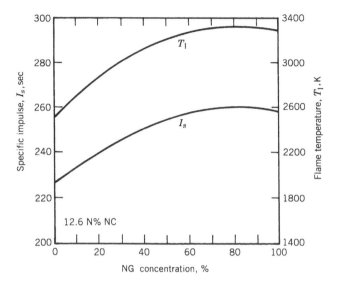

FIGURE 13-5. Specific impulse and flame temperature versus nitroglycerine (NG) concentration of double-base propellants. Reproduced from Ref. 13-4 with permission of the AIAA.

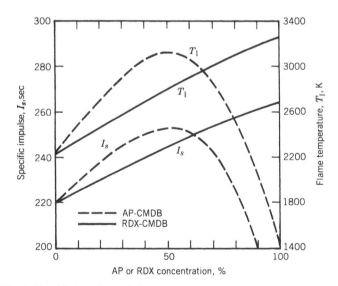

FIGURE 13-6. Specific impulse and flame temperature versus AP or RDX concentration of AP–CMDB propellants. Reproduced from Ref. 13-4 with permission of the AIAA.

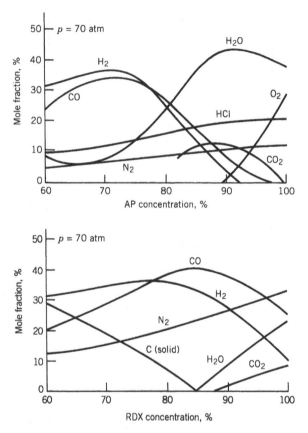

FIGURE 13–7. Calculated combustion products of composite propellant with varying amounts of AP or RDX. Adapted from Chapter 1 of Ref. 13–4 with permission of the AIAA.

thus make the effects of heat transfer more critical. The maximum values of I_s occur at about 50% AP and at 100% RDX (an impractical propellant concentration because it cannot be manufactured and will not have reasonable physical properties). At high concentrations of AP or RDX, the exhaust gases contain considerable H_2O and O_2 (as shown in Fig. 13–7); these enhance erosion rates in carbon-containing insulators or nozzle materials. In Fig. 13–7, the toxic HCl gas is present in concentrations between 10 and 20%, but in practical propellants it can seldom exceed 14%.

Nitramines such as RDX or HMX contain relatively few oxidizing radicals and the binder surrounding the nitramine crystals cannot be fully oxidized. As the binder decomposes at the combustion temperature, it releases gases rich in hydrogen and carbon monoxide (which reduce the molecular mass), cooling the gas mixture to lower combustion temperatures. The exhaust gases of AP-based and RDX-based CMDB propellant are shown in Fig. 13–7 where it can be seen that solid carbon particles disappear when the RDX content is above 85%.

13.3. HAZARDS

With proper precautions, training and equipment, all common propellants can be manufactured, handled, and fired safely. It is necessary and imperative to fully understand all hazards and methods for preventing dangerous situations from arising. Each material has its own set of hazards; some of the more common ones are described briefly below and also in Refs. 13–6 and 13–7. Not all apply to every propellant.

Inadvertent Ignition

If a rocket motor is unexpectedly ignited and starts combustion, the very hot exhaust gases may cause burns and local fires, or ignition of adjacent rocket motors. Unless the motor is constrained or fastened down, its thrust will accelerate it to unanticipated high velocities or erratic flight paths that can cause much severe damage. Its exhaust cloud can be toxic and corrosive. Inadvertent ignitions may be caused by the following effects:

Stray or induced currents that activate the igniter.

Electrostatic charging causing unintended sparks or arc discharges.

Fires excessively heating rocket motor exteriors, raising the solid propellant temperature above its ignition point.

Impacts (bullet penetration or dropping the rocket motor onto a hard surface).

Energy absorption from prolonged mechanical vibrations that overheat the propellant (e.g., transport over rough roads).

Radiation from nuclear explosions.

An electromechanical system called *safe and arm system* is usually included to prevent stray currents from activating the igniter. It averts ignition induced by currents in other wires of the vehicle, radar- or radio-frequency-induced currents, electromagnetic surges, or pulses from a nuclear bomb explosion. It prevents any electric currents from reaching the igniter circuit during its "unarmed" condition. When in the "arm" position, it accepts and transmits a start signal to the igniter.

Electrostatic discharges (ESD) may be caused by lightning, friction in insulating materials, or by the moving separation of two insulators. The buildup of high electrostatic potentials (thousands of volts) may, upon discharge, allow rapid increases in electric current, which in turn may lead to arcing or exothermic reactions along the current's path. For this reason all propellants, liners, or insulators need to have sufficient electric conductivity to prevent any buildup of such an electrostatic charge. The well-known inadvertent ignition of a Pershing ground-to-ground missile is believed to have been caused by an electrostatic discharge while in the transporter-erector vehicle. ESD capabilities depend on materials, their surface and volume resistivities, dielectric constants, and the breakdown voltages.

Viscoelastic propellants are excellent absorbers of *vibration energy* and can become locally hot when oscillating for extensive periods at particular frequencies.

This can happen in designs where a segment of the grain is not well supported and is free to vibrate at its natural frequencies. Because propellants can be accidentally ignited by extraneous means (such as mechanical friction or bullet impacts, or accidentally dropping a rocket motor) standard tests have been developed to measure the propellant's resistance to such energy inputs. Considerable effort is spent in developing new propellants resistant to these energy inputs.

Aging and Useful Life

This topic is briefly discussed in the section on Structural Design in the previous chapter. The *aging* of a propellant can be measured with test motors and propellant sample tests when the loading history during the life of the motor can be correctly anticipated. It has then been possible to estimate and predict the useful *shelf or storage life* of a rocket motor (see Refs. 13–7 and 13–8). When changes in physical properties, caused by estimated thermal or mechanical load cycles (cumulative damage), reduce the safety margin on stresses and/or strains to a danger point, the rocket motor is no longer considered to be safe to be ignited or operated. Once this age limit or its predicted, weakened condition is reached, the rocket motor has a high probability of failure and should be removed from any ready inventory and the aged propellant removed and replaced.

The *life* of a particular motor depends on its propellant composition, the frequency and magnitude of imposed loads or strains, and the design among other factors. Typical life values range from 5 to 25 years. Shelf life can usually be increased by increasing the physical strength of the propellants (e.g., by increasing the amount of binder), selecting chemically compatible, stable ingredients with minimal long-term degradation, and/or by minimizing the vibration loads, temperature limits, or number of cycles (i.e., controlling the storage and transport environments).

Case Overpressure and Failure

A rocket motor case will break and/or explode during operation when the chamber pressure exceeds the case's burst pressure. The release of high-pressure gases can cause explosions where motor pieces are thrown out into adjacent areas. Any sudden depressurization from chamber pressure to ambient pressure (which is usually below the deflagration limit) would normally stop the burning in a class 1.3 propellant (see Hazard Classification). Large pieces of unburned propellant are often found after violent case bursts. Case overpressure rocket motor failure may be caused by one of the following:

1. The grain is overaged, porous, or severely cracked and/or has major unbonded areas due to severe accumulated damage.
2. There have been a significant chemical changes in the propellant due to migration or slow, low-order chemical reactions; these can reduce the allowable physical properties, weakening the grain, so that it will crack or cause unfavorable increases in the burning rate. In some cases chemical reactions create

gaseous products, which produce many small voids and raise the pressure in sealed stored rocket motors.

3. The rocket motor has not properly been manufactured. Obviously, careful fabrication and inspection are a must.

4. The motor has been damaged. For example, a nick or dent in the case caused by improper handling will reduce the case strength. This can be prevented by careful handling and repeated inspections.

5. An obstruction plugs the nozzle (e.g., a loose large piece of insulation) causing a rapid increase in chamber pressure.

6. In propellants that contain hygroscopic ingredients, moisture absorption can degrade strength and strain capabilities by factors of 3 to 10. Rocket motors are routinely sealed to prevent humid air access.

Detonation versus Deflagration. When the burning rocket motor propellant is overpressurized, it may either continue to deflagrate (burn) or detonate (explode violently), as described in Table 13–4. In a detonation, the chemical reaction energy of the entire grain is released in a very short time (microseconds), and in effect it

TABLE 13–4. Comparison of Burning (Deflagration) and Detonation

Characteristic	Burning with Air	Deflagration within Rocket Motors	Detonation of Rocket Motor
Typical material	Coal dust and air	Propellant, no air	Rocket propellant or explosives
Common means of initiating reaction	Heat	Heat	Shock wave; sudden pressure rise plus heat
Linear reaction rate (m/sec)	10^{-6} (subsonic)	0.2 to 5×10^{-2} (subsonic)	2 to 9×10^3 (supersonic)
Shock waves	No	No	Yes
Time for completing reaction (sec)	10^{-1}	10^{-2}–10^{-3}	10^{-6}
Maximum pressure [MPa (psi)]	0.07–0.14 (10–20)	0.7–100 (100–14,500)	7000–70,000 (10^6–10^7)
Process limitation	By heat transfer at burning surface	Strength of case	By physical and chemical properties of material, (e.g., density, composition)
Increase in burning rate can result in:	Potential furnace failure	Overpressure and explosive failure of motor case	Motor case failure and violent rapid explosion of all the propellant
Remainder	Unburnt coal dust pockets may continue to burn	After case failure, unused propellant pieces will usually stop burning	No propellant remainder to burn
Classification	None	Class 1.3	Class 1.1

becomes an exploding bomb. Detonations happen with some propellants that contain certain ingredients (e.g., nitroglycerine or HMX as described later in this chapter). Proper designs, correct manufacture, and safe handling and operating procedures are essential in order to minimize or totally avoid detonations.

Any given propellant material may either burn or detonate depending on its chemical formulation, physical properties (such as density or porosity) the type and intensity of the initiation, the degree of confinement, and the geometric characteristics of the rocket motor. It is also possible for certain burning propellants to change suddenly from an orderly deflagration to a detonation. A simplified explanation of this transition is as follows: normal burning starts at the rated chamber pressure; then hot gases penetrate some existing but unknown pores or small cracks in the unburned propellant, where gas confinement causes the pressure to become locally very high; the combustion front then attains shock wave speeds with a low-pressure differential and then it accelerates further to the strong, fast, high-pressure shock wave, characteristic of detonations. The degree and rigidity of the motor geometric confinement and some scale factor (e.g., larger-diameter grain) influence the severity and occurrence of detonations.

Hazard Classification. Propellants that may transition from deflagration to detonation are considered more hazardous and are usually designated as class 1.1-type propellants. Class 1.3 propellants will not detonate even if while burning the case bursts as the chamber pressure becomes too high. The required tests and rules for determining this hazard category are treated in Ref. 13–9. Propellant samples are subjected to various tests, including impact tests (dropped weight) and card gap tests (which determine the force needed to initiate a propellant detonation when a sample is subjected to a blast from a known booster explosive). Even when the case bursts violently with a class 1.3 propellant, much of the remaining unburnt propellant is ejected and then usually stops burning. With a class 1.1 propellant, a powerful detonation can sometimes ensue, which rapidly gasifies all the remaining propellant, and is much more powerful and destructive than the bursting of the case under high pressure. Unfortunately, the term *explosion* has been used to describe both case bursting with its remaining unburned propellant fragmentation, and also the higher rate of energy release in a detonation which leads to more rapid and much more energetic fragmentation of the rocket motor.

The Department of Defense (DOD) classification of 1.1 or 1.3 determines the method of labeling and the cost of shipping rocket propellants, loaded military missiles, explosives, or ammunitions; it also defines required limits on propellant amounts that may be stored or manufactured in any one site and the minimum separation distance of that site to the next building or site. The DOD system (Ref. 13–9) is the same as that which has been used by the United Nations.

Insensitive Munitions

Any accidental ignition or otherwise unplanned rocket or military missile operation or explosion may cause severe damage to equipment and injure or kill personnel.

TABLE 13–5. Typical Testing for Insensitivity of Rockets and Missiles

Test	Description	Criteria for Passing
Fast cook-off	Build a fire (of jet fuel or wood) underneath the missile or its rocket motor	No reaction more severe than burning propellant
Slow cook-off	Gradual heating (6°F/hr) to failure	Same as above
Bullet impact	One to three 50-caliber bullets fired at short intervals	Same as above
Fragment impact	Small high-speed steel fragment	Same as above
Sympathetic detonation	Detonation from an adjacent similar motor or a nearby specific munition	No detonation of test motor
Shaped explosive charge impact	Blast from specified shaped charge in specified location	No detonation
Spall impact	Several high-speed spalled fragments from a steel plate which is subjected to a shaped charge	Fire, but no explosion or detonation

This may be avoided or minimized by making rocket motor designs and their propellants insensitive to a variety of energetic stimuli. The worst scenario is propellant detonation, releasing all the propellant's energy explosively, and this scenario must be avoided at all cost. Missiles and their rocket motors must undergo a series of prescribed tests to determine their resistance to inadvertent ignition using the most likely energy inputs during any possible battle situation. Table 13–5 describes a series of tests called out in military specifications, which are detailed in Refs. 13–3 and 13–10. Other tests besides those listed in Table 13–5 are sometimes needed, such as friction tests and drop tests. A *threat hazard assessment* must be made prior to any such tests, to evaluate the logistic and operational threats during the missile's life cycle. This may result in modifications to the test setups, changes to acceptance criteria, and/or the skipping of some such tests.

In all prescribed tests, the missiles together with their rocket motors are destroyed. If the rocket motor should detonate (an unacceptable result), the motor must be redesigned and/or undergo a change in propellant. There are some newer propellants that are more resistant to external stimuli and are therefore preferred for tactical missile applications, even though there may be a penalty in propulsion performance. If explosions (not detonations) occur, it may be possible to redesign the rocket motor and mitigate the explosion effects (make it less violent). For example, motor cases can have a provision for venting prior to explosion. Changes to shipping containers can also mitigate some of these effects. If the result is a fire (a sometimes acceptable result), it should be confined to the particular grain or rocket motor. Under some circumstances a burst failure of the case may also be acceptable. In the past decade a new class of insensitive propellants has been developed with new types of binders. Their purpose remains to minimize any drastic consequences (e.g., detonations) when the rocket motor is exposed to various unexpected energetic stimuli, such as external fires, and impact or pressure waves (see Table 13–5). Recently, insensitive propellants have been selected for some military missions, reflecting an important milestone in their development. The leading example is a recent composite propellant with HTPE (hydroxyl-terminated polyether) as the binder, see Ref. 13–11. HTPE has been qualified and applied to the MK 134 rocket motor for the ship-launched Evolved Sea Sparrow Missile (ESSM). This motor is co-manufactured by ATK in the United States and NAMMO Raufoss in Norway for member nations of the NATO Sea Sparrow Consortium. A relatively insensitive propellant has also been developed in France, namely, a nonmigrating ferrocene-grafted HTPB binder called Butacene. It has been qualified for a few systems in the United States and other countries; see Ref. 13–11.

Upper Pressure Limit

When the pressure-rise rate and the absolute pressure become sufficiently high (as in some impact tests or in the high acceleration of a gun barrel), some propellants will detonate. For many propellants these pressures are above approximately 1500 MPa or 225,000 psi, but for others they can be lower (as low as 300 MPa or 45,000 psi). The values quoted represent defined *upper pressure limits* beyond which such a propellant should not operate.

Toxicity

Many solid propellants do not have any significant toxicity problem. A number of propellant ingredients (e.g., some crosslinking agents and burning rate catalysts) and a few of the plastics used in fiber-reinforced cases can be dermatological or respiratory toxins; a few are carcinogens or suspected carcinogens. They, and the mixed uncured propellant containing these materials, must be handled carefully to prevent operator exposure. This means using gloves, face shields, good ventilation, and, with some high-vapor-pressure ingredients, gas masks. Usually, the finished or cured grain or motor is not toxic.

Exhaust plume gases can be very toxic if they contain beryllium or beryllium oxide particles, chlorine gas, hydrochloric acid gas, hydrofluoric acid gas, or some other fluorine compounds. When an ammonium perchlorate oxidizer is used, the exhaust gas may contain up to about 14% hydrochloric acid, which is a toxic gas. For large rocket motors this can mean many tons of highly toxic gas. Test and launch facilities for rockets with toxic plumes require very special precautions and occasionally decontamination processes, as explained in Chapter 21.

Safety Rules

Among the most effective ways to control hazards and prevent accidents are: (1) to acquaint personnel of the hazards of each propellant being handled by teaching them how to properly avoid hazardous conditions and to prevent accidents and how to recover from them; (2) to design the rocket motors, their fabrication as well as test facilities and equipment to be as safe as possible; and (3) to institute and enforce rigid safety rules during design, manufacture, and operation. There are many such rules. Examples include avoiding smoking (and matches) in areas where there are propellants or loaded motors, wearing spark-proof shoes and using spark-proof tools, shielding all electrical equipment, providing a water-deluge fire extinguishing system in test facilities to cool motors or extinguish burning, and/or proper grounding of all electrical equipment and items that could build up static electrical charges.

13.4. PROPELLANT INGREDIENTS

A number of relatively common propellant ingredients are listed in Table 13−6 for double-base propellants and for composite-type solid propellants in Table 13−7.

TABLE 13−6. Typical Ingredients of Double-Base (DB) Propellants and Composite-Modified Double-Base (CMDB) Propellants

Type	Percent	Acronym	Typical Chemicals
Binder	30–50	NC	Nitrocellulose (solid), usually plasticized with 20–50% nitroglycerine
Reactive plasticizer (liquid explosive)	20–50	NG	Nitroglycerine
		DEGDN	Diethylene glycol dinitrate
		TEGDN	Triethylene glycol dinitrate
		PDN	Propanediol-dinitrate
		TMETN	Trimethylolethane trinitrate
Plasticizer (organic liquid fuel)	0–10	DEP	Diethyl phthalate
		TA	Triacetin
		DMP	Dimethyl phthalate
			Dioctyl phthalate
		EC	Ethyl centralite
		DBP	Dibutyl phthalate
Burn Rate Modifier	Up to 3	PbSa	Lead salicylate
		PbSt	Lead stearate
		CuSa	Copper salicylate
		CuSt	Copper stearate
Coolant		OXM	Oxamide
Opacifier		C	Carbon black (powder or graphite powder)
Stabilizer and or antioxidant.	>1	DED	Diethyl diphenyl
		EC	Ethyl centralite
		DPA	Diphenyl amine
Visible flame Suppressant	Up to 2	KNO_3	Potassium nitrate
		K_2SO_4	Potassium sulfate
Lubricant (for extruded propellant only)	> 0.3	C	Graphite Wax
Metal fuel[a]	0–15	Al	Aluminum, fine powder (solid)
Crystalline oxidizer[a]	0–15	AP	Ammonium perchlorate
		AN	Ammonium nitrate
Solid explosive crystals[a]	0–20	HMX	Cyclotetramethylenetetranitramine
		RDX	Cyclotrimethylenetrinitramine
		NQ	Nitroguanidine

[a]Several of these, but not all, are added to CMDB propellant.

TABLE 13–7. Typical Ingredients of Composite Solid Propellants

Type	Percent	Acronym	Typical Chemicals
Oxidizer (crystalline)	0–70	AP	Ammonium perchlorate
		AN	Ammonium nitrate
		KP	Potassium perchlorate
		KN	Potassium nitrate
		ADN	Ammonium dinitramine
Metal fuel (also acts as a combustion stabilizer)	0–30	Al	Aluminum
		Be	Beryllium (experimental propellant only)
		Zr	Zirconium (also acts as burn rate modier)
Fuel/binder, polybutadiene type	5–18	HTPB	Hydroxyl-terminated polybutadiene
		CTPB	Carboxyl-terminated polybutadiene
		PBAN	Polybutadiene acrylonitrile acrylic acid
		PBAA	Polybutadiene acrylic acid
Fuel/binder, polyether and polyester type	0–15	PEG	Polyethylene glycol
		PCP	Polycaprolactone polyol
		PGA	Polyglycol adipate
		PPG	Polypropylene glycol
		HTPE	Hydroxyl-terminated polyether
		PU	Polyurethane polyester or polyether
Curing agent or crosslinker, which reacts with polymer binder	0.2–3.5	MAPO	Methyl aziridinyl phosphine oxide
		IPDI	Isophorone diisocyanate
		TDI	Toluene-2,4-diisocyanate
		HMDI	Hexamethylene diisocyanide
		DDI	Dimeryl diisocyanate
		TMP	Trimethylol propane
		BITA	Trimesoyl-1(2-ethyl)-aziridine
Burn Rate Modifier	0.2–3	FeO	Ferric oxide
		nBF	n-Butyl ferrocene
			Oxides of Cu, Pb, Zr, Fe
			Alkaline earth carbonates
			Alkaline earth sulfates
			Metallo-organic compounds
Explosive filler (solid)	0–40	HMX	Cyclotetramethylenetetranitramine
		RDX	Cyclotrimethylenetrinitramine
		NQ	Nitroguanadine
		CL-20	Hexanitrohexaazaisowurtzitane
Plasticizer/pot life control (organic liquid)	0–7	DOP	Dioctyl phthalate
		DOA	Dioctyl adipate
		DOS	Dioctyl sebacate
		DMP	Dimethyl phthalate
		IDP	Isodecyl pelargonate
Energetic plasticizer (liquid)	0–14	GAP	Glycidyl azide polymer
		NG	Nitroglycerine
		DEGDN	Diethylene glycol dinitrate
		BTTN	Butanetriol trinitrate
		TEGDN	Triethylene glycol dinitrate
		TMETN	Trimethylolethane trinitrate
		PCP	Polycaprolactone polymer

TABLE 13–7. (*Continued*)

Type	Percent	Acronym	Typical Chemicals
Energetic fuel/ binder	0–35	GAP	Glycidyl azide polymer
		PGN	Propylglycidyl nitrate
		BAMO/ AMMO	Bis-azidomethyloxetane/Azidomethyl- methyloxetane copolymer
		BAMO/ NMMO	Bis-azidomethyloxetane/3-Nitramethyl-3- methyloxetane copolymer
Bonding agent	> 0:1	MT-4	MAPO–tartaric acid–adipic acid condensate
(improves bond to solid particles)		HX-752	1,1'-Isophthaloyl-bis-(2-methylaziridine)
Stabilizer (reduces chemical deterioration)	> 0:5	DPA	Diphenylamine
		—	Phenylnaphthylamine
		NMA	*N*-methyl-*p*-nitroaniline
		—	Dinitrodiphenylanine
Processing aid	> 0:5	—	Lecithin
		—	Sodium lauryl sulfate

They are categorized by major *function*, such as *oxidizer, fuel, binder, plasticizer, curing agent*, and so on, and each category is briefly described later in this section. However, several ingredients can have *more than one function*. These lists are not complete because over 200 other ingredients have been tried in experimental rocket motors. Most ingredients in Table 13–6 have a shelf life of over 30 years.

Table 13–8 shows a classification for modern propellants (including some new types that are still in the experimental phase) according to their binders, plasticizers, and solid ingredients; these solids may be an oxidizer, a solid fuel, or a combination or a compound of both.

Ingredient properties and impurities both may have a profound effect on propellant characteristics. A seemingly minor change in one ingredient can cause measurable changes in ballistic properties, physical properties, migration, aging, and/or ease of manufacture. When the propellant's performance or ballistic characteristics have tight tolerances, ingredient purity and properties must conform to equally tight tolerances and careful handling (e.g., no exposure to moisture). In the remainder of this section, a number of the important ingredients, grouped by function, are briefly discussed.

Inorganic Oxidizers

Some thermochemical properties of several oxidizers and radical-containing oxygen compounds are listed in Table 13–9. Listed values depend on the chemical nature of each ingredient.

Ammonium perchlorate (NH_4ClO_4) is the most widely used crystalline oxidizer in solid propellants. It dominates the solid oxidizer field because of its desirable characteristics that include compatibility with other propellant materials, good performance, satisfactory quality and uniformity, low impact and friction

TABLE 13–8. Classification of Solid Rocket Propellants Used in Flying Vehicles According to Their Binders, Plasticizers, and Solid Ingredients

Designation	Binder	Plasticizer	Solid Oxidizer and/or Fuel	Propellant Application
Double base, DB	Plasticized NC	NG, TA, etc.	None	Minimum signature and smoke
CMDB[a]	Plasticized NC	NG, TMETN, TA, BTTN, etc.	Al, AP, KP	Booster, sustainer, and spacecraft
	Same	Same	HMX, RDX, AP	Reduced smoke
	Same	Same	HMX, RDX, azides	Minimum signature, gas generator
EMCDB[a]	Plasticized NC + elastomeric polymer	Same	Like CMDB above, but generally superior mechanical properties with elastomer added as binder	
Polybutadiene	HTPB	DOA, IDP, DOP, DOA, etc.	Al, AP, KP	Booster, sustainer, or spacecraft; used extensively in many applications
	HTPB	DOA, IDP, DOP, DOA, etc.	AN, HMX, RDX, some AP	Reduced smoke, gas generator
	CTPB, PBAN, PBAA	All like HTPB above, but somewhat lower performance due to higher processing viscosity and consequent lower solids content. Still used in applications with older designs.		
Polybutadyene with HMX or RDX	HTPB	DOA, IDP, DOP, DOA, etc.	AP, Al, HMX, or RDX	High energy vehicles
Polyether and polyesters	PEG, PPG, PCP, PGA, HTPE,[b] and mixtures	DOA, IDP, TMETN, DEGDN, etc.	Al, AP, KP, HMX, BiO$_3$	Booster, sustainer, or spacecraft
Energetic binder (other than NC)	GAP, PGN, BAMO/NMMO, BAMO/AMMO	TMETN, BTTN, etc. GAP-azide, GAP-nitrate, NG	Like polyether/polyester propellants above, but with slightly higher performance. Experimental propellant.	

[a]CMDB, composite-modified double-base; EMCDB, elastomer-modified cast double-base. For definition of acronyms and abbreviations of propellant ingredients see Tables 13–6 and 13–7.
[b]HTPE, hydroxyl-terminated polyether, binder for propellant developed for Insensitive Munitions by Orbital ATK.

TABLE 13–9. Comparison of Crystalline Oxidizers

Oxidizer	Chemical Symbol	Molecular Mass (kg/kg-mol)	Density (kg/m^3)	Total Oxygen Content (mass%)	Available Oxygen Content (mass%)	Remarks
Ammonium perchlorate	NH_4ClO_4	117.49	1949	54.5	34.0	Low n, low cost, readily available, high performance
Potassium perchlorate	$KClO_4$	138.55	2519	46.2	40.4	Low burning rate, medium performance
Sodium perchlorate	$NaClO_4$	122.44	2018	52.3		Hygroscopic, high performance, bright flame
Ammonium nitrate	NH_4NO_3	80.0	1730	60.0	20.0	Smokeless, medium performance, low cost

sensitivities, and good availability. Other solid oxidizers, particularly ammonium nitrate and potassium perchlorate, have occasionally been used in production rockets but are now replaced by more modern propellants containing ammonium perchlorate. None of the many other oxidizer compounds that were investigated during the 1970s have reached production status.

The oxidizing potential of the perchlorates is generally high, which makes this material suited for high-specific-impulse propellants. Both ammonium and potassium perchlorate are only slightly soluble in water, a favorable propellant trait. All the perchlorate oxidizers generate gaseous hydrogen chloride (HCl) and other toxic and corrosive chlorine compounds in their reaction with fuels. In the firing of rockets, care is required, particularly for very large rockets, to safeguard operating personnel and/or communities in the path of exhaust gas clouds. Ammonium perchlorate (AP) is supplied in the form of small white crystals. Particle size and shape influences their manufacturing process and propellant burning rate. Therefore, close control of crystal sizes and size distributions present in a given quantity or batch is required. AP crystals are rounded (to nearly ball shape) to make for easier mixing than with sharp, fractured crystals. From the factory, they come in sizes ranging from about 600 μm (1 μm = 10^{-6} m) diameter to about 80 μm. Diameters below about 40 μm are considered more hazardous (e.g., can easily be ignited and sometimes detonate) and are not shipped; instead, propellant manufacturers take the larger crystals and grind them (at a motor factory) to smaller sizes (down to 2 μm) just before they are incorporated into a propellant.

Inorganic nitrates are relatively low-performance oxidizers compared with perchlorates. However, *ammonium nitrate* (AN) is used in some applications because of its very low cost as well as its smokeless and relatively nontoxic exhaust. Its principal use is in low-burning-rate, low-performance rocket and gas generator applications.

Ammonium nitrate changes its crystal structure at several phase transformation temperatures. These changes cause slight changes in volume. One phase transformation at 32°C causes about a 3.4% change in volume. Repeated temperature cycling through this transition temperature creates tiny voids in the propellant, resulting in growth in the grain and a change in physical and/or ballistic properties. The addition of a small amount of stabilizer such as nickel oxide (NiO) or potassium nitrate (KNO_3) appears to change this transition temperature to above 60°C, a high enough value so that normal ambient temperature cycling will no longer cause recrystallization (Refs. 13–12 and 13–13). With such an additive, AN is known as *phase-stabilized ammonium nitrate* (PSAN). Moreover, AN is hygroscopic and any moisture absorption will degrade and destabilize propellants containing AN.

Fuels

This section discusses solid fuels of which *powdered spherical aluminum* is the most common. It consists of small spherical particles (5 to 60 μm diameter) and is used with a wide variety of composite and composite-modified double-base propellant formulations, usually constituting 14 to 20% of the propellant by weight. Small aluminum particles can burn in air and aluminum powder is mildly toxic if inhaled. During rocket combustion this fuel is oxidized to aluminum oxide. Such oxide particles tend to agglomerate and form larger particles. Aluminum increases the heat of combustion, the propellant density, the combustion temperature, and thus the specific impulse. The oxide starts in liquid droplet form during combustion but solidifies in the nozzle as the gas temperature drops. When in the liquid state, the oxide can form a molten slag, which can accumulate in pockets (e.g., around an improperly designed submerged nozzle), thus adversely affecting the vehicle's mass ratio. It also can deposit on walls inside the combustion chamber, as described in Refs. 13–14 and 13–15. Reference 13–16 addresses important issues related to adding aluminum as fuel to solid propellants,

Boron is a high-energy fuel that is lighter than aluminum and has a high melting point (2304°C). It is difficult to burn with high efficiency in combustion chambers of practical lengths. However, it can be efficiently oxidized if the boron particle size is sufficiently small. Boron has been used advantageously as a propellant in a rocket combined with an air-burning engine, where there is adequate combustion volume and atmospheric oxygen.

Beryllium burns much more easily than boron, improving the specific impulse of a solid propellant motor by about 15 sec, but both beryllium and its highly toxic oxide powders are absorbed by animals and humans when inhaled. The technology with composite propellants using powdered beryllium fuel has been experimentally proven, but its severe toxicity makes any earth-bound application unlikely.

Binders

In composite propellants, binders provide the structural matrix or glue with which the solid granular ingredients are held together. In raw form, these materials are liquid prepolymers or monomers. Polyethers, polyesters, and poly-butadienes have been used (see Tables 13–6 and 13–7). After they are mixed with their solid ingredients, cast and cured, they form a hard rubber-like material that constitutes the grain. Polyvinylchloride (PVC) and polyurethane (PU) (Table 13–1) were used 50 years ago and are still used in a few rocket motors, mostly of old design. Binder materials also act as fuels for solid propellant rockets and are oxidized in the combustion process. The binding ingredient, typically a polymer of one type or another, has a primary effect on motor reliability and its mechanical properties, propellant processing complexity, storability, aging, and costs. Some polymers undergo complex chemical reactions, crosslinking, and branch chaining during curing of the propellant. HTPB has been a favorite binder in recent years, because it allows somewhat higher solids fraction (88 to 90% of AP and Al), a small performance improvement, and relatively good physical properties at the temperature limits. Several common binders are listed in Tables 13–1, 13–6, and 13–7. Elastomeric binders can be added to plasticized double-base-type nitrocellulose to improve its physical properties. Polymerization occurs when the binder monomer and its crosslinking agent react (beginning in the mixing process) to form long chains and complex three-dimensional polymers. Other types of binders, such as PVC, cure or plasticize without molecular reactions (see Refs.13–4, 13–5, and 13–15). Often called *plastisol-type binders*, these form very viscous dispersions of powdered polymerized resins in nonvolatile liquids; they polymerize slowly.

Burning-Rate Modifiers

A burning-rate *catalyst* or burning-rate *modifier* helps to accelerate or decelerate combustion at the burning surface and thus increases or decreases the propellant burning rate. It permits tailoring of the burning rate to fit a specific grain design and thrust–time curve. Several are listed in Tables 13–6 and 13–7. Some, like iron oxide or lead stearate, increase the burning rate; however, others, like lithium fluoride, will reduce the burning rate of some composite propellants. Inorganic catalysts do not contribute to the combustion energy, but consume energy by being heated to combustion temperatures. These modifiers are effective because they can change combustion mechanisms, these are mentioned in Chapter 14 (examples of modifiers that change the burning rate of composite propellants are given in Chapter 2 of Ref. 13–4).

Burning rate is defined in Section 12.1; it is a strong function of the propellant composition. For composite propellants it may be increased by changing the propellant characteristics as follows:

1. Introduce a burning rate *catalyst*, or burning rate *modifier* (0.1 to 3.0% of propellant) or increase percentage of existing catalyst.
2. *Decrease* the *oxidizer particle size.*
3. *Increase oxidizer percentage.*
4. Increase the *heat of combustion* of the binder and/or the plasticizer.
5. Imbed high-conductivity *wires* or *metal staples* in the propellant.

Plasticizers

A plasticizer is usually a relatively low-viscosity, liquid organic ingredient, which also acts as fuel. It is added to improve the elongation of the propellant at low temperatures and to improve its processing properties, such as lower viscosity for casting or longer pot life of the mixed but uncured propellants. The plasticizers listed in Tables 13–6, 13–7 and 13–8 represent several examples.

Curing Agents or Crosslinkers

A curing agent or crosslinker causes prepolymers to form longer chains of larger molecular mass and interlocks between chains. Even though these curing agents are present in small amounts (0.2 to 3%), a minor change in their percentage can have a major effect on the propellant physical properties, manufacturability, and aging. They are used primarily with composite propellants and cause the binder to solidify and become hard. Several curing agents are listed in Table 13–7.

Energetic Binders and Plasticizers

Energetic binders and/or plasticizers are used in lieu of the conventional organic materials. They contain oxidizing species (such as azides or organic nitrates) as well as other organic species. They add some energy to the propellant causing a modest increase in performance. They serve also as binders to hold other ingredients, or as energetic plasticizers liquid. They can self-react exothermally and burn without a separate oxidizer. Glycidyl azide polymer (GAP) is an example of an energetic, thermally stable, hydroxyl-terminated prepolymer that can be polymerized. It has been used in experimental propellants. Other energetic binder or plasticizer materials are listed in Tables 13–6, 13–7 and 13–8.

Organic Oxidizers or Explosives

Organic oxidizers are highly energetic compounds with $-NO_2$ radicals or other oxidizing fractions incorporated into their molecular structure. References 13–4 and 13–15 describe their properties, manufacture, and applications. They are used with high-energy propellants and/or with smokeless propellants. They can be crystalline solids, such as the *nitramines* HMX or RDX, fibrous solids such as NC, or energetic plasticizer liquids such as DEGDN or NG. These materials can react or burn by themselves when initiated with enough activating energy and all may detonate under certain conditions. Both HMX and RDX are stoichiometrically balanced materials and their addition into either fuel or oxidizer reduces the values of T_1 and I_s. Therefore, when binder fuels are added to hold the HMX or RDX crystals in a viscoelastic matrix, it is also necessary to add an oxidizer such as AP or AN.

RDX and HMX are quite similar in structure and properties. Both are white crystalline solids that can be made in different sizes. For safety, they are shipped in a desensitizing liquid, which is removed prior to propellant processing. HMX has a higher density, a higher detonation rate, yields more energy per unit volume, and has a higher melting point than RDX. Also extensively used in military and commercial explosives are NG, NC, HMX, and RDX. To achieve higher performance or other desirable characteristics, HMX or RDX can be included in DB, CMDB, or composite propellants. The amount added can range up to 60% of the propellant. Processing propellant with these or similar ingredients can be hazardous and the necessary extra safety precautions make the processing more expensive.

Liquid *nitroglycerine* (NG) by itself is very sensitive to shock, impact, or friction. It is an excellent plasticizer for propellants when desensitized by the addition of other liquids (like triacetin or dibutyl phthalate) or by compounding it with nitrocellulose. It readily dissolves in many organic solvents, and in turn it acts as a solvent for NC and other solid ingredients (Ref. 13–15).

Nitrocellulose (NC) is a key ingredient in DB and CMDB propellants. It is made by the acid nitration of natural cellulose fibers from wood or cotton and is a mixture of several organic nitrates. Although crystalline, it retains the fiber structure of the original cellulose (see Ref. 13–15). The nitrogen content is important in defining the significant properties of nitrocellulose, which can range from 8 to 14%, but the grades used for propellant are usually between 12.2 and 13.1%. Since it is impossible to make NC from natural products with an exact nitrogen content, the required properties are achieved by careful blending. Since the solid-fiber-like NC material is difficult to make into a grain, it is usually mixed with NG, DEGDN, or other plasticizers to gelatinize or solvate it when used with DB and CMDB propellants.

A newly identified organic oxidizer, hexanitrohexaazaiso-wurtzitane (also known as HNIW or CL-20), is being extensively investigated for its potential as a practical propellant, see Refs. 13–1 and 13–17. It may yield the highest specific impulse to date, slightly higher than currently modified solid propellants with HMX, and can act as a most powerful explosive (20% more power than HMX). To date (in 2015), it has only been produced in small laboratory quantities. Two major factors have restrained the full implementation of Cl-20, namely, its high cost (up to $ 570/pound in 2013) and its high sensitivity—impact and friction tests indicate that CL-20 is less stable than HMX. CL-20's density is somewhat higher than any other explosive ingredients (2.04 g/cm^3). Investigations on its potential uses are continuing.

Additives

Additives perform many functions, including accelerating or lengthening *curing times*, improving *rheological properties* (easier casting of viscous raw mixed propellant), improving some *physical properties*, adding *opaqueness* to a transparent propellant to prevent radiation heating at places other than the burning surface, limiting *migration of chemical species* from the propellant to the binder or vice versa, minimizing any slow oxidations or *chemical deterioration* during storage, and improving *aging* characteristics or moisture resistance. *Bonding agents* are additives that enhance adhesion between the solid ingredients (AP or Al) and the binder. *Stabilizers* are intended to minimize slow chemical or physical reactions that may occur in propellants. *Catalysts* are sometimes added to crosslinker or curing agents to slow down curing rates. Lubricants are an aid the extrusion process. Desensitizing agents help to make propellants more resistant to inadvertent energy stimuli. Such additives are usually included in very small quantities.

Particle-Size Parameters

The size, shape, and size distribution of solid particles like AP, Al, or HMX within the propellant can have a major influence on composite propellant characteristics. These particles are made spherical in shape to allow for easier mixing and to attain higher solid percentages in the propellant than shapes of sharp-edged natural crystals. Normally, ground AP oxidizer crystals are rated according to particle size ranges as follows:

Coarse	400 to 600 μm (1 μm = 10^{-6} m)
Medium	50 to 200 μm
Fine	5 to 15 μm
Ultrafine	sub micrometer to 5 μm

Coarse and medium-grade AP crystals are handled as class 1.3 materials, whereas the fine and ultrafine grades are considered as class 1.1 high explosives and are usually manufactured on-site from medium or coarse grades, (see Section 13.3 for a definition of these explosive hazard classifications). Most propellants use a multimodal blend of oxidizer particle sizes to maximize the amount of oxidizer per unit volume of propellant, with the small particles filling part of the voids between the larger particles. A *monomodal* propellant has one size of solid oxidizer particles, a bimodal has two sizes (say, 20 and 200 μm), and a trimodal propellant has three sizes; multiple modes allow a larger mass of solids to be placed into a given volume. Problem 13–1 has a sketch depicting how voids between the largest particles are filled with smaller particles.

Figure 13–8 shows the influence of varying the ratio of coarse to fine oxidizer particle sizes on propellant burning rate together with the influence of a burning rate additive. Figure 13–9 shows that the effect of particle size of aluminum fuels

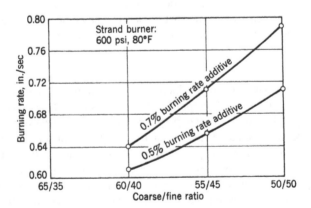

FIGURE 13–8. Typical effect of oxidizer (ammonium perchlorate) particle size mixture and burning rate additive on the burning rate of a composite propellant. From NASA report: 260-SL-3 Motor program, Volume 2, 260-SL-3 motor propellant development, NASA-CR-72262, AGC-7096, Jul 1967, 110 pp. Accession Number: N68-16051. http://ntrs.nasa.gov/archive/nasa/casi.ntrs.nasa.gov/19680006582_1968006582.pdf.

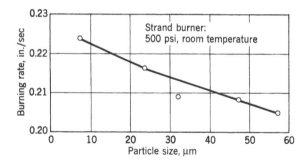

FIGURE 13-9. Typical effect of aluminum particle size on propellant burning rate for a composite propellant. From NASA report: Solid propellant processing factor in rocket motor design, NASA-SP-8075, 82 pp. (Oct 1972); N72-31767. http://hdl.handle.net/2060/19720024117; NTRS Document ID: 19720024117.

on propellant burning rate is much less pronounced than that of oxidizer particle size (Fig. 13–8 also shows an effect of particle size). Particle size, range, and shape for both the oxidizer (usually ammonium perchlorate AP]) and solid fuel (usually aluminum) have significant effects on solid packing fractions and on rheological properties (associated with the flowing or pouring of viscous liquids) of uncured composite propellants. By definition, *packing fraction* is the volume fraction of all solids when packed to minimum volume (a theoretical condition). High packing fractions make mixing, casting, and handling during propellant fabrication more difficult. Figure 13–10 shows a resulting distribution of AP particle size using a blend of sizes; the shape of this curve can be altered drastically by controlling size ranges and ratios. Also, solid particle size, range and shape affect the *solids loading*

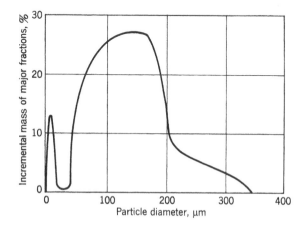

FIGURE 13-10. The oxidizer (AP) particle size distribution is a blend of two or more different particle sizes; this particular composite propellant consists of a narrow cut at about 10 μm and a broad region from 50 to 200 μm.

ratio, which is the mass ratio of solid to total ingredients in the uncured propellants. Computer-optimizing methods exist for adjusting particle-size distributions to improve the loading of solids, which can be as high as 90% in some composite propellants. Even though high solids loadings are desirable for high performance, they often introduce complexity and higher costs into the processing of propellant. Trade-offs among ballistic (performance) requirements, processability, mechanical strength, rejection rates, and facility costs are an ever-present concern with many high-specific-impulse composite propellants. References 13–4 and 13–15 report on the influence of particle size on motor performance.

13.5. OTHER PROPELLANT CATEGORIES

Gas Generator Propellants

Gas generator propellants are used to produce hot gases, not thrust. They generally have a low combustion temperature (800 to 1600 K), and most do not require internal insulators when used with metal cases. Typical applications of gas generators are listed in Table 12–1. Of the large variety of propellants utilized to for gas generators only a few will be mentioned.

Stabilized AN-*based propellants* have been used for many years with various binder ingredients. They give a clean, essentially smokeless exhaust and operate at a low combustion temperature. Because of their low burning rate they are useful for long-duration gas generator applications, say 30 to 300 sec. Typical compositions are given in Ref. 13–13 and Table 13–10 describes a propellant representative of early gas generators. It is often of interest to add some fuel to AN that acts as a coolant.

One method of reducing flame temperatures is to burn conventional AP propellants hot and then to add water to bring combustion gas temperatures down to where uncooled metals can contain them. This is used on the MX missile launcher tube gas generator (Ref. 13–18). Another formulation uses HMX or RDX with an excess of polyether- or polyester-type polyurethane.

For the inflation of automobile collision safety bags the combusted gas must be nontoxic, smoke free, have a low temperature (so as not to burn people), be quickly initiated; the propellant must be storable for rather long times without degrading and reliably available. One solution is to use alkali metal azides (e.g., NaN_3 or KN_3) with an oxide and an oxidizer. The resulting nitrates or oxides are solid materials that are removed by filtering, and the gas is clean largely composed of moderately hot nitrogen. In one model, air can be aspirated into the air bag by the hot, high-pressure gas (see Ref. 13–19). One particular composition uses 65 to 75% NaN_3, 10 to 28% Fe_2O_3, and 5 to 16% $NaNO_3$ as an oxidizer, a burn rate modifier, and a small amount of SiO_2 for moisture absorption. The resultant solid nitride slag is caught in a filter.

The ideal power P delivered by a gas generator can be expressed as (see Chapters 3 and 5)

$$P = \dot{m}(h_1 - h_2) = [\dot{m}T_1Rk/(k-1)][1 - (p_2/p_1)^{(k-1)/k}] \qquad (13-1)$$

TABLE 13–10. Typical Gas Generator Propellant Using Ammonium Nitrate Oxidizer

Ballistic Properties	
Calculated flame temperature (K)	1370
Burning rate at 6.89 MPa and 20°C (mm/sec)	2.1
Pressure exponent n (dimensionless)	0.37
Temperature sensitivity σ_p (%/K)	0.22
Theoretical characteristic velocity, c^* (m/sec)	1205
Ratio of specific heats	1.28
Molecular mass of exhaust gas	19
Composition (Mass Fraction)	
Ammonium nitrate (%)	78
Polymer binder plus curing agent (%)	17
Additives (processing aid, stabilizer, antioxidant) (%)	5
Oxidizer particle size, (μm)	150
Exhaust Gas Composition (Molar %)	
Water (steam)	26
Carbon monoxide	19
Carbon dioxide	7
Nitrogen	21
Hydrogen	27
Methane	Trace
Physical Properties at 25°C or 298 K	
Tensile strength (MPa)	1.24
Elongation (%)	5.4
Modulus of elasticity in tension (N/m²)	34.5
Specific gravity	1.48

where $\dot{m}$ is the mass flow rate, h_1 and h_2 are the enthalpies per unit mass respectively (at the gas generator chamber and exhaust pressure conditions), T_1 is the flame temperature in the gas generator chamber, R is the gas constant, p_2/p_1 is the reciprocal of the pressure ratio through which these gases are expanded, and k the specific heat ratio. Because flame temperatures are relatively low, there is no appreciable dissociation and frozen equilibrium calculations are usually adequate.

Smokeless or Low-Smoke Propellant

Several types of DB propellant, DB modified with HMX, nitramine (HMX or RDX) based composites, AN composites, and/or combinations of these have very few or no solid particles in their exhaust gases. They do not contain aluminum or AP (generally resulting in lower specific impulses than comparable propellants with AP) and have very little primary smoke, but they may produce secondary smoke in unfavorable weather. Several of these propellants have been used in tactical missiles. For certain military applications smokeless propellants are needed as discussed in Chapter 20.

It is difficult to make a solid propellant that produces truly smokeless exhaust gases. A distinction must be made, therefore, between *low-smoke* also called *minimum-smoke* (almost smokeless) and *reduced-smoke propellants,* which have a faintly visible plume. Visible smoke trails typically originate from solid metal oxide particles in the plume, such as aluminum oxide. With enough of these, the exhaust plume will scatter and/or absorb light and become as visible as *primary smoke* sources. Exhaust particles can also act as nuclei for moisture condensation, which occurs in saturated air or under high-humidity and low-temperature conditions. Moreover, vaporized plume molecules such as water or hydrochloric acid may condense in cold air to form droplets and thus a cloud trail. These processes create a *vapor trail* or *secondary smoke.*

As stated, *minimum-smoke propellants* are not a special class with a peculiar formulation but a variation of one of the classes mentioned previously. Propellants containing Al, Zr, Fe_2O_3 (a burn rate modifier), and/or other metallic species will form visible and often undesirable clouds in the exhaust. *Reduced-smoke propellants* are usually composite propellants with low concentrations of aluminum (1 to 6%) that results in having low percentages of aluminum oxide in their exhaust plume; they are faintly visible as primary smoke but may precipitate heavy secondary smoke in unfavorable weather. Their specific impulse is much better than that of minimum-smoke propellants, as seen in Fig. 13–1.

Igniter Propellants

The process of propellant ignition is discussed in Section 14.2, and several types of igniter hardware are discussed in Section 15.3. Propellants for igniters, a specialized field of propellant technology, are briefly described here. Requirements for igniter propellants include the following:

Fast high heat release and high gas evolution per unit propellant mass to allow rapid filling of grain cavity with hot gas and to partially pressurize the chamber.

Stable initiation and operation over a wide range of pressures (from subatmospheric to the high chamber pressures) and smooth burning at low pressures with no ignition overpressure surges.

Rapid initiation of igniter propellant burning and low ignition time delays.

Low sensitivity of burn rate to ambient temperature changes and low burning rate pressure exponent.

Proper start, operation and storage over the required ambient temperature ranges.

Safe and easy to manufacture, and safe to ship and handle.

Satisfactory aging characteristics and long life.

Minimal moisture absorption or degradation with time.

Low cost of ingredients and fabrication.

Low or no toxicity and low corrosive effects.

Some igniters not only generate hot combustion gases but also can produce hot solid particles or hot liquid droplets, which radiate heat and impinge on the propellant surface in the chamber where they embed themselves and assist in propellant burning on the exposed grain surface.

There is a wide variety of igniter propellants and their development has been largely empirical. Black powder, which was used in early rocket motors, is no longer favored because its properties are difficult to duplicate. Extruded double-base propellants are now frequently utilized, usually in the form of a large number of small cylindrical pellets. In some instances, rocket propellants used in the main grain are also used for the igniter grain, sometimes slightly modified. They are made in the form of a small rocket motor within the larger motor that is to be ignited. A common igniter formulation uses 20 to 35% boron and 65 to 80% potassium nitrate with 1 to 5% binder. Binders typically include epoxy resins, graphite, nitrocellulose, vegetable oils, polyisobutylene, and other binders listed in Table 13–7. Another formulation contains magnesium with a fluorocarbon (Teflon); this gives hot particles and hot gases (Refs. 13–20 and 13–21). Other igniter propellants are listed in Ref. 13–11.

13.6. LINERS, INSULATORS, AND INHIBITORS

Liners, insulators, and inhibitors represent three layer types that reside at grain interfaces and are defined in Section 12.3. Their materials do not contain any oxidizing ingredients but they may ablate, cook, char, vaporize, or disintegrate in the presence of hot gases. Many will actually burn if the hot combustion gases contain even small amounts of oxidizing species, but they will not usually burn by themselves. Liners, internal insulators, or inhibitors must be *chemically compatible* with the propellant and with each other to avoid migration (described below) and/or changes in material composition; they must have *good adhesive strength* so that they stay bonded to the propellant, or to each other. The *temperatures* at which they experience damage or large *surface regressions* should be adequately high. They should all have low densities in order to reduce inert mass. Typical materials are neoprene (specific gravity 1.23), butyl rubber (0.93), a synthetic rubber called ethylene propylene diene or EPDM (0.86), or a propellant binder such as polybutadiene (0.9 to 1.0); these values are low relative to propellant specific gravities of 1.6 to 1.8. For low-smoke propellant these three rubber-like materials should only give off some gases but few, if any, solid particles (see Ref. 13–22).

The *liners* principal purpose is to provide a proper bond between the grain and the case, or between the case and any internal thermal insulation. In addition to the desired characteristics listed in the previous paragraph, the *liner* should be a soft stretchable rubber-type thin material (typically 0.02 to 0.04 in. thick with 200 to 450% elongation) to allow relative movement along the bond line between the grain and the case. This differential expansion is needed because the thermal coefficient of expansion of the grain is typically an order of magnitude higher than that of the case. A liner will also seal fiber-wound cases (particularly thin cases), which are often porous, so that high-pressure hot gases cannot escape. Typical liners for tactical guided missiles have been made from polypropylene glycol (about 57%) with a titanium oxide filler (about 20%), a di-isocyanate crosslinker (about 20%), and minor ingredients such as an antioxidant. Rocket motor cases have to be preheated to about 82°C prior to liner application. Often used as polymer for liners is ethylene propylene diene monomer (EPDM) linked into ethylene propylene diene terpolymer to form a synthetic rubber; it adheres and elongates nicely.

In some present-day motors, *internal insulators* not only provide for thermal protection of the case (from the hot combustion gases) but also often serve as a *liner* providing good bonding between propellant and insulator or insulator and case. Most motors today still have a separate liner and an insulating layer. Internal thermal insulation should also fulfill these additional requirements:

1. It must be erosion resistant, particularly at the motor aft-end or blast tube. This may be achieved in part by using tough elastomeric materials, such as neoprenes or butyl rubbers that are chemically resistant to the hot gases and to the impact of particulates. Such surface integrity is also achieved when a porous black-carbon layer formed on a heated surface (called a porous char layer) remains after some interstitial materials have been decomposed and vaporized.

2. It must provide good thermal resistance and have low thermal conductivity to limit heat transfer to the case and thus keep the case below its maximum allowable temperature, which is usually between 160 and 350°C for the plastic in composite material cases and about 550 and 950°C for most steel cases. These are accomplished by filling the insulator either with silicon oxide, graphite, Kevlar, or ceramic particles. Asbestos, an excellent filler material, is no longer used because of health hazards.

3. It should allow large-deformations or strains to accommodate grain deflections upon pressurization or temperature cycling, and should transfer loads between the grain and the case.

4. Surface regression should be minimal so as to retain much of its original geometric surface contour and allow for thin insulation.

A simple relationship for the internal insulation thickness d at any location in the rocket motor depends on the exposure time t_e, the erosion rate r_e (obtained from erosion tests at the likely gas velocity and temperature), and the safety factor f which can range from 1.2 to 2.0:

$$d = t_e r_e f \qquad (13\text{--}2)$$

For small test rocket motors, some designers use the rule that the insulation depth should be twice the charred depth of the insulation.

Usually, the thickness of insulators cannot be uniform, varying by factors up to 20. It has to be thicker at locations near the nozzle where it is exposed to longer intervals and higher scrubbing velocities than at insulator locations protected by bonded propellant. Before making any material selection, it is necessary to evaluate the burned-propellant flow field and thermal environment (combustion temperature, gas composition, pressure, exposure duration, and internal ballistics) in order to carry out proper thermal analyses (erosion predictions and estimated thickness of insulator). Evaluation of loads and deflections under loads at different motor locations are also needed to estimate shear and compression stresses. When high stresses or relief flaps are involved, structural analyses are also needed. Various software codes, such as those mentioned in Refs. 13–23 and 13–24, have been used.

Inhibitors are usually made of the same materials as internal insulators. They are applied (bonded, molded, glued, or sprayed) to grain surfaces that should not burn. In a segmented rocket motor, for example (see Fig. 15–2), where burning is allowed only on the internal port area, the faces of the cylindrical grain sections may be inhibited.

Migration describes the transfer of mobile (liquid) chemical species from the solid propellant to the liner, insulator, or inhibitor, and vice versa. Liquid plasticizers such as NG or DEGDN or unreacted monomers or liquid catalysts are known to migrate. This migratory transfer occurs very slowly but may cause dramatic changes in physical properties (e.g., the propellant next to the liner becomes brittle or weak), and there are several instances where nitroglycerine that migrated into an insulator made it flammable. Migration can be prevented or inhibited by using (1) propellants without plasticizers, (2) insulators or binders with plasticizers identical to those used in propellants, (3) a thin layer of an impervious material or a migration barrier (such as PU or a thin metal film), and (4) an insulator material that will not allow migration (e.g., PU) (see Ref. 13–25).

The graphite–epoxy case of the launch-assist rocket motors used to boost the Delta launch vehicle utilizes a three-layer *liner:* EPDM (ethylene propylene diene terpolymer) as a thin primer to enhance bond strength, a polyurethane barrier to prevent migration of the plasticizer into the EPDM liner, and a plasticized HTPB-rich liner to prevent burning next to the case–bond interface. Composite AP–Al propellants also use the same HTPB binder.

Liners, insulators, and/or inhibitors may be applied to the grain in several ways: by painting, coating, dipping, spraying, or by gluing sheets or strips to the case or the grain. Often automated, robotic machines are used to achieve uniform thicknesses and high quality. Reference 13–22 describes the manufacture of particular insulators.

External insulation is often added to the outside of the motor case when it is part of the vehicle's structure, particularly in tactical missiles or high-acceleration launch boosters. Such insulation reduces the heat flow from any hot boundary layers outside the vehicle surface (aerodynamic heating) to the case, and then to the propellant. It may thus prevent fiber-reinforced plastic cases from becoming weak or propellants from becoming soft or, in extreme situations, from being ignited. Such insulator must withstand oxidation from hot air flows, have good adhesion, have structural integrity to loads imposed by the flight or launch, and must have reasonable cure temperatures. Materials ordinarily used as internal insulators are unsatisfactory because they burn in the atmosphere generating additional heat. The best is a nonpyrolyzing, low-thermal-conductivity refractory material (Ref. 13–26) such as certain high-temperature paints. Any internal and external insulation also helps to reduce grain temperature fluctuations and thus thermal stresses imposed by thermal cycling, such as day–night variations or high- and low-altitude temperature variations for airborne missiles.

13.7. PROPELLANT PROCESSING AND MANUFACTURE

The manufacture of solid propellant involves many complex physical and chemical processes. In the past, propellants have been produced by several different processes such as compaction or pressing of powder charges, extrusion of propellant through dies under pressure using heavy presses, and mixing with a solvent which is later evaporated. Even for the same type of propellant (e.g., double-base, composite, or composite double-base), fabrication processes are usually not identical for different manufacturers, motor types, sizes, or propellant formulation, and no single simple generalized process flowsheet or fabrication technique prevails. Most rocket motors in production today use composite-type propellants and therefore more emphasis on this process is given here.

Figure 13–11 is a representative flowsheet for the manufacture of a complete solid rocket motor with a composite propellant made by batch processes. Processes marked with an asterisk are potentially hazardous, are usually operated or controlled remotely, and are usually performed in buildings designed to withstand potential fires or explosions. Mixing and casting processes are the most complex, being more critical than other processes in determining quality, performance, burn rate, and physical properties of the resulting propellant.

The rheological properties of uncured propellants (e.g., their flow properties in terms of shear rate, stress, and time) are all-important to their processability and these properties usually change substantially throughout the path of the processing line. Batch-type processing of propellants, including the casting (pouring) of propellants into motors that serve as their own molds, is the most common method. For very large motors several days are needed for casting up to perhaps 40 batches into a single case to form a single grain. Vacuum is almost always imposed on the propellant during the mixing and casting operations to remove air and other dispersed gases and to avoid air bubbles in the grain. Viscosity measurements of the mixed propellant (in the range of 10,000 to 20,000 poise) are made for quality control. Vacuum, temperature, vibration, energy input to the mixer, and time are some factors affecting the viscosity of the uncured propellant. Time is important in terms of *pot life*, that is, that period of time the uncured propellant remains reasonably fluid after mixing before it cures and hardens. Short pot lives (a few hours) require fast operations in emptying mixers, measuring for quality control, transporting, and casting into motors. Some binder systems, such as those using PVC, present a very long pot life and avoid the urgency or haste in the processing line. References 13–5, 13–11, and 13–27 give details on propellant processing techniques and equipment.

Double-base propellants and modified double-base propellants are manufactured by a different set of processes. The key goal here is the diffusion of liquid nitroglycerine into the fibrous solid matrix of nitrocellulose, thus forming, by means of solvation, a fairly homogeneous, well-dispersed, relatively strong solid material. Several processes for making double-base rocket propellant are in use today, including extrusion and slurry casting. In the slurry casting process the case (or the mold) is filled with solid casting powder (a series of small solid pellets of nitrocellulose with a small amount of nitroglycerine) and the case is then flooded

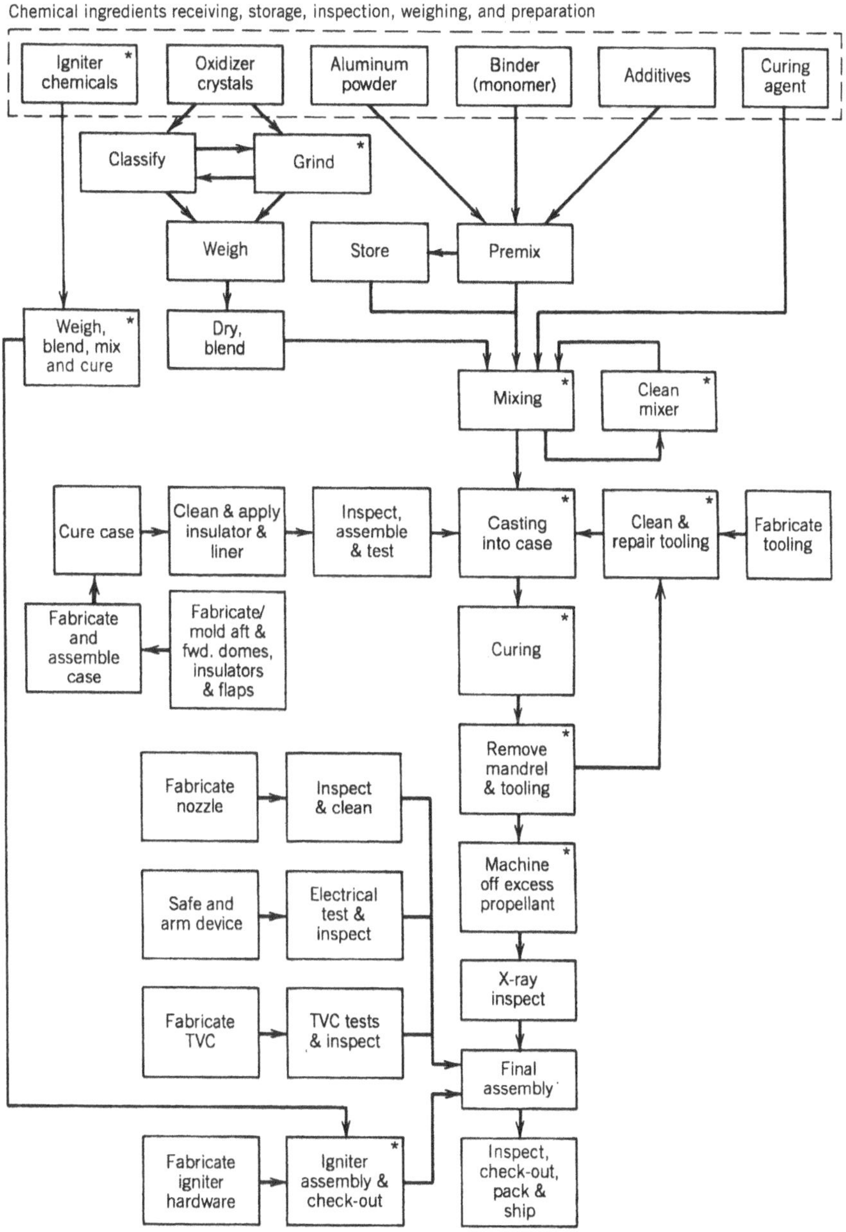

Chemical ingredients receiving, storage, inspection, weighing, and preparation

FIGURE 13–11. Simplified manufacturing process flow diagram for a rocket motor and its composite solid propellant. An asterisk means it is a more hazardous operation.

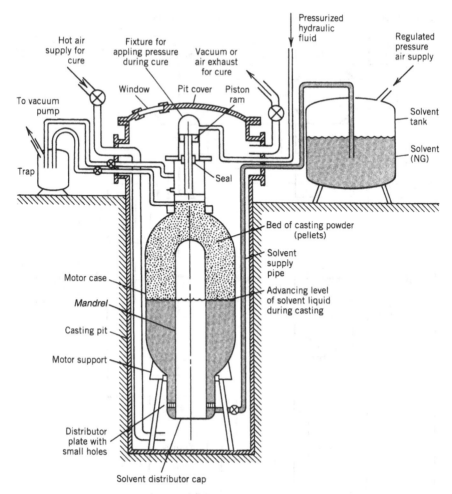

FIGURE 13–12. Basic diagram of one system for slurry casting and initial curing of a double-base solid propellant.

with liquid nitroglycerine, which then solvates the pellets. Figure 13–12 shows a simplified diagram of a typical setup for a slurry cast process. Double-base propellant manufacturing details are shown in Refs. 13–5 and 13–15.

Mandrels are used during casting and curing to assure the proper internal cavity or perforation pattern. They are made of metal in the shape of the internal bore (e.g., star or dogbone) and are often slightly tapered and coated with a nonbonding material, such as Teflon, to facilitate the withdrawal of the mandrel after curing without tearing the grain. For complicated internal passages, such as a conocyl, complex built-up mandrels, which can be withdrawn through the nozzle flange opening in smaller pieces or which can be collapsed are necessary. Some manufacturers have had success in making permanent mandrels (that are not withdrawn but stay with the motor) out of a lightweight foamed propellant that burn very quickly once ignited.

An important objective in processing is to produce a propellant grain free of cracks, low-density areas, voids, or other flaws. In general, voids and other flaws degrade the ballistic and mechanical properties of the propellant grain. Many small dispersed gas bubbles in a propellant grain may result in an abnormally high burning rate, one so high as to cause excessive pressure and catastrophic case failure.

The finished grain (or rocket motor) is usually inspected for defects (cracks, voids, and debonds) using X-rays, ultrasonics, heat conductivity probes, and/or other nondestructive inspection techniques. Propellant samples are taken from each batch, tested for rheological properties, and cast into physical property specimens and/or small rocket motors, which are then cured and subsequently tested. Any determination of the sensitivity of rocket motor performance, including possible failure, to propellant voids and other flaws often requires the test firing of motors with known defects. Data from such tests are important in establishing inspection criteria for accepting and rejecting production rocket motors.

Special processing equipment is required in the manufacture of propellants. For composite propellants this includes mechanical mixers (usually with two or three blades rotating on vertical shafts agitating propellant ingredients in a mixer bowl under vacuum), casting equipment, curing ovens, and/or machines for automatically applying the liner or insulation to the case, see Ref. 13–11. Double-base processing requires equipment for mechanically working the propellant (rollers, presses) or special tooling for allowing a slurry cast process. Computer-aided filament winding machines are used for laying the fibers of fiber-reinforced plastic cases and nozzles.

Casting of propellant into large boosters and strap-ons very near the launch site and the vertical keeping of rocket motors has the potential to decrease problems associated with handling and transportation, as well as with propellant slump. This decreases the time from manufacturing to launch but is not recommended for long storage times in any vertical position.

PROBLEMS

1. Ideally, solid oxidizer particles in a propellant may be treated as spheres of uniform size. Three sizes of particles are available: Coarse at 500 μm, medium at 50 μm, and fine at 5 μm, each at a specific gravity of 1.95, and a viscoelastic fuel binder is available at a specific gravity of 1.01. Assume that these materials can be mixed and vibrated so that the solid particles will touch each other so that there are no voids in the binder, and that the particles occupy a minimum of space similar to the sketch of the cross section shown here. It is desired to put 94 wt-% of oxidizer into the propellant mix, as this will give maximum performance. **(a)** Determine the maximum weight percentage of oxidizer if only coarse crystals are used or if only medium-sized crystals are used. **(b)** Determine the maximum weight of oxidizer if both coarse and fine crystals are used, with the fine crystals filling the voids between the coarse particles. What is the optimum relative proportion of coarse and fine particles to give a maximum of oxidizer? **(c)** Same as part **(b)**, but use coarse and medium crystals only. Is this better and, if so, why? **(d)** Using all three sizes, what is

the ideal weight mixture ratio and what is the maximum oxidizer content possible and the theoretical maximum specific gravity of the propellant? (*Hint*: The centers of four adjacent coarse crystals form a tetrahedron whose side length is equal to the diameter.)

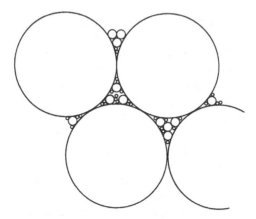

2. Suggest one or two specific applications (e.g., anti-aircraft, intercontinental missile, space launch vehicle upper stage, etc.) for each of the propellant categories listed in Table 13–2 and explain why it was selected in comparison to other propellants.

3. Prepare a detailed outline of a procedure to be followed by a crew operating a propellant mixer. This 1-m³ vertical solid propellant mixer has two rotating blades, a mixing bowl, a vacuum pump system to allow mix operations under vacuum, feed chutes or pipes with valves to supply the ingredients, and variable-speed electric motor drive, a provision for removing some propellant for laboratory samples, and a double-wall jacket around the mixing bowl to allow heating or cooling. It is known that composite propellant properties are affected by mix time, small deviations from the exact composition, temperature of the mix, the mechanical energy added by the blades, the blade speed, and the sequence in which the ingredients are added. It is also known that bad propellant would result if there are leaks that destroy the vacuum, if the bowl, mixing blades, feed chutes, and so on, are not clean but contain deposits of old propellant residue from a prior mix on their walls, if they are not mixed at 80°C, or if the viscosity of the mix becomes excessive. Usually, the sequence of loading ingredients needs to be: (1) prepolymer binder, (2) plasticizer, (3) minor liquid additives, (4) solid consisting of first powdered aluminum followed by mixed bimodal or trimodal AP crystals, and (5) finally the polymerizing agent or crosslinker. Refer to Fig. 13–11. Samples of the final liquid mix are taken to check for viscosity and density. Please list all the sequential steps that the crew should undertake before, during, and after the mixing operation. Each major step in this outline of the procedure for mixing a composite propellant has a purpose or objective that should be stated. Any specific data for each step (e.g., mass, duration, temperature, blade speed, etc.) is not required to be listed, but if known can be stated. Mention all instruments (e.g., thermometers, wattmeter, etc.) that the crew should have and identify those that they must monitor closely. Assume that all ingredients were found to be of the desired composition, purity, and quality.

4. Determine the longitudinal growth of a 24-in.-long freestanding grain with a linear thermal coefficient of expansion of $7.5 \times 10^{-5}/°F$ for temperature limits of -40 to $140°F$. *Answer*: 0.32 in.

5. The following data are given for an internally burning solid propellant grain with inhibited end faces and a small initial port area:

Length	40 in.
Port area	27 in.^{2}
Propellant weight	240 lbf
Initial pressure at front end of chamber	1608 psi
Initial pressure at nozzle end of chamber	1412 psi
Propellant density	0.060 lbm/in.^{3}
Vehicle acceleration	$21.2g_0$

Determine the initial forces on the propellant supports produced by the pressure differential and by the vehicle acceleration.

Answer: 19,600 lbf, 5090 lbf.

6. A solid propellant rocket motor with an end-burning grain has a thrust of 4700 N and a 14 sec duration. Four different burning rate propellants are available, all with approximately the same performance and the same specific gravity, but different AP mix and sizes and different burning rate enhancing ingredients. They are 5.0, 7.0, 10, and 13 mm/sec. The preferred L/D is 2.60, but values of 2.2 to 3.5 are acceptable. The impulse-to-initial-weight ratio is 96 at an L/D of 2.5. Assume optimum nozzle expansion at sea level. Chamber pressure is 6.894 MPa or 1000 psia and the operating temperature is 20°C or 68°F. Determine grain geometry, propellant mass, approximate hardware mass, and initial mass.

7. For the rocket in Problem 6 determine the approximate chamber pressure, thrust, and duration at 245 and 328 K. Assume the temperature sensitivity (at a constant value of A_b/A_t) of 0.01%/K does not change with temperature.

8. A fuel-rich solid propellant for a gas generator drives a turbine of a liquid propellant turbopump. Determine its mass flow rate. The following data are given:

Chamber pressure	$p_1 = 5$ MPa
Combustion temperature	$T_1 = 1500$ K
Specific heat ratio	$k = 1.25$
Required pump input power	970 KW
Turbine outlet pressure	0 psia
Turbine efficiency	65%
Molecular mass of gas	22 kg/kg-mol
Pressure drop between gas generator and turbine nozzle inlet	0.10 MPa

Windage and bearing friction is 10 kW. Neglect start transients.

Answer: $\dot{m} = 0.288$ kg/sec.

9. A gas generator propellant the following characteristics:

Burn rate at standard conditions	4.0 mm/sec
Burn time	110 sec
Chamber pressure	5.1 MPa
Pressure exponent n	0.55
Propellant specific gravity	1.47

Determine the size of an end-burning cylindrical grain.

Answer: Single end-burning grain 27.2 cm in diameter and 31.9 cm long, or two end-burning opposed grains (each 19.6 cm diameter × 37.3 cm long) in a single chamber with ignition of both grains in the middle of the case.

REFERENCES

13–1. O. Bolton et al., "High Power Explosive with Good Sensitivity: A 2:1 Cocrystal of CL-20:HMX," *Crystal Growth, Design*, Vol. 12, No. 9, August 2012, pp. 4311–4314. doi:10.1021/cg3010882.

13–2. A. Davenas, "Solid Rocket Motor Design," Chapter 4 of G. E. Jensen and D. W. Netzer (Eds.), *Tactical Missile Propulsion*, Vol. 170, *Progress in Astronautics and Aeronautics*, AIAA, Reston, VA, 1996.

13–3. "Hazards Assessment Tests for Non-Nuclear Ordnance," Military Standard MIL-STD-2105B (government-issued specification), 1994; M. L. Chan et al., "Advances in Solid Propellant Formulations," Chapter 1.7 in V. Yang, T. B. Brill, and W.-Z. Ren (Eds.), *Solid Propellant Chemistry, Combustion, and Motor Interior Ballistics, Progress in Astronautics and Aeronautics*, Vol. 185, AIAA, Reston VA, 2000, pp. 185–206.

13–4. N. Kubota, Chapter 1, "Survey of Rocket Propellants and Their Combustion Characteristics," in K. K. Kuo and M. Summerfield (Eds.), *Fundamentals of Solid-Propellant Combustion, Progress in Astronautics and Aeronautics*, Vol. 90, American Institute of Aeronautics and Astronautics, New York, 1984; see also Chapters 1.1 to 1.8 in V. Yang, T. B. Brill, and W.-Z. Ren (Eds.), *Solid Propellant Chemistry, Combustion, and Motor Interior Ballistics, Progress in Astronautics and Aeronautics*, Vol. 185, AIAA, Reston VA, 2000.

13–5. C. Boyars and K. Klager, *Propellants: Manufacture, Hazards and Testing*, Advances in Chemistry Series 88, American Chemical Society, Washington, DC, 1969.

13–6. Chemical Propulsion Information Agency, *Hazards of Chemical Rockets and Propellants*, Vol. II, *Solid Rocket Propellant Processing, Handling, Storage and Transportation*, NTIS AD-870258, May 1972.

13–7. H. S. Zibdeh and R. A. Heller, "Rocket Motor Service Life Calculations Based on First Passage Method," *Journal of Spacecraft and Rockets*, Vol. 26, No. 4, July–Aug. 1989, pp. 279–284; doi: 10.2514/3.26067.

13–8. D. I. Thrasher, Chapter 9, "State of the Art of Solid Propellant Rocket Motor Grain Design in the United States," in *Design Methods in Solid Rocket Motors*, Lecture Series LS 150, AGARD/NATO, Apr. 1988.

13–9. "Explosive Hazard Classification Procedures," DOD, U.S. Army Technical Bulletin TB 700-2, updated 1989; T. L. Boggs et al., "Hazards Associated with Solid Propellants," Chapter 1.9 in V. Yang, T. B. Brill, and W.-Z. Ren (Eds.), *Solid Propellant Chemistry, Combustion, and Motor Interior Ballistics, Progress in Astronautics and Aeronautics*, Vol. 185, AIAA, Reston VA, 2000. http://oai.dtic.mil/oai/oai?& verb=getRecord&metadataPrefix=html&identifier=ADA407332

13–10. "Ammunition and Explosive Hazard Classification Procedures," U.S. Department of Defense, U.S. Army TB 700-2, U.S. Navy NAVSEAINST 8020.8, U.S. Air Force TO 11A-1-47, Defense Logistics Agency DLAR 8220.1, Jan. 1998 rev.

13–11. A. Davenas, *Solid Rocket Propulsion Technology*, Elsevier Science, New York, 1992 (originally published in French); see also A. Davenas, "Development of Modern Solid

Propellants," *Journal of Propulsion and Power*, Vol. 19, No. 6, Nov.–Dec. 2003, pp. 1108–1128; doi: 10.2514/2.6947.

13–12. G. M. Clark and C. A. Zimmerman, "Phase Stabilized Ammonium Nitrate Selection and Development," *JANNAF Propellant Char. Subc. Mtg., CPIA Publication 435*, Oct. 1985, pp. 65–75; AD-B099756L.

13–13. J. Li and Y. Xu, "Some Recent Investigations in Solid Propellant Technology for Gas Generators," AIAA Paper 90-2335, Jul. 1990; doi: 10.2514/6.1990-2335.

13–14. S. Boraas, "Modeling Slag Deposition in the Space Shuttle Solid Motor," *Journal of Spacecraft and Rockets*, Vol. 21, No. 1, Jan.–Feb. 1984, pp. 47–54.

13–15. V. Lindner, "Explosives and Propellants," Kirk-Othmer, *Encyclopedia of Chemical Technology*, Vol. 9, pp. 561–671, John Wiley & Sons, New York, 1980.

13–16. R. Geisler, "A Global View of the Use of Aluminum Fuel in Solid Rocket Motors," AIAA Paper 2002-3748, Jul. 2002.

13–17. U, R, Nair et al., "Hexanitrohexaazaisowurtzitane (CL-20) and CL-20 Based Formulations," *Combustion Explosives and Shock Waves*, Vol. 41, No. 2, 2005, pp. 121–132.

13–18. J. A. McKinnis and A. R. O'Connell, "MX Launcher Gas Generator Development," *Journal of Spacecraft and Rockets*, Vol. 20, No. 3, May–Jun. 1983; doi: 10.2514/3.25590.

13–19. F. McCullough, W. F. Thorn, L. B. Katter, and E. W. Schmidt, "Gas Generator and Aspirator for Automatic Occupant Restraint Systems," SAE Technical Paper 720413, 1972, doi: 10.4271/720413.

13–20. A. Peretz, "Investigation of Pyrotechnic MTV Compositions for Rocket Motor Igniters," *Journal of Spacecraft and Rockets*, Vol. 21, No. 2, Mar.–Apr. 1984, pp. 222–224; doi: 10.2514/3.8639.

13–21. P. Gillard and F. Opdebeck, "Laser Diode Ignition of B/KNO3 Pyrotechnic Mixture," *Combustion Science and Technology*, Vol. 179, No. 8, Aug. 2007, pp. 1667–1699; doi: 10.1080/00102200701259833; G. Frut, "Mistral Missile Propulsion System," AIAA Paper 89–2428, Jul. 1989; doi: 10.2514/6.1989-2428.

13–22. J. L. Laird and R. J. Becker, "A Novel Smokeless Non-flaking Solid Propellant Inhibitor," *Journal of Propulsion and Power*, Vol. 2, No. 4, Jul.–Aug. 1986, pp. 378–379; doi: 10.2514/3.22898.

13–23. M. Q. Brewster, "Radiation–Stagnation Flow Model of Aluminized Solid Rocket Motor Insulation Heat Transfer," *Journal of Thermophysics and Heat Transfers*, Vol. 3, No. 2, Apr. 1989, pp. 132–139; doi: 10.2514/3.139.

13–24. A. Truchot, Chapter 10, "Design of Solid Rocket Motor Internal Insulation," in *Design Methods in Solid Rocket Motors*, Lecture Series LS 150, AGARD/NATO, Brussels, Apr. 1988.

13–25. M. Probster and R. H. Schmucker, "Ballistic Anomalies in Solid Propellant Motors Due to Migration Effects," *Acta Astronautica*, Vol. 13, No. 10, 1986, pp. 599–605; doi: 10.1016/0094-5765(86)90050-0.

13–26. L. Chow and P. S. Shadlesky, "External Insulation for Tactical Missile Motor Propulsion Systems," AIAA Paper 89–2425, Jul. 1989; doi: 10.2514/6.1989-2425.

13–27. W. W. Sobol, "Low Cost Manufacture of Tactical Rocket Motors," *Proceedings of 1984 JANNAF Propulsion Meeting*, Vol. II, Chemical Propulsion Information Agency CPIA Publ. 390, Vol. 2, Johns Hopkins University, Columbia, MD, 1984, pp. 219–226; AD-A143025.

CHAPTER 14

SOLID PROPELLANT COMBUSTION AND ITS STABILITY

In this the third of four chapters on solid propellant rocket motors we discuss combustion of solid propellants, physical and chemical processes of burning, ignition or start-up process, extinction of burning, and combustion instability.

When compared to other power plants, the combustion process in rocket propulsion systems is very efficient because gas composition in solid propellant reactions is essentially uniform and combustion temperatures are relatively very high; these two accelerate the rate of chemical reaction, helping to achieve nearly complete combustion. As mentioned in Chapter 2, the energy released in combustion can be between 94 and 99.5% of the total available. This is difficult to improve, so rocket motor designers have been concerned not so much with the burning process as with controlling combustion (start, stop, and heating effects) and with preventing the occurrence of combustion instabilities.

14.1. PHYSICAL AND CHEMICAL PROCESSES

Combustion in solid propellant motors involves exceedingly complex reactions taking place in the solid, liquid, and gas phases of heterogeneous mixtures. Not only are the physical and chemical processes occurring during solid propellant combustion not fully understood, but analytical combustion models have remained oversimplified and unreliable. Experimental observations of burning propellants show complicated three-dimensional microstructures and other flame structures, intermediate products in the liquid and gaseous phases, spatially and temporally variant processes, aluminum agglomerations, nonlinear response behavior, formations of carbon particles, and other complexities yet to be adequately reflected in most mathematical models.

Some insight into the solid propellant combustion process can be gained by understanding the behavior of one major oxidizer ingredient, such as ammonium perchlorate, which has been fairly well studied. This oxidizer is capable of self-deflagration with a low-pressure combustion limit at approximately 2 MPa. It has at least four distinct "froth" zones of combustion between 2 and 70 MPa, with the existence of a liquid froth on the surface of the crystal during deflagration between 2 and 6 MPa, and with a change in the energy transfer mechanism (particularly at about 14 MPa). Its influence on combustion is critically dependent on oxidizer purity. Surface regression rates for ammonium perchlorate range from 3 mm/sec at 299 K and 2 MPa to 10 mm/sec at 423 K and 1.4 MPa.

Many of the various polymeric binders used in composite propellants are less well characterized, and their combustion properties may vary, depending on the binder type, heating rate, and combustion chamber pressure.

The addition of powdered aluminum (sized from 2 to 400 μm) is known to favorably influence specific impulse and combustion stability. Photographs of burning aluminum particles show that they usually collect into relatively large agglomerates (100 or more particles) during combustion. The combustion behavior of powdered aluminum depends on several variables, including particle size and shape, surface oxides, binders, and the combustion wave environment. In Ref. 14–1, such solid propellant combustion aspects are extensively described.

Visual observations and simple experimental flame measurements, from *strand burner* tests, give some insight into the combustion process. For double-base propellants the combustion flame structure appears to be homogeneous and one-dimensional along the burning direction, as shown in Fig. 14–1. As heat from the combustion melts, decomposes and vaporizes the solid propellant at its burning surface, the resulting gases emerge already premixed. A brilliantly radiating bright flame zone can be seen where most of the chemical reaction is expected to occur together with a dark zone between the bright flame and the burning surface. Thus, the bright hot reaction zone appears detached from the propellant surface. Combustion does occur inside the dark zone but it is emitting in the infrared. The dark zone thickness decreases with increasing chamber pressure, and any higher heat transfer to the burning surface causes the burning rate to increase. Experiments on strand burners in an inert nitrogen atmosphere, reported in Ref. 14–1 (Kubota, Chapter 2), show this rather dramatically: for pressures of 10, 20, and 30 atm the dark zone thickness is 12, 3.3, and 1.4 mm, respectively, and the corresponding burning rates are 2.2, 3.1, and 4.0 mm/sec. The overall length of the visible flame shortens as the chamber pressure increases and the heat release per unit volume near the surface also increases. In DB propellants, directly over the burning surface in a bright but thin fizz or combustion zone some burning and heat release occurs. Beneath that is a zone of liquefied/bubbling propellant that is thought to be very thin (less than 1 μm) and that has been labeled the *foam* or *degradation zone*. Here, the temperature becomes high enough for the solid to vaporize and break up or to degrade into smaller molecules, such as nitrogen dioxide (NO_2), aldehydes, or nitric oxide (NO), which exit from the foaming surface. The solid propellant layer underneath the burning surface is being heated by conduction from above.

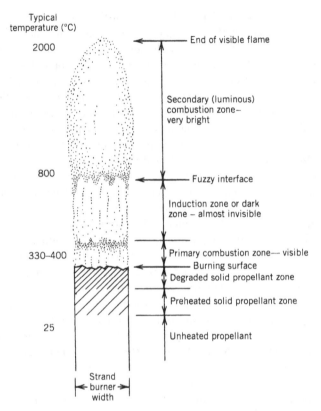

FIGURE 14–1. Schematic diagram of the combustion flame structure of a *double-base propellant* as seen with a strand burner in an inert atmosphere. Adapted from Ref. 14–1 with permission of the AIAA.

The presence of burn rate catalysts in the propellant appears to affect the primary combustion zone rather than processes in the condensed phase. These catalyze the reaction at or near the surface, increase or decrease the heat input into the burning surface, and thereby change the amount of propellant that is burning.

A typical flame from an AP/Al/HTPB* propellant looks very different than that from double-base propellants, as seen in Fig. 14–2. Here, the luminous flame appears attached to the burning surface, even at low pressures. There is no dark zone. Oxidizer-rich decomposed gases from the AP particles diffuse into the fuel-rich decomposed gases from the fuel ingredients, and vice versa. Some solid particles (aluminum, AP crystals, small pieces of binder, or combinations of these) break loose from the surface and they continue to react and degrade while in the gas flow. Thus, the burning gases may contain liquid particles of hot aluminum oxides, which radiate intensively. Both the propellant material and the burning surface are far from homogeneous. The flame structure is unsteady (flickers), three dimensional, and not

---

*Acronyms are explained in Tables 13–6 and 13–7.

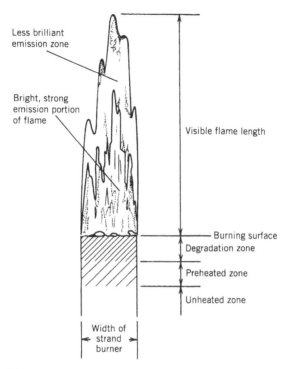

FIGURE 14–2. Diagram of the flickering, irregular combustion flame of a *composite propellant* (69% AP, 19% Al, plus binder and additives) in a strand burner with a neutral atmosphere. Adapted from Ref. 14–1 with permission of AIAA.

truly axisymmetric. Flame structures and burning rates from composite-modified cast double-base (CMDB) propellants with AP and Al seem to approach those of composite propellants, particularly when their AP content is high. Again there is no dark zone here and the flame structure is unsteady and not axi-symmetric, exhibiting the more complicated three-dimensional flame structures.

According to Ref. 14–1, the flame structure for double-base propellants with added nitramine shows only a thin dark zone with a slightly luminous degradation zone at the burning surface. The dark zone also decreases with increased pressure. Any gases issuing from RDX or HMX are essentially neutral (not oxidizing) when decomposing as pure ingredients. For a CMDB/RDX propellant, the degradation products of RDX solid crystals interdiffuse with the gases from the DB matrix just above the burning surface, before the RDX particles can produce monopropellant flamelets. Thus, an essentially homogeneous premixed gas flame is being formed, even though the solid propellant itself is heterogeneous. The flame structure appears to be one-dimensional. The burning rate of this propellant decreases when the RDX percentage is increased but seems almost unaffected by changes in RDX particle size. Much work has been done to characterize the burning behavior of different solid propellants. See Ref. 14–1 (Chapters 2, 3, and 4 by Kishore and Gayathri, Boggs, and Fifer, respectively) and Refs. 14–2 to 14–8.

The burning rate of all propellants is influenced by chamber pressure (see Section 12.1 and Eq. 12-5), by the initial ambient solid propellant temperature, by any burn rate catalyst present in the propellant, by particle size and the size distribution of solid additives like aluminum, and to a lesser extent by other ingredients and manufacturing process variables. Erosive burning has been discussed in Chapter 12. Combustion instability treated later in this chapter.

14.2. IGNITION PROCESS

In this section we discuss the necessary mechanisms for initiating solid propellant grain combustion. Individual propellants successfully used for igniters are mentioned in Section 13.5. Hardware, types, design, and integration of igniters into the motor are described in Section 15.3. Reference 14–1 contains reviews the state of the art on ignition, data from experiments, and on analytical models.

Solid propellant ignition comprises a series of complicated rapid events that start upon the receipt of a signal (usually electric) that ignites the propellant and this hot gas flows from the igniter to the motor grain surface, spreading of a flame over the entire burning surface area, filling the chamber cavity with gases, and elevating the chamber pressure without serious abnormalities (such as overpressures, combustion oscillations, damaging shock waves, hangfires [delayed ignition], flame extinguishment, and chuffing). The *igniter* in a solid rocket motor is the component that heats and generates the gases necessary for motor ignition.

Rocket motor ignition must usually be completed in fractions of a second for all but the very large motors (see Ref. 14–9). The pressure inside the motor rises to its equilibrium state in a very short time, as shown in Fig. 14–3. Conventionally, the ignition process is divided into three phases for analytical purposes:

Phase I, *Ignition time lag*: The period from the instant the igniter receives its signal until a portion of the grain surface burns and produces hot gases.

Phase II, *Flame-spreading interval*: The time from first ignition of the grain surface until the complete grain burning area has been ignited.

Phase III, *Chamber-filling interval*: The time for completing the chamber gas-filling process and for reaching equilibrium chamber pressure and mass flow.

Ignition will be successful once sufficient grain surface is ignited and burning, so that the rocket motor will continue to raise its own pressure up to the operating chamber pressure. If the igniter is not powerful enough, some grain surfaces may burn only for a short time and the flame will be extinguished; the relevant critical processes seem to be located in a gas-phase reaction above the burning surface, where propellant vapors and/or decomposition products interact with each other and with the igniter gas products.

Satisfactory attainment of equilibrium chamber pressures with full gas flows depends on (1) the characteristics of the igniter and its gas temperature, composition and issuing flow, (2) the motor propellant composition and grain surface ignitability,

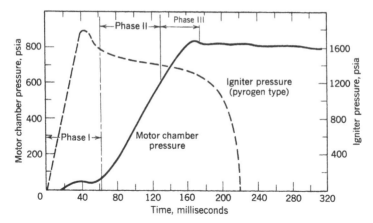

FIGURE 14–3. Typical ignition pressures as a function of time. Shown are the time traces of the igniter propellant and of its motor-chamber propellant. An electric signal is received a few milliseconds before time zero.

(3) the radiation and convection heat transfer rates between the igniter gas and grain surface, (4) the grain flame spreading rate, and (5) the dynamics of filling the motor free volume with hot gas (see Ref. 14–10). The quantity and type of caloric energy needed to ignite a particular motor grain in its prevailing environment has a direct bearing on most of the igniters' design parameters—particularly those affecting its required heat output. The ignitability of propellants at any given pressure and temperature is normally shown as a plot of ignition time versus heat flux received by the propellant surface, as in Fig. 14–4; these data were obtained from laboratory open-air tests. Propellant ignitability is affected by several factors, including (1) propellant formulation, (2) propellant grain surface initial temperature, (3) surrounding pressure, (4) modes of heat transfer, (5) grain surface roughness, (6) propellant age, (7) composition and hot-solid-particle content of igniter gases, (8) igniter propellant and its initial temperature, (9) velocity of hot igniter gases relative to the grain surface, and (10) cavity volume and configuration. Figure 14–4 and other data in Chapter 15 both show that the ignition time becomes shorter with increases in both heat flux and chamber pressure. If a shorter ignition delay is required, then a more powerful igniter will be needed. Radiation rates significantly affect ignition transients as described in Ref. 14–11. In Section 15.3 we treat the analysis and design of igniters.

14.3. EXTINCTION OR THRUST TERMINATION

Sometimes it is necessary to stop or extinguish burning in a solid propellant motor before all the propellant has been consumed. For example:

1. When a flight vehicle has reached the desired flight velocity (a ballistic missile attaining its predetermined velocity or a satellite achieving its desired orbit height), or when a precise total impulse cutoff is needed.

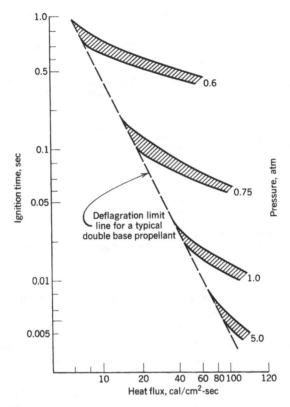

FIGURE 14–4. Propellant ignitability curves: effect of heat flux on ignition time for a specific rocket motor.

2. As a safety measure, when it appears that a flight test vehicle will unexpectedly fly out of the safe boundaries of a flight test range facility.

3. To avoid collisions of stages during stage separation maneuvers (requiring a thrust reversal) for multistage flight vehicles.

4. During research and development, when the test requires examining a partially burned rocket motor.

Common mechanisms for achieving extinction are listed below and are described in Ref. 14–1.

1. Very *rapid depressurization*, usually by a sudden and large increase of the nozzle throat area or by the fast opening of additional gas escape areas or ports. A most common technique neutralizes the net thrust or reverses its direction by suddenly opening exhaust ports in the forward end of the motor case. Such thrust reversals, using ports located on the forward bulkhead of the case, are achieved in the upper stages of Minuteman and Poseidon missiles. They are

accomplished by highly predictable and reproducible explosive devices that suddenly open additional gas escape areas (thus causing pressure reduction) and neutralize the thrust by exhausting gases in a direction opposite to that of the motor nozzle. To balance side forces, thrust termination blowout devices and their ducts are always designed as symmetrically opposed sets (two or more pairs). The motor in Fig. 1–5 has four symmetrically placed openings that are blown into the forward dome of the case by circular explosive cords; two of the sheathed circular cord assemblies are seen on the outside of the forward dome wall. Ducts that lead hot gases from such openings to the outside of the vehicle are not shown in Fig. 1–5. Any forward gas flows occur only for a very brief period of time, during which the thrust actually reverses. The rapid depressurization stops combustion suddenly at the propellant burning surfaces. Explosive cords must be properly designed so that they do not cause any detonation or explosion of any remaining unburned propellant.

2. During some motor development projects, it can be helpful to examine a partially consumed grain. Motor operation can be stopped by quenching its flames with an inhibiting liquid, such as water. Reference 14–12 shows that adding a detergent to the water allows better contact with the burning surface and reduces the amount of water needed for quenching.

3. *Lowering the combustion pressure* below the deflagration limit pressure. Compared to item 1, this depressurization occurs quite slowly. Nearly all solid propellants have a low-pressure combustion limit of 0.05 to 0.15 MPa. This means that some propellants will not extinguish when vented during a static sea-level test at 1 atm (0.1013 MPa) but will stop burning if vented at higher altitudes.

Sudden depressurizations are effective because the primary combustion zone at the propellant surface has a longer time lag than the gaseous combustion zone, which, at the lower pressures, more quickly develops a lower reaction rate moving farther away from the burning surface. Any gases created by vaporization and pyrolysis of the hot solid propellant cannot all be consumed in gaseous reactions close to the surface, and some will not burn completely. As a result, heat transfer to the propellant surface will be quickly reduced by several orders of magnitude and reactions at the propellant surface will diminish and stop. Experimental results, see Ref. 14–1, show that higher initial combustion pressures require faster depressurization rates (dp/dt) to achieve extinction.

14.4. COMBUSTION INSTABILITY

There are two basic types of combustion instability: (1) sets of acoustic resonances or pressure oscillations, which can occur in any rocket motor, and (2) vortex shedding phenomena, which occur only in relatively large segmented grains or in grains with circular slots.

Acoustic Instabilities

When solid propellant rocket motors experience unstable combustion, pressures in interior gaseous cavities (made up by the volume of the port or perforations, fins, slots, conical or radial groves) oscillate by at least 5% and often by more than 30% of their nominal value. When such instability occurs, any heat transfer to the burning surfaces, the nozzle, and the insulated case walls greatly increases; the burning rate, chamber pressure, and thrust usually also increase causing total burning duration to decrease. Changes in the thrust–time profile may cause significant changes in a vehicle's flight path, which at times can lead to mission failure. If prolonged and with high vibration energy levels, an instability may cause hardware damage, such as overheating that leads to case and/or nozzle failure. Unstable conditions should be avoided and must be carefully investigated and remedied when they do occur during the motor's development program. Final motor designs must be free of any instability.

There are fundamental differences between solid and liquid propellant combustion transient behavior. In liquid propellants chamber geometries are fixed; liquids in feed systems and in injectors are physically not part of the oscillating gases in the combustion chamber but may interact strongly with pressure fluctuations. In solid propellant motors, the geometry of the oscillating cavity increases in size as burning proceeds and there are other strong damping factors, such as solid particles and energy-absorbing viscoelastic materials. In general, combustion instability problems do not occur frequently or in every motor development and, when they do occur, they are rarely the cause of motor failures or disintegrations. Nevertheless, dire failures have happened.

While acoustically "softer" than a liquid rocket combustion chamber, the combustion cavity of a solid propellant rocket is still a low-loss acoustical cavity containing a very large acoustical energy source—the combustion process itself. Even small fractions of the energy released by combustion are more than sufficient to drive pressure vibrations to an unacceptable level. Undesirable oscillations in the combustion cavity of solid propellant rocket motors have been a continuing problem in the design, development, production, and even long-term (10-year) retention and proper aging, of some solid rocket motors for missiles.

Several combustion instabilities can occur spontaneously, often at some particular time during the motor burn period, a phenomenon usually repeatable in identical motors. Both longitudinal and transverse waves (radial and tangential) can occur. Figure 14–5 shows a pressure–time profile with typical instabilities. As pressure oscillations increase in magnitude, thrust and burning rate also increase. Oscillating frequencies seem to be a function of cavity geometry, propellant composition, pressure, and internal flame field. As the internal grain cavity enlarges and (some) oscillation frequencies and local velocities during the normal burning process change, oscillations often abate and disappear. The time and severity of vibrations due to combustion tend to change with ambient grain temperature prior to motor operation.

For simple grains within cylindrical port areas, the resonant transverse mode oscillations (tangential and radial) correspond roughly to those shown in Fig. 9–4 for liquid propellant thrust chambers. The longitudinal or axial modes, usually at a lower

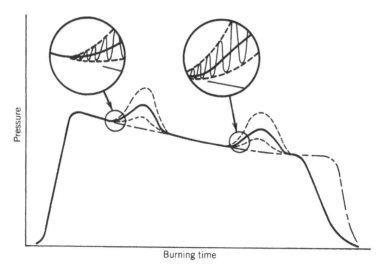

FIGURE 14–5. Simplified diagram showing two periods of combustion instability in the pressure–time history, with enlargements of two sections of this curve. The dashed lines show the upper and lower boundaries of the high-frequency pressure oscillations, and the dot-dash curve is the behavior without instability after a slight change in propellant formulation. The vibration period shows a rise in the mean pressure. With vibration, the effective burning time is reduced and the average thrust is higher but total impulse remains unchanged.

frequency, represent an acoustic wave traveling parallel to the motor axis between the forward end of the perforation and the convergent nozzle section. Harmonic frequencies of these basic vibration modes may also be excited. Internal cavities can become very complex since they may include igniter cases, movable as well as submerged nozzles, fins, cones, slots, star-shaped perforations, and/or other shapes, as described in the section on grain geometry in Chapter 12—determination of the resonant frequencies of complicated cavities is always challenging. Furthermore, internal geometries of resonating cavities continually increase as burning propellant surfaces recede reducing transverse oscillation frequencies.

The instability known as either *bulk mode*, or *Helmholtz mode*, L^* *mode*, or *chuffing mode*, is not a wave mode as described above. It occurs at relatively low frequencies (typically below 150 Hz and sometimes below 1 Hz), and pressures remain essentially uniform throughout the volume. The unsteady velocity is close to zero, but pressures do rise and fall. The gas motion (in and out of the nozzle) corresponds to the classical Helmholtz resonator mode, similar to exciting a tone when blowing across the open mouth of a bottle (see Fig. 9–7). It occurs at low values of L^* (see Eq. 8-9), sometimes during the ignition period, and disappears as the motor internal volume enlarges or the chamber pressure increases. *Chuffing* relates to the periodic low-frequency discharge of a bushy, unsteady flame of short duration (typically less than 1 sec) followed by periods of no visible flame, during which slow outgassing and vaporization of the solid propellant accumulates hot gas in the chamber. This often occurs near the end of burning. The motor experiences spurts of combustion and consequent pressure buildup followed by periods of essentially no flow with nearly ambient pressure. Such dormant periods can extend for a fraction of a second to a few seconds (Ref. 14–13 and Chapter 6 by Price in Ref. 14–1).

A useful method of visualizing unstable pressure waves is shown in Figs. 9–5 and 14–6 and presented in Ref. 14–14. It consists of Fourier- analysis sets of the measured pressure vibration spectrum, each taken at a different burning time and displayed at successive vertical positions on a time scale, thus providing a map of amplitude versus frequency versus burning time. Figure 14–6 shows a low-frequency longitudinal or axial mode and two tangential modes, whose frequencies are reduced in time by the enlargement of the cavity; it also shows the timing of different vibrations, and their onset and demise.

Initiation or triggering conditions for any particular vibration mode is still not well understood but has to do with the energetics of combustion at the propellant surface. A sudden change in pressure can be a trigger, such as when a piece of broken-off

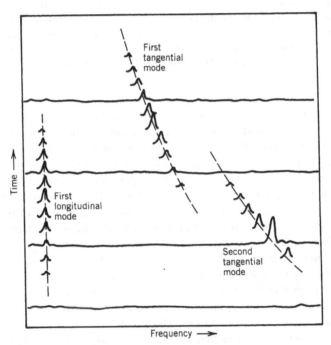

FIGURE 14–6. Simplified example of mode frequency display also called a "waterfall" diagram of a rocket motor firing. Only a few complete time–frequency curves are shown; for ease of visualization the other time lines are partly omitted except near the resonating frequencies. The height of each wave is proportional to pressure. As the cavity volume increases, the frequencies of the transverse modes decrease. The frequency of the longitudinal mode (aligned with the cavity center-line) does not change with time. Adapted from Chapter 13 of Ref. 14–1, with permission of AIAA.

insulation or unburned propellant flows through the nozzle, temporarily blocking all or a part of its flow area (causing a momentary pressure rise). Structural vibrations are known to induce instability (see Ref. 12–13) caused by the physical movement of the grain in and out of the flame zone; this effect is similar to the way acoustic waves couple with the propellant response function.

The balance between amplifying and damping factors shifts during burning and this may cause either growth or abatement of specific modes of vibration. The *response* in a solid propellant relates to changes in gaseous mass production and in energy release at the burning surface when stimulated by pressure perturbations. When a momentary high-pressure peak occurs at a burning surface, it increases the instantaneous heat transfer and thus the burning rate, causing the mass flow from that surface to also increase. Velocity perturbations along the burning surface are also believed to cause changes in mass flow. The following are phenomena that contribute to amplifying the vibrations or to gains in acoustic energy, see Ref. 14–1 (Price, Chapter 6):

1. The combustion process' dynamic response to flow disturbances or oscillations in the burning rate. This response can be determined from tests with T-burners as described on the following pages. The response function depends on the frequency of these perturbations and on propellant formulation. Combustion response may not be in phase with the disturbance. Effects of boundary layers on velocity perturbations have been reported in Ref. 14–15.
2. Interactions of flow oscillations with the main flow, similar to the operation of musical wind instruments or sirens (see Ref. 14–16).
3. The fluid-dynamic influence of vortexes.

Phenomena that contribute to diminishing vibrations or damping are energy-absorbing processes; they include the following:

1. Viscous damping in the boundary layers at the walls or propellant surfaces.
2. Damping by particles or droplets flowing in oscillating gas/vapor flows, which is often substantial. These particles accelerate and decelerate, being "dragged" along by the motion of the gas, a viscous flow process that absorbs energy. The attenuation for each particular vibration frequency has an optimum at a particular particle size; high damping for low-frequency oscillation (large motors) occurs with relatively large solid particles (8 to 20 μm); for small motors or high-frequency waves the highest damping occurs with small particles (2 to 6 μm). Attenuation drops off sharply when the particle size distribution in the combustion gas is not concentrated near the optimum for damping.
3. Energy from longitudinal and mixed transverse/longitudinal waves is lost out via the nozzle exhaust. Energy from purely transverse waves does not appear to dampen by this mechanism.
4. Acoustic energy is absorbed by the viscoelastic solid propellant, insulator, and at the motor case; its magnitude is difficult to estimate.

All solid propellants may experience instabilities. *Propellant characteristics* have a strong effect on its susceptibility to instabilities. Changes in the binder, particle-size distribution, ratio of oxidizer to fuel, ingredient types, and burn-rate catalysts can all affect stability, often in unpredictable ways. As a part of characterizing a new or modified propellants (e.g., determining their ballistic, mechanical, aging, and performance characteristics), many rocket companies now also evaluate propellants for stability behavior, as described below.

Analytical Models and Simulation of Combustion Stability

Several investigations have been aimed at mathematical models that simulate the combustion behavior of solid propellants. This has been reviewed by T'ien in Chapter 13 of Ref 14–1. Several aspects of *combustion stability* are treated in some depth in Refs. 14–17 and 14–18.

With the aid of computers it has been possible to successfully simulate combustion in some limited cases, such as in validating or extrapolating experimental results or making limited predictions of the stability of motor designs. Such simulations apply to well-characterized propellants, where empirical constants (such as propellant response or particle-size distribution) have been determined and where the range of operating parameters, internal geometries, or sizes can be narrow. It is unlikely that a reliable but simple analysis will be found for predicting the occurrence, severity, nature, and location of instability for any arbitrary propellant and motor design. The physical and chemical phenomena involved are too complex (e.g., multidimensional, unsteady, nonlinear, and influenced by many variables) and difficult to emulate mathematically without a large number of simplifying assumptions. However, analyses may give insight into physical behavior and be a valuable contributor to solving instability problems and hence they have been routinely used for preliminary design evaluation of grain cavities.

Combustion Stability Assessment, Remedy, and Design

In contrast to liquid rocket technology, a universally accepted combustion stability rating procedure does not presently exist for full-scale solid rocket motors. Undertaking stability tests on large full-scale flight-hardware rocket motors is expensive, and therefore lower-cost methods, such as subscale motors, T-burners, and other test equipment, have been used to assess motor stability.

The best known and most widely used method of gathering combustion stability-related data is the use of a *T-burner*, an indirect and limited method that does not use a full-scale motor. Figure 14–7 is a sketch of a standard T-burner; typically, it has a 1.5-in. internal diameter double-ended cylindrical burner vented at its midpoint (see Ref. 14–19). Venting can be through a sonic nozzle to the atmosphere or by a pipe connected to a surge tank, which maintains a constant level of pressure in the burner cavity. T-burner designs and usage usually concentrate on that portion of the frequency spectrum dealing with expected transverse oscillations in a full-scale motor. The desired acoustical frequency, to be imposed on the

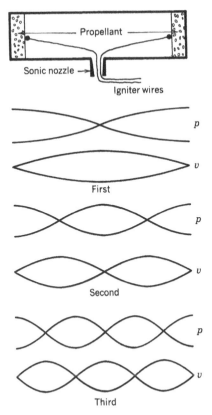

FIGURE 14–7. Schematic of a standard T-burner and its three longitudinal mode standing waves (pressure p and velocity v).

propellant charge as it burns, determines the burner length (i.e., distance between closed ends).

The nozzle location, midway between the ends of the T-burner, minimizes the attenuation of fundamental longitudinal mode oscillations (in the propellant grain cavity). Theoretically, an acoustic pressure node exists at the center and antinodes occur at the ends of the cavity. Acoustic velocity nodes are out of phase with pressure waves and occur at the ends of the burner. Propellant charges are often in the shape of disks or cups cemented to the end faces of the burner. The gas velocity in the burner cavity is kept intentionally low (Mach 0.2 or less) compared with velocities in a full-scale motor. This practice minimizes the influence of velocity-coupled energy waves and allows the influence of pressure-coupled waves to be more clearly recognized.

Use of T-burners for assessing the stability of full-scale solid rocket motors presupposes the existence of valid theoretical models for phenomena occurring in both the T-burner and the actual rocket motor; such theories are still not fully validated. In addition to assessing solid rocket motor combustion stability, the T-burner also is used to evaluate new propellant formulations including the importance of seemingly small changes in ingredients, such as a change in aluminum powder particle size and oxidizer grind method. Tests with T-burners are the current standard method of measuring a propellant's *pressure-coupled response*, which is defined as the ratio of the oscillating burning rate over the mean burning rates to the oscillatory pressure over the mean pressure (Ref. 14–20). Pressure coupling, the coupling between acoustic pressure oscillations at the surface of a burning solid propellant with the combustion processes of the propellant, is a most dominant driving mechanism. It is a function of frequency, pressure, and propellant formulation. Pressure-coupled responses represent a key input to rocket motor stability prediction programs.

Once an instability has been observed or predicted in a given rocket motor, designers proceed to fix the problem but there are no standard methods for selecting the remedy. Each of the three remedies below has been successful in at least one application:

1. Changing the grain geometry to shift the frequencies away from undesirable values. Sometimes, changing fin locations, port cross-sectional profile, or number of slots can be successful.

2. Changing propellant composition. Using aluminum as an additive has been most effective in curing transverse instabilities, provided that the particle-size distribution of the aluminum oxide is favorable to optimum damping at the distributed frequency. Changing size distribution and using other particulates (Zr, Al_2O_3, or carbon particles) has also been effective in some cases. Sometimes changes in the binder can work.

3. Adding a mechanical device for attenuating unsteady gas motions or changing the natural frequency of cavities. Various inert resonance rods, baffles, or paddles have been successfully added, mostly as a fix to an existing motor with observed instability. These can change the resonance frequencies of cavities, but introduce additional viscous surface losses, add extra inert mass, and cause potential problems with heat transfer and/or erosion.

Combustion instability is usually addressed during the design process, often through a combination of mathematical simulations, understanding similar problems in other motors, studies of possible changes, and with supporting experimental work (e.g., T-burners, measuring particle-size distribution). Most solid propellant rocket companies have in-house two- and three-dimensional computer programs to calculate likely acoustic modes (axial, tangential, radial, and combinations of these) for a given grain/motor, initial and intermediate cavity geometries, and combustion gas properties using thermochemical analyses. Data on combustion responses (dynamic burn rate behavior) and damping can be obtained from T-burner tests. Effects of particle size can be estimated from prior experience, observation through windows, or plume measurements (Ref. 14–21). Estimates of nozzle losses, friction, or other damping losses need to be included. Depending on the balance between gain and damping, it may be possible to arrive at conclusions on a grain's propensity to instability for each specific instability mode that is analyzed. If unfavorable, either the grain geometry or the propellant has to be modified. If favorable, full-scale motors are built and tested to validate the predicted burning characteristic stability. There is always a practical trade-off between the amount of work spent on extensive analysis, subscale experiments, and computer programs (which will not always guarantee a stable motor) and taking a chance that a retrofit will be needed after full-scale motors have been tested. When the instability is not discovered until after the motor is in production, it is more difficult, time consuming, and thus more expensive to fix the problem.

Vortex-Shedding Instability

This instability is associated with surface burning of the grain at inner-slots. Large segmented rocket motors have slots between segments, and some grain configurations have slots that intersect the centerline of the grain. Figure 14–8 shows hot gases

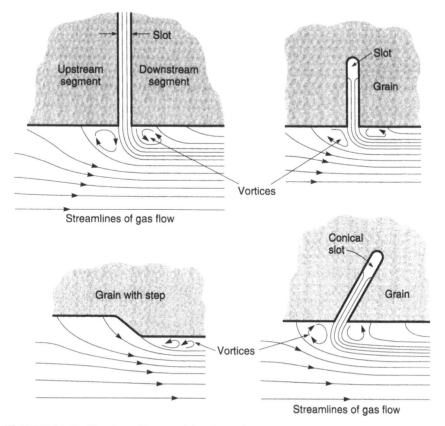

FIGURE 14–8. Sketches of four partial grain sections each with a slot or a step. Heavy lines identify the burning surfaces. The flow patterns result in the formation of vortices. The shedding of these vortices can induce flow oscillations and pressure instabilities.

from burning slot or step surfaces entering the main flow in the perforation or central cavity of the grain, see Fig. 15–2. Hot gases from slots then turn into a direction toward the nozzle. Flow from these side streams restricts the flow emanating from the upstream side of the perforation and, in effect, reduces the port area. Such restrictions cause the upstream port pressures to rise, sometimes substantially. This interaction of two subsonic gas flows causes turbulence; vortices form and are periodically shed or allowed to flow downstream, thereby causing unstable flow patterns. These shed vortices can also interact with existing acoustic instabilities. Reference 14–22 gives a description and Ref. 14–23 a method for analyzing such vortex-shedding phenomena. The remedy is usually to apply inhibitors to some of the burning surfaces or to change the grain geometry—for example, increasing the width of a slot reduces the local velocities and the vortices become less pronounced.

Several types of circumferential slots are illustrated in Fig. 14–8; when improperly designed, they produce noticeable pressure drops across the slot, causing propellant deflection, which may lead to propellant breakup and motor failure due to overpressures. A set of design rules for downstream slot tailoring, as found in Ref. 14–24, have led to successful redesigns.

PROBLEMS

1. **a.** Calculate the length of a T-burner to give a first natural oscillation of 2000 Hz using a propellant that has a combustion temperature of 2410 K, a specific heat ratio of 1.25, a molecular mass of 25 kg/kg-mol, and a burning rate of 10.0 mm/sec at a pressure of 68 atm. The T-burner is connected to a large surge tank and prepressurized with nitrogen gas to 68 atm. The propellant disks are 20 mm thick. Make a sketch to indicate the T-burner dimensions, including the disks.

   b. If the target frequencies are reached when the propellant is 50% burned, what will be the frequency at propellant burnout?

   Answer: (a) Length before applying propellant = 0.270 m (b) frequency at burnout = 1854 Hz

2. An igniter is needed for a rocket motor similar to one shown in Fig. 12–1. Igniters have been designed by various oversimplified design rules such as Fig. 14–3. The motor has an initial internal grain cavity volume of 0.055 m^3 and an initial burning surface of 0.72 m^2. The proposed igniter propellant has these characteristics: combustion temperature 2500 K and an energy release of about 40 J/kg-sec. Calculate the minimum required igniter propellant mass **(a)** if the cavity has to be pressurized to about 2 atm (ignore heat losses); **(b)** if only 6% of the igniter gas energy is absorbed at the burning surface, and it requires about 20 cal/cm^2-sec to ignite in about 0.13 sec.

3. Using the data from Fig. 14–4, plot the total heat flux absorbed per unit area versus pressure to achieve ignition with the energy needed to ignite being just above the deflagration limit. Then, for 0.75 atm, plot the total energy needed versus ignition time. Give an interpretation of the results and trend for each of the two curves.

REFERENCES

14–1. N. Kubota, Chapter 1, "Survey of Rocket Propellants and Their Combustion Characteristics"; K. Kishore and V. Gayathri, Chapter 2, "Chemistry of Ignition and Combustion of Ammonium-Perchlorate-Based Propellants"; T. L. Boggs, Chapter 3, "The Thermal Behavior of Cyclotrimethylene Trinitrate (RDX) and Cyclotetramethylene Tetranitrate (HMX)"; R. A. Fifer, Chapter 4, "Chemistry of Nitrate Ester and Nitramine Propellants"; C. E. Hermance, Chapter 5, "Solid Propellant Ignition Theories and Experiments"; M. Kumar and K. K. Kuo, Chapter 6, "Flame Spreading and Overall Ignition Transient"; E. W. Price, Chapter 13, "Experimental Observations of Combustion Instability"; J. S. T'ien, Chapter 14, "Theoretical Analysis of Combustion Instability"; all in K. K. Kuo and M. Summerfield (Eds.), *Fundamentals of Solid-Propellant Combustion*, Vol. 90, *Progress in Astronautics and Aeronautics*, American Institute of Aeronautics and Astronautics, New York, 1984.

14–2. G. Lengelle, J. Duterque, and J. F. Trubert, Chapter 2.2, "Physico-Chemical Mechanisms of Solid Propellant Combustion," in V. Yang, T. B. Brill, and W.-Z. Ren (Eds.), *Solid Propellant Chemistry, Combustion, and Motor Interior Ballistics*, Vol. 185, *Progress in Astronautics and Aeronautics*, AIAA, Reston VA, 2000.

14–3. C. Youfang, "Combustion Mechanism of Double-Base Propellants with Lead Burning Rate Catalyst," *Propellants, Explosives, Pyrotechnics*, Vol. 12, 1987, pp. 209–214.

14–4. N. Kubota et al., "Combustion Wave Structures of Ammonium Perchlorate Composite Propellants," *Journal of Propulsion and Power*, Vol. 2, No. 4, Jul.–Aug. 1986, pp. 296–300.

14–5. T. Boggs, D. E. Zurn, H. F. Cordes, and J. Covino, "Combustion of Ammonium Perchlorate and Various Inorganic Additives," *Journal of Propulsion and Power*, Vol. 4, No. 1, Jan.–Feb. 1988, pp. 27–39.

14–6. T. Kuwahara and N. Kubota, "Combustion of RDX/AP Composite Propellants at Low Pressure," *Journal of Spacecraft and Rockets*, Vol. 21, No. 5, Sept.–Oct. 1984, pp. 502–507.

14–7. P. A. O. G. Korting, F. W. M. Zee, and J. J. Meulenbrugge, "Combustion Characteristics of Low Flame Temperature, Chlorine-Free Composite Propellants," *Journal of Propulsion and Power*, Vol. 6, No. 3, May–Jun. 1990, pp. 250–255.

14–8. N. Kubota and S. Sakamoto, "Combustion Mechanism of HMX," *Propellants, Explosives, Pyrotechnics*, Vol. 14, 1989, pp. 6–11.

14–9. L. H. Caveny, K. K. Kuo, and B. J. Shackleford, "Thrust and Ignition Transients of the Space Shuttle Solid Rocket Booster Motor," *Journal of Spacecraft and Rockets*, Vol. 17, No. 6, Nov.–Dec. 1980, pp. 489–494.

14–10. "Solid Rocket Motor Igniters," NASA SP-8051, March 1971 (N71-30346).

14–11. I. H. Cho and S. W. Baek, "Numerical Simulation of Axisymmetric Solid Rocket Motor Ignition with Radiation Effect," *Journal of Propulsion and Power*, Vol. 16, No. 4, Jul.–Aug. 2000, pp. 725–728.

14–12. J. Yin and B. Zhang, "Experimental Study of Liquid Quenching of Solid Rocket Motors," AIAA Paper 90-2091.

14–13. B. N. Raghunandam and P. Bhaskariah, "Some New Results of Chuffing in Composite Solid Propellant Rockets," *Journal of Spacecraft and Rockets*, Vol. 22, No. 2, Mar.–Apr. 1985, pp. 218–220.

14–14. P. M. J. Hughes and E. Cerny, "Measurement and Analysis of High Frequency Pressure Oscillations in Solid Rocket Motors," *Journal of Spacecraft and Rockets*, Vol. 21, No. 3, May–Jun. 1984, pp. 261–265.

14–15. R. A. Beddini and T. A. Roberts, "Response of Propellant Combustion to a Turbulent Acoustic Boundary Layer," *Journal of Propulsion and Power*, Vol. 8, No. 2, Mar.–Apr. 1992, pp. 290–296.

14–16. F. Vuillot and G. Avalon, "Acoustic Boundary Layers in Solid Propellant Rocket Motors Using Navier-Stokes Equations," *Journal of Propulsion and Power*, Vol. 7, No. 2, Mar.–Apr. 1991, pp. 231–239.

14–17. L. De Luca, E. W. Price, and M. Summerfield (Eds.), *Nonsteady Burning and Combustion Stability of Solid Propellants*, Vol. 143, *Progress in Astronautics and Aeronautics*, AIAA, Washington DC, 1992.

14–18. V. Yang, T. B. Brill, and W.-Z. Ren (Eds.), *Solid Propellant Chemistry, Combustion, and Motor Interior Ballistics*, Vol. 185, *Progress in Astronautics and Aeronautics*, AIAA, Reston VA, 2000.

14–19. R. L. Coates, "Application of the T-Burner to Ballistic Evaluation of New Propellants," *Journal of Spacecraft and Rockets*, Vol. 3, No. 12, Dec. 1966, pp. 1793–1796; M. Barrere, Chapter 2, "Introduction to Nonsteady Burning and Combustion Stability," pp. 17–58, and L. D. Strand and R. S. Brown, Chapter 17, "Laboratory Test Methods for Combustion Stability Properties of Solid Propellants," pp. 689–718, in L. De Luca, E. W. Price, and M. Summerfield (Eds.), *Nonsteady Burning and Combustion Stability of Solid Propellants*, Vol. 143, *Progress in Astronautics and Aeronautics*, AIAA, Washington DC, 1992.

14–20. F. E. Culick, A Review of Calculations for Unsteady Burning of a Solid Propellant, *AIAA Journal*, Vol. 6, No. 12, Dec. 1968, pp. 2241–2255.

14–21. E. D. Youngborg, J. E. Pruitt, M. J. Smith, and D. W. Netzer, "Light-Diffraction Particle Size Measurements in Small Solid Propellant Rockets," *Journal of Propulsion and Power*, Vol. 6, No. 3, May–Jun. 1990, pp. 243–249.

14–22. F. Vuillot, "Vortex Shedding Phenomena in Solid Rocket Motors," *Journal of Propulsion and Power*, Vol. 11, No. 4, 1995.

14–23. A. Kourta, "Computation of Vortex Shedding in Solid Rocket Motors using a Time-Dependent Turbulence Model," *Journal of Propulsion and Power*, Vol. 15, No. 3, May–Jun. 1999.

14–24. L. Glick and L. H. Caveny, "Comment on Method of Reducing Stagnation Pressure Losses in Segmented Solid Rocket Motors," *Journal of Propulsion and Power,* Vol. 10, No. 2, Mar.–Apr. 1994, pp. 295–296. (Technical comment on W. A. Johnston, *JPP*, Vol. 8, No. 3, May–Jun 1992, pp. 720–721).

CHAPTER 15

SOLID ROCKET MOTOR COMPONENTS AND DESIGN

This is the last of four chapters on solid propellant rocket motors. Here, we describe key components such as the motor case, nozzle, and igniter; subsequently we elaborate on the design of these motors. Although thrust vector control mechanisms are also components of many rocket motors, they are described separately in Chapter 18.

15.1. ROCKET MOTOR CASE

Solid propellant rocket motor cases enclose the propellant grain and also serve as highly loaded pressure vessels, often as part of the flight vehicle structure. Case design and fabrication technology has progressed to where efficient and reliable motor cases can now be produced consistently for many solid rocket applications. Problems arise, however, when the established technology is used improperly; incorrect design analysis or understating of requirements may lead to improper materials and process controls, and to omitting necessary nondestructive tests at critical points in the fabrication process. Besides constituting the structural body of the rocket motor including its nozzle and propellant grain, the case frequently serves also as the primary structure of the missile or launch vehicle. Thus, optimizing case designs frequently entail trade-offs between case design parameters and vehicle design parameters. Often, case design is further influenced by assembly and fabrication requirements.

Table 15–1 lists many types of case loads with their sources; only some of these must be considered at the beginning of case design and they depend on rocket motor application. Additionally, environmental conditions peculiar to each specific motor usage must be carefully considered. Typically, the scrutinized conditions include the following: (1) temperature changes leading to thermal stresses and strains

TABLE 15–1. Rocket Motor Case Loads

Origin of Load	Type of Load/Stress
Internal pressure	Tension biaxial, vibration
Axial thrust	Axial, vibration
Rocket motor nozzle	Axial, bending, shear, thermal
Thrust vector control actuators	Axial, bending, shear
Thrust termination equipment	Biaxial, bending
Aerodynamic control surfaces or wings mounted to case	Tension, compression, bending, shear, torsion
Staging	Bending, shear
Flight maneuvering	Axial, bending, shear, torsion
Vehicle mass and wind forces on launch pad	Axial, bending, shear
Dynamic loads from vehicle oscillations	Axial, bending, shear
Start pressure surge	Biaxial tension
Ground handling, including lifting	Tension, compression, bending, shear, torsion
Ground transport	Tension, compression, shear, vibration
Earthquakes (large motors)	Axial, bending, shear

(internal heating, external or aerodynamic heating, and ambient temperature cycling during storage); (2) stress corrosion (moisture/chemical, galvanic, and hydrogen embrittlement); and (3) space environment stressors (vacuum or radiation).

Three classes of materials have traditionally been used: high-strength metals (steels, alloys of aluminum or titanium), wound-filament reinforced plastics, and a combination of these where metal cases have externally wound filaments for extra strength. Table 15–2 compares several typical materials. For filament-reinforced materials it gives data not only for the composite material, but also for several strong filaments and typical binders. Since the strength-to-density ratio is higher for composite materials, they have less inert mass but there are some important disadvantages; filament-wound cases with a plastic binder are usually superior on a vehicle performance basis. Metal cases combined with external filament-wound reinforcements and spiral-wound metal ribbons glued together with plastics have also been successful.

The shape of motor cases is usually determined from their grain configuration and/or from geometric vehicle constraints on length or diameter. Case configurations range from long, thin cylinders (length to diameter ratio, L/D, of 10) to spherical or near-spherical geometries (see Figs. 1-5, 12-1 to 12-4, and 12-17). Spherical shapes provide the lowest case mass per unit of enclosed volume. Because the case is often a key structural element of the vehicle, it has to allow for the mounting of other components. Propellant mass fractions of motors are usually strongly influenced by the mass of the case and they typically range from 0.70 to 0.94. The higher values apply to upper-stage motors. For small-diameter motors the mass fraction is lower, because of practical wall thickness constraints and because the ratio of the wall surface area (which varies roughly as the square of the diameter) to chamber volume (which varies

TABLE 15–2. Physical Properties of Selected Solid Propellant Motor Materials at 20 °C

Material	Tensile Strength, N/mm² (10³ psi)	Modulus of Elasticity, N/mm² (10⁶ psi)	Density, g/cm³ (lbm/in.³)	Strength to Density Ratio (1000)
Filaments				
Glass	1930–3100	72,000	2.5	1040
	(280–450)	(10)	(0.090)	
Aramid (Kevlar 49)	3050–3760	124,000	1.44	
	(370–540)	(18.0)	(0.052)	2300
Carbon fiber	3500–6900	230,000–300,000	1.53–1.80	2800
	(500–1000)	(33–43)	(0.055–0.065)	
Binder (by itself)				
Epoxy	83	2800	1.19	70
	(12)	(0.4)	(0.043)	
Filament-Reinforced Composite Material				
Glass	1030	35,000	1.94	500
	(150–170)	(4.6–5.0)	(0.070)	
Kevlar 49	1310	58,000	1.38	950
	(190)	(8)	(0.050)	
Graphite IM	2300	102,000	1.55	1400
	(250–340)	(14.8)	(0.056)	
Metals				
Titanium alloy	1240	110,000	4.60	270
	(155–160)	(16)	(0.166)	
Alloy steel (heat treated)	1400–2000	207,000	7.84	205
	(200–280)	(30)	(0.289)	
Aluminum alloy 2024 (heat treated)	455	72,000	2.79	165
	(66)	(10)	(0.101)	

Source: Data adapted in part from Chapter 4 by Evans and Chapter 7 by Scippa Ref. 12–1.

roughly as the cube of diameter) is less favorable in small sizes. Minimum case thicknesses always need to be higher than would be determined from simple stress analyses because they have to include two layers of filament strands for fiber composites, and because certain minimum metal thickness' are dictated by manufacturing and handling considerations.

Simplified membrane theory can be used to predict approximate stresses in solid propellant rocket chamber cases; it assumes no bending of the case walls and that all the loads are taken in tension. For a cylinder of radius R and thickness d, with chamber pressure p, the longitudinal stress σ_l is one-half of the tangential or hoop stress σ_θ:

$$\sigma_\theta = 2\sigma_l = pR/d \qquad (15\text{--}1)$$

For a cylindrical case with hemispherical ends, the cylinder walls have to be twice as thick as the walls of the end closures.

Combined stresses should never exceed the working stresses of the wall material. As a rocket motor begins to operate, an increasing internal pressure p causes case growth in both the longitudinal and circumferential directions, and these deformations must be considered in designing supports for the rocket motor and/or propellant grain. Let E be Young's modulus of elasticity, v be Poisson's ratio (0.3 for steel), and d be the wall thickness; then, the growth in length L and in diameter D due to pressure can be expressed as

$$\Delta L = \frac{pLD}{4Ed}(1 - 2v) = \frac{\sigma_l L}{E}(1 - 2v) \qquad (15\text{--}2)$$

$$\Delta D = \frac{pD^2}{4Ed}\left(1 - \frac{v}{2}\right) = \frac{\sigma_\theta D}{2E}\left(1 - \frac{v}{2}\right) \qquad (15\text{--}3)$$

Details can be found in regular texts on thin shells and membranes. For a hemispherical chamber end, the stress in each of two directions at right angles to each other is equal to the longitudinal stress of a cylinder of identical radius. For ellipsoidal end-chamber closures, the local stress varies with position along the surface and maximum stresses are larger than that in hemispheres. Radial displacements of a cylinder end are not the same as those for hemispherical or ellipsoidal closures when computed by thin-shell theory. Thus existing discontinuities may result in some shearing and bending stresses. Similarly, the presence of an igniter boss attachment, or a pressure gauge, or a nozzle can superimpose bending and shear stresses on the simple tension stresses in the case. In these locations it is necessary to locally reinforce or thicken the case wall.

During rocket motor operation, heat transfer from the hot gases causes a continuous rise of temperature at the case walls. This heating is reduced at locations where there is internal thermal insulation. In locations where the propellant in bonded directly to the case walls (or to a thin liner layer, which in turn is bonded to the case walls), there can also be essentially little or no heat transfer to these walls (except for a modest amount just before burnout). When the rocket motor case is also the outer skin in a flying vehicle, for certain speeds and altitudes hot boundary layers develop that produce external heating at the case walls. In order to reduce this external heating, an outer layer of oxidation resistant insulation has successfully been applied for certain atmospheric flight applications.

Heat transfer in solid propellant rocket motors seldom reaches a steady state and local temperatures for all nozzle and case components continuously increase during motor operation. Analyses that result in temperature-time histories have been useful for determining thicknesses and other thermal insulation component parameters and are also useful for estimating thermal stresses. Finite-element computer stress analyses are used by motor design companies to determine case design configurations with desirable stress values. Such computations must be done simultaneously with corresponding stress analyses on the grain (since it imposes loads on the case), and with any other coupled thermal analysis that determines thermal stresses and deformations.

Case designs need to provide a means for attaching the case to the vehicle, as well as for attaching nozzles (rarely more than one nozzle) and igniters, together with provisions for loading the grain. Sometimes there are also attached aerodynamic surfaces (fins), sensing instruments, a raceway (external conduit for electrical wires), handling hooks, and thrust vector control actuators with their power supply. For upper stages of ballistic missiles the case structure can also include blow-out ports or thrust termination devices, as described in Chapter 14. Typical methods for attaching these items include multiple pins (tapered or straight), snap rings, and/or bolts. Gaskets and/or O-ring seals are needed to prevent gas leaks.

Metal Cases

Metal cases have several advantages compared to filament-reinforced plastic cases: they are more rugged and will take the considerable rough handling required in many tactical missile applications, are usually ductile and yield before failure, and may be heated to relatively high temperatures (700 to 1000 °C or 1292 to 1832 °F, and higher with refractory metals) thus requiring less insulation. They will not deteriorate significantly with time or weather exposure and are easily adapted to take concentrated loads, if made thicker at a flange or boss. Since a metal case has much higher density and less insulation, it occupies less volume than a fiber-reinforced plastic case; therefore, for the same external envelope, it may contain somewhat more propellant.

Figure 15–1 shows various sections of a typical large solid rocket motor case made of welded steel pieces. The shape of the case, particularly the length-to-diameter ratio

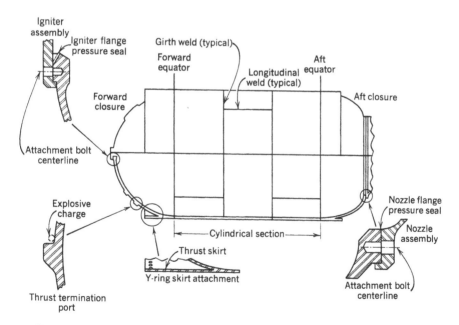

FIGURE 15–1. Typical large solid rocket motor case made of welded alloy steel.

in cylindrical cases, influences not only the stresses withstood by the case but the quantity of case material required to enclose a given amount of propellant. For very large and long motors both the propellant grain and the motor case have been made in sections; these *case segments* are then mechanically attached and sealed to each other at the launch site. The segmented solid rocket booster used in the Space Shuttle is shown in Fig. 15–2 and discussed in Ref. 15–1. Although this rocket motor is now out of production, it represents a good example of a segmented design; other similarly segmented rocket motors are still in use today. For critical seals between segments, a multiple-O-ring type of joint is often used as shown in Fig. 15–3 and discussed in Ref. 15–2. Segments are needed when an unsegmented rocket motor would be too large and too heavy to be transported over ordinary roads (i.e., cannot make road turns) or railways (will not go through some tunnels or under some bridges) and/or are too difficult to fabricate.

Small metal *cases for tactical missile rocket motors* can be extruded or forged (and subsequently machined), or made in three pieces as shown in Fig. 12–4. Such motor case is designed for loading a freestanding grain where the case, nozzle, and blast tube are sealed by O-rings (see Chapter 6 of Ref. 15–4 and Chapter 7 of Ref. 15–5). Since mission velocities for most tactical missiles are relatively low (100 to 1500 m/sec), their propellant mass fractions are also relatively low (0.5 to 0.8) and the percentage of inert motor mass is high. Safety factors for tactical missile cases are often higher to allow for rough handling and cumulative damage. The emphasis in selecting motor cases (and other hardware components) for tactical missiles is therefore not on highest performance (lowest inert motor mass), but on reliability, long life, low cost, safety, ruggedness, and/or survivability.

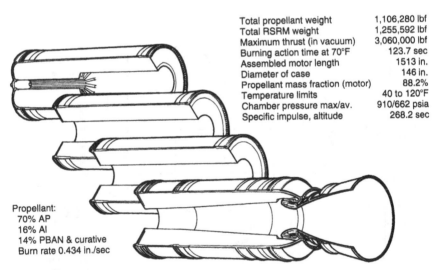

Total propellant weight	1,106,280 lbf
Total RSRM weight	1,255,592 lbf
Maximum thrust (in vacuum)	3,060,000 lbf
Burning action time at 70°F	123.7 sec
Assembled motor length	1513 in.
Diameter of case	146 in.
Propellant mass fraction (motor)	88.2%
Temperature limits	40 to 120°F
Chamber pressure max/av.	910/662 psia
Specific impulse, altitude	268.2 sec

Propellant:
70% AP
16% Al
14% PBAN & curative
Burn rate 0.434 in./sec

FIGURE 15–2. Simplified diagram of the four segments of the historic Space Shuttle solid rocket booster motor. Details of the thrust vector actuating mechanism or the ignition system are not shown. Courtesy of NASA and Orbital ATK

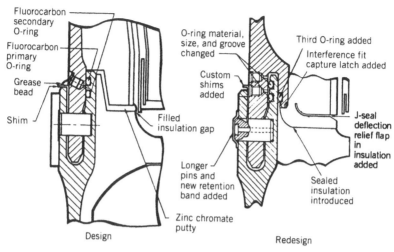

FIGURE 15–3. The joints between segments of the Shuttle Solid Rocket Booster (SRB) were redesigned after its dramatic failure. The improvements were not only in a third O-ring, the mechanical joint, and its locking mechanism, but also featured a redesign of the insulation between propellant segments. See Ref. 15–3. Courtesy of NASA

High-strength alloy steels have been the most common case metals, but others, like alloys of aluminum, titanium, and nickel, have also been used. Table 15–2 gives a comparison of rocket motor case material properties. Extensive knowledge exists today for designing and fabricating motor cases with low-alloy steels with strength levels of over 240,000 psi.

Maraging steels have strengths up to about 300,000 psi in combination with high fracture toughness. The term *maraging* is derived from the fact that these alloys exist as relative soft low-carbon martensites in their annealed condition and attain higher strengths from aging at relatively low temperatures.

Stress-corrosion cracking in certain metals presents unique problems that may result in spontaneous failure without any visual evidence of an impending catastrophe. Emphasis given to lightweight thin metal cases aggravates stress corrosion and crack propagation that often starts from a flaw in the metal, with failures occurring at a stress levels below the documented yield strength of the original metal.

Wound-Filament-Reinforced Plastic Cases

Filament-reinforced cases use continuous filaments of strong fibers wound in precise patterns and bonded together with a plastic. Their principal advantage is lower weight. Most plastics soften when they are heated above about 180 °C or 355 °F; they need inserts or reinforcements to allow fastening or assembly of other components and to accept concentrated loads. Thermal expansion of reinforced plastics is often higher than that of metal and the thermal conductivity is much lower, causing higher temperature gradients. References 15-4 and 15-5 explain the design and winding of these composite cases, and Ref. 15–6 discusses their damage tolerance.

Typical fiber materials are, ordered by increasing strength, glass, aramids (Kevlar) and carbon, as listed in Table 15–2. Typically, for the same load the inert mass of a case made of carbon fiber is about 50% of one made with glass fibers and around 67% that of a case mass made with Kevlar fibers.

Individual fibers are strong only in tension (2400 to 6800 MPa or 350,000 to 1,000,000 psi). Fibers are held in place by plastic binders of relatively low density; this prevents fibers from slipping and thus weakening in shear or bending. In a filament-wound composite case (with existing tension, hoop, and bending stresses) filaments are not always oriented along the direction of maximum stress and the materials may include low-strength plastics; therefore, the strength of composites may be a factor of 3 to 5 less than the strength of the filament itself. The traditional plastic binder is a thermosetting epoxy material, which limits maximum temperatures to between 100 and 180 °C or about 212 and 355 °F. Although resins with higher temperature limits are available (295 °C or 563 °F), their fiber adhesion has not been sufficiently strong. Typical safety factors used (in deterministic structural analysis) are for failure to occur at 1.4 to 1.6 times the maximum operating stresses, and proof testing is done up to 1.15 to 1.25 times operating pressures.

A typical case design is shown schematically in Fig. 15–4. The forward and aft ends and the cylindrical portion are wound on a *preform* or mold which already contains the forward and aft rings. The direction in which the bands are laid onto the mold and the tension that is applied to the bands is critical in obtaining proper case designs. Curing is done in an oven and may be done under pressure to assure high density and minimum voids in the composite material. The *preform* is then removed.

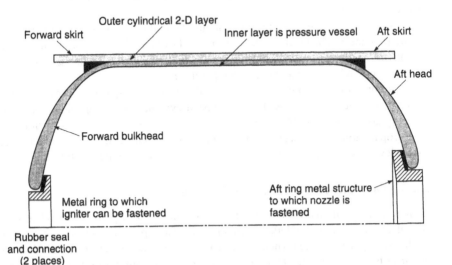

FIGURE 15–4. Simplified half-section of a typical design of a filament-wound composite material case. Elastomeric adhesive seals are shown in black. The outer layer reinforces the cylinder portion and provides attachment skirts. The thickness of the inner case increases at smaller diameter.

One way is to use sand with a water-soluble binder for the preform; after curing the case, the preform is washed out with water. Since filament-wound case walls can be porous, they must be sealed. The liner between the case and the grain can be the only seal that prevents hot gases from seeping through the case walls. Scratches, dents, and moisture absorption degrade case strengths.

In some designs insulator material is placed on the preform before winding and the case is cured simultaneously with the insulator, as described in Ref. 15–7. In another design the propellant grain with its forward and aft closures is used as the preform. A liner is applied to this grain, then an insulator, and the high-strength fibers of the case are wound in layers directly over the insulated live propellant. Here, curing has to be done at relatively low temperatures so that the propellant will not be adversely affected. This process works well with extruded cylindrical grains. There are also cylindrical cases made of steel with an overwrap layer of filament-wound composite material, as described in Ref. 15–8.

Allowable stresses are usually determined from tensile tests with roving or band clusters and rupture tests on subscale composite cases made by an essentially identical filament winding process. Some companies further reduce the allowable strength value to account for degradation due to moisture, manufacturing imperfections, or nonuniform densities.

In a wound motor case the filaments must be oriented in the direction of the principal stress and must be proportioned in number to the magnitude of that stress. Compromise occurs around parts needed for nozzles, igniters, and so on, where orientations are kept as close to the ideal as practicable. Filaments are customarily clustered in *yarns, rovings*, or *bands*, as indicated in Fig. 15–5. By using two or more winding angles (i.e., helicals and circumferentials) and by calculating the proportion of filaments in each direction, a balanced stress structure may be achieved. The ideal balance is for each fiber in each direction to carry an equal tension load. Realistically, filaments supported by an epoxy resin must also absorb stress compressions, bending loads, cross-laminar shears, and interlaminar shears. Even though the latter stresses are small compared to the tensile ones, each must be analyzed since they can lead to case failure before any filament fails in tension. In a proper design, failure should occur only when filaments reach their ultimate tensile strength, rather than from stresses in other directions. Figure 18-5 shows the cross section, made of ablative materials, of a Kevlar filament motor case and detail of its flexible nozzle.

15.2. NOZZLES[*]

Supersonic nozzles provide for the efficient expansion of hot chamber gases and have to withstand the severe environment of high heat transfer and erosion. Advances in material technology presently allow substantial mass reductions and performance improvements. Nozzles range in size from 0.05 in. throat diameter to about 90 in.

---

[*]In the Seventh edition of this book this nozzle section was revised and partly rewritten by Terry A. Boardman of a predecessor organization to Orbital ATK.

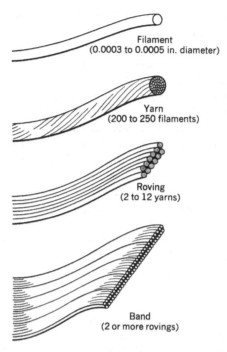

FIGURE 15–5. Filament winding terminology (each sketch is drawn to a different scale).

(to date only experimental) , with operating durations of a fraction of a second (small thrusts) to several minutes (large thrusts) (see Chapters 2 and 3 of Ref. 15–4 and Chapter 6 in Ref. 15–5).

Classification

Nozzles for solid propellant rocket motors can be classified into five categories as listed below and as shown in Fig. 15–6.

1. *Fixed nozzle.* Simple and frequently used in tactical weapon propulsion systems for short-range air-, ground-, and sea-launched missiles, also as strap-on propulsion for space launch vehicles such as Atlas and Delta, and in spacecraft rocket motors for orbital transfer. Typical throat diameters are between 0.25 and 5 in. for most tactical missile nozzles, but some are larger.

2. *Movable nozzle.* Provides thrust vector control to a flight vehicle. As explained in Chapter 18, one movable nozzle can provide pitch and yaw control and two are needed for roll control. Movable nozzles are typically submerged and use a flexible seal or a ball-and-socket joint with two actuators 90° apart to achieve omniaxial motion. Movable nozzles are primarily used in long-range strategic propulsion ground- and sea-launched systems (typical throat diameters are 7 to 15 in. for the first stage and 4 to 5 in. for the third stage) and in large space

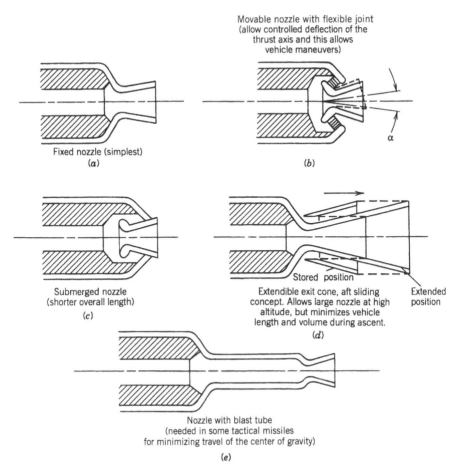

FIGURE 15–6. Simplified diagrams of five common nozzle configurations. Here all are shown with a simple conical nozzle, but actual nozzle exits may curve.

launch boosters. Titan's boost rocket motor and Ariane's V solid rocket booster have throat diameters in the 30- to about 90-in. range.

3. *Submerged nozzles.* A significant portion of the nozzle structure is submerged within the combustion chamber or case, as shown in Figs. 15–2 and 15–6 *b, c.* Submerging the nozzle reduces the overall motor length somewhat, which in turn reduces the vehicle length and its inert mass. It is important for length-limited applications such as silo- and submarine-launched strategic missiles as well as their upper stages, and for space propulsion systems. Within a fixed motor length, it also allows an increase of propellant mass inside the case. The *splitline* between fixed and movable sections is more favorably located than with external subsonic splitline movable nozzles (see Ref. 15–9). Reference 15–10 describes the sloshing of trapped molten aluminum oxide that can accumulate in the groove around a submerged nozzle; this accumulation is undesirable, but can be minimized by proper design.

4. *Extendible nozzle.* Commonly referred to as an extendible exit cone (EEC), although it is not always exactly conical. It is used on strategic missile propulsion upper-stage systems and upper stages on space launch vehicles to maximize motor-delivered specific impulse. As shown in Fig. 12-3, this nozzle has a fixed low-area-ratio section, which is enlarged to a higher area ratio by adding one or more nozzle cone extension pieces. The extended nozzle improves specific impulse by doubling or tripling the initial expansion ratio, thereby significantly increasing the nozzle thrust coefficient. This system thus allows very high expansion ratio nozzles to be packaged in a relatively short length, thereby reducing vehicle inert mass. The nozzle cone extensions reside in their retracted position during the boost phase of the flight and are moved into place before their rocket motor is started but after separation from the lower stage. Typically, electromechanical ball screw actuators deploy exit cone extensions.

5. *Blast-tube-mounted nozzle.* These are used with tactical air- and ground-launched missiles. The blast tube allows the rocket motor's center of gravity (CG) to be close to or ahead of the vehicle CG. This limits the CG travel during motor burn and makes flight stabilization much easier. See Fig. 12-4.

As stated in Chapter 12 nozzle throat areas often enlarge slightly during hot firing. An enlargement of more than 10% is usually considered unacceptable. With slight enlargements of the throat area, both the chamber pressure and the specific impulse will slightly decrease. When single value for specific impulse applicable during their entire operation is used in rocket motor specifications it is named the *effective specific impulse*.

Design and Construction

Almost all solid rocket motor nozzles are cooled by heat absorption in ablative materials. In general, construction of a rocket nozzle features steel or aluminum shells (or housings) designed to carry structural loads (where motor operating pressures and nozzle TVC actuator loads are usually the biggest), and composite ablative insulators (with or without liners) which are bonded to the cases or to the housings. These ablative liners are designed to insulate metal housings, provide the internal aerodynamic contour necessary for efficient combustion gas expansion to generate thrust, and ablate and char in a controlled and predictable manner in order to prevent heat buildups that could substantially weaken the structural housings or the bonding materials. Solid rocket motor nozzles are designed to ensure that the thickness of ablative liners is sufficient to maintain the liner-to-housing adhesive bond line below temperatures that would degrade the adhesive structural properties during rocket motor operation. Nozzle designs are detailed in Figs. 1–5, 12–1 to 12–4, and 15–7.

The construction of nozzles ranges from simple, single-piece nonmovable graphite nozzles to complex multipiece nozzles capable of controlling the direction of the thrust vector. The simpler, smaller nozzles are typically of applications with low chamber pressure, short durations (perhaps less than 10 sec), low area ratios, and/or low thrust. Two typical small, simple built-up nozzles are shown in Fig. 15–7. Complex

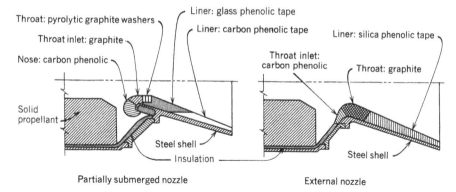

FIGURE 15–7. Half-section sketches for two nozzle designs for small solid propellant rocket motors that employ ablative heat sink wall pieces and graphite throat inserts resistant to high temperatures, erosion, and oxidation. The pyrolytic washers or disks are so oriented that their high conductivity direction is perpendicular to the nozzle axis.

nozzles are necessary to meet more difficult design requirements such as providing thrust vector control, operating at higher chamber pressures (and thus at higher heat transfer rates) and/or higher altitudes (large nozzle expansion ratios), producing very high thrust levels, and surviving longer motor burn durations (above 30 sec).

Figures 15–8 and 15–9 illustrate design features of a large and relatively complex solid rocket motor with thrust vector control. It was used as the Reusable Solid Motor rocket (RSRM) booster of the Space Shuttle Launch Vehicle, which was retired in

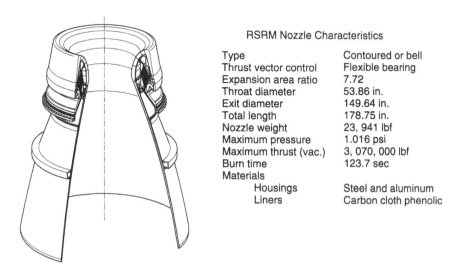

RSRM Nozzle Characteristics

Type	Contoured or bell
Thrust vector control	Flexible bearing
Expansion area ratio	7.72
Throat diameter	53.86 in.
Exit diameter	149.64 in.
Total length	178.75 in.
Nozzle weight	23, 941 lbf
Maximum pressure	1.016 psi
Maximum thrust (vac.)	3, 070, 000 lbf
Burn time	123.7 sec
Materials	
Housings	Steel and aluminum
Liners	Carbon cloth phenolic

FIGURE 15–8. External quarter section view of nozzle configuration of the historic Space Shuttle reusable solid rocket motor (RSRM). Courtesy of Orbital ATK

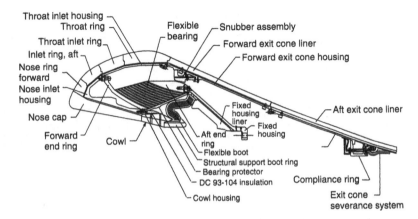

FIGURE 15–9. Section through movable nozzle shown in Fig. 15–8 with component identification. Courtesy of Orbital ATK

2011. It is typical of other large flexible nozzles. The two RSRMs provided 71.4% of the lift-off thrust of the Space Shuttle launch vehicle shown in Figs. 1-14 and 15-2. The nozzle was designed to provide structural and thermal safety margins during the Shuttle booster's 2-min burn time and consisted of nine carbon-cloth phenolic ablative liners bonded to six steel and aluminum housings. The housings were bolted together to form the structure of the nozzle. A flexible bearing (described further in Chapter 18), made of thin rubber sheets vulcanized to spherically shaped steel shims, enabled the nozzle to vector omni-axially up to 8 degrees from centerline to provide thrust vector control. Since these metal housings were recovered and reused after each flight, an exit cone severance system (a circumferential linear shaped charge) was used to cut off a major section of the aft exit cone just below the aft exit cone aluminum housing to minimize splashdown loading on the remaining components. NASA's new first-stage boosters (segmented solid propellant rocket motors) are being patterned after the Space Shuttle's SRM design.

From a performance perspective, any nozzle should efficiently expand the gases originating from the motor's combustion chamber to produce thrust. Simple nozzles with noncontoured conical exit cones can be designed using the basic thermodynamic relationships presented in Chapter 3 to establish the throat area, nozzle half angle, and expansion ratio. More complex contoured (bell-shaped) nozzles are used to reduce divergence losses, improve the specific impulse, and reduce nozzle length and mass. Section 3.4 gives data on designing bell-shaped nozzles with optimum wall contours (that avoid shock waves and minimize impact of particulates in the exhaust gas).

Two-dimensional, two-phase, reacting-gas method-of-characteristics flow codes are used to analyze gas-particle flows in nozzles for the determination of optimal nozzle contours and of shortened nozzle lengths that maximize specific impulse while yielding acceptable erosion characteristics. Such codes provide results for all identified specific impulse loss mechanisms (in units of secs), which produce a less than ideal performance. An example is given in Table 15–3.

TABLE 15–3. Calculated Losses in the Space Shuttle Booster RSRM Nozzle

Theoretical specific impulse (vacuum conditions)	278.1 sec
Delivered average specific impulse (vacuum conditions)	268.2 sec
Losses (*calculated*):	(9.9 sec total)
Two-dimensional two-phase flow (includes divergence angle loss)	7.4 sec
Throat erosion (reduces nozzle area ratio)	0.9 sec
Boundary layer (wall friction)	0.7 sec
Submergence (flow turning)	0.7 sec
Finite rate chemistry (chemical equilibrium)	0.2 sec
Impingement (of Al_2O_3 particles on nozzle wall)	0.0 sec
Shock (if turnback angle is too high or nozzle length too low)	0.0 sec
Combustion efficiency (incomplete burning)	0.0 sec

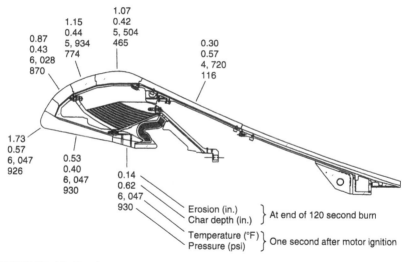

FIGURE 15–10. Erosion measurements and char depth data of the carbon fiber phenolic material of the nozzle of the Space Shuttle Reusable Solid Rocket Motor. Courtesy of Orbital ATK

Figure 15–10 illustrates amounts of carbon cloth phenolic liner removed by chemical reactions and by particle impingement erosion, the liner char depth, and gas temperature and pressure at selected locations in the RSRM nozzle. Erosion on the nose cap (1.73 in.) is high primarily as a result of impingement by Al_2O_3 particles traveling down the motor bore. The mechanical impact of these particles removes charred liner material. In contrast, the radial throat erosion of 1.07 in. results primarily from the carbon liner material reacting chemically with oxidizing species in the combustion gas flow at the region of greatest heat transfer. At the throat location, impingement erosion is low because any Al_2O_3 particles present are traveling parallel to the nozzle surface.

The often used acronym ITE means *integral throat/entrance* and refers to a single-piece nozzle throat insert that also includes a part of the converging entry section. ITE nozzle inserts can be seen in Figs. 1-5, 12-1, 12-3, and 12-4. Most ITEs are made from carbon-carbon (either 3 or 4 directionally reinforced). The introduction of ITEs has improved reliability and simplified designs (see Ref. 15–11).

As already stated, *nozzle throat erosion* causes throat diameters to enlarge during operation and is one problem encountered in nozzle design. Usually, a throat area increase larger than about 10% is considered unacceptable for most applications, since it causes a significant reduction in thrust and chamber pressure. Erosion occurs not only at the throat region (typically, at 0.01 to 0.25 mm/sec or 0.004 to 0.010 in./sec), but also at the sections immediately upstream and downstream of it, as shown in Fig. 15–10. Nozzle assemblies typically lose 3 to 12% of their initial inert mass during operation. Erosion is caused by complex interactions between the high-temperature, high-velocity gas flow, the chemically aggressive species in the gas, and the mechanical abrasion by small particles. Carbon in nozzle materials reacts with hot species like O_2, O, OH, or H_2O becoming oxidized; the cumulative concentration of these species is an indication of the likely erosion. Tables 5-6 and 5-7 give chemical concentrations for the exhaust species from aluminized propellants. Fuel-rich propellants (which contain little free O_2 or O) and propellants where some of the gaseous oxygen is removed by aluminum oxidation show fewer erosive tendencies. Uneven erosion in a nozzle causes thrust misalignment. Some tactical designs use a particular grain design that burns very progressively but counter this with a high erosion-rate material to control the maximum system pressure.

For bell-shaped nozzles finding optimum nozzle wall contours requires analysis (computer codes using the method of characteristics are mentioned in Section 3.4) to determine wall contours, which effectively turn the gas to near axial flow without introducing strong shock waves. Results given in Section 3.4 are for gaseous exhaust jets (without entrained solid particles), as may occur with liquids or double-base solid propellants. Figure 3-13 illustrates key parameters, which describe such bell shaped nozzles. The initial angle θ_i (angle through which supersonic flow is turned immediately downstream of the nozzle throat) is typically 30 to 50° and the exit angle θ_e is 10 to 15°. Such bell-shaped contour allows a 1 to 2% increase in specific impulse, when compared with an equivalent conical nozzle exit having a 15° half angle. For some applications, designers have used a parabolic curve or a circular arc without performance penalties.

With solid particles in the exhaust gas bell-shaped nozzles (designed for a purely gaseous exhaust) quickly erode, particularly at the nozzle exit region where a diverging flow is turned back to a near axial direction. With the insulation eroded, the nozzle exit metal housing quickly burns out. All solid composite propellants have some amount of small solids (mostly Al_2O_3) in their exhaust. A solution for not losing excessive nozzle insulation materials is to go to a smaller value of the initial angle θ_i (20 to 32°) on a gentler bell-shaped contour. Compared to a 15° conical nozzle, the specific impulse gain of such contour is no more than 0.7%. In general, nozzle throat inlet contours are not critical and can be based on any hyperbolic curve that uniformly accelerates the combustion gas flow to sonic velocity at the throat. In the supersonic region the task is more complicated, see Section 3.4. for details.

Heat Absorption and Nozzle Materials

Since solid rocket motors do not reach thermal equilibrium, the temperature of all components exposed to heat flow increases continuously during operation. In proper thermal designs, critical locations only reach their maximum allowable temperature a short time after the rocket motor stops operation. Nozzle components rely on their heat-absorbing capacity (high specific heat and high energy required for material decomposition) and slow heat transfer (good insulation with low thermal conductivity) to withstand the stresses and strains imposed by thermal gradients and loads. The maximum allowable temperature for motor materials is taken as just below the temperature at which excessive degradation occurs (e.g., the material loses strength, melts, becomes too soft, cracks, pyrolyses, unglues, or oxidizes too rapidly). Operating durations are thus affected by nozzle design and by the amount of heat-absorbing and insulating material present. Stated in a different way, one principal design objective here is to arrive at a nozzle with just sufficient heat-absorbing-material and insulation so that its structures and joints will do the job for the duration of the application under all likely operating conditions, while complying with mandated margins or factors of safety.

Selection and application of proper materials is the key to successful designs of solid rocket nozzles. Table 15–4 groups various typical nozzle materials according to their usage. High-temperature exhausts from solid rockets present an unusually severe environment for nozzle materials, especially when metalized propellants are employed.

About 80 years ago nozzles were manufactured from a single piece of *molded polycrystalline graphite* and some were supported by metal housing structures. They eroded easily, but were low in cost. We still use them today for short duration, low chamber pressure, low altitude flight applications with low thrust, such as in certain tactical missiles. For more severe conditions a throat insert or ITE is placed into the graphite piece; this insert is a denser, better grade of graphite; later *pyrolytic graphite* washers and fiber-reinforced carbon materials came into use. Figure 15–7 shows a set of non-isotropic pyrolytic graphite washer in the throat insert of small nozzles (they are not used now). For a period of time tungsten inserts were used; they had very good erosion resistance, but were heavy and their melting point was eventually exceeded as higher motor pressures and hotter propellants came into use. The introduction of high-strength carbon fibers in a carbon matrix has been a major advance in high-temperature materials. For small and medium-sized nozzles, ITE pieces have been made of *carbon–carbon*, the present abbreviation for carbon fibers in a carbon matrix (see Ref. 15–12). The orientation of the fibers can be two-directional (2D) or three-directional (3D), as described below. Some properties of these materials are listed in Tables 15–4 and 15–5. For large nozzles the then existing technology did not allow the fabrication of large 3D carbon–carbon ITE pieces, so layups of carbon fiber (or silicon fiber) cloth in a phenolic matrix were used. An example of a self-supporting ablative exit cone is the Naxeco-phenolic structure used in the first stage of the Vega launch vehicle (Ref. 15–13).

TABLE 15–4. Typical Motor Nozzle Materials and Their Functions

Function	Material	Remarks
Structure and pressure container (housing)	Aluminum	Limited to 515 °C (959 °F)
	Low-carbon steel, high-strength steels, and special alloys	Good between 625 and 1200 °C (1157 and 2192 °F), depending on material; rigid and strong
Heat sink and heat-resistant material at inlet and throat section; severe thermal environment and high-velocity gas with erosion	Carbon-carbon	Three- or four-directional interwoven filaments, strong, expensive, limited to 3300 °C (5972 °F)
	Carbon or Kevlar fiber cloth with phenolic or plastic resins	
	Molded graphite	Used with large throats
	Tungsten, molybdenum	For low chamber temperatures and low pressures only; low cost
		Heavy, expensive, subject to cracking; resist erosion. Rarely used in current designs
Insulator (behind heat sink or flame barrier); not exposed to flowing gas	Ablative plastics, with fillers of silica or Kevlar, phenolic resins	Want low conductivity, good adhesion, ruggedness, erosion resistance; can be filament wound or impregnated cloth layup with subsequent machining
Flame barrier (exposed to hot low-velocity gas)	Ablative plastics (same as insulators but with less filler and tough rubber matrix)	Lower cost than carbon–carbon; better erosion resistance than many insulators
	Carbon, or silica fibers with phenolic resin	Cloth or ribbon layups; woven and compressed, glued to housing
	Carbon–carbon	Higher temperature than others, three-directional weave or layup
Nozzle exit cone	Ablative plastic, self-supporting or with metal or structural plastic structure	Heavy, limited duration; cloth or woven ribbon layups, glued to housing
	Refractory metal (niobium alloy, Cb-103, rarely tantalum or molybdenum)	Section near nozzle exit radiation cooled. Strong, needs coating for oxidation resistance; can be thin but heavy
	Carbon–carbon, needs gas seal	Radiation cooled, higher allowable temperature than metals; two- or three-directional weave, strong, often porous

TABLE 15–5. Comparison of Properties of Molded and Pyrolytic Graphite, Carbon–Carbon, Carbon Cloth, and Silica Cloth Phenolic

	ATJ Modern Graphite	Pyrolytic Graphite	Three-Directional Carbon Fibers in a Carbon Matrix	Carbon Cloth Phenolic	Silica Cloth Phenolic
Density (lbm/in.^{3})	0.0556	0.079	0.062–0.072	0.053	0.062
Thermal expansion (in./in.-°F)	0.005–0.007	0.00144 (warp) 0.0432 (fill)	$1 - 9 \times 10^{-6}$	8.02×10^{-6}	7.6×10^{-6}
Thermal conductivity (Btu/in.-sec/°F) at room temperature	1.2×10^{-3} 1.5×10^{-3}	4.9×10^{-5} (warp)[a] 4.2×10^{-5} (fill)[a]	$2 \text{ to } 21 \times 10^{-5}$ (warp)[a] $8 \text{ to } 50 \times 10^{-5}$ (fill)[a]	2.2×10^{-3} (warp)[a]	1.11×10^{-3} (warp)[a]
Modulus of elasticity (psi) at room temperature	1.5×10^6 (warp)[a] 1.2×10^6 (fill)[a]	4.5×10^6 (warp)^2 1.5×10^6 (fill)[a]	$35 \text{ to } 80 \times 10^6$	2.86×10^{-6} (warp)[a] 2.91×10^{-6} (fill)[a]	3.17×10^{-6} (warp)[a] 2.86×10^{-6} (fill)[a]
Shear modulus (psi)	—	0.2×10^6 (warp)[a] 2.7×10^6 (fill)[a]	—	0.81×10^6	0.80×10^6
Erosion rate (typical)[b] (in./sec)	0.004–0.006	0.001–0.002	0.0015–0.012	0.005–0.010	0.010–0.020

[a]*Warp* is in direction of principal fibers. *Fill* is at right angles to warp.
[b]Surface wear rates at which a test article recedes under agreed upon operating conditions—used for comparison only.

Regions immediately downstream of the throat have less heat transfer, less erosion, and lower temperatures than the throat, and less expensive materials are usually satisfactory here. This includes various grades of graphite, or ablative materials, strong high-temperature fibers (carbon or silica) in a matrix of phenolic or epoxy resins, which are described later in this section. Figure 18-5 shows a movable nozzle with multilayer *insulators* behind the graphite nozzle pieces directly exposed to heat. These insulators (between the very hot throat piece and housing) limit the heat transfer and prevent excessive housing temperatures.

In the *diverging exit section* the heat transfer and temperatures are even lower and similar but less capable ordinary materials can be used there. This segment can be built integral with the nozzle throat (as it is in most small nozzles), or it can be a separate one- or two-piece subassembly, which is then fastened to the smaller diameter throat segment. Ablative materials without oriented fibers as in cloth or ribbons, but with short fibers or insulating ceramic particles, can be utilized. For large area ratios (upper stages and space transfer vehicles), the nozzle will often protrude beyond the vehicle's boattail surface. This allows for efficient radiation cooling since any exposed exit cone can radiate directly to space. High-temperature metals (niobium, titanium, stainless steel, or a thin carbon–carbon shell) have been used for radiation cooling in some upper-stage or spacecraft exit cone applications. Since such cooled nozzle sections can quickly reach thermal equilibrium, their duration can be unlimited.

The *housing* or *structural support of the nozzle* uses the same material as the metal case, such as steel or aluminum. Housings are never allowed to become very hot. Some of the simpler, smaller nozzles (with one, two, or three pieces of mostly graphite) do not have a separate housing structure, but use the ITE for the structure.

Estimates of nozzle internal temperatures and temperature distributions with time can be made using two-dimensional finite element software for *transient heat transfer analyses*. These are similar in principle to the transient heat transfer method described in Section 8.5 where some results are shown in Fig. 8-20. After firing, nozzles cool by conducting heat from hotter inner parts outer sections. Sometimes these outer pieces will end up exceeding their temperature limit and suffer damage. *Structural analyses* (of stresses and strains) of key nozzle components are highly interdependent with heat transfer analyses, which determine the component temperatures. Design must also allow for thermal growth and for differential expansion of adjacent parts.

Typical materials used for ITEs or nozzle throat inserts are listed in Table 15–5. These materials are exposed to the most severe heat transfer conditions and undergo high thermal stresses and temperatures. Their physical properties are often anisotropic, that is, vary with the orientation or direction of the crystal structure or of reinforcing fibers. *Polycrystalline graphites* are extruded or molded units. Different grades with different densities and capabilities are available. As already mentioned, they are used extensively for simple nozzles and for ITE parts. *Pyrolytic graphite* is strongly anisotropic and has excellent conductivity in one preferred direction. A nozzle using pyrolytic graphite is shown in Fig. 15–7. It is fabricated by depositing graphite crystals on a substrate in a furnace containing decomposing methane gas. While pyrolytic graphite use is declining, it is still installed in current rocket motors of older design.

Carbon–carbon materials are made from carefully oriented sets of *carbon fibers* (woven, knitted, threaded, needled, or laid up in patterns) in a *carbon matrix*. Two-directional (2D) materials have fibers in two directions, 3D have fibers oriented in three directions (usually at right angle to each other), and 4D has fibers oriented in one of two ways, often called "tetrahedral" and "hexagonal" (Ref.15-14). In the tetrahedral construction, the carbon bundles are the same diameter in all four directions represented by the four diagonals of a cube. In the hexagonal construction, three of the four directions of reinforcement are in a plane perpendicular to the axial direction. The three directions of yarn bundles in the perpendicular plane are at 60° to each other and are usually smaller in diameter. Highly densified materials are superior in high heat transfer regions, such as the throat. Multidirectional fiber reinforcements allow them to better withstand the high thermal stresses introduced by the steep temperature gradients within the component. There are two methods in use for matrix deposition: chemical vapor deposition (CVD) and liquid impregnation followed by further pyrolyzation. Carbon-carbon densification is discussed in Ref. 15–9.

The French (Herakles) Naxeco Sepcarb carbon-carbon and the Naxeco 3-D reinforced phenolic materials are presently in production and have flown in Vega launches. Carbon-silicon carbide and refractory metal nozzle components have reached production in the last decade.

Ablative Materials. These commonly used materials in rocket motors nozzles are also used in some insulators. They are usually made from a composite material with high-temperature organic or inorganic high-strength fibers, namely, high silica glass, aramids (Kevlar), or carbon fibers, impregnated with organic plastic materials such as phenolic or epoxy resins. The fibers may be individual strands or bands (applied in a geometric pattern on a winding machine), or come as a woven cloth or ribbon, all impregnated with resin.

Ablation may result from a combination of surface melting, sublimation, charring, decomposition in depth, and from film cooling. As shown in Fig. 15–11, progressive layers of ablative material undergo an endothermic degradation, that is, physical and chemical changes occur that absorb heat. While some ablative material evaporates

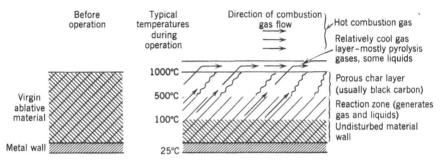

FIGURE 15–11. Zones in an ablative material during rocket operation with two sets of fibers (seen as multiple lines) at 45° to the flow.

(and some types also have a viscous liquid phase), enough charred and porous solid material should remain on and below the surface to preserve the basic geometry and surface integrity. Upon rocket start, ablative materials act as a thermal heat sink, but their poor conductivity causes their surface temperature to rise rapidly. At temperatures of 650 to 800 K some of resins start to decompose endothermically into a porous carbonaceous char and into pyrolysed gases. As the char depth increases, these gases undergo an endothermic cracking process as they percolate through the char in a counterflow direction to the heat flux. These gases then form an artificial fuel-rich, protective, relatively cool but flimsy, boundary layer over the char.

Since char is almost all carbon and can withstand 3500 K or 6000 R, porous char layers allow the original surface to be maintained (but with a rough surface texture) and provides geometric integrity. However, the char's carbon may be oxidized by certain combustion gas species in which case there is a slow regression of the charred surface. Char is structurally weak and can be damaged or abraded by direct impingement of solid particles from the flowing gas. Ablative material construction is used for some or all parts of the chambers and/or nozzles shown in Figs. 1–5, 6–9a, 12–1 to 12–4, and 15–10.

Ablative parts are formed either by high-pressure molding (~55 to 69 MPa or 8000 to 10,000 psi at 149 °C or 300 °F) or by tapewrapping on a shaped mandrel followed by either hydroclaving, typically at 1000 psi and about 300 °F, or by autoclaving at about 250 psi and 330 °F. *Tapewrapping* has been a common method of forming in very large nozzles. The wrapping procedure normally includes heating a shaped mandrel (~54 °C or 130 °F), heating the tape and resin (66 to 121°C or 150 to 250 °F), pressurizing the tape of fiber material and the injected resin in place while rolling (~35,000 N/m or 200 lbf/in. per inch of width), and maintaining the proper rolling speed, tape tension, wrap orientation, and resin flow rate. Experience has shown that the as-wrapped density is an important indicator of procedural acceptability, with the desired criterion being near 90% of the autoclaved density. Resin content usually ranges between 25 and 35%, depending on the fabric-reinforcing material and the particular resin and its filler material. Normally, the mechanical properties of the cured ablative material, and also the durability of the material during rocket operation, correlate closely with the cured material density. Within an optimal density range, a low density usually means poor bonding of the reinforcing layers, high porosity, low strength, and high erosion rates.

We note here that in liquid propellant rocket engines ablatives have been used effectively in very small thrust chambers (where there is insufficient regenerative cooling capacity), in pulsing, restartable spacecraft control rocket engines, in the exit region of large nozzle and in variable-thrust (throttled) rocket engines. Figure 6–9a shows an ablative nozzle extension for a large liquid propellant rocket engine.

Heat transfer properties of many available ablative and other fiber-based materials are highly dependent on their design, composition, and construction. Figure 15–12 shows sketches of several common fiber orientation approaches. The orientation of *fibrous reinforcements*, whether in the form of tape, cloth, filaments, or random short fibers, has a marked impact on the erosion resistance of composite nozzles (Fig. 15–10 shows erosion data). When perpendicular to the gas flow, erosion

Hot gas ⟶

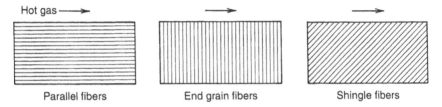

Parallel fibers   End grain fibers   Shingle fibers

FIGURE 15–12. Simplified sketches of three different types of fiber-reinforced ablative materials.

resistance is the greatest but heat transfer to outboard surrounding materials is also the highest. Good results have been obtained at throat and entrance locations when the fibers are at 40 to 60° relative to the gas flow, as the taped-wrapped ply at these angles can still provide high erosion resistance. However, at the exit cone, shallower ply angles (0 to 20°) are used as they reduce the heat transfer while simultaneously providing erosion protection and reducing nozzle weight (Ref. 15–14.)

15.3. IGNITER HARDWARE

The process of ignition is described in Section 14.2 and in Section 13.5 some propellants used in igniters are mentioned. In this section we discuss specific igniter types, their locations, and their hardware (see Ref. 15–15).

Since igniter propellant mass is small (often less than 1% of the motor propellant) and burns mostly at low chamber pressure (hence low I_s), it has a negligible contribution to the motor's overall total impulse. It is the designer's aim to reduce the igniter propellant mass and the igniter inert hardware mass to a minimum—just big enough to assure ignition of the grain under all operating conditions.

Figure 15–13 shows several proven locations for the igniter. When mounted at the forward end, the gas flow over the propellant surface helps to achieve ignition.

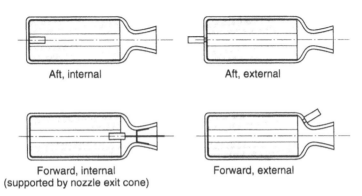

Aft, internal   Aft, external

Forward, internal   Forward, external
(supported by nozzle exit cone)

FIGURE 15–13. Simplified diagrams of mounting options for igniters. Grain configurations details are not shown.

With aft mounting there is little gas motion, particularly near the forward end; here, ignition must rely on the temperature, pressure, and heat transfer from the igniter gas. When mounted internally through the nozzle, igniter hardware and supports are discarded shortly after use and there is no inert mass penalty for the igniter case. A unique annular igniter, sleeve-shaped around a submerged nozzle throat region, has been used successfully with a few of the larger rocket motors. The two basic most common types, *pyrotechnic igniters* and *pyrogen igniters*, are discussed below.

Pyrotechnic Igniters

In industrial practice, pyrotechnic igniters are defined as igniters (other than pyrogen-type igniters) that use solid explosives or energetic propellant-like chemical formulations (usually small pellets of propellant that provide large burning surfaces and short burning times) as the heat-producing materials. This definition fits a wide variety of designs, known in the trade as bag and carbon igniters, powder can, plastic case, pellet basket, perforated tube, combustible case, jellyroll, string, or sheet igniters. A common pellet-basket design is shown in Fig. 15–14, being typical of the pyrotechnic igniters. Ignition of the main charge (in this case pellets consisting of 24% boron–71% potassium perchlorate–5% binder) is accomplished by stages; first, on receipt of an electrical signal the initiator releases the energy of a small amount of sensitive powdered pyrotechnic housed within the initiator, commonly called the squib or the primer charge; next, the booster charge is ignited by heat released from the squib; and finally, the main ignition charge propellants are ignited.

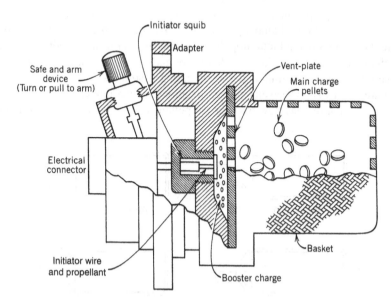

FIGURE 15–14. Typical pyrotechnic igniter with three different propellant charges that ignite in sequence. Only a few pellets are shown. The larger surface of the pellets gives very short igniter burning times.

A special form of pyrotechnic igniter is the *surface-bonded* or *grain-mounted igniter*. It has its initiator included within a sandwich of flat sheets; the layer touching the grain has the main pyrotechnic charge. This form of igniter is used with multipulse rocket motors with two or more end-burning grains. The ignition from the second and successive pulses of these motors presents unusual requirements for available space, compatibility with grain materials, life, and for the resulting pressures and temperatures from booster grain operation. Advantages of the sheet igniter include light weight, low volume, and high heat flux at the grain surface. Any inert material employed (such as wires and electric ceramic insulators) is usually blown out of the motor during ignition and resulting impacts have at times either damaged or plugged the nozzle, particularly if the blown material has not been intentionally broken up into small pieces.

Pyrogen Igniters

A pyrogen igniter is basically a small unit, containing all the elements of a rocket motor, used to ignite a larger rocket motor, see Ref. 15–11, and is not designed to produce thrust. They all consist of one or more nozzle orifices (both sonic and supersonic types) and most use conventional rocket motor grain formulations (sometimes the same as the main propellant grain) and design technology. Heat transfer from the pyrogen gas to the motor grain is largely convective, with hot gases contacting the inner grain surface, in contrast to pyrotechnic igniters that transfer heat by high-energy radiation. Figures 12–1, 12–2, and 12–20 illustrate rocket motors with a typical pyrogen igniter; in Fig. 18–5 the igniter has three nozzles and a cylindrical grain with high-burn-rate propellant. For pyrogen igniters, initiator and booster charge designs are often very similar to those used in pyrotechnic igniters. Reaction products from the main charge impinge on the inside surface of the rocket motor grain, producing motor ignition. Common practice on very large motors is to mount them externally, with the pyrogen igniter pointing its jet up through the large motor nozzle, in which case the igniter becomes a piece of ground-support equipment.

Two approaches are commonly used to safeguard against rocket motor misfires, or inadvertent ignition; one is to use of the classical *safe and arm device* and the second is to *design safeguards* into the initiator. Energy sources causing unintentional ignitions (usually disastrous when they happen) can be (1) static electricity, (2) induced electrical currents from electromagnetic radiation—such as radar, (3) induced electrical currents from ground test equipment, communication apparatus, or nearby electrical circuits in the flight vehicle, and (4) heat, vibration, or shock from motor handling and operations. Functionally, the *safe and arm* device serves as an electrical switch to keep the igniter circuit grounded and interrupted when not operating; in some designs it also mechanically misaligns or blocks the igniter's gas flow passage so that unwanted ignition is precluded even when the initiator charge may fire. As the device is moved into the *arm* position, the electric ignition circuit is no longer blocked and an ignition flame can reliably propagate from the igniter's booster and main charges to the propellant surface.

Electric initiators in motor igniters are also called squibs, glow plugs, primers, and headers; they always constitute the initial element in the ignition train and, if properly designed, can act as a safeguard against unintended ignition. Three typical designs of initiators are shown in Fig. 15–15. Both (a) and (b) types form structurally a part of the rocket motor case and generically act as headers. In the integral diaphragm type (a), the initial ignition energy is passed in the form of a shock wave through the diaphragm activating the acceptor charge, with the diaphragm remaining integral. The same principle is also used to transmit shock waves through metal case walls or metal inserts in filament-wound cases; here, the combustion case would not need to be penetrated and can remain sealed. The header in type (b) resembles a simple glow plug with two high-resistance bridgewires buried in the initiator propellant charge. The design in type (c) employs a small bridgewire (0.02 to 0.10 mm) of low-resistance material, usually platinum or gold, that explodes by the application of a high-voltage discharge.

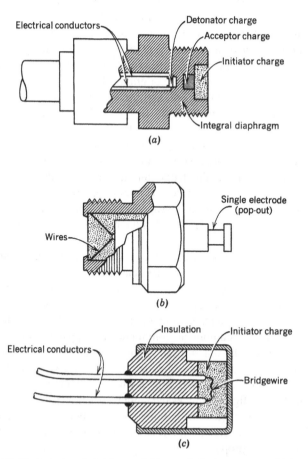

FIGURE 15–15. Typical electric initiators; (a) integral diaphragm type; (b) header type with double bridgewire; (c) exploding bridgewire type.

Initiator safeguard aspects appear as a basic design feature in the form of (1) a minimum threshold electrical energy required for activation, (2) voltage-blockage provisions (usually, air gaps or semiconductors in the electrical circuit), and/or (3) responsiveness only to a specific energy pulse or frequency band. Invariably, such safeguards may compromise to some degree the safety provided by the classical safe and arm device.

A more recent method of initiating igniter action of is to use a laser as an energy source to start the combustion in an initiator charge. Properly designed, there are no problems with induced currents or other inadvertent electrical initiations. The energy from a small neodymium/YAG laser, external to the motor, travels in fiber-optical glass cables to the pyrotechnic initiator charge (Ref. 15–16). Sometimes an optical window in the case or closure wall allows the initiator charge to be inside the case.

Igniter Analysis and Design

The basics of initiating ignition are common to all designs and applications of pyrotechnic and pyrogen igniters. In general, current analytical models of physical and chemical processes relevant to igniter design (including heat transfer, propellant decomposition, deflagration, flame spreading, and chamber filling) are far from complete and accurate. See Chapter 5 by Hermance and Chapter 6 by Kumar and Kuo in Ref. 14–1, and Ref. 15–15. Analyses and design of igniters, regardless of the type, depend heavily on experimental results. When data on past successes and failures with full-scale motors is included, the effects of some important parameters have become quite predictable. For example, Fig. 15–16 can be useful in estimating the mass of the igniter main charge for rocket motors of various sizes (i.e., motor free volumes). From these data,

$$m = 0.12(V_F)^{0.7} \qquad\qquad (15{-}4)$$

where m is the igniter charge mass in grams and V_F is the motor free volume in cubic inches (the void in the case not occupied by propellant). A larger igniter mass flow means a shorter ignition delay. Ignition time events are shown in Fig.14-3.

15.4. ROCKET MOTOR DESIGN APPROACH

Although there are many common elements in the design of all solid propellant rocket motors, there is as yet no universal, well-defined set of procedures or design methods. Each application presents different requirements. Individual designers and their organizations have different approaches, background, step sequences, or emphasis. Approaches also vary depending on the amount of available data on design issues, propellants, grains, hardware, materials, with the degree of novelty (many "new" motors are actually modifications of proven existing motors), and/or on available/validated computer software.

The following items are usually part of the preliminary design process: they start with requirements for the flight vehicle and rocket motor, such as those listed in

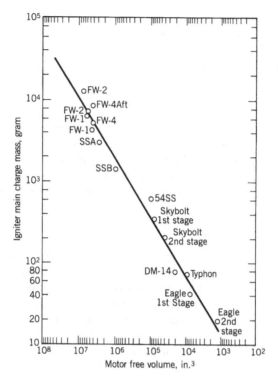

FIGURE 15–16. Igniter charge or propellant mass versus motor free volume, based on data from various-sized rocket motors that use AP/Al composite propellant. Data with permission from Ref. 15–17

Table 15–6. When the motor under design has enough similarity to proven existing motors, their parameters and flight experiences will be helpful in reducing the design effort and in enhancing confidence in the new design. Selection of propellant and grain configuration are usually made early in the design process (propellant selection is discussed in Chapter 13 and grains in Chapter 12). It is uncommon for any one propellant to satisfy all of the following key requirements, namely, necessary rocket performance (I_s), burning rate to suit the desired thrust–time curve, and required strength (maximum stress and strain). A well-characterized propellant with a proven grain configuration or a well-tested piece of hardware suitably modified to fit the new application will usually be preferred.

Analysis is a key ingredient in design; for example, calculations of structural integrity should be undertaken at places where stresses or strains likely exceed those that can be tolerated by the grain, or other key components, at loading limits in the expected range of environmental conditions. An evaluation of thermal growth and thermal stresses is usually performed for every heated component or assembly. An analysis of the nozzle as described in an earlier section of this chapter should be done, particularly if the nozzle is complex or includes thrust vector control. When gas-flow

TABLE 15–6. Typical Requirements and Constraints for Solid Rocket Motors

Requirement Category	Examples
Application	Definition of mission, vehicle and propulsion requirements, flight paths, maneuvers, environment, number required
Functional	Total impulse, thrust–time curve, ignition delay, initial motor mass, specific impulse, TVC angles and angular accelerations, propellant fraction, class 1.1 or 1.3, burn time, and tolerances on all of these parameters
Interfaces	Attachments to vehicle, fins, TVC system, power supply, instruments, lifting and transport features, grain inspection, control signals, shipping container
Operation	Storage, launch, flight environment, temperature limits, transport loads or vibrations, plume characteristics (smoke, toxic gas, radiation), life, reliability, safe and arm device functions, field inspections
Structure	Loads and accelerations imposed by thrusting and by vehicle (flight maneuvers), stiffness to resist vehicle oscillations, safety factors
Insensitive munitions (military application)	Response to slow and fast cook-off, bullet impact, sympathetic detonation, shock tests
Cost and schedule	Stay within the allocated time and money
Deactivation	Method of removing/recycling of propellants, safe disposal of over-age motors
Constraints	Limits on volume, length, or diameter; minimum acceptable performance, maximum cost, maximum inert mass
Schedule	Design completion, test articles completion, qualification, delivery

considerations show that erosive burning that cannot be tolerated is likely to occur during any portion of the burning duration, modifications to the propellant, the nozzle material, or the grain geometry may need to be made. Usually, preliminary evaluation is also done of acoustic resonances in the grain cavity to identify possible combustion instability modes (see Chapter 14). Motor performance calculations that include heat transfer and stresses at critical locations will usually need to be undertaken.

Since there is considerable interdependence and feedback between parameters (propellant formulation, grain geometry/design, stress analysis, thermal analysis, major hardware component designs, and their manufacturing processes), it is difficult to finalize any design without going through several iterations. Data from tests of laboratory samples, subscale motors, and full-scale motors will strongly influence each step.

Preliminary layout drawings or CAD (computer-aided design) images of the rocket motor with its key components will have to be made in sufficient detail to provide sizes and reasonably accurate dimensions. For example, preliminary design of the thermal insulation (often from a heat transfer analysis) will provide dimensions for that insulator. Such layouts are then used to estimate volumes, inert masses, propellant masses, propellant mass fraction or center of gravity.

A simplified flow diagram of one particular approach to motor preliminary design and development steps is shown in Fig. 15–17. Not shown in this diagram are many side steps, such as igniter design and tests, liner/insulating selection, thrust vector

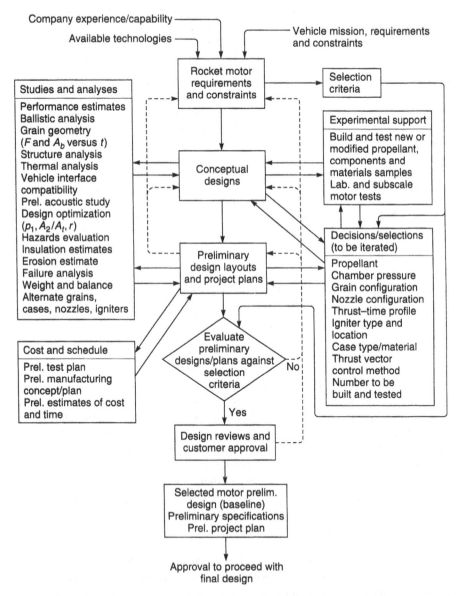

FIGURE 15–17. Diagram of one approach to preliminary design activity sequences and interrelations. Dashed lines indicate feedback paths. Some items listed here only apply to specific types of rocket motors.

control design and test, reliability analyses, evaluation of alternative designs, material specifications, inspection/quality control steps, safety provision, special test equipment, special test instrumentation, and others.

If performance requirements are narrow and ambitious, it will be necessary to carefully study cumulative performance tolerances together with various other parameters. For example, practical tolerances may be assigned to the propellant density, nozzle throat diameter (for erosion), burn rate scale factor, initial burning surface area, propellant mass, or pressure exponent. These, in turn, will appear as tolerances in process specifications, specific inspections, dimensional tolerances, or accuracy of propellant ingredient weighing. Cost (see Section 19-4) is always a major factor and a portion of the design effort will be spent looking for lower-cost materials, simpler manufacturing processes, fewer assembly steps, and/or lower-cost component designs. For example, tooling for casting, mandrels for case winding and tooling for insulator molding can be expensive. Time needed for completing a design can be shortened when there is effective communication and a cooperative spirit between all design-individuals (propellant specialists, analysts, customer representatives, manufacturing and test personnel, and/or vendors) concerned with the project. Reference 15–18 deals with some of the uncertainties in a particular booster motor design, and Ref. 15–19 discusses design optimization.

A *preliminary project plan* is usually formulated simultaneously with the preliminary-design work. Two decades or more ago the project plan was made after preliminary design was completed. Presently, with a strong emphasis on affordability, personnel working on preliminary designs also have to focus on reducing costs on all components and processes. The project plan reflects such decisions and defines the number of rocket motors and key components to be built, the availability and lead time of critical materials or components, the type and number of tests (including aging or qualification tests); also, project plans identify the manufacturing, inspection, and test facilities to be used, the number and kind of personnel (and when they will be needed), and any special tooling or fixtures. These decisions and these data are needed to make realistic *estimates of cost* and realistic *preliminary schedules*. When estimates exceed allowable costs or desired delivery schedules, then some changes are required. For example, there may be changes in the number of units to be built, the number and types of tests, or redesigns for easier, less costly assemblies. However, such changes must not compromise reliability or performance. It is particularly difficult to make good plans and good cost or time estimates when the rocket motor has not been well defined or designed in sufficient detail. Such plans and estimates must therefore be based largely on experience with prior successful similar rocket motors.

Final result of any preliminary design will consist of layout drawings or CAD images of the selected configuration, a prediction of performance, an estimate of rocket motor mass (and, if needed, also the travel of the center of gravity), an identification of the propellant, grain, geometry, insulation, and of several key materials for the hardware components. Any estimate of predicted reliability and motor life needs to be accompanied by supporting data. All information would then be presented for review of the selected preliminary design. This review would be undertaken with a diverse group of rocket motor experts including the vehicle designer, safety engineers, specialists in manufacturing, assembly, or inspection, customer representatives, analysts, and others. Here, the preliminary design team explains why they selected that particular design and how it meets the requirements. With competent reviewers there usually will be suggestions for changes or further improvements. The project plan, preliminary cost estimates, and a preliminary schedule are sometimes included in the design review, but more often these are presented to a separate group of experts, or just to the customer's experts.

After the design review and approval of the selected preliminary design, a detailed or *final design* of all parts and components and the writing of certain specifications can begin. During manufacture and development testing, further design changes may become necessary to improve the manufacture, reduce cost, or remedy technical problems that have arisen. In many organizations the final, detailed design is again submitted for design review before manufacturing can begin. The new motor will then start its development testing. With the larger, expensive rocket motors that have a lot of heritage from prior proven motors, the development may include a single rocket motor firing with extensive instrumentation. For motors which are to be built in large quantities and for motors with major new features, the development and qualification may involve testing as few as 2 to 4 motors. Final design ends when all detail drawings or CAD images and a final parts list are completed, and specifications for motor testing, certain manufacturing operations, and/or materials/component acceptance have been prepared. Detailed designs are considered complete only after the motor successfully passes all its development and qualification tests and begins production.

Example 15–1. This example gives one particular method for determining key design parameters of a solid rocket motor that uses a composite propellant. These results may then be used as the basis for preliminary design. The rocket is to be launched at altitude and it will fly at constant altitude. The following data are given:

Specific impulse (actual)	$I_s = 240$ sec at altitude and with $p_1 = 1000$ psia
Burning rate	$r = 0.8$ in./sec at 1000 psia and 60 °F grain temperature
Propellant density	$\rho_b = 0.066$ lbm/in.^{3}
Specific heat ratio	$k = 1.25$
Chamber pressure, nominal	$p_1 = 1000$ psi
Desired average thrust	$F = 20,000$ lbf (at altitude)
Maximum vehicle diameter	$D = 16$ in.
Desired burn duration	$t_b = 5.0$ sec
Ambient pressure	3.0 psi (at altitude)
Vehicle payload	5010 lbm (includes structure, controls, power supply)

Approximately neutral burning is desired. Assume no erosion of nozzle material, no loss of insulation during burning, and sea-level gravity.

SOLUTION

a. **Basic design.** The total impulse I_t and propellant weight at sea level w_b are obtained from Eqs. 2-2 and 2-5. $I_t = Ft_b = I_s w_b = 20{,}000 \times 5.0 = 100{,}000$ lbf−sec. The propellant weight is $100{,}000/240 = 417$ lbf. Allowing for a loss of 2% for manufacturing tolerances and slivers, the total propellant weight becomes $1.02 \times 417 = 425$ lbf. The volume occupied by this propellant V_b is given by $V_b = w_b/\rho_b = 425/0.066 = 6439$ in.^{3}. The web thickness $b = rt_b = 0.80 \times 5 = 4.0$ in.

b. **Case dimensions.** The outside diameter is fixed at 16.0 in. Heat-treated steel with an ultimate tensile strength of 220,000 psi is to be used. Its wall thickness d can be determined from Eq.15-1 for simple circumferential stress as $d = (p_1 D)/(2\sigma)$. The value of p_1 to be used here depends on the selection of a safety factor, which in turn depends on the heating of the wall, prior experience with the material, and so on; a safety factor of 2.0 is recommended to allow for surface scratches, combined stresses and welds, and rough field handling (below we use 1.98). When the value of D is the average diameter to the center of the tube wall, then d becomes

$$d = 1.98pD/[2\sigma] = 1.98 \times 1000 \times 16.0/(2 \times 220{,}000) = 0.072 \text{ in.}$$

A spherical head end and a spherical segment at the nozzle end similar to Fig.12-1 are assumed.

c. **Grain configuration.** The grain is to be cast into the case but will be thermally isolated from the case with an elastomeric insulator plus a very thin liner of combined average thickness of 0.100 in. inside the case; the actual thickness will be less than 0.10 in. inside the cylindrical and forward closure regions, but thicker in the nozzle entry area. The outside diameter D_o for the grain is determined from the case thickness and liner to be $16.0 - 2 \times 0.072 - 2 \times 0.10 = 15.66$ in. The inside diameter D_i of a simple hollow cylinder grain would be the outside diameter D_o minus twice the web thickness or $D_i = 15.66 - 2 \times 4.0 = 7.66$ in. For a simple cylindrical grain, the volume determines its effective length, which can be determined from the geometry, namely,

$$V_b = \frac{\pi}{4}L(D_o^2 - D_i^2)$$

$$L = \frac{6439 \times 4}{\pi(15.66^2 - 7.66^2)} = 43.95 \text{ in.}$$

The web fraction would be $2b/D_o = 8/15.66 = 0.511$. The L/D_o of the grain is (approximately) $43.95/15.66 = 2.81$.

For grains with this web fraction and this L/D_o ratio, Table 12-4 suggests the use of an internal burning tube with fins or cones. A conocyl configuration is selected, although a slotted tube or fins would also be satisfactory. These grain shapes are shown in Figs. 12–1 and 12–17. The initial or average burning area will be found from Eqs. 12-1 and 2-5: namely $F = \dot{w}I_s = \rho_b A_b r I_s$ (at the Earth's surface)

$$A_b = \frac{F}{\rho_b r I_s} = \frac{20{,}000}{0.066 \times 0.8 \times 240} = 1578.3 \text{ in.}^2$$

The actual grain now has to be designed into the case with spherical ends, so it will not be a simple cylindrical grain. An approximate total volume occupied by the grain is found by subtracting the perforation volume from the chamber volume. There is a full hemisphere at the head end and a partial hemisphere of propellant at the nozzle end (estimated here at 0.6 the volume of a full hemisphere):

$$V_b = \frac{1}{2}(\pi/6)D_o^3(1 + 0.6) + (\pi/4)D_o^2 L$$
$$- (\pi/4)D_i^2(L + D_i/2 + 0.3D_i/2) = 6439 \text{ in.}^3$$

This is solved for L, with $D = 15.66$ in. and inside diameter $D_i = 7.66$ in. The answer is $L = 38.24$ in. The initial internal hollow-tube burn area is about

$$\pi D_i(L + D_i/2 + 0.3D_i/2) = 1040 \text{ in.}^2$$

The desired burn area of 1578 in.^{2} is larger than 1040 in.^{2} by 538 in.^{2} Therefore, an additional burn surface area of 538 in.^{2} will have to be designed into the cones of a conocyl configuration or as slots in a slotted tube design. Actually, a detailed geometrical study should be made analyzing the instantaneous burn surface after arbitrary short time intervals and selecting a detailed grain configuration where A_b stays approximately constant. This example does not go through a preliminary stress and elongation analysis, but they should also be done.

d. **Nozzle design.** Nozzle parameters can be determined from Chapter 3. The thrust coefficient C_F can be found from curves of Figs. 3-5, 3-6, and 3-7 or Eq. 3-30 for $k = 1.25$ and a pressure ratio of $p_1/p_2 = 1000/3 = 333$, as $C_F = 1.73$. The throat area is from Eq. 3-31:

$$A_t = F/(p_1 C_F) = 20,000/(1000 \times 1.73) = 11.56 \text{ in.}^2$$

The nozzle throat diameter is $D_t = 3.836$ in. The nozzle area ratio for optimum expansion (Fig. 3-5) A_2/A_t is about 27. The exit area and diameter are therefore about $A_e = 312$ in.^{2} and $D_e = 19.93$ in. However, this is larger than the maximum vehicle diameter of 16.0 in. ($A_2 = 201$ in.^{2}), which governs the maximum for the outside of the nozzle exit. Allowing for an exit cone plus local thermal insulation thickness adding up to 0.10 in., the internal nozzle exit diameter D_2 is 15.80 in. and A_2 is 196 in.^{2}. This would allow only a maximum area ratio of 196/11.56 or 16.95. Since C_F values do not change appreciably for this new area ratio, it can be assumed that the nozzle throat area may stay unchanged. This nozzle can have a thin wall in its exit cone, but requires heavy ablative materials, probably in several layers near the throat and convergent nozzle regions. Thermal and structural analyses of the nozzle are not shown here.

e. **Weight estimate.** The steel case weight (assume a cylinder with two spherical ends and that steel weight density is 0.3 lbf/in.^{3}) is

$$d\pi DL\rho + (\pi/4)dD^2\rho = 0.072\pi \times 15.83 \times 38.24 \times 0.3$$
$$+ 0.785 \times 0.072 \times 15.83^2 \times 0.3$$
$$= 45.3 \text{ lbf}$$

With attachments, flanges, igniter fitting, and pressure tap bosses, this weight increases to 51 lbf. The nozzle weight is composed of the weights of the individual parts, estimated for their densities and geometries. This example does not go through such detailed calculations, but merely gives the result of 30.2 lbf. Assume further an expended igniter propellant weight of 2.0 lbf and a full igniter weight of 5.0 lbf. The weights are then

Case weight at sea level	51.0 lbf
Liner/insulator	14.2 lbf
Nozzle, including fasteners	30.2 lbf
Igniter case and wires	2.0 lbf
Total inert hardware weight	97.4 lbf
Igniter propellant	3 lbf
Propellant (effective)	417 lbf
Unusable propellant (2%)	8 lbf
Total weight	525.4 lbf
Propellant and igniter powder	428.0 lbf

f. **Performance.** Since the total impulse is $I_t = F\Delta t = 20,000 \times 5 = 100,000$ lbf-sec, the total impulse-to-weight ratio is $100,000/525.4 = 190.3$. Comparison with I_s shows this to be an acceptable value, indicating good performance. The total launch weight is $5010 + 525.4 = 5535$ lbf, and the weight at burnout or thrust termination is $5535 - 428 = 5107$ lbf. The initial and final thrust-to-weight ratios and accelerations become

$$F_i/w = 20,000/5535 = 3.61$$
$$F_f/w = 20,000/5107 = 3.92$$

The acceleration in the direction of thrust is 3.61 times the sea-level gravitational acceleration at start and 3.92 at burnout.

g. **Erosive Burning.** The ratio of the port area to the nozzle throat area at start is $(7.66/3.836)^2 = 3.97$. This is close to the limit of 4.0, and erosive burning is not likely to be significant. A simple analysis of erosive burning in the conical cavity should also be made, but not shown here.

With abrasion of nozzle material and insulation, the actual total mass flow and thrust will be enhanced slightly. As the nozzle area increases from erosion, the chamber pressure will decrease somewhat and the specific impulse will be slightly lower. For this problem, there is now enough information to make preliminary design drawings, project plans, an initial cost estimate, and a tentative schedule.

PROBLEMS

1. In Figs. 14-3 and 14-4, it can be seen that higher case pressures and higher heat transfer rates promote faster ignition. One way to promote this more rapid ignition is for the nozzle to remain plugged until a certain minimum initial pressure has been reached, at which time the nozzle plug will be ejected. Analyze the time saving achieved by such a device, assuming that the igniter gas evolution follows Eqs. 12-3 and 12-5. Under what circumstances is this an effective method? State any relevant assumptions about cavity volume, propellant density, and so on.

2. Compare a bare simple cylindrical case with hemispheric ends (ignore nozzle entry or igniter flanges) for an alloy steel metal and two reinforced fiber (glass and carbon)-wound filament case. Use the properties in Table 15–2 and thin shell structure theory. Given:

Length of cylindrical portion	370 mm
Outside cylinder diameter	200 mm
Internal pressure	6 MPa
Web fraction	0.52
Insulator thickness (average)	
for metal case	1.2 mm
for reinforced plastic case	3.0 mm
Volumetric propellant loading	88%
Propellant specific gravity	1.80
Specific impulse (actual)	248 sec
Nozzle igniter and mounting provisions	0.20 kg

Calculate and compare the theoretical propulsion system flight velocity (without payload) in a gravity-free vacuum for these three cases.

3. The following data are given for a case that can be made of either alloy steel or fiber-reinforced plastic.

Type	Metal	Reinforced Plastic
Material	D6aC	Organic filament composite (Kevlar)
Physical properties	See Table 15-2	See Table 15-2
Poisson ratio	0.27	0.38
Coefficient of thermal expansions, m/m-K $\times 10^{-6}$	8	45
Outside diameter (m)	0.30	0.30
Length of cylindrical section (m)	0.48	0.48
Hemispherical ends		
Nozzle flange diameter (m)	0.16	0.16
Average temperature rise of case material during operation (°F)	55	45

Stating any assumptions, determine the growth in diameter and length of the case due to pressurization, heating, and the combined growth and interpret the results.

4. A high-pressure helium gas tank at 8000 psi maximum storage pressure and 1.5 ft internal diameter is proposed. Use a safety factor of 1.5 on the ultimate strength. The following candidate materials are to be considered:

Kevlar fibers in an epoxy matrix (see Table 15–2).

Carbon fibers in an epoxy matrix.

Heat-treated welded titanium alloy with an ultimate strength of 150,000 psi and a weight density of 0.165 lbf/in.[3]

Determine the dimensions and sea-level weights of these three tanks and discuss their relative merits. To contain the high-pressure gas in a composite material that is porous, it is also necessary to include a thin metal inner liner (such as 0.016-in.-thick aluminum) to prevent loss of gas; this liner will not really carry structural loads, but its weight and volume need to be considered.

5. Make a simple sketch and determine the mass or sea-level weight of a rocket motor case that is made of alloy steel and is cylindrical with hemispherical ends. State any assumptions you make about the method of attachment of the nozzle assembly and the igniter at the forward end.

Outer case and vehicle diameters	20.0 in.
Length of cylinder portion of case	19.30 in.
Ultimate tensile strength	172,000 psi
Yield strength	151,300 psi
Safety factor on ultimate strength	1.65
Safety factor on yield strength	1.40
Nozzle bolt circle diameter	12.0 in.
Igniter case diameter (forward end)	3.00 in.
Chamber pressure, maximum	1520 psi

6. Design a solid propellant rocket motor with insulation and liner. Use the AP/Al-HTPB propellant from Table 12-3 for Orbus 6. The average thrust is 3600 lbf and the average burn time is 25.0 sec. State all the assumptions and rules used in your solution and give your reasons for them. Make simple sketches of a cross section and a half section with overall dimensions (length and diameter), and determine the approximate loaded propellant mass.

7. The STAR 27 rocket motor (Fig. 12-1 and Table 12-3) has an average erosion rate of 0.0011 in./sec. (a) Determine the change in nozzle area, thrust, chamber pressure, burn time, and mass flow at cutoff. (b) Also determine those same parameters for a condition when, somehow, a poor grade of ITE material was used that had three times the usual erosion rate. Comment on the difference and acceptability.

Answer: Nozzle area increases by about (a) 5.3% and (b) 14.7% and decrease of chamber pressure at cutoff decreases by approximately the same percentage.

REFERENCES

15–1. NASA, *National Space Transportation System*, Vols. 1 and 2, U.S. Government Printing Office, Washington, DC, Jun. 1988 (Space Shuttle Description).

15–2. M. Salita, "Simple Finite Element Analysis Model of O-Ring Deformation and Activation during Squeeze and Pressurization," *Journal of Propulsion and Power*, Vol. 4, No. 6, Nov.–Dec. 1988.

15–3. A. J. McDonald and J. R. Hansen, *Truths, Lies and O-Rings, Inside the Space Shuttle Challenger Disaster*, Florida University Press, Gainesville FL, 2009.

15–4. J. H. Hildreth, Chapter 2, "Advances in Solid Rocket Motor Nozzle Design and Analysis Technology since 1970"; A. Truchot, Chapter 3, "Design and Analysis of Solid Rocket Motor Nozzle"; P. R. Evans, Chapter 4, "Composite Motor Case Design"; A. J. P. Denost, Chapter 5, "Design of Filament Wound Rocket Cases,"; H. Baham and G. P. Thorp, Chapter 6, "Consideration for Designers of Cases for Small Solid Propellant Rocket Motors,"; all in *Design Methods in Solid Rocket Motors*, AGARD Lecture Series LS 150, Advisory Group for Aerospace Research and Development, NATO, revised 1988.

15–5. B. H. Prescott and M. Macocha, Chapter 6, "Nozzle Design"; M. Chase and G. P. Thorp, Chapter 7, "Solid Rocket Motor Case Design"; in G. E. Jensen and D. W. Netzer (Eds.), Vol. 170, *Progress in Astronautics and Aeronautics*, American Institute of Aeronautics and Astronautics, 1996.

15–6. A. de Rouvray, E. Haug, and C. Stavrindis, "Analytical Computations for Damage Tolerance Evaluations of Composite Laminate Structures," *Acta Astronautica*, Vol. 15, No. 11, 1987, pp. 921–930.

15–7. D. Beziers and J. P. Denost, "Composite Curing: A New Process," AIAA Paper 89–2868, Jul. 1989.

15–8. A. Groves, J. Margetson, and P. Stanley, "Design Nomograms for Metallic Rocket Cases Reinforced with a Visco-elastic Fiber Over-wind," *Journal of Spacecraft and Rockets*," Vol. 24, No. 5, Sept.–Oct. 1987, pp. 411–415.

15–9. Monograph Space Vehicle Design Criteria (Chemical Propulsion), *Solid Rocket Motor Nozzles*, NASA-SP-8115, Jun. 1975.

15–10. S. Boraas, "Modeling Slag Deposition in Space Shuttle Solid Rocket Motor," *Journal of Spacecraft and Rockets*, Vol. 21, No. 1, Jan.–Feb. 1984.

15–11. L. H. Caveny, R. L. Geisler, R. A. Ellis, and T. L. Moore, "Solid Rocket Enabling Technologies and Milestones in the United States," *Journal of Propulsion and Power*, Vol. 19, No. 6, Nov.–Dec. 2003, pp. 1038–1066.

15–12. B. H. Broguere, "Carbon/Carbon Nozzle Exit Cones: SEP's Experience and New Developments," AIAA Paper 97–2674, Jul. 1997.

15–13. M. Berdoyes, M. Dauchier, C. Just, "New Ablative Material Offering SRM Nozzle Design Breakthroughs," AIAA paper 2011-6052, Jul. 2011. doi: 10.2514/6 .2011-6052.

15–14. R. A. Ellis, Personal Communication, 2013; R. A. Ellis, J. C. Lee, and F. M. Payne, CSD, USA; and A. Lacombe, M. Lacoste, and P. Joyez, SEP, France, "Development of a Carbon-Carbon Translating Nozzle Extension for the RL10B-2 Liquid Rocket Engine." Presented at 33rd AIAA/ASME/SAE/ASEE Joint Propulsion Conference, Seattle, WA, AIAA paper 97-2672, Jul. 6–9, 1997.

15–15. "Solid Rocket Motor Igniters," NASA SP–8051, March 1971 (N71–30346).

15–16. R. Baunchalk, "High Mass Fraction Booster Demonstration," AIAA Paper 90–2326, July 1990.

15–17. L. LoFiego, *Practical Aspects of Igniter Design*, Combustion Institute, Western States Section, Menlo Park, CA, 1968 (AD 69–18361).

15–18. R. Fabrizi and A. Annovazzi, "Ariane 5 P230 Booster Grain Design and Performance Study," AIAA Paper 89–2420, Jul. 1989.

15–19. A. Truchot, Chapter 11, "Overall Optimization of Solid Rocket Motors," in *Design Methods in Solid Rocket Motors*, AGARD Lecture Series LS 150, Advisory Group for Aerospace Research and Development, NATO, revised 1988.

CHAPTER 16

HYBRID PROPELLANTS ROCKET PROPULSION*

Rocket propulsion bipropellant concepts in which one component propellant is stored in the liquid phase and the other as a solid are called *hybrid propulsion systems* or *hybrid rocket engines* (there is no universally accepted designation). See Fig. 1–6. Hybrid propulsion is of considerable interest to commercial rocket endeavors primarily for space applications. The main advantages of hybrids are: (1) more safety and ruggedness than conventional chemical propulsion systems—vehicles using hybrid propellants may survive unharmed effects of bullet impacts, fires under the vehicle, inadvertent droppings, and of explosions from adjacent munitions or other rockets; (2) start–stop–restart capabilities; (3) relative simplicity compared to liquids which may translate into low overall system cost; (4) higher specific impulse than solid rocket motors and higher density-specific impulse (see Eq. 7–3) than many common liquid bipropellant engines; and (5) capability to smoothly change thrust on demand over wide ranges. Even though hybrids have many unique features of their own, they do benefit from developments in liquid and solid rockets. Early hybrid motors were produced for target drones and for one small tactical missile. Experimental hybrid motors in sizes between 2 and 250,000 lbf thrust have been developed and ground tested. A few of medium-sized and small hybrid motors have flown in experimental vehicles. However, as of 2015 none of the larger systems have been designated for production applications. Several different manned suborbital space vehicles have flown with a hybrid propulsion system, as described later in this chapter; these vehicles are launched at altitude (typically 40,000 feet) from a carrier airplane.

Among the disadvantages of hybrid systems are: (1) mixture ratio and hence specific impulse may vary during "steady flow operation" (as well as during throttling)

---

*This chapter was contributed originally for the Sixth edition by Terry A. Boardman and modified versions appeared in the Seventh and Eighth editions. It has been further revised for this edition.

of the nonsolid component; (2) relatively complicated solid geometries are needed with significant and unavoidable fuel residues (slivers) that reduce the mass fraction and may vary unpredictably with random throttling; (3) prone to large-amplitude, low-frequency pressure fluctuations (termed chugging); and (4) present descriptions are incomplete, both for solid-fuel regression rates and for large hybrid systems motor-scaling effects (resulting in part from their relatively complicated internal grain design configurations). While cracks in the solid component of the hybrid may not be as catastrophic as those in solid motors, total surface area does influence propellant release, and exposed burning areas can change unpredictably during rocket operation (in some of the more complicated designs).

There are presently three distinct configurations of hybrid propulsion systems (Ref. 16–1). By far the most common arrangement, termed the *classical* or *typical configuration*, has the fuel in the solid phase and the oxidizer stored as a liquid, and many such arrangements have been experimentally evaluated. The *inverse* or *reverse configuration* has the oxidizer as a solid and the fuel in liquid form; comparatively little work has been done on this arrangement. A third one, called the *mixed hybrid configuration*, includes a small amount of solid oxidizer implanted with the solid fuel, and the fuel-rich mixture is then burned with additional liquid oxidizer injected either in an afterburner chamber, or simultaneously at the head and end of the grain cavity. This last configuration results from attempts to enhance fuel regression rates and thus increase thrust and chamber pressure (Ref. 16–2). Increased regression rates can also be obtained in the classical configuration with metallized fuel added to the grain. Designs with special chambers upstream and downstream of the solid component region are common with all three configurations. Figure 16–1 shows a proposed classical configuration design consisting of liquid-oxygen burning hydroxyl-terminated-polybutadiene (HTPB) solid fuel originally intended for Space Shuttle lift-off booster type of applications. The oxidizer can be either a noncryogenic (storable) or a cryogenic liquid depending on application and required specific impulse. For this hybrid application, the oxidizer is injected into a precombustion or vaporization chamber upstream of the primary fuel region. The fuel grain cavity contains numerous axial combustion ports that generate fuel vapor that then reacts with the flowing oxidizer. An aftermixing chamber is provided to ensure that all fuel and oxidizer are essentially completely burned before entering the nozzle.

16.1. APPLICATIONS AND PROPELLANTS

Hybrid propulsion units are well suited to applications or missions requiring throttling, command shutdown and restart, and long-duration missions requiring storable propellants associated with infrastructure nontoxic operations (manufacturing and launch) that benefit from non-self-deflagrating propulsion systems. Such applications might include primary boost propulsion for space launch vehicles, upper stage propulsion, and satellite maneuvering thrusters. Hybrids have been used to boost winged suborbital manned spaceships.

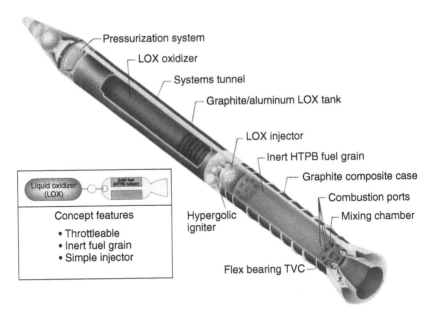

FIGURE 16–1. Depiction of a hybrid rocket preliminary concept aimed at boosting a large space vehicle. It has an inert solid fuel grain, pressurized liquid oxygen feed system, and can be throttled. Multiple ports are required to achieve the substantial fuel surface necessary for high flow rates.

Many early hybrid rocket motor developments involved target missiles and low-cost small tactical missile applications (Refs. 16–3 and 16–4). Other development efforts focused on high-energy upper-stage motors. In one program (Ref. 16–3), a hybrid motor was developed for high-performance upper-stage applications with design requirements that included a nominal thrust level of 22,240 N and an 8:1 throttling range; here oxygen difluoride was selected as the oxidizer with a lithium hydride/polybutadiene fuel grain (both highly toxic). In recent years development efforts have concentrated on prototypes for space launch applications.

A more practical, although lower energy, upper-stage hybrid propellant system uses 90 to 95% or "high-test" hydrogen peroxide oxidizer combined with HTPB fuel. Hydrogen peroxide (Ref. 16–5) is considered storable for time periods typical of upper-stage missions (to mission completion on the order of several months) and is relatively inexpensive. In solid rocket motors, HTPB is used as the binder to consolidate aluminum fuel with an ammonium perchlorate oxidizer matrix. In hybrids, HTPB becomes the entire fuel constituent. HTPB is low cost, processes easily, and will not self-deflagrate under any known conditions.

A common propellant combination for large hybrid booster applications has been liquid oxygen (LOX) oxidizer with HTPB fuel. Liquid oxygen is a widely used cryogenic oxidizer in the space launch industry; it is relatively safe and delivers high performance at relatively low cost. This hybrid propellant combination produces a nontoxic, reasonably smoke-free exhaust and is favored for future booster applications because it is chemically and performance-wise equivalent to LOX–kerosene bipropellant systems. In general, metallized solid fuels may increase performance, but beryllium is very toxic, boron is hard to ignite, lithium has a low heat of combustion, and aluminum oxides often increase the molecular mass of the combustion products often beyond the temperature gain. Work with hydrides and slurries containing these metals (to offset their intrinsic disadvantages) remains in its research stages. On the other hand, hydrogen peroxide (H_2O_2) and HAN (hydroxyl ammonium nitrate) have proven desirable themochemical properties, non-toxic exhausts and attractive density-specific impulses (Ref.16–5). Their regression rates and combustion efficiencies are only comparable to those obtained using LOX, but they have storage advantages and comparable degrees of "greenness."

Hybrid propellants may benefit from the addition of powdered aluminum to the fuel for some applications where a smoky or intensely radiative exhaust is no detriment. This addition increases the combustion temperature, reduces the stoichiometric mixture ratio, and increases fuel density as well as overall density-specific impulse. Although density-specific impulse ($\rho_f I_s$) is increased, addition of aluminum to the fuel may reduce the actual specific impulse. Figure 16–2 illustrates theoretical vacuum-specific impulse levels (calculated at 1000 psia chamber pressure and a 10:1 nozzle expansion area ratio) for a variety of cryogenic and storable oxidizers used in

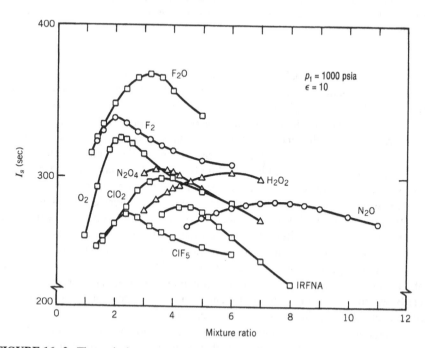

FIGURE 16–2. Theoretical vacuum-specific impulse of selected oxidizers reacted with hydroxyl-terminated polybutadiene fuel. The I_s of the O_2/HTPB propellant is comparable to that of a LOX/kerosene bipropellant engine. IRFNA is red fuming nitric acid.

TABLE 16–1. Thermochemical Properties of Selected Oxidizers Reacted with HTPB Fuel

Oxidizer	Type	Boiling Point (°C)	Density (g/cm^3)	$\Delta_f H^a$ (kcal/mol)
O_2	Cryogenic	−183	1.149	−3.1
O_3	Cryogenic	−112	1.614	+30.9
N_2O	Cryogenic	−88	1.226	+15.5
N_2O_4	Storable	+21	1.449	+2.3
IRFNA[b]	Storable	+80 to +120	1.583	−41.0
H_2O_2	Storable	+150	1.463	−44.8

[a]$\Delta_f H$ is the heat of formation at standard conditions as defined in Chapter 5.
[b]Inhibited red fuming nitric acid.

conjunction with HTPB fuel. Table 16–1 tabulates the heat of formation for HTPB reacted with various oxidizers.

A suborbital sounding rocket was successfully launched in 2002 by Lockheed Martin using a large hybrid rocket motor. The American Rocket Company (AMROC) before its demise tested many hybrid motors ranging from 100 to 250,000 lbf in thrust. In 1998, SpaceDev (Ref. 16–6) obtained the technical rights, proprietary data, and patents produced by AMROC. A smaller commercial version of the Space Shuttle has been made by Scaled Composites with a hybrid propulsion system (developed by SpaceDev) which, in a well-publicized accomplishment, won the Ansari X-Prize when it's SpaceShipOne (see Fig. 16–3) reached space in 2004 with two suborbital

FIGURE 16–3. SpaceShipOne gliding down for landing. This air ship uses a hybrid propulsion system to produce its main thrust. The propellant is N_2O-HTPB. (Photo by scaled composites. SpaceShipOne is a Paul G. Allen Project © Mojave Aerospace Ventures, LLC.)

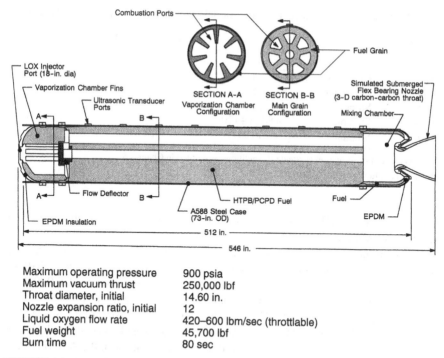

Maximum operating pressure	900 psia
Maximum vacuum thrust	250,000 lbf
Throat diameter, initial	14.60 in.
Nozzle expansion ratio, initial	12
Liquid oxygen flow rate	420–600 lbm/sec (throttlable)
Fuel weight	45,700 lbf
Burn time	80 sec

FIGURE 16–4. Simplified depiction of an experimental 250,000 lbf thrust, hybrid booster with a vaporization chamber, aft mixing chamber and two different solid fuel grains. The vaporization chamber fins and flow deflector are designed to promote flame holding in combustion ports. The fuel ingredient PCPD is polycyclopentadaine.

manned flights. In 1999, a consortium of aerospace companies also tested several 250,000-lbf thrust LOX/HTPB hybrid prototypes as a candidate strap-on booster for space launch vehicles (see Ref. 16–7). In these motors, polycyclopentadiene (PCPD) was added to the HTPB fuel to increase fuel density by about 10% over HTPB alone. The motors were designed to operate for 80 sec at a LOX flow rate of 600 lbm/sec with a maximum chamber pressure of 900 psi. Figure 16–4 illustrates a cross section of one motor configuration. Test results indicated that additional work is necessary to develop large hybrid motor configurations that exhibit stable combustion throughout the motor burn, and for a better understanding of fuel regression-rate scale-up factors.

Hybrid fuel grains are commonly ignited by providing a source of heat, which initiates gasification of the solid fuel grain at the head end of the motor. Subsequent initiation of oxidizer flow provides the required flame spreading to fully ignite the rocket motor. Ignition is typically accomplished by injection of a hypergolic fluid combination into the motor combustion chamber. Using the motor described in Fig. 16–4 as an example, a mixture of triethyl aluminum (TEA) and triethyl borane (TEB) is injected into the vaporization chamber. This TEA–TEB mixture ignites spontaneously on contact with air or with fuel in the combustion chamber, vaporizing fuel in the dome region. Subsequent injection of liquid oxygen completes motor ignition. TEA–TEB mixtures are currently used for motor ignition in the Atlas and Delta commercial launch vehicles. References 16–8 and 16–9 provide descriptions of solid fuels that will ignite spontaneously at ambient conditions when sprayed with specific oxidizers other than LOX. Small hybrid rocket motors, such as those used in laboratory environments with gaseous oxygen, can be electrically ignited by passing current through a resistor (such as steel wool) placed in the combustion port, or by use of separate propane or hydrogen ignition systems.

16.2. INTERIOR HYBRID MOTOR BALLISTICS

As the fuel grain in the classical hybrid configuration contains no oxidizer, combustion processes occur only in the gaseous phase and hence fuel surface regression rates are markedly different from those of a solid rocket motor. Because the solid fuel must vaporize before combustion, fuel surface regression is intrinsically related to the coupling of combustion port fluid dynamics and the heat transfer to the fuel grain surface. The primary combustion region is known to be contained within a relatively narrow flame zone located within the boundary layer region that develops and grows over the fuel grain surface (Ref. 16–10). Heat is transferred to fuel grain surfaces by convection and radiation. Because the study of hybrids is largely empirical, it is well understood that any given motor characteristics will strongly depend on the propellant system and on the scale and configuration of the combustion chamber being described.

Figure 16–5 shows a simplified model of the hybrid combustion process for a nonmetallized (or low radiation) fuel system. Fuel vaporized as a result of flame zone

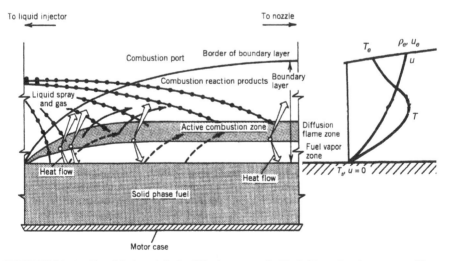

FIGURE 16–5. Simplified model of a diffusion-controlled hybrid combustion process, illustrating a flame zone embedded within the fuel boundary layer. T is temperature and u is velocity, the subscript e denotes external to boundary layer.

heating flows away from the surface toward the flame region, while oxidizer convects from the free stream (core flow) to the flame zone by turbulent diffusion. The flame establishes itself at a location within the boundary layer determined solely by the stoichiometry under which combustion can occur. The thickness of this flame zone is governed primarily by the rate at which oxidation reactions occur. Such rates are largely dependent on local pressures and typically follow an exponential dependence on temperature.

Factors beyond pressure and gas temperature affecting the development of the fuel-grain boundary layer, and hence fuel regression characteristics, include grain composition, combustion port oxidizer mass flow rate, and combustion port length and cross-sectional area. Existing boundary layers are more complex and more robust than those introduced in Chapter 3 inside supersonic nozzles. Heat transfer relationships between the gas and the solid phase strongly depend on whether the boundary layer is laminar or turbulent. In a typical hybrid using oxygen, the Reynolds number per unit axial length is on the order of 1 to 2×10^5 per inch of grain length for flux levels between 0.3 and 0.67 lbm/sec-in.^{2} Thus, fully turbulent boundary layers govern convective heat transfer (to nonmetallized fuel grains) and empirical equations must often be utilized. Figure 16–6 summarizes overall behavior of many oxidizer/solid-fuel combinations. As shown in this figure, there are three distinct regimes as a function of increasing mass–flow–velocity in the free stream ($G = \rho v$, total or oxidizer mass flow rate). At the low mass flux regime, radiative heat transfer phenomena are manifest in the form of pressure and diameter effects on the optical transmissivity of the propellant gas (Refs. 16–11 and 16–12), effects which may also arise from any metal loading present; at the "melting limit" the fuel grain may melt, char, or undergo subsurface decomposition (Ref. 16–11). The intermediate range represents fully turbulent heat and mass transfer, and the regression rate dependence on $G^{0.8}$ is consistent with the traditional turbulent diffusion results. At the high mass flow rates effects from gas-phase kinetics on chemical reactions are apparent and a different type of pressure dependence appears (the "flooding limit" results when the flame is extinguished by the high oxidizer flow rates, and depends on pressure and chemical

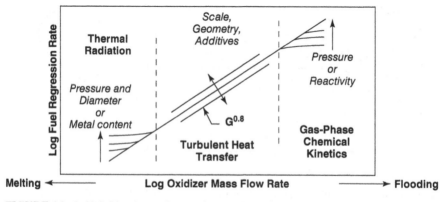

FIGURE 16–6. Hybrid regimes of regression rate dependencies. (Adapted from Ref. 16–11.)

reactivity of the propellants). Because solid fuel regression rates are a function of many parameters, the reader is encouraged to consult the relevant literature before using any of the correlations included in this chapter. A thorough summary of classical hybrid regression rate equations is given in Ref. 16–11, which clearly exemplifies the complexity of this subject.

With nonmetallized fuel grains, at the pressures and flux levels of interest for propulsion applications, heat transfer by convection is considered to be much larger than that transferred by gas-phase radiation or radiation from soot particles in the flow. As a result the basic characteristics of fuel grain regression may be explored via analyses of convective heat transfer in a turbulent boundary layer. Thus, for the central regime of Fig. 16–6, the following equation may be used for the fuel regression rate, $\dot{r}$:

$$\dot{r} = 0.036\frac{G^{0.8}}{\rho_f}\left(\frac{\mu}{x}\right)^{0.2}\beta^{0.23} \qquad (16\text{--}1)$$

Here, G is the total free-stream propellant mass velocity (oxidizer and fuel flow per unit area) at any given axial combustion port location x, ρ_f is the solid-phase fuel density, μ is the combustion gas viscosity, and β is a nondimensional fuel mass flux parameter reflecting fuel vaporization and evaluated at the fuel surface. All dimensions in Eq. (16–1) need to be in the English Engineering system of units to match the numerical constant. The parameter β is frequently called the *blowing coefficient* and can be shown to also represent a nondimensional enthalpy difference between the fuel surface and flame zone (i.e., $\Delta h/h_v$ where h_v is the heat of gasification). Equation (16–1) indicates that hybrid fuel regression rates in this nonradiative regime are strongly dependent on G and rather weakly dependent on both axial location (x) and fuel blowing characteristics (β). It should also be noted that these regression rates are not explicitly dependent on chamber pressure. This is not the case for the other regimes shown in Fig. 16–6 because metallized hybrid fuel systems may exhibit pronounced pressure dependences (see Ref. 16–13).

As propellants move down the combustion port, the gasified fuel being added in the port passage increases the total mass flux. In locations operating at low mixture ratios, this fuel mass increase may be of the same order as the oxidizer mass flow initially entering the port. Given the weak dependence of regression rate on x in Eq. (16–1), it would be expected that fuel regression would increase along the flow direction with increases of G. While this generally turns out to be the case, fuel regression has been observed to both increase and decrease with increasing x, depending on the details of the rocket motor configuration. In practice, axial fuel regression characteristics are strongly influenced by the method of oxidizer injection and by existing precombustion/vaporization chamber design characteristics. Some general trends observed with increasing x are: total mass flux increases, boundary layer thickness growths, flame-standoff distance from the surface increases, combustion port average gas density increases, and oxidizer concentration decreases.

Since the blowing coefficient β is not only an aerodynamic parameter but also a thermochemical parameter, and since its x dependence is of the same order, Eq. (16–1) is often simplified for purposes of preliminary engineering design by lumping the effects of x, β, fuel density, and gas viscosity into one parameter, usually given the symbol a. Moreover, in practice, some deviations from the theoretical 0.8-power of the mass velocity exponent are often noted. The resulting simplified form of Eq. (16–1) is therefore written as

$$\dot{r} = a(G_o)^n \qquad (16\text{–}2)$$

Here, G_o is the oxidizer mass velocity (i.e., the oxidizer mass flow rate divided by the combustion port cross-sectional area). Also, the parameters a and n are empirically fitted constants. In the central range of Fig. 16–6, $\dot{r}$ has been observed to have values between 0.05 and 0.2 in./sec and n has been observed to range between 0.4 and 0.7.

In the low mass flux end of Fig. 16–6 or when metallized fuels are utilized, radiation heat transfer cannot be neglected. Recent studies of this region have yielded accurate but elaborate empirical fits, which are reported in Ref. 16–11. In essence, a radiation heat flux term must be added to the convective heat arriving at the fuel surface, and this results in modifications to the turbulent diffusion relations applicable here. A simpler empirical correlation that somewhat represents modifications in this end range behavior (Ref. 16–14) may be written as follows:

$$\dot{r} \approx 2.50\,\dot{r}_{\text{ref}}(1 - e^{-p_1/p_{\text{ref}}})(1 - e^{-D/D_{\text{ref}}}) \qquad (16\text{–}3)$$

where p_{ref} and D_{ref} are empirically selected reference pressures and diameters, and the form of $\dot{r}_{\text{ref}}$ is given by Eq. (16–2) (multiplied by 2.50 to normalize it).

At the high mass flux end, chemical kinetic effects dominate, and here a strong pressure dependence of the form $(p_1)^{0.5}$ is both predicted and observed. A useful empirical fit in the region where heterogeneous reaction kinetics plays a strong role is of the form

$$\dot{r} = (\text{constant})(p_1)^{0.5}G^n x^{-m} \qquad (16\text{–}4)$$

where the exponent n varies between 0.3 and 0.4 (lower values than 0.8 from the middle domain), and the exponent m varies between 0.1 and 0.2 reflecting a consistently weak dependence on the axial location (see Ref. 16–11 for details).

Whenever there is deep reliance on empirical information, as in this field, further guidelines are needed before using any given formulation; otherwise, applying these equations must be restricted to their range of experimental observation. *Dimensional analysis and similarity theories* are designed to generalize experimental results by casting relevant variables in suitable dimensionless forms (e.g., see Eq. 8–20 and Problem 16–2) but the most satisfactory approach is the one where theory and experiment go hand in hand. The interested reader should consult Refs. 16–12 and 16–15.

16.3. PERFORMANCE ANALYSIS AND GRAIN CONFIGURATION

A characteristic operating feature of classical hybrids configurations is that fuel regression rates are usually less than one-third those for composite solid rocket propellants. It is very difficult to obtain fuel regression rates comparable to burn rates in solid rocket motors. Consequently, practical high-thrust hybrid motor designs must have multiple perforations (combustion ports) in the fuel grain to produce the relatively large fuel surface area required. The performance of a hybrid rocket motor (defined in terms of delivered specific impulse) depends critically on the degree of flow mixing attained in the combustion chamber. High performance stems from high combustion efficiency, which is a direct function of the thoroughness with which unburned oxidizer mixes with unburned fuel. Downstream of the fuel grain, extra chamber combustion ports can serve to promote higher combustion efficiencies in the turbulent mixing environment of unreacted fuel and oxidizer.

A cross section of a typical high-thrust hybrid fuel grain is shown in Fig. 16–4. The number of combustion ports required is a motor optimization problem that must account for the desired thrust level, acceptable shifts in mixture ratio during burn, motor length and diameter constraints, and desired oxidizer mass velocity. Hybrid rocket motor designs typically begin with the specification of a desired thrust level and propellant combination. Subsequently, selection of the desired operating oxidizer-to-fuel mixture ratio (O/F ratio) determines propellant characteristic velocity. Once the characteristic velocity and mixture ratio are specified, total propellant flow rate and the subsequent split between oxidizer and fuel flow rates necessary to produce the required thrust level can be computed. The necessary fuel flow rates in a hybrid are determined by total fuel surface areas (perimeter and length of the combustion ports) and by fuel regression rates. As will be shown in subsequent sections, fuel regression rates are primarily determined by the oxidizer mass velocity, also called the oxidizer flux which is equal to the mass flow rate of oxidizer in a combustion port divided by the port cross-sectional area. Thus, fuel flow rates are intrinsically linked to oxidizer flow rates and cannot be independently specified, as in liquid rocket engines.

Much but not all of the technology from liquid and solid propellant rockets can be directly applied to hybrid rockets; the main differences lie in the driving mechanisms for solid component burning and hybrid fuel regression. In solid propellants, the oxidizer and fuel ingredients are well mixed during propellant manufacturing process; combustion occurs as a result of heterogeneous chemical reactions on or very near the surface of the solid propellant; the solid propellant burning rate is controlled by chamber pressure and follows the well-established burning rate of Eq. 12–5, with empirical coefficients derived experimentally for specific propellant formulations. Since the rate of propellant gasification per unit area in a solid rocket motor (at a given propellant bulk temperature and in the absence of erosive burning) is determined only by chamber pressure, motor thrust is predetermined by the initial propellant grain surface area and grain geometrical characteristics.

In hybrids with metallized fuel grains, radiation from the meta- oxide particle cloud in the combustion port contributes a major portion of the total heat flux to the fuel grain. The local regression rate of the fuel is also quite sensitive to the general turbulence level of the combustion port gas flow (Refs. 16–16 and 16–17). Localized combustion gas eddies or recirculation zones adjacent to the fuel surface act to significantly enhance the regression rate in these areas. Hybrid fuel regression rate is considered to be insensitive to fuel grain bulk temperatures over the range in which solid rocket motors may operate (−65 to +165 °F). This is due to the absence of heterogeneous fuel/oxidizer reactions at the fuel surface (in which the reaction rates are temperature dependent) and because, over the above temperature range, any change in heat content of the solid fuel is small compared to the heat necessary to initiate vaporization of the fuel surface.

Selection of fuel ingredients can also have a significant impact on grain regression rates, which are largely a function of the energy required to convert the fuel from solid to vapor phase (h_v). This energy is called the heat of gasification and, for polymeric fuels, includes the energy required to break polymer chains (heat of depolymeriza-tion) and the heat required to convert the polymer fragments to gaseous phase (heat of vaporization). The term "heat of vaporization" is often used as a catchall phrase to include all decomposition mechanisms in hybrid fuels. In nonmetallized fuels, low heats of gasification tend to produce higher regression rates. In metallized fuels, the addition of ultra-fine aluminum (UFAl) powder (particle sizes on the order of 0.05 to 0.1 μm) to HTPB has been noted to significantly increase fuel regression rates rel-ative to a pure HTPB baseline (see Ref. 16–18 and Fig. 16–7). Hybrid propellants containing aluminum particles with diameters typical of those used in solid rocket propellants (40 to 400 μm) do not exhibit this effect.

An alternative form of Eq. (16–2), to account for an observed pressure and/or port diameter dependency, is written as

$$\dot{r} = aG_o^n p_1^m D_p^l \qquad (16\text{--}5)$$

where m has been observed to vary between zero and 0.25 and l between zero and 0.7.

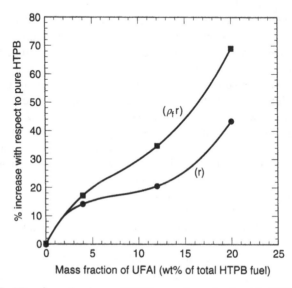

FIGURE 16–7. Ultra-fine aluminum (UFAl) powder mixed with HTPB significantly increases the fuel regression rate.

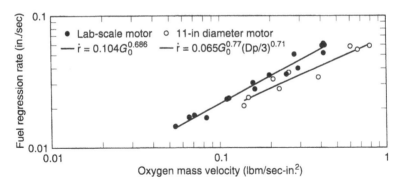

FIGURE 16–8. Regression-rate decrease with hybrid motor scale (combustion port diameter) increases.

Figure 16–8 illustrates surface regression rate data obtained for the combustion of HTPB fuel grains and gaseous oxygen in rocket motor tests at two different scales. The first data set were obtained by testing fuel grains in a small laboratory-scale (2-in. motor diameter with a 0.43-in. combustion port diameter) rocket at varying gaseous oxygen flux levels. A least-squares regression analysis, performed to determine the constants in Eq. (16–2), indicates that, at this scale, the following relationship best describes the regression rate characteristics of HTPB as a function of oxygen mass flux (English Engineering units):

$$\dot{r}_{HTPB} = 0.104G_o^{0.681} \tag{16–6}$$

Data obtained with the same propellant system in a larger 11-in.-diameter hybrid motor with combustion port diameters ranging between 3 and 6 in. exhibited a relatively stronger dependence on combustion port diameter. Data from this testing was best matched with an expression in the form of Eq. (16–5) (English Engineering units):

$$\dot{r}_{HTPB} = 0.065G_o^{0.77}(D_p/3)^{0.71} \tag{16–7}$$

Such difference in fuel regression characteristics between the two motor scales illustrates one of the central difficulties in hybrid motor design, that is, that of scaling ballistic performance. Scaling issues in hybrid motors are currently not well understood (in part because of the lack of sufficient valid data for different motor sizes) while the literature abounds with empirical regression-rate scaling relationships. Computational fluid dynamic approaches that resolve the hybrid flow field and calculate fuel surface heating appear to offer the best hope for analytically evaluating scale effects (Ref. 16–19).

Dynamic Behavior

Dynamic behavior in hybrid propulsion systems is important because, as stated earlier, mixture ratio always varies even during steady-state oxidizer flow. When mass storage term changes (i.e., the propellant gas that remains to fill the increasing chamber volume as in solid motors, Section 12.1) may be neglected in the combustor cavity, then Eqs. 6-2 and 3-32 apply, namely,

$$\dot{m} = \dot{m}_o + \dot{m}_f = \frac{p_1 A_t}{c^*} \tag{16-8}$$

The steady thrust of a hybrid rocket motor can then be expressed as

$$F = \dot{m} I_s g_0 = (\dot{m}_o + \dot{m}_f) I_s g_0 \tag{16-9}$$

Changing the thrust or throttling a hybrid motor (in the classical configuration) is achieved by changing the oxidizer flow rate, usually by means of a throttling valve in the oxidizer feed line. The fuel flow is a function of the oxidizer flow, but not necessarily a linear function. For circular port geometries with inner radius R, Eq. (16–2) may be recast as

$$\dot{r} = a \left(\frac{\dot{m}_o}{\pi R^2} \right)^n \tag{16-10}$$

The mass flow rate of fuel is given by

$$\dot{m}_f = \rho_f A_b \dot{r} = 2\pi \rho_f R L \dot{r} \tag{16-11}$$

where A_b is the combustion port surface area and L is the port length. Combining Eqs. (16–10) and (16–11), we obtain the fuel production rate in terms of port radius and oxidizer mass flow rate:

$$\dot{m}_f = 2\pi^{1-n} \rho_f L a \dot{m}_o^n R^{1-2n} \tag{16-12}$$

From this expression we note that, for $n = \frac{1}{2}$, the fuel mass flow rate becomes independent of combustion port radius and varies as the square root of oxidizer mass flow rate. For such situation, if the oxidizer flow is reduced to one-half of its rated value, the fuel flow will be reduced by a factor of 0.707 and the rocket motor thrust, which depends on the total propellant flow ($\dot{m}_f + \dot{m}_o$), will not vary linearly with the change in oxidizer flow. Usually, as thrust is decreased by reducing oxidizer flow, the mixture ratio ($\dot{m}_o / \dot{m}_f$) is also reduced, becoming increasingly fuel rich. In some hybrid motor concepts, a portion of the oxidizer is injected in a mixing chamber downstream of the fuel grain in order to maintain a more constant mixture ratio. However, for most applications, system design may be optimized over the range of mixture ratios encountered with very little degradation of average specific impulse due to throttling.

Equation (16–12) also indicates that, for constant oxidizer flow, fuel flow will increase with increasing port radius if $n < \frac{1}{2}$. For $n > \frac{1}{2}$, fuel flow will decrease with increasing port radius.

For a fuel grain incorporating N circular combustion ports, Eq. (16–10) can be simply integrated to give combustion port radius, instantaneous fuel flow rate, instantaneous mixture ratio, and total fuel consumed as functions of burn time (t):

Combustion port radius $R(t)$ as a function of time and oxidizer flow rate:

$$R(t) = \left\{ a(2n + 1)\left(\frac{\dot{m}_o}{\pi N}\right)^n t + R_i^{2n+1} \right\}^{1/(2n+1)} \tag{16-13}$$

Instantaneous fuel flow rate:

$$\dot{m}_f(t) = 2\pi N\rho_f La\left(\frac{\dot{m}_o}{\pi N}\right)^n \left\{ a(2n + 1)\left(\frac{\dot{m}_o}{\pi N}\right)^n t + R_i^{2n+1} \right\}^{(1-2n)/(1+2n)} \tag{16-14}$$

Instantaneous mixture ratio:

$$\frac{\dot{m}_o}{\dot{m}_f}(t) = \frac{1}{2\rho_f La}\left(\frac{\dot{m}_o}{\pi N}\right)^{1-n} \left\{ a(2n + 1)\left(\frac{\dot{m}_o}{\pi N}\right)^n t + R_i^{2n+1} \right\}^{(2n-1)/(2n+1)} \tag{16-15}$$

Total fuel consumed:

$$m_f(t) = \pi N\rho_f L\left[\left\{ a(2n + 1)\left(\frac{\dot{m}_o}{\pi N}\right)^n t + R_i^{2n+1} \right\}^{2/(2n+1)} - R_i^2 \right] \tag{16-16}$$

where L is the fuel grain length, R_i the initial port radius, N the number of combustion ports of radius R_i in the fuel grain, and $\dot{m}_o$ and $\dot{m}_f$ are the total oxidizer and fuel flow rates, respectively. Although the above equations are valid strictly for circular combustion ports, they may be used to give qualitative understanding for hybrid motor behavior applicable to the burning in noncircular ports as well.

16.4. DESIGN EXAMPLE

The preliminary design parameters for a large hybrid booster are to be determined as per the following example.

Example 16–1. Determine several key hybrid RPS design parameters for a proposed large hybrid rocket booster stage of a space flight vehicle which is to be tested at sea level.

Fuel	HTPB
Oxidizer	Liquid oxygen
Required booster initial thrust (vacuum)	3.1×10^6 lbf
Burn time	120 sec
Fuel grain outside diameter	150 in.
Initial chamber pressure	700 psia
Initial mixture ratio (Eq. 6-1)	2.0
Initial nozzle area ratio	7.72
Ambient pressure	14.696 psia (sea level)

TABLE 16–2. Theoretical Characteristic Velocity c* and Ratio of Specific Heats k for Reaction Gases of Liquid Oxygen–HTPB Fuel

Mass Mixture Ratio	c^*(ft/sec)	k
1.0	4825	1.308
1.2	5180	1.282
1.4	5543	1.239
1.6	5767	1.201
1.8	5882	1.171
2.0	5912	1.152
2.2	5885	1.143
2.4	5831	1.138
2.6	5768	1.135
2.8	5703	1.133
3.0	5639	1.132

Data is based on thermochemical rocket propellant analyses similar to those described in Chapter 5.

SOLUTION. Using the ratio of specific heats from Table 16–2 shown below and the given initial nozzle exit area ratio, the sea-level thrust coefficient is determined from Eq. 3–30 to be 1.572. Initial nozzle throat area and throat diameter are determined from Eq. 3–31

$$A_t = \frac{F_v}{C_{F_v}p_1} = \frac{3.1 \times 10^6 \text{lbf}}{(1.572)(700 \text{lbf}/\text{in.}^2)} = 2817.16 \text{ in.}^2$$

then $D_t = 59.89$ in. From Table 16–2, c* is listed for initial mixture ratio of 2.0 as 5912 ft/sec. Theoretical c^* values are typically degraded to account for combustion inefficiencies due to incomplete oxidizer/fuel mixing. Using a factor of 95%, the delivered c^* is 5616 ft/sec. Total initial propellant flow rate can now be determined from Eq. 3–32 as (g_0 is a conversion factor)

$$\dot{m} = \frac{g_0 p_1 A_t}{c^*} = \frac{\left(32.174 \frac{\text{lbm-ft}}{\text{lbf-sec}^2}\right)(700 \text{lbf}/\text{in.}^2)(2817.16 \text{ in.}^2)}{(0.95)(5912 \text{ft}/\text{sec})} = 11{,}297 \text{ lbm/sec}$$

Mixture ratio is defined as in Eq. 6-1,

$$r = \dot{m}_o/\dot{m}_f$$

Initial fuel and oxidizer flow rates follow at the initial mixture ratio of 2.0:

$$\dot{m} = \dot{m}_o + \dot{m}_f = \dot{m}_f(r + 1)$$

$$\dot{m}_f = \frac{11{,}297 \text{ lbm/sec}}{3} = 3765.6 \text{ lbm/sec}$$

$$\dot{m}_o = 11{,}297 - 3766 = 7531.2 \text{ lbm/sec}$$

Figure 16–9a illustrates a candidate seven-circular-port symmetric fuel grain configuration. The dotted lines represent the diameters to which the combustion ports burn at the end

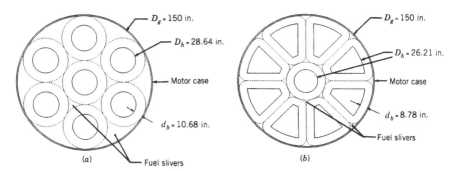

FIGURE 16–9. (a) Circular fuel grain combustion ports are volumetrically inefficient and leave large slivers at burnout. (b) Quadrilateral port hybrid grain configuration minimizes residual fuel sliver at burnout.

of 120 sec. The problem is to determine the initial port diameter such that, at the end of the specified 120-sec burn time, the grain diameter constraint of 150 in. is satisfied. The unknown quantity in this problem is the initial combustion port radius, R_i, and the fuel burn distance, d_b. In terms of initial port radius, the burn distance can be expressed via Eq. (16–13) as

$$d_b = R(t, R_i)|_{t=120} - R_i$$

The fuel grain diameter requirement of 150 in. is satisfied by the following relation:

$$150 \text{ in.} = 6R_i + 6d_b$$

Subscale motor test data indicate that one expression for the fuel surface regression rate can be described by Eq. (16–6). Assuming that these data are valid for the flux levels for circular ports and port diameters under consideration (ignoring potential regression rate scaling issues), the above two relations can be combined to solve for the initial port radius and distance burned, yielding

$$R_i = 14.32 \text{ in.} \qquad d_b = 10.68 \text{ in.}$$

Knowing the initial port radius, the oxidizer mass velocity can be determined:

$$G_o = \frac{\dot{m}_o}{NA_p} = \frac{7531 \text{lbm/ sec}}{7\pi(14.32 \text{ in.})^2} = 1.67 \text{ lbm/in.}^2\text{- sec}$$

The initial fuel regression rate may be explicitly determined from Eq. (16–6):

$$\dot{r}_i = 0.104 G_{oi}^{0.681} = 0.104(1.67 \text{ lbm/ft}^2\text{- sec})^{0.681} = 0.1475 \text{ in./ sec}$$

From the initial fuel mass flow rate, determined to be 3765.6 lbm/sec, the fuel grain length required for a seven-circular-port design may be found from Eq. (16–11):

$$L = \frac{\dot{m}_f/N}{2\pi R_i \rho_f \dot{r}_i} = \frac{(3765.6 \text{ lbm/ sec})/7}{\pi(28.65 \text{ in.})(0.0331 \text{ bm/in.}^3)(0.1475 \text{ in./ sec})} = 1224.6 \text{ in.}$$

Using Eqs. (16–8), 16-9, 16-14, 16-15, and 16-16, while neglecting effects of throat erosion, the general operating characteristics of the booster may be computed with respect to time. The total fuel and oxidizer required for a 120-sec burn time are determined to be 451,872 and 903,720 lbm, respectively. The total propellant mass required is therefore 1,355,592 lbm.

Selection of circular fuel ports is not an efficient way of designing a hybrid grain since large fuel slivers will remain at the end of burn and it is heavier than other designs. In the preceding example, a sliver fraction (1 minus fuel consumed divided by fuel loaded) of about 30% applies. Recognizing that uniform burn distances around each port, as well as between combustion ports and the case wall, will minimize residual fuel sliver, the outer ring of circular ports may be replaced with quadrilateral-shaped ports. Such a grain is illustrated in 16-9b. If, as before, the grain diameter is constrained to be 150 in., the grain geometry is uniquely determined by specification of the initial fuel and oxidizer flow rates, number of ports, burn time, and the requirement that the burn distance around each port be equal. Additionally, the hydraulic diameter D_h (four times port area divided by port perimeter) of all ports should be equal to assure that all ports have the same mass flow rate.

For the example in Fig. 16–9b, the nine-port grain configuration results in a theoretical fuel sliver fraction of about 4%. In reality, the sliver fraction for both designs will be somewhat greater than theoretical values since some web must be designed to remain between ports at the end of the burn duration (to prevent slivers from being expelled out of the nozzle). Table 16–3 compares key features of the circular port grain design (Fig. 16–9a) and the quadrilateral grain design (Fig. 16–9b).

TABLE 16–3. Comparison of Circular Port and Quadrilateral Port Grain Designs

Design Parameter	Circular Port	Quadrilateral Port
Oxidizer flow rate (lbm/sec)	7531	7531
Initial fuel flow rate (lbm/sec)	3766	3766
Burn time (sec)	120	120
Grain outer diameter (in.)	150	150
Number of combustion ports	7	9
Oxidizer flux (lbm/sec/in.^{2})	1.67	1.07 (est.)
Fuel regression rate (in./sec)	0.1475	0.109 (est.)
Grain length (in.)	1,225	976.1 (est.)
Fuel consumed (lbm)	451,872	348,584 (est.)
Theoretical sliver fraction (%)	30	4

(est.) = estimated value

In this example, the fuel consumed by the quadrilateral port design is estimated to be less than that consumed by the circular port design. Therefore, the total impulse of the two designs could differ. If fuel consumed were constrained to be the same in each design, we would find that, as the number of quadrilateral fuel ports would increase, the grain length would decrease and grain diameter would increase. In practice, the hybrid motor designer must carefully balance vehicle system requirements, such as total impulse and envelope constraints, with available grain design options to arrive at an optimum motor configuration. Total propellant and propellant contingency necessary to accomplish a specific mission will depend upon such factors as residual fuel and oxidizer allowances at motor cutoff, ascent trajectory throttling requirements, which impact overall mixture ratio and oxidizer utilization, and additional propellant if a Δu (vehicle velocity necessary to achieve mission objectives) contingency reserve is required.

Using Table 16–2 to obtain c^*, the initial sea-level delivered specific impulse for the circular port booster design may be calculated as

$$I_{s_v} = \frac{(C_F)_v c^*}{g_0} = \frac{(1.572)(0.95)(5912 \text{ft/ sec})}{32.174 \frac{\text{lbm-ft}}{\text{lbf} - \text{sec}^2}} = 274 \text{ sec}$$

In general, the effect of throat erosion in ablative nozzles on overall rocket motor performance depends on the initial throat diameter. For the booster design under consideration, an assumed 0.010-in./sec erosion rate acting only at the throat will reduce the nozzle expansion area ratio from 7.72 to 7.11 over the 120-sec burn time. Using an estimated end-of-burn mixture ratio and if throat erosion is accounted for, a reduction of specific impulse of about 1.0% compared with the noneroding throat assumption would be expected. The throat material erosion rate in a hybrid is generally significantly greater than that of a solid propellant system and is a strong function of chamber pressure and mixture ratio. Erosion of carbonaceous throat materials (carbon cloth phenolic, graphite, etc.) is primarily governed by heterogeneous surface chemical reactions involving the reaction of carbon with oxidizing species present in the flow of combustion gases such as O_2, O, H_2O, OH, and CO_2 to form CO. Hybrid motor operation at oxygen-rich mixture ratios and high pressure will result in comparatively high throat erosion rates. Operation at fuel-rich mixture ratios and pressures below 400 psi will result in very low throat erosion rates.

Current practice for preliminary design of hybrid booster concepts is to couple a fuel regression rate model, a grain design model, and booster component design models in a numerical preliminary design procedure. Using numerical optimization algorithms, such a computer model looks for the optimum booster design that maximizes selected optimization variables, such as booster ideal velocity or total impulse, while minimizing booster propellant and inert weight. New manufacturing techniques (additive manufacturing) and green propellants are presently being investigated, see Ref. 16–23.

16.5. COMBUSTION INSTABILITY

Some hybrid combustion processes tend to produce rough pressure-time characteristics. However, a well-designed hybrid can typically limit combustion roughness to tolerable levels, approximately 2 to 3% of mean chamber pressure. In any combustion device, pressure fluctuations will tend to organize themselves around the natural acoustic frequencies of the combustion chamber or the oxidizer feed system. While significant combustion pressure oscillations at chamber natural-mode acoustic frequencies have been observed in numerous hybrid motor tests, such oscillations have not proved to be an insurmountable design problem. When pressure oscillations have occurred in hybrid motors, they have been observed to grow only to a limiting amplitude that depends on factors such as oxidizer feed system and injector characteristics, fuel-grain geometric characteristics, mean chamber pressure levels, and oxidizer mass velocities. Unbounded growth of pressure oscillations, such as may occur in solid and liquid rocket motors, has not been observed in hybrid motors.

In static tests, hybrid motors appear to exhibit two basic types of instability: oxidizer-feed-system induced instability (nonacoustic), and flame holding instability (acoustic). Oxidizer feed system instability is essentially a chugging type as described in Chapter 9 and arises when the feed system is sufficiently pliable or "soft." In cryogenic systems, this implies a high level of compressibility from sources such as vapor cavities or two-phase flow in feed lines combined with insufficient isolation from motor combustion processes. Figure 16–10 (a) illustrates feed system induced instability in a 24-in.-diameter hybrid rocket motor that operated at a LOX flow rate of 20 lbm/sec with HTPB fuel. This instability is manifested by high-amplitude, periodic oscillations well below the first longitudinal (1-L) acoustic mode of the

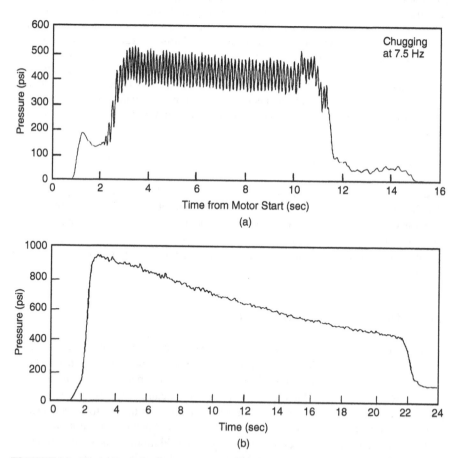

FIGURE 16–10. (a) Periodic, large-amplitude, low-frequency combustion pressure oscillations are an example of oxidizer feed system induced "chug"-type combustion instability in a 24-in.-diameter LOX/HTBP motor. (b) Example of stable combustion in the same 24-in.-diameter LOX/HTPB motor, exhibiting an overall combustion roughness level of 1.3%.

combustor. In this example the oscillation frequency is 7.5 Hz whereas the 1-L mode frequency is approximately 60 Hz. Stiffening the feed/injection system (e.g., making it more resistant to vibration) can eliminate such oscillation. This is accomplished by increasing the injector pressure drop (thus making propagation of motor pressure disturbances upstream through the feed system more difficult) and eliminating sources of compressibility in the feed system. Chugging-type instabilities in hybrid motors have proven amenable to analysis in terms of prediction and prevention (Ref. 16–20). For comparison, Fig. 16–10(b) shows a pressure–time trace from the same 24-in.-diameter hybrid motor exhibiting under stable combustion while being operated at a LOX flow rate of 40 lbm/sec and at a maximum chamber pressure of 900 psi.

Flame-holding instabilities relevant to hybrid motors were first observed during the development of solid fuel ramjets (Ref. 16–21). A solid fuel ramjet is essentially a hybrid motor operating on atmospheric oxygen available from rammed air. Flame-holding instabilities in hybrids typically manifest at acoustic frequencies and appear in longitudinal modes. No acoustic instabilities in hybrid motors have been observed in the higher frequency tangential or radial modes that appear in solid rocket motors or in liquid rocket engines. Flame-holding instabilities arise from inadequate flame stabilization inside boundary layers (Ref. 16–22) and are not associated with feed system flow perturbations. In one flame-holding instability example, an 11-in.-diameter hybrid motor was tested with gaseous oxygen (GOX) oxidizer and HTPB fuel, using an injector producing a conical flow field; here, oxygen flow was initiated through the hybrid motor at a pressure of 90 psi for 2 sec prior to motor ignition; the motor was ignited using a hydrogen torch that continued to operate for approximately 1 sec following motor ignition. During the first second of motor operation, the hydrogen igniter flame stabilized the motor but when the igniter flame was extinguished, the motor became unstable. In this case stable combustion was achieved by changing the flow field within the motor, using an injector producing an axial flow field. Figure 16–11 shows the results, decomposing the pressure versus time signal for this unstable example into its frequency components via fast Fourier transform techniques. The 1-L (longitudinal) acoustic oscillation mode is clearly visible at around 150 Hz.

It is now apparent that flame-holding instabilities can be eliminated by several means, all of which act to stabilize combustion in the boundary layers. The first method is to use a "pilot flame" derived from injecting a combustible fluid such as hydrogen or propane to provide sufficient oxidizer preheating at the leading edge region of the boundary layer flame zone. With this technique, motor stability characteristics are relatively insensitive to the nature of the injector flow field. In the example above, the hydrogen torch igniter acted as a pilot during its period of operation. A second method involves changing the injector flow field to ensure that a sufficiently large hot gas recirculation zone is present at the head end of the fuel grain. Such a zone can be created by forcing the upstream flow over a rearward-facing step or by strong axial injection of oxidizer (see Fig. 16–12). Axial injection in the correct configuration produces a strong counter-flowing hot gas recirculation zone, similar to that in a rearward-facing step, at the head end of the diffusion flame (conical injection produces a much smaller and usually ineffective recirculation zone).

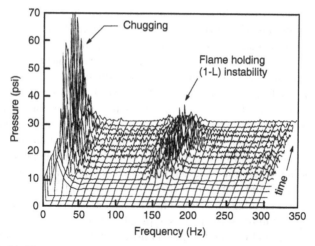

FIGURE 16–11. Frequency-versus-amplitude plot at successive time intervals for an 11-in.-diameter GOX/HTPB motor test shows pressure oscillations in the motor first longitudinal (1-L) acoustic mode at 150 Hz due to flame-holding instability.

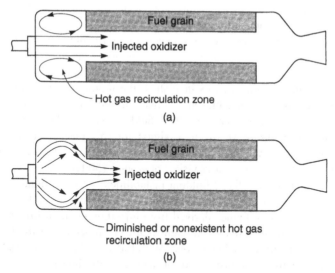

FIGURE 16–12. (a) Sketch of axial injection into a vaporization chamber of oxidizer that results in a strong hot gas flow recirculation zone at the fuel grain leading edge, producing stable combustion. (b) Conical oxidizer injection may produce a weak or nonexistent hot gas flow recirculation zone at the fuel grain leading edge, resulting in unstable combustion.

These techniques produce flow field results very similar to those produced by bluff body flame stabilizers, used in jet engine afterburners and solid fuel ramjets to prevent flame blowoff. The recirculation zone acts to entrain hot gas from the core flow, which provides sufficient oxidizer preheating for the leading edge of the boundary layer diffusion flame to stabilize combustion.

Comparison of the average pressure levels in the above-mentioned instability illustrates an interesting phenomenon. For the same motor operating conditions (oxidizer flow rate, grain geometry and composition, and throat diameter) the average pressure level in the unstable motor is significantly greater than that in the stable motor. This same phenomenon has been noted in solid propellant motors and results from intensification of heat transfer to the fuel surface due to gas convection at the fuel surface oscillating at high frequency. The high heating rate results in the vaporization of more fuel than would otherwise occur in equilibrium conditions, thus producing a higher average chamber pressure.

Despite recent advances in understanding causes of and solutions for combustion instability in hybrid motors, development of a comprehensive, predictive theory of combustion stability remains one of the major challenges in hybrid technology development (see Ref. 16–24).

SYMBOLS

a	burning or regression rate coefficient (units of a depend on value of oxidizer flux exponent)	variable units
A_p	combustion port area	m^2 (in.^{2})
A_s	fuel grain surface area	m^2 (in.^{2})
A_t	nozzle throat area	m^2 (ft^2)
c^*	characteristic velocity	m/sec (ft/sec)
C_{F_v}	vacuum thrust coefficient	dimensionless
c_p	heat capacity	J/kg-K (Btu/lbm-°R)
d_b	fuel grain burn distance	m (in.)
D_h	hydraulic diameter ($4A_p/P$)	m (in.)
D_p	combustion port diameter	m (in.)
D_t	nozzle throat diameter	m (in.)
F_v	vacuum thrust	N (lbf)
G	mass velocity	kg/m^2-sec (lbm/ft^2-sec)
G_o	oxidizer mass velocity	kg/m^2-sec (lbm/ft^2-sec)
g_0	acceleration of gravity—conversion factor	m/sec^2—32.174 lbm-ft/lbf-sec^2
h	convective heat transfer coefficient	J/m^2-sec/K (Btu/ft^2-sec/°R)
h_v	heat of gasification	J/kg (Btu/lbm)
Δh	flame zone–fuel surface enthalpy difference	J/kg (Btu/lbm)

H_f	heat of formation	J/kg-mol (kcal/mol)
I_s	specific impulse	Sec
k	specific heat ratio	dimensionless
L	combustion port length	m (in.)
$\dot{m}$	propellant flow rate	kg/sec (lbm/sec)
$\dot{m}_f$	fuel flow rate	kg/sec (lbm/sec)
$\dot{m}_o$	oxidizer flow rate	kg/sec (lbm/sec)
n, m, l	burning or regression rate pressure exponent	dimensionless
P	combustion port perimeter	m (in.)
p_1	chamber pressure	MPa (lbf/in.^{2})
R	combustion port radius	m (in.)
R_i	initial combustion port radius	m (in.)
Re	Reynolds number	dimensionless
$\dot{r}$	fuel regression rate	mm/sec (in./sec)
r	oxidizer to fuel mixture ratio	dimensionless
T	temperature	°C (°F)
t	time	Sec
u_e	gas free-stream velocity in axial direction	m/sec (ft/sec)
v	gas velocity normal to fuel surface	m/sec (ft/sec)
x	axial distance from leading edge of fuel grain	m (in.)
y	length coordinate normal to fuel surface	m (in.)

Greek Letters

α	fuel surface absorptivity	Dimensionless
β	boundary layer blowing coefficient	dimensionless
μ	gas viscosity	N−sec/m^2 (lbf−sec/ft^2)
ρ_1	combustion chamber gas density	kg/m^3 (lbm/in.^{3})
ρ_e	free stream gas density	kg/m^3 (lbm/in.^{3})
ρ_f	fuel density	kg/m^3 (lbm/in.^{3})

Subscripts

e	boundary layer edge conditions	
f	fuel	
i	initial conditions	
o	oxidizer	
s	surface conditions	
x	axial distance from leading edge of fuel grain	m (in.)
ref	reference conditions	

PROBLEMS

1. Consider a classical hybrid motor with a solid fuel grain stored in a single cylindrical enclosure. The combustion port radius is R and the length L. Using the fuel burning rate given by Eq. (16–2), namely, $\dot{r} = a(G_o)^n$ with $n = 0.7$:

   a. What happens to the thrust with time for a constant oxidizer mass flow rate during the burn?

   b. What happens to the mixture ratio if the oxidizer rate is decreased?

   Answers: (**a**) and (**b**) decreases.

2. **a.** Write Eq. (16–1) in terms of a *Reynolds number* based on the free-stream gas flow and the axial location x. What does this say about the Reynolds number dependence of surface regression rate?

   b. It is know that the blowing coefficient is itself roughly proportional to $x^{0.2}$. What additional assumptions might be necessary to transform Eq. (16–1) into 16–2?

3. Calculate the ideal density-specific impulse (in units of kg–sec/m³) for the propellant combination LOX/HTPB at an oxidizer-to-fuel ratio of 2.3 at a pressure ratio across the nozzle of 1000/14.7 and an optimum expansion. Use the information in Table 16–2 and a fuel density of 920 kg/m³. Compare your result with the value of LOX/RP-1 bipropellant listed in Table 5–5.

   Answer: 3×10^5 kg-sec/m³

4. Plot the instantaneous mixture ratio given by Eq. (16–15) as a function of the index n and time. Use values of n greater and less than ½ (at $n = \frac{1}{2}$ the time dependence vanishes). In order to focus only on the effects of n, work with the modified variables ψ (related to mixture ratio) and τ (related to time) defined as follows:

$$\psi = [(2n + 1)\tau + 1]^{(2n-1)/(2n+1)}$$

$$\psi \equiv \frac{\dot{m}_o}{\dot{m}_f} \frac{2\rho_f La}{R_i \left(\dfrac{\dot{m}_o}{\pi N R_i}\right)^{1-n}} \quad \text{and} \quad \tau \equiv a \left(\frac{\dot{m}_o}{\pi N R_i^2}\right)^n \frac{t}{R_i}$$

5. A classical hybrid propulsion system has the following characteristics:

Propellants	LOX/HTPB
Nominal burn time	20 sec
Initial chamber pressure	600 psi
Initial mixture ratio	2.0
Initial nozzle exit area ratio	10.0
Fuel grain outside diameter	12 in
Grain geometry	Single cylinder case bonded
Initial HTPB temperature	65 °F

   Determine initial and final thrust, specific impulse, propellant masses, and useful propellant.

REFERENCES

16–1. K. K. Kuo, Chapter 15, "Challenges of Hybrid Rocket Propulsion in the 21st Century," in M. A. Chiaverini and K. K. Kuo (Eds.), *Fundamentals of Hybrid Rocket Combustion and Propulsion*, Vol. 218, *Progress in Astronautics and Aeronautics*, AIAA, Reston, VA, 2007.

16–2. R. A. Frederick Jr., J. J. Whitehead, L. R. Knox, and M. D. Moser, "Regression Rates of Mixed Hybrid Propellants," *Journal of Propulsion and Power*, Vol. 23, No. 1, Jan.–Feb. 2007.

16–3. D. Altman and A. Holzman, "Overview and History of Hybrid Rocket Propulsion," in M. A. Chiaverini and K. K. Kuo (Eds.), *Fundamentals of Hybrid Rocket Combustion and Propulsion*, Vol. 218, *Progress in Astronautics and Aeronautics*, AIAA, Reston, VA, 2007, Chapter 1.

16–4. H. R. Lips, "Experimental Investigation of Hybrid Rocket Engines Using Highly Aluminized Fuels," *Journal of Spacecraft and Rockets*, Vol. 14, No. 9, Sept. 1977, pp. 539–545.

16–5. S. Heister and E. Wernimont, Chapter 11, "Hydrogen Peroxide, Hydroxil Ammonium Nitrate, and Other Storable Oxidizers," in M. A. Chiaverini and K. K. Kuo (Eds.), *Fundamentals of Hybrid Rocket Combustion and Propulsion*, Vol. 218, *Progress in Astronautics and Aeronautics*, AIAA, Reston, VA, 2007.

16–6. SpaceDev website: http://www.spacedev.com.

16–7. T. A. Boardman, T. M. Abel, S. E. Chaflin, and C. W. Shaeffer, "Design and Test Planning for a 250-klbf-Thrust Hybrid Rocket Motor under the Hybrid Demonstration Program," AIAA Paper 97-2804, Jul. 1997.

16–8. S. R. Jain and G. Rajencran, "Performance Parameters of Some New Hybrid Hypergols," *Journal of Propulsion and Power*, Vol. 1, No. 6, Nov.–Dec. 1985, pp. 500–501.

16–9. U. C. Durgapal and A. K. Chakrabarti, "Regression Rate Studies of Aniline-Formaldehyde-Red Fuming Nitric Acid Hybrid System," *Journal of Spacecraft and Rockets*, Vol. 2, No. 6, 1974, pp. 447–448.

16–10. G. A. Marxman, "Combustion in the Turbulent Boundary Layer on a Vaporizing Surface," Tenth Symposium on Combustion, The Combustion Institute, 1965, pp. 1337–1349.

16–11. M. A. Chiaverini, Chapter 2, "Review of Solid-Fuel Regression Rate Behavior in Classical and Nonclassical Hybrid Rocket Motors," in M. A. Chiaverini and K. K. Kuo (Eds.), *Fundamentals of Hybrid Rocket Combustion and Propulsion*, Vol. 218, *Progress in Astronautics and Aeronautics*, AIAA, Reston, VA, 2007.

16–12. F. P. Incropera and D. P. DeWitt, *Fundamentals of Heat and Mass Transfer*, 5th ed., John Wiley & Sons, Hoboken NJ, 2002.

16–13. L. D. Smooth and J. C. Price, "Regression Rates of Metallized Hybrid Fuel Systems," *AIAA Journal*, Vol. 4, No. 5, September 1965, pp. 910–915.

16–14. D. Altman and R. Humble, Chapter 7, "Hybrid Rocket Propulsion Systems," in R. W. Humble, G. N. Henry, and W. J. Larson, *Space Propulsion Analysis and Design*, McGraw-Hill, New York, 1995.

16–15. A. Gany, Chapter 12, "Similarity and Scaling Effects in Hybrid Rocket Motors," in M. A. Chiaverini and K. K. Kuo (Eds.), *Fundamentals of Hybrid Rocket Combustion and Propulsion*, Vol. 218, *Progress in Astronautics and Aeronautics*, AIAA, Reston, VA, 2007.

16–16. P. A. O. G. Korting, H. F. R. Schoyer, and Y. M. Timnat, "Advanced Hybrid Rocket Motor Experiments," *Acta Astronautica*, Vol. 15, No. 2, 1987, pp. 97–104.

16–17. W. Waidmann, "Thrust Modulation in Hybrid Rocket Engines," *Journal of Propulsion and Power*, Vol. 4, No. 5, Sept.–Oct. 1988, pp. 421–427.

16–18. M. A. Chiaverini et al., "Thermal Pyrolysis and Combustion of HTPB-Based Fuels for Hybrid Rocket Motor Applications," AIAA paper 96-2845, Jul. 1996; "M. A. Chiaverini, K. K. Kuo, A. Peretz, and G. C. Harting, Regression-Rate and Heat-Transfer Correlations for Hybrid Rocket Combustion, *Journal of Propulsion and Power*, Vol. 17, No. 1, Jan.–Feb. 2001, pp. 99–110.

16–19. V. Sankaran, Chapter 8, "Computational Fluid Dynamics Modeling of Hybrid Rocket Flowfields," in M. A. Chiaverini and K. K. Kuo (Eds.), *Fundamentals of Hybrid Rocket Combustion and Propulsion*, Vol. 218, *Progress in Astronautics and Aeronautics*, AIAA, Reston, VA, 2007.

16–20. T. A. Boardman, K. K. Hawkins, R. S. Wasson, and S. E. Claflin, "Non-Acoustic Feed System Coupled Combustion Instability in Hybrid Rocket Motors," AIAA Hybrid Rocket Technical Committee Combustion Stability Workshop, 31st Joint Propulsion Conference, 1995.

16–21. B. L. Iwanciow, A. L. Holzman, and R. Dunlap, "Combustion Stabilization in a Solid Fuel Ramjet," 10th JANNAF Combustion Meeting, 1973.

16–22. T. A. Boardman, D. H. Brinton, R. L. Carpenter, and T. F. Zoladz, "An Experimental Investigation of Pressure Oscillations and Their Suppression in Subscale Hybrid Rocket Motors," AIAA Paper 95-2689, Jul. 1995.

16–23. S. A. Whitmore et al., "Survey of Selected Additively Manufactured Propellants for Arc-Ignition of Hybrid Rockets," paper AIAA 2015-4034, Orlando FL, 2015.

16–24. A. Karabeyoglu, Chapter 9, "Combustion Instability and Transient Behavior in Hybrid Motors," in M. A. Chiaverini and K. K. Kuo (Eds.), *Fundamentals of Hybrid Rocket Combustion and Propulsion*, Vol. 218, *Progress in Astronautics and Aeronautics*, AIAA, Reston, VA, 2007.

CHAPTER 17

ELECTRIC PROPULSION

Chapters 1 and 2 present information on rocket propulsion devices that use electrical energy for heating and/or directly ejecting propellant thus utilizing an energy source that is separate from the propellant itself. The purpose of this chapter is to provide a more complete presentation of the various thrusters, power supplies, applications, and flight performance. Vector notation is used in several background equations. At present, most electric propulsion concepts are not suitable for earth liftoffs.

In electric propulsion the term *thruster* is used the same way as *engine* is in liquid propellant and *motor* in solid propellant rockets. In addition to a separate *energy source*, such as solar or nuclear with its auxiliaries (concentrators, heat conductors, pumps, panels, and radiators), the basic subsystems of a typical space electric propulsion system are: (1) *conversion devices* to transform the spacecraft's electrical power to voltages, frequencies, pulse rates, and currents suitable for particular electrical propulsion systems; and (2) one or more *thrusters* to convert the electric energy into kinetic energy of the propellant exhaust. Additionally needed are: (3) a *propellant system* for storing, metering, and delivering the propellant and/or propellant fill provisions; (4) several controls for starting and stopping power and propellant flow; and some also need (5) thrust vector control units (also called TGAs—thrust/gimbal assemblies).

Electric propulsion is unique in that it encompasses both thermal and nonthermal systems as classified in Chapter 1. Also, since the energy source is divorced from the propellant, the choice of propellant is guided by factors considerably different to those in chemical propulsion. In Chapter 3, ideal relations that apply to all thermal thrusters are developed and they apply also to thermal-electric (or electrothermal) systems. Concepts and equations for nonthermal electric systems are defined in this chapter.

From among the many ideas and designs of electric propulsion devices reported to date, one can distinguish the following three fundamental types:

1. *Electrothermal.* Propellant is heated electrically and expanded thermo-dynamically; that is, gas is accelerated to supersonic speeds through a converging/diverging nozzle, as in chemical rocket propulsion systems.
2. *Electrostatic.* Acceleration is achieved by the interaction of electrostatic fields with nonneutral or charged propellant particles such as atomic ions, charged droplets, or colloids.
3. *Electromagnetic.* Acceleration is achieved by the interaction of electric and magnetic fields within a plasma. Moderately dense plasmas, found in high-temperature and/or nonequilibrium gases, are electrically neutral overall and reasonably good conductors of electricity. Some devices add a nozzle to enhance performance.

A general description of these three types was given in Chapter 1 and in Figs. 1–8 to 1–10. Figure 17–1 and Tables 2–1 and 17–1 show approximate power and performance values for several types of electric propulsion units. Note that thrust

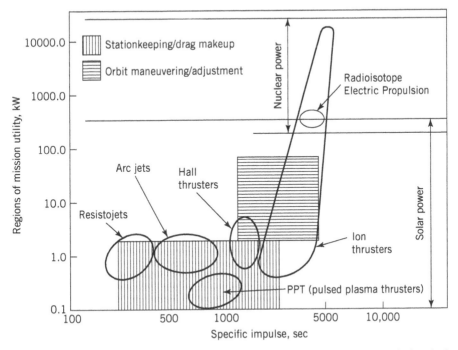

FIGURE 17–1. Overview of the approximate regions of application of several electrical propulsion systems in terms of power and specific impulse.

TABLE 17-1. Typical Performance Parameters of Various Types of Electrical Propulsion Systems

Type	Thrust Range (mN)	Specific Impulse (sec)	Thruster Efficiency[a] (%)	Thrust Duration	Typical Propellants	Kinetic Power per Unit Thrust (w/mN)
Resistojet (thermal)	200–300	200–350	65–90	Months	NH_3, N_2H_4, H_2	0.5–6
Arcjet (thermal)	200–1000	400–800	30–50	Months	H_2, N_2, N_2H_4, NH_3	2–3
Ion thruster	0.01–500	1500–8000	60–80	Years	Xe,Kr,Ar,Bi	10–70
Solid pulsed plasma (PPT)	0.05–10	600–2000	10	Years	Teflon	10–50
Magnetoplasma dynamic (MPD)	0.001–2000	2000–5000	30–50	Weeks	Ar,Xe,H_2,Li	100
Hall thruster	0.01–2000	1500–2000	30–50	Months	Xe,Ar	100
Monopropellant rocket[b]	30–500,000	200–250	87–97	Hours or minutes	N_2H_4	

[a]See Eq. 17–2.
[b]Listed for comparison only.

levels are small relative to those of chemical rocket propulsion systems and that tiny valves for metering low flows are a challenge, but that values of specific impulse can be substantially higher; this translates into a longer operational life for those satellites whose life is propellant limited. Presently, electric thruster gives accelerations too low for overcoming the high-gravity fields of planetary launches. They operate exclusively in space, which also matches the near-vacuum exhaust pressures required for electrostatic and electromagnetic systems. Since all flights envisioned with electric propulsion operate in a reduced gravity or gravity-free space, they must be launched from Earth by chemical rocket systems. Launching from sufficiently low-gravity bodies such as moons and asteroids, however, is currently feasible without chemical assist. Recent interest in very small spacecraft has given rise to a rocket subfield called *micropropulsion*, see Ref. 17–1; here power levels below 500 W and thrusts below 1 mN are required because total vehicle mass (m_0) is less than 100 kg.

Present types of electric propulsion system depend on some vehicle-borne power source—based on solar, chemical, or nuclear energy—and on internal power conversion and conditioning equipment. The mass of the electric generating equipment, even when solar energy is employed, is much larger than that of the thrusters, particularly when thruster efficiency is low, and this translates into appreciable increases in inert-vehicle mass (or dry mass). Modern satellites and spacecraft have substantial communications and other electrical requirements; typically, these satellites share their electrical power source, and thus avoid assigning that extra mass to the propulsion system. The power source is often a completely separate subsystem. What remains to be tagged to the propulsion system is the specialized power-conditioning unit (PCU or power-processing unit, PPU), except in cases where it is also shared with other spacecraft components.

Electric propulsion has been considered for space applications since the inception of the space program in the 1950s but only began to make a widespread impact in the mid-1990s. This has primarily been the result of the availability of sufficiently large amounts of electrical power in spacecraft. Basic principles on electric propulsion devices are given in References 17–2 to 17–4, along with applications, although that information relates to early versions of such devices. Table 17–2 gives a comparison of advantages and disadvantages of several types of electric propulsion. Pulsed devices differ from continuous ones in that startup and shutdown transients degrade their effective performance. Pulsed devices, however, are of practical importance, as detailed later in this chapter.

Applications for electric propulsion fall into several broad mission categories (these have already been introduced in Chapter 4):

1. *Overcoming translational and rotational perturbations* in satellite orbits, such as north–south station keeping (NSSK) of satellites in geosynchronous orbits (GEO) or aligning telescopes or antennas or drag compensation of satellites in low (LEO) and medium earth orbits (MEO). For a typical north–south station-keeping task in a 350-km orbit, a velocity increment of about 50 m/sec every year or 500 m/sec for 10 years might be needed. Several different electric propulsion systems have actually flown in this type of mission.

TABLE 17–2. Comparison of Electrical Propulsion Systems

Type	Advantages	Disadvantages	Comments[a]
Resistojet (electrothermal)	Simple device; easy to control; simple power conditioning; low cost; relatively high thrust and efficiency; can use many propellants, hydrazine augmentation	Lowest I_s; heat loss; gas dissociation; indirect heating of gas; nozzle erosion	Operational
Arcjet (electrothermal and electromagnetic)	Direct heating of gas; low voltage; relatively high thrust; can use catalytic hydrazine augmentation; inert propellant	Low efficiency; erosion at high power; low I_s; high current; heavy wiring; heat loss; more complex power conditioning	Relatively high thrust/power. Operational up to 2 kW
Ion propulsion (electrostatic)	High specific impulse; high efficiency; throttleable; inert propellant (Xenon)	High voltages; low thrust per unit area; massive power supply	Operational in GEO satellites (Boeing 702HP). Space probes (DS1, DAWN, Artemis, GOCE, EURECA, Hayabusha)
Pulsed plasma (PPT) (electromagnetic)	Simple device; low power; solid propellant; no gas or liquid feed system; no zero-g effects on propellant	Low thrust; Teflon reaction products are toxic, may be corrosive or condensable; inefficient	Teflon PPT flown on EO-1 Operational
MPD steady-state plasma (electromagnetic)	Scalable; high I_s; high thrust per unit area	Difficult to simulate analytically; high specific power; heavy power supply; lifetime validation required.	Few have flown
Hall thruster	Desirable LEO I_s range; compact; inert propellant (Xenon)	Single inert-gas propellant; high beam divergence; erosion	Operational SMART-1, AEHF

[a]The abbreviations listed under Comments refer to specific electric propulsion systems.

2. Increasing satellite speed while overcoming the relatively weak gravitational field some distance away from the earth, such as *orbit raising* from LEO to a higher orbit or even to GEO. Circularizing an elliptical orbit may require a vehicle velocity increase of 2000 m/sec and going from LEO to GEO typically might require up to 6000 m/sec. All electric upper stages are being developed for orbit raising, but when transit times are unacceptably long combinations of chemical and electrical thrusters have been used (see Section 17.1).

3. Missions such as *interplanetary travel* and *deep space probes* are also candidates for electric propulsion. A return to the moon, missions to Mars and Jupiter, and missions to comets and asteroids are of interest. A few such electric thruster missions are presently under way such as NASA's DAWN.

4. A number of newer missions look at electric propulsion for either precision attitude/position control or formation-flying relative position control needed for multi-satellite communications. Several electric propulsion units have been developed for these and similar types of mission like the Boeing 702SP and Loral's "all electric satellite" and the Lockheed-Martin's AEHF (Advanced Extremely High Frequency) satellite.

As an illustration of the benefit in applying electric propulsion, consider a typical geosynchronous communications satellite with a 15-year lifetime and with a mass of 2600 kg. For NSSK the satellite might need an annual velocity increase of some 50 m/sec; this requires about 750 kg of chemical propellant for the entire period, which is more than one-quarter of the satellite mass. Using an electric propulsion system with a specific impulse of 2800 sec (about nine times higher than a chemical rocket), the propellant mass can be reduced to perhaps less than 100 kg. A power supply and electric thrusters would have to be added, but the inert mass of the chemical system can be deleted. Such an electric system would save perhaps 450 kg or at least 18% of the satellite mass. With launch costs estimated at $50,000 per kilogram delivered to GEO, this is a potential saving of $22,500,000 per satellite; lighter satellites may qualify for smaller launch vehicles allowing additional savings. Alternatively, more propellant could be stored in the satellite, thus extending its useful life. Additional savings may be realized when electric propulsion is used for both station keeping and orbit rising.

The propulsive output or *kinetic power of the jet* P_j originates from the energy rate supplied by the available power source (P_e) diminished by: (1) losses in the power conversion, such as from solar or nuclear into electrical energy; (2) conversions into the forms of electric power (voltage, frequency, etc.) required by the thrusters; and (3) losses of the conversion of electric energy delivered to the thruster into propulsive jet energy (see thruster efficiency η_t below). The kinetic power (P_j) per unit thrust (F) may be expressed with the following relation, assuming no significant pressure thrust (i.e., $c = v_2$) and no appreciable exit flow divergence:

$$\frac{P_j}{F} = \frac{\frac{1}{2}\dot{m}v^2}{\dot{m}v} = \frac{1}{2}v = \frac{1}{2}g_0 I_s \qquad (17\text{--}1)$$

where $\dot{m}$ is propellant mass flow rate, v mass-average jet discharge velocity (v_2 or c in Chapters 2 and 3), and I_s specific impulse. This jet power-to-thrust ratio is therefore proportional to the effective exhaust velocity or equivalently the specific impulse. It is sometimes concluded here that thrusters with substantially high values of I_s will require more power and therefore bigger power supplies, but this is generally not so. It is shown in this chapter that systems with high specific impulse tend to operate at much longer *propulsive times* (t_p) and thus at smaller thrust levels (for the same total impulse) so that power requirements may actually be comparable.

Thruster efficiency η_t is defined as the ratio of the thrust producing kinetic power of the exhaust beam (axial component) to the total electrical power applied to the thruster ($\sum(IV)$, where I is current and V voltage), including any power used in evaporating and/or ionizing propellant:

$$\eta_t = \frac{\text{power of the jet}}{\text{electrical power input}} = \frac{\frac{1}{2}\dot{m}v^2}{\sum(IV)} \qquad (17\text{--}2)$$

Then, from the fundamentals in Chapter 2 (Eqs. 2–18 and 2–21)

$$\frac{F}{P_e} = \frac{2\eta_t}{g_0 I_s} \qquad (17\text{--}3)$$

where P_e represents the total electric power input to the thruster in watts and given by the product of the electrical current and all associated voltages (hence the summation sign, $\sum$). The power required from the natural source is found through the inclusion of the first two additional conversion efficiencies outlined above Eq. 17–1.

In summary, thruster efficiency accounts for all energy losses that do not result in propellant kinetic energy, including (1) the wasted electrical power (stray currents, ohmic resistances, etc.); (2) unaffected or improperly activated propellant particles (propellant utilization); (3) loss of thrust resulting from dispersion (direction and magnitude) of the exhaust; and (4) heat losses. Thus, η_t measures how effectively electric power and propellant are used in the production of thrust. When electrical energy is not the only input energy, Eq. 17–2 needs to be modified; for example, certain chemical monopropellants may release energy, as in hydrazine decomposition within a resistojet.

17.1. IDEAL FLIGHT PERFORMANCE

Because of their low thrust, flight regimes for space vehicles propelled by electric thrusters are quite different from those using chemical rockets. Accelerations tend to be relatively low (10^{-4} to $10^{-6}g_0$), thrusting times are typically long, and spiral trajectories are often more advantageous for electrically propelled spacecraft. Figure 17–2 shows three schemes for going from LEO to GEO including an increasing spiral (using multiple thrusters and lasting several months), a Hohmann ellipse (see Section 4.4 and Fig. 4–9 on the *Hohmann orbit*, which is optimum with chemical propulsion and lasts hours to perhaps days) as well as a "supersynchronous"

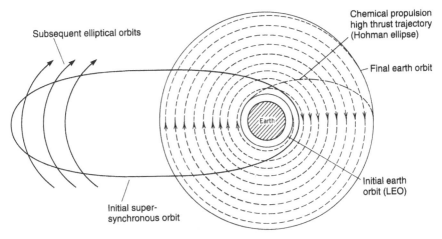

FIGURE 17–2. Simplified diagram of trajectories going from a low earth orbit (LEO) to higher earth orbits using chemical propulsion (short duration), electric propulsion alone with a multiple spiral trajectory (long duration) and with a mixed chemical orbit approach as an alternate from LEO (intermediate duration). From an initial elliptical orbit, continuous thrusting with electric propulsion at a fixed inertial attitude lowers orbit apogee and raises perigee until reaching the final high circular orbit. See Ref. 17–5.

orbit transfer (Ref. 17–5). Because long spiral transfer orbit durations may be impractical, shorter time trajectories have been implemented (of a few weeks duration) such as those using chemical propulsion to achieve the very eccentric elliptical orbits; from there, electric propulsion is continuously and effectively used to attain GEO.

It is instructive to analyze flight performance with electrical thrusters in terms of power and relevant masses (Ref. 17–6). Let m_0 be the total initial mass of the vehicle stage, m_p the total mass of the propellant to be expelled, m_{pl} the payload mass to be carried by the particular stage under consideration, and m_{pp} the mass of the power plant (which is relatively more substantial than in chemical rockets) – m_{pp} consists of the empty propulsion system and includes the thruster, propellant storage and feed system. The power source, with its conversion system and auxiliaries and all associated structure, is considered here as part of m_{pl} as it is most often shared with payload operations (Ref. 17–7). The initial mass thus becomes

$$m_0 = m_p + m_{pl} + m_{pp} \qquad (17\text{–}4)$$

The energy source input to the power supply is always larger than its electrical power output; raw energy converted into electrical power at the desired voltages, frequencies, and power levels is modified by the conversion efficiency (presently exceeding 24% for photovoltaic and up to 30% for rotating machinery). This converted electrical output P_e supplies the propulsion system. The ratio of electrical power P_e to power plant mass m_{pp} is defined as α, an important new term often referred to as the *specific power* (or by its inverse, the *specific mass*) of the power plant or entire propulsion system. This specific power must be defined for each design because, even with the same type of thruster, α somewhat depends on the engine–module configuration (this includes the number of engines that share the same power conditioner, redundancies, valving, etc.):

$$\alpha \equiv P_e/m_{pp} \tag{17-5}$$

Specific power is considered to be proportional to thruster power and reasonably independent of m_p. Its value depends strongly on the type of electric thruster and somewhat on the engine module configuration design. Presently, typical values of α in US designs range between 1.0 and 600 W/kg. With technological advances, it is expected that certain thrusters will exceed these α values (see for example Ref. 17–8). Electrical power is converted by the thruster into kinetic energy of the exhaust propellant; accounting for losses through the thruster efficiency η_t, defined in Eqs. 17–2, the mass of the power plant now becomes

$$m_{pp} = m_p v^2/(2\alpha t_p \eta_t) \tag{17-6}$$

where m_p is the total useful propellant mass, v the effective exhaust velocity, and t_p the time of operation or propulsive time when propellant is being ejected at some uniform rate.

Using Eqs. 17–4, 17–5, and 17–6 together with Eq. 4–7, we obtain a relation for the "reciprocal payload mass fraction" (see Ref. 17–6):

$$\frac{m_0}{m_{pl}} = \frac{e^{\Delta u/v}}{1 - (e^{\Delta u/v} - 1)v^2/(2\alpha t_p \eta_t)} \tag{17-7}$$

This result assumes a gravity-free and drag-free flight. The change of vehicle velocity Δu which results from the propellant being exhausted at a speed v is plotted in Fig. 17–3 as a function of payload mass fraction. The specific power α and the thruster efficiency η_t together with the propulsive time t_p are combined into a *characteristic speed* v_c:

$$v_c = \sqrt{2\alpha t_p \eta_t} \tag{17-8}$$

This characteristic speed does not represent a physical quantity but rather a grouping of parameters that has units of speed; it can be thought of as the speed a power plant's inert mass m_{pp} would attain if its full power output were converted into kinetic energy. Equation 17–8 includes the propulsive time t_p, which is typically the actual mission time (mission time cannot be smaller than thrusting time). From Fig. 17–3 it can be seen that, for a given payload fraction (m_{pl}/m_0) and characteristic speed (v_c), there is an optimum value of v represented by the peak in vehicle velocity increment; this is later shown (in Section 17.4) to signify that there exists a particular set of most desirable flight operating conditions (also see Ref. 17–6).

A peak in the curves in Fig. 17–3 occurs because the inert mass of the power plant m_{pp} increases as the specific impulse increases whereas propellant mass decreases.

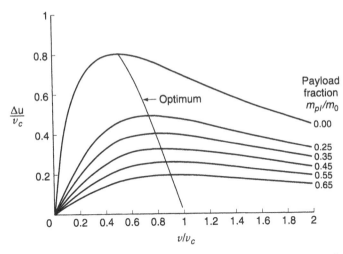

FIGURE 17–3. Normalized vehicle velocity increment as a function of normalized exhaust velocity for various payload fractions with negligible inert mass of propellant tanks. The optima of each curve are connected by a line that represents Eq. 17–9.

As indicated in Chapter 19 and elsewhere, this trend is generally true for all rocket propulsion systems and leads to the statement that, *for a given mission, theoretically there is an optimum range of specific impulse that maximizes* $\Delta u/v_c$ *and thus a most favorable propulsion system design*. The peak of each curve in Fig. 17–3 is nicely bracketed by the ranges $\Delta u/v_c \leq 0.805$ and $0.505 \leq v/v_c \leq 1.0$. This means that for any given electric propulsion system an *optimum operating time* t_p^* will be nearly proportional to the square of the total required change in vehicle velocity and thus large $\Delta u's$ would correspond to very long mission times. Moreover, the *optimum specific impulse* I_s^* is nearly proportional to Δu so that large vehicle velocity changes would necessitate proportionately high specific impulses.

The peak in the curves of Fig. 17–3 may be found from Eq. 17–7 as

$$\left(\frac{v}{\Delta u}\right)(e^{\Delta u/v} - 1) - \frac{1}{2}\left(\frac{v_c}{v}\right)^2 - \frac{1}{2} = 0 \qquad (17\text{–}9)$$

This relates Δu, v, and v_c to maximum payload fraction (see Ref. 17–2).

All equations quoted so far apply equally to the three fundamental types of electric rocket systems. The only engine parameters necessary are the overall efficiency, which ranges from 0.4 to 0.8 in well-designed electric propulsion units, and α, which varies more broadly.

One difficulty with the above formulation is that the equations are underconstrained in that in traveling down the optimum curve too many parameters need to be independently assigned. Additionally, results need to be validated with respect to overall mission constraints. We will return to this topic in Section 17.4 where we extend and refine the optimization results shown above.

Example 17–1. Determine the flight characteristics of an electric propulsion thruster for raising a low satellite orbit. The following typical values are given:

$$I_s = 2000 \text{ sec}; F = 0.20 \text{N}$$

$$\text{Duration} = 4 \text{ weeks} = 2.42 \times 10^6 \text{ sec}$$

$$\text{Payload mass} = 100 \text{ kg}$$

$$\alpha = 100 \text{ W/kg}; \eta_t = 0.5$$

SOLUTION. The propellant flow is, from Eq. 2–6 or 2–12 ($c = v_2 = v$),

$$\dot{m} = F/(I_s g_0) = 0.2/(2000 \times 9.81) = 1.02 \times 10^{-5} \text{kg/sec}$$

The total required propellant is

$$m_p = \dot{m} t_p = 1.02 \times 10^{-5} \times 2.42 \times 10^6 = 24.69 \text{ kg}$$

The required electrical power is

$$P_e = \frac{1}{2} \dot{m} v^2 / \eta_t = \frac{1}{2}(1.02 \times 10^{-5} \times 2000^2 \times 9.81^2)/0.5 = 3.92 \text{ kW}$$

The mass of the power supply system is, from Eq. 17–5,

$$m_{pp} = P_e/\alpha = 3.92/0.1 = 39.2 \text{ kg}$$

The vehicle mass before and after engine operation (see Eq. 17–4) is

$$m_1 = 100 + 24.7 + 39.2 = 163.9 \text{ kg} \quad \text{and} \quad m_2 = 139.2 \text{ kg}$$

The velocity increase of the vehicle under ideal vacuum and zero-g conditions (Eq. 4–6) is

$$\Delta u = v \ln[m_0/(m_0 - m_p)] = 2000 \times 9.81 \ln(163.9/139.2) = 3200 \text{ m/sec}$$

The acceleration of the vehicle is

$$a = \Delta u/t = 3200/2.42 \times 10^6 = 1.32 \times 10^{-3} \text{m/sec}^2 = 1.35 \times 10^{-4} g_0$$

The flight's energy increase after 4 weeks of continuous thrust-producing operation is not enough to get from LEO to GEO (which would have required a change of vehicle velocity of about 4700 m/sec with continuous low thrust as compared to the 3200 m/sec above). During its travel the satellite would have made about 158 revolutions around the earth and raised the orbit by about 13,000 km. Moreover, this case does not represent an optimum. In order to satisfy Eq. 17–9 it would be necessary to increase the burn duration (operating time) or change the thrust, or both. The above approach may be compared to the Hohmann method in Example 4–2.

17.2. ELECTROTHERMAL THRUSTERS

In this category, electric energy is used to heat the propellant, which is then thermo-dynamically expanded through a supersonic nozzle. There are two generic types in use today:

1. The *resistojet*, in which solid components with high electrical resistance dissi-pate power and in turn heat the propellant, largely by convection.
2. The *arcjet*, in which current flows through the bulk of the propellant gas ioniz-ing it with an electrical discharge. Compared to the resistojet, the arcjet is less governed by material limitations; this method introduces heat directly into the gas (which can reach local gas temperatures of 20,000 K or more). The electrothermal arcjet is a device where magnetic fields (either external or self-induced by the current) are not as essential for producing thrust as is the nozzle. As shown in Section 17.3 (and Fig. 17–10), arcjets can also operate as electromagnetic thrusters, but there magnetic fields are externally imposed for acceleration and propellant densities are much lower. Thus, there are some arc–thruster configurations that may be classified as either electrothermal or electromagnetic.

In a recent concept called VASIMR (Variable Specific Impulse Magnetoplasma Rocket, Ref. 17–9), the propellant is heated and ionized using radio waves and then the resulting plasma is expanded through a *magnetically generated supersonic nozzle*. This concept, which is still under development as of 2015, may also be categorized as electrothermal.

Resistojets

These devices represent the simplest type of electrical thruster. As the propellant flows, it contacts one or more ohmically heated refractory-metal surfaces, such as (1) coils of heated wire, (2) heated hollow tubes, (3) heated knife blades, and (4) heated cylinders. Power requirements range between 1 W and several kilowatts; here a broad range of terminal voltages, AC or DC, can be designed, and there are no special requirements for power conditioning. Thrust can be steady or intermittent as programmed in the power and propellant flow.

Material limitations presently cap the operating temperatures to under 2700 K, yielding maximum possible specific impulses of about 300 sec. The highest specific impulse has been achieved with hydrogen (because of its lowest molecular mass), but its low density makes propellant storage too bulky (cryogenic storage is unreal-istic for most space missions). Since virtually any propellant is appropriate, a large variety of different gases have been used, such as O_2, H_2O, CO_2, NH_3, CH_4, and N_2. Also, hot gases resulting from the catalytic decomposition of hydrazine (which produces approximately 1 volume of NH_3 and 2 volumes of H_2 [see Chapter 7]) have been successfully operated. A system using liquid hydrazine (Ref. 17–10) has the advantage of being compact and the catalytic decomposition preheats the mixed gases to about 700 °C (1400 °F) prior to their being heated electrically to higher temperatures; this reduces the required electric power while taking advantage of a well-proven space chemical propulsion concept. Figure 17–4 shows details of such a hybridized resistojet, which is fed downstream from a catalyst bed where hydrazine is decomposed. Performance increases range between 10 and 20% and Table 17–3 shows these performance values.

Resistojets have been proposed for manned long-duration deep space missions, where the spacecraft's waste products (e.g., H_2O or CO_2) could then be used as propellants. Unlike ion engines or Hall thrusters, the same resistojet hardware can be used with different propellants.

In common with nearly every electric propulsion systems, resistojets have a propellant feed system that supplies either a gas from high-pressure storage tank or a liquid under zero gravity conditions. Liquids require positive tank expulsion mechanisms, which are discussed in Chapter 6, and pure hydrazine needs heaters to keep it from freezing in space.

High-temperature materials used for resistor elements include rhenium and refractory metals and their alloys (e.g., tungsten, tantalum, molybdenum), platinum

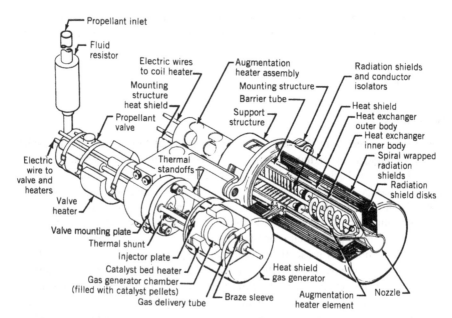

FIGURE 17–4. Resistojet augmented by hot gas from catalytically decomposed hydrazine; two main assemblies are present: (1) a small catalyst bed with its electromagnetically operated propellant valve with heaters to prevent hydrazine from freezing, and (2) an electrical resistance spiral-shaped heater surrounded by thin radiation shields, a refractory metal exhaust nozzle, and high-temperature electrical insulation supporting the power leads. Courtesy of Aerojet Rocketdyne.

TABLE 17–3. Selected Performance Values of a Typical Resistojet with Augmentation

Propellant for resistojet	Hydrazine liquid, decomposed by catalysis
Inlet pressure (MPa)	0.689–2.41
Catalyst outlet temperature (K)	1144
Resistojet outlet temperature (K)	1922
Thrust (N)	0.18–0.33
Flow rate (kg/sec)	$5.9 \times 10^{-5} - 1.3 \times 10^{-4}$
Specific impulse in vacuum (sec)	280–304
Power for heater (W)	350–510
Power for valve (max.) (W)	9
Thruster mass (kg)	0.816
Total impulse (N-sec)	311,000
Number of pulses	500,000
Status	Operational

Source: Data sheet for model MR-501, Aerojet Rocketdyne.

(stabilized with yttrium and zirconia), as well as cermets. For high-temperature electrical (but not thermal) insulation, boron nitride has been used effectively.

A design objective in resitojets is to keep heat losses in the chamber low relative to the power consumed. This can be done by (1) the use of external insulation, (2) internally located radiation shields, and (3) entrant flow layers or cascades. Another important reason for heat-isolation is to keep any stored propellant from overheating under all operating conditions, including thrust termination (liquid hydrazine may detonate if heated above 480 K and in some cases at temperatures as low as 370 K).

The choice of chamber pressure is influenced by several factors. For a given mass flow rate, in molecular gases high pressures reduce molecular gas dissociation losses in the chamber, increase the rate of recombination in nozzle exhausts, improve heat exchanger performance, and reduce the size of both the chamber and the nozzle. However, high pressures cause higher heat transfer losses, higher stresses at the chamber walls, and may accelerate the rate of nozzle throat erosion. The lifetime of resistojet hardware is often dictated by the nozzle throat life. Good design practice, admittedly a compromise, sets the chamber pressure in the range of 15 to 200 psia.

Thruster efficiencies of resistojets range between 65 and 85%, depending on propellant composition and exhaust gas temperature. In Table 17–3, the specific impulse and thrust increase as the electric power of the heater is increased. An increase in flow rate (at constant specific power) results in an actual decrease in performance. The highest specific power (power over mass flow rate) is achieved at relatively low flow rates, low thrusts, and modest heater augmentation. At sufficiently high temperatures, dissociation of molecular gases noticeably reduces the energy that is available for thermodynamic expansion.

Even with its comparatively lower value of specific impulse, the resistojet's superior efficiency contributes to far higher values of thrust/power (see Eq. 17–3) than any of its competitors. Additionally, these engines possess the lowest overall system empty mass since they do not require power processors and their plumes are uncharged (thus avoiding the additional equipment that ion engines require). Resistojets have been employed in Intelsat V, Satcom 1-R, GOMS, Meteor 3–1, Gstar-3, and Iridium spacecraft. They are most attractive for low to modest levels of mission velocity increments, where power limits, thrusting levels and times, and plume effects are mission drivers.

Arcjets

The basic elements of an arcjet thruster are shown in Fig. 1–8 where the relative simplicity of the physical components gives no hint to its rather complicated phenomenology. The arcjet overcomes gas temperature limitations of the resistojet with an electric discharge that directly heats the propellant gases to temperatures much higher than those of the surrounding walls. The arc stretches from the tip of a central cathode and the anode, which is part of a coaxial nozzle that accelerates the propellant flow as it is being heated. Arcjet components must be electrically insulated from each other and must be designed to withstand high temperature gas environments. At the nozzle it is desirable that the arc attach itself as an annulus in the divergent portion just downstream of the throat (see Figs. 1–8 and 17–5). The region of attachment is known to move around depending on the magnitude of the arc voltage and on the mass flow rate. In reality, arcs tend to be highly filamentary and heat only a small portion of the flowing gases unless the nozzle throat dimension is sufficiently small; bulk heating is done by mixing, often with the aid of vortex flows and turbulence.

Arcs are inherently unstable, often forming pinches and wiggles; they can only be stabilized with external electric fields and/or by swirling vortex motions at the outer gas layers. The arc-current's flow configuration at the nozzle throat can be quite nonuniform and arc instabilities and erosion at the throat usually become life limiting. Proper mixing of cooler outer gas layers with arc-heated inner gases tends to stabilize the arc while lowering its conductivity, which in turn requires higher voltages of operation. In some designs the arc is made longer by lengthening the throat.

The analysis of arcjets is based on *plasma physics* applied to a moving partially ionized fluid. Any conduction of electricity through a gaseous medium requires that a certain level of ionization be present. This ionization is obtained from an electrical discharge, that is, from the electric breakdown of the cold gas (resembling a lightning

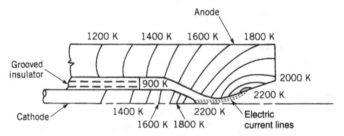

FIGURE 17–5. Simplified half-section with an approximate typical temperature distribution in the electrodes of an arcjet.

discharge in the atmosphere but, unlike it, a power supply feeds the current in either a continuous or pulsed fashion). Gaseous conductors of electricity follow a generalized version of *Ohm's law*; in an ordinary *uniform medium* where an electrical current I is flowing across an area A through a distance d by virtue of a voltage drop V, we may interpret Ohm's law (where R is the resistance) as

$$V = IR = (I/A)(AR/d)(d) \tag{17-10}$$

For the given uniform medium, we may define the electric field as $E = V/d$, the current density as $j = I/A$, and lastly introduce the *electrical conductivity* as $\sigma = d/AR$. Thus, we may now rewrite the basic Ohm's law as simply $j = \sigma E$. The (scalar) electrical conductivity is directly proportional to the density of unattached or *free electrons* that, under equilibrium conditions, may be found from Saha's equation (Ref. 17–11). Strictly speaking, Saha's equation applies to thermal ionization only (and not necessarily to electrical discharges). For most gases, either high temperatures or low ionization energies or both are required for any useful ionization. However, since only about one free electron per million atoms/molecules is sufficient for workable conductivity levels, an inert gas may be "seeded" with low ionization potential vapors, as has been demonstrated in other plasmas. The magnitude of σ, the plasma's electrical conductivity, may be calculated based on the motion of free electrons from

$$\sigma = e^2 n_e \tau / \mu_e \tag{17-11}$$

Here e is the electron charge, n_e the free-electron number density, τ a mean time between collisions, and μ_e the electron mass.

Actually, arc currents are nearly always influenced by magnetic fields, external or self-induced, and a *generalized Ohm's law* (Ref. 17–12) for moving gases is needed, such as the following vector form (this equation is given in scalar forms in the section on electromagnetic devices):

$$\mathbf{j} = \sigma[\mathbf{E} + \boldsymbol{v} \times \mathbf{B} - (\beta/\sigma B)(\mathbf{j} \times \mathbf{B})] \tag{17-12}$$

The vector motion of the gas containing charged particles is represented by a velocity $\boldsymbol{v}$; a magnetic induction field is given as $\mathbf{B}$ (a scalar B in the above equation is required in the last term) and an electric field as $\mathbf{E}$. In Eq. 17–12, both the current density $\mathbf{j}$ and the conductivity are understood to relate only to the free electrons as does β, the Hall parameter. This Hall parameter consists of the electron cyclotron frequency (ω) multiplied by the mean time it takes an electron to lose its momentum by collisions with the heavier particles (τ). The second term in Eq. 17–12 ($\boldsymbol{v} \times \mathbf{B}$) is the *induced electric field* due to any plasma motion normal to the magnetic field, and the last term represents the Hall electric field which is perpendicular to both the current vector and the applied magnetic field vector as the cross product (i.e., the "×") implies (for simplicity ionslip and electron pressure gradients have been omitted here). Magnetic fields are responsible for most of the peculiarities observed in arc behavior, such as pinching (a constriction arising from the current interacting with its own magnetic field), and play a central role in nonthermal electromagnetic types of thrusters, as discussed in a following section.

To initiate an arcjet, a much higher voltage than necessary for operation needs to be momentarily applied in order to 'break down' the cold gas to produce a plasma. Some arcjets require an extended initial burn-in period before consistent running ensues. Because the conduction of electricity through a gas is inherently unstable, arcs use a conventional external or 'ballast' resistance to allow steady-state operation. The cathode must run hot and is usually made of tungsten with 1 or 2% thorium (suitable up to about 3000 K). Boron nitride, an easily shaped high-temperature electrical insulator, is commonly used.

Presently, most arcjets are rather inefficient since less than half of the electrical energy goes into kinetic energy of the jet; the nonkinetic part of the exhaust plume (residual internal energy and ionization) is the largest loss. About 10 to 20% of the electric power input is usually dissipated and radiated as heat to space or transferred by conduction from the hot nozzle to other parts of the system. Arcjets, however, are potentially scalable to larger thrust levels than any other electric propulsion systems. Generally, arcjets exhibit about one-sixth the thrust-to-power ratio of resistojets because of their increased specific impulse coupled with relatively low values of efficiency. Moreover, arcjets have another disadvantage in that the required power processing units need to be more complex than those for resistojets due to the intricacy of arc phenomena.

The life of an arcjet is often severely limited by electrode erosion and vaporization, which arise at the arc attachment or foot-point, and because of the high operating temperatures in general. The rate of erosion is influenced by the particular propellant in combination with the electrode materials (argon and nitrogen give higher erosion rates than hydrogen), and by pressure gradients, which are much higher during start or pulsing transients (often by a factor of 100) than during steady-state operation. A variety of propellants have been used in arcjet devices, but the industry has now settled on the catalytic decomposition products of hydrazine (N_2H_4); see Section 7.4.

An arcjet downstream from a catalytic hydrazine decomposition chamber looks similar to the resistojet of Fig. 17–4, except that the resistor is replaced by a smaller diameter chamber where an arc heats the gases, see Fig. 17–6. Also, larger cables are needed to supply the higher operating currents. Decomposed hydrazine would enter the arc at a temperature of about 760 °C. Liquid hydrazine is relatively storable and provides a low-volume, light-weight propellant supply system when compared to gaseous propellants, but its use requires appropriate thermal control of tanks to prevent freezing and thermal tank isolation to prevent detonations from hydrazine overheated by external sources. Table 17–4 shows on-orbit performance of a system of a 2-kW hydrazine arcjet used for N-S-stationkeeping. Specific impulses range up to 600 sec for hydrazine arcjets (Ref. 17–13). A 30-kW ammonia arcjet program (ESEX) flew in 1999 (Ref. 17–14) with 786 sec specific impulse and 2 N thrust. NASA and NOAA presently operate hydrazine arcjet thrusters in the GOES-R spacecraft with an I_s of 600 sec. The performance of hydrazine augmented arcjets is roughly twice that of hydrazine resistojets.

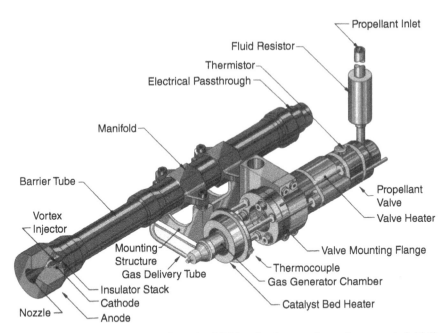

FIGURE 17–6. Perspective drawing of a 2-kW arcjet thruster. Its performance is initially augmented by the catalytic-decomposition of hydrazine into moderately hot gases, which are in turn fed through an electric arc and further heated. The arc is located at the centerline of the flow passage at the throat region of a converging-diverging nozzle. Courtesy of Aerojet Rocketdyne, Redmond Operations

TABLE 17–4. On-Orbit Data for the 2-kW MR-510 Hydrazine Arcjet System

Propellant	Hydrazine
Steady thrust	222–258 mN
Mass flow rate	40 mg/sec
Feed pressure	200–260 psia
Power control unit (PCU) input	4.4 kW (two thrusters)
System input voltage	70 V DC
PCU efficiency	91%
Specific impulse	570–600 sec
Dimensions	
Arcjet (approx.)	$237 \times 125 \times 91$ mm^3
PCU	$631 \times 359 \times 108$ mm^3
Mass	
Arcjet (4) and cable	6.3 kg
PCU	15.8 kg
Total impulse	1,450,900 N-sec

Source: From Ref 17–13.

17.3. NONTHERMAL ELECTRICAL THRUSTERS

The acceleration of hot propellant gases using a supersonic nozzle is the most conspic-uous feature of thermal thrusting. Now we turn our study to propellant acceleration by electrical forces where no enclosure area changes are essential for direct gas acceler-ation. Electrostatic (or Coulomb) forces and electromagnetic (or Lorentz) forces can accelerate a suitably ionized propellant to speeds ultimately limited by the speed of light (note that thermal thrusting is essentially limited by the speed of sound in the plenum chamber). The microscopic vector force $\mathbf{f}_e$ on a *singly charged particle* may be written as

$$\mathbf{f}_e = e\mathbf{E} + e\boldsymbol{v}_e \times \mathbf{B} \qquad (17\text{–}13)$$

where e is the electron charge magnitude, $\mathbf{E}$ the electric field vector, $\boldsymbol{v}_e$ the velocity of the charged particle, and $\mathbf{B}$ the magnetic field vector. Summing these electromagnetic forces on all charges gives the total force per unit volume vector $\tilde{\mathbf{F}}_e$ (scalar forms of this equation follow):

$$\tilde{\mathbf{F}}_e = \rho_e\mathbf{E} + \mathbf{j} \times \mathbf{B} \qquad (17\text{–}14)$$

Here ρ_e is the *net charge density* and $\mathbf{j}$ the vector electric current density. With plas-mas, which by definition have an equal mixture of positively and negatively charged particles within a volume of interest, this net charge density vanishes. On the other hand, the current due to an electric field does not vanish in plasmas because pos-itive ions move opposite to electrons, thus adding to the current (but in plasmas with free electrons this ion current is very small). From Eq. 17–14, we can surmise that an electrostatic accelerator should have a nonzero net charge density, commonly referred to as the *space–charge density*. A common electrostatic accelerator is the ion thruster, which operates with positive ions; here magnetic fields are unimportant in the accelerator region. Electromagnetic accelerators (e.g., the MPD and PPT) oper-ating only with plasmas rely solely on the Lorentz force to accelerate the propellant. The Hall accelerator may be thought of as a crosslink between ion and electromag-netic thrusters. These three types of accelerator are discussed next. Presently, exten-sive research and development efforts with nonthermal thrusters continue to be truly international.

Analyses of electrostatic and electromagnetic thruster are based on the basic laws of electricity and magnetism which are found in Maxwell's equations complemented by the electromagnetic force relation (Eq. 17–14) and the generalized Ohm's law (Eq. 17–12). In addition, various phenomena peculiar to ionization and to gaseous conduction need to be considered. These subjects form the basis of the discipline of magnetohydrodynamics or MHD, a proper treatment of which is beyond the scope of this book.

Electrostatic Devices

Electrostatic thrusters rely on Coulomb forces to accelerate propellant gases comprised of nonneutral charged particles. They operate only in a near vacuum where internal particle collisions are few. The electric force is proportional to the space charge density; since all charged particles are of the same "sign," they move in the same direction. Electrons are easy to produce and readily accelerated, but their extremely light mass makes them impractical for electric propulsion. From thermal propulsion fundamentals (see Chapter 3) we showed that "the lighter the exhaust particle the better." However, the momentum carried by electrons is relatively negligible even at nonrelativistic high velocities, and the thrust per unit area derived from any electron flow remains negligible even when the effective exhaust velocity or specific impulse is high (see Problem 17–11). Accordingly, electrostatic thrusters use singly charged high-molecular-mass atoms as *positive ions* (a proton has 1836 times the mass of the electron and typical ions of interest contain hundreds of proton-equivalent particles). There has been some research work with tiny liquid droplets or *charged colloids* which can in turn be some 10,000 times more massive than atomic particles. In terms of power sources and internal equipment, the use of colloids permits more desirable characteristics for electrostatic thrusters—for example, high voltages and low currents in contrast to the conventional low voltages and high currents with their associated massive wiring and switching requirements.

Electrostatic (ion) thrusters may be categorized by the charged particle source. Note that exit beam neutralization is required in all these schemes.

1. *Electron bombardment thrusters.* Positive ions are produced by bombarding a gas or vapor, such as xenon or mercury, with electrons usually emitted from a heated cathode in a suitable plenum chamber. Ionization voltages can be either DC or RF. In these ion thrusters (see Fig. 1–9), acceleration is accomplished with a separate electrical source applied through a series of suitably manufactured and positioned electrically conducting grids. This method is the oldest and presently the most common.

2. *Field emission thrusters.* With the field emission electric propulsion (FEEP) concept, positive ions are obtained from a liquid metal source flowing through capillary tubes and several geometrical arrangements are possible (Ref. 17–15). Liquid metals such as indium (Ref. 17–16) or cesium when subjected to high enough electric fields ($> 10^6$V/cm) produce molecular ions that flow into an accelerating region. The injector, ionizer, and accelerator are all part of the same voltage circuit which operates typically at values over 10 kV; I_s values are around 8000 to 9000 sec. This is a robust concept being considered for micropropulsion (thrust levels below 1 mN, Ref. 17–1) applications. Some FEEPs have been space qualified.

3. *Colloids.* Here charged liquid droplets produced by an electrospray (or field evaporation process) are used. The propellants are typically liquid metals (or low volatility ionic liquids). These are presently under development and also of interest for micropropulsion, Ref. 17–1.

Examples of space flights where electrostatic units have provided the primary propulsion are the NSTAR (a DC-electron-bombardment xenon ion thruster, Refs. 17–3 and 17–17) used in NASA's DS1 and the ongoing DAWN missions, and

RITA (radio frequency ion thrusters) flown in several European missions and Japan's Hayabusa spacecraft with four microwave ion engines is another mission which included landing and taking off from an asteroid, Refs. 17–3 and 17–18. Boeing has extensively used "L-3 Communications" XIPS 25-cm ion thruster in its geosynchronous communications satellites.

Basic Relationships for Electrostatic Thrusters

An electrostatic thruster, regardless of type, consists of the same series of basic ingredients, namely, a propellant source, several forms of electric power, an ionizing chamber, an accelerator region, and a means of neutralizing the exhaust (see Fig. 1–9). While all Coulomb-force accelerators require a net charge density of unipolar ions, the exhaust beam must be neutralized to avoid a space–charge buildup outside of the craft, which easily nullifies the operation of the thruster. Neutralization is achieved by the injection of electrons downstream of the accelerator. The ion exhaust velocity is a function of the voltage V_{acc} imposed across an accelerating chamber consisting of ion-permeable grids, and the mass of the charged particle μ of electrical charge e. In the conservation of energy equation the kinetic energy of a charged particle must equal the electrical energy gained from the field, provided that there are no collisional or other losses. For charges injected with a velocity v_i,

$$\frac{1}{2}\mu(v_x^2 - v_i^2) = eV_x \tag{17–15}$$

This equation describes one-dimensional transit along the accelerator coordinate x. The total ion speed issuing from the accelerator becomes,

$$v = \sqrt{2eV_{acc}/\mu + v_i^2} \tag{17–16}$$

When e is ion charge in coulombs, μ is the ion mass in kilograms, and V_{acc} is in volts, then v is in meters per second. Using $\mathfrak{M}$ to represent the molecular mass of the ion ($\mathfrak{M} = 1$ kg/kg-mole for a proton), then, for singly charged ions, the equation above becomes v(m/sec) $= 13,800 \sqrt{V_{acc}/\mathfrak{M}}$ when v_i may be neglected. References 17–3 and 17–4 contain more detailed treatments of the applicable theory.

In an ideal ion thruster, the electric current I across the accelerator consists of all the propellant mass flow rate (fully but singly ionized and purely unidirectional), i.e.,

$$I = \dot{m}(e/\mu) \tag{17–17}$$

The total ideal thrust from the accelerated particles is given by Eq.2–13 (without the pressure thrust term, as gas pressures are extremely low):

$$F = \dot{m}v = I\sqrt{2\mu V_{acc}/e + (\mu v_i/e)^2} \tag{17–18}$$

For a given current and accelerator voltage, the thrust is proportional to the mass-to-charge ratio (μ/e) of the charged particles and high molecular mass ions are favored because they yield high thrust per unit volume. Both the thrust and the power absorbed by the electrons in the neutralizing region are small (about 1%) and can thus be neglected.

The current density j that can flow within a nonneutral charged particle beam has a theoretical limit (unconnected to any source-current saturation) which depends on the beam's geometry and on the voltage applied (see Refs. 17–2, 17–3 and 17–19). This fundamental constraint is caused by an internal electric field associated with the ion cloud that *opposes* the imposed electric field so that only a certain maximum number of charges of the same sign can pass simultaneously through the accelerator region. This *space-charge limited current* is found from Poisson's equation as traditionally applied to a one-dimensional planar electrode region. We next define of the current density in terms of the space–charge density (ρ_e) as:

$$j = \rho_e v \qquad (17\text{–}19)$$

Injection velocities can never be zero, but when the injected charges have a negligible kinetic energy compared to that gained within the accelerator operating at V_{acc}, (i.e., dropping out v_i), it is possible to solve Eqs. 17–16, 17–19, and 17–20 directly and obtain the classic relation known as *Child–Langmuir's law*:

$$j = \frac{4\varepsilon_0}{9}\sqrt{\frac{2e}{\mu}}\frac{(V_{acc})^{3/2}}{d^2} \qquad (17\text{–}20)$$

Otherwise, when the injected ion kinetic energy is sufficiently large relative to the energy gained in the accelerator, Eq. 17–16 remains unmodified and an enhanced current beyond its Child-Langmuir value may result. As long as the current emitted from the source has not saturated, analysis shows that the current increase may be as high as j_{max}, given below as Eq.17–21 (where $\kappa \geq 0$ is the ratio of the initial ion energy to the energy gained in the accelerator, assuming monoenergetic beam injection), see Ref. 17–20,

$$j_{max} = j(\kappa^{1/2} + (1+\kappa)^{1/2})^3 \qquad (17\text{–}21)$$

Returning now to Eq. 17–20 where $v_i \approx 0$, d represents the accelerator interelectrode distance and ε_0 the *permittivity of free space* which, in SI units, becomes 8.854×10^{-12} farads/meter. In SI units Eq. 17–20, the classical saturation current density may be expressed (for atomic or molecular ions) as

$$j = 5.44 \times 10^{-8} V_{acc}^{3/2}/(\mathfrak{M}^{1/2}d^2) \qquad (17\text{–}22)$$

Here the current density is in A/m², the voltage is in volts, and the distance in meters. For xenon with electron bombardment ionization, values of j vary from 2 to about 10 mA/cm². In practice, acceleration is seldom one-dimensional and the current density and cross sectional area depend on accelerator voltage as well as on electrode configuration and spacing.

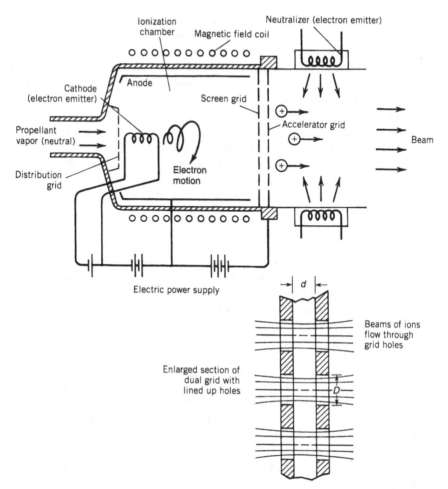

FIGURE 17–7. Simplified schematic diagram of an electron bombardment ion thruster showing, enlarged, a two-screen accelerator grid section. Presently thrusters use a three-grid acceleration unit and permanent magnets instead of coils.

Using Eqs. 17–18 (with $v_i \approx 0$) and 17–22 and letting the beam cross section be circular so that the current through each grid hole (see Fig.17–7) becomes $I = (\pi D^2/4)j$, the corresponding thrust may be rewritten as

$$F = (2/9)\pi\varepsilon_0 D^2 V_{acc}^2/d^2 \tag{17–23}$$

In SI units, this classical space-charge limited relation becomes

$$F = 6.18 \times 10^{-12} V_{acc}^2 (D/d)^2 \tag{17–24}$$

The ratio of a beam's diameter D to the accelerator–electrode grid spacing d is regarded as an *aspect ratio* for the ion accelerator region. For multiple grids with equal holes (see Figs. 17–7 and 17–8) the diameter D is that of the individual perforation

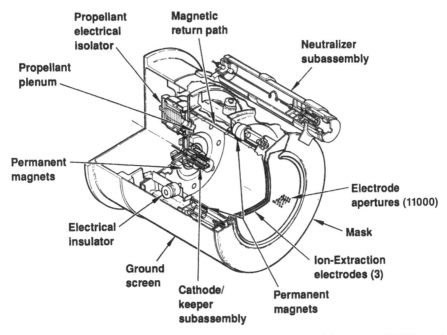

FIGURE 17–8. External view and section of a 500-watt ion propulsion system (XIPS), rated at 18 mN and 2800 sec. Also shown are hollow-cathodes for ionization and for beam neutralization. Xenon gas is delivered to the ionizer, then accelerated through the extraction electrodes with an added "screen electrode," after this section the ion beam is neutralized. Drawing courtesy of L-3 Communications Electron Technologies, Inc., and the American Physical Society

hole and the distance d is the mean spacing between grids. Because of space–charge limitations, D/d can have values no higher than about one for simple, single-ion beams. This implies a rather stubby thruster design with many perforations and the need for large numbers or multiple parallel ion beamlets for high thrust requirements, but other practical considerations also need to be considered.

Using Eqs. 17–1, 17–2 and 17–17, and allowing for losses in the conversion of potential energy to kinetic energy, the power needed for the electrostatic accelerator region becomes

$$P_e = IV_{acc} = \left(\frac{1}{2}\right) \dot{m} v^2 / \eta \qquad (17\text{–}25)$$

Here, the electrostatic thruster efficiency (η) is part of the overall thruster efficiency η_t (see Eq. 17–2) and among other losses includes the ionization-energy expenditure. Ionization energies represent an input necessary to make the propellant respond to the electrostatic force and are nonrecoverable. The ionization energy is found from the ionization potential (ε_I) of the atom or molecule times the current flow, as the Example 17–2 demonstrates. Table 17–5 shows molecular masses and ionization potentials for different gaseous propellants. In actual practice, considerably higher voltages than the ionization potential are required to operate the ionization chamber.

TABLE 17–5. Ionization Potentials for Various Gases

Gas	Ionization Potential (eV)	Molecular or Atomic Mass (kg/kg-mol)
Cesium vapor	3.9	132.9
Bismuth	7.3	209
Mercury vapor	10.4	200.59
Xenon	12.08	131.30
Krypton	14.0	83.80
Hydrogen, molecular	15.4	2.014
Argon	15.8	39.948

Example 17–2. For an electron-bombardment ion thruster that uses a two-grid accelerator section and xenon gas the following data are given:

Working fluid	xenon (131.3 kg/kg-mol, 12.08 eV)
Net accelerator voltage	750 V
Distance d between accelerator grids	2.5 mm
Diameter D of each grid opening	2.0 mm
Number of holes per grid	25,000

Determine the thrust, exhaust velocity, specific impulse, mass flow rate, propellant needed for 91 days' operation, the power of the exhaust jets, and the thruster efficiency including ionization losses. Assume $v_i \approx 0$.

SOLUTION. The ideal thrust from each beamlet is obtained from Eq. 17–24:

$$F = 6.18 \times 10^{-12} \times (750)^2 \times (2/2.5)^2 = 2.23 \times 10^{-6} \text{N per beam}$$

The total ideal thrust is obtained by multiplying by the number of holes in the accelerator grid:

$$F = 25,000 \times 2.23 \times 10^{-6} = 55.67 \text{ mN}$$

The exhaust velocity and specific impulse are obtained from Eq. 17–16:

$$v = 13,800\sqrt{750/131.3} = 32,982 \text{ m/sec}$$

$$I_s = 32,982/9.81 = 3362 \text{ sec}$$

The overall mass flow rate, obtained from Eq. 2–6, is

$$\dot{m} = F/v = 55.67 \times 10^{-3}/32,982 = 1.68 \times 10^{-6} \text{ kg/sec}$$

For a cumulative period of 91 days or 7.8624×10^6 sec of operation, the amount of xenon propellant needed (assuming no losses) is

$$m_p = \dot{m}t_p = 1.68 \times 7.8624 = 13.27 \text{ kg}$$

The kinetic energy rate in the jet is

$$\frac{1}{2}\dot{m}v^2 = 0.5 \times 1.68 \times 10^{-6} \times (32,982)^2 = 914 \text{ W}$$

The ionization loss (l_l) represents a nonrecoverable ionization expense and is related to the ionization potential of the atom ($\varepsilon_I = 12.08$) times the total ion current I or, equivalently, the number of coulombs produced per second (see Table 17–5 and Eq. 17–17):

$$l_l = (12.08) \times (1.68 \times 10^{-6} \times 1.602 \times 10^{-19})/(1.67 \times 10^{-27} \times 131.3) = 14.9 \text{ W}$$

As can be seen, the rate of ionization energy for this ideal case is about 2% of the accelerator energy rate. The current is found from Eq. 17–17 to be just under 10 mA. Discharge losses and others detract from the high ideal efficiency of this device, which would be 98.3%; in fact, the energy needed per ion produced in the XIPS is reported to be above 200 eV/ion bringing the overall efficiency down to 60%, see Ref. 17–3, even though discharge ionization losses can be as low as 140 eV/ion.

Ionization Schemes. Ordinary gaseous propellants must be ionized before they can be electrostatically accelerated. Even though all ion acceleration schemes are fundamentally the same, several ionization schemes are available. Most devices ionize using direct current discharges (DC) but some use high-frequency alternating currents (RF). The ionization chamber is responsible for most of the size, mass, and internal efficiency of these thrusters.

Ionization of a gas by electron bombardment is a well-established technology (Refs. 17–2, 17–3 and 17–19). Electrons emitted from a thermionic (hot) cathode or the more efficient "hollow cathode" are made to interact with a gaseous propellant flow inside a suitable ionization chamber. The chamber pressures are low, about 10^{-3} torr or 0.134 Pa. Figure 17–7 depicts a typical electron-bombardment ionizer, which contains neutral atoms, positive ions, and electrons. Emitted electrons, attracted toward the chamber's cylindrical anode, are forced to spiral by the axial magnetic field, thus enabling the numerous collisions with propellant atoms needed for ionization; the more contemporary devices incorporate "ring-cusp magnetic circuits," which rely on a "magnetic mirror effect" to control and filter the discharge electrons. A radial electric field removes electrons from the chamber and an axial electric field moves positive ions toward the accelerator grids. These grids are designed to act as "porous electrodes," where only the positive ions are accelerated. Electrons losses are minimized by maintaining the cathode potential negatively biased at both the inner grid electrode and at the opposite wall of the chamber. An external circuit routs the extracted electrons from the cylindrical anode and re-introduces them at the exhaust beam in order to neutralize it.

Figure 17–8 shows the cross section of a xenon ion propulsion thruster with *three perforated electrically charged grids* or 'ion extraction electrodes'. Here the inner one is charged to the cathode potential (typically 1000 V with respect to the spacecraft ground plasma potential), the second or 'accelerator electrode' is typically charged to – 200 V, and the third or 'decelerator electrode' is tied to the neutralizer (so as to reduce sputter-eroded products and to improve the beam focusing in the near field).

Hence the saturation current density is given the potential difference between the first two electrodes and the extraction velocity is given by the potential difference between the screen and accelerator electrodes. Each grid hole is suitably aligned with a similar opening in other grids and the ion beamlets flow through these holes.

Other key thruster components are (1) cathode heaters, (2) a propellant feed system, (3) electrical insulators, and (4) permanent magnets. Reference 17–17 describes a 500-W xenon thruster. Hollow cathodes are efficient electron emitters but, because of their size and complexity, carbon-nanotube electron-field emitters are presently being explored. Xenon, the highest molecular mass stable inert gas, has been the propellant of choice. Xenon is a minor component of air, in a concentration of about 9 parts in 100 million, so it is relatively rare and expensive. It is easily stored below its critical temperature as a liquid and does not pose any problems of condensation or toxicity. Pressure regulators for xenon need to be quite sophisticated because no leakages can be tolerated as flows are quite small.

Electromagnetic Thrusters

This third category of electric propulsion devices accelerates propellant gases that have been heated up to a plasma state. Plasmas are neutral mixtures of electrons and positive ions (often including un-ionized atoms/molecules) that can readily conduct electricity, existing at temperatures usually above 5000 K or 9000 °R. According to electromagnetic theory, whenever a conductor carries a current perpendicular to a magnetic field, a body force is exerted on that conductor in a direction at right angles to both the current and the magnetic field. Unlike the ion thruster, this acceleration process yields a neutral exhaust beam. Another advantage is the relatively high thrust density, or thrust per unit area, which can be 10 to 100 times that of electrostatic thrusters.

Basically, designs of electromagnetic thrusters consist of a region of electrically conductive gas where a high current produced by an applied electric field accelerates the propellant though the action of either an external or a self-generated magnetic field. Many conceptual arrangements have undergone laboratory study, some with external and some with self-generated magnetic fields, some suited to continuous thrusting and some limited to pulsed thrusting. Table 17–6 shows ways in which electromagnetic thrusters are categorized. Because there is a wide variety of devices with a correspondingly wide array of names, we will use the term *Lorentz-force accelerators* when referring to their principle of operation. For all of these devices the plasma current must be part of the electrical circuit and most accelerator geometries are constant area. Motion of the propellant, a moderate-density plasma—usually a combination of ionized and cooler gas particles, is due to a complex set of interactions. This is particularly true of short duration (3 to 10 μsec) pulsed-plasma thrusters where nothing reaches a thermal equilibrium state.

Conventional Thrusters—MPD and PPT. Any description of *magneto-plasma-dynamic* (MPD) and *pulsed-plasma* (PPT) electromagnetic thrusters is based on plasma conduction in the direction of the applied electric field but perpendicular

TABLE 17-6. Characterization of Electromagnetic Thrusters

	Thrust Mode	
	Steady State	Pulsed (Transient)
Magnetic field source	External coils or permanent magnets	Self-induced
Electric current source	Direct-current supply	Capacitor bank and fast switches
Working fluid	Pure gas, mixtures, seeded gas, or vaporized liquid	Pure gas or stored as solid
Geometry of path of working fluid	Axisymmetric (coaxial) rectangular, cylindrical, constant or variable cross section	Ablating plug, axisymmetric, other
Special features	Using Hall current or Faraday current	Simple requirement for propellant storage

to the magnetic field, with both of these vectors in turn normal to the direction of plasma acceleration (see Ref. 17–12). Equation 17–12 may be specialized here to a Cartesian coordinate system where the plasma's "mass-mean velocity" is in the x direction, the external electric field is in the y direction (E_y), and the magnetic field acts in the z direction (B_z). A simple manipulation of Eq. 17–12, with negligible Hall parameter β, yields a scalar equation for the current, noting that only j_y (termed the Faraday current), E_y, and B_z remain present:

$$j_y = \sigma(E_y - v_x B_z) \tag{17-26}$$

and the Lorentz force (from Eq. 17–14) becomes

$$\tilde{F}_x = j_y B_z = \sigma(E_y - v_x B_z)B_z = \sigma B_z^2(E_y/B_z - v_x) \tag{17-27}$$

Here $\tilde{F}_x$ represents the *force "density" within the accelerator* and should not be confused with F the total thrust force; $\tilde{F}_x$ has units of force per unit volume (e.g., N/m^3). The axial velocity v_x is a mass-mean velocity that increases internally along the accelerator length; the actual thrust equals the exit value (v_{max} or v_2) multiplied by the mass flow rate. It is noteworthy that, as long as E_y and B_z (or E/B) remain somewhat constant, both the current and the force decrease along the accelerator length due to the *induced field* $v_x B_z$, which subtracts from the impressed value E_y. Such plasma velocity behavior translates into a diminishing force along Faraday accelerators, with eventual limits on the final axial velocity. Although not practical, it would seem desirable to design for increasing E/B along the channel in order to maintain substantial accelerating forces throughout. But it is not necessarily of interest to design for peak exit velocities because these might translate into unrealistically long accelerators (see Problem 17–8). It can be shown that practical considerations would restrict these exit velocities to below one-tenth of the maximum value of E_y/B_z.

A "gas-dynamic approximation" (essentially an extension of the classical concepts of Chapter 3 to plasmas in electromagnetic fields) by Resler and Sears (Ref. 17–21) indicates that further complications are possible, namely, that a constant area accelerator channel would *choke* if the plasma velocity does not have the very specific value of $[(k-1)/k](E/B)$ at the sonic location of the accelerator. This *plasma tunnel velocity* would have to be equal to 40% of the value of E/B for inert gases, since k (their ratio of specific heats) equals 1.67. Thus, constant area, constant E/B accelerators could be severely constrained because Mach 1 corresponds only to about 1000 m/sec in typical inert gas plasmas; constant-area choking in real systems, where the properties E, B, and σ are actually quite variable, is more likely to manifest itself as one or more instabilities. Another problem here is that values of the conductivity and electric field are usually difficult to determine and a combination of analysis and measurement is often required to evaluate, for example, Eq. 17–12. Fortunately, most plasmas are reasonably good conductors even when less than 10% of the gas particles are ionized.

Figure 17–9 shows the simplest plasma accelerator, which employs a self-induced magnetic field. This is the *pulsed plasma thruster* (PPT) where an accelerating plasma burst or "*bit*," created by a spark discharge between the accelerator electrodes, is powered by a capacitor, which in turn is charged from the spacecraft's power supply. The flow of current through the plasma rapidly discharges the capacitor and hence such mass-flow-rate pulse must be synchronized with a "discharge schedule." The discharge current closes a "current loop," one which induces a significant magnetic field perpendicular to the plane of the rails. Analogous to a metal conductor in an electric motor, the Lorentz force acts on the movable plasma segment, accelerating it along the rails. Hence, no area changes (i.e., nozzles) are necessary to accelerate the propellant.

As indicated in Fig. 17–9, the system (though containing a plasma along with other more ordinary resistances) may be modeled with an equivalent *L-R-C pulsed circuit* where L is the lumped-parameter total circuit inductance, R is the total Ohmic resistance, and C is an effective capacitance. It is undesirable to let the current reverse

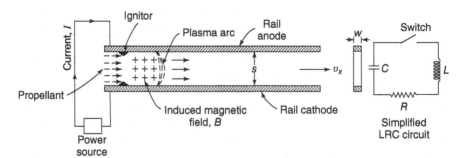

FIGURE 17–9. Simple rail accelerator for self-induced magnetic field acceleration of current-carrying plasma. This concept illustrates the basic physical interactions; it suffers from propellant losses, resulting in low efficiencies.

during each pulsing as this diminishes the overall thrust; a nonoscillatory pulse may be realized through the $R > 2\sqrt{L/C}$ "impedance-balancing criterion" (when this is not physically possible, a "quenching diode" is inserted, see Ref. 17–22). In some designs, the hot plasma pulse has sufficient gas-dynamic expansion capabilities to contribute to the thrust (needing the addition a nozzle for thermal expansion). In many typical configurations, the rate of increase of circuit inductance with the distance the arc travels (x) is the sole propulsive contributor $(F = \frac{1}{2}I^2(dL/dx)$ where I is the current in the loop); since the induced magnetic field is a function of accelerator geometry, the axial thrust obtained may be written for the two configurations listed below as

Linear:
$$F = \frac{1}{2}\mu_0 I^2(s/w) \tag{17–28}$$

Coaxial:
$$F = \left(\frac{1}{4}\pi\right)\mu_0 I^2 \ln\left(r_o/r_i + \frac{3}{4}\right) \tag{17–29}$$

The *linear* rail *spacing* is s and the rail *width* is w as shown in Fig. 17–9; for the *coaxial* accelerator the equation above describes an electrode geometry of inner radius r_i and constant outer radius r_o (space qualification for the coaxial micro-PPT has not been reported to date). Because plasmas are nonmagnetic, the value of μ_0 remains the same as free space, namely, $4\pi \times 10^{-7}$ H/m.

A practical first version of the PPT that flew in 1968 is shown in Chapter 1 as Fig. 1–10. The propellant is stored as a solid Teflon bar that is pushed by a spring against two linear rails where a pulsed discharge is initiated across surface momentarily ablating a small portion of propellant. Teflon stores well in space, is easy to handle, and ablates without significant charring. There are no tanks, valves, synchronizing controls, or zero gravity feed requirements. A rechargeable capacitor fed from a power processing unit in the spacecraft provides the power input. Thrust results only as rapid pulses unlike most other electric propulsion devices; these pulsed thrusters are very compatible with precise control and positioning maneuvers where mean thrust can be varied by changing the pulsing rate or the total number of pulses N. If a "total impulse per pulse" or *impulse bit* (in units of N-sec or lbf-sec) is denoted by I_{bit}, then the overall vehicle change of velocity Δu would be the sum of N tiny but equal pulses; using Eqs. 2–4 and 4–6 a much simplified result equivalent to Eq. 4–36 may be obtained provided that $NI_{bit}/cm_0 << 1$ where cm_0 is the product of the effective exhaust velocity c with the initial mass, as would relate to the PPT:

$$\Delta u = c\sum_{n=1}^{N} \ln\left[\frac{m_0 - (n-1)I_{bit}/c}{m_0 - nI_{bit}/c}\right] \approx \sum_{n=1}^{N} \frac{I_{bit}/m_0}{1 - nI_{bit}/cm_0}$$

$$\approx \frac{I_{bit}}{m_0}\left[1 + \frac{N}{2}(N+1)\frac{I_{bit}}{cm_0}\right] \tag{17–30}$$

Much electrical energy is lost in PPT circuits on top of the ionization energy, which cannot be recovered; moreover, the bits of mass being accelerated do not typically exit well collimated and propellant utilization has been poor in earlier designs. Besides its very low efficiency, a disadvantage of the PPT has been the size and

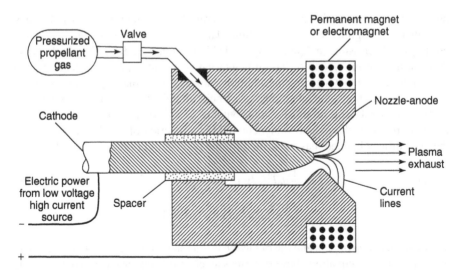

FIGURE 17–10. Simplified diagram of a magnetoplasma dynamic (MPD) arcjet thruster. Similar in construction to the thermal arcjet shown in Fig. 1–8, it has a strong magnetic field to produce propellant acceleration.

mass of the early capacitors and associated power conditioning equipment. Interest in PPTs has been renewed for micropropulsion (small satellite propulsion) using better designs.

Figure 17–10 shows a hybrid electrothermal–electromagnetic arcjet concept, which is of interest because unlike the PPT it has a high-thrust capability. It can produce continuous large thrusts and Russia and Japan have flown some versions. Compared to electrothermal arcjets, these devices operate at relatively lower propellant pressures and much higher electric and magnetic fields. Hydrogen and argon are common propellants used for such MPD arcjets. Unlike ion thrusters, exhaust beam neutralization is unnecessary. Problems of high electrode erosion, massive electrical components, and low efficiencies (with their associated heat dissipation needs) have slowed implementation of these devices, which also depend on relatively large, more complex, and expensive quantities of electric power.

Hall-Effect Thrusters. When plasma densities are low enough and/or magnetic fields sufficiently high, the Hall-effect electric field becomes significant. This is the same phenomenon that is observed as the semiconductor Hall effect where a voltage appears transverse to the applied electric field. Hall currents can be understood to represent the motion of an electron "guiding center" (Ref. 17–11) in a crossed electric and magnetic field arrangement where electron collisions must be relatively insignificant. Hall thrusters are of interest because they represent a very practical operating region for space propulsion. Russian scientists were the first to successfully deploy them in many vehicles, a design originally called the *stationary plasma thruster* or SPT. Hall thrusters are described in Ref. 17–3.

In order to understand the Hall thruster principle, it is necessary to rewrite in scalar form the generalized Ohm's law, Eq. 17–12. Because the *electron Hall parameter β* (introduced in Eq. 17–12 and described below Eq.17–34) is no longer negligible, we arrive at two equations (shown in Cartesian form):

$$j_x = \frac{\sigma}{1 + \beta^2}[E_x - \beta(E_y - v_x B_z)] \qquad (17\text{–}31)$$

$$j_y = \frac{\sigma}{1 + \beta^2}[(E_y - v_x B_z) + \beta E_x] \qquad (17\text{–}32)$$

In typical designs, imposing a longitudinal electric field E_x causes a current density j_x to flow in the applied field direction together with a Hall current density j_y which flows in the direction transverse to E_x. The associated induced Hall electric field E_y is externally shorted to maximize the Hall current, and the electrodes are "segmented" in order not to short out the axial electric field E_x. It is necessary that $\beta E_x > v_x B_z$. Such arrangement unavoidably results in a very complicated design. For space propulsion, engineers prefer the cylindrical geometry because it yields a simpler, more practical design; here the applied magnetic field (B_r) is radial and the applied electric field is axial (E_x); the thrust-producing Hall current j_θ is azimuthal and clockwise and, because it closes on itself, it automatically shorts out its associated Hall electric field. The relevant geometry is shown in Fig. 17–11, and the equations for the orthogonal current density and the Hall current density now become

$$j_x = \frac{\sigma}{1 + \beta^2}[E_x + \beta v_x B_r] \qquad (17\text{–}33)$$

$$j_\theta = \frac{\sigma}{1 + \beta^2}[\beta E_x - v_x B_r] \qquad (17\text{–}34)$$

where, for acceleration to take place, $\beta E_x > v_x B_r$.

The current density j_x is necessary for ionization (by electron collision) because here the discharge chamber coincides with a portion of the accelerator region.

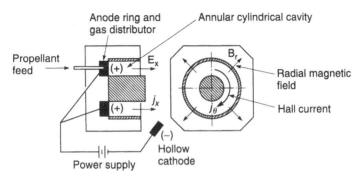

FIGURE 17–11. Simplified diagram of a cylindrical Hall accelerator configuration showing how an applied axial field results in a transverse current that accelerates the plasma. A significant axial current density j_x represents an inefficiency for Hall devices.

The Hall current j_θ performs the acceleration through the Lorentz force $j_\theta B_r$. The electron Hall parameter β is calculated from the product of the *electron cyclotron frequency* (Ref. 17–11) $\omega = eB/\mu_e$ and the *collision time* τ for electrons with the heavier particles, which is part of the electrical conductivity in Eq. 17–11. In order for a Hall generator to be of interest, β must be much greater than one (in fact, Ref. 17–23 indicates that it should be at least 100), whereas ion motion must proceed relatively unaffected by magnetic effects. Large β's are obtained most readily at low plasma densities, which yield large times between electron collisions. Figure 17–12 shows a cutout of the original SPT design with a redundant set of hollow cathodes (presently only one is used because of their greater reliability) and a solenoid magnetic pair for producing the magnetic field. In Hall thrusters, the propellant gas, xenon or argon, is fed in the vicinity of the anode; some gas is also provided through the cathode for better cathode operation. While the discharge chamber is not physically separate from the accelerator region, the absence of ions in the first portion of the chamber effectively differentiates the ionization region from the rest of the accelerator. The local charge mass and density of the ions and electrons, together with the magnetic field profiles, need to be tailored such that the ion motion is mostly axial, while the electron motion mostly spiral; this makes any given fixed physical design inflexible to changes of propellant and unsuitable for robust throttling. A variation of the original nonconducting accelerator wall SPT design is a smaller channel with metallic walls; this "thruster with an anode layer" (TAL) has comparable performance with a higher thrust density.

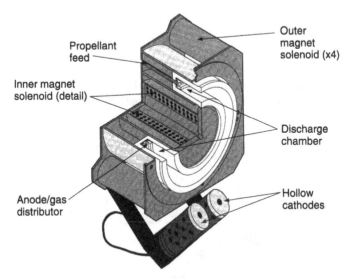

FIGURE 17–12. External view and quarter section of a 1350-watt Hall accelerator (SPT-100). It is rated at a thrust of 83 mN at a specific impulse of 1600 sec. The radial magnetic field is produced by an inner solenoid and four external solenoids. Ionization takes place at the beginning of the insulated annular channel. Modern Hall thrusters use only one cathode positioned like the neutralizer in Fig. 17–8. Drawing courtesy of FAKEL

Hall thrusters may be classified as either electromagnetic devices (as above) or electrostatic devices where the space–charge in the ion acceleration region is neutralized by an electron current transverse to the ion flow (Refs. 17–23 and 17–24). If we can mentally separate the process of ionization from that of acceleration, then it is easy to see that electrons swirling within the accelerator act to neutralize the ion space–charge as it moves from anode to cathode. This, in effect, decreases the magnitude of the accelerating fields and removes most of the beam-focusing requirements. In reality, there is some small interaction between the azimuthal electron current and the ion current, but it diminishes in proportion to the magnitude of the Hall parameter β.

The Hall thruster yields the best β-efficiency (η_H as defined below) when β is very large. The high β limit is found, from Eqs. 17–14, 17–33 and 17–34 and the definition of the plasma conductivity σ (Eq. 17–11), as

$$j_x \rightarrow \sigma v_x B_r/\beta = \rho_e v_x \qquad \text{and} \qquad j_\theta \rightarrow \sigma E_x/\beta \qquad (17\text{–}35)$$

$$\tilde{F} = j_\theta B_r \rightarrow \rho_e E_x \qquad (17\text{–}36)$$

$$\eta_H \equiv \tilde{F} v_x/j_x E_x \rightarrow 1.0 \qquad (17\text{–}37)$$

The magnetic flux density B_r is shown in Fig. 17–11. Equation 17–36, the accelerating force at this high Hall parameter limit, is the electrostatic force and, since the exit ionization levels are about 90%, this corresponds in principle to an ion engine without any of its severe space–charge current limitations. Even though electron densities are on the order of 10^{15} to $10^{17}/\text{m}^3$, the effective space–charge densities (ρ_e) are considerably lower because of charge neutralization, and they approach zero at the exit. Note that the β-efficiency η_H as defined above strictly reflects the influence of the electron Hall parameter β; this efficiency is ideal, representing the loss that arises from the total current vector not being perfectly normal to the flow direction. The overall efficiency is still given by Eq. 17–2.

Example 17–3. The BPT-4000 Hall thruster is an advanced electric propulsion system used in GEO satellites (Ref. 17–25). Some reported values include an overall efficiency of 59% at a specific impulse of 2000 sec with an input power of 4.5 kW, while delivering a thrust of 280 mN. During $\approx$ 5800 hr. of operation, the system is reported to have utilized $\approx$ 250 kg of xenon throughput (which represents the magnitude of the flow rate). Calculate the mass flow rate and the overall efficiency and compare to the values given.

SOLUTION. The mass flow rate may be calculated from Eq. 2–17 as

$$\dot{m} = \frac{F}{I_s g_0} = \frac{0.28}{(2000)(9.81)} = 14.3 \times 10^{-6} \text{kg/ sec}$$

And the thruster efficiency from Eq. 17–3

$$\eta_t = \frac{F I_s g_0}{2P_e} = \frac{(0.28)(2000)(9.81)}{2(4500)} = 0.61$$

This propellant mass flow rate is very low which is desirable for long mission durations; the calculated mass flow rate is somewhat higher than implied by the given data (250/5800), likely because of a lack of specificity. The calculated efficiency is somewhat greater that the quoted 59% because minor system losses have not been taken into account. Furthermore, the internal or Hall efficiency (Eq. 17–37) cannot be determined here because additional information is necessary (for very high values of β the Hall efficiency will always approach one).

Hall thrusters have flown in many Russian satellites and in the European Space Agency's (ESA) SMART-1 mission. The BPT-4000 in Example 17–3 has been used by US Air Force in several missions and Hall thrusters are presently being implemented for orbit raising and station keeping on geostationary US satellites. Higher specific impulses and some *throttling* capability are deemed desirable for other applications.

17.4. OPTIMUM FLIGHT PERFORMANCE

Now we return to the discussion of flight performance. In Section 17.1 the fundamental background for the design of an optimum propulsion system was introduced. That discussion remained incomplete because the specific power and the efficiency of individual thrusters, among other things, need to be known for further analysis. In any given mission, the payload m_{pl} and velocity increment Δu need to be specified along with upper limits on electric power available (Ref. 17–26). In the analysis of Section 17.1, for any desired $\Delta u/v_c$, one can find an optimum v/v_c given a payload ratio (v_c is the characteristic speed, Eq. 17–8); however, even when the choice of an electric propulsion system has been made, thrust time t_p is unspecified and thus total initial mass m_0 also remains unspecified. Thrust time or "burn time" has the smallest value at zero payload and continuously increases with increasing payload ratio. Concurrently, the required specific impulse changes, making the problem underconstrained.

Given the payload mass m_{pl} and the vehicle velocity increment Δu, the following spacecraft design procedure might be implemented for attaining the optimum results of Section 17.1:

1. Select a payload mass fraction—from Fig. 17–3 this yields an optimum $\Delta u/v_c$.
2. From the given Δu, deduce the value of the characteristic speed v_c.
3. From the optimum value of v/v_c in Fig. 17–3 at the given mass fraction, or Eq. 17–9, calculate the corresponding value of v or I_s.
4. Select an engine that can deliver this optimum I_s and from its properties (i.e., α and η_t) find the thrusting time t_p from Eq. 17–8.
5. Calculate m_p from Section 17.1, including Eq. 4–7 and the given payload ratio.
6. Check that the available vehicle electrical power (from Eq. 17–6), vehicle volume, and the desirable mission time and total cost are not exceeded.

As may be evident, a unique criterion for the choice of the assumed payload mass fraction is still missing above. One possible approach to this problem is to look for some "dual optimum," namely, to seek the shortest burn time consistent with the highest payload mass fraction of the flight vehicle. A maximum for the product of m_{pl}/m_0 with $\Delta u/v_c$ does exist as a function of v/v_c. Such dual optimum defines a minimum overall mass for a specified payload consistent with minimum transfer time (for comparable values of efficiency and specific mass). Table 17–7 gives estimated values of the specific power α along with a corresponding range of specific impulse and efficiency for electric propulsion systems in present engine inventories.

The optimum formulation in Section 17.1, however, needs to be modified to account for the portion of tank-mass which derives from propellant loading. With few exceptions, an additional 10% (the *tankage mass fraction*) of the propellant mass shows up as tank or container mass (this could be further refined to include reserve propellant). Reference 17–27 includes information on this *tankage mass fraction* for various thrusters. Fortunately, the analysis presented earlier is little modified and it turns out that the optima are driven toward higher specific impulses and longer times of operation. For any arbitrary tankage fraction allowance, φ,

$$\frac{\Delta u}{v_c} = \frac{v}{v_c} \ln \left[\frac{(1+\varphi) + (v/v_c)^2}{(m_{pl}/m_0 + \varphi) + (v/v_c)^2} \right] \qquad (17-38)$$

When $\varphi = 0.1$ the actual value for the jointly optimized payload ratio can be shown to be 0.46, with corresponding ratios of vehicle velocity increment as 0.299 and propellant exhaust velocity as 0.892. Further scrutiny indicates that this peak is rather broad and that payload ratios between 0.34 and 0.58 are within 6% of the mathematical optimum. Since engine parameters are rather "inelastic," and since spacecraft designers deal with numerous constraints that are not propulsion related, using this wider range of optima is a practical necessity.

Given the desirable 0.34 to 0.58 optimum payload-ratio range, under this method of optimization we may select one or more thrusters within the range $0.2268 \leq (I_s^*/\Delta u) \leq 0.4263$, where the optimized specific impulse (I_s^*) is in seconds and the velocity change in m/sec. Since the vehicle's change in velocity is known, this criterion yields the resulting desirable limits in specific impulse. Figure 17–13 shows curves depicting the parameters in Eq. 17–38 for $\varphi = 0.1$ in an expanded dual-optimum neighborhood. The oval insert encloses an expanded region of interest, one bounded by a contour curve, which represents values 10% below the mathematical peak.

The utility of this approach hinges on the validity of the employed engine information. In particular, the specific power should represent all the inert components of the propulsion system that can be identified to depend on the power level. Payload mass must reflect all the mass that is neither proportional to the electrical power nor propellant related in addition to the actual "payload." The tankage fraction must be proportional to the total propellant mass as required for Eq. 17–38. It has been assumed that there is available a source of electricity (typically from 28 to 300 VDC for solar-powered craft) which is not tagged to the propulsion system. Our analysis also assumes that the efficiency is not a function of specific impulse (in contrast to Ref. 17–29); this implies the use an average or effective efficiency value at the relevant specific impulses. Since each individual thruster inside the oval of Fig. 17–13

TABLE 17–7. Summary of Current Technology in Typical Electric Propulsion Systems

Engine Type	Identification (Reference)	Specific Power, α (W/kg) (estimated)	Thruster Efficiency, η_t	Specific Impulse, I_s (sec)	Power (W)	Thrust (N)	Lifetime (hr)	Status
Resistojet	N_2H_4 (17–19, 17–28)	333–500	0.8–0.9	280–310	500–1500	0.2–0.8	>390	Operational
	NH_3 (17–19)		0.8	350	500			Operational
	Primex MR-501B (17–28)			303–294	350–510	0.369–0.182	>389	Operational
Arcjet	N_2H_4 (17–28)	313	0.33–0.35	450–600	300–2000	0.2–0.25	>830–1000	Operational
	NH_3 (17–19)	270–320	0.27–0.36	500–800	500–30 k	0.2–0.25	1500	Operational
	MR-509 (17–28) (c)	115.3	>0.31	>502 (545)	1800	0.213–0.254	>1575	Operational
	MR-510 (17–28) (c)	150	>0.31	>570–600	2170	0.222–0.258	>2595	Operational
Ion propulsion	Busek CMNT Colloid	1.7		150–275	24	5–30 μN	>2200	Qualified
	Alta FEEP-150	588	>0.9	5000–8000	20	1–150 μN		Qualified
	XIPS-13 (17–28) (a)		0.46, 0.54	2585, 2720	427, 439	0.0178, 0.018	12,000	Qualified
	XIPS-25 (17–28) (a)		0.65, 0.67	2800	1400	0.0635	>4350	Operational
	NSTAR/DS1/DAWN (a)	278	0.62	3100	2300–2500	0.093	>30,000	Operational
	NEXT NASA GRC	110	0.7	1400–4300	560–6900	0.25–0.235	>30,000	Operational
	ETS-VI IES (Jap.)(17–28)		0.4	3000	730	0.02		Operational
	DASA RIT-10 (Ger.)(17–28) (a)		0.38	3000–3150	585	0.015		Operational
Hall	Snecma—PPS 1350	283	0.55	1650	1200–1600	0.088	9500	Operational
	SPT (XE) (17–28)		0.48	1600	150–1500	0.04–0.2	>4000	Operational
	ARC/Fakel SPT-100 (17–19) (b)	169.8	0.48	1600	1350	0.083	>7424	Operational
	Fakel SPT-70 (17–3) (b)		0.46, 0.50	1510, 1600	640–660	0.04	9000	Operational
	TAL D-55 (Russia) (17–28)	<50.9	0.48, 0.50–0.60	950–1950	600–1500	0.082	>5000	Operational
	BPT-4000 Hall (c) (17–25)	366	0.59	2000	4500	0.28	>8000	Operational
MPD–Pulsed	Teflon PPT (17–19)	1	0.07	1000	1–200	4000 N-sec	>10^7 pulses	Operational
	LES 8/9 PPT (17–28)		0.0068, 0.009	836, 1000	25, 30	0.0003	>10^7 pulses	Operational
	NASA/Primex EO-1 (c)	<20	0.098	1150	up to 100	3000 N-sec		Operational
	PRS-101 (c)			1150		1.4 mN, 2 Hz		Operational
	EPEX arcjet (Jap.) (17–28)		0.16	600	430	0.023		Operational

Manufacturers: (a): L3 Communications, (b): Fakel (Russia), (c): Aerojet Rocketdyne.

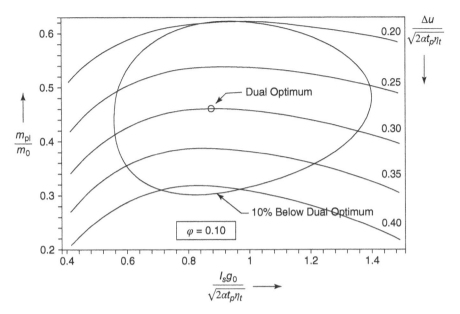

FIGURE 17–13. Payload fractions in the neighborhood of the dual optimum. *The oval* represents a contour 10% below the dual-optimum peak. The parameters v/v_c and $\Delta u/v_c$ in Fig. 17–3 are shown here in their equivalent form.

only spans a limited range of specific impulse, this assumption is not too restrictive. During the continuous thrust schedules required by electric propulsion systems, thrust time represents mission time.

Example 17–4. To change orbit from LEO to GEO a velocity increment of 4700 m/sec is required with continuous low thrust (see Example 17–1). Pick three engines types that would conform to the dual optimum criterion shown in Fig. 17–13, and determine their corresponding payload fractions (m_{pl}/m_0) and travel times (t_p). Use relevant information from Table 17–7 and note from Fig. 17–13 that the largest range of specific impulse (i.e., the widest part of the oval) lies just above the curve $\Delta u/v_c = 0.3$ (which also contains the actual dual optimum point).

SOLUTION. From the problem statement we can directly solve for the common $v_c = 4700/0.3 = 15{,}667$ m/sec and this in turn provides from Fig. 17–13 at $\Delta u/v_c = 0.3$ a specific impulse range within the oval of 894 to 2156 sec. The table below shows relevant data for three thrusters that meet this I_s range somewhat closely:

	Arcjet(NH$_3$)	Hall(BPT-4000)	Ion(NSTAR)
I_s (sec)	800	1983	2800
α (W/kg)	295	366	278
η_t	0.32	0.57	0.50
φ	0.10	0.12	0.12

We proceed to calculate from $\Delta u = 4700$ m/sec and $\Delta u/v_c = 0.3$, the range $v/v_c = 0.56$ to 1.35 – note that the tankage mass fractions shown above are all close to $\varphi \approx 0.1$. The resulting values shown below come from Fig. 17–13 or Eq. 17–38 (m_{pl}/m_0) and Eq. 17–8 (t_p). Note $v = c = I_s g_0$.

Arcjet (NH$_3$): $m_{pl}/m_0 = 0.391$ and $t_p = 15$ days (the lower range of I_s).
Hall (BPT-4000): $m_{pl}/m_0 = 0.429$ and $t_p = 6.8$ days.
Ion (NSTAR): $m_{pl}/m_0 = 0.34$ and $t_p = 10$ days (the higher range of I_s) ($I_s = 2800$ sec; being throttled down, it has an efficiency lower than 0.62).

From the results above, the BPT-400 thruster appears to the most desirable being closest to the "dual-optimum point," thus confirming our earlier statement of "highest payload fraction consistent with shortest time." What is needed now is to specify the payload so as to be able size-up the thrusters needed (power configuration, number of units), which may add further discriminators within the various possibilities. Note that both the payload fraction and the travel time depend on thruster properties, namely, specific impulse, efficiency and specific power. Cost, lifetime, availability, and other mission constraints need consideration next. These may drive the design away from the dual optimum because higher payload fractions can be obtained with $\Delta u/v_c < 0.30$ but with longer travel times whereas with $\Delta u/v_c > 0.30$ we get shorter travel times at the expense of lower payload fractions.

17.5. MISSION APPLICATIONS

Three principal application areas have been mentioned at the introduction to this chapter. The selection of a particular electric propulsion system for any given flight application depends not only on the characteristics of the propulsion system but also on the propulsive requirements of the particular flight mission and on the proven performance of the specific candidate propulsion system along with vehicle interfaces and the power conversion and storage systems. In general, the following criteria can be enumerated:

1. For very precise low-thrust station-keeping and attitude control applications, *pulsed thrusters* are best suited. Some systems, however, utilize continuous thrust because pulsing generates noise or vibrations that cannot be tolerated.

2. For deep-space missions where the vehicle velocity increment is substantial, systems with *very high specific impulse* will give better performance. This is seen from the ratio $\Delta u/v \approx 0.335$ at the dual optimum location (data from Fig.17–13).

3. While very high spacecraft velocity increments optimize at the higher specific impulses, propulsive times also increase and more rapidly (as shown in Section 17.1, the optimum specific impulse is nearly proportional to the square root of the thrust operating time, from $0.505 \leq v/v_c \leq 1.0$ and Eq. 17–8). This makes the optimum thrust and corresponding power generally lower which is favorable to electrostatic and electromagnetic units that operate exclusively in the specific impulse range above 1000 sec. See Example 17–4.

4. Since most missions of interest require long life, *system reliability* is a key selection criterion. *Extensive testing* under all likely environmental conditions (temperatures, accelerations, vibration, and radiation) under zero pressure and zero-gravity is required for high reliability. Ground testing and qualification of electric engines needs to be as thorough as that of their chemical counterparts. Earth bound simulation of the low pressures in space requires large vacuum test chambers.

5. There is a premium gain for *high thruster efficiencies* and *high power-conversion efficiencies*. These will reduce the propellant mass, the inert mass of the power supply system and reduce thermal control requirements, all of which translate into lower total mass and higher vehicle performance. Technology advances in specific power (α) and decreases in tankage fraction (φ) are also desirable toward minimizing inert mass.

6. For every propulsion mission there is a theoretically *optimum range of specific impulse* (see Fig. 17–3) and thus an optimum electrical propulsion scheme. While this optimum could be superseded by other conflicting system constraints (e.g., flight time or maximum power or size constraints or cost), the present inventory of proven electrical propulsion systems should meet most goals after some modifications.

7. The present *state of the art in electrical power sources* may limit the type and size of electric propulsion systems that can be used for missions to the outer planets, unless nuclear energy power generation on board the spacecraft is further developed and its launch safety concerns become more acceptable to society.

8. Practical factors, such as the storing and feeding of liquids in zero gravity, the availability of propellant (in the case of xenon), the conditioning of power to the desired voltage, frequency, and pulse duration, as well as redundancies in key system elements, the survival of sensors or controllers in long flights, and the inclusion of automatic self-checking devices along with cost, all influence the selection and application of specific types of electric propulsion rockets.

9. In addition to tankage and other internal considerations, *propellant selection* will also be governed by certain interface criteria such as plume noninterference with communication signals. Plumes must also be thermally benign and noncondensing on sensitive surfaces of the spacecraft such as optical windows, mirrors, and solar cells. See Chapter 20.

10. Selection methods may evolve as one or more flight-proven electrical propulsion systems become available. Also, it may be possible and more practical to modify a less than optimum existing proven unit to meet a desired new application.

Synchronous or geostationary satellites are of particular interest for communications and earth observation; their long life requires extensive station-keeping propulsion requirements. Until recently, the main limitation to any such life increase had been the propellant mass requirement. There are also electric propulsion opportunities for orbit raising from LEO to GEO and there is interest in including propulsion system that can do both orbit rising and station keeping. "All-Electric" is a recently minted term that refers to satellites that do not use conventional chemical propellants to reach their operational orbit; although such orbit-raising process may take somewhat longer, the satellites themselves are much lighter than those with chemical propellant of similar means and they retain the capability to change orbital planes after launch. Earth satellites in inclined orbits with precise time–trajectory position requirements need propulsion units to maintain such orbits for counteracting certain perturbing natural forces, as described in Chapter 4. Drag compensation in low earth orbit (atmospheric skimming) satellites can be accomplished with electric thrusters.

Increasing life trends in earth-orbit satellites from a minimum of 8 years to at least 15 years significantly increases their total impulse and the durability requirements of their propulsion system. For example, the north–south station-keeping (NSSK) function of a typical geosynchronous satellite requires about 40,000 to 45,000 N-sec or 9000 to 10,000 lbf-sec of impulse per year. Table 17–8 shows some of the characteristics required of small and large electric thrusters for various propulsion functions in space. A good review of the state of the art up to 1997 is found in Ref. 17–30.

TABLE 17–8. Space Propulsion Application and Characteristics for Three Thrust Levels of Electric Propulsion Thrusters[a]

Thrust Class	Application (Life)	Characteristics	Status
Micronewtons (μN)	E–W station keeping Attitude control Momentum wheel unloading (15–20 years) Formation Flying[b]	10–500 W power Precise impulse bits of $\sim 2 \times 10^{-5}$ N-sec	Operational & in development (micropropulsion)
Millinewtons (mN)	N–S station keeping Orbit changes Drag cancellation Vector positioning (20 years)	Kilowatts of power Impulse bits $< 2 \times 10^{-3}$ N-sec for N–S, impulse/year of 46,000 N-sec/100 kg spacecraft mass	Operational
0.01–10 N	Orbit raising Interplanetary travel Solar system exploration (1–5 years)	Long duration 1 kW–300 kW of power Intermittent and continuous operation	Operational

[a]Presently, nearly 95% of mission's applications of electric propulsion are for auxiliary propulsion such as orbit maintenance.
[b]Formation Flying is used in several communication LEO satellites. By having exactly equal spans between, they can provide continuous coverage in the region under them.

17.6. ELECTRIC SPACE-POWER SUPPLIES AND POWER-CONDITIONING SYSTEMS

The availability of substantial amounts of electrical power in space is considered to be a key requirement for high thrust electrical propulsion. Several combinations of energy sources and conversion methods have reached prototype stages, but only solar cells (photovoltaic), isotope thermoelectric generation units (nuclear), and fuel cells (chemical) have advanced to the point of routine space flight operation. Power output capacity of larger operational systems has been increasing from the low one-kilowatt range to the medium tens of kilowatts required for some missions. The high end of a hundred or more kilowatts is still pending some technological (and political) breakthroughs.

Space power level requirements have increased with the increased capacity of earth-orbit communications satellites and with planned missions, both manned and robotic, to moons and nearby planets. Payload requirements and thrust duration dictate power levels. Commercial communications system power satellites can temporarily reduce their communications volume during orbit maintenance so that electric power supplies do not need a dedicated unit for the propulsion system, but larger power demands may need enhanced solar cell capabilities. Many communications satellites actually share part or all of the power-conditioning equipment with their electric thrusters.

Power Generation Units

Electric power generation units are classified as either *direct* (no moving mechanical parts) or *dynamic*. When the primary driver is reliability and the total power is small, direct conversion has been preferred but, at the advent of the Space Shuttle era and with the manned International Space Station (ISS), dynamic systems are being reconsidered. Many diverse concepts have been evaluated for meeting the electrical power demands of spacecraft, including their electric propulsion needs. Direct energy conversion methods considered include photovoltaic, thermoelectric, thermionic, and electrochemical, while methods with moving parts include the Brayton, Rankine, and Stirling cycles.

Batteries. Batteries may basically be classified as either primary or secondary. *Primary batteries* consume their active materials and convert chemical energy into electrical. *Secondary batteries* store electricity by utilizing a reversible chemical reaction and are designed to be recharged many times. There are both dry-cell and wet-cell primary batteries. The importance of primary batteries passed with the short-lived satellites of the early 1960s. Secondary batteries with recharging provisions afford electrical power at higher output levels and longer lifetimes than primary batteries. All batteries must be sealed against the space vacuum or housed inside pressurized compartments. Secondary batteries are a critical component of solar cell systems for power augmentation and emergency backup and for periods when the satellite is in the earth's shadow. Lithium-ion batteries have been used in space applications since 2003 because they offer higher energy/power density and operating voltage than conventional ones.

Fuel Cells. *Chemical fuel cells* are conversion devices used to supply space-power needs for two to four weeks and for power levels up to 40 kW in manned missions. A catalyzer controls the chemical reaction to yield electricity directly; there is also some heat evolved which must be removed to maintain a desirable fuel cell temperature. Fuel cells are too massive for both robotic and long-duration missions, having also had some reliability problems, but technological improvements have considerably advanced their performance and attractiveness.

Solar Cell Arrays. Solar cells rely on the photovoltaic effect to convert the sun's electromagnetic radiation into electricity. In silicon cells, sunlight generates a voltage and a photocurrent across a *p-n junction* running as an inverse photodiode. Commercially available devices have reached 20% efficiency, but single-bandgap (i.e., single-junction) semiconductors are restricted by a theoretical upper limit of 33.7%. Newer multijunction solar cells can circumvent this efficiency limit because each layer is made to absorb a separate part of the spectrum as the light passes though the stack; presently, arrays with greater than 24% efficiencies are in earth orbit with multijunction semiconductors (up to 46% laboratory efficiencies have been reported by Fraunhofer ISE/Soltec). Another aspect of typical solar cell arrays is that many had been originally designed for only a 28 VDC output, although the present designs such as those for the ISS bus operate at 160 VDC and the NSTAR/DAWN missions uses 80 to 160 VDC. Today, there are designs (termed *high-voltage* or *direct drives*, Ref. 17–31) that output increased voltages from their solar cell arrays, up to 300 VDC and beyond. As they become more operational, their higher voltages will considerably simplify power-conditioning equipment in many existing electric-propulsion systems and that may result in valuable inert mass savings. But care must be taken to avoid electrostatic discharges that can more easily occur at some earth orbits.

Solar cells have supplied electrical power in most of long-duration space missions. The first solar-cell unit was launched in March 1958 on Vanguard I and successfully energized data transmission for six years. Solar arrays exist in sizes up to 10 kW and could grow to 10 MW sizes in earth orbits. Typically, solar cell arrays are designed with a 20% overcapacity to allow for material degradation toward "end of life." Losses in performance are due to radiation and particle impact damage, particularly in the radiation belts around the Earth. There have been continuous improvements in cell efficiency, reliability, and power per unit mass; for example, standard silicon cells deliver 180 W/m^2 with arrays of 40 W/kg. Newer gallium arsenide cells produce 220 W/m^2 and are more radiation resistant than silicon cells; these cell arrays are presently space qualified and together with parabolic concentrators can reach 100 W/kg (Ref. 17–27).

Factors that affect the specific mass of a solar array, besides its conversion efficiency, include *the solar constant* (which varies inversely as the square of the distance from the sun) and the manufactured thinness of the cell. Orientation to the sun is a more critical factor when solar concentrators are being used. Cell output is a function of cell temperature; performance in present designs can drop as much as 20% for a 100 °F increase in operating temperature so that thermal control has been critical. Solar cell panel configurations can be (1) fixed and body mounted to the spacecraft, (2) rigid and deployable (protected during launch and positioned in space), (3) flexible panels that are deployed (rolled out or unfolded), and (4) deployable with solar concentrator capabilities.

In addition to the solar arrays, their structure, deployment, and orientation equipment, other required items that include batteries, power-conditioning and distribution systems, must be assigned to the power source mass and volume. Despite their apparent bulkiness and battery dependence, solar-cell electrical systems have emerged as the dominant generating-power system for unmanned spacecraft.

Nuclear Thermoelectric and Thermionic Systems. Nuclear energy from long-decay radioisotopes and in some cases from fission reactors has played some role in the production of electricity in space. Both thermoelectric (based on the Seebeck effect) and thermionic (based on the Edison effect) devices have been investigated. These generators have no moving parts and can be made of materials reasonably resistant to radioactive environments. But their specific power is relatively low and cost, availability, and efficiency have been a limiting factor. The Voyager 2 spacecraft has been powered by three radioisotope thermoelectric generators which collectively delivered 420 watts at launch (in 1977); this spacecraft has been in operation for 37 years (as of 2014).

Throughout the 1950s and 1960s nuclear fission reactors were regarded as the most promising way to meet the high power demands of space missions, particularly trips to the outer planets involving months and perhaps years of travel. Radioisotope thermoelectric power has been embodied in a series of SNAP (Systems for Nuclear Auxiliary Power) electrical generating units that were designed and tested, ranging from 50 W to 300 kW of electrical output. More recent space nuclear reactor programs include NASA's Project Prometheus and the Russian TOPAZ that has been space tested up to nearly 6 kW. The latter consists of sets of nuclear rods each surrounded by a thermionic generator. Usage of direct-conversion nuclear power generation, like in NASA's Multi-Mission Radioisotope Thermoelectric Generator (MMRTG), include powering the Mars Science Laboratory (Curiosity) currently exploring the planet.

Thermionic converters have a significant mass advantage over thermoelectric ones, based on their higher effective radiator temperatures. Since thermal efficiencies for both thermoelectric and thermionic conversion have been below 10% and since all unconverted heat must be radiated, at higher operating temperatures thermionic radiators can be less massive. Moreover, cooling must be present at times when no electricity is generated since the heat source cannot be "turned off." Depending on the location of the waste heat, designs involving heat pipes or recirculating cooling fluids are needed.

Long-Duration High-Output Dynamic Systems. On earth, designs of electric power generation with outputs of 10 to 1000 kW have been based on Stirling or Rankine heat engine cycles with nuclear, chemical, and even solar power sources. Overall efficiencies can be between 10 and 40%, but the associated hardware remains complex, including bearings, pumps, reactors, control rods, shielding, compressors, turbines, valves, and heat exchangers. Superconducting magnets together with advances in the state-of-the-art of seals, bearings, and flywheel energy storage have made some dynamic units relatively more attractive. There remain development issues about high-temperature materials that will withstand intense nuclear radiation fluxes over several years and there are still some concerns about achieving the required reliability in such complex systems in the space environment. While limited small-scale experiments have been conducted, the development of these systems remains a challenge. An advanced Stirling radioisotope generator (labeled Radioisotope Electric Propulsion, see Fig. 17–1) is under development to replace radioisotope thermoelectric generators (RTGs) for future NASA missions.

Power-Conditioning Equipment (PCU or PPU)

Power-conditioning units or equipment are a necessary component of electric propulsion systems because of inevitable mismatches in voltage, frequency, power rate, and other electrical properties between the space-power generating unit and the electric thruster. Power-conditioning equipment has been more expensive, more massive, and more difficult to qualify than the thruster itself. If the thrust is pulsed, as in the PPT, the power-conditioning unit has to provide pulse-forming networks for momentary high currents, exact timing of different outputs, and control and recharging of condensers. Electrostatic engines commonly require up to 1800 VDC (except for FEEPs); the output of solar-cell arrays is typically 28 to 300 VDC, so there is a need for DC-to-DC inverters and step-up transformers to accomplish this task. Often this equipment is housed in a single "black box," termed the *power conditioner*. Modern conditioning equipment contains all the internal logic required to start, safely operate, and stop the thruster; it is controlled by on–off commands sent by the spacecraft control processor. Besides the above functions that are specific to each engine, power-conditioning equipment may have to provide circuit protection and propellant flow control as well as necessary redundancies.

As may be apparent from Table 17–7, one of the largest contributors to the specific mass (α) of any electric-propulsion system can be the power-conditioning equipment. Here, electrothermal units have the simplest and lightest conditioning equipment, some needing none. Ion engines, on the other hand, have the heaviest equipment, with Hall thrusters somewhere in between (Ref. 17–24). PPTs have tended to have a large mass, but advances in energy storage capacitors have improved this situation. In fact, advances in solid-state electronic pulse circuits together with lighter, more efficient, and higher temperature power-conditioning hardware are areas of great interest in electric propulsion. Equipment efficiency tends to be high, about 90% or more, but here heat is generated at low temperatures and must be radiated to maintain their required moderately low temperatures of operation. While most present electric propulsion systems share the spacecraft electrical "bus system," in some applications where EPS acts as the primary propulsion a special electrical bus called a "Direct Drive" (Ref. 17–31) is provided. When feasible, direct drives may allow simplification or elimination of some conditioning equipment, but a low-pass filter is still needed for electromagnetic interference (EMI) control (more information in Ref. 17–29).

SYMBOLS

a	acceleration, m/sec^2(ft/sec^2)
A	area, cm^2 or m^2
$\boldsymbol{B}, B_z$	magnetic flux density, web/m^2 or tesla
B_r	radial magnetic flux density, web/m^2 or tesla
c	effective exhaust velocity, m/sec
C	circuit capacitance, farad/m
c_p	specific heat, J/kg-K
d	accelerator grid spacing, cm (in.)
D	hole or beam diameter, cm (in.)
e	electronic charge, 1.602×10^{-19} coulomb
E	electric field, V/m
E_x	longitudinal electric field, V/m
E_y	transverse electric field, V/m
$\boldsymbol{f}$	microscopic force on a particle
F	thrust force, N or mN (lbf or mlbf)
$\tilde{F}_x$	accelerating force density inside channel, N/m^3(lbf/ft^3), see Eq. 17–14
g_0	constant converting propellant ejection velocity units to sec, 9.81 m/sec^2 or 32.2 ft/sec^2
I	total current, A
I_{bit}	impulse bit, N-sec or lbf-sec
I_s	specific impulse, sec [(I_s^*) optimum]
$\boldsymbol{j}, j_{max}$	current density, A/m^2
j_x, j_y	orthogonal current density components
j_θ	Hall current density, A/m^2
k	specific heat ratio
l_I	ionization loss, W
L	circuit inductance, henry
m_p	propellant mass, kg (lbm)
m_{pp}	power plant mass, kg (lbm)
m_{pl}	payload mass, kg (lbm)
m_0	initial total vehicle mass, kg (lbm)
$\dot{m}$	mass flow rate, kg/sec (lbm/sec)
$\mathfrak{M}$	atomic or molecular mass, kg/kg-mol (lbm/lb-mol)
n_e	electron number density, m^{-3}(ft^{-3})
N	number of pulses

P	power, W
P_e	electrical power, W
P_{jet}	kinetic power of jet, W
r_i	inner radius, m
r_o	outer radius, m
R	plasma resistance, ohms
S	distance, cm (in.)
t	time or duration, sec
t_p	propulsive time, sec [t_p^* optimum]
T	absolute temperature, K (°R)
Δu	vehicle velocity change, m/sec (ft/sec)
v, v_e	propellant or charged particle exhaust velocity, m/sec (ft/sec)
v_x, v_i	plasma velocity or injection velocity along accelerator, m/sec
v_c	characteristic speed, m/sec
V	voltage, V
V_x, V_{acc}	Local or total accelerator voltage, V
w	rail width, m
x	linear dimension, m (ft)

Greek Letters

α	specific power, W/kg (W/lbm)
β	electron Hall parameter (dimensionless)
ε_0	permittivity of free space, 8.85×10^{-12} farad/m
ε_I	ionization potential, eV
η_H	Hall thruster β-efficiency
η_t, η	thruster efficiency
κ	ion-ratio of injected to gained accelerator energy ≥ 0
μ	ion mass, kg
μ_e	electron mass, 9.11×10^{-31} kg
μ_0	permeability of free space, $4\pi \times 10^{-7}$ henry/m
φ	tankage mass fraction
ρ_e	space–charge, coulomb/m^3
σ	plasma electrical conductivity, mho/m
τ	mean collision time, sec (also characteristic time, sec)
ω	electron cyclotron frequency, (sec)$^{-1}$

PROBLEMS

1. The characteristic velocity $v_c = \sqrt{2t_p \alpha \eta}$ is used to achieve a dimensionless representation of flight performance analysis. Derive Eq. 17–38 without any tankage fraction allowance (i.e., $\varphi = 0$). Also, plot the payload fraction against v/v_c for several values of $\Delta u/v_c$. Discuss your results with respect to the optimum performance.

2. For the special case of zero payload in Problem 1, determine the maximized values of $\Delta u/v_c$, v/v_c, m_p/m_0, and m_{pp}/m_0 in terms of this characteristic velocity.

   Answer: $\Delta u/v_c = 0.805$, $v/v_c = 0.505$, $m_p/m_0 = 0.796$, $m_{pp}/m_0 = 0.204$.

3. For a space mission with an incremental vehicle velocity of 85,000 ft/sec and a specific power of $\alpha = 100\,\text{W/kg}$, determine the optimum values of I_s and t_p for two maximum payload fractions, namely 0.35 and 0.55. Take the thruster efficiency as 100% and $\varphi = 0$.

   Answer: For 0.35: $I_s = 5.11 \times 10^3$ sec; $t_p = 2.06 \times 10^7$ sec; for 0.55: $I_s = 8.88 \times 10^3$ sec; $t_p = 5.08 \times 10^7$ sec.

4. Derive Eq. 17–7 using $m_{pp} = m_p[(v/v_c)^2 + (v/v_c)^3]$ instead of Eq. 17–6 (take $\varphi = 0$); this form penalizes the high I_s and/or short t_p missions. Plot and compare to the results shown on Fig. 17–3.

5. An ion thruster uses heavy positively charged particles with a charge-to-mass ratio of 500 coulombs per kilogram, producing a specific impulse of 3000 sec. **(a)** What two-screen grid acceleration voltage would be required for this specific impulse? **(b)** If the accelerator spacing is 6 mm, what would be the diameter of an ion beam producing 0.5 N of thrust at this accelerator voltage?

   Answer: **(a)** 8.66×10^5 V; **(b)** $D = 1.97$ mm.

6. An argon ion thruster has the following characteristics and operating conditions:

Voltage across ionizer = 400 V	Voltage across accelerator = 3×10^4 V
Diameter of ion source = 5 cm	Accelerator electrode spacing = 1.2 cm

   Calculate the mass flow rate of the propellant, the thrust, and the thruster overall efficiency (including ionizer and accelerator). Assume singly charged ions.

   Answer: $\dot{m} = 2.56 \times 10^{-7}\,\text{kg/sec}$; $F = 9.65 \times 10^{-2}\,\text{N}$; $\eta_t = 98.7\%$.

7. For a given power source of 300 kW electrical output, a propellant mass of 6000 lbm, $\alpha = 450\,\text{W/kg}$, and a payload of 4000 lbm, determine the thrust, ideal velocity increment, and duration of powered flight for the following three cases:

   a. Arcjet: $I_s = 500$ sec $\eta_t = 0.35$
   b. Ion engine: $I_s = 3000$ sec $\eta_t = 0.75$
   c. Hall engine: $I_s = 1500$ sec $\eta_t = 0.50$

   Answers:

   a. $t_p = 3.12 \times 10^5$ sec; $\Delta u = 3.63 \times 10^3$ m/sec; $F = 42.8$ N.
   b. $t_p = 5.24 \times 10^6$ sec; $\Delta u = 2.18 \times 10^4$ m/sec; $F = 15.29$ N.
   c. $t_p = 1.96 \times 10^6$ sec; $\Delta u = 1.09 \times 104$ m/sec; $F = 20.4$ N.

8. A formulation for the exit velocity in an MPD that allows for a simple estimate of the accelerator length is shown below; these equations relate the accelerator distance to the velocity implicitly through the acceleration time t. Considering a flow at a constant plasma of density ρ_m (which does not choke), solve Newton's second law first for the speed $v(t)$ and then for the distance $x(t)$ and show that

$$v(t) = (E_y/B_z)[1 - e^{-t/\tau}] + v(0)e^{-t/\tau}$$

$$x(t) = (E_y/B_z)[t + \tau e^{-t/\tau} - \tau] + x(0)$$

where $\tau = \rho_m / \sigma B_z^2$ and has units of seconds. For this simplified plasma model of an MPD accelerator, calculate the distance needed to accelerate the plasma from rest up to $v = 0.01(E/B)$ and the time involved. Take the plasma conductivity as $\sigma = 100\,\text{mho/m}$, $B_z = 10^{-3}$ tesla web/m^2), $\rho_m = 10^{-3}\,\text{kg/m}^3$, and $E_y = 1000\,\text{V/m}$.

Answer: 503 m, 0.1005 sec.

9. Assume that a materials breakthrough makes it possible to increase the operating temperature in the plenum chamber of an *electrothermal engine* from 3000 to 4000 K. Nitrogen gas is the propellant which is available from tanks at 250 K. Neglecting dissociation, and taking $\alpha = 200\,\text{W/kg}$ and $\dot{m} = 3 \times 10^{-4}\,\text{kg/sec}$, calculate the old and new Δu corresponding to the two temperatures. Operating or thrust time is 10 days, payload mass is 1000 kg, and $k = 1.3$ for the hot diatomic molecule.

Answer: 610 m/sec old, 711 new.

10. An arcjet delivers 0.26 N of thrust. Calculate the vehicle velocity increase under gravitationless, dragless flight for a 28-day thrust duration with a payload mass of 100 kg. Take thruster efficiency as 50%, specific impulse as 2600 sec, and specific power as 200 W/kg. This is not an optimum payload fraction; estimate an I_s which would maximize the payload fraction with all other factors remaining the same.

Answer: $\Delta u = 4.34 \times 103$ m/sec; $I_s = 2020$ sec (decrease).

11. A patent application describes an electrostatic thruster that accelerates electrons as the propellant. The inventor points out that the space–charge limited thrust is independent of the propellant mass and that electrons are very easy to produce (by cathode surface emission) and much easier to accelerate than atomic ions. Show using the basic relationships for electrostatic thrusters given in this chapter that electron acceleration is impractical for electrostatic thrusters. Assume that the required thrust is 10^{-5} N per accelerator hole, that there are several thousand holes of "aspect ratio" $D/d = 1.0$ in the accelerator, and that the neutralizer operates with protons (which have a mass 1836 times that of the electron).

12. For each of the three thrusters in Example 17–4, calculate the thrust F and input power P_e that would apply for a payload mass m_{pl} of 100 kg. What would result if the spacecraft power supply is limited to 30 kW but mission time could extend up to 100 days?

Answer: Arcjet: 0.754 N, 9.24 kW; Hall: 1.73 N, 29.6 kW; Ion: 1.49 N, 40.9 kW.

REFERENCES

17–1. M. M. Micci and A. D. Ketsdever (Eds.), *Micropropulsion for Small Spacecraft*, Progress in Astronautics and Aeronautics, Vol. 187, AIAA, Reston, VA, 2000; J. Mueller et al., "Survey of Propulsion options for Cubesats," Jet Propulsion Laboratory Caltech, CPIAC JSD CD-62 (Abstract No. 2010-9915BP).

17–2. R. G. Jahn, *Physics of Electric Propulsion*, McGraw-Hill Book Company, New York, 1968, pp. 103–110. See also http://alfven.princeton.edu/papers/Encyclopedia.pdf

17–3. D. M. Goebel and I. Katz, *Fundamentals of Electric Propulsion—Ion and Hall Thrusters*, John Wiley and Sons, Hoboken, NJ, 2008.

17–4. P. J. Turchi, Chapter 9, "Electric Rocket Propulsion Systems," in R. W. Humble, G. N. Henry, and W. J. Larson (Eds.), *Space Propulsion Analysis and Design*, McGraw-Hill, New York, 1995, pp. 509–598.

17–5. A. Spitzer, "Near Optimal Transfer Orbit Trajectory Using Electric Propulsion," AAS Paper 95–215, American Astronautical Society Meeting, Albuquerque, NM, Feb. 13–16, 1995.

17–6. D. B. Langmuir, Chapter 9, "Low-Thrust Flight: Constant Exhaust Velocity in Field-Free Space," in H. Seifert (Ed.), *Space Technology*, John Wiley & Sons, New York, 1959.

17–7. D. C. Byers and J. W. Dankanich, "Geosynchronous Earth Orbit Communication Satellite Deliveries with Integrated Electrical Propulsion," *Journal of Propulsion and Power*, Vol. 24, No. 6, 2008, pp. 1369–1375.

17–8. M. J. Patterson, L. Pinero, and J. S. Sovey, "Near-Term High Power Ion Propulsion Options for Earth-Orbital Applications," AIAA Paper 2009–4819, Denver, Colorado, Aug. 2009.

17–9. F. R. Chang Diaz, et al, VASIMR Engine: Project Status and Recent Accomplishments, AIAA Paper 2004-0149, Reno Nevada, Jan. 2004; see also Wikipedia entry: Variable Specific Impulse Magnetoplasma Rocket.

17–10. C. D. Brown, *Spacecraft Propulsion*, AIAA Education Series, Washington, DC, 1996.

17–11. F. F. Chen, *Introduction to Plasma Physics*, Plenum Press, New York, 1974.

17–12. G. W. Sutton and A. Sherman, *Engineering Magnetohydrodynamics*, McGraw-Hill Book Company, New York, 1965.

17–13. D. M. Zube et al., "History and Recent Developments of Aerojet Rocketdyne's MR-510 Hydrazine Arcjet Systems," Space Propulsion Conference 2014, paper SP2014_2966753, Cologne, Germany.

17–14. D. R. Bromaghim et al., "Review of the Electric Propulsion Space Experiment (ESEX) Program," *Journal of Propulsion and Power*, Vol. 18, No. 4, July–August 2002, pp. 723–730.

17–15. J. Mueller, Chapter 3, in Ref. 17–1.

17–16. M. Tajmar, A. Genovese, and W. Steiger, "Indium Field Emission Electric Propulsion Microthruster Experimental Characterization," *Journal of Propulsion and Power*, Vol. 20, No. 2, Mar.–Apr. 2004.

17–17. J. R. Beattie, XIPS Keeps Satellites on Track, *The Industrial Physicist*, Vol. 4, No. 2, Jun. 1998; C. E. Garner and M. D. Rayman, "In Flight Operation of the Dawn Ion Propulsion System Through Survey Science Orbit at Ceres," paper AIAA 2015-3717, Orlando, FL, 2015.

17–18. K. J. Groh and H. W. Loeb, State of the Art of Radio Frequency Ion Thrusters, *J. Propulsion*, Vol. 7 No. 4, Jul.–Aug. 1991, pp. 573–579; N. Nishiyama et al., "In-Flight Operation of the Haybusa2 Ion Engine System in the EDVEGA Phase," paper AIAA 2015-3718, Orlando, FL, 2015.

17–19. P. G. Hill and C. R. Peterson, *Mechanics and Thermodynamics of Propulsion*, Addison-Wesley Publishing Company, Reading, MA, 1992.

17–20. S. Liu and R. A. Dougal, "Initial Velocity Effect on Space-Charge-Limited Currents," *Journal of Applied Physics*, Vol. 78, No. 10, Nov. 16, 1995, pp. 5919–59–25; O. Biblarz, "Ion Accelerator Currents Beyond the Child-Langmuir Limit," AIAA Paper 2013–4109, San Jose, CA, Jul. 2013.

17–21. E. L. Resler, Jr., and W. R. Sears, "Prospects of Magneto-Aerodynamics," *Journal of Aeronautical Sciences*, Vol. 24, No. 4, Apr. 1958, pp. 235–246.

17–22. S. S. Bushman and R. L. Burton, "Heating and Plasma Properties in a Coaxial Gas-dynamic Pulsed Plasma Thruster," *Journal of Propulsion and Power*, Vol. 17, No. 5, Sept.–Oct. 2001, pp. 959–966.

17–23. V. Kim, "Main Physical Features and Processes Determining the Performance of Stationary Plasma Thrusters," *Journal of Propulsion and Power*, Vol. 14, No. 5, Sept.–Oct. 1998, pp. 736–743.

17–24. C. H. McLean, J. B. McVey, and D. T. Schappell, "Testing of a U.S.-Built HET System for Orbit Transfer Applications," AIAA Paper 99-2574, Jun. 1999.

17–25. B. Welander et al., "Life and Operating Range Extension of the BPT 4000 Qualification Model Hall Thruster," AIAA Paper 2006–5263, Sacramento, CA, Jul. 2006.

17–26. D. Baker, Chapter 10, "Mission Design Case Study," in R. W. Humble, G. N. Henry, and W. J. Larson (Eds.), *Space Propulsion Analysis and Design*, McGraw-Hill, New York, 1995.

17–27. M. Martinez-Sanchez and J. E. Pollard, "Spacecraft Electric Propulsion—An Overview," *Journal of Propulsion and Power*, Vol. 14, No. 5, Sept.–Oct. 1998, pp. 688–699.

17–28. J. D. Filliben, "Electric Propulsion for Spacecraft Applications," *Chemical Propulsion Information Agency Report CPTR 96-64*, The Johns Hopkins University, Dec. 1996.

17–29. M. A. Kurtz, H. L. Kurtz, and H. O. Schrade, "Optimization of Electric Propulsion Systems Considering Specific Power as a Function of Specific Impulse," *Journal of Propulsion and Power*, Vol. 4, No. 2, 1988, pp. 512–519.

17–30. J. D. Filliben, "Electric Thruster Systems," *Chemical Propulsion Information Agency Report CPTR-97-65*, The Johns Hopkins University, Jun. 1997.

17–31. T. W. Kerslake, "Effect of Voltage Level on Power System Design for Solar Electric Propulsion Missions," *Journal of Solar Energy Engineering*, Vol. 126, No. 3, Aug. 2004, pp. 936–944.

CHAPTER 18

THRUST VECTOR CONTROL

Controlling the flight path and the attitude of a rocket-propelled vehicle enables it to reach a precise flight destination. Rocket propulsion systems always provide a "push" toward an intended destination, but they also can be made to provide torques that rotate the vehicle in conjunction with the propulsive force. By controlling the direction of thrust vectors through mechanisms described in this chapter, it is possible to influence a vehicle's pitch, yaw, and roll rotations. Thrust vector control units integrated with the principal propulsion system are only effective while it is operating and producing an exhaust jet. For periods of free flight, when the main rocket propulsion system is off, separate propulsion units are needed for achieving control over attitude or flight path. In space, many vehicles utilize dedicated attitude-control systems with multiple independent thrust-producing units (see Figs. 1–14 and 4–14). Examples of other dedicated attitude control arrangements are shown in Figs. 3–16, 6–14, 12–27, and 12–28. A related history of flight trajectory control with liquid propellant rocket engines is given in Ref. 18–1.

All chemical propulsion systems may have one of several types of thrust vector control (TVC) mechanisms. Some may apply to solid, hybrid, and liquid propellant rocket propulsion systems, but most are specific to only one of these propulsion categories. In this chapter we describe mechanisms that consist of a single-nozzle (Section 18.1) and those that use two or more nozzles (Section 18.2).

Aerodynamic fins (fixed or movable) are very effective for controlling vehicle flight within the atmosphere, and almost all weather rockets, anti-aircraft missiles, and air-to-surface missiles use them. Even though aerodynamic control surfaces add some drag, their effectiveness in terms of vehicle mass, turning moment, and actuating power consumption is difficult to surpass with any other method. Vehicle flight control is also achieved with a separate attitude control propulsion system as described in Sections 4.5 and 6.7. Here six or more small liquid propellant thrusters (with separate feed systems and controls) provide moment forces to vehicles in flight during, before, or after operation of the main rocket propulsion system. Figure 4–14 indicates that it requires 12 nozzles to obtain pure torques about 3 perpendicular axes. Rockets are the principal means available for TVC in space.

Motives for TVC may be stated as: (1) willful changes to a flight path or trajectory (e.g., changing flight-path direction of a target-seeking missile); (2) vehicle rotation or changes in attitude during powered flight; (3) deviation corrections from intended trajectory or attitude during powered flight; or (4) thrust corrections for misalignments of fixed nozzles in main propulsion system during its operation (when the main thrust vector misses the vehicle's center of mass). In all prior analyses, we have implicitly assumed that the trust vector passes through the center of mass and is perfectly aligned with the direction of flight (see Fig. 4–6). For TVC, moments are purposely generated about the *center of mass* to control flight trajectories. The location of the side force (or its perpendicular component) along the vehicle's axis determines its moment arm and thus the magnitude of the force needed for any given application; this implies that locations as far away as possible from the center of mass (i.e., the longest moment arms) are most desirable to minimize the needed forces and thus the propellant usage.

Pitch moments are those that raise or lower the nose of a vehicle; *yaw moments* turn the nose sideways; and *roll moments* are applied about the main axis of the flying vehicle (Fig. 18–1). Most often, the thrust vector of the main rocket nozzle is in the direction of the vehicle axis and goes through its center of mass. Thus it is possible to obtain pitch and yaw control moments by simple deflections of the main rocket thrust vector. Roll control, however, requires the use of two or more rotary vanes or two or more separately hinged propulsion system nozzles. Figure 18–2 shows how a pitch moment is obtained from a hinged thrust chamber or nozzle. Side force and pitch moment vary as the *sine* of the effective angle of thrust vector deflection.

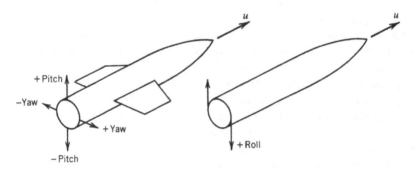

FIGURE 18–1. Moments applied to a flying vehicle.

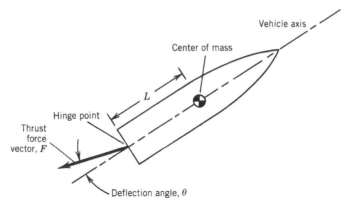

FIGURE 18–2. In the arrangement depicted, the pitch moment applied to the vehicle is $FL \sin\theta$.

18.1. TVC MECHANISMS WITH A SINGLE NOZZLE

Many different mechanisms have been successfully used. Several are illustrated in Refs. 18–2 and 18–3. They can be classified into four categories:

1. Mechanical deflection of the main nozzle or thrust chamber.
2. Insertion of heat-resistant movable bodies into the exhaust jet; these experience aerodynamic forces and cause deflection of part of the exhaust gas flow.
3. Injection of fluid into a side portion of the diverging nozzle section, causing an asymmetrical distortion of the supersonic exhaust flow.
4. Separate thrust-producing unit that is not part of the main propulsion system providing flow through one or more of its own nozzles.

Each category is described briefly below and in Table 18–1, where the four categories are separated by horizontal lines. Figure 18–3 illustrates several TVC mechanisms. All of the TVC schemes shown here have been used in production vehicles.

In the *gimbal* or *hinge* scheme (a hinge permits rotation about one axis only, whereas a gimbal is essentially a universal joint), the whole engine is pivoted on a bearing and thus the thrust vector rotates. For small angles this scheme has negligible losses in specific impulse and is used in many vehicles. It requires a flexible propellant-piping unit or bellows to send the propellant from the fixed vehicle tanks to the movable engine. The Space Shuttle (Fig. 1–14) has two gimballed orbit maneuver engines, and three gimballed main engines. Figures 6–1 and 8–17 show gimbaled engines. Some Soviet launch vehicles use multiple thrusters and hinges (Figs. 11–1 and later in 18–10 indicate four hinges), in contrast many U.S. vehicles use gimbals.

Jet vanes as shown in Fig. 18–3 are pairs of heat-resistant, wing-shaped surfaces submerged in the hot exhaust jet of a fixed rocket nozzle. They were first

TABLE 18-1. Thrust Vector Control Mechanisms

Type	L/S[a]	Advantages	Disadvantages
Gimbal or hinge	L	Simple, proven technology; low torques, low power; up to $\pm 12°$; duration limited only by propellant supply; very small thrust loss	Requires flexible piping; high inertia; large actuators and a power supply for high slew rate
Movable nozzle (flexible bearing)	S	Proven technology; no sliding, moving seals; predictable actuation power; up to $\pm 12°$	High actuation forces; high torque at low temperatures; variable actuation force
Movable nozzle (rotary ball with gas seal)	S	Proven technology; no thrust loss if entire nozzle is moved; $\pm 20°$ possible	Sliding, moving hot gas spherical seal; highly variable actuation power; limited duration; needs continuous load to maintain seal
Jet vanes	L/S	Proven technology; low actuation power; high slew rate; roll control with single nozzle; $\pm 9°$	Thrust loss of 0.5–3%; erosion of jet vanes; limited duration; extends missile length
Jet tabs	S	Proven technology; high slew rate; low actuation power; compact package	Erosion of tabs; thrust loss, but only when tab is in the jet; limited duration
Jetavator	S	Proven on Polaris missile; low actuation power; can be lightweight	Erosion and thrust loss; induces vehicle base hot gas recirculation; limited duration
Liquid-side injection	S/L	Proven technology; specific impulse of injectant nearly offsets weight penalty; high slew rate; easy to adapt to various motors; can check out before flight; components are reusable; duration limited by liquid supply; $\pm 6°$	Toxic liquids are needed for high performance; often difficult packaging for tanks and feed system; sometimes requires excessive maintenance; potential spills and toxic fumes with some propellants; limited to low vector angle applications
Hot-gas-side injection	S/L	Lightweight; low actuation power; high slew rate; low volume/compact; low performance loss, such as turbine gas exhaust	Multiple hot sliding contacts and seals in hot gas valve; hot piping expansion; limited duration; requires special hot gas valves
Hinged auxiliary thrust chambers for high thrust engine	L	Proven technology; feed from main turbopump; low performance loss; compact; low actuation power; no hot moving surfaces; unlimited duration	Additional components and complexity; moments applied to vehicle are small; not used for 20 years in United States
Turbine exhaust gas swivel for large engine	L	Swivel joint is at low pressure, low performance loss; lightweight; proven technology	Limited side forces; moderately hot swivel joint; used mostly for roll control

[a]L, used with liquid propellant engines; S, used with solid propellant motors.

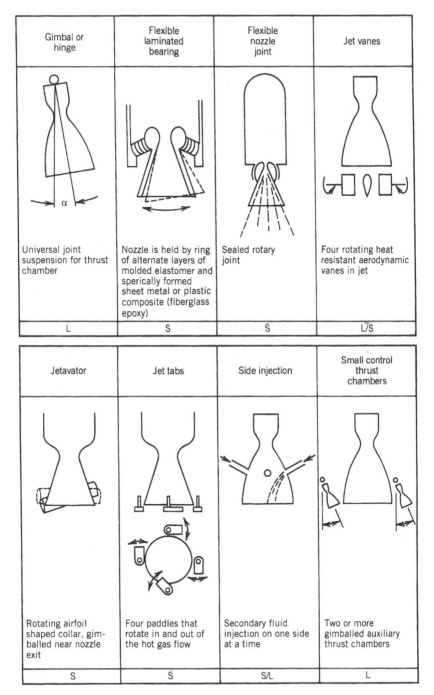

FIGURE 18–3. Schematic diagrams for eight different TVC mechanisms. Actuators and structural details are not shown. The letter L means it is used with liquid propellant rocket engines and S means used with solid propellant motors.

used in the 1940s. They cause extra drag (up to 2 to 5% less I_s; drag increases with larger vane deflections) and vane materials erode. Graphite jet vanes were used in German V-2 missiles in World War II and in Scud missiles fired by Iraq in 1991. The advantage of having roll control with a single nozzle often outweighs other performance penalties. A recent application is discussed in Ref. 18–4.

Small auxiliary thrust chambers were used in the Thor and early version of Atlas missiles. They provided roll control while the principal rocket engine operated. They are fed propellant from the same feed system as the main rocket engine. Such a scheme is still used on some Russian booster rocket vehicles.

Secondary fluid injection through a nozzle wall into the main gas stream induces oblique shocks in the nozzle's divergent section, causing an unsymmetrical distribution of the main gas flow, which produces the side force. Secondary fluids can originate from a stored liquid or from a separate hot gas generator (the gas would need to be sufficiently cool to be piped), a direct bleed from the chamber, or the injection of a catalyzed monopropellant. When the needed deflections are small, this is a low-loss scheme, but for large enough moments (large side forces) the amount of secondary fluid can become excessive. This scheme has found application in a few large solid propellant rockets, such as Titan IIIC and one version of Minuteman.

Of all the mechanical deflection types, *movable nozzles* and *gimballed engines* are the most efficient. They do not significantly reduce thrust or specific impulse and are weight-competitive with other mechanical types. Flexible nozzles (shown in Figs. 18–3 and 18–4) are commonly used with solid propellant motors for TVC. Molded, multilayer bearing packs act as seals, load transfer bearings, and viscoelastic flexures. The deformation of a stacked set of doubly curved elastomeric (rubbery)

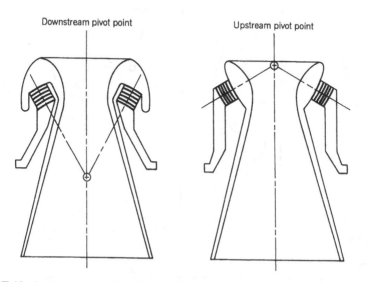

FIGURE 18–4. Two methods for using flexible nozzle bearings with different locations for the center of rotation. The bearing support ring is made of metal or plastic composite shims formed into rings with spherical contours (white) bonded together by layers of molded elastomer or rubber (black stripes). Although only five elastomeric layers are shown for clarity, many flexible bearings have 10 to 20 layers. Used with permission from Ref. 18–2.

layers between spherical metal sheets or spherically formed plastic composites made from fiberglass–epoxy is used to carry loads and allow angular deflections of the nozzle axis. Flexible seal nozzles have been used in launch vehicles and large strategic missiles, where environmental temperature extremes are modest. At low temperatures, elastomers become stiff and their actuation torques increase substantially, requiring a much more powerful actuation system. Figure 18–5 depicts a different type of flexible nozzle; it uses a movable joint with a toroidal hydraulic bag to

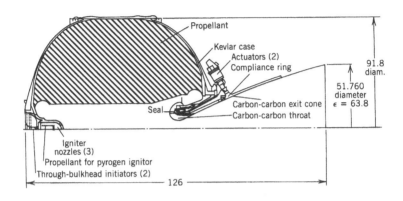

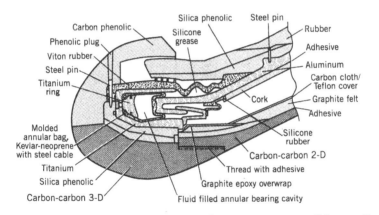

FIGURE 18–5. Simplified cross section of an upper-stage solid propellant rocket motor (IUS) using an insulated carbon-fiber/carbon-matrix nozzle, an insulated Kevlar filament-wound case, a pyrogen igniter, forward and aft stress-relieving boots, a fluid-filled bearing, and an elastomeric seal assembly in the nozzle to allow 4 ½° of thrust vector deflection. This motor has a loaded weight of 22,874 lbf, a propellant with hydroxyl-terminated polybutadiene binder, a weight of 21,400 lbf, a burnout weight of 1360 lbf, a motor mass fraction of 0.941, a nozzle throat diameter of 6.48 in., and a nozzle exit area ratio of 63.8. The motor burns for 146 sec at an average pressure of 651 psi (886 psi maximum) and an average thrust of 44,000 lbf (60,200 lbf maximum), with an effective altitude specific impulse of 295 sec. Top drawing is cross section of motor; bottom drawing is enlarged cross section of nozzle package assembly. The motor is an enlarged version of Orbus-6 described in Fig. 12–3. From C. A. Chase, "IUS Solid Motor Overview," JANNAF Conference, Monterey, California, 1983; courtesy of former Chemical Systems Div., United Technologies Corp.

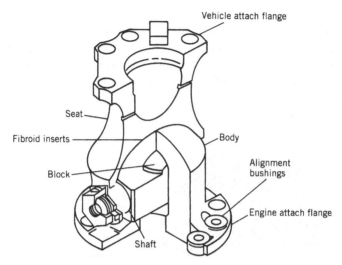

FIGURE 18–6. Gimbal bearing for the Space Shuttle main engine; it is typical of thrust carrying mounts on top of an engine. Courtesy of Aerojet Rocketdyne, Inc.

transfer loads; there are double seals to prevent leaks of hot gas and various insulators to keep the structure below 200 °F or 93 °C.

Two gimbals mechanisms will now be described in more detail. Figure 18–6 shows the gimbal bearing assembly used in the Space Shuttle's main engine. It supported the weight of the engine and transmitted the thrust force. It consists of a ball-and-socket universal joint with contact and intermeshing spherical (concave and convex) surfaces. Sliding can occur on these surfaces as the gimbal assembly is rotated. When assembling the engine to the vehicle, some offset bushings are used to align the thrust vector. Some of the design features and performance requirements of this gimbal are listed in Table 18–2. The maximum angular motion can actually be larger than the deflection angle during operation so as to allow for various tolerances and alignments. Actual deflections, alignment tolerances, friction coefficients, angular speeds, and accelerations during operation are usually much smaller than the maximum values listed in the table.

A scheme only suitable for liquid propellant rocket engines operating with a gas generator cycle is listed at the bottom of Table 18–1. The exhaust gas from the turbine is fed into one or more nozzles and maximum torque is limited by the amount of hot gas available. The Russian RD-119 upper-stage engine used six hot-gas valves to control the flow of turbine exhaust gases to four pitch and yaw fixed nozzles and four smaller roll control fixed nozzles. Clearly visible in Fig. 6–9a, the RS-68 engine has only one continuously flowing nozzle for the turbine exhaust without valves. This exhaust flow augments the thrust. It controls roll using a swivel.

Table 18–3 and Ref. 18–5 give design *actuator systems* requirements for TVC with flexible bearings in the IUS solid rocket motor nozzle. This system is shown in Figs. 12–3 and 18–5 and detailed in Table 12–3. One version of this nozzle can

TABLE 18–2. Characteristics and Performance Requirements of the Gimbal Bearing Assembly of the Space Shuttle Main Engine

Engine weight to be supported (lbf)	Approx. 7000
Thrust to be transmitted, (lbf)	512,000
Gimbal assembly weight (lbf)	105
Material is titanium alloy	6Al-6V-2Sn
Dimensions (approximate) (in.)	11 dia. × 14
Angular motion (deg)	
Operational requirement (max.)	±10.5
Snubbing allowance in actuators	0.5
Angular alignment	0.5
Gimbal attach point tolerance	0.7
Overtravel vector adjustment	0.1
Maximum angular capability	±12.5
Angular acceleration (max.) (rad/sec^2)	30
Angular velocity (max.) (deg/sec)	20
Angular velocity (min.) (deg/sec)	10
Lateral adjustment (in.)	±0.25
Gimbal duty cycle about each axis	
Number of operational cycles to 10.5°	200
Nonoperational cycles to 10.5°	1400
Coefficient of friction (over a temperature range of 88–340 K)	0.01–0.2

Source: Courtesy of Aerojet-Rocketdyne.

TABLE 18–3. Design Requirements for TVC Actuation System of an IUS Solid Rocket Motor

Item	Requirement
Performance parameter	
Input power	31 A/axis maximum at 24–32 VDC; > 900 W (peak)
Stroke	10.2 cm (4.140 in.) minimum
Stall force	1.9 kN (430 lbf) minimum
Accuracy	±1.6mm (±0.063 in.) maximum
Frequency response	>3.2 Hz at 100° phase lag
No load speed	8.13 cm/sec (3.2 in./sec) minimum
Stiffness	28.9 kN/cm (16,600 lbf/in.) minimum
Backlash	±0.18 mm (0.007 in.) maximum
Reliability	>0.99988 redundant drive train, > 0.999972 single thread element
Weights	
Controller	5.9 kg (13 lbf) maximum, each
Actuator	7.04 kg (15.5 lbf) maximum, each
Potentiometer	1.23 kg (2.7 lbf) maximum, each
System	22.44 kg (49.4 lbf) maximum

Source: Reproduced from Ref. 18–5 with permission of former Chemical Systems Div., United Technologies Corp.

deflect 4° maximum plus 0.5° for margin and another is rated at 7.5°. This system has two electrically redundant electromechanical actuators using ball screws, two potentiometers for position indication, and one controller that provides both the power drive and the signal control electronics for each actuator. A variable-frequency, pulse-width-modulated (PWM) electric motor drive of small size and low weight is used for the power and forces involved. It has a pair of mechanisms that will lock the nozzle in a fixed pitch-and-yaw position as a fail-safe device.

Thrust vector *alignment* is a necessary task during rocket hardware assembly. Usually, the neutral position (no deflection, in many vehicles the thrust axis coincides with the vehicle axis) of the thrust vector should go through the center of gravity of the vehicle. Any TVC mechanism has to allow for alignment or adjustments in angle as well as position of the TVC center point with the intended vehicle axis. The geometric centerline of the diverging section of the nozzle is generally considered to be the thrust direction. Alignment bushings are shown in Fig. 18–6. In small-sized nozzles, alignment accuracies of one-quarter of a degree and axis offsets of 0.020 in. have been achieved with proper measuring fixtures.

The *jet tab TVC system* depicted in Fig. 18–3 produces low torques and is appropriate for flight vehicles with low exit area ratio nozzle expansions. Thrust losses can be high when tabs are rotated at full angle into the jet, but negligible when the tabs are in their neutral position outside of the jet. On most flights, as the time-averaged position of the tab is at a very small angle, the average thrust loss is small. Jet tabs can consist of very compact mechanisms and have been used successfully on tactical missiles with solid propellant rocket motors. An example of the jet tab assembly for the booster rocket motor of the Tomahawk cruise missile is shown in Fig. 18–7. Four tabs, independently actuated, are rotated in and out of the motor's exhaust jet during its 15-sec duration of rocket operation. A tab that blocks 16% of the nozzle exit area is equivalent to a thrust vector angle deflection of 9°. The maximum attainable angle is 12° and the slew rate is fast (100°/sec). The vanes are driven by four linear small push–pull hydraulic actuators with two servo valves and an automatic integral controller. Power is supplied by compressed nitrogen stored at 3000 psi. An explosive valve releases the gas to pressurize an oil accumulator in a blowdown mode. The vanes are made of tungsten to minimize the erosion from solid particles in the exhaust gas.

Jetavators have been used in tactical submarine-launched missiles. Their thrust loss is roughly proportional to the vector angle. This mechanism is shown in Fig. 18–3 and mentioned in Table 18–1.

The concept of TVC by *secondary fluid injection* into the exhaust stream dates back to 1949 and can be credited to A. E. Wetherbee Jr. (U.S. Patent 2,943,821). Application of *liquid injection thrust vector control* (LITVC) to production vehicles began in the early 1960s. Both inert (water) and reactive fluids (such as hydrazine or nitrogen tetroxide) have been used. Side injection of reactive liquids is still used on some of the older vehicles, although it requires a pressurized propellant tank and feed system. The high-density injection of a liquid is preferred because its tank will be relatively small and its pressurization will require less mass. Because other schemes have better performance, liquid injection TVC will probably not be selected for new applications.

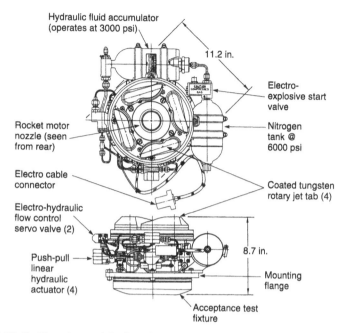

FIGURE 18–7. Two views of the jet tab assembly, packaged in a doughnut shape volume around the nozzle of the Tomahawk cruise missile's solid propellant booster rocket motor. Hydraulic actuators, located just beyond the nozzle exit, rotate the tabs in and out of the nozzle exhaust jet. Courtesy of Space and Electronics Group, Northrop Grumman.

Hot gas injection (HGITVC) of the combustion products from solid or liquid propellants is inherently attractive from a performance and packaging viewpoint. In the past, there had been no production applications of HGITVC because of unacceptable material erosion in the hot gas valves. However, two factors now make hot-gas-side injection feasible. First, hot gas valves are made with the newer carbon–carbon structural parts and with modern insulators. A hot gas system with a limited duration hot gas carbon valve is described in Ref. 18–6. Also, advances in metallurgy have allowed the development of hot valves made of rhenium alloy, a high-temperature metal suitable for hot gas valve applications. The second factor is the development of solid propellants that are less aggressive (less AP, Al_2O_3, and/or fewer oxidizing gas ingredients) thus reducing nozzle and valve erosion; this helps hot gas valves and insulated hot gas plumbing to better survive for limited durations, but often at the expense of propulsion system performance. The Russians have a production HGITVC system with the gas generators located within the combustion chamber, see Ref. 18–7.

With either liquid or solid propellants, hot gases can be bled off the main combustion chamber or may be generated in a separate gas generator. Hot gas valves can be used to (1) control side injection of hot gas into a large nozzle, or (2) control a pulsing flow through a series of small fixed nozzles similar to small attitude control thrusters described in Chapters 4, 6, and 12. In liquid propellant engines it is feasible to withdraw gas from the thrust chamber at any location where there is an intentional fuel-rich mixture ratio; the gas temperature should be low enough (about 1100 °C or 2000 °F) so that uncooled metal hardware can be used in HGITVC valves and piping.

The total side force resulting from secondary fluid injection into the main stream of the supersonic nozzle may be expressed with two force components: (1) the force associated with the momentum of the injectant and (2) pressure unbalances acting over areas of the internal nozzle wall. The second term results from unbalanced wall pressures within the nozzle caused by shock formation, boundary layer separation, differences between injectant and undisturbed nozzle stream pressures, and primary-secondary combustion reactions (for chemically active injectants). The strength of any shock pattern and of the pressure unbalance created between opposite walls in the nozzle depend on many variables, including the properties of the injectant and whether it is liquid or gas. In the case of injecting a reactive fluid, the combustion occurring downstream of the injection port(s) usually produces larger pressure unbalances than those obtained by liquid vaporization only. However, benefits from combustion depend on chemical reaction rates being high enough to keep the reaction zone close to the injection port. TVC performances that are typical of inert and reactive liquids and hot gas (solid propellant combustion products) are indicated in Fig. 18–8. This plot, showing force and mass flow ratios, is a parametric representation commonly used in performance comparisons.

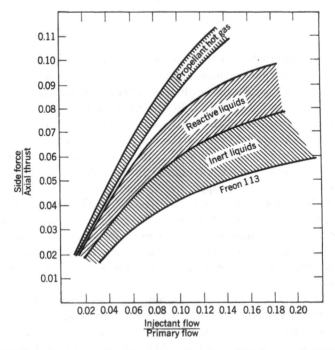

FIGURE 18–8. Typical performance regions of various side injectants in TVC nozzles.

18.2. TVC WITH MULTIPLE THRUST CHAMBERS OR NOZZLES

Each of various concepts shown in Fig. 18–3 can provide pitch and yaw moments to a vehicle. Roll control, however, can only be obtained when there are at least two separate vectorable nozzles, four fixed pulsing or throttled flow nozzles, or two rotary jet vanes submerged in the exhaust gas from a single nozzle.

Several concepts have been developed and flown that use two or more rocket engines or a single engine or motor with two or more actuated nozzles. Two fully gimballed thrust chambers or motor nozzles can provide roll control using very slight differential angular deflections. For pitch and yaw control deflections need to be larger, of the same angle and direction for both nozzles, and the deflection magnitude needs to be the same for both nozzles. All TVC motions can also be achieved with four hinged (see Fig. 11–1) or gimballed nozzles (as depicted later in Fig.18–12). Figure 18–9 shows the rocket motor of an early version of the Minuteman I missile booster (first stage) with four hinge-mounted nozzles. This motor is described in Table 12–3.

The Russian RD-0124A liquid propellant rocket engine is an example of TVC with four hinged thrust chambers. It is used in the upper stages of the new Angara family of space launch vehicles, and also its predecessor the RD-0124 is used in the third stage of the Soyuz 1–1B. See Figs. 18–10, 18–11 and Table 18–4. The RD-0124A

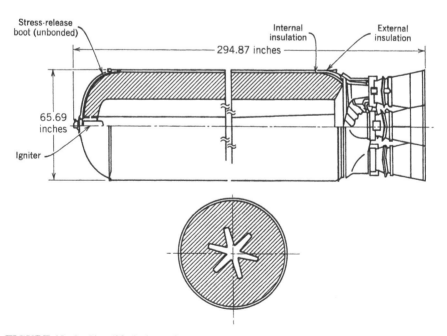

FIGURE 18–9. Simplified view of an early version of the first-stage Minuteman I missile motor using composite-type propellant bonded to the motor case. Four hinged nozzles provide pitch, yaw, and roll control. Courtesy of U.S. Air Force.

has 2 booster or feed pumps in addition to the main turbopump and is operated on a staged combustion engine cycle, Ref. 18–8.

Compared to its older version, the RD-0124A engine has a longer operating duration, together with certain design simplifications and a smaller dry-engine mass (however, these two engines are very similar, having 60% common parts). They are representative of Russian liquid propellant rocket engines that use a single large turbopump to feed multiple thrust chambers. A small amount of high-pressure fuel is tapped off at the main pump discharge for supplying hydraulic power to each TVC actuator for the 4 hinged thrust chambers. The hinge mechanism/structure is located around the nozzle-throat's cooling jacket; this reduces the moment arm and the required effective engine diameter at the nozzle exit plane, when compared to injector-mounted hinge mechanisms.

Figure 18–10 shows a *blow-back plume gas shield* also known as *flame barrier* or *thermal protection diaphragm*. It is intended to prevent low pressure hot gases (from the impingement of the 4 plumes with each other) from reaching the vehicle's engine compartment. With such a sheet, made of heat resistant material, the blow back gases flow out in between adjacent nozzles. Today these flame shields are used

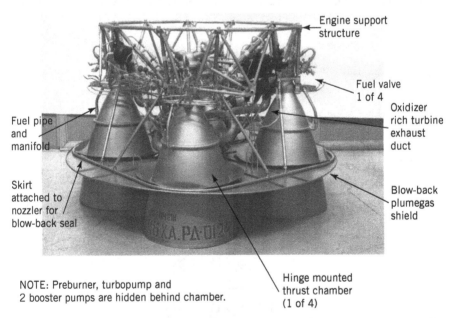

NOTE: Preburner, turbopump and
2 booster pumps are hidden behind chamber.

FIGURE 18–10. This Russian RD-0124 (similar to the newer RD-124A) upper stage engine has four hinge-mounted thrust chambers. Its characteristics are given in Table 18–4 and flow diagram shown in Fig. 18–11. The large circular bulkhead near the bottom prevents hot-gas blowbacks (discussed later in this section). In addition to thrust, this engine provides pitch, yaw, and roll control to the upper stage of its launch vehicle. Hidden behind the front thrust chamber are a preburner, two booster pumps and a single vertically mounted turbopump. Courtesy of J. Morehart and The Aerospace Corporation.

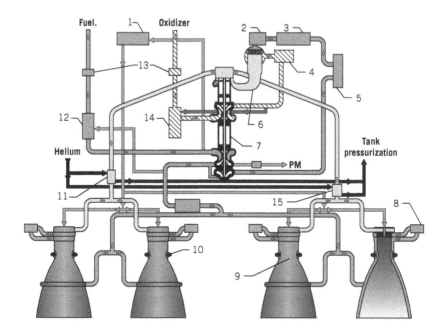

1 – start-up ampoule; 2 – fuel valve; 3 – start-up ampoule;
4 – oxidizer valve; 5 – regulator; 6 – preburner;
7 – turbopump assembly; 8 – fuel valve; 9 – thrust chamber;
10 – hinge fitting; 11 – heat exchanger; 12 – booster pump;
13 – starter valve; 14 – booster pump; 15 – heat exchanger.

FIGURE 18–11. Simplified flow diagram of the RD-0124 liquid propellant upper stage rocket engine with four hinged thrust chambers. Two starter ampules contain a toxic hypergolic mixture that readily ignites with the propellant or air. The call out "PM" (power module) refers to the distribution and control of high-pressure fuel (tapped off at the fuel pump discharge) that is used to drive the thrust vector control actuators of the thrust chambers (these items are not shown). Courtesy of J. Morehart and The Aerospace Corporation.

TABLE 18–4. Selected Characteristics of the RD-0124A Rocket Engine (4 thrust chambers)

Propellants	Liquid oxygen and kerosene
Vacuum thrust, kN (lbf)	254.3 (57,170)
Vacuum specific impulse, sec	359.1
Chamber pressure, MPa	15.7
Dry engine mass, kg	500
Operating duration, sec	420
Engine height/diameter, m	1.573/2.400
Nozzle exit area ratio (estimated)	165

Data from several Russian sources.

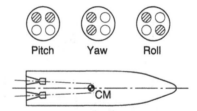

FIGURE 18–12. Differential throttling with four fixed-position thrust chambers can provide flight maneuvers. In this simplified diagram, the shaded nozzle exits indicate a throttled condition or reduced thrust. The larger forces from the unthrottled engines impose turning moments on the vehicle. For roll control the nozzles are slightly inclined and their individual thrust vectors do not go through the center of mass of the vehicle.

in the majority of large multiple rocket nozzles or with rocket engine clusters in several countries.

The differential throttling concept as shown in Fig. 18–12 has no gimbal and does not use any single nozzles methods as described in Fig. 18–3. It has four non-moveable thrust chambers and their axes are almost parallel to and set off from the vehicle's centerline. Two of the four thrust chambers are selectively throttled (typically thrust is reduced by only 2 to 15%). The four nozzles may be supplied from the same feed system or they may belong to four separate but identical rocket engines. This differential throttling system has been used in an experimental Aerospike rocket engine described in Chapter 8 of the Eighth edition of this book and in one Russian launch vehicle.

18.3. TESTING

Testing of thrust vector control systems often includes actuation of the system after assembly on the propulsion system or on the vehicle (usually without propellant and before flight). For example, the Space Shuttle main engine was put through certain gimbal motions (without rocket firing) prior to a flight. A typical acceptance test series for TVC systems (prior to the delivery to an engine manufacturer) may include the determination of input power, accuracy of deflected positions, angular speeds or accelerations, signal response characteristics, and/or validation of overtravel stops. The ability to operate under extreme thermal environments, under various vehicle or propulsion system generated vibrations, temperature cycling, and ignition shocks (high momentary acceleration) would usually be part of qualification tests.

Side forces and roll torques are usually relatively small compared to the main thrust and pitch or yaw torques. Their accurate static test measurement can be difficult, particularly at low vector angles. Elaborate, multicomponent test stands employing multiple load cells and isolation flexures are needed to assure valid measurements.

18.4. INTEGRATION WITH VEHICLE

The vehicle's guidance and control system directs actuations or movements of TVC devices during flight (see Ref. 18–9). This system measures the three-dimensional position, velocity vectors, and rotational rates of the vehicle and compares them with the desired position, velocity, and rates. Any error signals between these two sets of parameters are transformed by computers in TVC controllers into control commands for actuating the TVC device until the error signals are reduced to an acceptably low value. The vehicle's computer control system determines the timing of the actuation, the direction, and magnitude of the deflection. These systems tend to become complex and include servomechanisms, power supplies, monitoring/failure detection devices, actuators with their controllers, and kinetic compensators.

Criteria governing selection and design of TVC systems stem from vehicle needs and account for steering-force moments, force rates of change, flight accelerations, duration, performance losses, dimensional and weight limitations, available vehicle power, reliability, delivery schedules, and costs. For the TVC designer these translate into such factors as duty cycle, deflection angle, angle slew rate, power requirement, kinematic position errors, and many other vehicle–TVC and motor–TVC interface details, besides the program aspects of costs and delivery schedules.

Interface details include electrical connections to and from the vehicle flight controller, power supplies, mechanical attachment with fasteners for actuators, and sensors to measure the position of the thrust axis and/or the actuators. Design features to facilitate testing of the TVC systems, easy access for checkouts or repairs, or to facilitate resistance to high-vibration environments are usually included. Because the TVC subsystem is usually physically connected to the vehicle and mounted to the rocket's nozzle, the designs of these components must be coordinated and integrated. Nozzle–TVC interfaces are discussed in Refs. 6–1 (TVC of liquid rocket engines and their control architecture) and 18–9.

The actuators can be hydraulic, pneumatic, or electromechemical (lead screw), and usually include position sensors to allow feedback to the controller. Proven power supplies include high-pressure cold stored gases, batteries, warm gases from gas generators, hydraulic fluids pressurized by cold or warm gas generators, electric or hydraulic power from the vehicle's power supply, and electric or hydraulic power from a separate turbogenerator (in turn driven by a gas generator). The last type is used for relatively long-duration high-power applications, such as the power package used in the Space Shuttle solid rocket booster TVC, explained in Ref. 18–10. Selection of actuation schemes and their power supplies depend on minimum weight, minimum performance loss, simple controls, ruggedness, reliability, ease of integration, linearity between actuating force and vehicle moments, cost, and other factors. The required frequency response is higher when the vehicle is small, such as in small tactical missiles. The frequency response listed in Table 18–3 is more typical of larger spacecraft applications. Sometimes the TVC system is integrated with a movable aerodynamic fin system, as reported in Ref. 18–11. Control aspects of electrome-chemical actuators are discussed in Ref. 18–12.

PROBLEMS

When solving problems in thrust vector control, information regarding the *center of mass* of the vehicle and either the *centroidal radius of gyration* or the appropriate *mass moment of inertia* should be inferred from the geometry of the rocket vehicle or should be given explicitly. Many elementary books in dynamics carry useful tables in this regard (see, for example, J. L. Meriam, L. G. Kraige, and J. N. Bolton, *Engineering Mechanics: Dynamics*, Eighth edition, John Wiley and Sons, Inc., Hoboken, NJ, 2015). Note that as propellant is consumed, in most propulsion systems the *center of mass* may move along the vehicle axis during operation.

1. A single-stage weather sounding rocket has a takeoff mass of 1020 kg, a sea-level initial acceleration of $2.00g_0$, 799 kg of useful propellant with an average specific gravity of 1.20, a burn duration of 42 sec, a vehicle body shaped like a cylinder with an L/D ratio of 5.00 and with a nose cone having a half angle of 12°. Assume the center of mass does not change during flight. The vehicle tumbled (rotated in an uncontrolled manner) during the flight and failed to reach its objective. Subsequent evaluation of its design and assembly showed that the maximum possible thrust misalignment of the main thrust chamber was 1.05° with a maximum lateral offset d of 1.85 mm but assembly records show it was 0.7° and 1.1 mm for this vehicle. Since the propellant flow rate was essentially constant, the thrust at altitude cutoff was 16.0% larger than at takeoff. Determine the maximum torque applied by the main thrust chamber at start and at cutoff, without any operating auxiliary propulsion systems. Then determine the approximate maximum angle through which the vehicle would rotate during powered flight, assuming no drag. Discuss these results.

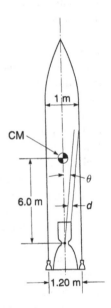

2. A propulsion system with a thrust of 400,000 N is expected to have a maximum thrust misalignment θ of $\pm\,0.50°$ and a horizontal offset d of the thrust vector of 0.125 in. as shown in the following sketch. One of four small reaction control thrust chambers will be used to counteract the disturbing torque. What should be its maximum thrust level and best orientation? Distance of vernier gimbal to center of mass (CM) is 7 m.

REFERENCES

18–1. G. P. Sutton, Section 4.9, "Steering or Flight Trajectory Control," *History of Liquid Propellant Rocket Engines*, AIAA, 2006, pp. 218–227.

18–2. A. Truchot, "Design and Analysis of Solid Rocket Motor Nozzles," in *Design Methods in Solid Rocket Motors*, AGARD Lecture Series 150, Advisory Group for Aerospace Research and Development, NATO, Revised Version, 1988, Chapter 3.

18–3. B. H. Prescott and M. Macocha, "Nozzle Design," in G. E. Jensen and D. W. Netzer (Eds.), *Tactical Missile Propulsion*, Vol. 170, *Progress in Astronautics and Aeronautics*, American Institute of Aeronautics and Astronautics, 1996, pp. 177–186.

18–4. A. B. Facciano et al., "Evolved SeaSparrow Missile Jet Vane Control System Prototype Hardware Development", *Journal of Spacecraft and Rockets*, Vol. 39, No. 4, July–August 2002, pp. 522–531.

18–5. G. E. Conner, R. L. Pollock, and M. R. Riola, "IUS Thrust Vector Control Servo System," paper presented at 1983 JANNAF Propulsion Meeting, Monterey, CA, February 1983.

18–6. M. Berdoyes, "Thrust Vector Control by Injection of Hot Gas Bleed from the Chamber Hot Gas Valve," AIAA Paper 89–2867, July 1989.

18–7. A. M. Lipanov, "Historical Survey of Solid-Propellant Rocket Development in Russia", *Journal of Propulsion and Power*, Vol. 19, No. 6, Nov.–Dec. 2003, pp. 1073.

18–8. Personal communication with Dr. J. Morehart of the Aerospace Corporation (2014, 2015).

18–9. J. H. Blakelock, *Automatic Control of Aircraft and Missiles*, 2nd ed., John Wiley & Sons, New York, 1991.

18–10. A. A. McCool, A. J. Verble, Jr., and J. H. Potter, "Space Transportation System's Rocket Booster Thrust Vector Control System," *Journal of Spacecraft and Rockets*, Vol. 17, No. 5, Sept.–Oct. 1980, pp. 407–412.

18–11. S. R. Wassom, L. C. Faupel, and T. Perley, "Integrated Aerofin/Thrust Vector Control for Tactical Missiles," *Journal of Propulsion and Power*, Vol. 7, No. 3, May–June 1991, pp. 374–381.

18–12. D. E. Schinstock, D. A. Scott, and T. T. Haskew, "Modelling and Estimation for Electromechanical Thrust Vector Control of Rocket Engines," *Journal of Propulsion and Power*, Vol. 14, No. 4, Jul.–Aug. 1998, pp. 440–446.

CHAPTER 19

SELECTION OF ROCKET PROPULSION SYSTEMS

This chapter covers a general presentation of the process of selecting rocket propulsion systems for a given mission. There are many factors, restraints, and analyses that need to be considered and evaluated before suitable selections can be properly made. The objectives are to operate at very high combustion efficiencies and prevent recurring or destructive combustion instabilities. See Refs. 19–1 to 19–6. Because design problems most often have several possible engineering solutions, this can be a complicated task. Three specific aspects of selection are covered here in some detail:

1. Comparison of merits and disadvantages of liquid propellant rocket engines with solid propellant rocket motors and hybrids.
2. Key factors used in evaluating particular rocket propulsion systems and in selecting from among several competing candidates.
3. Necessary interfaces between the propulsion system and the flight vehicle and/or the overall system.

A propulsion system is really one of several subsystems of a flight vehicle. The vehicle, in turn, can be part of a larger overall system. An example of an overall system would be a communications network with ground stations, transmitters, and several satellites; each satellite may require a multi-stage flight vehicle during launch with all associated propulsion systems as well as on-orbit attitude-control propulsion systems (each with specific propulsion requirements). Additionally, the length of active time in orbit is a system parameter that affects both satellite size and the total impulse propulsion requirements.

Vehicle subsystems (such as its structure, power supplies, propulsion, guidance, control, communications, ground support, and/or thermal control) often pose conflicting requirements. *Only through careful analyses and proper system-engineering studies is it possible to find compromises that allow all subsystems to operate satisfactorily and in harmony with each other.* The "systems engineering" approach is now utilized routinely, see Ref. 19–1, and engineering design has advanced considerably in recent times with computer-aided design (CAD) now commonly used. Certain publications address specifically the design of space systems (e.g., Refs. 19–1 to 19–3) and the design of liquid propellant engines (e.g., Ref. 19–4). In our book, Sections 11.6, "System Integration and Engine Optimization," and 15.4, "Rocket Motor Design Approach," are preludes to this chapter and may contain some duplicate content.

Systems engineering is a useful discipline for rocket propulsion systems selection. This topic can be defined in several ways; one being (adapted from Ref. 19–5) "a logical process of activities, analyses and engineering designs, that transforms a set of requirements arising from specific mission objectives in an optimum way. It ensures that all likely aspects of a project or engineering system have been considered and integrated into a consistent whole." Such studies comprise all elements of a propulsion system and its ground support, all interfaces with other vehicle subsystems or ground-based equipment, and consider safety aspects, risks, and costs, together with the orderly selection of the most suitable propulsion system.

There are commonly *three levels of requirements* from which *propulsion system requirements* can be derived, (see Ref. 19–5). At each level, certain activities, studies, and/or trade-off evaluations are involved.

1. At the top level are *mission* defining requirements, such as space communications or missile defense. Here, analyses and optimizations are usually carried out by the mission responsible organizations. They define rocket propulsion parameters like trajectories, orbits, payloads, number of vehicles, life, and the like, and document them as mission requirements specifications.

2. From the mission requirements document, definitions and *specifications for the flight vehicle* are derived. These may include vehicle size, vehicle mass, number and size of vehicle stages, types of propulsion, required mass fraction, minimum specific impulse, thrust vector and/or attitude control needs, allowable accelerations, acceptable propellants, restraints on engine size and/or inert mass, cost restraints, allowable start delays, desired thrust–time profiles, preferred propulsion system types (liquid, solid, hybrid, electrical), ambient temperature limits, and so forth. Engineering analyses together with preliminary designs and additional specifications are usually done by the organization responsible for the flight vehicle.

3. *Propulsion requirements* form the basis for the final design and development of the propulsion system, and they are derived from the above two levels. These may include, for example, an optimum thrust–time profile, restart requirements, liquid propellant engine cycle, pulsing schedules, solid rocket motor case L/D (length/diameter), grain configuration, restraints on travel of the center of mass, details of thrust vector and/or attitude control mechanisms, reliability, cost and/or schedule restraints, and expected number of propulsion

systems and spare parts to be delivered. These requirements are often done by the vehicle development organization with inputs from one or more selected propulsion organizations.

Much of the design (i.e., three-dimensional modeling of the propulsion system), most of the analyses (i.e., stress analysis), testing, keeping track of all hardware assemblies and part inventories, and inspection records or NC programming are done today using computers. Interfacing these computer programs with each other and with organizational sectors that deal with some or all aspects of the propulsion system becomes an important objective.

All mission, vehicle, and propulsion system *requirements* can be related to performance, cost, and/or reliability. In general, if cost criteria have the highest priority then the propulsion system will be different from one that has performance as its top priority. For any given mission, one criterion is usually more important than the others. There is a strong interdependence between the three levels of requirements and the criteria categories mentioned. Some propulsion system requirements (which is usually a second-tier subsystem) strongly influence the vehicle and vice versa. An improvement in propulsion performance, for example, can affect vehicle size, overall system cost, and/or life (which normally translate into reliability and cost).

19.1. SELECTION PROCESS

The selection process for the vehicle and its associated rocket propulsion systems is an integral part of any overall vehicle design effort. Selection is based on a *series of criteria* largely derived from the *requirements,* which serve to evaluate and compare alternate propulsion system proposals. In addition to the chosen application, determining the most suitable rocket propulsion system depends on the experience of those making the selection and their ability to express many propulsion system characteristics quantitatively, the quality and amount of applicable data available, and on available time and resources for examining alternate propulsion systems. We describe here a somewhat idealized selection process, the one depicted in Fig. 19–1, but there are other sequences to do the job of selection.

Because total vehicle's performance, flight control, operation, and/or maintenance critically depend on the performance, control, and operation of the rocket propulsion system (and vice versa), the selection process will normally consist of several iterations in defining both the vehicle and propulsion requirements that satisfy the given mission. This iterative process involves both the system's organization (and vehicle/system contractor) and one or more propulsion organizations (or rocket propulsion contractors). Documentation and reporting can take many forms; electronic storage has greatly expanded capabilities to network, record, and retrieve documents.

Several competing candidate systems are usually evaluated. They may originate from different rocket propulsion organizations, perhaps on the basis of modifications

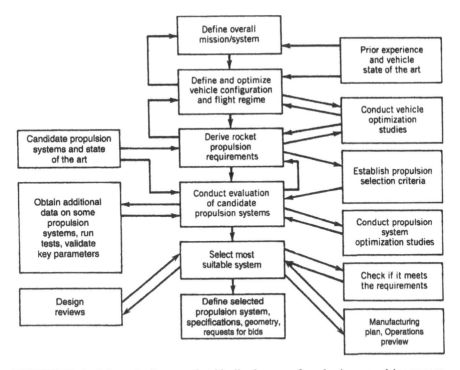

FIGURE 19–1. Schematic diagram of an idealized process for selecting propulsion systems.

of some existing rocket propulsion system, or may include some novel technology, or may be new types of systems specifically configured to fit the vehicle or mission needs. In making evaluations it will be necessary to compare the candidate propulsion systems with each other and to rank-order them (in accordance with some selection criteria on how well they meet each requirement). This necessitates studies for each candidate system and also, sometimes, additional testing. For example, statistical analyses of the functions, failure modes, and safety factors of all key components may lead to quantifiable reliability estimates. For some criteria (such as safety or prior related experience) it may not be possible to compare candidate systems quantitatively but only subjectively.

For any given mission, several rocket parameters often need to be optimized. *Trade-off studies* can be used to optimize these parameters in accordance with overall requirements. Such trade-off studies have often been performed on the following items: number of thrust chambers, optimum chamber pressure (considering performance and inert propulsion system mass), selected mixture ratio (performance, tank volume), solid propellant case configuration (L/D), cost versus propulsion system performance, chamber pressure versus size and/or mass of propulsion system, chamber pressure versus heat transfer or cooling method, alternate grain configurations, minimum center of mass travel during operation, thrust magnitude/profile versus powered flight duration, nozzle exit area ratio versus performance, nozzle mass, and/or nozzle dimensions (length, diameter), alternate TVC (thrust vector control) concepts versus mass and power, chamber pressure versus heat transfer (cooling method or insulation) and inert mass. See Ref. 19–5.

Early in the selection process a tentative recommendation is made as to whether the propulsion system should be a solid propellant motor, a liquid propellant engine, an electrical thruster, a solar-thermal system, or some combination of these. Each type has its own thrust regime, specific impulse, thrust-to-weight ratio (acceleration), and/or likely duration, as indicated in Table 2–1 and Fig. 2–4, where these factors have been listed for several chemical rocket propulsion devices and several types of non-chemical engines. Liquid engines and solid motors are covered in Chapters 6 to 15, hybrids in Chapter 16, and electric propulsion in Chapter 17.

If an existing vehicle is to be upgraded or modified, the question arises if its propulsion system should also be changed (e.g., higher thrust, more total impulse, or faster thrust vector control). While there might be some trade-off studies leaning toward modifying propulsion parameters, in upgrades one normally does not consider an entirely different propulsion system, as is often done in entirely new vehicles or missions. Also, it is rare that an identical rocket propulsion system will satisfy two different applications; frequently, design changes, interface modifications, and requalification will be necessary to adapt an existing rocket propulsion system to another application. Most often, proven qualified propulsion systems that fit the desired requirements have an advantage in cost, scheduling, and reliability.

Electric propulsion systems address a set of unique space applications with low thrusts, low acceleration trajectories, high specific impulses, long operating times, and relatively massive power supplies. They perform competitively for certain space transfer and orbit maintenance missions. With more electric propulsion systems flying than ever before, the choice of proven electric propulsion thruster types has widened. At the time of this writing, there are plans for all-electric-propulsion upper launch stages. These systems, together with their design approaches, are described in Chapter 17.

When chemical propulsion systems are deemed most suitable for a particular application, selection has to be made between liquid propellant engines, solid propellant motors, or hybrid rocket propulsion systems. Some of the major advantages and disadvantages of liquid propellant engines and solid propellant motors are presented in Tables 19–1 through 19–4. These lists may not be complete, some items need further information, some items may be controversial, and a number are restricted to particular applications. Most of the entries from these lists may be converted to evaluation criteria; for any specific mission, relevant entries would be rank-ordered in accordance with their relative importance. A quantification of many of these items would be needed.

The question is often asked: which are better, solid or liquid propellant rocket propulsion systems? A clear statement of strongly favoring one or the other can only be made when referring to a specific set of flight vehicle missions. Today, solid propellant motors (SPRMs) seem to be preferred for tactical missiles (air-to-air, air-to-surface, surface-to-air, or short-range surface-to-surface) and ballistic missiles

TABLE 19–1. Solid Propellant Rocket Motor Advantages

Simple design (few or no moving parts).

Easy to operate (little or no preflight checkout).

Ready to operate quickly (push button readiness). Can start in zero gravity.

Will not leak, spill, or slosh.

Sometimes less overall weight for low total impulse application.

Can be throttled or stopped and restarted (usually only once) if preprogrammed and predesigned.

Can provide TVC (thrust vector control) but at increased complexity.

Can be stored for 10 to 30 years.

Usually, higher overall density; more compact package, smaller vehicle (less drag) for boosting lower stages.

Some propellants have nontoxic, clean exhaust gases, often at a performance penalty.

Thrust termination devices permit control over total impulse.

Ablation, erosion, and gasification of insulator, nozzle wall, and liner materials contribute to mass flow and thus to total impulse, but increase exhaust molecular mass.

Detonation avoided when a tactical motor is subjected to energetic stimulant (e.g., bullet impact or external five) by judicious designs and special propellants.

Some tactical missile motors were produced in large quantities (over 200,000 per year).

Some designed for recovery, refurbishing, and reuse (Space Shuttle solid rocket motor).

TABLE 19–2. Liquid Propellant Rocket Engine Advantages

The highest chemical specific impulses obtained with certain liquid propellants. This increases the payload carrying ability, vehicle velocity increment and the attainable mission velocity.

Can be randomly throttled and randomly stopped and restarted; can be efficiently pulsed (some small thrust sizes over 10^6 times). Thrust–time profiles can be randomly controlled; this allows a reproducible flight trajectory.

Cutoff impulse controllable with thrust termination device (better control of vehicle terminal velocity).

Some controls can be checked out just prior to operation. Can be tested at full thrust on ground or in vehicle at launch pad prior to flight.

Can be designed for reuse after field services and checkout.

Thrust chamber (or some part of the vehicle) cooled by propellant and made lightweight.

Storable liquid propellants have been kept in vehicle for more than 20 years and engine can be ready to operate quickly.

With pumped propulsion feed systems and large total impulse, the inert propulsion system mass (including tanks) can be relatively very low (thin tank walls and low tank pressure), allowing a high propellant mass fraction.

Some propellants have nontoxic or environmentally acceptable exhausts.

Some propellant feed systems supply several thrust chambers in different parts of vehicle.

Can provide component redundancy (e.g., dual check valves) to enhance reliability.

With multiple engines, can operate with one or more shut off (engine out capability).

The geometry of low-pressure tanks can be designed to fit most vehicles' space constraints (i.e., mounted inside wing or nose cone).

Can withstand many ambient temperature cycles without deterioration.

Placement of propellant tanks within the vehicle can minimize the travel of the center of gravity during powered flight. This enhances the vehicle's flight stability and reduces flight control forces.

Plume radiation and smoke can be low with certain liquid propellants.

TABLE 19–3. Solid Propellant Rocket Motor Disadvantages

Explosion and fire potential is high; failures can be catastrophic; many cannot take bullet impacts or being dropped onto a hard surface without self-ignition.

Many require environmental permits and safety features for transport on public conveyances.

Under certain conditions some propellants and grains can detonate.

Ambient temperature variations during storage or in repeated flight (under the wing of a military aircraft) cause progressive grain-crack formation increasing the burning area, causing excessive combustion pressures and limiting the life of safe operation.

If designed for reuse, the motor requires factory rework and new propellants.

Requires an ignition system.

Each restart requires a separate ignition system and additional insulation—in practice, one restart.

Limited firing duration for typical single pulse motor compared to liquid engines.

Exhaust gases are toxic for composite propellants containing ammonium perchlorate.

Some propellants or propellant ingredients can deteriorate (self-decompose) in storage.

Most solid propellant plumes cause more radio-frequency attenuation than liquid propellant plumes.

Once ignited, cannot change predetermined thrust or duration.

If combustion gases contain more than a few percent particulate carbon, aluminum, aluminum oxide, or other metal, the vehicle's exhaust will be smoky and the plume radiation will be greatly increased.

Thrust and operating duration will vary with initial ambient grain temperature and cannot be easily controlled. Thus, the vehicle's flight path, velocity, altitude, or range can vary.

Large boosters can in some designs take a few seconds to start.

Thermal insulation is required in almost all rocket motors.

Cannot be hot fire tested prior to use.

Need a safety provision (electrical grounding) to prevent inadvertent self-ignition, which would lead to an unplanned motor firing. This can cause a disaster.

Rough handling or transport on bumpy roads or exceeding temperature limits can cause cumulative grain damage and rocket motor may no longer be safe to ignite.

TABLE 19–4. Liquid Propellant Rocket Engine Disadvantages

Relatively complex design, more moving parts and components, more things to go wrong.

Cryogenic propellants cannot be stored for long periods except when tanks are well insulated and escaping vapors are recondensed. Propellant loading occurs at the launch stand or the test facility and requires cryogenic propellant storage facilities.

Spills or leaks of certain propellants can be hazardous, corrosive, toxic, and cause fires.

More overall engine weight for most short-duration, low-total-impulse applications (resulting in low propellant mass fraction).

Nonhypergolic propellants require an ignition system.

Tanks need to be pressurized by a separate pressurization subsystem. This can require heavy high-pressure inert gas storage (2000–10,000 psi) often for long periods of time.

More difficult to control combustion instabilities.

Bullet impact cause leaks, sometimes fires and explosions, but usually no detonations.

A few propellants (e.g., red fuming nitric acid, nitrogen tetroxide) give off very toxic vapors or fumes.

Usually require more volume due to lower average propellant density, separate pressurizing gas storage, and the relatively inefficient packaging of engine components.

When a vehicle falls over and breaks up at the launch stand and fuel and oxidizer are spilled and intimately mixed, it is possible for an explosive mixture to be created and ignited.

Sloshing in tank may cause flight stability problem and heavy side loads on tanks; this can be minimized with baffles.

Some monopropellants can explode with excessive temperature rises (e.g., in cooling jackets).

Aspirated gas can cause combustion interruption or instability if tank outlet is uncovered.

Smoky exhaust (soot) plume can occur with some hydrocarbon fuels.

TABLE 19–4. (*Continued*)

Needs special design provisions for start in zero gravity.

With cryogenic liquid propellants there is a start delay caused due to the time needed to cool the system flow passage hardware to cryogenic temperatures.

Life of cooled large thrust chambers may be limited to perhaps 100 or more starts.

High-thrust unit requires several seconds to start.

If tank outlet is uncovered during sloshing or vortexing, then aspirated pressurizing gas will flow to the thrust chamber and it can cause combustion interruption or instability.

(long- and short-range surface-to-surface) because of their instant readiness and compactness; the absence of spills or leaks of hazardous liquids are also important criteria in these applications. Liquid propellant rocket engines (LPREs) seem to be preferred for space-launched main propulsion units and upper stages, because of their higher specific impulse, relatively clean exhaust gases, and random throttling capability. They are also favored for postboost control systems and attitude control systems, because of their random multiple pulsing capabilities with precise impulse cutoff, and for pulsed axial and lateral thrust propulsion on hit-to-kill interceptors. There are, however, always exceptions to these preferences.

When selecting rocket propulsion systems for a major new multiyear high-cost project, considerable time and effort are spent in evaluation and in developing rationales for quantitative comparisons. In part, this is in response to government policies as well as to international competition. Multiple studies are often done by competing systems and rocket propulsion organizations; formal reviews are then used to assist in arriving at a proper selection.

When submitting a proposal for a new or modified unit, it is necessary to have a well-defined project plan, a proper cost estimate, and a realistic schedule for the proposed work. These can only be prepared after the following become available: (1) design layouts of the geometry with functions and features of the proposed system, (2) descriptions of planned tests, data handling, special test equipment, and safe test facilities, (3) plans for manufacturing that include fabrication steps, tooling fixtures, available suitable factory space, and special manufacturing equipment, (4) lists of qualified parts suppliers and other vendors, (5) descriptions and availability of all needed materials, and (6) a list of experienced personnel.

19.2. CRITERIA FOR SELECTION

Criteria used in selecting a particular rocket propulsion system are unique to the particular mission or vehicle application. However, some selection factors may apply to several applications. Typical performance criteria include propellant composition, thrust profile, minimum specific impulse, operating duration, thrust variation (% throttling), life, system mass or weight, nozzle exit area ratio, solid propellant parameters (strength, elongation, or storage temperature limits), maximum allowed residual liquid propellant or solid propellant grain slivers, and ignition parameters;

for reusable vehicles these criteria should include number of restarts, provisions for draining residual liquid propellants and for cleaning and drying out tank feed systems. Most tactical missiles with SPRMs include criteria to account for tolerating external fires, bullet impacts, or nearby explosions that do not cause propellant detonation. For storage or transport over public conveyances, rules and safety standards regarding rocket propulsion systems loaded with solid propellants must be well understood. Allowable tolerances/variations for each of the items above need to be well defined. For a more complete list of these criteria see Ref. 19–6.

Any evaluation/selection groups will include experienced technical personnel from the vehicle development/fabrication organization, and such groups might be augmented by personnel from Government Laboratories, consultants, and/or other outside experts. Proposal selection is done in accordance with well-established and prioritized criteria by personnel from the organization that will do the contracting of the propulsion system. Actual selection will depend on the balancing of various relevant factors in accordance with their importance, benefits, and/or potential impact on the system, and on quantifying as many of these selection factors as possible through analysis, extrapolation of prior experience/data, cost estimates, weights, and/or separate tests. See Ref. 19–5. Layouts, inert mass estimates, center-of-gravity analyses, vendor cost estimates, preliminary stress or thermal analyses, and other preliminary design efforts are necessary to quantify the selection parameters. A comparative examination of the interfaces of alternate propulsion systems can also be a part of the process.

For a spacecraft that contains optical instruments (e.g., telescopes, horizon seekers, star trackers, or infrared radiation seekers) the exhaust plume must be free of possible contaminants that may deposit or condense on optical windows, mirrors, lenses, photovoltaic cells, or radiators, and degrade their performance, and also be free of particulates that scatter sunlight into the aperture of optical instruments, thus causing erroneous signals. Here, conventional composite solid propellants and pulsing storable bipropellants are *not* satisfactory, but cold or heated clean gas jets (H_2, Ar, N_2, etc.) and monopropellant hydrazine reaction gases are usually acceptable. This topic is discussed in Chapter 20. Another example is the need for smokeless propellant-exhaust plumes for avoiding any visual detection of smoke or vapor trails. This applies particularly to tactical missiles. Only a few solid propellants and several liquid propellants can be nearly smokeless, essentially free of vapor trails under most weather conditions.

Sometimes selection criteria conflict with each other. For example, some propellants with a very high specific impulse are more likely to experience combustion instabilities; higher chamber pressures increase the pressure ratio across the nozzle and will give an increase in specific impulse, improving vehicle performance, but the combustion chamber (and in LPREs also the propellant feed system) will require thicker, heavier walls and this inert mass increase will reduce the vehicle's mass ratio, which will in turn reduce the vehicle velocity (see Sec. 4–6); in electric propulsion, their high specific impulse is usually accompanied by low thrust and massive power-generating and conditioning equipment. Compromises must to be made when propulsion requirements are incompatible with each other. For example, the monitoring by extra sensors can prevent occurrence of certain types of failures by promoting remedial action and thus enhancing propulsion system reliability, yet these extra sensors and control components contribute to system mass and complexity, and their possible failure can reduce overall reliability. The selection process may also include feedbacks for when the stated propulsion requirements cannot be met or do not make sense, that leads to revisions or redefinitions of initial mission or vehicle requirements.

Once cost, performance, and reliability drivers have been identified and quantified, priorities for the system's criteria can be identified and the selection of the best propulsion system for a specified mission can proceed. The final propulsion requirements may result after several iterations take place, and will usually be documented, for example, in a *propulsion requirement specification*. A substantial number of records will be required here (such as rocket engine or rocket motor acceptance documents, CAD (computer-aided design) images, parts lists, inspection records, laboratory test data, etc.). Usually, there are many specifications associated with design and manufacturing as well as with vendors, materials, and so on. There must also be disciplined procedures for making and approving design and manufacturing changes. This stage now becomes the starting point for the design and development of the selected propulsion system.

19.3. INTERFACES

An *interface* can be considered to be a surface (usually irregular) forming the boundary between two adjacent bodies or assemblies. Three interfaces types are of concern to rocket propulsion systems (RPSs). The first interface is between the system and its vehicle—these two are usually designed and fabricated separately but must form when assembled a single viable structure which transmits (at the interface) electrical power and/or electrical signals (e.g., hydrazine heaters, command and control/start signals, measuring instruments electric outputs, TVC power/signals, etc.), and in some cases transfer fluids (pressurized gas and loading liquid propellants). For large vehicles, a second interface exists between the RPS and the launch stand on the ground; here the same interface connections apply as above, but there are additional provisions to allow separation of the vehicle from the launch stand at liftoff. This includes explosive bolts to release the vehicle structurally from the launch stand, electrical disconnect devices (often wire cutters), and fluid disconnect fittings (valves) in the fill pipes designed for minimum spill losses of gas or liquid propellant. The third interface is between any two stages in a multistage flight vehicle; for stage separation all structural interconnect links and all electric connections between the two stages need to be disconnected or severed just prior to separation. This may include using any remaining high-pressure gas (or small separation rocket motors) for reverse thrust on the lower stage and these actions need to be controlled during the separation maneuvers.

In Section 19.2 interfaces between the propulsion system and the vehicle and/or overall system were identified as criteria to be considered for study in the selection of propulsion systems. See also Ref. 19–5. Only a few rocket propulsion systems are straightforward to integrate and interface with their vehicles. Furthermore, interfaces are an important aspect of any design and development discipline. Interfaces assure system functionality and compatibility between the propulsion system and the vehicle's other subsystems under all likely operating conditions and mission options. Usually, an interface document or specification is prepared for use by designers and operating and/or maintenance personnel.

A list of systems in increasing order of complexity follows. Along with cold gas systems, simple solid propellant rocket motors have the fewest and least complex set of interfaces. Monopropellant liquid rocket engines also have relatively few and simple interfaces. Solid propellant motors with TVC and thrust termination capabilities have additional interfaces compared to simple motors. Bipropellant rocket engines are more complex than monopropellant LPREs. The number and difficulty of interfaces increase when they have turbopump feed systems, throttling features, TVC, and/or pulsing capabilities. In electric propulsion systems the number and complexity of interfaces is highest for electrostatic thrusters with pulsing capabilities, when compared to steady-state electrothermal systems; in general, the more complex electric propulsion systems produce higher values of specific impulse. When the mission includes recovery and reuse of the propulsion system or in manned vehicles (where the crew can monitor and/or override propulsion system commands), several more additional interfaces, safety features, and requirements need to be introduced.

19.4. COST REDUCTION

For the past 15 years there has been increased emphasis on *cost reduction* in rocket propulsion systems for space flight. Table 19–5 lists some cost reduction examples that have been achieved to date. This list is neither complete nor comprehensive and only some items may apply to any one particular system, but it does represent many relevant issues. *A reduction in cost is acceptable only when it does not diminish or compromise the intended function or flight performance or environmental compatibility of the affected components, subassemblies or systems; it must always allow the vehicle to fly and complete the intended mission within all its overall constraints.*

TABLE 19–5. Examples of Actual Cost Reduction in Rocket Propulsion Systems

1. Replace toxic liquid propellants and/or cryogenic propellants with less toxic and/or storable liquid propellants. The cost savings come from fewer safety precautions, reduced or no thermal insulation, shorter launch preparations and launch operations, less personnel protection and less protective equipment, and possibly negligible losses of propellant by vaporization.

2. Change existing construction materials with others that are less costly or easier to process and fabricate. An example can be found in Ref. 19–7.

3. When available, replace custom designed parts with carefully selected standard parts. Examples are standard fasteners (bolts, clamps, and screws), common cleaning liquids, standard thickness of commercially available sheet metal, O-rings, pipes or tube sizes or fittings, etc.

4. Combine two or more separate parts into a single unit to allow less handling, fewer setups, and often lower manufacturing costs. This can also help to avoid errors. New manufacturing techniques can simplify certain complicated parts.

5. Standard commercial manufacturing equipment should be used whenever feasible. This may include ordinary machine tools, simple welders, ordinary tube bending equipment, simpler tooling and simpler fixtures. This may avoid the development of some specialized machinery, costing less and being easier to maintain.

6. Reduce:
   a. Tight dimensional tolerances if it will reduce fabrication costs. High precision requires extra steps and setup times (and should be reserved only for the most critical components).
   b. The number of manufacturing steps and set-ups in processing components or assemblies or inspections.
   c. The spare parts inventory to an acceptable minimum.
   d. The organization's overhead costs.
   e. The number of component tests and complete system tests during development, flight rating, qualification, and/or production.

7. Change class 1.1 solid propellants for class 1.3 propellants, which have fewer safety requirements and are less detonation sensitive.

8. Where feasible, minimize the requirements, number and time for inspections and for tests during manufacturing. This can encompass factory pressure tests, electrical continuity tests, surface hardness tests, dimensional and geometry measurements, material, composition and impurities, etc.

9. Simplify and reduce the number of receiving operations or receiving inspections for purchases such as propellants, components, materials, and subassemblies.

10. Collect and sell unused materials, remaining after the propulsion system parts have been fabricated and assembled (e.g., chips from machining, scraps of sheet metal pieces, excess or unused solid propellant ingredient materials, etc.).

11. When feasible, modify designs with ease of fabrication and inspection in mind.

12. When a sufficient number of flights are planned, then reusable vehicles and reusable propulsion systems (e.g., using new composite materials, Ref. 19–7) may allow cost savings. For example, the Space Shuttle Main Engines were reconditioned (drained, cleaned, inspected, and retested) for reuse.

REFERENCES

19–1. W. J. Larson and J. R. Wertz (Eds.), *Space Mission Analysis and Design*, 2nd ed., Microcosm, Inc., and Kluver Academic Publishers, Boston, 1992.

19–2. J. C. Blair and R. S. Ryan, "Role of Criteria in Design and Management of Space Systems", *Journal of Spacecraft and Rockets*, Vol. 31, No. 2, Mar.–Apr. 1994, pp. 323–329.

19–3. R. W. Humble, G. N. Henry, and W. J. Larson, *Space Propulsion Analysis and Design*, McGraw-Hill, New York, 1995.

19–4. D. K. Huzel and D. H. Huang, *Modern Engineering for Design of Liquid Propellant Rocket Engines*, rev. ed. *Progress in Astronautics and Aeronautics*, Vol. 147, AIAA, Washington, DC, 1992.

19–5. P. Fortescue, J. Stark, and G. Swinerd, *Spacecraft Systems Engineering*, 3rd ed., John Wiley & Sons, Hoboken, NJ, 2003.

19–6. Table 19–6 (pgs. 703 to 705) and Table 19–7 (pgs. 706 to 708) of G. P. Sutton and O. Biblarz, *Rocket Propulsion Elements*, 8th Ed., John Wiley & Sons, Hoboken, NJ, 2010.

19–7. S. Schmidt et al., "Advanced ceramic matrix composite materials for current and future propulsion technology applications," *Acta Astronautica*, Vol. 55, Nos. 3–9, Aug.–Nov. 2004, pp. 409–420.

CHAPTER 20

ROCKET EXHAUST PLUMES

This chapter is an introduction to rocket exhaust plumes providing a general background, description of various plume phenomena and their effects, and references for further study. The plume consists of moving hot exhaust gases (often with entrained small particles) issuing from rocket nozzles. Such gas formations are not uniform in structure, velocity, or composition. They contain several different flow regions with supersonic shock waves. Plumes are usually visible as a brilliant flame, emitting intense radiation energy in the infrared, visible, and ultraviolet, and are a strong noise source. Many plumes leave trails of smoke or vapor and some contain toxic gases. At high altitudes, plume gases spread over large regions, and some portion of the plume can flow backwards outside of the nozzle and reach components of the flight vehicle.

Plume characteristics (size, shape, structure, emission intensities of photons and/or sound pressure waves, visibility, electrical interference, and/or smokiness) depend not only on the characteristics of the particular rocket propulsion system and its propellants but also on the flight path, flight velocity, altitude, local weather conditions (such as wind, humidity, and/or clouds), and vehicle configuration. See Refs. 20–1 to 20–3. In recent decades, progress has been made in understanding the complex interacting physical, chemical, optical, aerodynamic, and combustion phenomena within plumes by means of laboratory experiments, computer simulations, plume measurements during static firing tests and flight tests, and/or simulated altitude tests in vacuum chambers; yet, some plumes are not fully understood or predictable. Shown in Table 20–1 are many applications where quantitative predictions of plume behavior are useful.

TABLE 20–1. Applications of Plume Technology

Design/Develop/Operate Flight Vehicles, Their Propulsion Systems, Launch Stands, or Launch Equipment

For any given propulsion system and operating conditions (altitudes, weather, speed, afterburning with atmospheric oxygen, etc.) determine or predict the plume dimensions, its temperature profiles, emissions, or other plume parameters.

Determine likely heat transfer to components of vehicle, test facility, propulsion system or launcher, and prevent damage by design changes. Include afterburning and recirculation.

Estimate the ability of vehicle and test facilities to withstand intensive plume noise.

Determine the aerodynamic interaction of the plume with the airflow around the vehicle, which can cause changes in drag and plume shape.

Reduce impingement on vehicle components (e.g., plumes from attitude control thrusters hitting a solar panel); this can cause excessive heating or impingement forces that may turn the vehicle.

Estimate and minimize erosion effects on vehicle or launcher components.

Prevent deposits of condensed species on spacecraft windows, optical surfaces, solar panels, or radiating heat emission surfaces.

Determine the backscatter of sunlight by plume particulates or condensed species, and minimize the scattered radiation that can reach into optical instruments on the vehicle, because this can give erroneous signals.

Protect personnel using a shoulder-fired rocket launcher from heat, blast, noise, smoke, and toxic gas.

Detect metal trace element vapors (Fe, Ni, Cu, Cr) in the plume as indicators of thrust chamber damage and potential failure during thrust chamber development.

Detect and Track Flight of Vehicles

Analysis and/or measurement of plume emission spectrum or signature.

Identify plumes of launch vehicles from a distance when observing from spacecraft, aircraft, or ground stations, using IR, UV, or visible radiations and/or radar reflections.

Distinguish their emissions from background signals.

Detect and identify smoke and vapor trails.

Track plumes and predict the flight path or determine the launch location.

Estimate amount of secondary smoke under foul weather conditions.

Reduce secondary smoke.

Transmit a vehicle motion vector to a missile defense system.

Provide an aim point on the body of a flight vehicle.

Improve Understanding of Plume Behavior

Improve theoretical approaches to plume phenomena.

Improve or create novel or more realistic computer simulations.

Provide further validation of theory by experimental results from flight tests, laboratory investigations, static tests, or tests in simulated altitude facilities.

Understand and minimize the generation of high-energy noise.

Understand the mechanisms of smoke, soot, or vapor formation, thus learning how to control them and minimize them.

Provide a better understanding of emission, absorption, and scatter within plume.

Provide a better prediction of chemiluminescence.

Understand the effect of shock waves, combustion vibration, or flight maneuvers on plume phenomena.

Understand the effects of plume gas remains on the stratosphere or ozone layer.

Develop a better algorithm for simulating turbulence in different parts of the plume.

TABLE 20–1. (*Continued*)

Minimize Radio-Frequency Interference

Determine the plume attenuation for specific antennas and antenna locations on the vehicle.
Estimate and reduce the attenuation of radio signals that have to pass through the plume, typically between an antenna on the vehicle and an antenna on the ground or on another vehicle.
Estimate and reduce radar reflections from plumes.
Estimate and reduce the electron density and electron collision frequency in the plume; for example, by reducing certain impurities, such as sodium.

20.1. PLUME APPEARANCE AND FLOW BEHAVIOR

Figure 20–1 shows a sketch of the front part of a nozzle exhaust plume from a large propulsion system operating at low earth altitudes. The plume begins at the nozzle exit plane of the rocket propulsion system, where it has its smallest cross section. It expands in diameter as the plume gases move supersonically away from the vehicle. Exit flows from nozzles are not uniform most often underexpanded as explained in Chapter 3 (see Fig. 3–10). The relative mean velocity between the exhaust gases of the plume (the difference between the exhaust and the vehicle velocities) and the ambient air diminishes along the length of the plume, eventually approaching zero near the plume's tail end. Low thrust propulsion systems have small plumes and thus emit much less radiation and acoustic energy.

A plume's inviscid core expands until the dynamic pressure of the external flow forces it to turn. The only way a supersonic flow can turn is through an oblique shock, also called a barrel shock, as will be explained later. This intercepting oblique shock wave curves toward the plume axis, where a strong normal shock wave, of a diameter

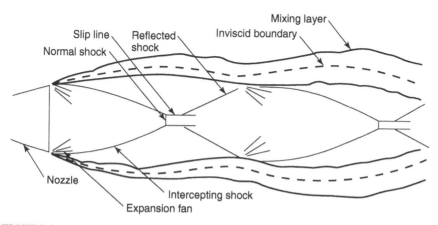

FIGURE 20–1. Simplified schematic diagram of the near-field section of the exhaust plume of a large rocket propulsion unit operating at low altitude. (Used with permission from Ref. 20–1.) Similar figures also are in Refs. 20–2 and 20–3.

smaller than the nozzle exit, forms. The inner flow is reflected at its boundary as an oblique expansion shock or reflection wave. The region immediately after the normal shock wave is subsonic; through normal shocks the kinetic energy of the incoming supersonic flow velocity is converted into thermal energy with major increases in local temperature and pressure. This normal wave and the hot subsonic region immediately downstream of it are clearly visible in many liquid propellant plumes. Shock waves are discussed later in this Chapter. After flowing through the hot subsonic zone, gases in the plume are recompressed. This series of supersonic oblique and normal shock waves repeat themselves to form a characteristic pattern often called Mach diamonds, because of the diamond-shaped visible appearance. Such Mach diamonds expand in scale with increasing altitude.

In the *mixing layer* (identified in Fig. 20–1), hot gases from the plume (usually fuel rich with extra H_2 and CO) mix and burn with the oxygen in ambient air. This secondary combustion creates extra heat and causes the temperature in the mixing layer to rise by several hundred degrees. Therefore, the mixing layer is usually a strong radiation emitter. With some propellants (O_2/H_2 or LOX/alcohol) the mixing layer is transparent but the Mach diamonds in the core can be clearly visible. With most solid propellants and with certain storable liquid propellants, this mixing layer is not transparent, in part because of the emission and absorption by suspended particulates. Much infrared radiation from the inner plume core is obscured or absorbed by the bright outer mixing layers.

When the axis of the plume is aligned with the vehicle axis, then a straight inner core and a symmetrical mixing layer are formed as seen in Fig. 20–1 and the sketches and information in Table 20–2. When the plume is at an angle to the vehicle axis, as is the case with target-homing upper stages of antimissile vehicles, the plume shape and the mixing layer are geometrically very different. In the case of a retrorocket (where the firing direction of the propulsion system opposes that of the vehicle) the plume shape is again changed, but it can be symmetrical (Ref. 20–4).

Table 20–2 shows three sketches for typical changes of plume configuration with altitude and lists relevant heat sources. In general, the length and diameter of plumes increase with altitude but their inner temperatures and pressures decrease along their length, and plumes typically grow to become larger than the source vehicle. All hot plume portions emit radiation. The first sketch is really a longer version of the plume in Fig. 20–1. It has a long inviscid core (with exhaust gases that have not yet been mixed with air) with many Mach diamonds; at low altitudes (under 5 km, where $p_2 \sim p_3$) the shape of the plume is nearly cylindrical and the mixing layer is either cylindrical or slightly conical. The second sketch identifies several plume extents. In the transient region, the shock waves and the radiation diminish with every Mach diamond and more of the exhaust gas is mixed with ambient air; at higher altitudes (10 to 25 km), the nozzle exit pressure p_2 far exceeds the atmospheric pressure p_3 and the plume spreads out further and becomes larger; at even higher altitudes (above approximately 35 km when $p_2 \gg p_3$), the atmospheric pressure is very low and the plume may extend to several kilometers in diameter with only one or two Mach diamonds visible. In the third sketch two relatively large diameter shock waves can

TABLE 20–2. Altitude Characteristics of Large Plumes

Plume Configuration (not to same scale)			

Altitude, km	0–20	20–50	80 and above
Heat transfer sources			
1. Shock waves	Minor source in central core	Major source	Major source
2. Afterburning mantle	Major source	Minor source	Almost zero
3. Base recirculation	Usually minor[a]	Major source[a]	Minor source
4. Particulates (with solid propellants)	Can be major source	Major source	Minor source
Plume size	Longer than vehicle	Larger diameter and longer yet	Shock waves can be over 5 km wide
Plume shape	Near cylinder	Nearly conical	Core is small
Dimensions, m	10–100 long	200–1000 long	1–15 km wide
Number of visible shock diamonds	6 or more	3–4	Often only first inviscid region, sometimes $1\frac{1}{2}$ shock diamonds

[a]Even higher intensity with two or more nozzles firing.

be seen; one is in the thin air just ahead of the flight vehicle and is known as the bow shock wave. It occurs when the vehicle is at supersonic velocities. The other shock wave is attached to the tail end of the flight vehicle and is called the exhaust gas shock wave (a smaller version of this shock also appears at lower altitudes). In the far field (tail end of plume) ambient air and the exhaust gases are fairly well mixed throughout the plume cross section and the local pressure is essentially that of the ambient.

Underexpanded supersonic flows emerging from a nozzle experience a Prandtl–Meyer expansion through waves that attach themselves to the nozzle exit lip. This expansion allows the outer streamlines just outside the nozzle to bend increasing the Mach number of the gases in the outer layers of the plume, and, at higher altitudes, may cause some small portion of the supersonic plume to bend by more than 90° from the nozzle axis. The theoretical limit for Prandtl–Meyer expansions from Mach one is about 129° for gases with $k = 1.4$ (air) and about 160° for gases with $k = 1.3$ (typical for a rocket exhaust mixture, see Ref. 20–5), but from higher nozzle exit Mach numbers this expansion is much less. See Fig. 20–2.

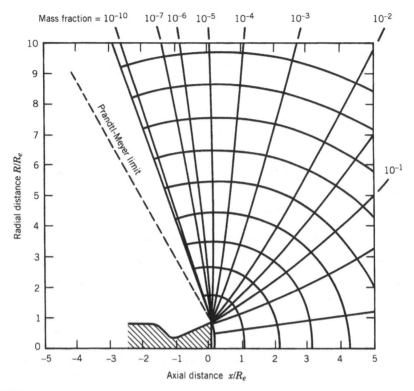

FIGURE 20–2. Density profile for vacuum plume expansion using a flow model for a small storable bipropellant thruster. The axial distance x and the plume radius R have been normalized with the nozzle exit radius R_e. Here $k = 1.25$, the Mach number of the nozzle exit is 4.0, and the nozzle cone half angle is $19°$.

This backward flow is analyzed to estimate possible heat, impingement, and contamination effects on vehicle components (see Ref. 20–6).

The boundary layers adjacent to the nozzle internal walls comprise viscous regions where the flow velocities are lower than in the core or inviscid region. The velocity decreases to zero right next to the wall. For large nozzles these boundary layer can be quite thick, say 2 cm or more. Figure 3–15 shows subsonic and supersonic regions within the boundary layer inside a nozzle's divergent section; it also shows temperature and velocity profiles. While supersonic flows are restricted in the angle through which they can deflect, subsonic boundary layer flows at the nozzle lip may deflect up to $180°$. Although these subsonic boundary layers represent only a small portion of the total mass flow, they nevertheless allow exhaust gases to flow backward outside of the nozzle, particularly at higher altitudes. This backflow has been known to cause heating of and sometimes chemical damage to exposed vehicle and propulsion system parts.

In the plume, the mass distribution or relative density is not uniform, as can be seen in Fig. 20–2, which is based on calculations for a high-altitude exhaust. Here 90%

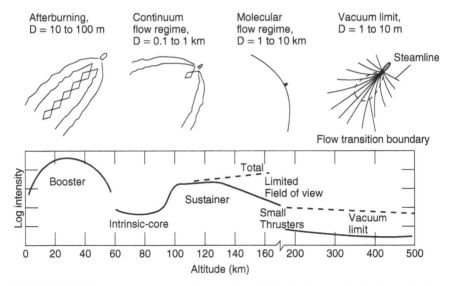

FIGURE 20–3. For a three-stage ascending rocket vehicle the plume radiation intensity from the three propulsion systems will vary with thrust level or mass flow, altitude, propellant combination, and gas temperatures within the plume. The four sketches are not drawn to the same scale. Copyright 2000 by The Aerospace Corporation.

of the flow is within $\pm\,44°$ of the nozzle axis and only one hundred thousandth, or 10^{-5}, of the total mass flow is bent by more than $90°$.

Figure 20–3 shows sketches of the radiation intensity and shape of plumes for three propulsion systems, each driving one of three stages—all part of a single flight vehicle. In the upper portion of Fig. 20–3 each sketch lists a key plume dimension (D), also depicting some shock waves and the plume's shape (each sketch is drawn to a different scale). The wiggly lines indicate effects of turbulence and wind on the mixing layer surrounding the core. Gaseous plumes exchange radiation only at specific narrow wavelengths or spectrum frequencies. The plume here does not contain any particulates. Each chemical species in the exhaust (H_2, H_2O, CO, N_2, etc.) has its own characteristic emission/absorption wave lengths. The lower part of the figure shows the plume radiation intensities of the various propulsion systems on a logarithmic scale as observed from a satellite. The dotted line is the estimated total radiation from the sustainer plume.

In Fig. 20–3, the rise in total emission intensity from the booster stage with altitude is due to an increase in plume diameter and a decrease in atmospheric absorption between the booster and the space-based sensor; the subsequent decrease of its total intensity can be explained as the gradual decrease in afterburning as the vehicle gains altitude. The gap in the curves is due to stage separation (termination of booster thrust and subsequent start of the sustainer propulsion system). The sustainer has lower thrust and a smaller plume and thus less total radiation intensity. Radiation intensity from the sustainer propulsion system rises as its plume expands and the radiating gas volume becomes larger. The decrease of intensity shown in the figure is due to the limited field of view and sensitivity of the specific sensor, which can only see a piece of the large plume (about 1 by 1 km). The lower part of the figure shows the plume radiation intensities of the various propulsion systems on a logarithmic scale as observed from a satellite.

The sustainer or second stage typically has a larger nozzle exit area ratio than the first or booster stage, and its plume gases at the nozzle exit have a substantially lower exit pressure. When the plume gas exit pressure is close to the ambient air pressure at altitude, the initial shape of the sustainer propulsion system plume will be nearly cylindrical and similar in shape to the initial booster plume shape at low altitude, but smaller. The initial sustainer plume size and radiation intensity are small compared to the final booster plume, and this is in part responsible for the initial trough in the curve of the sustainer propulsion system in Fig. 20–3. As the mixing of air and plume diminishes with altitude, emissions due to afterburning also diminish, and this is also partly responsible for the trough. As altitude increases further, the plume of the sustainer spreads out and forms a larger body of radiating gas—hence more overall intensity of radiation.

At high altitude (above about 170 km) the plume expands further and eventually only the intrinsic core (one half or one and one half Mach diamonds) has a strong enough radiation to be measured by a remote sensor (Ref. 20–4).

The multiple thrust chamber propulsion systems (with typically 6 to 12 small thrusters) of the third stage (it could be a satellite or a multiple warhead upper stage) have much lower thrusts and even much smaller plumes than the other two stages, and thus their radiation intensities are the lowest. The only portion of the plumes hot enough to emit significant radiation is the inviscid core just downstream of the nozzle exit. Also radiation may come from the very hot gases inside the small combustion chambers, but these can only be observed within a narrow angle through its nozzle. From space, radiation from small thrusters is often difficult to distinguish from the earth's background radiation.

The booster and sustainer vehicles both operate in that part of the atmosphere where *continuum flow* prevails; that is, the mean free path of molecules is small relative to vehicle dimensions, frequent collisions between molecules occur, gases follow the basic gas laws and they exhibit compression or expansion waves. As higher altitude is reached, the continuum regime changes into a *free molecular flow* regime, where there are fewer molecules per unit volume and the mean free path becomes larger than the key dimension of the vehicle (e.g., length). Here plumes spread out substantially, some reaching diameters in excess of 10 km. Only the exhaust gases close to the nozzle exit experience continuum flow conditions that allow the flow streamlines to spread out by means of successive Prandtl–Meyer expansion waves; once the gas reaches the boundary shown by the elliptical dashed line in the last sketch on the right in Fig. 20–3, the flow outside the dashed line will be in the free molecular regime and molecules will continue to spread out in straight lines into space. The phenomenology of rocket exhaust plumes as seen from a space-based surveillance system is described in Ref. 20–1.

Spectral Distribution of Radiation

As already stated, gas species radiate in specific narrow bands of the spectrum. The primary radiation emissions from most plume gases lie usually in narrow bands of the infrared, and to a lesser extent in narrow bands in the ultraviolet, with relatively little energy in the visible spectrum. Radiant emissions depend on size, particular propellants, their mixture ratio, and their respective exhaust gas compositions. For example, the exhaust from the liquid hydrogen–liquid oxygen combination contains mostly water vapor and hydrogen, with a minor percentage of oxygen and dissociated species (see Table 5–4). Its radiation is strong in specific wavelength bands, such as 2.7 and 6.3 µm for water (infrared region) and 122 nanometers for hydrogen (ultraviolet region). As shown in Fig. 20–4, a hydrogen–oxygen plume is essentially transparent or colorless; visible radiation from the inside white Mach diamonds is due to chemical reactions of the minor species O_2, OH, H, or O in the hot center region. Another propellant combination, nitrogen tetroxide with methylhydrazine fuel, gives strong emissions in the infrared region; in addition to the strong H_2O and hydrogen emissions mentioned previously; there are intense emissions for CO_2 at 4.7 µm,

FIGURE 20–4. Visible plume created by the oxygen–hydrogen propellants of the Vulcain 60 thrust chamber, which has a specific impulse of 439 sec (at altitude), a nozzle expansion area ratio of 45, and a mixture ratio of 5.6. The upstream portions of the diamonds of the multiple shock wave patterns are visible in the core of the plume because of emissions from reacting minor species. Courtesy of Airbus Defence and Space GmbH 2016, Germany.

CO at 4.3 μm and weaker emissions in the ultraviolet and visible (due to bands of CN, CO, N_2, NH_3, and other intermediate and/or final gaseous reaction products). These give the plume a pink orange–yellow color, but it is still partly transparent.

The exhaust of many solid propellants and some liquid propellants contains *solid particles* and may also contain *condensed liquid droplets*. In this book the term "particle" covers both types. In Tables 5–8 and 5–9 examples of solid propellant are given that contain in their incandescent white exhaust plumes about 10% small liquid droplets or solid particles of Al_2O_3; some kerosene-burning liquid propellants and many solid propellants have a small percentage of soot or small carbon particles in their exhaust. Receding, eroding, or charring thermal insulation in solid propellant rocket motors and ablative materials from nozzles can also contribute solid carbon particles to the exhaust. The radiation spectrum from hot solids is continuous, peaking in the infrared region, but it also has substantial emissions in the visible and ultraviolet region; such continuous spectrum is usually more intense in the visible than the narrow-band gaseous emissions in the plume. Afterburning in the mixing layer increases the temperature of the particles by several hundred degrees and thus intensifies their radiation emission. Even with only 2 to 5% solid particles, the plumes radiate brilliantly and are therefore very visible. However, outer layer particles in the plume may obscure the central core so that the shock wave patterns can no longer be observed.

The distribution of small particles is far from uniform. Figure 20–5 shows a split-image sketch of two plumes one with solid particles shown in the lower half and another with condensed small liquid droplets in the upper half. Continuum radiation emission energy from hot solid particles in rocket exhausts is usually higher and often more visible than that of all gaseous species. These small particulates (encompassing condensed droplets and solid particles) are accelerated or dragged along by the gas flow, but their velocities always lag the higher gas velocities. Large solid particles have more inertia to changes in flow direction than the smaller ones. As the plume spreads out, portions of the gas may turn to flow at more than a right angle to the plume axis, but the solid particles can only turn through small angles (perhaps 5° to 30°). They do not reach the outer boundaries of the plume as shown in Fig. 20–5. The central area of the plume, which contains essentially all the larger solid particles is surrounded by an annular region with contains a few larger particles, mostly medium and small sized particles. Most of the smaller solid particles are in a next annular flow region. The outer regions of the plume contain few or no solid particles.

Mechanisms involving condensed liquid droplets are somewhat different. Certain gases condense to form liquid droplets at a particular liquefying temperature, irrespective of location or stream line within the plume. As the gases expand in the plume, they cool dramatically but they may also absorb heat from surrounding hot particulates. Therefore, the distribution of condensed droplets becomes relatively uniform in the plume.

In the upper half of Fig. 20–5, it can be seen that any super-cooled water vapor actually condenses downstream from the nozzle exit, presumably by homogeneous nucleation. The principal observable effect would be the scattering of sunlight.

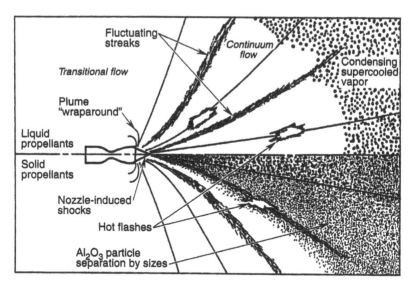

FIGURE 20–5. Simplified diagram of particulates in a plume flow at altitude. The upper half shows the condensation of gases into small liquid droplets and the lower half shows the plume from a solid motor with small aluminum-oxide solid particles near the centerline of the plume. Copyright 2000 by The Aerospace Corporation.

The lower half of this figure shows solid alumina particles. Some of the larger particles persist as small liquid droplets from the nozzle into the plume before solidifying by radiation cooling (Refs. 20–1 and 20–4).

Plumes with particulates remain visible by scattered sunlight for a relatively long time even after the vehicle has gone. They can be removed by local winds. Such smoke or vapor trail marks the path of the vehicle that has just passed. Eventually the plume cools down, diffuses, and will no longer be visible as it is assimilated by the atmosphere.

Plumes frequently display some fluctuating visible hot streaks along their stream lines. These are hot moving flashes, pockets of more intense radiation, or locally hotter mixtures of gases as seen in Fig. 20–5. These hot moving gas local anomalies originate from unsteady combustion and injection processes in the combustion chamber and are also caused by turbulence. See Ref. 20–1.

The total radiation from the plume also heats adjacent vehicle or propulsion system components. Predictions of radiative emission require an understanding of plume components and of the radiant absorption by intervening atmospheric or plume gases (see Refs. 20–1 and 20–7). Any heat transferred from the plume to vehicle components (by radiation and convection) will depend on propellant combination, nozzle configuration, vehicle geometry, size, number of nozzles, trajectory, altitude, and the secondary turbulent flow around the nozzles and on the tail section of the vehicle. Blowback plume gas shields are installed to prevent damage to the vehicle as shown in Figure 18–10.

Observed or measured values of radiant emissions need to be corrected for various factors. Signal strength diminishes as the square of the distance between the plume and the observation station, and observed magnitudes can change significantly during flight. Attenuation is a function of wavelength, atmospheric conditions (rain, fog or clouds), the air mass and gas contents between the hot part of the plume and the observing location, and depends on the flight vector direction relative to the line of sight. The total radiant emission is a maximum when seen at right angles to the plume (see Refs. 20–1, 20–3, and 20–7). Radiation measurements can be biased by background radiation (important for observation from satellites) or by Doppler shifts.

Multiple Nozzles

It is common to have more than one propulsion system operating simultaneously, or more than one nozzle sending out hot exhaust gas plumes. For example, several large launch vehicles have in its first stage multiple liquid propellant rocket engines and two or more solid propellant motors as strap-on boosters, all running simultaneously. The interference and impingement of these plumes with one another causes regions of high temperature in the combined plumes and therefore larger emissions, but the emissions will no longer be axisymmetrical. Also, multiple nozzles can cause distortions in the airflow near the rear end of the vehicle and influence the vehicle drag and augment hot backflows from the plume locally. With multiple propulsion systems plumes impinge with each other and can create local hot spots. At high altitudes the plumes of multiple propulsion systems appear to merge; analytically, this can at times be approximated as flow from an equivalent single nozzle providing the total mass flow, but with the properties and characteristics of the individual nozzles (Ref. 20–4).

With either liquid or solid propellants, the impingement of two or more hot plumes on each other redistributes the hot gases when compared to a single nozzle exhaust. Plume spreading, capable of exceeding 90°, and impingement patterns change with altitude. Created by such impingement, some of the hot low-pressure gases flow up and then sideways between thrust chambers or between nozzles from the center to the outside below the boattail of the vehicle. A blowback protection sheet or "Thermal Shield" is installed to protect engine components and vehicle equipment from the hot low pressure gases and from the intense plume radiation. This thermal shield is made of a temperature resistant material; it can include features to allow for the angular motion of nozzles during thrust vector control. An example of such a thermal shield for the RD-124 engine can be seen in Fig. 18–10. A majority of space vehicles use such thermal shields.

Plume Signature

This is the term used for all plume characteristics in the infrared, visible, and ultraviolet, and for its electron density, smoke, or vapors, relating to any particular vehicle, mission, rocket propulsion system, and propellant (see Refs. 20–3, 20–8 to 20–10). In many military applications it is important to reduce the plume signature in order to minimize detection and/or tracking. Fractions of H_2 and CO in the nozzle exit gases are perhaps the most significant factors influencing plume signature. As plume temperatures increase, higher levels of radiation and radio-frequency interaction will occur. Emissions can be reduced by selecting a propellant combination or mixture ratio with a lower combustion temperature; unfortunately, this usually gives a lower performance. Ways to reduce smoke are described later in this Chapter. Often, the specifics of plume signatures are requirements for any new or modified rocket-propelled military flight vehicles, as are the limits imposed on spectral emissions in certain regions of the spectrum and the amount of acceptable smoke.

The atmosphere absorbs energy in certain regions of the spectrum, for example, by its carbon dioxide and water vapor contents. These molecules absorb and attenuate radiation in the frequency bands peculiar to these two species. Since many plume gases contain a lot of CO_2 or H_2O, attenuation within the plume itself can be significant. Plume energy or intensity, as measured by spectrographic instruments, has to be corrected for attenuation of intervening air or plume gases.

Vehicle Base Geometry and Recirculation

Nozzle exit geometries and flight vehicle's tail (or aft base) configurations influence the plume. Figure 20–6 shows a nozzle exit whose diameter is almost the same as the

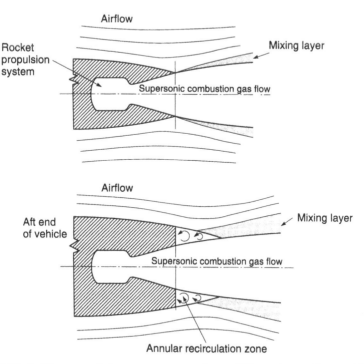

FIGURE 20–6. Diagrams of flow patterns around two different boat tails or vehicle aft configurations, with and without hot gas recirculation.

base or tail diameter of the vehicle bodies. When these two diameters are not close to each other, a flat doughnut-shaped base or tail region will form. In this region the high-speed-combustion gases velocity is usually larger than the air speed of the vehicle (which is about the same as the local air velocity relative to the vehicle) and unsteady flow vortex-type recirculation zones will be induced. This greatly augments the afterburning, the heat release to the base, and usually creates a low pressure on this base. Such lower pressure in effect increases the vehicle's drag.

Airflow patterns at the vehicle's tail, which helps to maintain the vehicle's aerodynamic stability, may be different with different tail geometries (e.g., straight cylinders, diminishing diameters, or increasing-diameter conical shapes). Airflow patterns and mixing layers change dramatically with angle of attack, tail fins, or wings, causing unsymmetrical plume shapes and possibly flow separation. In some cases the recirculation of fuel-rich exhaust gases mixing with air will ignite and burn; this dramatically increases the heat transfer to the base surfaces and causes changing plume characteristics.

Compression and Expansion Waves

Shock waves are surface of discontinuities in supersonic flows. In rocket plumes they represent very rapid transformations of kinetic energy to potential and thermal energies within a very thin wave surface. Fluid crossing a stationary shock rises suddenly and irreversibly in pressure and decreases in velocity. When flow passes through a *normal shock wave* there is no change in flow direction. Such normal shocks produce the largest increase in pressure (and local downstream temperature) and the flow velocity behind normal shocks is subsonic. When the incoming flow is at an angle less than 90° to the shock wave surface, it is known as a *weak compression wave* or as an *oblique shock wave*. Figure 20–7 illustrates these flow relationships and shows the angle of flow change. Gas temperatures immediately behind a normal shock wave approach the stagnation temperature in the combustion chamber and hence radiation increases greatly. Also, here (and in other hot plume regions) dissociation of gas species and chemical luminescence (emission of visible light) may occur, as seen (downstream of strong shock waves) in Fig. 20–4.

All gas expansions in supersonic flows behave with fairly similar geometric relationships. The opposite of a shock occurs at a surface where the flow undergoes a Prandtl–Meyer *expansion wave*, which is a surface where pressures are reduced and velocities are increased. Often, there is a series or fan of weak expansion waves next to each other; these occur outside the nozzle exit section where the static gas pressure is higher than the ambient pressure (underexpansion), as shown in Figs. 20–1 and 20–2.

Plumes from hydrogen–oxygen liquid propellant combustion consist mostly of superheated water vapor and hydrogen gas, which are largely invisible. However, they may become faintly visible from chemical reactions of minor species in hot zones that are believed to produce observed pale pink-orange-and-white skeletal wave patterns. These patterns are seen in Table 20–2 and Fig. 20–4.

Supersonic gas flow patterns out of the nozzle exit remain undisturbed until they change direction in a wave front or go through a normal shock. Diamond-shaped

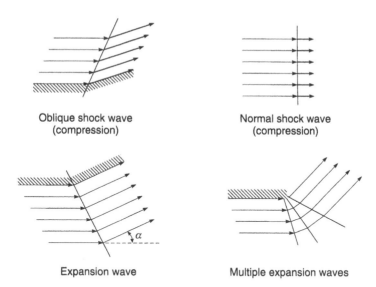

Oblique shock wave
(compression)

Normal shock wave
(compression)

Expansion wave

Multiple expansion waves

FIGURE 20–7. Simplified diagrams of oblique shock waves or compression waves, normal shocks, and expansion waves. The change in the length of the arrows is an indication of the change in gas velocity as the flow crosses the wave front.

patterns result from compression and expansion wave surfaces. These patterns (shown in Table 20–2 and Fig. 20–4) repeat themselves and are clearly visible even in largely transparent plumes, as seen from hydrogen–oxygen or alcohol–oxygen propellant combinations. These diamond patterns become weaker with each succeeding wave. Mixing layers act as wave reflectors because an expansion wave is reflected as a compression wave.

As seen in Fig. 20–1, the interface surface between the rocket exhaust plume gases and the air flowing over the vehicle (or the air aspirated by the high-velocity plume) acts as a "free boundary." Oblique shock waves are reflected at a free boundary as an opposite wave. For example, an oblique compression wave is reflected as an oblique expansion wave. This boundary is not usually a simple surface of revolution, but an annular layer, sometimes called a slipstream or mixing layer, which is distorted by turbulence.

20.2. PLUME EFFECTS

Smoke and Vapor Trails

Smoke is unacceptable in a number of military missile applications. It interferes with optical signal transmissions needed with line-of-sight or electro-optical guidance systems. Smoke also hampers the vision of a soldier guiding a wire-controlled antitank weapon. A visible trail allows easy and rapid detection of a missile being fired, and the tracking of a flight path may reveal a covert launch site. Smoke is produced not only during rocket operation, but also during the irregular combustion of solid propellant remainders after rocket cutoff (or chuffing), which, as described in Chapter 14, produces small puffs of flame and smoke at frequencies between 2 and 100 Hz.

Primary smoke consists of a suspension of many very small solid particles in a gas, whereas secondary smoke originates from condensed small liquid droplets, such as condensed vapor trails, fog, or clouds. Secondary smoke may form in all-gas plumes and also in plumes that contain primary smoke. Many propellants leave visible trails of smoke and/or vapor from their plumes during powered rocket flight (see Refs. 20–3, 20–8, and 20–9). These trails are shifted by local winds after the vehicle has passed. They are most visible in the daytime, because they originate from reflected or scattered sunlight. Solid particles that form primary smoke are mainly aluminum oxide (Al_2O_3, typically 0.1 to 3 μm diameter) from composite solid propellants. Other solid particles in solid propellant exhausts come from unburned aluminum, zirconium or zirconium oxide (a combustion stabilizer), or iron or lead oxides (a burn-rate catalyst). Carbon particles (soot) arise from the thermal insulation used in solid propellants motors and in some liquid propellants from the hydrocarbon fuels themselves.

During expansion, rocket exhaust plumes may cool by radiation, gas expansion, and by mixing with colder ambient air (if below its dewpoint temperature, water vapor will condense). Of course, all these effects depend on local weather conditions. If the ambient temperature is low enough (e.g., at high altitude) and/or if the gas expands to low temperatures, water droplets can freeze to form small ice crystals or snow. At high altitude, CO_2, HCl, and other gases may also condense. Most rocket exhaust gases contain between 5 and 35% water, and the exhaust from the liquid hydrogen/liquid oxygen propellant combination may contain as much as 80%. When the exhaust contains tiny solid particles, these then act as condensation nuclei for water vapor thus increasing the amount of nongaseous material in the plume making the plume even more visible.

When reducing smoke in the plume is important to the mission, then a *reduced-smoke solid propellant or a minimum-smoke propellant* must be used. Alternatively a liquid propellant combination, known not to form small carbon particles, may be selected. These are described in Chapters 7 and 13. Even then, secondary smoke trails can form under coldweather and high-humidity conditions, though these may be difficult to see under some weather conditions.

Toxicity

Exhaust gases from many rocket propulsion systems contain highly toxic and/or corrosive gas species, which may cause severe health hazards and potential environmental damage around launch or test sites. Accidental spills of liquid oxidizers, such as nitrogen tetroxide or red fuming nitric acid, can generate toxic, corrosive gas clouds, which have higher density than air and will stay close to the ground. Exhaust gases such as CO or CO_2 present a health hazard if inhaled in concentrated doses. Gases such as hydrogen chloride (HCl) from solid propellants using a perchlorate oxidizer, nitrogen dioxide (NO_2), nitrogen tetroxide (N_2O_4), or vapors of nitric acid (HNO_3) have relatively low levels of allowable inhalation concentration before health damage can occur. Chapter 7 lists some safe exposure limits. Potential damage increases with the amount of toxic species in the exhaust, the mass flow rate or thrust level, and the duration of the rocket firing in regions at or near the test/launch site.

Dispersion by wind and diffusion/dilution with air can reduce concentrations of toxic materials to tolerable levels within a few minutes, but this depends on their quantity and on the local weather conditions, as explained in Chapter 21. Careful attention must therefore be given to scheduling launch or test operations for times when the wind will disperse these gases to nearby areas. For very highly toxic exhaust gases (e.g., those containing beryllium oxide or certain fluorine compounds), and usually for all relatively small thrust levels, exhaust gases from static test facilities must be captured, chemically treated, detoxified, and purified before release into the atmosphere.

Noise

Acoustical noise is an unavoidable by-product of thrust; it is particularly important in large launch vehicles with large propulsion systems. It is a primary design consideration in the vehicle and in much of the ground-support equipment, particularly electronic components. Besides being an operational hazard to personnel in and around rocket-propelled vehicles, it can be a severe annoyance to communities near rocket test sites. The *acoustic power* emitted by the Saturn V space flight vehicle at launch was about 2×10^8 W, enough to light up about 200,000 average homes if it were available as electricity.

Acoustic energy emission is mainly a function of exhaust gas velocity and density, nozzle configuration, and the velocity of sound in the surrounding medium. In these terms, the chemical rocket is the noisiest of all aircraft and missile propulsion devices. Sound intensity is highest near the nozzle exit and diminishes with the square of the distance from the source. Analytical models of noise from a rocket exhaust usually divide the plume into two primary areas, one upstream of the shock waves and the other one downstream (subsonic), with high-frequency sound coming from the first and low frequency from the second. The shock wave itself generates sound, as does the strong turbulent mixing of the high-velocity exhaust with its relatively more quiescent surroundings. Sound emission is normally measured in terms of microbars (μbar) of *sound pressure*, but is also expressed as *sound power* (W), *sound intensity* (W/ft^2), or *sound level* (decibels, dB). The relationship that exists among a decibel scale, the power, and intensity scales is difficult to estimate intuitively since the decibel is a logarithm of a ratio of two power quantities or two intensity quantities. Further, expression of a decibel quantity must also be accompanied by a decibel scale reference, for example, the quantity of watts corresponding to 0 dB. In the United States, the most common decibel scale references 10^{-13} W power, whereas the European scale references 10^{-12} W.

A large rocket can generate sound levels of about 200 dB (reference 10^{-13} W), corresponding to 10^7 W of sound power, compared to 140 dB for a 75-piece orchestra generating 10 W. Reducing the sound power by 50% reduces the value by only about

3 dB. In terms of human sensitivity, a 10-dB change usually doubles or halves the noise for the average person. Sound levels above 140 dB frequently introduce pain to the ear and levels above 160 become intolerable. In small propulsion systems the noise level is somewhat lower and often can be ignored. There is a remarkable similarity between the spectra of infrared radiation and acoustic wave propagation (Ref. 20–4).

Spacecraft Surface Contamination

Contamination of sensitive surfaces of a spacecraft by rocket exhaust products can be a problem to vehicle designers and users. When deposited on a surface these products can noticeably degrade the function of surfaces, such as solar cells, optical lenses, radiators, windows, thermal-control coatings, and mirrored surfaces. Propellants that have condensed liquids and/or solid particles in their exhaust appear to be more troublesome than propellants with mostly gaseous products, such as oxygen–hydrogen or monopropellant hydrazine; plumes from most solid propellants contain contaminating species. Practically all studies here have been concerned with small storable liquid propellant attitude control pulse motors in the thrust range 1.0 to 500 N (the type commonly used for controlling vehicle attitude and orbit positioning over long periods of time). Deposits of hydrazinium nitrate and other substances have been found. The amount of accumulated exhaust products on surfaces is a function of many variables, including propellant composition, combustion efficiency, combustion pressure, nozzle expansion ratio, surface temperature, and rocket–vehicle interface geometries. The prediction of exhaust contamination of spacecraft surfaces is only partly possible through analytical calculations. Reference 20–10 provides a comprehensive analytical model and computer program for liquid bipropellant rockets.

Another effect of clouds of condensed species (either tiny liquid droplets or solid particles) is to scatter sunlight and cause solar radiation to be diverted toward optical instruments on the spacecraft, such as cameras, telescopes, IR trackers, or star trackers; this diversion causes erroneous instrument measurements. Scatter depends on propellant composition, density, size, and changes of particulates as well as on plume location relative to the instruments, their sensitivity to optical frequencies, and their surface temperature.

Radio Signal Attenuation

In general, rocket exhaust plumes interfere with the transmission of radio-frequency signals that must pass through the plume in the process of guiding the vehicle, radar detecting, or communicating with it. Here, solid propellant exhaust plumes usually cause more interference than liquid rocket engine plumes. Signal attenuation is a function of free electron density and of electron collision frequency. Given these two parameters for the entire plume, the amount of attenuation a signal experiences in passing through the plume can be calculated. Figure 20–8 shows a minimum plume model containing contours of constant electron density and electron collision frequency for momentum transfer, a model sufficient for predicting signal

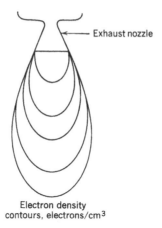

Exhaust nozzle

Electron density
contours, electrons/cm³

FIGURE 20–8. Exhaust plume model for predicting attenuation of radio communications signals. The contours shown are for either equal electron density or electron collision frequency; the highest values are near the nozzle exit.

attenuation. Free-electron density and activity in the exhaust plume are influenced by many factors, including the propellant formulation (particularly alkali contents), exhaust temperature, motor size, chamber pressure, flight speed and altitude, and distance downstream from the nozzle exit. Computer methods have been developed for analyzing physical and chemical compositions, including electron densities, and the attenuation characteristics of exhaust plumes (Refs. 20–11 to 20–14).

The relationship between several influential rocket motor and vehicle design factors may be summarized from experience with typical solid propellant rockets as follows:

1. The presence of alkali-metal molecules increases attenuation; changing the concentration level of potassium from 10 to 100 ppm increases the relative attenuation some 10-fold at low altitudes. Both potassium ($\approx$150 ppm) and sodium ($\approx$50 ppm) are considered to be impurities in commercial-grade nitrocellulose and ammonium perchlorate.

2. The concentration of aluminum fuel has a major influence; increasing its percentage from 10 to 20% increases the attenuation fivefold at sea level and three- to fourfold at 7500 m altitude.

3. Increasing the chamber pressure for any given aluminized propellant from 100 to 2000 psi reduces the relative attenuation by about 50%.

4. Attenuation varies with the distance downstream from the nozzle exit plane and can become four to five times as great as the nozzle exit plane, depending on the flight altitude, nozzle geometry, oxidizer-to-fuel ratio, flight velocity, altitude, and other parameters.

For many solid rocket applications, attenuation of radio or radar signal strength by the exhaust plume is little consequence. When it is, attenuation can usually be kept at acceptable levels by controlling the level of alkali impurities in propellant ingredients and by using nonmetal fuels or low percentages (<5%) of aluminum. Rocket motors with high expansion ratio nozzles help bring down electron concentrations since they recombine faster with positive ions as the exhaust temperature falls.

Electrons in the plume greatly increase its radar cross section, and hot plumes are readily detected with radar. The plume is usually a stronger radar reflector than its flight vehicle. A radar homing missile seeker would hence focus on the plume and not the vehicle. A reduction of plume cross section is desirable (lower gas temperature, less sodium impurities) in combat situations.

Plume Impingement on Structures

In most reaction control systems there are several small thrusters pointed in different directions. Although small thrusters have small plumes and the intensity of their radiation emissions is relatively low, they may cause severe damage to the vehicle. There have been cases where small thruster plumes have impinged upon some critical surfaces of a space vehicle, such as extended solar cell panels, radiation heat rejection surfaces, or aerodynamic control surfaces. This is more likely to happen at high altitudes, where plume diameters are large. Such impingement can lead to surface overheating and to unexpected turning moments.

Also, during stage separation, there have been occasions where the plume of the upper stage impinges on the lower vehicle stage. This may cause overheating and impact damage not only to the lower stage (being separated), but by reflection or recirculation, also to the upper stage. Other impingement examples are upon docking maneuvers or the launching of multiple rockets (nearly simultaneously) from a military barrage launcher. When the plume of a rocket missile impacts on another while flying, it may cause the latter to experience a change in flight path, often not hitting the intended target.

Heat Transfer to Clusters of Liquid Propellant Rocket Engines

According to Ref. 20–15, all multiengine space launch vehicles (e.g., Figs. 1–12 and 1–13) experience strong heating during liftoff and ascent. This mainly affects the lower portions of the launch vehicle and arises from both plume radiation and hot-exhaust-gases convection. Such heating changes in intensity and location during the various ascent phases, which can be conveniently divided into four stages: in the first one, at the start of the launch, heat radiation comes from each individual rocket plume and from atmospheric flows cooling the vehicle's base; phase two occurs at intermediate altitudes where air currents moving across the base compete with updraft jets created by the multiple plumes interacting with each other—during these first two stages heating remains relatively low. At higher altitudes, plumes spread out and the radiation source increases in size. There is little or no mixing (see Fig. 20–1). Peak base heating occurs at high altitudes where low atmospheric pressures cause the individual rocket plumes to expand and combine into a single plume; air flows over the base decrease significantly and hot gases may come in contact with the vehicle's lower structures (see Table 20–2 and Fig. 20–3). Temperatures level off during the fourth stage as the rocket vehicle reaches the edge of space and where engine plumes remain combined with hot gases recirculating around the rocket's base. The liquid in cooling jackets is usually able to adequately cool the thrust chamber itself; however, uncooled components (such as pressure switches, vehicle structures, antennas, temperature sensors, stage separation devices, etc.) can overheat and may fail. Knowing estimated temperatures at each stage allows designers to build hardware that can handle the heat loads.

As stated earlier, many large liquid propellant rocket engines with multiple chambers have a *flame barrier* or *blow back prevention shield* near the vehicle's bottom as seen in Fig. 18–10 for the RD-124 engine. This heat shield is positioned at right angles to the vehicle axis, fastened to the vehicle's perimeter and to the exit section of the individual thrust chambers. Such a barrier prevents the failure of heat sensitive engine and vehicle components (above the barrier) from plume radiation and from hot gas blowbacks from plume mixing layers. With multi-thrust chamber systems, low-pressure gas blowbacks can issue from certain portions of the plumes as they impinge into each other, particularly at the higher altitudes. The heat resistant barrier is designed to prevent any such damage. If the thrust chambers are gimballed or hinged for thrust vector control, the barrier has to have special features to allow nozzle angular movements without opening any holes in it.

20.3. ANALYSIS AND MATHEMATICAL SIMULATION

The complicated plume structure, behavior, and many of its physical attributes have been simulated by mathematical algorithms, and a number of sophisticated computer programs exist (see Refs. 20–1 to 20–3 and 20–16 to 20–20). Although there has been much progress in using mathematical simulations of plume phenomena, results of such analyses are not always reliable or useful for making predictions of many plume characteristics; however, the models do help in the understanding of plumes and in extrapolating test results to different conditions within narrow limits. Some of the physics of plumes is not yet fully understood.

All simulations require simplifying assumptions to make solutions possible and their field of application is usually limited. Analysis is aimed at predicting plume parameters such as temperature, velocity, pressure profile, radar cross section, heat transfer, radiations, condensation, deposits on optical surfaces, impact forces, and/or chemical species. These analyses are usually focus on separate spatial segments of the plume (e.g., core, mantle, supersonic versus subsonic regions, continuum versus free molecular flow, near or far field), and need different assumptions about the dynamics of the flow (many neglect turbulence effects and/or the interaction between boundary layers and shock waves); most are suitable only for steady-state conditions. The algorithms also treat differently the chemical reactions, solids content, energy releases, composition changes within the plume, flight and altitude regimes, airflow and vehicle interactions, and/or selected regions of the spectrum (e.g., IR only). They may require assumptions about particle sizes and their amounts, spatial and size distributions, gas velocity distributions, the geometry and boundaries of the mixing layer, and/or the turbulence behavior. The more complex mathematical models use multidimensional models. Plume analyses using several different models are shown in Refs. 20–16 to 20–20. Many solutions are based in part on extrapolating measured data from actual plumes to guide the analyses.

Actual measurements on plumes during static and flight tests are used to verify mathematical models requiring highly specialized instrumentation, careful calibrations and characterization, skilled personnel, and a sophisticated application of various correction factors. Extrapolating computer results to regions or parameters that have not been fully validated has often given poor results. There are validated computer programs for plume phenomena predictions but most are proprietary or limited in distribution.

PROBLEMS

1. List at least two parameters that are likely to increase the total radiant emission from plumes, and explain how they accomplish this. For example, increasing the thrust increases the radiating mass and size of the plume.

2. Look up the term *chemiluminescence* in a technical dictionary or chemical encyclopedia; provide a definition and explain how it can affect plume radiation.

3. Describe the changes of total radiation intensity from a plume of a vertically ascending rocket vehicle launched at sea level as seen by an observer also at sea level, but at a distance of 50 km away from the launch location.

4. A rocket vehicle flies horizontally at 25 km altitude. Its plume radiation emission is measured by an instrument in the gondola of a balloon 50 km away, but at the same altitude of 25 km. At this location the air has a very low density and plume radiation absorption by the air is small and may be neglected. Describe any corrections that need to be applied to the measurements in order to determine the plume's radiant emission.

REFERENCES

20–1. F. S. Simmons, *Rocket Exhaust Plume Phenomenology*, Aerospace Press, The Aerospace Corporation, El Segundo, California, 2000.

20–2. S. M. Dash, "Analysis of Exhaust Plumes and Their Interaction with Missile Airframes," in M. J. Hemsch and J. N. Nielson (Eds.), *Tactical Missile Aerodynamics*, a book in the series of *Progress in Astronautics and Aeronautics*, Vol. 104, AIAA, Washington, DC, 1986.

20–3. A. C. Victor, Chapter 8, "Solid Rocket Plumes," G. E. Jensen and D. W. Netzer (Eds.), *Tactical Missile Propulsion*, a book in the series of *Progress in Astronautics and Aeronautics*, Vol. 170, AIAA, 1996.

20–4. Personal communication from Frederick S. Simmons of The Aerospace Corporation, 2008 and 2015.

20–5. S. M. Yahya, *Fundamentals of Compressible Flow*, 2nd revised printing, Wiley Eastern Limited, New Delhi, 1986.

20–6. R. D. McGregor, P. D. Lohn, and D. E. Haflinger, "Plume Impingement Study for Reaction Control Systems of the Orbital Maneuvering Vehicle," AIAA Paper 90-1708, June 1990.

20–7. R. B. Lyons, J. Wormhoudt, and C. E. Kolb, "Calculation of Visible Radiations from Missile Plumes," in *Spacecraft Radiative Heat Transfer and Temperature Control*, a book in the series of *Progress in Astronautics and Aeronautics*, Vol. 83, AIAA, Washington, DC, June 1981, pp. 128–148.

20–8. *Rocket Motor Plume Technology*, AGARD Lecture Series 188, NATO, Jun. 1993.

20–9. Terminology and Assessment Methods of Solid Propellant Rocket Exhaust Signatures, AGARD Advisory Report 287, NATO, Feb. 1993.

20–10. R. J. Hoffman, W. D. English, R. G. Oeding, and W. T. Webber, "Plume Contamination Effects Prediction," Air Force Rocket Propulsion Laboratory, December 1971.

20–11. L. D. Smoot and D. L. Underwood, Prediction of Microwave Attenuation Characteristics of Rocket Exhausts, *Journal of Spacecraft and Rockets*, Vol. 3, No. 3, Mar. 1966, pp. 302–309.

20–12. K. Kinefuchi et al., "Prediction of In-Flight Radio Frequency Attenuation by a Rocket Plume by Applying CFD/FDTD Coupling," AIAA Paper 2013-3790, July 2013.

20–13. A. Mathor, "Rocket Plume Attenuation Model," AIAA Paper 2006–5323, Jun. 2006.

20–14. K. Kinefuchi, K. Okita, I. Funaki, and T. Abe, "Prediction of In-Flight Radio Frequency Attenuation by a Rocket Plume by Applying CFD/FDTD Coupling," AIAA Paper 2013-3790, Jul. 2013.

20–15. H. Kenyon, "Modelling heat off rocket engines," *Aerospace America*, Vol. 53, No. 4, Apr. 2015, pp. 16–18.

20–16. I. Boyd, *Modeling of Satellite Control Thruster Plumes*, PhD thesis, Southampton University, England, 1988.

20–17. S. M. Dash and D. E. Wolf, "Interactive Phenomena in Supersonic Jet Mixing Plumes, Part I: Phenomenology and Numerical Modeling Technique," *AIAA Journal*, Vol. 22, No. 7, Jul. 1984, pp. 905–913.

20–18. S. M. Dash, D. E. Wolf, R. A. Beddini, and H. S. Pergament, "Analysis of Two Phase Flow Processes in Rocket Exhaust Plumes," *Journal of Spacecraft*, Vol. 22, No. 3, May–Jun. 1985, pp. 367–380.

20–19. C. B. Ludwig, W. Malkmus, G. N. Freemen, M. Slack, and R. Reed, "A Theoretical Model for Absorbing, Emitting and Scattering Plume Radiations," in *Spacecraft Radiative Transfer and Temperature Control*, a book of the series of *Progress in Astronautics and Aeronautics*, Vol. 83, AIAA, Washington, DC, 1981, pp. 111–127.

20–20. A. M. Kawasaki, D. F. Coats, and D. R. Berker, "A Two-Phase, Two-Dimensional, Reacting, Parabolized Navier-Stokes Solver for the Prediction of Solid Rocket Motor Flow Fields," AIAA Paper 92–3600, Jul. 1992.

CHAPTER 21

ROCKET TESTING

21.1. TYPES OF TESTS

Before any rocket propulsion systems are put into operational use, they are subjected to several different types of tests, many of which are outlined below in the approximate sequence in which they are normally performed.

1. Manufacturing inspection and fabrication tests on individual parts (dimensional inspection, pressure tests, X-raying, leak checks, electric continuity tests, electromechanical checks, etc.). These tests are usually done at the factory.
2. Component tests (functional and operational on igniters, nozzles, insulation, valves, controls, injectors, structures, tanks, motor cases, thrust chambers, turbopumps, thrust-vector control, etc.). May need special equipment/facilities.
3. Static rocket propulsion system tests (with complete propulsion system) on a test stand: (*a*) complete propulsion system tests (under rated conditions, off-design conditions, with intentional variations in environment or calibration); (*b*) with liquid propellants a partial or simulated rocket operation (for proper function, calibration, ignition, operation—often without establishing full thrust or operating for the full duration); for reusable or restartable rocket propulsion systems this can include several starts, long-duration endurance tests, and postoperational inspections and reconditioning; (c) some tests on chemical propulsion systems and nearly all tests on electrical propulsion systems are performed in large vacuum facilities that simulate the high-altitude rarified atmosphere conditions.

4. Static vehicle tests (when rocket propulsion system is installed in a restrained, nonflying vehicle or stage) on a vehicle test stand.
5. Flight tests: (*a*) with a specially instrumented or new propulsion system in a developmental flight test vehicle; (*b*) with a production vehicle.

Each of these five types of tests can be performed on at least three basic program types:

1. Research on and development or improvement of a new (or modified) rocket propulsion system, its propellants, materials, or components.
2. Evaluation of the suitability of a new (or modified) rocket engine or novel rocket motor for an alternate specified application or for flight readiness.
3. Production proof tests and quality assurance of existing operational rocket propulsion systems.

The first two program types are usually concerned with novel devices and often involve the testing and measurement of new concepts or conditions using experimental rocket propulsion systems. Here examples are the testing of a new solid propellant grain, the development of a novel control valve assembly, and the measurement of the thermal expansion of a nozzle exhaust cone during firing operation.

Production tests are concerned with measurements of a few basic parameters on propulsion systems to assure that their performance, reliability, and operation are within specified tolerance limits. If the number of units is large, test equipment and instrumentation used for these tests are usually partly or fully automated and designed to permit the testing, measurement, recording, and evaluation in a minimum amount of time.

During early development phases of a program, many special or unusual tests are performed on components and on complete rocket systems to prove specific design features and performance characteristics. Special facilities and instrumentation or modification of existing test equipment are used. Examples are large centrifuges to operate propulsion systems under acceleration, or variable frequency vibration equipment for shaking complete systems and/or components. During the second type of program, special tests are usually conducted to establish statistically the performance and reliability of a rocket system by operating a number of units of the same design. During this phase, tests are also made to demonstrate the ability of the rocket system to withstand extreme operating conditions limits, such as high and low ambient temperatures, variations in fuel composition, changes in the vibration environment, or exposure to moisture, rain, vacuum, or rough handling during storage or transport. To demonstrate safety, sometimes intentional malfunctions, spurious signals, or known manufacturing flaws are introduced into the propulsion system to determine the capability of its control system or the safety devices to handle and prevent potential failures.

Before an experimental rocket system can be flown in a vehicle it has to pass a set of *preliminary flight rating tests* aimed at demonstrating its safety, reliability, and performance. It is not a single test, but a series of tests under various specified conditions, operating limits, and performance tolerances, simulated environments, and intentional malfunctions. Only thereafter may the rocket system be used in experimental flights. However, before they can be put into production, rocket systems have to pass another a series of tests under a variety of rigorous specified conditions, known as the *qualification test* or *preproduction test*. Once a particular propulsion system has been *qualified*, or passed such a qualification test, it is usually forbidden to make any changes in design, fabrication, or materials without going through a careful review, extensive documentation, and often also a partial or full set of requalification tests.

The amount and expense of component and complete propulsion systems testing has greatly decreased in the last few decades. The reasons here are more experience with prior similar systems, less hardware development, better sensors, better analytical model simulations, better prediction of system parameters, use of health monitoring systems to identify incipient damage (which saves hardware), and more confidence in predicting certain failure modes and their location. Validated computer software have removed many uncertainties and obviated needs for some tests. In some applications the number of firing tests has decreased by a factor of 10 or more.

21.2. TEST FACILITIES AND SAFEGUARDS

For chemical rocket propulsion systems, each test facility usually has the following major systems or components:

1. A *test cell* or *test bay* where the article (a rocket propulsion system or a thrust chamber) to be tested is mounted, usually in a special test fixture. If the test is hazardous, the test facility must have provisions to protect operating personnel and to limit damage in case of accidents or explosions.

2. An *instrumentation system* with associated computers for sensing, maintaining, measuring, analyzing, correcting, and recording the various physical and chemical parameters. It usually includes calibration systems and timers to accurately synchronize the measurements.

3. A *control system* for starting, stopping, and changing the operating conditions during tests.

4. Systems for handling and lifting heavy or awkward assemblies, for supplying (or removing) liquid propellant, and for providing maintenance, security, and safety.

5. For highly *toxic propellants* and *toxic plume gases* the capture of hazardous gases or vapors has been required (by firing inside a closed duct system) to remove most or all of the hazardous ingredients (e.g., by wet scrubbing and/or chemical treatment), allowing the release of the nontoxic portion of the exhaust gases and the safe disposal of the toxic solid or liquid residues. With exhaust gases containing fluorine, for example, the toxic gas removal

can be achieved by scrubbers with water that contains dissolved calcium; this will then form calcium fluoride, which is essentially insoluble and can be precipitated and removed.

6. In some tests specialized equipment and unique facilities are needed to conduct static testing under various environmental conditions or under simulated emergency conditions. For example, high and low ambient temperature tests of large rocket motors may require a temperature-controlled enclosure around the motor; a rugged explosion-resistant facility is needed for bullet impact tests of propellant-loaded missile systems and also for cook-off tests, where gasoline or rocket fuel is burned with air below a stored missile. Similarly, special equipment is needed for vibration testing, measuring thrust vector side forces and moments in three dimensions, and/or determining total impulse for very short pulse durations at low thrust.

Most rocket propulsion testing is accomplished in sophisticated facilities under closely controlled conditions. Modern rocket propulsion test facilities are frequently located several miles away from the nearest community to prevent or minimize effects of excessive noise, vibration, explosions, and toxic exhaust clouds. Figure 21–1 shows one type of an open-air test stand for vertically down-firing large liquid propellant thrust chambers, medium-sized SPRMs or LPREs (100,000 to 2 million pounds thrust). It is best to fire the propulsion system in a direction (vertical or horizontal) similar to the actual flight condition. Figure 21–2 shows a simulated altitude test facility for rockets of about 10.5 metric tons thrust force (46,000 lbf). It requires a vacuum chamber in which to mount the engine, a set of steam ejectors to create the vacuum, water sprays to reduce the gas temperature, and a cooled diffuser. With flow magnitudes of chemical rocket propellant combustion gases, however, it is impossible to maintain a high vacuum in these kinds of facilities; typically, between 15 to 4 torr (equivalent to 20 to 35 km altitude) can be maintained. This type of test facility allows the operation of rocket propulsion systems with high-nozzle-area ratios that would normally experience flow separation at sea-level ambient pressures.

Prior to performing any test, it is common practice to educate the test crew by going through repeated dry runs to familiarize each person with his or her responsibilities and procedures, including all emergency procedures.

Typical personnel and plant security and/or safety provisions in a modern test facility include the following:

1. Concrete-walled blockhouses or control stations for the protection of personnel, instruments and ground-based computer terminals (see Fig. 21–3), located remotely from the actual rocket propulsion test stand.

2. Remote controls, indications, and recordings of all hazardous operations and measurements; isolation of propellants and exhaust gases from all instrumentation and control rooms.

3. Automatic or manual emergency water deluge and fire-extinguishing systems, where applicable.

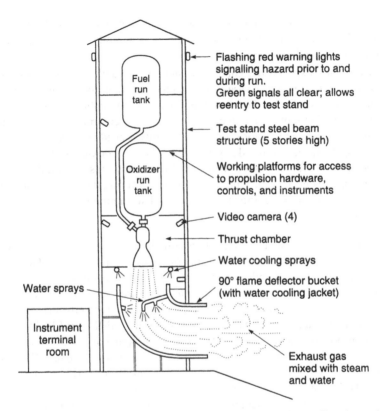

Flashing red warning lights signalling hazard prior to and during run.
Green signals all clear; allows reentry to test stand

Test stand steel beam structure (5 stories high)

Working platforms for access to propulsion hardware, controls, and instruments

Video camera (4)

Thrust chamber

Water cooling sprays

90° flame deflector bucket (with water cooling jacket)

Exhaust gas mixed with steam and water

Fuel run tank

Oxidizer run tank

Water sprays

Instrument terminal room

FIGURE 21–1. Sketch of a typical static test stand for a large liquid propellant thrust chamber firing vertically downward. Only a small part of the exhaust plume (between the nozzle exit and flame bucket entrance) is visible. The cooled flame bucket turns the exhaust gas plume horizontally thus preventing from digging a hole in the ground. Not shown here are cranes, equipment for installing and/or removing the thrust chamber, safety railings, high-pressure gas tanks, propellant tank pressurization systems, as well as separate storage tanks for fuel, oxidizer, and/or cooling water with their feed systems, or any small workshops.

4. Closed circuit television systems for remotely viewing the test.

5. Warning signals (sirens, bells, horns, lights, speakers) to notify personnel to clear the test area prior to testing, and an all-clear signal when the conditions are no longer hazardous.

6. Quantity and distance restrictions on the storage of liquid propellants, some solid propellant ingredients, and solid propellants to minimize damage in the event of explosions; robust separation of fuels and oxidizers.

7. Barricades around hazardous test articles to reduce shrapnel damage in the event of a blast.

8. Explosion-proof electrical systems, spark-proof shoes, and nonspark hand tools to prevent ignition of flammable materials, such as certain SPRM ingredients or vapors.

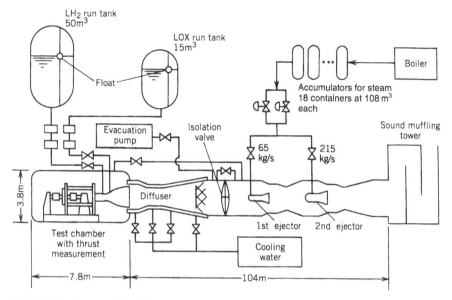

FIGURE 21–2. Simplified diagram of a simulated altitude, horizontal firing test facility for the LE-5 Japanese-designed thrust chamber (liquid oxygen–liquid hydrogen propellants) showing the method of creating a vacuum (6 torr during operation and 13 torr prior to start). The operating duration is limited to about 10 min by the capacity of the steam ejector system. Reproduced from Ref. 21–1 with permission of the AIAA.

9. For certain hazardous propellants also safety clothing (see Fig. 21–4), including propellant- and fire-resistant suits, face masks and shields, gloves, special shoes, and hard hats.

10. Rigid enforcement of rules governing area access, smoking, safety inspections, and so forth.

11. Limitations on the number and type of personnel that may be in a hazardous area at any time.

Monitoring the Environment and Controlling Toxic Materials

Open-air testing of chemical rocket propulsion systems frequently requires measurement and control of exhaust cloud concentrations and gas movement in the surrounding areas for safeguarding personnel and the environment. Most test and launch facilities have several stations (both inside and outside the facilities) for collecting and measuring air samples before, during, and after testing. Toxic clouds of exhaust gases (some with particulates) can result from normal rocket operation or from vapors or reaction gases from unintentional propellant spills, or from fire borne gases, explosions, or from the intentional destruction of vehicles in flight or rockets on the launch stand. Environmental government regulations usually limit the maximum local concentration or the total quantity of toxic gas or particulates that

FIGURE 21–3. Photograph of an actual RS-68A acceptance test. The test conductor is in the middle back row; to his right is the engine test article engineer and to his left is the controls and vehicle simulation engineer (the row in front primarily controls facility stations, like the oxygen and hydrogen systems). Up on the top are units monitoring engine firing. Present day testing is nearly all controlled remotely with computers, which also display read-out meters and act as data recorders. Reproduced by courtesy of Aerojet Rocketdyne.

can be released to the atmosphere. The toxic nature of some of these liquids, vapors, and gases has been mentioned in Chapters 7 and 13. One method of control is for tests with discharges of moderately toxic gases or products to be postponed until favorable weather and wind conditions are present.

Extensive analytical studies and measurements of the environmental exposure from explosions, industrial smoke, and exhausts from missile and space vehicle launchings provide useful background for predicting the atmospheric diffusion and downwind concentrations of rocket exhaust clouds. In ground-test analyses, the toxic cloud source is modelled as a *point source* and in flight tests as a *ribbon source*. Diffusion rates of exhaust clouds are influenced by many propulsion system variables, including propellant types, vehicle size, exhaust gas temperatures, and thrust duration; also by many atmospheric variables, including wind velocity, direction, turbulence, humidity, vertical stability or lapse rate (see definition below), and by the surrounding terrain. Reference 21–2 describes hazards related to toxic gas cloud concentrations and dispersals. Reference 21–3 evaluates the environmental impact of rocket exhausts from large units on the ozone layer in the stratosphere as well as on the ground weather near the test site (it concludes that the impacts are generally small and temporary). Reference 21–4 describes a test area atmospheric measuring network.

FIGURE 21–4. Plastic safety suit, gloves, boots, and hood used by test personnel in handling hazardous and/or corrosive liquid propellants. The safety shower, which starts automatically when a person steps onto the platform, washes away splashed or spilled propellant. Official U.S. Air Force photograph.

A widely used relationship for predicting atmospheric diffusion of gas clouds has been formulated by O. G. Sutton (Ref. 21–5). Many modern equations and models relating to downwind concentrations of toxic clouds are extensions of O. G. Sutton's theory. Given below are equations from Ref. 21–5 of primary interest to rocket and missile operators.

For instantaneous ground-level point-source nonisotropic conditions,

$$\chi_{(x,y,z,t)} = \frac{Q}{\pi^{3/2} C_x C_y C_z (\bar{u}t)^{3(2-n)/2}} \exp \left[(\bar{u}t)^{n-2} \left(\frac{x^2}{C_x^2} + \frac{y^2}{C_y^2} + \frac{z^2}{C_z^2} \right) \right] \qquad (21-1)$$

For continuous ground-level point-source nonisotropic conditions,

$$\chi = \frac{2Q}{\pi C_y C_z \bar{u} x^{2-n}} \exp \left[-x^{n-2} \left(\frac{y^2}{C_y^2} + \frac{z^2}{C_z^2} \right) \right] \qquad (21-2)$$

where χ is concentration in grams per cubic meter, Q is source strength (grams for instantaneous, grams per second for continuous); $C_{x,y,z}$ are diffusion coefficients in the x, y, z planes, respectively; $\bar{u}$ is average wind velocity in meters per second, t is time in seconds, and the coordinates x, y, z are in meters measured from the center of the moving cloud in the instantaneous case and from a ground point beneath the plume axis in the continuous case. The exponent n is a stability or turbulence coefficient, ranging from almost zero for highly turbulent conditions to 1 as a limit for extremely stable conditions, and usually falling between 0.10 and 0.50.

A few basic definitions relevant to the study of atmospheric diffusion of exhaust clouds are as follows:

1. *Micrometeorology.* The study and forecasting of atmospheric phenomena restricted to a region approximately 300 m above the Earth's surface and to a horizontal distance of approximately 5 miles.

2. *Lapse rate.* The rate of decrease in temperature with increasing height above the Earth's surface. The standard atmosphere (Appendix 2) has a lapse rate of about 6.4 °C per 1000 m. Lapse rates are also affected by altitude, wind, and humidity.

3. *Inversion or Inversion Layer.* Condition of negative lapse rates (temperature increases with increasing height). Usually forming near the ground at night.

The following are some general rules and observations derived from experience with atmospheric diffusion of rocket exhaust clouds:

1. Inversion presents a very stable layer and greatly reduces any vertical dispersion (the higher the lapse rate, the greater the vertical dispersion).

2. A highly stable atmospheric condition tends to keep the exhaust plume or cloud intact and away from the Earth's surface, except when the exhaust products are much heavier than the surrounding air.

3. High winds increase the rate of diffusion and reduce the thermal effects.

4. For short firings (<500 sec) approximate dosages downwind are about the same as from an instantaneous point source.

5. When the plume reaches about one-fourth the distance to a given point before emission is stopped, peak concentrations will be about three-fourths of those from a continuous source of equal strength.

6. The presence of an inversion layer significantly restricts the mixing or diffusion capacity of the propulsion system exhaust gases with the atmosphere; the effective air mass is that mass existing between the Earth's surface and the inversion layer.

7. Penetrations of the inversion layer arising from buoyant forces in the hot exhaust cloud seldom occur.

8. Earth surface cloud dosages drop rapidly when missiles or space launch vehicles are destroyed in flight above a height of 1500 m when compared to lower altitudes of 600 to 1000 m.

Interpretation of the hazards that exist once the concentration of the toxic agent is known also requires knowledge of their effects on the human body, and the environment. Government tolerance limits for humans are given in Chapter 7 and in Ref. 7–5. There are usually three limits of interest: one for the short-time exposure of the general public, one for an 8-hr exposure limit of workers, and one for concentrations needing evacuation. Depending on the toxic chemical, the 8-hr limit may vary from 5000 ppm for a gas such as carbon dioxide, to less than 1 ppm for more toxic substances such as fluorine. Human poisoning by rocket exhaust products usually occurs from gas and fine solid particle inhalation, but also any solid residuals that may remain around a test facility for weeks or months following a test firing can enter the body through cuts and other avenues. Moreover, certain liquid propellants may cause burns and skin rashes or are poisonous when ingested, as explained in Chapter 7.

21.3. INSTRUMENTATION AND DATA MANAGEMENT

In the last few decades, considerable progress has been made in instrumentation and data management. For further study the reader is referred to standard textbooks on instruments and computers used in testing, such as Ref. 21–6. Some of the physical quantities commonly measured during rocket propulsion testing are:

1. Forces (thrust, thrust vector control side forces, short thrusting pulses).
2. Flows (hot and cold gases, liquid fuels, liquid oxidizers, leaks).
3. Pressures (chamber, propellant, pump inlet/outlet, tank, etc.).
4. Temperatures (chamber or case walls, propellant, structure, nozzle).
5. Timing and command sequencing of valves, switches, igniters, full pressure attainments, and others.
6. Stresses, strains, and vibrations (combustion chamber, structures, solid propellants, liquid propellant lines, local accelerations of vibrating parts) (Ref. 21–7). Also, thermal growths.
7. Movements and positions of parts (valve stems, gimbal position, deflection of parts under load or heat).
8. Voltages, frequencies, and currents in electrical or control subsystems.
9. Visual observations (flame configuration and color, test article failures, explosions) using high-speed cameras or video cameras.
10. Special quantities such as propellant strains, turbopump shaft speeds, liquid levels in propellant tanks, burning rates, flame luminosities, sound pressures and/or exhaust gas composition.
11. Ambient air conditions (pressure, temperature wind speed and direction, toxic gas content) at stations in the test facility area and also at stations downwind of the test area.

During any one test of a production RPS, only some of the 11 measurements listed above are made and then automatically compared (by computer) with the same measurements on a prior identical rocket propulsion system, which performed properly in previous tests. Any deviation between the two that is identified by the computer unit is usually investigated; if appropriate, an adjustments or hardware modifications are then made to bring the rocket propulsion system being tested to the desired performance.

Reference 21–8 describes some specialized diagnostic techniques used in propulsion systems, such as using nonintrusive optical methods, microwaves, and ultrasound for measurements of temperatures, velocities, particle sizes, or burn rates in solid propellant grains. Many of these sensors incorporate specialized technologies and, often, unique software. Each measured parameter can be obtained with different types of instruments, sensors, and analyzers, as indicated in Ref. 21–9.

Measurement System Terminology

Each measurement or each measuring system may require one or more *sensing elements* (often called transducers or pickups), a device for *recording, displaying*, and/or *indicating* the sensed information, and often also another device for *conditioning, amplifying, correcting*, or *transforming* the sensed signal into the form suitable for *recording, indicating, display*, or *analysis*. Recording of rocket propulsion test data has been performed in several ways, such as on *chart recorders* or in digital form on *memory devices*, such as on magnetic tapes or disks. Definitions of several significant terms are given below and in Ref. 21–6.

Range refers to a region extending from the minimum to the maximum rated value over which the measurement system will give a true and linear response. Usually an additional margin is provided to permit temporary overloads without damage to the instrument or need for recalibration.

Errors in measurements are usually of two types: (1) *human errors* (improperly reading or incorrectly adjusting the instrument, chart, or record and improperly interpreting or correcting these data), and (2) *instrument or system errors*, which usually fall into four classifications: static errors, dynamic response errors, drift errors, and hysteresis errors (see Refs. 21–6 and 21–10). *Static errors* are usually fixed errors due to fabrication and installation variations; these errors can usually only be detected by careful calibration, and an appropriate correction can then be applied to the reading. *Drift error* is the change in output over a period of time, usually caused by random wander and environmental conditions. To avoid drift error the measuring system has to be calibrated at frequent intervals at standard environmental conditions against known standard reference values over its whole range. *Dynamic response errors* occur when the measuring system fails to register the true value of the measured quantity while this quantity is changing, particularly when it is changing rapidly. For example, thrust force has an embedded dynamic component due to vibrations, combustion oscillations, interactions with the support structure, and the like. These dynamic changes can distort or amplify thrust reading unless the test stand structure, the rocket mounting structure, and the thrust measuring and recording system are properly designed to avoid any harmonic excitation or excessive energy damping. To obtain good dynamic responses requires careful analysis and design of the total system.

A *maximum frequency response* refers to the maximum frequency at which the instrument system will measure true values. The natural frequency of the measuring system should be above the limiting response frequency. Generally, a high-frequency response requires more complex and expensive instrumentation. The instrument system (sensing elements, modulators, and recorders) must be capable of fast responses. Most of the measurements in rocket testing are made with one of two types of instruments: those made under nearly steady static conditions, where only relatively gradual changes in the quantities occur, and those made with fast transient conditions, such as rocket starting, stopping, or vibrations (see Ref. 21–11). This latter type of instrument has frequency responses above 200 Hz, sometimes as high as 20,000 Hz. Such fast measurements are necessary to evaluate physical phenomena in rapid transients.

Linearity of the instrument refers to the ratio of the input (usually pressure, temperature, force, etc.) to the output (usually voltage, output display change, etc.) over the range of the instrument. Very often static calibration errors indicate deviations from a truly linear response; a nonlinear response can cause appreciable errors in dynamic measurements. *Resolution* refers to the minimum change in the measured quantity that can be detected with a given instrument. *Dead zone* or *hysteresis* errors are often caused by the absorption of energy within the instrument system or by instrument-mechanism play; each may limit the resolution of the instrument.

Sensitivity refers to the change in response or reading caused by particular influences. For example, *temperature sensitivity* and *acceleration sensitivity* refer to the change in measured value caused by ambient temperature and flight acceleration respectively. These are commonly expressed in percent change of measured value per unit of temperature or acceleration. Such information can serve to correct readings to reference or standard conditions.

Errors in measurements can arise from many sources. Reference 21–12 gives a standardized method, including mathematical models, for estimating errors, component by component, as well as any cumulative effect in the instrumentation and recording systems.

Electrical interference or "noise" within an instrumentation system, including the power supply, transmission lines, amplifiers, and recorders, can affect the accuracy of the recorded data, especially when low-output transducers are in use. Methods for measuring and eliminating objectionable electrical noise are given in Ref. 21–13.

Test Measurements

For every test of a rocket propulsion system or one of its components there should be one or more objectives, an identification of the article to be tested, a description of the test, an identification of the test facility, a list of specific measurements to be performed, instructions on storing, analyzing or displaying of the data, an interpretation of the results, and a conclusion.

For many tests, especially with large liquid propellant rocket engines, the amount of data generated can be very large. Manual data analysis with skilled individuals was the original method, but it was too cumbersome; today computers have become commonplace. They can be programmed to correct raw measured data, store it, analyze it, correct it, and display it. Some control systems can be programed to control the engine being tested and minimize failures. This approach is really a version of a health monitoring system (HMS) described at the end of Section 11.5 and also discussed in the next section.

Often, only a portion of recorded data is actually analyzed or reviewed during and/or after testing. In complex rocket propulsion system development tests, sometimes between 100 and 400 different instrument measurements are made and recorded. Some data need to be sampled frequently (e.g., some transients may be sampled at rates higher than 1000 times per second), whereas other data need to be taken at lower intervals (e.g., temperature of mounting structures may be needed only every 1 to 5 sec). Multiplexing of data is commonly practiced to simplify data transmission. For most rocket tests, computer systems contain a configuration file to indicate data characteristics for each channel, such as range, gain, data references, type of averaging, parameter characteristics, and/or data correction algorithms. Most data are not analyzed or printed; a detailed analysis occurs only if there is reason for scrutinizing particular test events in more detail. Such analysis may occur months after the actual tests and may not even be done on the same computer.

Proper *selection* of type, number of measurements, and/or frequency of data collection is crucial in the sensing of the health status of any propulsion system to be tested. This selection is usually performed by the engineers while developing the particular propulsion system, by identifying critical parameters that influence performance and possibly also likely impending failures.

Actual *test measurements* are performed with a variety of sensors, modified to fit standard interfaces so they can be connected to a common computer or network. See Ref. 21–14. Measured outputs from each instrument or sensor are corrected for changes in ambient temperature, instrument calibration factors, nonlinear outputs, conversions of analog data into digital, and/or filtering of data signals to eliminate signals outside ranges of interest. *Recording* of data can be done by a computer located within the engine, on the ground, or in the flight vehicle. *Manipulation* of data may include changes of scale or providing displays (table, plots, or curves) of selected engine parameters in support of analysis of test results.

Health Monitoring System (HMS)

All HMSs rely on receiving corrected data inputs from selected sensors, whose measurements were described in the prior paragraphs. HMSs are also discussed in Section 11.5. From *simulated and validated analytical descriptions of the propulsion system* it is possible to derive the nominal (desired) values for each parameter at a particular time during the test, and these values are entered into the HMS computer. They can include the nominal chamber pressure, thrust, nozzle wall temperature, voltage, as well as many others for both steady-state and transient conditions—all at various times of engine operation. For each measured parameter and time, programmed analytical models provide nominal values together with upper and lower limits (also known as red line values) that identify the safe operating range of that parameter. These analytical models usually also include transient conditions (start/stop). Values for each parameter are often validated (and sometimes corrected) by actual test data from prior successful test-firings from equivalent types of rocket propulsion systems.

Then, *automatic comparisons* are made by the HMS computer between the *actual measured corrected data* and equivalent *data from analytical simulations*. These comparisons will show if any measured parameter is satisfactory (falls between the two limit lines) or unsatisfactory exceeding safe limits. Since damage to a chemical rocket system can occur very quickly (in less than a second), immediate action is necessary. In an impending failure, other measurements will usually also exceed their safe limits. For example, if the turbine inlet temperature of an LPRE becomes too hot, turbine blades may be damaged. If the turbine inlet gas temperature exceeds its safe limit, then usually the turbopump's shaft speed and the turbine inlet manifold temperature will also exceed their safe limits. By sensing any impending critical failure with more than one red line measurement, likely incipient engine failures are thus confirmed; also, any occasional single sensor malfunction can be eliminated as an insufficient cause for starting a remedial action.

Decisions on possible remedial actions are done very quickly on a computer (certainly much faster than human decisions). If the computer receives three simultaneous related measurements that exceed their safe limit values, and if all three are related to the same potential failure, then remedial action is allowed; the computer will automatically query a table of remedies, and corrective actions, which fit the three overlimit values, are quickly identified and selected, and this remedy is automatically initiated. For the turbine inlet temperature example given above, this action may include throttling the total flow to the gas generator or changing the mixture ratio of the gas generator flow (provided the gas generator valves have a throttling capability) or shutting the engine down safely. Such fast actions with decisions made by a computer have saved much hardware in development testing. They represent a relatively new feature of rocket engine testing.

For some component tests *programmable logic controllers* are used to control test operations instead of general-purpose computers, and this application usually requires some software development.

21.4. FLIGHT TESTING

Flight testing of the larger rocket propulsion systems is always conducted in conjunction with flight tests of their vehicles and other subsystems such as guidance, vehicle controls, or ground support. These flights usually occur along missile and space launch ranges, often over an ocean. If a flight test vehicle deviates from its intended path and appears to be headed for a populated area, a range safety official (or a computer) will either have to destroy the vehicle, abort the flight or cause it to correct its course. Many propulsion systems therefore include devices that will either terminate operation (shut off the rocket engine or open thrust termination openings into rocket motor cases as described in Section 14.3) or trigger explosive devices that will cause the vehicle (and therefore also the propulsion system) to disintegrate in flight.

Flight testing requires special launch test range support equipment together with means for observing, monitoring, and recording data (cameras, radar, telemetering, etc.), equipment for assuring range safety and for reducing data and evaluating flight test performance, and specially trained personnel. Different equipment is needed for different kinds of vehicles. This may include launch tubes for shoulder-held infantry support missile launchers, movable turret-type mounted multiple launchers installed on an army truck or a navy ship, transporters for larger missiles, and track-propelled launch platforms or fixed complex launch pads for spacecraft launch vehicles. Launch equipment has to have provisions for loading or placing vehicles into a launch position, for allowing access of various parts and connections to launch support equipment (checkout, monitoring, fueling, etc.), for aligning or aiming vehicles, or for withstanding the exposure to hot rocket plumes at launch.

During experimental flights extensive measurements need to be made on the behavior of the various vehicle subsystems; for example, rocket propulsion parameters, such as chamber pressure, feed pressures, temperatures, and so on, are measured and these data are telemetered and transmitted to a ground receiving station for recording, monitoring and analyzing. Some flight tests also rely on salvaging and examining the test vehicle after the flight.

21.5. POSTACCIDENT PROCEDURES

In the testing of any rocket propulsion system there will invariably be failures, particularly when some of the operating parameters are close to their limit. With each failure comes an opportunity to learn more about the design, materials, propulsion performance, fabrication methods, and/or the test procedures. A careful and thorough investigation of each failure is needed to learn the likely causes in order to identify remedies or fixes to prevent similar failures in the future. The lessons to be learned from these failures are perhaps some of the most important benefits of development testing. A formalized postaccident approach is often used, particularly if the failure had a major impact, such as high cost, major damage, or personnel injury. Any major failure (e.g., the loss of a space launch vehicle or severe damage to a test facility) often causes the program to be stopped and further testing or flights put on hold until failure causes are determined and remedial actions have been taken to prevent a recurrence.

Of utmost concern immediately after a major failure are the needed steps to respond to the emergency. These may include giving first aid to injured personnel, bringing the propulsion system and/or the test facilities to a safe/stable condition, limiting further damage from chemical hazards to the facility or the environment, working with local fire departments, medical or emergency maintenance staff or ambulance personnel, and debris clearing crews, and quickly providing factual statements to the management, the employees, the news media, and the public. It also includes controlling access to the facility where the failure has occurred and preserving evidence for the subsequent investigation. All test personnel, particularly the supervisory people, need to be trained not only in preventing accidents and minimizing any impact of a potential failure, but also on how to best respond to the emergency. Reference 21–15 suggests postaccident procedures involving rocket propellants.

REFERENCES

21–1. K. Yanagawa, T. Fujita, H. Miyajima, and K. Kishimoto, "High Altitude Simulation Tests of LOX-LH2 Engine LE-5," *Journal of Propulsion and Power*, Vol. 1, No. 3, May–Jun. 1985, pp. 180–186.

21–2. *Handbook for Estimating Toxic Fuel Hazards*, NASA Report CR-61326, Apr. 1970.

21–3. R. R. Bennett and A. J. McDonald, "Recent Activities and Studies on the Environmental Impact of Rocket Effluents," AIAA Paper 98-3850, Jul. 1998.

21–4. R. J. Grosch, "Micro-Meteorological System," Report TR-68-37, Air Force Rocket Propulsion Laboratory, Nov. 1968 (AD 678856).

21–5. O. G. Sutton, *Micrometeorology*, Chapter 8, McGraw-Hill, New York, 1973.

21–6. D. Ramsey, *Principles of Engineering Instrumentation*, University of Michigan, Ann Arbor, Ml, 1996.

21–7. K. G. McConnell and P. S. Varoto, *Vibration Testing: Theory and Practice*, 2nd ed., John Wiley & Sons, Hoboken, NJ, 2008.

21–8. Y. M. Timnat, "Diagnostic Techniques for Propulsion Systems," *Progress in Aerospace Sciences (Series)*, Vol. 26, No. 2, 1989, pp. 153–168.

21–9. R. S. Figliola and D. B. Beasley, *Theory and Design for Mechanical Measurements*, 4th ed., John Wiley & Sons, Hoboken, NJ, 2005.

21–10. R. Cerri, *Sources of Measurement Error in Instrumentation Systems*, Preprint 19-LA-61, Instrument Society of America, Research Triangle Park, NC.

21–11. P. M. J. Hughes and E. Cerny, "Measurement and Analysis of High-Frequency Pressure Oscillations in Solid Rocket Motors," *Journal of Spacecraft and Rockets*, Vol. 21, No. 3, May–Jun. 1984, pp. 261–265.

21–12. *Handbook for Estimating the Uncertainty in Measurements Made with Liquid Propellant Rocket Engine Systems*, Handbook 180, Chemical Propulsion Information Agency, Apr. 30, 1969 (AD 855130).

21–13. "Grounding Techniques for the Minimization of Instrumentation Noise Problems," Report TR-65-8, Air Force Rocket Propulsion Laboratory, Jan. 1965 (AD 458129).

21–14. S. Schmalzel, F. Figuero, J. Morris, S. Mandayam, and R. Polikar, "An Architecture for Intelligent Systems Based on Smart Sensors," *IEEE Transactions on Instrumentation and Measurement*, Vol. 34, No. 4, Aug. 2005.

21–15. D. K. Shaver and R. L. Berkowitz, *Post-accident Procedures for Chemicals and Propellants*, Noyes Publications, Park Ridge, NJ, 1984.

APPENDIX 1

CONVERSION FACTORS
AND CONSTANTS

Conversion Factors (arranged alphabetically)

Acceleration $(L\ t^{-2})^*$

1 m/sec^2 = 3.2808 ft/sec^2 = 39.3701 in./sec^2
1 ft/sec^2 = 0.3048 m/sec^2 = 12.0 in./sec^2
g_0 = 9.80665 m/sec^2 = 32.174 ft/sec^2 (standard gravity constant at Earth's surface)

Area (L^2)

1 ft^2 = 144.0 in.^{2} = 0.092903 m^2
1 m^2 = 1550.0 in.^{2} = 10.7639 ft^2
1 in.^{2} = 6.4516 × 10^{-4} m^2

Density $(M\ L^3)$

Specific gravity is dimensionless, but has the same numerical value as *density* expressed in g/cm^3; 1 liter = 10^{-3} m^3

1 kg/m^3 = 6.24279 × 10^{-2} lbm/ft^3 = 3.61273 × 10^{-5} lbm/in.^{3}
1 lbm/ft^3 = 16.0184 kg/m^3
1 lbm/in.^{3} = 2.76799 × 10^4 kg/m^3

---

*The letters in parentheses after each heading indicate the dimensional parameters (L = length, M = mass, t = time, and T = temperature).

Energy, also Work or Heat $(M\,L^2t^{-2})$

1.0 Btu = 1055.056 J (joule)
1.0 kW-hr = 3.600×10^6 J
1.0 ft-lbf = 1.355817 J
1.0 cal = 4.1868 J
1.0 kcal = 4186.8 J

Force $(M\,Lt^{-2})$

1.0 lbf = 4.448221 N
1 dyne = 10^{-5} N
1.0 kg (force) [used in Europe] = 9.80665 N
1.0 ton (force) [used in Europe] = 1000 kg (force)
1.0 N = 0.2248089 lbf
1.0 millinewton (mN) = 10^{-3} N

Weight is the *force* acting on a *mass* being accelerated by gravity (g_0 applies at the Earth's surface)

Length (L)

1 m = 3.2808 ft = 39.3701 in.
1 ft = 0.3048 m = 12.0 in.
1 in. = 2.540 cm = 0.0254 m
1 mile = 1.609344 km = 1609.344 m = 5280.0 ft
1 nautical mile = 1852.00 m
1 mil = 0.0000254 m = 1.00×10^{-3} in.
1 micron (μm) = 10^{-6} m
1 astronomical unit (au) = 1.49600×10^{11} m

Mass (M)

1 slug = 32.174 lbm
1 kg = 2.205 lbm = 1000 g
1 lbm = 16 ounces = 0.4536 kg

Power $(M\,L^2t^{-3})$

1 Btu/sec = 0.2924 W (watt)
1 J/sec = 1.0 W = 0.001 kW

1 cal/sec = 4.186 W
1 horsepower = 550 ft-lbf/sec = 745.6998 W
1 ft-lbf/sec = 1.35581 W

Pressure $(M\,L^{-1}\,t^{-2})$

1 bar = $10^5\,N/m^2$ = 0.10 MPa
1 atm = 0.101325 MPa = 14.696 psia
1 mm of mercury = 133.322 N/m^2 (or Pa)
1 MPa = $10^6\,N/m^2$
1 psi or lbf/in.^{2} = 6894.757 N/m^2

Speed (or linear velocity) $(L\,t^{-1})$

1 ft/sec = 0.3048 m/sec = 12.00 in./sec
1 m/sec = 3.2808 ft/sec = 39.3701 in./sec
1 knot = 0.5144 m/sec
1 mile/hr = 0.4770 m/sec

Specific Heat $(L^2\,t^{-2}\,T^{-1})$

1 g-cal/g-°C = 1 kg-cal/kg-K = 1 Btu/lbm-°F = 4.186 J/g-°C = 1.163 × 10^{-3} kW-hr/kg-K

Temperature (T)

1 K = (9/5)°R = 1.80°R
0°C = 273.15 K
0°F = 459.67°R
°C = (5/9)(°F − 32) and °F = (9/5)°C + 32

Time (t)

1 mean solar day = 24 hr = 1440 min = 86, 400 sec
1 calendar year = 365 days = 3.1536 × 10^7 sec

Viscosity $(M\,L^{-1}\,t^{-1})$

1 centistoke = 1.00 × $10^{-6}\,m^2/sec$ (kinematic viscosity)
1 centipoise = 1.00 × 10^{-3} kg/m sec
1 lbf-sec/ft^2 = 47.88025 kg/m sec

Constants

J	Mechanical equivalent of heat = 4.186 joule/cal = 777.9 ft-lbf/Btu = 1055 joule/Btu
R'	Universal gas constant = 8314.3 J/kg-mol-K = 1545 ft-lbf/lbm-mol-°R
V_{mole}	Molecular volume of an ideal gas = 22.41 liter/kg-mol at standard conditions
e	Electron charge = 1.60218×10^{-19} coulomb
	Permittivity of vacuum = 8.854187×10^{-12} farad/m
	Gravitational constant = 6.674×10^{-11} m³/kg-sec²
	Boltzmann's constant $\qquad$ $1.38065003 \times 10^{-23}$ J/K
	Electron mass $\qquad$ 9.109381×10^{-31} kg
	Avogadro's number $\qquad$ 6.02252×10^{26}/kg-mol
σ	Stefan–Boltzmann constant $\qquad$ 5.6696×10^{-8} W/m²-K⁻⁴

PROPERTIES OF THE EARTH'S STANDARD ATMOSPHERE

Sea-level pressure is 0.101325 MPa (or 14.696 psia or 1.000 atm).

Altitude (m)	Temperature (K)	Pressure Ratio	Density (kg/m^3)
0 (sea level)	288.150	1.0000	1.2250
1,000	281.651	8.8700×10^{-1}	1.1117
3,000	268.650	6.6919×10^{-1}	9.0912×10^{-1}
5,000	255.650	5.3313×10^{-1}	7.6312×10^{-1}
10,000	223.252	2.6151×10^{-1}	4.1351×10^{-1}
25,000	221.552	2.5158×10^{-2}	4.0084×10^{-2}
50,000	270.650	7.8735×10^{-4}	1.0269×10^{-3}
75,000	206.650	2.0408×10^{-5}	3.4861×10^{-5}
100,000	195.08	3.1593×10^{-7}	5.604×10^{-7}
130,000	469.27	1.2341×10^{-8}	8.152×10^{-9}
160,000	696.29	2.9997×10^{-9}	1.233×10^{-9}
200,000	845.56	8.3628×10^{-10}	2.541×10^{-10}
300,000	976.01	8.6557×10^{-11}	1.916×10^{-11}
400,000	995.83	1.4328×10^{-11}	2.803×10^{-12}
600,000	999.85	8.1056×10^{-13}	2.137×10^{-13}
1,000,000	1000.00	7.4155×10^{-14}	3.561×10^{-15}

Source: *U.S. Standard Atmosphere*, National Oceanic and Atmospheric Administration, National Aeronautics and Space Administration, and U.S. Air Force, Washington, DC, 1976 (NOAA-S/T-1562).

SUMMARY OF KEY EQUATIONS FOR IDEAL CHEMICAL ROCKETS

Parameter	Equations	Equation Numbers
Average exhaust velocity, v_2 (m/sec or ft/sec) (assume that $v_1 = 0$)	$v_2 = c - (p_2 - p_3)A_2/\dot{m}$	2–15
	When $p_2 = p_3$, $v_2 = c$	
	$v_2 = \sqrt{[2k/(k-1)]RT_1[1 - (p_2/p_1)^{(k-1)/k}]}$	3–16
	$= \sqrt{2(h_1 - h_2)}$ with $v_1 \approx 0$	3–15
Effective exhaust velocity, c (m/sec or ft/sec)	$c = c^* C_F = F/\dot{m} = I_s g_0$	3–32, 3–33
	$c = v_2 + (p_2 - p_3)A_2/\dot{m}$	2–15
Thrust, F (N or lbf)	$F = c\dot{m} = cm_p/t_p$	2–16, 4–9
	$F = C_F p_1 A_t$	3–31
	$F = \dot{m}v_2 + (p_2 - p_3)A_2$	2–13
	$F = \dot{m}I_s g_0$	2–6
Characteristic velocity, c^* (m/sec or ft/sec)	$c^* = c/C_F = p_1 A_t/\dot{m}$	3–32
	$c^* = \dfrac{\sqrt{kRT_1}}{k\sqrt{[2/(k+1)]^{(k+1)/(k-1)}}}$	3–32
	$c^* = I_s g_0/C_F = F/(\dot{m}C_F)$	3–32, 3–33
Thrust coefficient, C_F (dimensionless)	$C_F = c/c^* = F/(p_1 A_t)$	3–31, 3–32
	$C_F = \sqrt{\dfrac{2k^2}{k-1}\left(\dfrac{2}{k+1}\right)^{(k+1)/(k-1)}\left[1 - \left(\dfrac{p_2}{p_1}\right)^{(k-1)/k}\right]}$ $+ \dfrac{p_2 - p_3}{p_1}\dfrac{A_2}{A_t}$	3–30
Total impulse	$I_t = \int F\,dt = Ft = I_s w$	2–1, 2–2, 2–5
Specific impulse, I_s (sec)	$I_s = c/g_0 = c*C_F/g_0$	2–16, 3–33
	$I_s = F/\dot{m}g_0 = F/\dot{w}$	2–5
	$I_s = v_2/g_0 + (p_2 - p_3)A_2/(\dot{m}g_0)$	2–15
	$I_s = I_t/(m_p g_0) = I_t/w$	2–4, 2–5

Parameter	Equations	Equation Numbers
Propellant mass fraction, ζ (dimensionless)	$\zeta = m_p/m_0 = \dfrac{m_0 - m_f}{m_0}$	2–8, 2–9
	$\zeta = 1 - \mathbf{MR}$	4–4
Mass ratio of vehicle or stage, $\mathbf{MR}$ (dimensionless)	$\mathbf{MR} = \dfrac{m_f}{m_0} = \dfrac{m_0 - m_p}{m_0}$	2–7, 2–10
	$= m_f/(m_f + m_p)$	
	$m_0 = m_f + m_p$	2–10
Vehicle velocity increase in gravity-free vacuum, Δu (m/sec or ft/sec) (assume that $u_0 = 0$)	$\Delta u = -c \ln \mathbf{MR} = c \ln \dfrac{m_0}{m_f}$	4–6
	$= c \ln[m_0/(m_0 - m_p)]$	4–5, 4–6
	$= c \ln[(m_p + m_f)/m_f]$	
Propellant mass flow rate, $\dot{m}$ (kg/sec or lb/sec)	$\dot{m} = Av/V = A_1 v_1/V_1$	3–3
	$= A_t v_t/V_t = A_2 v_2/V_2$	3–24
	$\dot{m} = F/c = p_1 A_t/c^*$	2–16, 3–32
	$\dot{m} = p_1 A_t k \dfrac{\sqrt{[2/(k+1)]^{(k+1)/(k-1)}}}{\sqrt{kRT_1}}$	3–24
Mach number, M (dimensionless)	$M = v/a = v/\sqrt{kRT}$	3–11
	At throat, $v = a$ and $M = 1.0$	
Nozzle area ratio, ϵ	$\epsilon = A_2/A_t$	3–19
	$\epsilon = \dfrac{1}{M_2} \sqrt{\left[\dfrac{1 + \dfrac{k-1}{2}M_2^2}{1 + \dfrac{k-1}{2}}\right]^{(k+1)/(k-1)}}$	3–14
Isentropic flow relationships for stagnation and free-stream conditions	$T_0/T = (p_0/p)^{(k-1)/k} = (V/V_0)^{(k-1)}$	3–7
	$T_x/T_y = (p_x/p_y)^{(k-1)/k} = (V_y/V_x)^{k-1}$	
Satellite velocity, u_s, in circular orbit (m/sec or ft/sec)	$u_s = R_0\sqrt{g_0/(R_0 + h)}$	4.26
Escape velocity, v_e (m/sec or ft/sec)	$v_e = R_0\sqrt{2g_0/(R_0 + h)}$	4.25
Liquid propellant engine mixture ratio r and propellant flow $\dot{m}$	$r = \dot{m}_o/\dot{m}_f$	6.1
	$\dot{m} = \dot{m}_o + \dot{m}_f$	6.2
	$m_f = \dot{m}/(r+1)$	6.4
	$m_o = r\dot{m}/(r+1)$	6.3
Average density ρ_{av} for (or average specific gravity)	$\rho_{av} = \dfrac{\rho_o \rho_f(r+1)}{r\rho_f + \rho_o}$	7.2
Characteristic chamber length L^*	$L^* = V_c/A_t$	8.9
Solid propellant mass flow rate $\dot{m}$	$\dot{m} = A_b r \rho_b$	12–1
Solid propellant burning rate r	$r = ap_1^n$	12–5
Ratio of burning area A_b to throat area A_t	$K = A_b/A_t$	12–4
Temperature sensitivity of burning rate at constant pressure	$\sigma_p = \dfrac{1}{r}\left(\dfrac{\partial r}{\partial T_b}\right)_{p_1}$	12–8
Temperature sensitivity of pressure at constant K	$\pi_K = \dfrac{1}{p_1}\left(\dfrac{\partial p}{\partial T_b}\right)_K$	12–9

INDEX